CARBON NANOTUBES – GROWTH AND APPLICATIONS

Edited by **Mohammad Naraghi**

Carbon Nanotubes – Growth and Applications
Edited by Mohammad Naraghi

Published by InTech
Janeza Trdine 9, 51000 Rijeka, Croatia

Publishing Process Manager Viktorija Zgela

Technical Editor Teodora Smiljanic

Cover Designer InTech Design Team

Additional hard copies can be obtained from orders@intechopen.com

Carbon Nanotubes – Growth and Applications, Edited by Mohammad Naraghi
p. cm.
ISBN 978-953-307-566-2

For EU product safety concerns:
IN TECH d.o.o., Prolaz Marije Krucifikse Kozulić 3, 51000 Rijeka, Croatia,
info@intechopen.com or visit our website at intechopen.com.

Meet the editor

Dr. Mohammad Naraghi received his PhD degree from the University of Illinois at Urbana Champaign, Department of Aerospace Engineering, in 2009. His PhD research was in the field of nanomechanics and the application of MEMS sensors and actuators to investigate the mechanical behavior of soft nanofibers. His thesis research received the "Roger A. Strehlow Memorial Award" for outstanding research accomplishments by the Aerospace Engineering Department at UIUC. Since then, Dr. Naraghi has been working as a post-doctorate research fellow at Northwestern University, in the Mechanical Engineering Department. Starting from January 2012, he will be an Assistant professor at Texas A and M University, Department of Aerospace Engineering. His main field of expertise is bio-inspired high performance light-weight nanocomposites, multiscale mechanics of carbon nanotube based materials, nanomechanics, mechanical characterization of soft nanostructures, and application of MEMS to nanomechanics.

Contents

Preface

This book contains a collection of major research accomplishments in the past decade or so in the area of nanotechnology and specifically carbon nanotubes. The book has been brought together by the efforts of scientists from all over the world, through the support of the multidisciplinary Open Access publisher of books and journals, InTech (http://www.intechweb.org/) with the aim of providing easy access to high quality research in the field.

"There's plenty of room at the bottom," said Richard Feynman, when in part he was referring to the arrangement of atoms in a "desirable arrangement". While a desirable choice will depend solely on the application of the material, in many ways one may think of carbon nanotubes (CNTs) as a desirable one. CNTs are allotropes of carbon arranged in cylindrical structures with diameters of as low as a fraction of nanometer to several tens of nanometers, which are bound together with sp2 bonds. From a mechanical point of view CNTs are among the strongest materials that have existed on the Earth, possessing strength and stiffness of as high as 100 GPa and 1 TPa, owing to the strong sp2 bonds between Carbon atoms. Moreover, the remarkable stability of the sp2 bonds and their delocalized electrons provides CNTs with excellent thermal stability and tunable electronic properties. In addition, the pristine structure of CNTs can be modified to make their surfaces for applications which require more active surfaces. Therefore, CNTs have been proposed as a primary material for many applications ranging from multifunctional nanocomposites, electronics, energy storage devices, to gas and other chemical detection tools, and for pharmaceutical applications.

While current scientific literature has focused significantly on the properties of individual CNTs and their laboratory made samples, one of the focuses of this book is on applications of CNTs. Various topics discussed in this book are as follows:

The opening part of the book is titled "**Growth of CNT Structures and Application Specific Modification of Their Structures**", which discusses the fabrication of CNTs and methods of their functionalization. Given the scalability of the method, chemical vapor deposition is the major method of fabrication of CNTs presented in this part. Moreover, in this part, the decoration of CNT surfaces with biological macromolecules to generate smart nanocomposites is discussed.

As mentioned earlier, a great portion of the book is devoted to the application of CNTs in parts 2, 3 and 5. In part 2, "**CNTs in Biological and Medical Applications**", the application of CNTs as biosensors in the detection of, for instance, DNA/proteins and metalloproteins, in labeling and molecular delivery platforms, and the application of CNTs in biological nanocomposites such as orthopedic bone cement are discussed.

Part 3, "**CNTs as Chemical Sensors**", presents another class of the application of CNTs which utilizes the interesting electrochemical properties of CNTs for the detection of analytical concentration of some compounds of the analyte solution, for instance in pharmaceutical applications.

The 4th part of the book, "**Health Hazard Potentials of CNTs**", deals with the potential health issues related to any application and handling of CNTs. These studies at the moment are not conclusive. In some extreme cases, some researchers have described the toxicity of CNTs to be similar to asbestos, which tangle destructively with chromosomes, due to their minute dimensions. Therefore, in general, special care should be taken in handing CNTs, while more studies are required to further establish relations between CNTs and toxicity.

The last part of the book, part 5, "**CNTs for Energy Related Applications**", is mostly focused on the application of CNTs in energy conversion devices and proton exchange membrane fuel cells and on development of highly efficient catalysts.

Finally, I would like to thank all the authors and the staff of the InTech for the support to make this book into reality.

Mohammad Naraghi
Mechanical Engineering Department
Northwestern University
USA

Part 1

Growth of CNT Structures and Application Specific Modifications of Their Structure

1

Carbon Nanotubes and Carbon Nanotubes/Metal Oxide Heterostructures: Synthesis, Characterization and Electrochemical Property

Yong Hu and Changfa Guo
Zhejiang Key Laboratory for Reactive Chemistry on Solid Surfaces and Institute of Physical Chemistry, Zhejiang Normal University, Jinhua, China

1. Introduction

Carbon nanotubes (CNTs), a novel carbonaceous material, have been widely studied since their discovery in 1991. CNTs have highly unique electronic, mechanical, catalytic, adsorption, and transport properties, making them interesting for a variety of applications. It is very important for the use of CNTs to prepare CNTs with various constructs in large scale and ideal quality. Up to now, CNTs have been produced by various kinds of strategies, mainly including Arc discharge, laser ablation, and chemical vapor deposition (CVD). Compared with the other two methods, CVD is considered as the most promising method for easily scaled-up to batch-scale production duo to simplicity and economy. At present, a lot of modified CVD were studied and explored for preparing pure CNTs in large scale, especially single-walled CNTs (SWCNTs), such as plasma-enhanced CVD (PECVD) and thermal CVD. A vapor–liquid–solid (VLS) mechanism is generally accepted for CNTs growth, and catalyst plays an important role in the CVD synthesis of CNTs, which is needed to be removed in follow purity of CNTs for further use. To avoid the problem, some modified CVD methods without metal particles as catalyst were also carried out for producing relative pure CNTs.

CNTs possess large specific surface areas due to their high aspect ratio, while their structural integrity and chemical inertness support relatively high oxidation stability in comparison with graphite. For these reasons, CNTs are promising building blocks for hybrid materials, which could endow CNTs more excellent performances for widely potential fields. CNTs-based hybrid materials, including CNTs-inorganic species and CNTs-organic species, were synthesized by numerous diverse strategies, which can be classed roughly into two kinds: ex situ approach (attaching nanoparticles to CNTs) and in situ synthesis (Directly on the CNTs surface). CNTs/metal oxide heterostructures are an important kind of CNTs-inorganic composites. The synergistic effect of CNTs and metal oxide makes CNTs/metal oxide heterostructures possess the properties of the both and have more widely potential application. The electrochemical properties of the CNTs/metal oxide heterostructures have attracted researchers' interest duo to their excellent performances on electrochemical catalyst, electrochemical sensor, supercapacitors and batteries. Over the past few years a great number of CNTs/metal oxide heterostructures, such as TiO_2/CNTs, Co_3O_4/CNTs, Au/CNTs, Au/TiO_2/CNTs, Co/CoO/Co_3O_4/CNTs (Fig. i) [1] etc, were synthesized and their electrochemical properties were investigated as well.

Numerous synthetic methodologies have been developed for the preparation of CNTs and their composites. In this chapter, besides the preparation of nanocomposites based on CNTs, we also focus on investigating the electrochemical applications of CNT/metal-oxide or hydroxide composites.

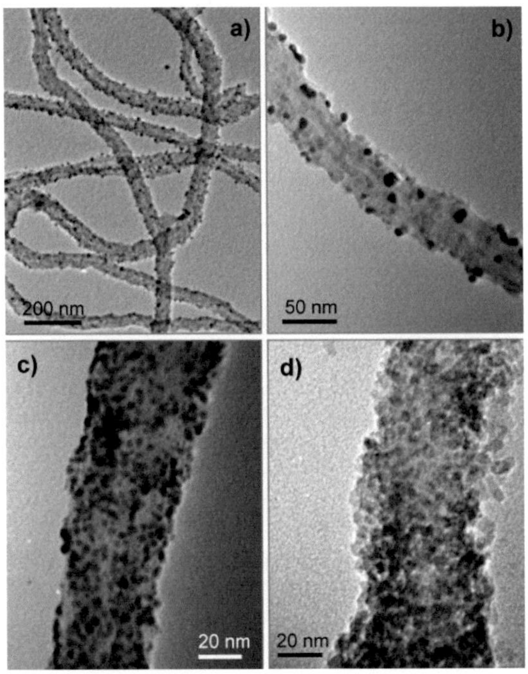

Fig. i. a) and b) a large view and a detailed view on the Au/TiO_2/CNTs nanocomposites prepared by the photo-assisted method, (c) a detailed view on the Au/TiO_2/CNTs prepared by the self-assembly method, and (d) a detailed view on the TiO_2/Co_3O_4/CNTs nanocomposite prepared by the self-assembly method. (Courtesy of J. Li et al. [1])

2. Synthesis of CNTs

CNTs with tubular structures are made entirely of rolled-up layers of interconnected carbon atoms [2,3], and with diameters ranging from about one nanometer to tens of nanometers and lengths up to centimeters. CNTs can be open-ended or closed by a hemispherical fullerene-type cap, depending on their synthesis method [4]. CNTs have highly unique electronic, mechanical, catalytic, adsorption, and transport properties, making them interesting for a variety of applications [5-13]. The scale-up preparation of CNTs with high purity and homogeneity is essential for the use of CNTs in reality. In several past decades, a lot of effort was made to investigate the synthesis strategies of CNTs and modify constantly the promising methods. By far, various strategies, maily including Arc discharge, laser ablation, and various chemical vapor depositions (CVD), have been found to produce SWCNTs and MWCNTs with different merit and demerit. In this section, a brief discussion about the progress in preparation of CNTs is divided into two parts (synthesis of SWCNTs and synthesis of MWCNTs) and made.

2.1 Synthesis of SWCNTs

SWCNT can be considered to be a unique molecule, with different physical properties, depending on their chirality, which determines that SWCNTs are primarily metallic or semiconducting. Although obtaining a single aliquot of one (n,m) nanotube is not necessary in some cases, it is still necessary to obtain nanotubes that are free of non-nanotube carbon impurities and metal catalyst particles (employed to make SWCNTs).

SWCNTs were first synthesized by the coevaporation of a cobalt catalyst and graphite in an electric arc [3,14]. In the early arc-generated nanotube material, the SWCNT fibers typically consisted of 7-14 bundled SWCNTs, with the individual tubes being 1.0-1.5 nm in diameter. The early arc-generated materials also contained relatively low weight percent densities of SWCNTs. An improvement in the nanotube yield was observed by changing the metal catalysts employed in the arc-technique [15]. Fig. 1a displays a transmission electron microscope (TEM) image of a typical arc SWCNT material generated with an iron catalyst [16]. Note that the nanotube bundles are relatively small (containing~3-5 nanotubes). SWCNTs were later produced at much higher yield by the method of laser vaporization [17]. Crystalline ropes micrometers in length containing 10-100 s of individual SWCNTs were easily obtained [18]. Fig. 1b shows a typical TEM image of crude laser-generated material produced with an Alexandrite laser operating at ~45 W/cm^2, with Co and Ni at 0.6 % each and an external furnace temperature of 1200 °C [19]. Note the large, very long bundles of SWCNTs present in Fig. 1b. Although there is a high density of SWCNTs (30-40 wt %), non-nanotube carbon impurities, as well as metal catalyst particles, are still clearly visible in the TEM image. Laser-generated SWCNTs typically have low defect concentrations [20], making them easier to purify, since they are not as likely to be destroyed by the acids generally employed to remove metal catalyst particles. For example, Laser-genrated SWCNTs were purified with dilute HNO_3 to remove the metal catalyst particles and then burned in air at 550 °C to remove the non-nanotube carbon [21]. However, the laser technique is often considered to be too expensive to be industrially scalable.

Consequently, numerous research groups have turned to the development of CVD as a potentially low-cost scalable technique for the production of SWCNTs. In 1996, SWCNT growth employing CVD on a supported catalyst was demonstrated as a promising route to carbon nanotube production [22]. Multiple reports quickly followed, further establishing CVD as a viable large-scale production process [23-32]. Typically, CVD materials contain more metal and often smaller and shorter SWCNT bundles than those produced in the laser processes. Fig. 1c shows a TEM image of commercially available CVD material produced by a high pressure carbon monoxide (HiPCO) process [29]. CVD production of isolated nanotubes has been achieved on oxidized silicon substrates using an iron catalyst [33]. Additionally, isolated SWCNTs have been generated in the gas phase by the technique of hot wire chemical vapor deposition (HWCVD) [34]. Fig. 1d shows high-resolution TEM images of isolated SWCNTs produced by this HWCVD process. For comparison, a high resolution TEM image of a bundle of purified laser-generated SWCNTs is also provided in

Fig. 1e [9]. A multistep process for the purification of HiPCO CVD-generated nanotubes was reported in 2001 [35]. Recently, K. E. Hurst reported a simple "laser cleaning" method than can be employed to purify a host of CNT materials [36]. The unavoidable metal species remaining in the SWCNT products would result in obvious disadvantages for both intrinsic property characterization and application exploration of SWCNTs. Despite sustained efforts, it has been until now an intractable problem to remove metal catalysts completely from SWCNT samples without introducing defects and contaminations. Surprisingly, in 2009, Liu et al. proposed a simple and effective method for growing SWCNTs via a metal-catalyst-free CVD process on a sputtering deposited SiO_2 film. Metal-free, pure, and dense SWCNTs were obtained [37]. The successful growth of SWCNTs using a nonmetal catalyst can provide valuable implications for understanding the growth mechanism of SWCNTs in-depth, which accordingly will facilitate the controllable synthesis and applications of carbon nanotubes.

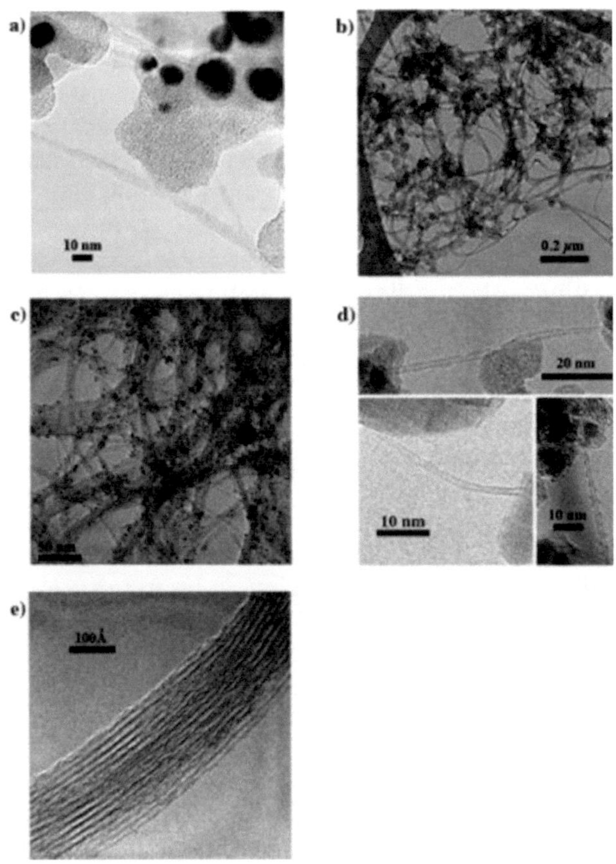

Fig. 1. TEM images of bulk SWCNT materials showing (a) typical arc-material generated with a Co/Ni catalyst mixture. (b) Alexandrite laser-generated crude material produced with ~45W/cm^2 and a Co/Ni catalyst mixture at 1200 °C. (c) commercially available CVD material produced by the HiPCO process. (d) Isolated SWCNTs produced by a continuous HWCVD process where ferrocene is employed to supply the metal catalyst. (e) Bundle of purified laser-generated SWCNTs. (Courtesy of A. C. Dillon et al. [5])

2.2 Synthesis of MWCNTs

MWCNTs were discovered by Iijima in 1991 while vaporizing carbon in an electric arc [2] and were then produced at higher yield by increasing the pressure of the helium gas atmosphere [38]. MWCNTs typically have inner diameters of ~3-20 nm and are then surrounded by concentric graphene sheets with an interstitial spacing between the sheets of ~3.4 Å. The number of concentric graphene sheets can range from 2 to ~100. High resolution TEM images of MWCNTs are provided in Fig. 2 [39]. In Fig. 2a the high-resolution TEM image clearly shows that the distance between layers of the MWCNTs measures ~3.4 Å. Fig. 2b depicts a MWCNT with ~20 concentric shells. MWCNTs have electronic properties similar to those of graphite and are thus semimetals. They are promising for multiple applications including strong composite materials, field emission displays, and adsorbents for gas separation or storage [40-47].

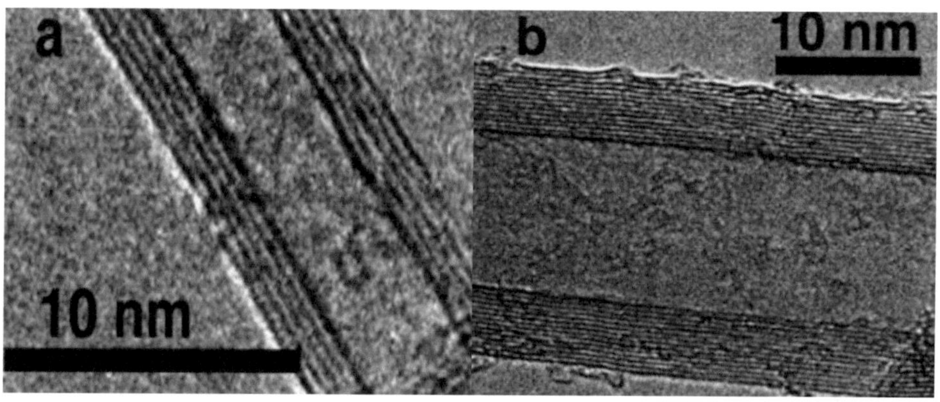

Fig. 2. TEM images of MWCNTs (a) showing that the distance between layers of the MWCNTs measures ~0.34 nm and (b) depicting a MWCNT with ~15 concentric shells. (Courtesy of A. C. Dillon et al. [39])

Similar to SWCNTs, a continuous low-cost production method producing MWCNTs that are easily purified is required for MWCNTs to be incorporated in emerging technologies. In 1997 [48] and 1998 [49], ferrocene was utilized as a gas-phase catalyst in a CVD process for continuous MWCNT formation from methane at 1150 °C. However, the 1997 study reported materials containing more amorphous carbon than arc-generated MWCNTs, presumably due to a lower synthesis temperature than that achieved in the arc process [48]. In the 1998 study, the outer layers of the tubes were not graphitic [49], making them more difficult to purify. Later, ferrocene and ethylene were employed in CVD of MWCNTs between 650 and 950 °C [50]. Again, carbon impurities were observed at high-density. The authors concluded that further work was necessary to improve the nanotube microstructure and yield [50]. High-purity aligned, graphitic MWCNTs were synthesized via decomposition of a ferrocene-xylene mixture at ~675 °C. However, although the catalyst was supplied in the gas-phase, nucleation of tube growth occurred only for iron species deposited on a quartz substrate, resulting in a surface growth mechanism and limiting yields to available surface area [51]. CVD techniques employing benzene pyrolysis [52] and the decomposition of ethylene [53] and acetylene [54] on supported metal catalysts have been demonstrated as viable large-scale production methods. MWCNTs have also been grown on supported metal particles or films via CVD [55-59], plasma-enhanced CVD [60-70], hot-wire chemical vapor deposition [71,72], and plasma-enhanced HWCVD methods [73,74]. One HWCVD report employed evaporation of the Fe-Cr filament to supply a gas-phase catalyst, resulting in MWCNTs with a high density of structural defects and significant carbon impurities [75]. Although more research was deemed necessary, this method had potential for large-scale production, since it was not substrate dependent. In 2003, the first continuous high-density MWCNT formation with minimal non-nanotube carbon impurites was demonstrated with HWCVD employing methane as the carbon source and ferrocene as a gas-phase catalyst [39]. The metal catalyst impurities were simply removed via refluxing in dilute HNO_3 [39]. Surprisingly, Multi-walled carbon nanotubes has been synthesized using acetylene as carbon sources with a metal-free mild chemical vapor deposition process by Du et al. [76]. The authors not only gave a simple and facile way to synthesize MWCNTs without metal but also provided valuable information for a deeper understanding of CNT formation in CVD.

3. Synthesis of CNTs/metal oxide heterostructures

The transition metal oxides are an important family of inorganic nanomaterials with abundant properties in optics, electronics, magnetics and catalysis. The property of metal oxide nanomaterials can be further tuned by varying their composition, structure and morphology [77].

Composite materials based on CNTs and metal oxide nanomaterials integrate the unique characters and functions of the two types of components and may also exhibit some new properties caused by the cooperative effects between the two kinds of materials [78–82]. Therefore, these composite materials have shown very attractive potential applications in many fields. This section summarizes the handling of CNTs and the preparation of CNT-based nanocomposites.

3.1 Purification and dispersion of CNTs

Generally, the carbon nanotubes as-prepared are grown as mixtures of carbon nanotubes and impurities such as amorphous carbon, metal catalyst particles and carbon nanoparticles. These impurities significantly influence the properties of CNTs and limit their applications. Consequently, development of economical purification methods has become an important issue to the development and practical applications of the CNTs. A commonly used purification approach consists of two steps. The thermal or acid oxidation treatment on raw CNTs can effectively remove the amorphous carbon, carbon nanoparticles and carbon layers coated on the metal particles [83,84]. The following acid refluxing treatment removes the residual naked metal-oxide particles [85]. This method is time consuming and has the disadvantage of low yield and damaging the nanotubes. Some extraordinary methods such as magnetic filtration, microwave irradiation, electrochemical oxidation, surfactant-assisted purification and $C_2H_2F_4$ or SF_6 treatment have shown higher efficiency in removing the contaminants in the CNT samples and are less damaging to the nanotube structure [86–89].

Due to the strong π-π stacking interactions between the neighboring CNTs, they tend to aggregate into bundles, making it very difficult for them to be dispersed in water and organic solvents. Their insolubility has become a great obstacle to the manipulation and application of CNTs. Dispersing nanotubes at the individual nanotube level is critical for a better performance of CNTs in most applications, especially the preparation of CNT-based composites. Therefore, many strategies have been explored to improve the solubility of CNTs in solvents. They can be classified mainly into two types, one is the sidewall covalent functionalization, and the other is the noncovalent modification using guest molecules as solubilizers. By sonication in mixtures of sulfuric and nitric acids or sulfuric acid and hydrogen peroxide, carbonyl and carboxylic groups can be introduced onto the nanotubes [90]. This is one of the most prevalently used covalent modification methods. The modified CNTs can be further functionalized by esterification or amidation reaction [91,92]. Other methods such as carbene cycloaddition, diazonium reaction and grafting of polymers have also been successfully used in the functionalization of CNTs [93,94]. The covalent methods convert the carbon atom hybridization from sp^2 to sp^3, leading to a disruption of electronic structure of CNTs. The noncovalent modification methods, such as polymer and DNA wrapping, π-π stacking interactions with aromatic molecules and surfactant-assisted dispersion, are based on van der Waals or π-π stacking between CNTs and solubilizer molecules [95–98]. The noncovalent strategy offers the advantage of remarkably improving the CNTs' solubility without disrupting the electronic structure of the tubes. Therefore it is more attractive than the covalent method for the maintenance of pristine structure and properties of CNTs. SWCNTs can be directly dispersed in the imidazolium-based ionic liquids simply by mechanical milling. The concentration of SWCNTs can be as high as 1wt%, which is remarkably higher than that of conventional covalent and noncovalent approaches [99]. The prosperities of SWCNTs are well preserved, since only weak van der Waals force exists between SWCNTs and ionic liquids [100,101]. Therefore, imidazolium-based ionic liquids are superior solvents for the handling of CNTs [100,101].

3.2 Filling CNTs

The earliest attempt toward CNT-inorganic hybrids (in 1993) was the filling of MWCNTs with metal oxides (PbO [102] and Bi_2O_5 [103]). Because of their larger inner diameter (5-50 nm) compared with SWCNTs (1-1.5 nm), most efforts had been spent on filling MWCNTs, and it was not until five years later that Sloan et al. reported the filling of SWCNTs with $RuCl_3$ [104]. Recently, a few of researchers reported that magnetic nanoparticles or nanowires, such as cobalt, iorn and corresponding oxides, filled carbon nanotubes by various technologies and their properties were investigated as well [105-107]. However, carbon nanotube is difficult to be filled fully without few impurities duo to capillarity of itself.
Initially, the motivation arose mainly from the prospect of forming encapsulated or (upon oxidation) freestanding inorganic nanowires with new crystal structures or novel properties. Although both MWCNTs and SWCNTs have been filled with a vast number of compounds, little is known about their properties and potential in applications. This has been frequently blamed on high impurity levels in the synthesized hybrids, the lack of bulk quantities, and the need for specifically designed measurement devices [108]. In general, the distinction should be made between the intrinsic properties of the filler, the altered properties of the encapsulated material due to the confined-space, and novel properties arising from interactions between the filler and CNTs.

3.3 Ex situ approach: attaching metal oxide nanoparticles to CNTs

In this ex situ or building block approach, metal oxide nanoparticles are attached to the CNTs via linking agents that utilize covalent, π-π Stacking, or electrostatic interactions. In this approach, either the metal oxide nanoparticles or the CNTs (or the both) require modification with functional groups. The type of functionalization and, thus, the strength of interaction determine the distribution and concentration of the metal oxide nanoparticles on the CNT surface.

3.3.1 Covalent Interactions

Carboxyl groups on the surface of acid-treated CNTs are often used to attach amine-terminated or mercapto-terminated metal oxide nanoparticles via amide bonds [109]. This can be achieved either by directly linking amine-terminated or mercapto-terminated nanoparticles with the carboxyl groups or by modifying these carboxyl groups into thiol groups, which then anchor to colloidal nanoparticles. In a similar way, QDs have been linked by first stabilizing them with a mixed monolayer of trioctylphosphine oxide (TOPO) and 2-aminoethanethiol [110]. The resulting amino-functionalized QDs then reacted with the carboxylic groups of the acid-treated CNTs to form amide bonds.

Metal oxides can be attached to the carboxyl groups without any linking agent due to their hydrophilic nature, as recently demonstrated for MnO_2 [111], MgO [112], and TiO_2 [113]. However, the authors observed only weak interactions between the oxides and the acid-terminated CNTs, resulting in rather nonuniform distributions of the nanoparticles. Better adhesion was observed when capping agents were used. For instance, Sainsbury and Fitzmaurice produced capped TiO_2 via a standard sol-gel process using titanium tetraisopropoxide (TTIP) as precursors with cetyltrimethyl ammoniumbromide (CTAB) as the capping agent [114]. MWCNTs were modified with 2-amino-ethylphosphoric acid and then mixed with the oxide nanoparticles. The authors showed that the phosphonic acid groups on the CNTs were well-distributed and provided an excellent driving force for the attachment of TiO_2 nanoparticles (Fig. 3).

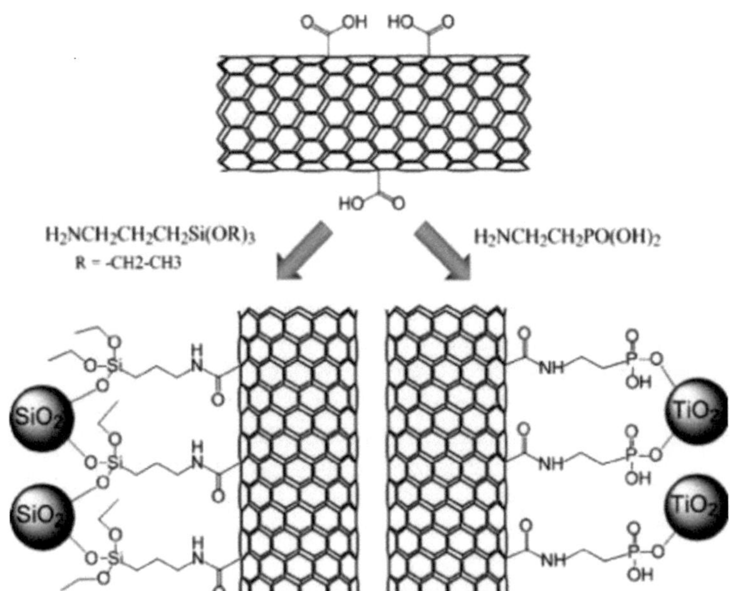

Fig. 3. Example of an ex situ attachment of SiO_2 and TiO_2 nanoparticles to functionalized CNTs via silane and phosphonic acid bonds, respectively. (Courtesy of T. Sainsbury et al. [114])

3.3.2 π-π stacking

This approach uses the moderately strong interactions between delocalized π-electrons of CNTs and those in aromatic organic compounds, such as derivatives of pyrene [115-117], porphyrins [118-120], phthalocyanines [121], or combinations thereof [122], as well as benzyl

alcohol or triphenylphosphine (Fig. 4) [123]. These molecules are often modified with long alkyl chains that are terminated with thiol, amine, or acid groups, which can then connect to metal oxide nanoparticles and enable their attachment to pristine CNTs via π-π stacking. For example, Li et al. used pyrene derivatives with a carboxylic termination to anchor magnetic nanoparticles such as Co or Fe_3O_4 [124].
One of the major advantages of this approach is that the pyrene compounds remain strongly adsorbed on the CNT surface after workup steps (e.g., washing, filtration) and thus provide enhanced solubility and allow continuous redispersing of the modified CNTs in various aqueous and organic solvents. Furthermore, spectroscopic experiments on CNT/Pt [125] and CNT/porphyrin hybrids[118,126] revealed an enhanced charge transfer from inorganic nanoparticles to the CNTs, mediated by the aromatic compound. This was also observed for attached Co and Fe_3O_4 nanoparticles [124], whose magnetic and electronic properties were altered due to a strong electron transfer. Moreover, this effect is tunable by the length of the chain.

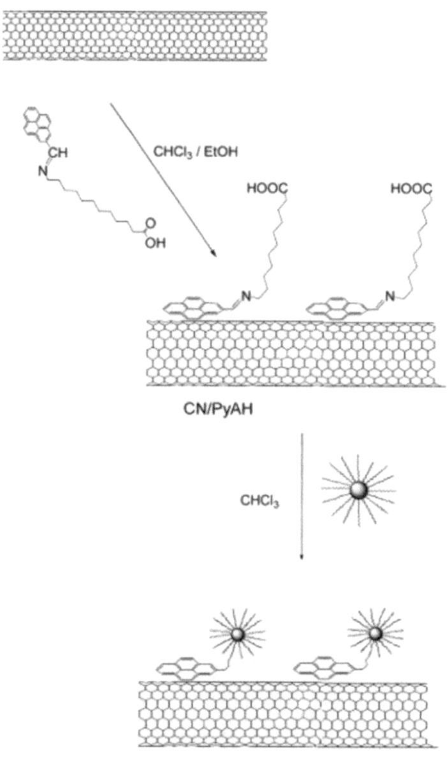

Fig. 4. Examples of linking agents and ligands used to attach metal (oxide) nanoparticles to pristine CNTs via π-π interactions: pyrene derivatives. (Courtesy of V. Georgakilas et al. [124])

3.3.3 Electrostatic interactions

The approach utilizes electrostatic interactions between modified CNTs and metal oxide nanoparticles. Among the known examples, the deposition of ionic polyelectrolytes to attract charged nanoparticles is the most common route [127-131]. These polyelectrolytes

typically bond covalently to the functional groups on the CNT, in contrast to polyethyleneimine (PEI), which interacts with CNTs via physisorption [132]. For instance, Sun et al. deposited Al_2O_3, ZrO_2, and TiO_2 nanoparticles on charged CNTs in a slightly modified way [133]. CNTs were pretreated in NH_3 at 600 °C to induce a positive surface charge. The addition of PEI increased the positive charges even further and enabled a better dispersion. Commercially available α-Al_2O_3 and 3Y-TZP were then dispersed in poly(acrylic acid) (PAA), which provided a negative surface potential over a wide range of pH values. Upon mixing, the Al_2O_3 and ZrO_2 nanoparticles formed strong electrostatic attractive interactions and covered the CNT surface completely.
Sun et al. attached nanocrystals of TiO_2 to acid-treated SWCNTs, also using PEI as a modifier (Fig. 5) [134]. In the first step, $TiCl_4$ was mixed with PEI, whose charged amino groups were protonated at a pH of 8. The positively charged amino groups of PEI accelerated the hydrolysis of $TiCl_4$ into TiO_2-nanoparticles and stabilized them electrosterically. These amine-terminated TiO_2-nanoparticles with positive charge then attached to acid-treated SWCNTs, either via amide linkage or through electrostatic interaction. Another route has been suggested by Jerome et al. [135], who grafted MWCNTs with poly-2-vinylpyridine (P2VP) to form carboxylate terminated alkyl chains, onto which positively charged magnetic Fe_3O_4 nanoparticles were anchored. The main advantage of the polymer route is that the polymers provide a very dense, uniform distribution of either negative or positive charges over the entire CNT surface, which enables very dense assemblies of metal oxide nanoparticles.

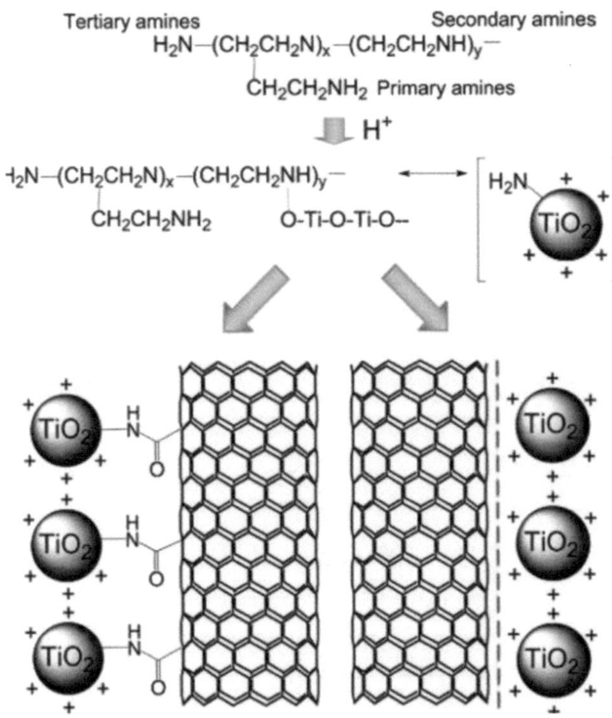

Fig. 5. Two-step process for the attachment of TiO_2 nanoparticles to CNTs: (1) modification of the nanoparticles with positively charged PEI, containing primary, secondary, and tertiary amines, and (2) attachment via amide linkage or electrostatic interactions. (Courtesy of J. Sun et al. [134])

In summary, these few examples demonstrate the simplicity and feasibility of the ex situ approach, which remains the method of choice for the deposition of metal oxide nanoparticles. The main advantage is the possibility of using prepared nanoparticles with controlled morphology, structure, shape, and size, and, therefore, a good structure property relationship. The downside of this route, however, is the need to chemically modify either the CNTs or the metal oxide. This process is often work-intensive, and functionalization alters both the surface chemistry of the CNTs and also, particularly for SWCNTs, their physical properties. Furthermore, the use of predefined building blocks restricts the synthesis of novel hybrid materials and, thus, the development of new physical properties.

3.4 In situ synthesis directly on the CNT surface

The metal oxides or hydroxides can also be formed directly on the surface of pristine or modified CNTs. The main advantage of this route is that the metal oxide or hydroxide can be deposited as a continuous amorphous or single-crystalline film with controlled thickness, or as discrete units in the form of nanoparticles, nanorods, or nanobeads. Furthermore, the CNTs may act as a support to stabilize uncommon or even novel crystal phases or prevent crystal growth during crystallization and phase-transformation processes. Finally, a variety of chemical and physical synthesis techniques can be applied. The deposition can be carried out either in solution via electrochemical reduction of metal salts, electro- or electroless deposition, sol-gel processing, and hydrothermal treatment with supercritical solvents, or from the gas phase using chemical deposition (CVD or ALD) or physical deposition (laser ablation, electron beam deposition, thermal evaporation, or sputtering).

3.4.1 Electrochemical techniques

Electrochemistry is a powerful technique for the deposition of various metal oxide nanoparticles, as it enables effective control over nucleation and growth [136,137]. Most research has been conducted on the deposition of metal oxides, as they are the metal oxides of choice for applications like heterogeneous electrocatalyst, sensors, supercapacitors and so on. In general, metal oxide nanoparticles are obtained via reduction of metal complexes by chemical agents (chemical reduction), or by electrons (electrodeposition). The size of the metal oxide nanoparticles and their coverage on the sidewalls of CNTs can be controlled by the concentration of the metal salt and various electrochemical deposition parameters, including nucleation potential and deposition time [138].

3.4.1.1 Chemical reduction and oxidation

These techniques involve reactions, in which the reduction of the precursor is carried out with liquid or gaseous reducing agents with the aid of heat, light, ultrasound, microwave, or supercritical CO_2 [139,140]. As an example, Sivvakkumar et al. deposited MnO_2 via chemical reduction of $KMnO_4$ [141]. The authors suspended the CNTs in Na-p-toluene sulfonate and pyrrole, which polymerized with the aid of ultrasonication. $KMnO_4$ was then slowly added and reduced with acetonitrile to form hydrous MnO_2. A very elegant variation of this process is shown in Fig. 6 and uses $KMnO_4$ as both the oxidizer and reactant [142]. In detail, pristine MWCNTs were first oxidized under reflux with $KMnO_4$, which introduced exclusively hydroxyl groups on the sidewalls of CNTs, in contrast to other oxidation treatments, e.g., with oxidizing acids. These hydroxyl groups then acted as anchors for Mn^{7+} ions, which subsequently were reduced to Mn^{4+} with citric acid to form a coating of ε-MnO_2. For comparison, a pretreatment of the CNTs in strong acids, which induces the formation of carboxyl groups, resulted instead in γ-MnO_2 [143]. Therefore, this work provides an interesting example of the effect of the CNT surface chemistry on the crystal structure of the metal oxide coating. In contrast to MnO_2, the deposition of other metal oxides typically requires oxidizing rather than reducing processes. For instance, Huang et al. added acid-treated CNTs to a solution containing ammonium metavanadate [144]. The adsorbed VO^{2+} ions were then oxidized with oxalic acid to V_2O_5. A different approach combined the reduction of cationic precursors by hydrogen with a water-in-oil microemulsion technique for the deposition of metal oxide nanoparticles [145]. Sun et al. used a slightly modified process [146]. The CNTs was mixed first with an aqueous solution of sodium dodecylbenzene sulfonate (NaD-DBS) and then with a mixture of Triton-X and cyclohexane, which resulted in very small water droplets on the CNT surface. Upon adding zinc acetate as the metal precursor, the Zn^{2+} ions concentrated in the aqueous phase and then reacted with NH_3 or LiOH to form spherical, hollow ZnOH nanoparticles. Subsequent calcinations oxidized them to create small and dense ZnO nanoparticles. In all cases, the microemulsion technique produced fine dispersions of small nanoparticles.

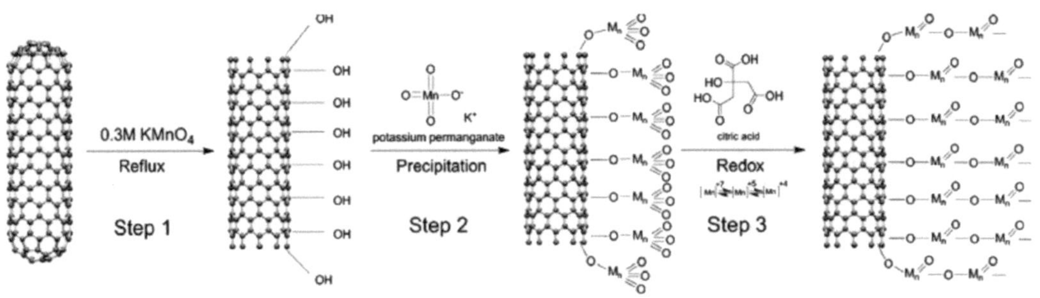

Fig. 6. Example of the deposition of MnO_2 on CNTs via chemical reduction. The multistep process involves (1) the oxidation of MWCNTs with $KMnO_4$ to form hydroxyl groups, (2) the precipitation of permanganate ions, and (3) their reduction with citric acid to MnO_2. (Courtesy of X. Xie et al. [142])

3.4.1.2 Electrodeposition

Many of the above-mentioned reduction/oxidation techniques are very time-consuming and as such allow impurities in the bath solution to be incorporated into the inorganic phase. In contrast, electrodeposited nanoparticles, exhibit higher purities as well as a good adhesion to the CNT surface [138].

In most cases, simple van der Waals interactions between the CNTs and the metal oxide nanoparticles seem to be sufficient to provide strong enough adhesion. Although most research currently concentrates on the electrodeposition of metal nanoparticles, there have also been a few reports on electrodeposited metal oxides. As an example, $TiCl_3$ was used as a precursor and electrolyte and was kept at pH 2.5 with HCl/Na_2CO_3. The deposition was then carried out via galvanostatic oxidation with 1 mA/cm^2 and resulted in a rather irregular and partial coating of a mixture of anatase and TiO_2-B [147]. The galvanostatic approach (3 mA/cm^2) has also been applied to codeposit Ni and Co oxides from their nitrates, with Pt and saturated calomel as the counter electrodes and reference electrodes, respectively [148]. This was also demonstrated by Kim et al., who produced a continuous 3 nm thick coating of RuO_2 via the potential cycling method while varying the potential from 200 to 1000 mV with a scan rate of 50 mV/s [149]. They observed that the gas atmosphere during the postannealing process had a significant effect on the morphology of RuOx [150].

One advantage of this technique is that the electrodeposition occurs to the same extent on both the sidewalls of the tubes and the tips [151]. Consequently, the presence of carboxyl or hydroxyl groups as activators is not required. However, the major drawback of electrodeposition is that it is difficult to produce bulk quantities.

3.4.2 Sol-gel process

The sol-gel process is a versatile, solution-based process for producing various ceramic and glass materials in the form of nanoparticles, thin-film coatings, fibers, or aerogels and involves the transition of a liquid, colloidal "sol" into a solid "gel" phase [152]. The sol-gel process is a cheap and low-temperature technique that allows fine control of chemical composition and the introduction of lowest concentrations of finely dispersed dopants. One of the major drawbacks is that the product typically consists of an amorphous phase rather than defined crystals and, thus, requires crystallization and postannealing steps.

In general, the sol-gel process has emerged as the most common technique to synthesize CNT/metal oxide hybrids. Early attempts concentrated on the dispersion of CNTs within a matrix of metal oxide nanoparticles. Vincent et al. synthesized TiO_2 nanoparticles using metal organic precursors and acetic acid as a gelator [153]. They observed that the dispersion of pristine CNTs was more stable when the TiO_2 nanoparticles had been produced in the presence of the CNTs (in situ) compared with the simple mixing of the two materials. Upon reducing the amount of TiO_2 with respect to CNTs, Jitianu et al. obtained a thin but rather irregular and partial coating of TiO_2 on CNTs [154,155]. Typically, the thickness of the coating can be controlled by various parameters, such as the reaction time [156], the reaction composition, and the choice of metal precursor [157]. For instance, in the case of TiO_2, the use of titanium tetraisopropoxide (TTIP) produced irregular coatings [154,158], while tetraethoxy orthotitanate (TEOT) [154] or tetrabutoxy orthotitanate (TBOT) [157,159] enabled a more uniform deposition. The sol-gel process was sometimes carried out under reflux [156], or with the aid of ultrasonication [160], microwave [161], or magnetic agitation [162], in order to enable faster and simultaneous nucleation resulting in a more homogeneous coating.

3.4.2.1 Covalent

These early works used pristine CNTs, whose hydrophobic nature provides little attractive interaction with the metal oxide and thus limits the quality of the coating. Similar to the ex situ approach, the most common approach to change the surface chemistry of CNTs is to treat them in strong oxidizing acids (H_2SO_4/HNO_3). This process introduces a variety of organic groups, with limited control over their number, type, and location, and causes surface etching and shortening of the tubes. Consequently, the metal oxide coatings on acid-treated CNTs were often nonuniform, although they provided better interaction in comparison with pristine CNTs. Despite these drawbacks, most researchers have used such acid-treated CNTs for various metal oxide coatings, including SnO_2 [160,161,163-165], TiO_2 [159,166-168], RuO_2 [169,170], CeO_2 [162], NiO [171], and mixed oxides [172].

3.4.2.2 Noncovalent

In a similar way to the ex situ approach, noncovalent attractions and π-π interactions can be used to grow the metal oxide on the surface of CNTs. Cao et al. modified CNTs with surfactants such as sodium dodecylsulfate (SDS) [173]. The hydrophobic aliphatic chain interacted with the surface of the CNTs, while the hydrophilic end attracted the metal ions of the $RuCl_3$ precursor, which then reacted to form RuO_2. Recently, this author has developed a nondestructive, simple process to coat pristine CNTs with TiO_2 by using benzyl alcohol as a surfactant [174]. Benzyl alcohol adsorbs onto the CNTs' surface via π-π interactions with the alcohol's benzene ring, while simultaneously providing hydrophilic hydroxyl groups for the hydrolysis of the titanium precursor (TBOT) (Fig. 7). In contrast to the sample without benzyl alcohol (Fig. 7A), the addition of small amounts of benzyl alcohol resulted in a very uniform coating that covered the whole CNT surface. The work further showed that benzyl alcohol strongly affected the phase transition from anatase to rutile, providing very small and uniform rutile nanocrystals with very high specific surface areas (60-100 m^2/g) without too great a hindrance of the anatase to rutile transformation.

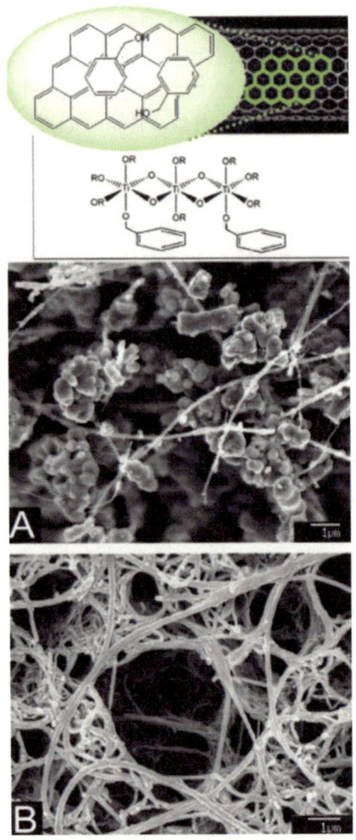

Fig. 7. (Top) Scheme of the beneficial role of benzyl alcohol in the in situ coating of pristine CNTs with TiO_2. The benzene rings of the alcohol adsorb onto the CNT surface via π-π interactions and at the same time provide a high density of hydroxyl groups for the hydrolysis of the titanium precursor directly on the CNT surface. (Bottom) SEM images of pristine MWCNTs (A) without the use of benzyl alcohol and (B) with a titanium-to-benzyl alcohol molar ratio of 5. (Courtesy of D. Eder et al. [174])

3.4.2.3 Electrostatic

In contrast to the ex situ approach, the use of electrostatic interactions for the in situ sol-gel route has been demonstrated for a few metal oxides. As an example, Hernadi et al. used CNTs that had been pretreated with SDS, dried, and redispersed in 2-propanol [175]. Using metal halides as precursors, the authors could successfully produce coatings of Al_2O_3 and TiO_2. Vietmeyer et al. used acid-treated CNTs to attract Zn^{2+} ions (using zinc acetate), which then reacted with LiOH in an ice-water bath to form ZnO nanoparticles [176]. While some metal oxide nanoparticles were attached to the CNT surface and tips, most of the nanoparticles formed large clusters rather than a continuous coating. For comparison, Jiang et al. enhanced the negative charge on the surface of acid-treated CNTs by simply depositing a thin layer of SDS and carried out the same reaction with zinc acetate and LiOH [177]. This surface modification significantly enhanced the attractive interactions with the metal ions and led to a more uniform coating with ZnO crystals.

3.4.3 Hydrothermal and aerosol techniques

In recent years, many organic-inorganic hybrids have been produced by hydrothermal techniques [178]. In contrast to standard sol-gel routes, the hydrothermal method typically enables the formation of crystalline particles or films without the need for postannealing and calcinations. Furthermore, the forced crystallization enables the formation of metal oxide nanowires and nanorods [179].

3.4.3.1 Vapor-assisted, polyol-assisted process

In this simplest case of hydrothermal synthesis, pristine or acid-treated CNTs were added to an aqueous solution of the precursor and treated in an autoclave at temperatures between 100 and 240 °C to produce crystalline films of ZnO [180], TiO_2 [181], or Fe_2O_3 [182]. These works consistently produced dense coatings of spherical or slightly elongated nanoparticles. Zhang et al. used aligned CNTs, which they precoated with a thin, amorphous layer of ZnO via magnetron sputtering [183]. Then they dissolved a fine ZnO powder in NaOH at a pH of 10-12, which provided a saturated solution of $Zn(OH)_4^{2-}$. The modified CNT carpet was then placed top-down in an autoclave, floating on the precursor solution. By keeping the reaction at a temperature of 100 °C for several hours, the Zn precursor nucleated on the CNT-ZnO film to grow ZnO nanowires perpendicular to the CNTs, with a thickness of 30-70 nm and lengths of up to 0.5 μm (Fig. 8).

In general, specific molecules provide efficient control of the crystal size of the nanoparticles, which hinder the crystal growth by steric configuration (capping agents). For example, Yu et al. dissolved copper acetate in water and diethylene glycol and added acid-treated CNTs [184,185]. Upon heating in an autoclave at 180 °C, the diethylene glycol-capped copper species condensed to CuOx and nucleated to form small Cu_2O crystals. After a reaction time of 2 h, the crystals were 5-10 nm in diameter and covered by an amorphous layer. Similarly, small crystals of Fe_3O_4 were produced by Jia et al., who used polyethylene glycol (PEG) and $FeCl_3$ and a reaction temperature of 200 °C [186]. However, the 5 nm crystals agglomerated to about 180 nm aggregates (nanobeads), which attached preferably to the carboxyl groups on the surface of acid-treated CNTs. Consequently, the magnetite nanobeads were concentrated at the tips of the CNTs and acted as a glue to form heterojunctions between the CNTs (Fig. 9).

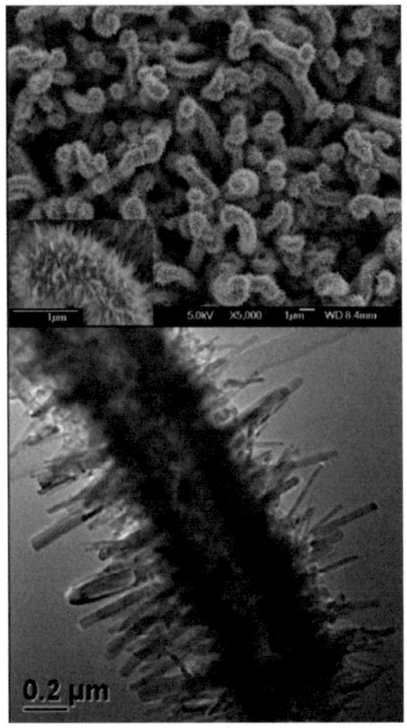

Fig. 8. Example of the hydrothermal synthesis: growth of ZnO nanowires perpendicular to MWCNTs. (Courtesy of W. D. Zhang et al. [183])

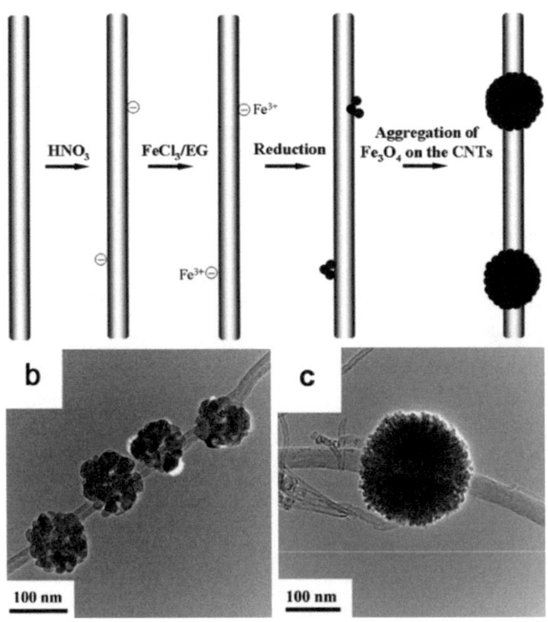

Fig. 9. Example of a polyol-assisted hydrothermal deposition of Fe_3O_4 on acid-treated MWCNTs. Polyethylene glycol (PEG) was used to reduce $FeCl_3$ and to control the size of the magnetite nanoparticles, which formed large aggregates (nanobeads) near the carboxyl groups on the CNT surface. (Courtesy of B. P. Jia et al. [186])

3.2.4.2 Supercritical solvent

A very important hydrothermal method involves the use of supercritical CO_2 as an antisolvent that reduces the solvent strength of ethanol, resulting in the precipitation of the oxide due to high saturation. Using metal nitrates or halides, this method has been applied to deposit Eu_2O_3 [187], CeO_2 [188], La_2O_3 [188], Al_2O_3 [188], SnO_2 [189,190], and Fe_2O_3 [191] onto pristine CNTs. Sun et al. used supercritical ethylenediamine as a solvent to produce thin coatings of RuO_2 [192]. They also observed various morphologies and structures of cerium oxide by simply changing the reaction temperature [188]. For instance, the authors could alter the composition of cerium oxide from preferentially Ce_2O_3 at 120 °C to CeO_2 at 150 °C. In contrast, a reaction temperature of 120 °C was needed to obtain a coating of SnO_2, while at 35 °C the oxide was only encapsulated [190].

3.4.4 Gas-phase deposition

Chemical and physical vapor deposition techniques are among the most common methods to produce metal oxide nanomaterials, as they provide excellent control over the size, shape, and uniformity of the oxide material. Furthermore, it is possible to deposit thin, continuous films on carbon substrates, without altering the 3D integrity of vertically aligned CNTs. This section provides examples of the synthesis of various CNT/metal oxide hybrids using physical techniques, such as evaporation, sputtering, and pulsed laser deposition (PLD), and chemical methods, including chemical vapor deposition (CVD) and atomic layer deposition (ALD).

3.4.4.1 Evaporation and sputtering

Physical vapor deposition involves the evaporation of material in a crucible under high vacuum, using either resistive heating (thermal evaporation) or electron bombardment (electron beam deposition). In contrast, sputtering (magnetron and radio frequency, RF) relies on plasma (typically argon) to bombard the target material, which can be kept at a relatively low temperature. Reactive sputtering involves a small amount of oxygen, which reacts with the sputtered material to deposit oxides. The deposition of metal oxides via thermal evaporation has been demonstrated by Kim et al., who mixed annealed CNTs with Zn powder in a ratio of 1:12 [193]. Depending on the reaction temperature, the Zn particles reacted with oxygen impurities in argon to form a coating on the CNTs consisting either of spherical particles (450 °C), nanowires (800 °C), or short nanorods (900 °C). RF or magnetron sputtering has been used to deposit RuOx, [194] and ZnO [195]. Furthermore, Jin et al. have cosputtered Ba and Sr in an oxygen atmosphere to obtain a BaO/SrO coating [196]. In most of these works, the coating around the CNTs was generally conformal, although in the case of vertically aligned CNTs ("carpet"), the metal oxide material was deposited predominantly along the top of the carpet. Interestingly, Fang et al. observed that the distribution of RuO_2 particles on vertically aligned CNTs was significantly better on nitrogen-doped CNTs compared with pure CNTs [197]. It is well-known that the incorporation of nitrogen into the sp^2-type walls of CNTs causes many structural defects (pyridine-like bonding) due to the different valence of nitrogen ions. It seems that these defects are more reactive toward the adsorption of RuO_2 than the graphitic walls of pure CNTs and allow a uniform dispersion along the sidewalls of nitrogen-doped CNTs (Fig. 10).

In summary, sputtered films typically have a better adhesion to the substrate than evaporated films and a composition closer to that of the source material. Sputtering also enables the deposition of materials with very high melting points and can be performed "top-down", while evaporation must be operated "bottom-up". On the other hand, evaporation techniques typically offer better structural and morphological control and more flexible deposition rates.

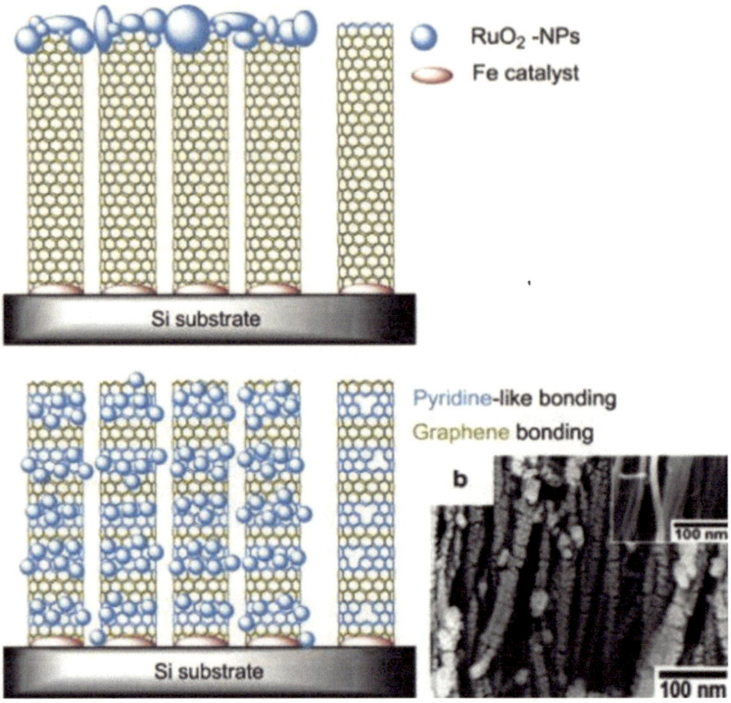

Fig. 10. Scheme of the deposition of RuO_2 nanoparticles on aligned CNTs (top) and nitrogen-doped CNTs (N-CNTs, bottom) via magnetron sputtering. The pyridine-like defects in the N-CNTs provide a stronger interaction with RuO_2 and a more uniform dispersion along the sidewalls. (Courtesy of W. C. Fang et al. [197])

3.4.4.2 Pulsed laser deposition

Pulsed laser deposition (PLD) is related to the evaporation techniques but utilizes a high-power pulsed laser beam. The directed laser pulse is absorbed by the target, vaporizes the material, and creates a plasma plume containing various energetic species, such as atoms, molecules, electrons, ions, clusters, particulates, and molten globules, which then expand into the vacuum and deposit on a typically hot substrate to nucleate and grow as a thin film. This process can occur in ultrahigh vacuum or in the presence of a background gas, such as oxygen, which is commonly used when depositing oxides. Ikuno et al. used individual MWCNTs (grown by arc discharge), which were attached to a molybdenum plate via electrophoresis. A pulsed Nd:YAG (yttrium aluminum garnet) laser with a wavelength of 355 nm was focused onto a target at a repetition rate of 10 Hz. A single-crystalline Si wafer and pellets of Zr, Hf, Al, ZnO, Au, and Ti were used as targets for the coating of SiOx, ZrOx, HfOx, AlOx, and ZnOx [198,199]. The advantages of this technique over other thin-film deposition methods include the relatively simple basic setup and the operation at room temperature. On the other hand, PLD has a lower average deposition rate than other deposition techniques, such as CVD or evaporation/sputtering techniques, but is faster than ALD.

3.4.4.3 Chemical vapor deposition

Chemical vapor deposition (CVD) is a versatile technique often used in the semiconductor industry that involves the growth of a solid material from the gas phase via chemical reaction at the surface of a substrate. In contrast to high-pressure/high-temperature synthesis, the CVD technique typically operates at medium temperatures (600-800 °C) and at slightly reduced atmospheric pressures. Because CVD utilizes reagents of very high purity, the technique is capable of synthesizing crystals with controlled purity and composition. Other advantages of CVD include the high deposition rate, good degree of control (purity and composition), and easy scalability. However, due to the fast deposition, it is difficult to achieve uniform and defect-free coatings when scaling down to a few nanometers. Several groups have used CVD to synthesize CNT-metal oxide hybrids with SnO_2 [200] and RuO_2 [201]. Kuang et al. deposited acid-treated MWCNTs on a Si waver and heated them to 550 °C in a SnH_4-containing N_2 atmosphere [200]. At this temperature, the precursor decomposed, attached to the functional groups of the CNTs, and reacted with the oxygen impurities to produce SnO_2. In a similar way, CNTs/RuO_2 [150], CNTs/Co_3O_4 [202] and CNTs/ZnO [203-205] have been synthesized by directly coating of prepared CNTs. On the other hand, An example of the in situ growth of CNTs is the work of Liu et al., who used a water-assisted CVD process to produce CNTs with a ZnO coating [206]. Most of the ZnO coating was observed at the tip of the CNTs. An interesting alternative to ZnO was demonstrated by Ho et al. [207]. The authors used Ni catalyst particles, supported on MgO, which were deposited on a Si wafer and placed inside a tube furnace. ZnO powder was put inside a thin ceramic tube, whose ends were covered with Al foil, and placed inside the reactor tube, close to the catalyst powder. Upon reaching a certain temperature (300-350 °C), assisted via a plasma-enhanced CVD technique, CNTs began to grow on the Ni particles and kept growing as the temperature was steadily increased to 700 °C over a period of about 6-7min. At this temperature, the Al foil melted and exposed the ZnO powder, which was instantaneously transformed to Zn vapor via carbothermal reduction and deposited on the still-growing CNTs as a thin coating of ZnO.

3.4.4.4 Atomic layer deposition

In contrast to CVD, the precursors for atomic layer deposition (ALD) are kept separate and exposed sequentially. Ideally, each precursor forms a monolayer on the substrate, and the excess vapor is removed before the next precursor is introduced. This process is then repeated until the deposited film reaches the desired thickness. Hence, ALD film growth is self-limiting and based on surface reactions, which enables deposition control on the atomic scale. As an example, Gomathi et al. used metal chloride precursors to coat acid-treated MWCNTs with TiO_2 and Al_2O_3 [208]. Similarly, Javey et al. demonstrated coatings of ZrO_2 as thin as 8 nm covering the top of horizontally attached SWCNTs for device applications [209]. ALD has unique advantages over other thin-film deposition techniques, as it can be operated at low temperatures and allows exact control over the thickness of the deposited coating. However, because of the sequential exposure of the precursors, the technique has the lowest deposition rate compared with CVD and PLD.

4. Properties and potential applications of CNTs/metal oxide hybrids

The improved performances of CNT/metal oxide hybrids are demonstrated in many applications, such as photocatalysis, electrocatalysis, and environmental catalysis, gas sensors and chemical sensors, supercapacitors and batteries, and field emission devices. In the section, some applications of electrocatalysis, electrochemical sensors, supercapacitors and batteries are highlighted and summarized as follow.

4.1 Electrocatalysis

Electrocatalyst is a new catalyst depending on electric energy. Which accelerate the reactions on the interface between electrode and electrolyte. It has significance in research work and also has important application value on organic wastewater treatment and degradation, electrolytic desulfurization, and electroreduction study. As an electrocatalyst, two basic conditions are necessary: (a) electrical conductivity and transfer free electron well; (b) efficient catalytic activity towards target substrate. Metal, especially noble metal, and some metal oxide or hydroxide are widely used as electrocatalysts [210]. In contrast, some common transition metal oxides or hydroxides, such as TiO_2, SnO_2, $Ni(OH)_2$, have many advantages including low cost, facile preparation, and thus are more interesting for wide study. CNTs endows CNT/metal oxide or hydroxide composites enhancement on electrocatalysis duo to electric conductivity and fast electron transfer.

For example, Qiao et al. reported that the MWCNTs/$Ni(OH)_2$-Ni electrode showed better catalytic effect on the oxidation of CH_4 in 1.0 M NaOH than $Ni(OH)_2$-Ni electrode [211]. ZnO-NWs/MWCNTs nanocomposite had been prepared via a hydrothermal process by Mo et al., and showed remarkable electrocatalytic activity towards H_2O_2 by comparing with bare MWCNTs [212]. Subseguently, Ma et al. discovered that ZnO-MWCNTs/Nafion film showed fast and excellent electrocatalytic activity to not only H_2O_2, also trichloroacetic acid [213]. Other metal oxide or hydroxide, such as FeOOH [214] and cobalt-coordinated polypyrrole [215], have been reported have obvious electrocatalysis towards oxygen reduction reaction. Recently, we synthesized CNTs/nickel glycolate polymer core-shell heterostructures and found that the composite have significant electrocatalysis towards glucose in 0.5 M NaOH, and catalytic activity dependent obviously on the thickeness of the nickel glycolate polymer on MWCNTs (Fig. 11).

The role of CNTs is often attributed to their ability to stabilize highly dispersed catalyst nanoclusters, resulting in higher specific surface areas. So, There is also important significant for fuel cell and biosensor to study further electrocatalysis of CNTs/metal oxide or hydroxide composites.

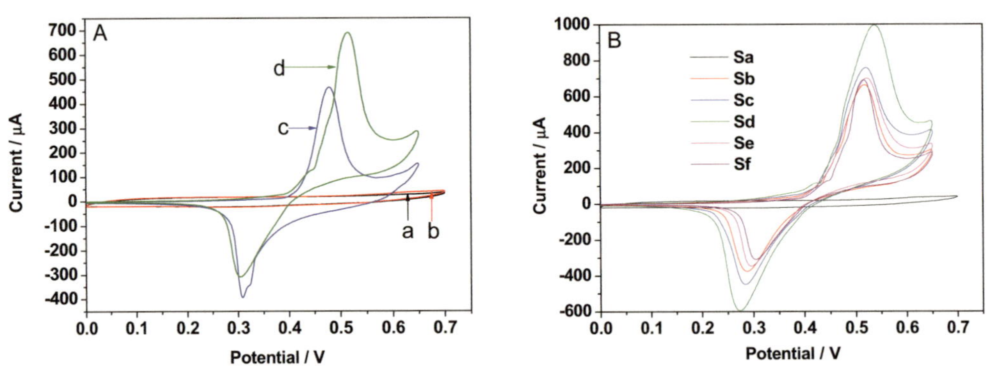

Fig. 11. A) CVs of MWCNTs (a,b) and MWCNT/nickel glycolate polymer core-shell heterostructures in 0.5 M NaOH with absence of (a,c) and presence of (b,d) 1 mM glucose. B) The electrocatalysis activity difference of the nanocomposites with different thickness of glycolate polymer shell towards glucose.

4.2 Electrochemical sensors

Metal oxide semiconductors (MOS) are prominent examples of sensing materials in gas sensors, as their electrical properties are highly affected by the surrounding gas environment. For instance, tungsten trioxide (WO_3) shows sensitivity to pollutants such as SO_2, H_2S, NO, and NH_3 [216,217], while SnO_2 is sensitive to NOx, CO, ethanol, and C_2H_4 [218]. However, their sensing properties often suffer degradation due to growth and aggregation. In contrast with metal oxide sensors, CNTs exhibit excellent adsorption properties due to their high specific surface area, which provides a large number of active surface sites [219-221]. As the CNTs' electric properties are effectively altered by very small amounts of adsorbed gas molecules, the CNT gas sensor can be operated at temperatures close to room temperature. However, as MWCNTs are not very sensitive to ambient gas, the use of CNTs as gas sensors is mainly restricted to SWCNTs. Furthermore, due to their long recovery times, CNT-based sensors typically need reactivation.

CNTs/MOS hybrid composite sensors have been fabricated and investigated widely. These sensors don't noly improve the efficiency of the metal oxide based sensors, such as enhanced sensitivity and selectivity to target gases, reduced response and recovery times, but also are operated at lowered temperature, even room temperature. Similar to core/shell heterostructures like MOS/CdS, CNT/MOS hybrids have shown improved photoluminescent quantum efficiencies and enhanced gas-sensing properties including reduced response and recovery times [163,222]. So far, CNT/SnO_2 has been tested for detection of CO [172], NO_2 [201,203], NH_3 [223], formaldehyde [224], and ethanol [225]. Wei et al. used a sol-gel process to coat pristine SWCNTs with SnO_2 and investigated the gas-sensing performance for NO_2 at room temperature [165]. They observed considerably enhanced sensitivities ($\Delta R/\Delta C$, Fig. 12) compared with the pure SnO_2 sensor. Because the morphology and surface area of the hybrid sensors were similar to those of the pure SnO_2, and the observed sensitivities increased with increasing CNT loading, the authors concluded that the advanced sensing behavior originated from a common interface with CNTs. In contrast to conventional SnO_2 sensors, which typically operate at temperatures between 200 and 500 °C, the SWCNT/SnO_2 hybrid gas sensors could indeed be operated at room temperature. When the NO_2 gas molecules adsorb on the surface of pure SnO_2, they extract electrons, leaving the oxide surface positively charged. This leads to the formation of a depletion zone and to an increase in the sensor resistance. In the CNT/SnO_2 hybrid sensor, the electric properties of the oxide are strongly enhanced by the highly conducting CNTs. Consequently, the sensor resistance is dominated by the Schottky barrier at the interface between the n-type SnO_2 grains and the p-type CNTs, causing the formation of additional depletion layers, which then amplifies the increase in resistance upon NO_2 adsorption and enables the operation of the gas sensor at room temperature.

As CNTs/metal oxide hybrid sensors emerges, some of which were investigated to applied in liquid sensing, especially biosensors including glucose sensor, hydrogen peroxide, hydrazine sensor and nitrite detection etc. The sensors have significant potential application in biology and medicine because of unavoidable drawbacks of enzyme sensor, such as temperature, PH, and unicity. More and more CNTs/metal oxide hybrids were reported to be used as various sensors in liquid.

For instance, Jiang et al. reported a highly sensitive nonenzymatic glucose sensor based on CuO nanoparticles-modified carbon nanotube electrode by sputtering deposition method [226]. The CuO/MWCNTs electrode exhibits an enhanced electrocatalytic property, low-working potential, high sensitivity, excellent selectivity, good stability, and fast amperometric sensing towards oxidation of glucose, thus is promising for the future development of nonenzymatic glucose sensors. Glucose was also detected with CNT/Fe_3O_4 hybrids electrochemically doped glucose oxidase reported by Liu et al. [227] The PA-Fe_3O_4-CNT-based glucose sensor shows much higher sensitivity and linearity than the PA-Fe_3O_4-based sensor, indicating that CNTs significantly enhance the performances of the biosensor. A novel hydrazine electrochemical sensor based on a carbon nanotube-wired ZnO nanoflower-modified electrode was reported via a simple process (ammonia-evaporation-induced synthetic method) by Fang et al. [228]. The authors observed that an optimized limit of detection of 0.18 μM at a signal-to-noise ratio of 3 and with a fast response time (within 3 s).

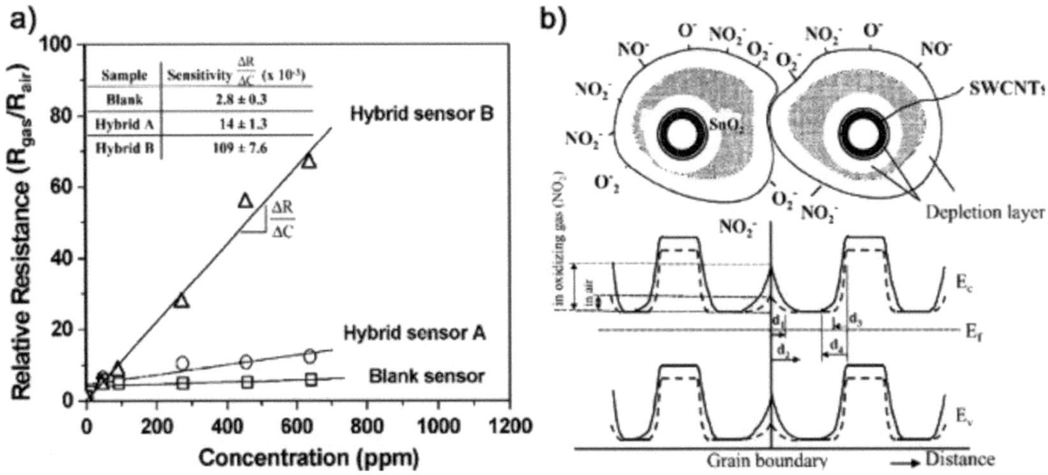

Fig. 12. Example of CNT-SnO_2 hybrid in gas sensors. (a) Relative resistances versus NO_2/air gas concentrations of SnO_2 and hybrids with low (A) and high (B) concentrations of CNTs at room temperature. The sensitivities (ΔR/ΔC) increase considerably with increasing CNT concentration. (b) Scheme of the presence of depletion zones near the CNT oxide interface. (Courtesy of B.Y. Wei et al. [165])

4.3 Supercapacitors and batteries

Reliable and affordable electricity storage is a prerequisite for optimizing the integration of renewable energy systems. Energy storage, therefore, has a pivotal role to play in the effort to combine a future, sustainable energy supply with the standards of technical services and products. For both stationary and transport applications, energy storage is of growing importance as it enables the smoothing of transient and/or intermittent loads and the downsizing of base-load capacity with substantial potential for energy and cost savings. The extended lifetime of batteries in handheld devices is credited not only to higher energy densities but also to a simultaneous reduction of energy consumption of the portable devices. In contrast, the electric vehicle has failed to become the accepted mode of transportation mainly because of the battery. Short distances between recharging and a limited service life of the battery are to blame, but also the incredible weight and volume of the batteries. Electrochemical capacitors (ECs) are energy-storage devices that possess higher energy and power density than conventional dielectric capacitors and batteries and are used in applications including electric vehicles, noninterruptible power supplies, dc power systems, lightweight electronic fuses, memory backups, and solar batteries [229]. The challenges for these applications concern limitations in volumetric/gravimetric power densities and RC time, long life, safety, simplicity of design, cost, and the possibility of recharging, and can only be accomplished by specially designed materials.

According to the mechanism of energy storage, ECs can be categorized into two classes [229]: (a) electrochemical double layer capacitors (EDLC), based on double-layer capacitance due to charge separation at the electrode/electrolyte interface, which thereby need materials with high specific surface area (e.g., activated carbon, CNTs), and (b) pseudocapacitors or supercapacitors, based on the pseudocapacitance of faradaic processes in active electrode materials such as transition metal oxides and conducting polymers.

Because of their exceptional electronic properties, which allow ballistic transport of electrons over long nanotube lengths, CNTs have been considered a most promising candidate for electrochemical capacitors [230,231]. However, pure CNTs possess a rather low specific capacitance, typically about 10-40 F/g, which depends on the microtexture, purity, and electrolyte [231]. A considerable enhancement can be expected from the combination of CNTs with an electroactive material, which provides the additional pseudocapacitance while each tube acts as a minute electrode. the electrochemical stability of transition metal oxides makes them a better choice, provided they are highly conducting. A wide range of oxides has been investigated for use in CNT hybrids, including NiO [171,232,233], MnO_2 [234-237], V_2O_5 [238,239], and RuO_2 [169,192,194]. In all of these studies, the synergistic effects of CNT oxide hybrids have shown a considerable improvement of the otherwise poor electric properties and deficient charge transfer channels of the pure oxide.

Among these transition metal oxides, ruthenium oxides are regarded as the most promising electrode material for supercapacitor applications due to their high specific capacitance, highly reversible redox reactions, wide potential window, and very long cycle life [192]. Park et al. reported a specific capacitance for CNT/RuO_2 (13 wt %) of 450 F/g at a potential scan rate of 20 mV s^{-1}, which could be further enhanced, by using hydrous ruthenium oxide and functionalized CNTs, to 800 F/g [170]. Reducing the thickness of the RuO_2 layer to 3 nm led to an even higher capacitance (1170F/g at a potential scan rate of 10 mV s^{-1}) [149]. All of these studies also demonstrated longer lifetimes and greater stabilities and rate capabilities, while the specific capacitances were in general considerably higher than those observed for pure MWCNTs (<80 F/g) or pure RuO_2 (<400 F/g).

In view of the rather high costs and environmental issues concerning RuO_2, and despite the lower specific capacitances, there has been a clear trend toward the use of hydrous manganese oxides in the past few years. Xie et al. produced a CNT/MnO_2 hybrid by in situ reduction of $KMnO_4$ on MWCNTs with citric acid and observed that the specific capacitance increased from 18 to 190 F/g [140]. Zhang et al. electrodeposited MnO_2 "nanoflowers" onto the surface of vertical CNT arrays and obtained capacitances of about 200 F/g, as well as excellent rate capabilities (50.8%, capacity retention at 77 A/g) and a long cycle life (3% capacity loss after 20 000 charge/discharge cycles) [236]. The authors proposed a model for the improved electrochemical storage characteristics based on high electronic conductivity in CNTs. In this model, the geometry of the MnO_2 nanoflowers and their hierarchical localization at CNT junctions limits the diffusion length of ions within the MnO_2 phase during the charge/discharge process and enhances the ionic conductivity of the hybrid.

To maximize the electrochemical utilization of the supercapacitor, it is desirable to (a) use a low loading of metal oxide and (b) increase the interfacial area between CNTs and metal oxide. Consequently, the preparation of a thin, uniform, connected coating on CNTs is expected to improve the specific capacitance significantly. This was confirmed by Nam et al.,

who observed increasing capacitances with decreasing NiO content and thus coating thickness [233].
A promising approach to enhance the capacitive performance has recently been reported by Sivakkumar et al., who synthesized a ternary hybrid of CNT/polypyrrole/hydrous MnO_2, which showed a significantly higher specific capacitance (280 F/g) compared to the binary CNT/MnO_2 (150F/g) [141]. The increase in capacitance is believed to be due to the better dispersion of hydrous MnO_2 in the composite electrode. However, the presence of the polymer also accounts for the poor cyclability behavior. Fang et al. chose a different approach by depositing RuO_2 on N-doped CNT [240], for which they measured considerably higher specific capacitances. The authors showed that the RuO_2 particles were considerably better dispersed on N-doped CNTs than on pure CNTs, due to the presence of structural defects in N-CNTs. Consequently, the hybrids with N-doped CNTs had an enhanced interface, which resulted in their better performance as a supercapacitor (Fig. 13).

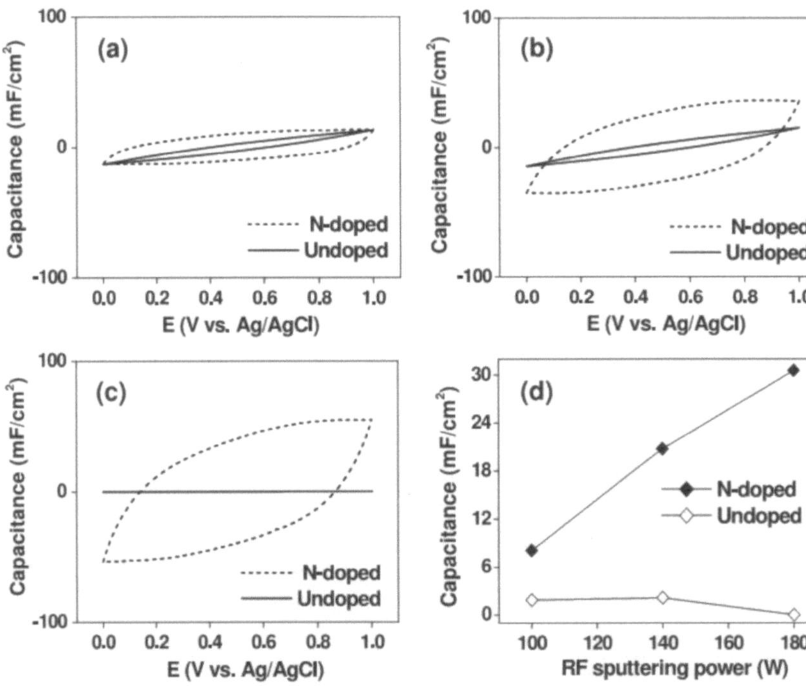

Fig. 13. Capacitance voltage diagrams of hybrids with RuO_2 using undoped and nitrogen-doped CNTs; (a) light (RF power 100 W), (b) moderate (RF power 140 W), and (c) heavy (RF power 180 W) loading of RuO_2. (d) Capacitance of hybrids as a function of RF sputtering power. In contrast to the undoped CNTs, the N-doped CNTs are coated completely with RuO_2, resulting in a larger interfacial area and an improved capacitance. (Courtesy of A. L. M. Reddy et al. [240])

A good cyclability is crucial to a successful application as an anode material for lithium-ion batteries, as demonstrated by Chen et al. for CNT/SnOx hybrids [164]. Pure CNTs typically show good cycle stability, but also a large and irreversible capacity loss in the first cycle. The authors showed that well-dispersed SnOx on CNTs enhanced the electrochemical performance significantly as the capacity fade of the hybrid electrode was strongly reduced compared to unsupported SnOx or pure CNT. The authors attributed the improvement in performance to the enhanced conductivity and ductility of the hybrid. This is in line with Zheng et al., who observed a 5-fold increase in electronic conductivity and thus enhanced cycling rate and stability in CNT/CuO hybrids compared to pure CuO.

In summary, electrochemical devices have been among the first applications for which CNT/metal oxide hybrids were tested. In the long run, CNT hybrids are expected to strongly enhance the batteries' intrinsic capacities, stabilities, and lifetimes, as well as the charging/discharging characteristics. Furthermore, the substitution of part of the heavy oxide material (e.g., in car batteries) with the lightweight CNTs will significantly reduce the weight of the batteries.

5. Conclusion and future directions

As reviewed in the former publications, numerous synthetic methodologies have been developed for the preparation of CNTs and composites. In this chapter, besides the preparation of nanocomposites based on CNTs, we also focus on investigating the electrochemical applications CNT/metal oxide or hydroxide composites. There is a wealth of opportunities for nanoscale electrochemical devices based on carbon materials. Furthermore, when the electrode dimension approaches the double layer thickness, interesting and unusual kinetic and electrostatic effects occur. It is likely that carbon nanotubes electrodes will be key players in the investigation of such phenomena.

6. Acknowledgments

Financial support from Zhejiang Provincial Natural Science Foundation of China (Y4110304), Zhejiang Qianjiang Talent Project (2010R10025), the Educational Commission of Zhejiang Province of China (Z200909406) are gratefully acknowledged.

7. References

[1] J. Li, S. B. Tang, L. Lu, H. C. Zeng. J. Am. Chem. Soc. 2007, 129, 9401
[2] S. Iijima. Nature. 1991, 354, 56
[3] S. Iijima. Nature. 1993, 363, 603
[4] P. J. F. Harris. Carbon Nanotubes and Related Structures; Cambridge University Press: Cambridge, U.K., 2003.
[5] A. C. Dillon, M. Yudasaka, M. S. Dresselhaus. J. Nanosci. Nanotechnol. 2004, 407, 691
[6] P. Avouris. MRS Bull. 2004, 29, 403
[7] A. Jorio, G. Dresselhaus, M. S. Dresselhaus. Carbon Nanotubes: ad anced topics in sythesis, structure and properties; Springer: New York, 2007
[8] A. C. Dillon, K. M. Jones, T. A. Bekkedahl, C. H. Kiang, D. S. Bethune, M. J. Heben. Nature. 1997, 386, 377
[9] A. C. Dillon, A. H. Mahan, J. L. Alleman, M. J. Heben, P. A. Parilla, K. M. Jones. Thin Solid Films. 2003, 430, 292
[10] J. H. Lehman, C. Engtrakul, T. Gennett, A. C. Dillon. Appl. Opt. 2005, 44, 483
[11] J. H. Lehman, R. Deshpande, P. Rice, A. C. Dillon. IR Phys. 2006, 47, 246
[12] A. C. Dillon, M. J. Heben. Appl. Phys. A: Mater. Sci. Process. 2001, 72, 133

[13] A. C. Dillon, A. H. Mahan, R. Deshpande, J. L. Alleman, J. L. Blackburn, P. A. Parillia, M. J. Heben, C. Engtrakul, K. E. H. Gilbert, K. M. Jones, R. To, S. H. Lee, J. H. Lehman. Thin Solid Films. 2006, 501, 216
[14] D. S. Bethune, C. H. Kiang, M. S. d. Vries, G. Gorman, R. Savoy, J. Vasquez, R. Beyers. Nature. 1993, 363, 605
[15] C. Journet, W. K. Maser, P. Bernier, A. Loiseau, M. Lamy de la Chapells, S. Lefrant, P. Deniard, R. Lee, J. E. Fischer. Nature. 1997, 388, 756
[16] S. Bandow. Private Communication.
[17] T. Guo, P. Nikolaev, A. Thess, D. T. Colbert, R. E. Smalley. Chem. Phys. Lett. 1995, 243, 49
[18] A. Thess, R. Lee, P. Nikolaev, H. Dai, P. Pitit, J. Robert, C. Xu, Y. H. Lee, S. G. Kim, A. G. Rinzler, D. T. Colbert, G. E. Scuseria, D. Tomanek, J. E. Fischer, R. E. Smalley. Science. 1996, 273, 483
[19] M. J. Heben, A. C. Dillon, K. E. H. Gilbert, P. A. Parilla, T. Gennett, J. L. Alleman, G. L. Hornyak, K. M. Jones. Hydrogen Mater. Vac. Syst. 2002, 671, 77
[20] A. C. Dillon, P. A. Parilla, J. L. Alleman, T. Gennett, K. M. Jones, M. J. Heben. Chem. Phys. Lett. 2005, 401, 522
[21] H. Kataura, Y. Maniwa, M. Abe, A. Fujiwara, T. Kodama, K. Kikuchi, H. Imahori, Y. Misaki, S. Suzuki, Y. Achiba. Appl. Phys. A: Mater. Sci. Process. 2002, 74, 349
[22] A. C. Dillon, T. Gennett, K. M. Jones, J. L. Alleman, P. A. Parilla, M. J. Heben. Adv. Mater. 1999, 11, 1354
[23] H. Dai, A. G. Rinzler, P. Nikolaev, A. Thess, D. T. Colbert, R. E. Smalley. Chem. Phys. Lett. 1996, 260, 471
[24] A. Peigney, C. Laurent, F. Dobigeon, A. J. Rousset. Mater. Res. 1997, 12, 613
[25] J. H. Hafner, M. J. Bronikowski, B. R. Azamian, P. Nikolev, A. G. Rinzler, D. T. Colbert, K. A. Smith, R. E. Smalley. Chem. Phys. Lett. 1998, 296, 195
[26] H. M. Cheng, F. Li, G. Su, H. Y. Pan, L. L. He, X. Sun, M. S. Dresselhaus. Appl. Phys. Lett. 1998, 72, 3282
[27] A. M. Cassell, J. A. Raymakers, J. Kong, H. J. Dai. Phys. Chem. B. 1999, 103, 6484
[28] J. -F. Colomer, C. Stephan, S. Lefrant, G. Van Tendeloo, I. Willems, Z. Konya, A. Fonseca, C. Laurent, J. B. Nagy. Chem. Phys. Lett. 2000, 317, 83
[29] M. Su, B. Zheng, J. Liu. Chem. Phys. Lett. 2000, 322, 321
[30] M. J. Bronikowski, P. A. Willis, D. T. Colbert, K. A. Smith, R. E. Smalley. J. Vac. Sci. Technol. A 2001, 19, 1800
[31] Q. Li, Y. Hao, C. Yan, Z. Jin, Z. Liu. J. Mater. Chem. 2002, 12, 1179
[32] G. L. Hornyak, L. Grigorian, A. C. Dillon, P. A. Parilla, K. M. Jones, M. J. Heben. J. Phys. Chem. B. 2002, 106, 2821
[33] L. M. Wagg, G. L. Hornyak, L. Grigorian, A. C. Dillon, K. M. Jones, J. Blackburn, P. A. Parilla, M. J. Heben. J. Phys. Chem. B. 2005, 109, 10435
[34] J. H. Hafner, C.-L. Cheung, T. H. Oosterkamp, C. M. Lieber. J. Phys. Chem. B. 2001, 105, 743
[35] A. H. Mahan, J. L. Alleman, M. J. Heben, P. A. Parilla, K. M. Jones, A. C. Dillon. Appl. Phys. Lett. 2002, 81, 4061
[36] I. W. Chiang, B. E. Brinson, A. Y. Huang, P. A. Willis, M. J. Bronikowski, J. L. Margrave, R. E. Smalley, R. H. Hauge. J. Phys. Chem. B. 2001, 105, 8297
[37] B. L. Liu, W. C. Ren, L. B. Gao, S. S. Li, S. F. Pei, C. Liu, C. B. Jiang, H. M. Cheng. J. Am. Chem. Soc. 2009, 131, 2082
[38] T. W. Ebbesen, P. M. Ajayan. Nature. 1992, 358, 220

[39] A. C. Dillon, A. H. Mahan, P. A. Parilla, J. L. Alleman, M. J. Heben, K. M. Jones, K. E. H. Gilbert. Nano Lett. 2003, 3, 1425
[40] W. D. Wang, P. Serp, P. Kalck, C. G. Silva, J. L. Faria. Mater. Res. Bull. 2008, 43, 958
[41] B. P. Ribaya, J. Leung, P. Brown, M. Rahman, C. V. Nguyen. Nanotechnology. 2008, 19
[42] L. Bokobza. Polymer 2007, 48, 4907
[43] R. P. Raffaelle, B. J. Landi, J. D. Harris, S. G. Bailey, A. F. Hepp. Materials Science and Engineering B-Solid State Materials for Advanced Technology. 2005, 233
[44] Q. Ngo, D. Petranovic, S. Krishnan, A. M. Cassell, Q. Ye, J. Li, M. Meyyappan, C. Y. Yang. IEEE Trans. Nanotechnol. 2004, 3, 311
[45] K. P. Gong, M. N. Zhang, Y. M. Yan, L. Su, L. Q. Mao, Z. X. Xiong, Y. Chen. Anal. Chem. 2004, 76, 6500
[46] H. T. Gomes, P. V. Samant, P. Serp, P. Kalck, J. L. Figueiredo, J. L. Faria. Appl. Catal., B. 2004, 54, 175
[47] A. Lueking, R. T. Yang. J. Catal. 2002, 206, 165
[48] L. C. Qin. J. Mater. Sci. Lett. 1997, 16, 457
[49] Y.-Y. Fan, F. Li, H.-M. Cheng, G. Su, Y.-D. Yu, Z.-H. Shen. J. Mater. Res. 1998, 13, 2342
[50] R. Marangoni, P. Serp, R. Feurer, Y. Kihn, P. Kalck, C. Vahlas. Carbon. 2001, 39, 443
[51] R. Andrews, D. Jacques, A. M. Rao, F. Derbyshire, D. Qian, X. Fan, E. C. Dickey, J. Chen. Chem. Phys. Lett. 1999, 303, 467
[52] M. Endo, K. Takeuchi, S. Igarashi, K. Kobori, M. Shiraishi, H. W. Kroto. J. Phys. Chem. Sol. 1993, 54, 1841
[53] M. Jose-Yacaman, M. Miki-Yoshida, L. Rendon, J. G. Santiesteban. Appl. Phys. Lett. 1993, 62, 202
[54] V. Ivanov, J. B. Nagy, P. Lambin, A. Lucas, X. B. Zhang, X. F. Zhang, D. Bernaerts, G. Van Tendeloo, S. Amelinckx, J. Van Landuyt. Chem. Phys. Lett. 1994, 223, 329
[55] J. T. L. Thong, C. H. Oon, W. K. Eng, W. D. Zhang, L. M. Gan. Appl. Phys. Lett. 2001, 79, 2811
[56] D. S. Tang, S. S. Xie, Z. W. Pan, L. F. Sun, Z. Q. Liu, X. P. Zou, Y. B. Li, L. J. Ci, W. Liu, B. S. Zou, W. Y. Zhou. Chem. Phys. Lett. 2002, 356, 563
[57] C. J. Lee, D. W. Kim, T. J. Lee, Y. C. Choi, Y. S. Park, W. S. Kim, Y. H. Lee, W. B. Choi, N. S. Lee, J. M. Kim, Y. G. Choi, S. C. Yu. Appl. Phys. Lett. 1999, 75, 1721
[58] S. Xie, W. Li, Z. Pan, B. Chang, L. Sun. Mater. Sci. Eng. A. 2000, 286, 11
[59] Y. J. Li, Z. Sun, S. P. Lau, G. Y. Chen, B. K. Tay. Appl. Phys. Lett. 2001, 79, 1670
[60] L. C. Qin, D. Zhou, A. R. Krauss, D. M. Gruen. Appl. Phys. Lett. 1998, 72, 3437
[61] M. Chen, C.-M. Chen, C.-F. Chen. J. Mater. Sci. 2002, 37, 3561
[62] Y. C. Choi, D. J. Bae, Y. H. Lee, B. S. Lee, G.-S. Park, W. B. Choi, N. S. Lee, J. M. Kim. J. Vac. Sci. Technol. A. 2000, 18, 1864
[63] S. J. Chung, S. H. Lim, C. H. Lee, J. Jang. Diamond Relat. Mater. 2001, 10, 248
[64] G. W. Ho, A. T. S. Wee, J. Lin, W. C. Tiju. Thin Solid Films. 2001, 388, 73
[65] H. Wang, J. Lin, C. H. A. Huan, P. Dong, J. He, S. H. Tang, W. K. Eng, T. L. Thong. J. Appl. Surf. Sci. 2001, 181, 248
[66] Q. Zhang, S. F. Yoon, J. Ahn, B. Gan, Rusli, M.-B. Yu. J. Mater. Res. 2000, 15, 1749
[67] H. J. Yoon, H. S. Kang, J. S. Shin, K. J. Son, C. H. Lee, C. O. Kim, J. P. Hong, S. N. Cha, B. G. Song, J. M. Kim, N. S. Lee. Physica B. 2002, 323, 344
[68] Y. S. Woo, D. Y. Jeon, I. T. Han, N. S. Lee, J. E. Jung, j. M. Kim. Diamond Relat. Mater. 2002, 11, 59
[69] H. Lee, Y.-S. Kang, P. S. Lee, J.-Y. Lee. J. Alloys Compd. 2002, 330, 569
[70] C. Bower, O. Zhou, W. Zhu, D. J. Werder, S. Jin. Appl. Phys. Lett. 2000, 77, 2767
[71] K. H. Park, K. M. Lee, S. Choi, S. Lee, K. H. Koh. J. Vac. Sci. Technol. B. 2001, 19, 946

[72] T. Ono, H. Miyashita, M. Esashi. Nanotechnology. 2002, 13, 62
[73] J.-h. Han, B.-S. Moon, W. S. Yang, J.-B. Yoo, C. Y. Park. Surf. Coat. Technol. 2000, 131, 93
[74] Z. F. Ren, Z. P. Huang, J. W. Xu, J. H. Wang, P. Bush, M. P. Siegal, P. N. Provincio. Science. 1998, 282, 1105
[75] C.-F. Chen, C.-L. Lin, C.-M. Jpn. Wang. J. Appl. Phys. 2002, 41, L67
[76] G. X. Du, Z. P. Zhu, L. D. Zhang, J. L. Song. Mater. Lett. 2010, 64, 1179
[77] C. N. R. Rao, C. A. K. Müller Achim, The Chemistry of Nanomaterials: Synthesis, Properties and Applications, JohnWiley, 2004
[78] G. G.Wildgoose, C. E. Banks, R. G. Compton. Small. 2006, 2, 182
[79] V. Georgakilas, D. Gournis, V. Tzitzios, L. Pasquato, D. M. Guldi, M. Prato. J. Mater. Chem. 2007, 17, 2679
[80] X. G. Hu, S. J. Dong. J. Mater. Chem. 2008, 18, 1279
[81] D. Vairavapandian, P. Vichchulada, M. D. Lay. Anal. Chim. Acta. 2008, 626, 119
[82] X. Peng, J. Chen, J. A. Misewich, S. S.Wong. Chem. Soc. Rev. 2009, 38, 1076
[83] I. W. Chiang, R. E. Smalley, J. L. Margrave, R. H. Hauge. J. Phys. Chem. B. 2001, 106, 1157
[84] T. Tohji, H. Takahashi, Y. Shinoda, N. Shimizu, B. Jeyadevan, I. Matsuoka, Y. Saito, A. Kasuya, T. Ohsuna, K. Hiraga, Y. Nishina, Nature. 1996, 383, 679
[85] E. E. Dujardin, A. Krishnan, M. M. J. Treacy. Adv. Mater. 1998, 10, 611
[86] H.Y. Xu, R. H. Hauge, R. E. Smalley, Nano Lett. 2005, 5, 163
[87] H. T. Fang, C. G. Liu, C. Liu, F. Li, M. Liu, H. M. Cheng. Chem. Mater. 2004, 16, 5744
[88] C. J. Murphy R., M. Cadek, B. McCarthy, M. Bent, A. Drury. J. Phys. Chem. B. 2002, 106, 3087
[89] K. Shen, S. Curran, H. F. Xu, S. Rogelj, Y. B. Jiang, J. Dewald, T. Pietrass. J. Phys. Chem. B. 2005, 109, 4455
[90] A. G. Liu, H. Dai, J. H. Hafner, R. K. Bradley, P. J. Boul, A. Lu, T. Iverson, K. She-limov, C. B. Huffman, F. Rodriguez-Macias, Y. -S. Shon, T. R. Lee, D. T. Colbert, R. E. Smalley. Science. 1998, 280, 1253
[91] M. A. Chen, H. Hu, Y. Chen, A. M. Rao, P. C. Eklund, R. C. Haddon. Science. 1998, 282, 95
[92] W. Fu, Y. Lin, L. A. Riddle, D. L. Carroll, Y. -P. Sun. Nano Lett. 2001, 1, 439
[93] M. E. Kamaras, H. Hu, B. Zhao, R. C. Haddon. Science. 2003, 301, 1501
[94] C. A. Strano, M. L. Usrey, P. W. Barone, M. J. Allen, H. Shan, C. Kittrell, R. H. Hauge, J. M. Tour, R. E. Smalley, T. W. Dujardin, A. Ebbesen, M. Krishnan, M. J. Treacy. Science. 2003, 301, 1519
[95] A. Zheng, E. D. Semke, B. A. Diner, R. S. McLean, S. R. Lustig, R. E. Richardson, N. G. Tassi. Nat. Mater. 2003, 2, 338
[96] Y. Chen, D.Wang, H. Dai. J. Am. Chem. Soc. 2002, 123, 3838
[97] J. -O. Besteman, F. G. M. Wiertz, H. A. Heering, C. Dekker. Nano Lett. 2003, 3, 727
[98] M. S. Moore, E. H. Haroz, R. H. Hauge, R. E. Smalley, J. Schmidt, Y. Talmon. Nano Lett. 2003, 3, 1379
[99] A. Fukushima, Y. Ishimura, T. Yamamoto, T. Takigawa, N. Ishii, T. Aida. Science. 2003, 300, 2072
[100] H. C. J. Wang, Y. Li. ACS Nano. 2008, 2, 2540
[101] J. Y. Wang, Y. Li. J. Am. Chem. Soc. 2009, 131, 5364
[102] P. M. Ajayan, S. Lijima. Nature. 1993, 361, 333
[103] P. M. Ajayan, T. W. Ebbesen, T. Ichihashi, S. Iijima, K. Tanigaki, H. Hiura. Nature. 1993, 362, 522
[104] J. Sloan, J. Hammer, M. Zwiefka-Sibley, M. L. H. Green. Chem. Commun. 1998, 3, 347

[105] C. Z. Loebick, M. Majewska, F. Ren, G. L. Haller, L. D. Pfefferle. J. Phys. Chem. C. 2010, 114, 11092
[106] D. Jain, R. Wilhelm. Carbon. 2007, 45, 602
[107] M. C. Schnitzler, M. M. Oliveira, D. Ugarte, A. J. G. Zarbin. Chem. Phys. Lett. 2003, 381, 541
[108] M. Monthioux, E. Flahaut, J. P. Cleuziou. J. Mater. Res. 2006, 21, 2774
[109] J. Chen, M. A. Hamon, H. Hu, Y. S. Chen, A. M. Rao, P. C. Ecklund, R. C. Haddon. Science. 1998, 282, 95
[110] J. M. Haremza, M. A. Hahn, T. D. Krauss, S. Chen, J. Calcines. Nano Lett. 2002, 2, 1253
[111] G.-X. Wang, B.-L. Zhang, Z.-L. Yu, M.-Z. Qu. Solid State Ionics. 2005, 176, 1169
[112] B. Liu, J. H. Chen, C. H. Xiao, K. Z. Cui, L. Yang, H. L. Pang, Y. F. Kuang. Energy Fuels. 2007, 21, 1365
[113] A. Kongkanand, R. M. Dominguez, P. V. Kamat. Nano Lett. 2007, 7, 676
[114] T. Sainsbury, D. Fitzmaurice. Chem. Mater. 2004, 16, 3780
[115] D. Q. Yang, B. Hennequin, E. Sacher. Chem. Mater. 2006, 18, 5033
[116] D. M. Guldi, G. M. A. Rahman, N. Jux, N. Tagmatarchis, M. Prato. Angew. Chem. Int. Ed. 2004, 43, 5526
[117] R. J. Chen, Y. Zhang, D. Wang, H. Dai. J. Am. Chem. Soc. 2001, 123, 3838
[118] H. Murakami, T. Nomura, N. Nakashima. Chem. Phys. Lett. 2003, 378, 481
[119] A. Satake, Y. Miyajima, Y. Kobuke. Chem. Mater. 2005, 17, 716
[120] G. M. A. Rahman, D. M. Guldi, S. Campidelli, M. Prato. J. Mater. Chem. 2006, 16, 62
[121] X. Wang, Y. Liu, W. Qiu, D. Zhu. J. Mater. Chem. 2002, 12, 1636
[122] F. D'Souza, R. Chitta, A. S. D. Sandanayaka, N. K. Subbaiyan, L. D'Souza, Y. Araki, O. Ito. Chem.sEur. J. 2007, 13, 8277
[123] Y. Mu, H. Liang, J. Hu, L. Jiang, L. Wan. J. Phys. Chem. B. 2005, 109, 22212
[124] V. Georgakilas, V. Tzitzios, D. Gournis, D. Petridis. Chem. Mater. 2005, 17, 1613
[125] J. Zhou, X. Zhou, X. Sun, R. Lib, M. Murphy, Z. Ding, X. Sun, T.-K. Sham. Chem. Phys. Lett. 2007, 437, 229
[126] D. M. Guldi, H. Taieb, G. M. A. Rahman, N. Tagmatarchis, M. Prato. Adv. Mater. 2005, 17, 871
[127] M. A. Correa-Duarte, L. M. Liz-Marza´n. J. Mater. Chem. 2006, 16, 22
[128] H. Chen, S. Donga. Talanta. 2006, 71, 1752
[129] J. W. Ostranger, A. A. Mamedov, N. A. Kotov. J. Am. Chem. Soc. 2001, 123, 1101
[130] K. Jiang, A. Eitan, L. S. Schadler, P. M. Ajayan, R. W. Siegel, N. Grobert, M. Mayne, M. Reyes-Reyes, H. Terrones, M. Terrones. Nano Lett. 2003, 3, 275
[131] M. A. Correa-Duarte, J. Perez-Juste, A. Sanchez-Iglesias, M. Giersig. Angew. Chem. Int. Ed. 2005, 44, 4375
[132] X. Hu, T. Wang, X. Qu, S. Dong. J. Phys. Chem. B. 2006, 110, 853
[133] J. Sun, L. Gao. J. Electroceram. 2006, 17, 91
[134] J. Sun, M. Iwasa, L. Gao, Q. Zhang. Carbon. 2004, 42, 895
[135] F. Stoffelbach, A. Aqil, C. Jerome, R. Jerome, C. Detrembleur. Chem. Commun. 2005, 4532
[136] D. J. Guo, H. L. Li. Electrochem. Commun. 2004, 6, 999
[137] T. M. Day, P. R. Unwin, N. R. Wilson, J. V. Macpherson. J. Am. Chem. Soc. 2005, 127, 10639
[138] D. J. Guo, H. L. Li. J. Electroanal. Chem. 2004, 573, 197
[139] Y. Xing. J. Phys. Chem. B. 2004, 108, 19255
[140] Y. Lin, X. Cui, C. Yen, C. M. Wai. J. Phys. Chem. B. 2005, 109, 14410

[141] S. R. Sivakkumar, J. M. Ko, D. Y. Kim, B. C. Kim, G. G. Wallace. Electrochim. Acta. 2007, 52, 7377
[142] X. Xie, L. Gao. Carbon. 2007, 45, 2365
[143] X. P. Huang, C. X. Pan, X. T. Huang. Mater. Lett. 2007, 61, 934
[144] B. Huang, R. Huang, D. Jin, D. Ye. Catal. Today. 2007, 126, 279
[145] I. Moriguchi, R. Hidaka, H. Yamada, T. Kudo, H. Murakami, N. Nakashima. Adv. Mater. 2006, 18, 69
[146] J. Sun, L. Gao, M. Iwasa. Chem. Commun. 2004, 832
[147] O. Frank, M. Kalbac, L. Kavan, M. Zukalova, J. Prochazka, M. Klementova, L. Dunsch. Phys. Status Solidi. 2007, 244, 4040
[148] K.-X. He, Q.-F. Wu, X.-G. Zhang, X.-L. Wang. J. Electrochem. Soc. 2006, 153, A1568
[149] I.-H. Kim, J.-H. Kim, Y.-H. Lee, K.-B. Kim. J. Electrochem. Soc. 2005, 152, A2170
[150] J. D. Kim, B. S. Kang, T. W. Noh, J.-G. Yoon, S. I. Baik, Y.-W. Kim. J. Electrochem. Soc. 2005, 152, D23
[151] C. Wang, M. Waje, X. Wang, J. M. Tang, R. C. Haddon, Y. Yan. Nano Lett. 2004, 4, 345
[152] C. J. Brinker, G. W. Scherer. Sol Gel SciencesThe Physics and Chemistry of Sol Gel Processing; Academic Press: New York, 1990
[153] P. Vincent, A. Brioude, C. Journet, S. Rabaste, S. T. Purcell, J. L. Brusq, J. C. Plenet. J. Non-Cryst. Solids. 2002, 311, 130
[154] A. Jitianu, T. Cacciaguerra, R. Benoit, S. Delpeux, F. Beguin, S. Bonnamy. Carbon. 2004, 42, 1147
[155] A. Jitianu, T. Cacciaguerra, M. H. Berger, R. Benoit, F. Beguin, S. Bonnamy. J. Non-Cryst. Solids. 2004, 34546, 596
[156] L. P. Zhao, L. Gao. Carbon. 2004, 42, 1858
[157] D. Eder, A. H. Windle. J. Mater. Chem. 2008, 18, 2036
[158] Y. Yu, J. C. Yu, J.-G. Yu, Y.-C. Kwok, Y.-K. Che, J.-C. Zhao, L. Ding, W.-K. Ge, P.-K. Wong. Appl. Catal., A. 2005, 289, 186
[159] H. Song, X. Qiu, F. Li, W. Zhu, L. Chen. Electrochem. Commun. 2007, 9, 1416
[160] W.-Q. Han, A. Zettl. Nano Lett. 2003, 3, 681
[161] J. Y. Bai, Z. D. Xu, Y. F. Zheng. Chem. Lett. 2006, 35, 96
[162] Y. Li, J. Ding, J. Chen, C. Xu, B. Wei, J. Liang, D. Wu. Mater. Res. Bull. 2002, 37, 313
[163] Y. X. Liang, Y. J. Chen, T. H. Wanga. Appl. Phys. Lett. 2004, 85, 666
[164] M. H. Chen, Z. C. Huang, G. T. Wu, G. M. Zhu, J. K. You, Z. G. Lin. Mater. Res. Bull. 2003, 38, 831
[165] B.-Y. Wei, M.-C. Hsu, P.-G. Su, H.-M. Lin, R.-J. Wu, H.-J. Lai. Sens. Actuators, B. 2004, 101, 81
[166] W. D. Wang, P. Serp, P. Kalck, C. G. Silva, J. L. Faria. Mater. Res. Bull. 2008, 43, 958
[167] W. D. Wang, P. Serp, P. Kalck, J. L. Faria. J. Mol. Catal. A: Chem. 2005, 235, 194
[168] W. D. Wang, P. Serp, P. Kalck, J. L. Faria. Appl. Catal., B. 2005, 56, 305
[169] G. Arabale, D. Wagh, M. Kulkarni, I. S. Mulla, S. P. Vernekar, K. Vijayamohanan, A. M. Rao. Chem. Phys. Lett. 2003, 376, 207
[170] J. H. Park, J. M. Ko, O. O. Parka. J. Electrochem. Soc. 2003, 150, A864
[171] J. Y. Lee, K. Liang, K. H. Ana, Y. H. Lee. Synth. Met. 2005, 150, 153
[172] R.-J. Wu, J.-G. Wu, M.-R. Yu, T.-K T, C.-T. Yeh. Sens. Actuators, B. 2008, 131, 306
[173] L. Cao, F. Scheiba, C. Roth, F. Schweiger, C. Cremers, U. Stimming, H. Fuess, L. Chen, W. Zhu, X. Qiu. Angew. Chem. Int. Ed. 2006, 45, 5315
[174] D. Eder, A. H. Windle. Adv. Mater. 2008, 20, 1787
[175] K. Hernadi, E. Ljubovic, J. W. Seo, L. Forro. Acta Mater. 2003, 51, 1447
[176] F. Vietmeyer, B. Seger, P. V. Kamat. Adv. Mater. 2007, 19, 2935

[177] L. Jiang, L. Gao. Mater. Chem. Phys. 2005, 91, 313
[178] M. Yoshimura, K. J. Byrappa. Mater. Sci. 2008, 43, 2085
[179] R. Menzel, A. M. Peiro´, J. R. Durrant, M. S. P. Shaffer. Chem. Mater. 2006, 18, 6059
[180] K. Byrappa, A. S. Dayananda, C. P. Sajan, B. Basavalingu, M. B. Shayan, K. Soga, M. Yoshimura. J. Mater. Sci. 2008, 43, 2348
[181] Y. Lee, H. J. Song, H. S. Shin, H. J. Shin, H. C. Choi. Small. 2005, 1, 975
[182] Y. Huang, J. Lin, X. X. Ding, C. Tang, C. Z. Gu, S. R. Qi. Mater. Lett. 2007, 61, 697
[183] W. D. Zhang. Nanotechnology. 2006, 17, 1036
[184] Y. Yu, L. L. Ma, W. Y. Huang, J. L. Li, P. K. Wong, J. C. Yu. J. Solid State Chem. 2005, 178, 1488
[185] Y. Yu, L. L. Ma, W. Y. Huang, F. P. Du, J. C. Yu, J. G. Yu, J. B. Wang, P. K. Wong. Carbon. 2005, 43, 670
[186] B. P. Jia, L. Gao, J. Sun. Carbon. 2007, 45, 1476
[187] L. Fu, Z. M. Liu, Y. Q. Liu, B. X. Han, J. Q. Wang, P. A. Hu, L. C. Cao, D. B. Zhu. Adv. Mater. 2004, 16, 350
[188] Z. Y. Sun, X. R. Zhang, B. X. Han, Y. Y. Wu, G. M. An, Z. M. Liu, S. D. Miao, Z. J. Miao. Carbon. 2007, 45, 2589
[189] X. Sun, Y. Chu, D. W. Wang, J. H. Du, B. Y. Zhang, F. P. Wang. Rare Metals. 2007, 26, 191
[190] Z. Sun, Z. Liu, B. Han, G. An. Mater. Lett. 2007, 61, 4565
[191] Z. Sun, H. Yuan, Z. Liu, B. Han, X. Zhang. Adv. Mater. 2005, 17, 2993
[192] Z. Sun, Z. Liu, B. Han, S. Miao, J. Du, Z. Miao. Carbon. 2006, 44, 888
[193] H. Kim, W. Sigmund. Appl. Phys. Lett. 2002, 81, 2085
[194] J. S. Ye, H. F. Cui, X. Liu, T. M. Lim, W. D. Zhang, F. S. Sheu. Small. 2005, 1, 560
[195] Y. W. Zhu, H. I. Elim, Y. L. Foo, T. Yu, Y. J. Liu, W. Ji, J. Y. Lee, Z. X. Shen, A. T. S. Wee, J. T. L. Thong, C. H. Sow. Adv. Mater. 2006, 18, 587
[196] F. Jin, Y. Liu, C. M. Day. Appl. Phys. Lett. 2007, 90, 143114
[197] W.-C. Fang, O. Chyan, C.-L. Sun, C.-T. Wu, C.-P. Chen, K.-H. Chen, L.-C. Chen, J.-H. Huang. Electrochem. Commun. 2007, 9, 239
[198] T. Ikuno, T. Yasuda, S. I. Honda, K. Oura, M. Katayama, J. G. Lee, H. J. Mori. Appl. Phys. 2005, 98, 114305
[199] T. Ikuno, M. Katayama, K. Kamada, S. Honda, J. G. Lee, H. Mori, K. Jpn. Oura. J. Appl. Phys. Lett. 2003, 42, L1356
[200] Q. Kuang, S. F. Li, Z. X. Xie, S. C. Lin, X. H. Zhang, S. Y. Xie, R. B. Huang, L. S. Zheng. Carbon. 2006, 44, 1166
[201] G. H. Deng, X. Xiao, J. H. Chen, X. B. Zeng, D. L. He, Y. F. Kuang. Carbon. 2005, 43, 1557
[202] Z. Fan, J. Chen, K. Cui, F. Sun, Y. Xu, Y. Kuang. Electrochim. Acta. 2007, 2007, 2959
[203] A. Chrissanthopoulos, S. Baskoutas, N. Bouropoulos, V. Dracopoulos, D. Tasis, S. N. Yannopoulos. Thin Solid Films. 2007, 515, 8524
[204] A. D. Lazareck, T.-F. Kuo, J. M. Xu, B. J. Taft, S. O. Kelley, S. G. Cloutier. Appl. Phys. Lett. 2006, 89, 103109
[205] A. D. Lazareck, S. G. Cloutier, T.-F. Kuo, B. J. Taft, S. O. Kelley, J. M. Xu. Nanotechnology. 2006, 17, 2661
[206] J. Liu, X. Li, L. Dai. Adv. Mater. 2006, 18, 1740
[207] Y. M. Ho, J. W. Liu, J. L. Qi, W. T. Zheng. J. Phys. D: Appl. Phys. 2008, 41, 665308
[208] A. Gomathi, S. R. C. Vivekchand, A. Govindaraj, C. N. R. Rao. Adv. Mater. 2005, 17, 2757

[209] A. Javey, H. Kim, M. Brink, Q. Wang, A. Ural, J. Guo, P. Mcintyre, P. McEuen, M. Lundstrom, H. Dai. Nat. Mater. 2002, 1, 241
[210] N. Ren, A. G. Dong, W. B. Cai, Y. H. Zhang, W. L. Yang, S. J. Huo, Y. Chen, S. H. Xie, Z. Gao, Y. Tang. J. Mater. Chem. 2004, 14 , 3548
[211] J. Qiao, S. N. Tang, Y. N. Tiana, S. M. Shuang, C. Dong, M. M. F. Choi. Sens. Actuators B. 2009, 138, 402
[212] G. Q. Mo, J. S. Ye, W. D. Zhang. Electrochimica Acta. 2009, 55, 511
[213] W. Ma, D. B. Tian. Bioelectrochemistry. 2010, 78, 106
[214] K. N. Zhu, H. Y. Qin, B. H. Liu, Z. P. Li. J. Power Sources. 2011, 196, 182
[215] T. Hirayama, T. Manako, H. Imai. e-J. Surf. Sci. Nanotech. 2008, 6, 237
[216] I. Jimenez, M. A. Centeno, R. Scotti, F. Morazzoni, A. Cornet, J. R. Morante. J. Electrochem. Soc. 2003, 150, H72
[217] B. Fru ¨hberger, M. Grunze, D. J. Dwyer. Sens. Actuators, B. 1996, 31, 167
[218] N. L. Wu, S. Y. Wang, I. A. Rusakova. Science. 1999, 285, 1375
[219] P. G. Collins, K. Bradley, M. Ishigami, A. Zettl. Science. 2000, 287, 1801
[220] N. Sinha, J. Z. Ma, J. T. W. Yeow. J. Nanosci. Nanotechnol. 2006, 6, 573
[221] J. Kong, N. R. Franklin, C. Zhou, M. G. Chapline, S. Peng, K. Cho, H. Dai. Science. 2000, 287, 622
[222] Z. Sun, X. Zhang, N. Na, Z. Liu, B. Han, G. An. J. Phys. Chem. B. 2006, 110, 13410
[223] N. Van Hieu, L. T. B. Thuy, N. D. Chien. Sens. Actuators, B. 2008, 129, 888
[224] J. Liu, Z. Guo, F. Meng, Y. Jia, J. Liu. J. Phys. Chem. C. 2008, 112, 6119
[225] Y. Chen, C. Zhu, T. Wang. Nanotechnology. 2006, 17, 3012
[226] L. C. Jiang, W. D. Zhang. Biosens. Bioelectron. 2010, 25, 1402
[227] Z. Liu, J. Wang, D. Xie, G. Chen. Small. 2008, 4, 462
[228] B. Fang, C. H. Zhang, W. Zhang, G. F. Wang, Electrochim. Acta. 2009, 55, 178
[229] B. E. Conway. J. Electrochem. Soc. 1991, 138, 1539
[230] C. Niu, E. K. Sichel, R. Hoch, D. Moy, H. Tennet. Appl. Phys. Lett. 1997, 70, 1480
[231] G. Lota, K. Lota, E. Frackowiak. Electrochem. Commun. 2007, 9, 1828
[232] W. Yi, S. G. Yu, W. T. Lee, I. T. Han, J. T. Jeong, Y. S. Woo, J. H. Lee, S. W. Jin, W. B. Choi, J. N. Heo, D. Jeon, J. M. Kim. J. Appl. Phys. 2001, 89, 4091
[233] K. W. Nam, K. H. Kim, E. S. Lee, W. S. Yoon, X. Q. Yang, K. B. Kim. J. Power Sources. 2008, 182, 642
[234] V. Subramanian, H. W. Zhu, B. Q. Wei. Electrochem. Commun. 2006, 8, 827
[235] M. G. Deng, B. C. Yang, Z. A. Zhang, Y. D. Hu. J. Mater. Sci. 2005, 40, 1017
[236] H. Zhang, G. Cao, Z. Wang, Y. Yang, Z. Shi. Nano Lett. 2008, 8, 2664
[237] A. L. M. Reddy, M. M. Shaijumon, S. R. Gowda, P. M. Ajayan. Nano Lett. 2009, 9, 1002
[238] G.-M. Wu, A.-R. Wang, M.-X. Zhang, H.-Y. Yang, B. Zhou, J. J. Shen. Sol Gel Sci. Technol. 2008, 46, 79
[239] W.-C. Fang. J. Phys. Chem. C. 2008, 112, 11552
[240] W.-C. Fang, K.-H. Chen, L.-C. Chen. Nanotechnology. 2007, 18, 485716

2

Synthesis of Carbon Nanomaterials in a Swirled Floating Catalytic Chemical Vapour Deposition Reactor for Continuous and Large Scale Production

Sunny E. Iyuke and Geoffrey S. Simate
University of the Witwatersrand, Johannesburg
South Africa

1. Introduction

Carbon nanotubes (CNTs), 'rediscovered' (Monthioux & Kuznetsov, 2006) by Iijima as a byproduct of fullerene synthesis (Iijima, 1991), have attracted enormous scientific and technological interest. Their myriad applications in various fields since their rediscovery are no longer debatable. However, their commercial applications still depend on large scale synthesis (several thousands of tons per year) and associated cost of production. Various methods have been developed for the production of CNTs (Dresselhaus et al., 2001; Agboola et al., 2007). However, the three very useful and widespread methodologies include arc discharge, laser ablation and chemical vapour deposition (CVD) (Robertson, 2004; Agboola et al., 2007). Two key requirements revealed in these methods are as follows, (i) a carbon source, and (ii) a heat source to achieve the desired operating temperature (See & Harris, 2007). In the arc discharge, CNTs are produced from carbon vapour generated by an electric arc discharge between two graphite electrodes (with or without catalysts), under an inert gas atmosphere (Journet et al., 1997; Lee et al., 2002; Agboola et al., 2007). In the laser ablation, a piece of graphite target is vapourised by laser irradiation under an inert atmosphere (Journet & Bernier, 1998; Paradise & Goswami, 2007). As for the technique of CVD , it involves the use of an energy source such as plasma, a resistive or inductive heater, or furnace to transfer energy to a gas phase carbon source in order to produce fullerenes, CNTs and other sp^2-like nanostructures (Meyyappan, 2004). As would be expected, some of these methods are more effective than others. The arc-discharge, though it produces CNTs of high quality with fewer structural defects, uses high temperature of up to 1500°C, which makes it difficult to be scaled up for commercial purposes. On the other hand, laser vaporisation method is an expensive technique because it involves high purity graphite rods and high power lasers. At the moment, the CVD methodology (or variations thereof) is the only promising process for the production of CNTs on a reasonably large-scale compared to arc-discharge and laser vaporization methods (Coleman, 2008). In addition, the process tends to produce nanotubes with fewer impurities (catalyst particles, amorphous carbon and non-tubular fullerenes) compared to other techniques (Esawi & Farag, 2007). The variants of the CVD are as a result of the means by which chemical reactions are initiated, the type of reactor used and the process conditions (Deshmukh et al., 2010). The CVD is simple, flexible and allows high specificity of single wall or multi wall nanotubes through appropriate selection of process parameters, e.g., metal catalysts, reaction temperature and flow rate of feed stock (Nolan et al., 1995; Agboola et al., 2007; Moisala et al., 2006). Recently, a swirled floating catalystic chemical vapour deposition (SFCCVD) reactor was developed with the aim of up-scaling the production capacity (Iyuke, 2007). The simplified schematic presentation of a SFCCVD is shown in Figure 1. Typically it consists of a vertical quartz or silica plug flow reactor inside a furnace. The upper end of the reactor is connected to a condenser which leads to two delivery cyclones where the CNTs produced are collected. Feed materials including carrier gases are uniformly blended with the aid of a swirled coiled mixer to give optimum catalyst-carbon source interaction.

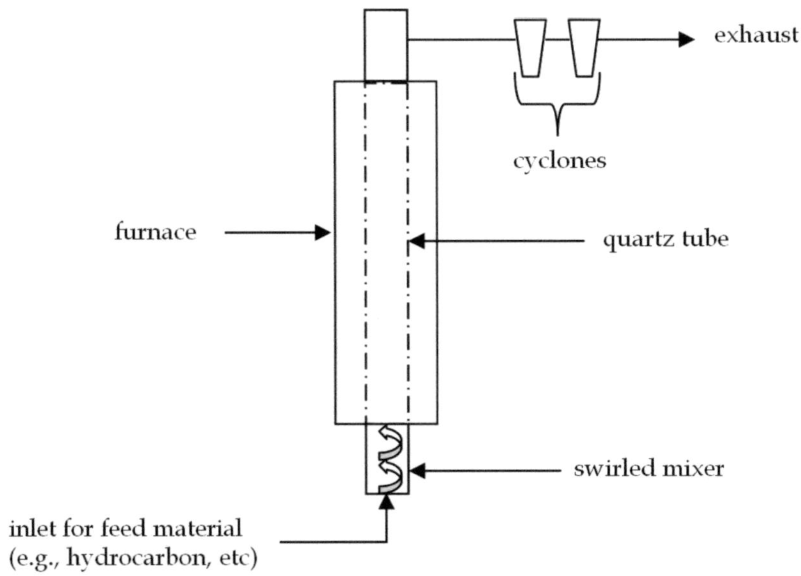

Fig. 1. Simplified schematic presentation of a swirled floating catalyst (or fluid) CVD

In this Chapter we discuss the continuous and large scale production of carbon nanomaterials in an SFCCVD reactor using various carbon sources. This reactor was found to be more successful than the microwave or the fixed-bed catalytic CVD modes (Iyuke et al., 2009). Furthermore, our large-scale synthesis method is simple and easily accessible to others with interest in this novel material. The chapter is divided into four themes covering, (1) the growth mechanisms and kinetics (section 2), (2) synthesis of both ordinary CNTs and carbon nanostructures (sections 3 and 4, respectively), (3) optimisation of the production (section 5), and (4) conclusions, challenges, and future prospects (section 6).

2. Growth mechanisms and kinetics

2.1 Growth mechanisms

Information on the mechanism of CNT growth, though important, is very scarce. As a result, more studies are needed to find out the characteristic mechanisms of CNT synthesis which may in turn guide the design and operation of continuous and large scale production of carbon nanomaterials. Presently, various growth models based on experimental and quantitative studies have been proposed (Agboola et al., 2007).

One very important factor in the formation of CNTs through the CVD is the catalyst. Usually the vapor-liquid-solid (VLS) model is used to explain the growth of CNTs using catalysts (Saito, 1995). The catalyst acts as a seed for nucleation and growth by controlling the overall reaction with the hydrocarbon source (Yoon and Baik, 2001). The catalytic graphitisation (CG) of carbon sources to carbon nanomaterials is a sequential process that involves hydrocarbon decomposition, carbon dissolution, diffusion, adsorption and precipitation of carbon atoms to produce graphitised material (Yoon and Baik, 2001; Iyuke et al., 2007). However, the most accepted growth model suggests that after the decomposition of the carbon source, carbon diffuses into the metal particles until the solution becomes saturated. Carbon saturation in the metal occurs either by reaching the carbon solubility limit in the metal at a given temperature or by lowering the solubility limit via temperature decrease (Moisala et al., 2003). Supersaturation of the saturated solution then results in precipitation of solid carbon from the metal surface (Moisala et al., 2003; See and Harris, 2007). In summary, the mechanism of the catalytic growth of CNT or carbon nanofibre (CNF) is generally accepted as consisting of three steps (de Jong and Geus, 2000): The first step is the decomposition of carbon containing gases on the metal surface, with carbon atoms deposited on the surface. In the second step the carbon atoms dissolve in and diffuse through the bulk of the metal particles. It must be noted, however, that it is still unclear whether carbon atoms diffuse on the particle bulk (Ducati et al., 2004), on the particle surface (Hofmann et al., 2005), or whether surface and bulk diffusion compete. The final step is the precipitation of the carbon in the form of CNTs or CNFs at the other side of the catalyst particle.

The driving force for carbon diffusion and for the global process of carbon nanomaterial formation is the difference in solubility of carbon at the gas/catalyst interface and the catalyst/carbon nanomaterial interface, which is determined by the affinity for carbon formation of the gas phase and the thermodynamic properties of the carbon nanomaterial, respectively (Snoeck et al., 1997). The above mechanism shows that the steady state growth of CNT is a delicate balance between carbon source dissociation, carbon diffusion through the particle, and the rate of nucleation and formation of graphitic layers (Yu et al., 2005). This is illustrated by a simplified model shown in Figure 2. Tubule formation such as CNTs is favoured over other forms of carbon such as graphitic sheets with open edges. This is because a tube contains no dangling bonds and, therefore, is in a low energy form (Dai, 2001).

As illustrated in Figure 3, for the deposition of CNTs on a substrate, two general growth modes are known; the nanotubes can either follow a base-growth mode (A) or tip-growth mode (B) (Dai, 2001). Base-growth occurs when the catalyst remains anchored to the substrate, while tip-growth mechanism happens when the particle lifts off the substrate and is observed at the top of the CNTs. These growth modes depend on the contact forces or adhesion forces between the catalyst particle and the substrate (Leonhardt et al., 2006). While a weak contact favours tip-growth mechanism, a strong interaction promotes base-growth (Bower et al., 2000; Song et al., 2004).

However, Gohier et al. (2008) dispute the notion of adhesion forces between the catalyst particle and the substrate as controlling growth mechanism. Instead, they propose the catalyst particle size, and the 'carbon diffusion paths' during CNT nucleation as the main determinants for controlling the growth mechanism. Their results showed that for the same substrate/catalyst couples, single or few-wall CNTs follow the base-growth mechanism while the tip-growth occurs only for the large multi-walled CNTs (MWCNTs). This behaviour was observed for three different catalysts commonly used in CNT synthesis, i.e., nickel, iron and cobalt. Besides the conventional explanation for growth mode change on catalyst/substrate interaction, Gohier and co-workers showed that growth mode can be explained by interaction between small carbon patches (polyaromatic carbon or reticulated carbon chains) and the catalyst. A strong interaction favours the formation of a graphene cap on the catalyst and leads to the single wall CNTs (SWCNTs) or few walls CNTs (FWCNT) growth via the base-growth mode. On the contrary, a weak interaction induces a diffusion of the graphitic section to the catalyst/substrate interface which drives the tip-growth mechanism.

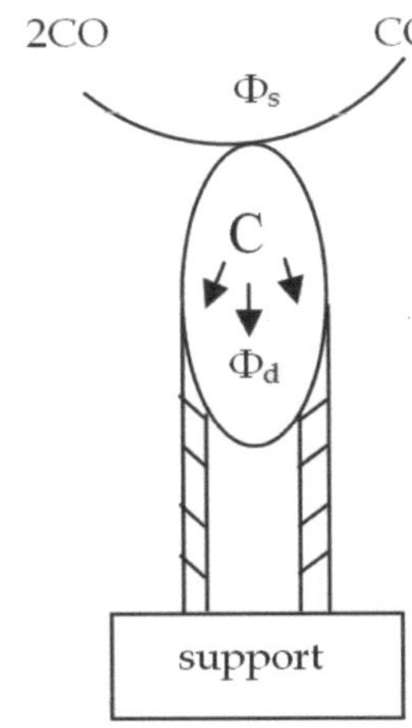

Fig. 2. Schematic representation of the CNT growth mechanism, where Φ_s represents surface decomposition rate, Φ_d means carbon diffusion rate. If $\Phi_s > \Phi_d$, encapsulating carbon forms on the catalyst surface; if $\Phi_s = \Phi_d$, the system is at balance and CNT grows at steady state (Yu et al., 2005).

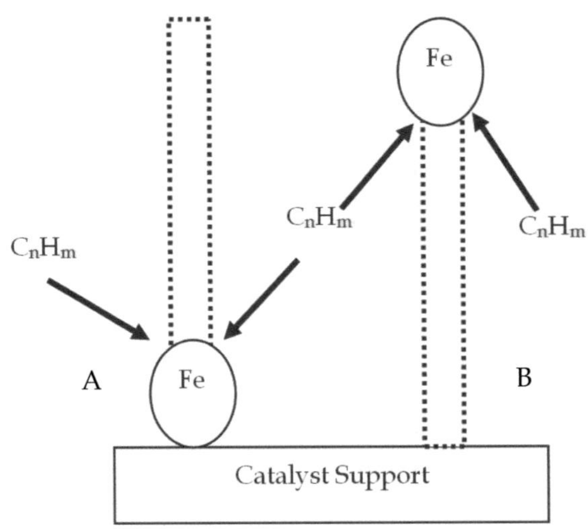

Fig. 3. General growth model (Dai, 2001)

Catalyst supports also play an important role. For example, in order to facilitate the large scale production of CNTs, support material should have a threefold action: minimises the formation of by- products (amorphous carbon, graphite nanospheres, etc), deters the aggregation of catalytic centres, and allows its removal from the resulting CNTs using simple chemical procedures (Tsoufis et al., 2007).

2.2 Kinetics

As already discussed mechanisms by which CVD occurs are very similar to those of heterogeneous catalysis; the reactant(s) adsorbs on the surface and then reacts on the surface to form a new surface (Fogler, 2006). Equation 1 represents the proposed mechanisms of catalytic graphitisation of a hydrocarbon (e.g., C_2H_2) to CNTs using a SFCCVD technique, whereby under heat the hydrocarbon and possibly the cracked fractions are dissociated into carbon atoms (Iyuke et al., 2007; Iyuke, 2007). The carbon atoms are deposited and adsorbed onto the catalyst (e.g., Fe) surface, which in turn react with each other to form C-C bonds thus producing graphene layers of CNTs produced.

$$\left.\begin{array}{l} C_2H_2 \rightarrow 2C + H_2 \\ C_2H_2 + 3H_2 \rightarrow 2CH_4 \\ 2CH_4 \rightarrow 2C + 4H_2 \\ \hline C_2H_2 \rightarrow 2C + H_2 \\ \qquad\qquad \text{heat} \end{array}\right\} nC \underset{\underset{\text{heat}}{k_d}}{\overset{\overset{k_a}{\text{Catalyst}}}{\underset{H_2\ \text{gas}}{\rightleftharpoons}}} C_n\text{-Fe} \xrightarrow[\text{heat}]{\text{C-C bond}\ k} n[CNT] + Fe + mH_2 \qquad (1)$$

Where the rate constants for the adsorption, desorption and reaction are k_a, k_d and k, respectively; m and n are stoichiometric coefficients.
The rate of CG (illustrated in Equation 1), can be written as a function of dissolution, adsorption and chemical reaction (Levenspiel, 1999; Iyuke et al., 2007).

$$r_{CG} = -r_{C_2H_2} = \text{fn (dissolution, adsorption, chemical reaction)} \qquad (2)$$

At appropriate temperature, where carbon nanomaterials grow, the dissolution of the hydrocarbon may be considered as complete, therefore, it becomes none rate limiting. As shown in Equation 1, the decomposed carbon atoms adsorb onto the catalyst surface, where adjacent carbon atoms react together to produce C-C bonds of graphene layers via a complex mechanism, which in turn produce carbon nanomaterials. Such complex mechanism can be expressed by rates of reactions catalysed by solid surfaces per unit mass as follows (Levenspiel, 1997; Iyuke and Ahmadun, 2002; Fogler, 2006):

$$-r_{C_2H_2} = \frac{1}{W_{cat}} \frac{dnC}{dt} = k\theta_C^n \qquad (3)$$

Where W_{cat} is the weight of catalyst used, nC is the number of moles of carbon, superscript n is the order of the reaction, θ_C is the fraction of catalyst surface covered by carbon atoms (i.e., surface concentrations of adsorbed carbon, C), k is the reaction rate constant, and C is either carbon or other graphitised carbon nanoparticles. Therefore, Equation 3 is the rate of reaction occurring on the catalyst surfaces which is proportional to the concentration of carbon (C) present on the surfaces or the fractions of surfaces covered by the carbon atoms of carbon nanomaterials (Iyuke et al., 2007). The commonly applied model, Langmuir-Hinshelwood, for reactions taking place between molecules or fragments of molecules adsorbed on the surface (Winterbottom and King, 1999) as shown in Equation 4 is adopted to obtain the reaction rate and equilibrium constant (Iyuke et al., 2007).

$$-r_{C_2H_2} = \frac{kC^n}{1+kC^n} \quad (4)$$

Equation 4 can be rearranged to give Equation 5.

$$\frac{1}{-r_{C_2H_2}} = \frac{1}{kC^n} + \frac{1}{k} \quad (5)$$

Using Equation 1, the exact model for the rate of reaction was developed below.

$$nC + Fe \rightleftarrows C_n\text{-}Fe \rightarrow CNT \quad (6)$$

Equation 6 was modeled as:

$$A + B \rightleftharpoons AB \rightarrow \text{Products} \quad (7)$$

Therefore, using the Langmuir-Hinshelwood mechanisms, the final model was obtained as,

$$r = \frac{kC_A^n}{k+1}\frac{\exp(1-\theta)}{kt} \quad (8)$$

Where C_A is the concentration of reactant (e.g., C_2H_2), θ is the surface coverage, k is the reaction rate constant which is proportional to the diffusion coefficient of carbon (Kim et al., 2005), n is the order of reaction, t is the time of reaction.

Using experimental data the value of n = 4 was obtained from a plot of $1/-rC_2H_2$ versus $1/C^n$ in Equation 5 (Iyuke et al., 2007). Consequently, the model in Equation 8 was used to predict the rate of production of CNTs at various temperatures and acetylene concentrations, and the results were compared to the experimental data (Table 1). Experimental data and results from the predictive model show reasonable agreement, with correlation coefficient ranging from 0.81 to 0.99 at various temperatures. The model equation is, therefore, able to represent data for SFCCVD technique in the production of CNTs. Both the experimental and predictive model results show that the concentration of acetylene and reaction temperatures affect the rate of CNT production.

Acetylene Concentration	CNT production rate (mg/s); experimental versus computed			
(ppm)	950°C	1000°C	1050°C	1100°C
1985	1.6(1.2)	2.7(1.9)	2.5(1.8)	2.1(1.9)
2728	2.8(3.0)	3.1(3.0)	4.3(3.0)	4.2(3.0)
3379	3.0(4.7)	4.0(4.7)	4.9(4.7)	3.8(4.7)
4058	4.5(5.8)	4.8(5.8)	4.2(5.8)	5.1(5.8)

Table 1. Experimental and computed production rate of CNTs at different temperatures (Iyuke et al., 2007).

3. Synthesis of ordinary carbon nanotubes

The production of CNTs in terms of the type and quality (for example) is controlled by several factors including carbon source. The commonly used gaseous carbon sources include carbon monoxide (CO) and hydrocarbon feed stocks such as methane, acetylene, ethylene, and n-hexane (Agboola et al., 2007). More recently, carbon dioxide (CO_2) has also been used in the synthesis of these carbon nanomaterials (Simate et al., 2010). There are typically two forms of CNTs according to the number of rolled up graphene layers that form the tube (as illustrated in Figure 4), i.e., SWCNTs and MWCNTs. A SWCNT is a graphene sheet rolled-over into a cylinder with typical diameter of the order of 1.2-1.4nm in magnitude (Journet and Bernier, 1998), while a MWCNT consists of concentric cylinders with an interlayer spacing of about 0.34nm (3.4Å) and a diameter typically of the order of 10-20 nm in magnitude (Dai, 2002).

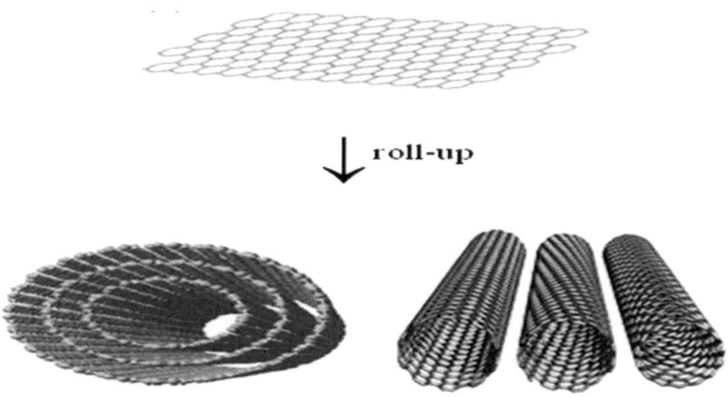

Fig. 4. Models and representation of multi-walled CNT and single-walled CNT (Merkoci, 2006; Liu, 2008).

In addition, there are two models which can be used to describe the structures of MWCNT. In the Russian doll model, a certain number of SWCNT with growing diameters are arranged in concentric cylinders. In the parachute model, a single graphite sheet is rolled in around itself. In MWCNTs, the nanotubes are typically bound together by strong van der Waals interaction forces and form tight bundles (Dai, 2002).

It seems likely that two different mechanisms operate during the growth of MWCNTs and SWCNTs, because the presence of a catalyst is absolutely necessary for the growth of the later (Ajayan, 2000). Furthermore, a key factor of CNT growth is the kinetics of carbon supply. An over-supply of carbon causes the formation of MWCNT rather than SWCNT because excess carbon activity allows the nucleation of extra nanotube walls (Wood et al., 2007). The stability of the double shell MWCNT is higher than SWCNTs on their own as the number of atoms increase (Sinnott et al., 1999). This is because the van der Waals interactions between the shells in the MWCNT increase as the nanotube gets longer. It is also observed that there is little difference in the stability of MWCNTs as a function of the helical arrangements of the shells (Sinnott et al., 1999). Compared to MWCNTs, one of the most significant features of SWCNT is the uniformity of size and the relative lack of any defects in the later. This explains the fact that methods to produce SWCNTs in significant large quantities developed rather quickly.

3.1 Use of acetylene

Acetylene has been the most widely used hydrocarbon in the production of CNTs by many researchers. Acetylene is more reactive than other hydrocarbons at the same reaction temperature, leading to CNTs of good quality and probably hinders the formation of carbon nanoshells which poison catalytic centres (Soneda et al., 2002). Furthermore, even molecular beam experiments have shown that acetylene is the most active growth precursor (Eres et al., 2005), and that it is the primary growth precursor in both hydrocarbon and alcohol feed stocks (Xiang et al., 2009; Zhong et al., 2009). Here, we discuss the results of using acetylene carbon source and organo-metallic ferrocene catalyst. Ferrocene was found to be a better catalyst precursor in the studied conditions since iron pentacarbonyl decomposes at lower temperatures resulting in the excessive growth of catalyst particles. Furthermore, its other advantages as an iron precursor include its innocuity and low cost (Barreiro et al., 2006).

The CNTs (and other nanomaterials to be discussed in section 4.1) were synthesized in a mixture of gases of acetylene, hydrogen and argon at the substrate temperature of 900-1050°C. Figure 5 shows typical transmission electron microscopy (TEM) images of CNTs grown in the temperature range of 1000 - 1050°C. The images reveal that each of the structures is hollow with inner diameter and length of several nanometers which confirm the presence of CNTs (Afolabi et al., 2007).

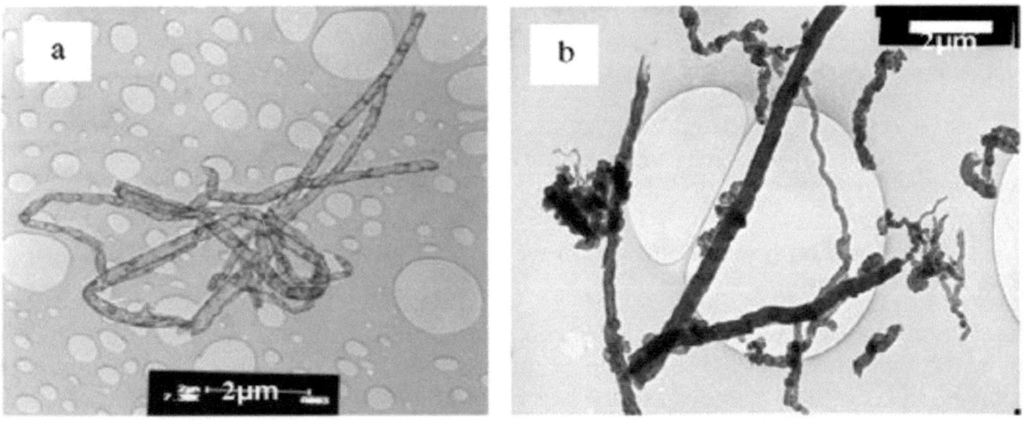

Fig. 5. TEM images of pure CNTs produced at (a) 1000°C and (b) 1050°C (Afolabi et al., 2007).

The diameter distribution of the CNTs based on the measurements from the sufficiently enlarged TEM images (results not shown in this Chapter) indicated that the diameter of the CNTs increased with increase in temperature (Afolabi et al., 2007). This was attributed to high rate of acetylene decomposition which maximised carbon generation leading to more CNT wall formation (Kumar and Ando, 2005; Afolabi et al., 2007).

The X-ray diffraction (XRD) pattern of the sample (Figure 6) reveals the characteristic pattern of graphitised carbon. The graphitic line (002) of this sample was observed at diffraction peak of 25.8° corresponding to inter-planner spacing of about 0.343 nm which is usually attributed to CNTs. This pattern also indicates high degree of crystalinity which suggest low content of amorphous carbon and impurities from catalysts (Afolabi et al., 2007). This observation is a marked difference from the results of many other CVD processes where impurities from catalyst employed in the pyrolysis process are always associated with the CNTs such that they have to be removed through various purification processes.

These multi-step purification processes can destroy the product. It is suggested that effective utilisation of the ferrocene catalyst during the decomposition process could have been responsible for the low content of iron impurities in the CNTs (Afolabi et al., 2007).

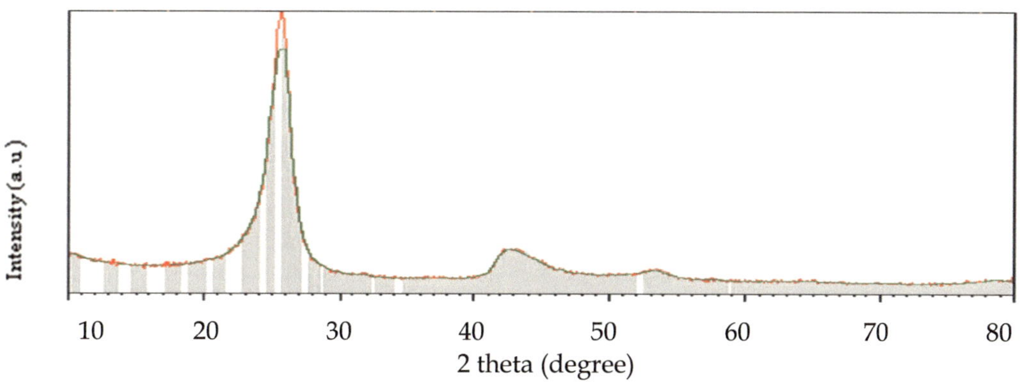

Fig. 6. XRD pattern of raw CNTs sample (Afolabi et al., 2007)

Table 2 shows the compositions of carbon nanomaterials at various conditions. These results indicate that accurate regulation of carbon source flow rate, hydrogen flow rate and temperature, which is a simple and versatile method, enables selective synthesis of various types of carbon nanomaterials. As can be seen from Table 2, CNTs were predominantly produced at very high temperatures (1000-1050°C). However, at temperatures above 1100°C there was decrease in production rate (Iyuke et al., 2007). The reduction in production at high temperature might have been due to the deactivation of the catalyst by carburization (Yamada et al., 2008) or sintering (Amama et al., 2009). Furthermore, low values of amorphous carbon were observed as the temperature increased as a result of amorphous carbon burning off at high temperatures (Iyuke et al., 2007).

Gas flow rate (mlmin^{-1})		Carbon nanomaterials at various temperatures (°C)			
Hydrogen	Acetylene	900	950	1000	1050
118	181	CNBas+CNTs	CNBas+CNTs	CNTs	CNTs
118	248	CNBas+CNTs	CNTs	CNTs	CNTs
118	308	CNBas+CNTs	CNTs	CNTs	CNTs
118	370	CNBas	CNTs	CNTs	CNTs
181	370	CNBas	CNTs	CNTs	CNTs
248	370	CNBas	CNTs	CNTs	CNTs

Table 2. Compositions of carbon nanomaterials at various conditions (Afolabi et al., 2007)

Similar studies on the SFCCVD reactor in the absence of hydrogen gas showed the production of CNTs and several other carbon nanomaterials at lower temperatures of 750°C to 900°C (Iyuke et al., 2009). For example, (1) a mixture of carbon nanoballs (CNBas), MWCNTs and CNFs at 750°C, (2) MWCNTs at 800°C, (3) a mixture of MWCNTs, SWCNTs and helical carbon nanofibres (HCNF) at 850°C, and (4) large quantities of HCNFs and strands of bamboo-like CNTs at 900°C. These results are different from the ones shown in Table 1 where high temperatures were needed for the production of carbon nanomaterials.

This can be attributed to the suppressing effect of hydrogen on the production of CNTs. The suppressing phenomenon was observed by several researchers as reviewed by Simate et al. (2010). The suppressing effect is due to the surface dehydrogenation of carbon to form methane. In fact, it has been observed that the presence of hydrogen can either accelerate or suppress the synthesis of CNTs (Yang and Yang, 1986; Endo, 1988; Ci et al., 2001) depending on the thermodynamics (Ci et al., 2001).

3.2 Use of ferrocene

This was an attempt to explore the use of organometallic complexes in the synthesis of CNT and other carbon nanomaterials. In particular, we explored the use of ferrocene both as a carbon source and the 'carrier' of elements to catalyst sites in CNT synthesis in our reactor. At a temperature above 500°C, ferrocene decomposes as given in Equation 9 (Leonhardt et al., 2006), and its unimolecular gas-phase decomposition is given in Equation 10 (Lewis and Smith, 1984).

$$Fe(C_5H_5)_2 \rightarrow Fe + H_2 + CH_4 + C_5H_6 + ... \quad (9)$$

$$Fe(C_5H_5)_2 \rightarrow Fe + C_5H_5 \quad (10)$$

This means that at a temperature higher than 500°C, solid or liquid-like iron particles and different kinds of hydrocarbons exist in the reaction zone of the SFCCVD equipment. Upon these particles, acting as catalyst nuclei, CNTs nucleate and grow with the carbon atoms provided solely by the ferrocene. This means that thermal decomposition of ferrocene provides both catalytic particles and carbon sources.

We investigated the production of CNTs from 800°C to 950°C by flowing a gas mixture consisting of argon and sublimated gaseous ferrocene through the reactor. Results showed that CNTs could be produced at all the temperatures with no significant difference in the amount of CNTs produced (i.e., $p > 0.05$). However, at higher reaction temperatures, there was less adherence of the product to the walls of the reactor. Therefore, for industrial processes, higher temperatures would need to be used.

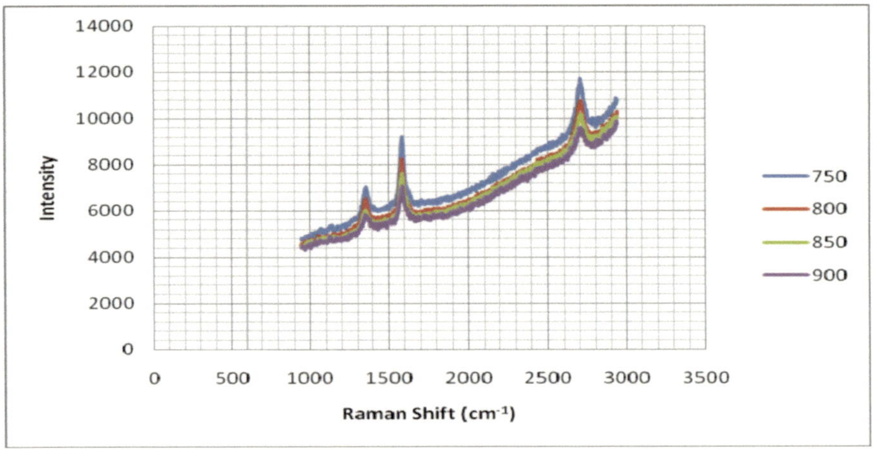

Fig. 7. Raman Shifts for the Products formed at different temperatures.

In contrast to results obtained by Barreiro and co-workers (Barreiro et al., 2006), where they predominantly produced SWCNT in a higher pressure horizontal CVD reactor using only ferrocene, the Raman spectroscopy of our samples showed that MWCNTs were produced as shown in Figure 7. As can be seen from the figure, the Raman spectra of CNTs produced show similar intensities for both the dispersive disorder-induced D-band (1353.1 cm^{-1}) and its second-order related harmonic G-band (1580.3 cm^{-1}). This strongly supports the fact that MWCNTs were produced.

3.3 Use of xylene/ferrocene mixture

This section reports the continuous and large scale production of MWCNTs from xylene/ferrocene mixture in the presence or absence of hydrogen (Yah et al., 2011). Xylene was used as the solvent for the catalyst and also acted as the main carbon source. In other words, xylene was used as a solvent to dissolve the ferrocene for easy transportation into the reactor and also functioned as a CNT growth promoter. It was selected as the hydrocarbon source since it boils (boiling point ≈ 140°C) below the decomposition temperature of ferrocene (≈ 190°C).

Just like other work already described, it was found that an increase in temperature increased the amount of CNTs produced (Table 3). While at 800°C there were no CNTs produced, (other than amorphous carbon) a very small amount of CNT was produced at 850°C. The non-production of CNTs at low temperatures (i.e., < 850°C), was due to the low diffusion rates of iron and carbon which prevented the CNTs from nucleating and growing from the iron particles, but resulted in the formation of amorphous carbon (Kunadian et al., 2008). This implies that below 850°C, the production of CNTs from xylene/ferrocene mixture is unfavourable. However, above 1000°C there was a decrease in the amount of CNTs produced as a result of short reaction time caused by the deactivation of catalyst sites which occurs at such high temperature (Yu et al., 2005).

The present findings showed that between 900 and 1000°C, with an optimal production at 950°C (Table 3), MWCNTs were well aligned. This differs from earlier reports by Jacques et al. (2000) where they also described the production of MWCNTs at 725-775°C, with an optimal production at 725°C. Their product had more nanoball-like (100 nm long) structures at 750°C. Our results were similar to those previously reported by Vivekchand et al. (2004) where they produced MWCNTs from the mixture of xylene and ferrocene by pyrolysis at furnace temperature in the range of 800-1000°C.

Reactor Conditions	800°C	850°C	900°C	950°C	1000°C	950°C (double reactor feed volume)	950°C (1:7 H_2:Ar carrier gas feed)
Observable number of CNTs per Cu grid square	0	2	8	16	13	18	47

Table 3. Effect of temperature on the amount of CNT formed (Yah et al., 2011).

With the introduction of hydrogen as one of the carrier gases, the amount of CNTs produced increased in the reactor by more than twice the amount produced at double reactor volume feed (see Table 3). Kunadian et al. (2008) also made similar observation when they prepared MWCNTs from xylene using the fixed bed CVD mode. This is because hydrogen has a strong influence in the gas-phase decomposition rate of ferrocene (Equation 11). This is an accelerating effect of hydrogen as stated in section 3.1.

$$Fe(C_2H_5)_2 \xrightarrow{H_2} Fe + 2C_2H_5 \qquad (11)$$

In the absence of hydrogen, ferrocene dissociation does not occur until about 1100-1200 K, while its reduction occurs at temperatures as a low as 673 K in the presence of hydrogen (Kuwana and Saito, 2007).

3.4 Use of acetylene and xylene/ferrocene mixture

This experimental work employed acetylene as the carbon source, hydrogen as a carrier gas, ferrocene dissolved in xylene as the catalyst precursor and the reactor operated from 900-1100°C in order to produce CNTs on a continuous basis. The results showed that CNTs produced at high ferrocene concentrations were riddled with high levels of iron impurities, while those at low ferrocene had only small traces of iron impurities. This was attributed to the limitation in the solubility of ferrocene in xylene at room temperature (Liu et al., 2002). In fact, Liu et al. (2002) observed that at low ferrocene concentrations, the catalysis efficiency of ferrocene is high. Though the quantity of CNTs produced at lower ferrocene concentrations was small, the low amorphous carbon and iron particles in the samples resulted in the formation of clean CNTs.

3.5 Use of carbon dioxide and methane

Though the pyrolysis of hydrocarbon precursors for synthesizing CNTs is very useful and is used widely, there are some disadvantages associated with these methods. Most hydrocarbons used in these methods are hazardous chemicals, and for most cases, the pyrolysis temperatures are around 1000°C, which are impractical for large scale industrial production (Qian et al., 2006). One approach to tackling this problem is by the use of CO_2 which is a cheap, non-toxic, low-energy, and abundant molecule on the earth (Qian et al., 2006). The CO_2 is easily formed by the oxidation of organic molecules during combustion or respiration. Furthermore, CO_2 can be acquired from natural reservoirs or recovered as a by-product of industrial chemical processes, so, no new production of CO_2 is necessary and there will be no addition to greenhouse gases (Young et al., 2000).
This work presents the successful production of CNTs and CNFs by catalytic decomposition of methane (CH_4) and CO_2 over an unsupported nickel alloy catalyst ($LaNi_5$) in a 'modified' SFCCVD reactor (Figure 8) at temperatures ranging from 650-850°C (Moothi, 2009; Maphutha, 2009).

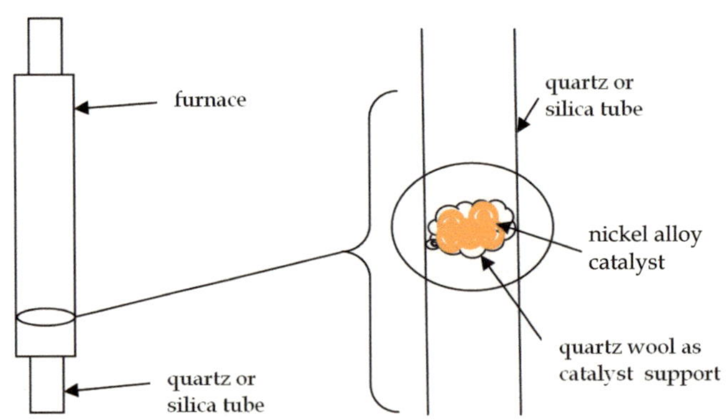

Fig. 8. Modified SFCCVD by attaching quartz wool as support for unsupported catalyst.

It was observed that CH_4 produced well defined CNTs at all the tested temperatures (650–850°C) as shown in Figure 9. A mixture of CNFs and CNTs was produced at lower temperatures (650–700°C) using CO_2, whereas at higher temperatures (750–850°C) only CNFs were produced as shown in Figure 10. The CNFs are cylindric nanostructures with graphene layers arranged as stacked cones, cups or plates. Furthermore, CNFs with graphene layers wrapped into perfect cylinders are called CNTs. In summary, this work has shown that the $LaNi_5$ catalyst is capable of decomposing CO_2 into CO and CNTs both of which are useful compounds. The CNTs produced are not entangled like other catalysts making it easier to separate them for further studies or used as nano-resistors.

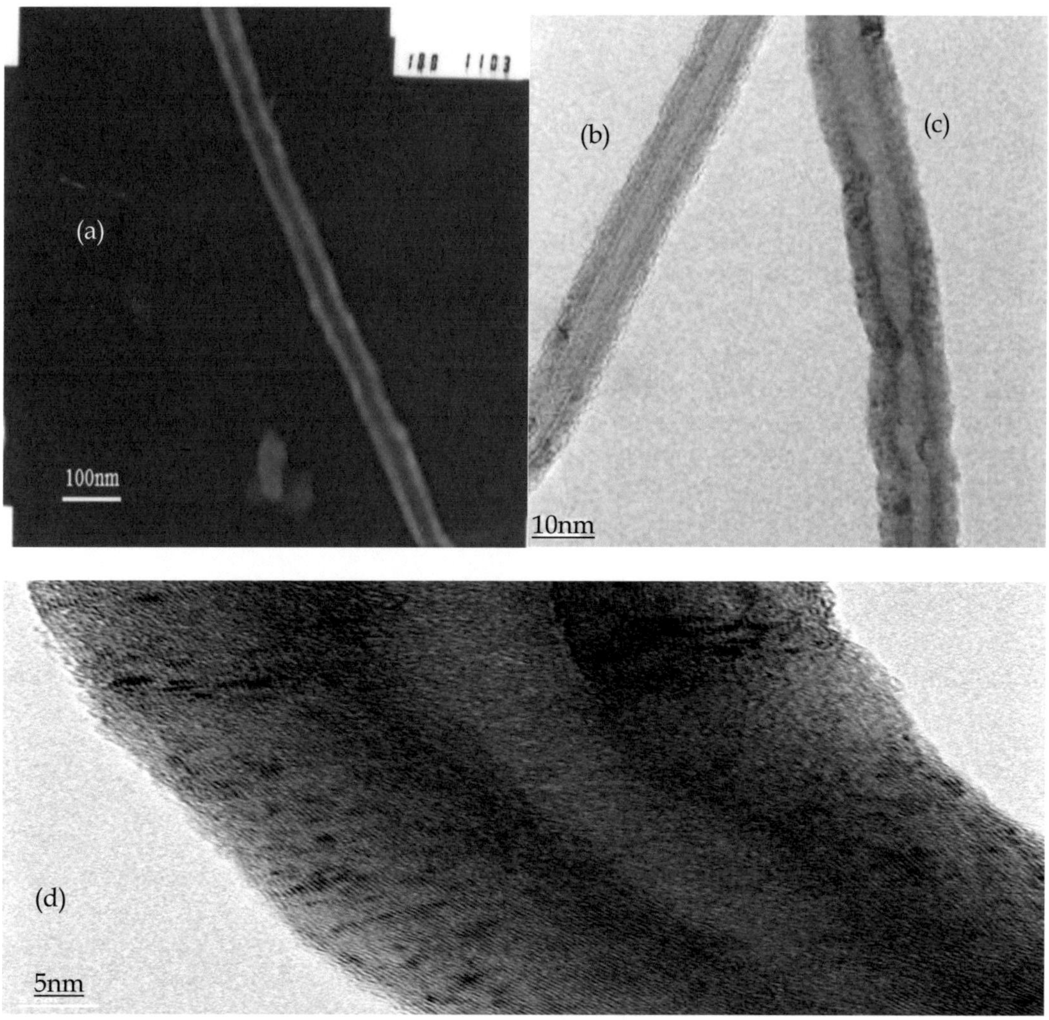

Fig. 9. (a) CNT produced at 650°C, (b) CNT produced at 750°C,with inside diameter and outside diameter of 6.6 nm and 16.6 nm, respectively (c) CNT produced at 750°C,with inside diameter and outside diameter of 10 nm and 23.2 nm, respectively, and (d) CNT produced at 850°C (Moothi, 2009; Maphutha, 2009)

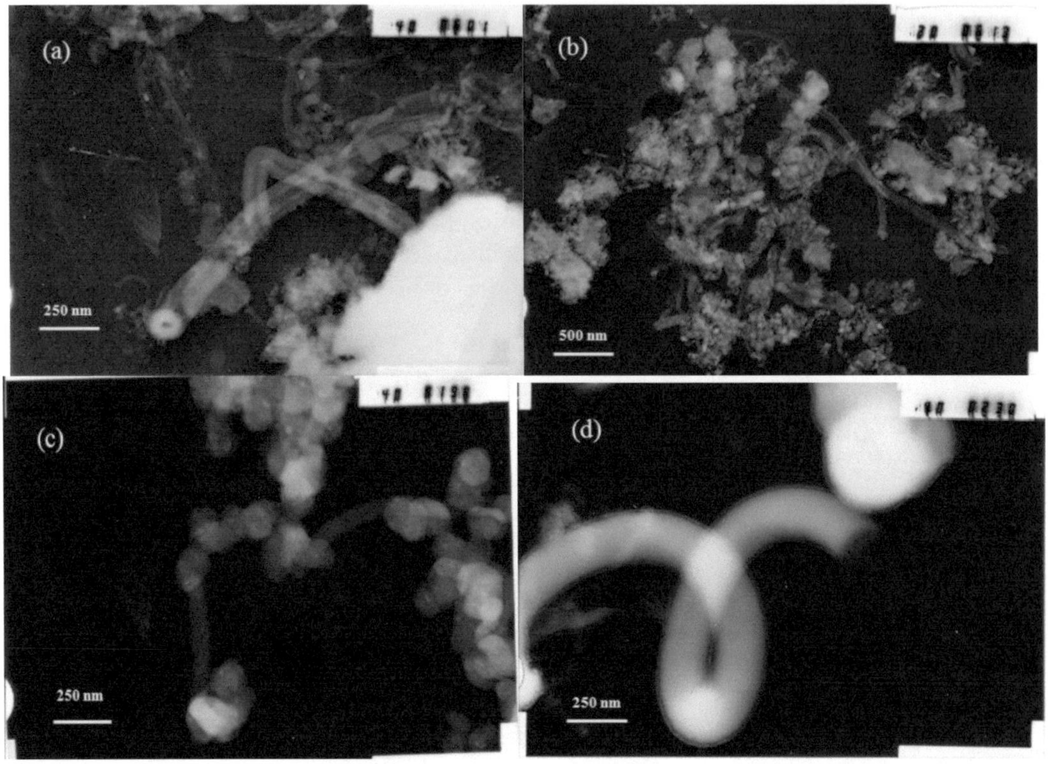

Fig. 10. (a) CNF/CNT produced at 650°C, growing from the large catalyst particles, (b) CNF/CNT produced at 700°C, (c) CNF produced at 750°C, with catalyst particles attached at ends, and (d) CNF produced at 850°C, with a circular structure and no catalyst embedded (Moothi, 2009; Maphutha, 2009)

4. Synthesis of other carbon nanostructures

As alluded to earlier, the nature of carbon nanomaterials produced by CVD method including SFCCVD depends on the working conditions such as temperature and pressure of operation, the volume and concentration of carbon source, the size and pretreatment of metallic catalysts, and the time of reaction (Paradise and Goswami, 2006). As a result, in the ambit of CNT production several other carbon nanomaterials may be produced in the CVD depending on the conditions. Some of these nanomaterials are CNBas (Zhong et al., 2000), CNFs (Endo et al., 2001), carbon nanorods (Liu et al., 2000) and even diamond particles (Zhang et al., 2006; Afolabi et al., 2009).

4.1 Carbon nanoballs

The CNBas are a group of shaped carbon materials with a spherical or near spherical shapes under 100 nm diameter. They include balls, spheres, microbeads, carbon blacks, onions and mesoporous micro-beads (Caldero-Moreno et al., 2005; Deshmukh et al., 2010). The spheres and balls can be hollow or solid and in this Chapter, spheres and balls will mean the same thing.

This is a continuation of the experimental work discussed earlier in section 3.1. As can be seen in Table 2, the higher acetylene flow rate and low temperature (900°C) tend to favour production of CNBas. Figures 11 shows a TEM image of CNBas obtained at hydrogen flow rate of 118 ml/min and above, and constant acetylene flow rate of 370 ml/min at the decomposition temperature of 900°C as shown in Table 2. The CNBas, instead of CNTs, were formed because of low residence time and scattered catalyst particles resulting from high gaseous flow rates. Scattered catalyst particles led to 'growth-sites starvation'. These CNBas were also found to contain negligible traces of amorphous carbon and catalyst.

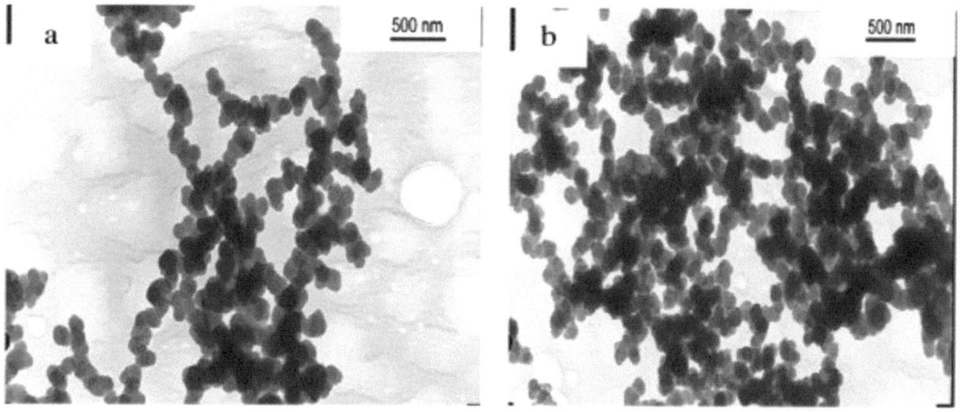

Fig. 11. TEM images of pure CNBas produced at (a) 900°C, C_2H_2 = 370 ml/min, H_2= 181 ml/min (b) 900°C, C_2H_2 = 370 ml/min, H_2 = 248 ml/min (Afolabi et al., 2007).

In other related work, CNBas were only formed at temperatures of 1000°C and above in the absence of catalysts (Iyuke et al., 2009). This shows that temperature was paramount in the production of CNBas without the use of catalysts. The sufficiently enlarged TEM images showed that the diameter of the CNBa depended on gas flow rate ratios (i.e., acetylene: argon). Higher ratios tended to synthesise CNBas with smaller diameters while lower ratios produced CNBas with larger diamters. In other words, as the flow rate of acetylene increased, the size of CNBas decreased, and vice-versa. This observation relates the size of the CNBas to the residence time of acetylene in the reactor. The residence time of acetylene will be higher at the lower flow rate of acetylene and this will allow the products to grow further before exiting with carrier gas into the cyclones. Results also showed that the CNBa size increases as the pyrolysis temperature increases (Mhlanga et al., 2010).
The CNBas were also observed in experimental test-works performed in section 3.4. The CNBas were obtained at very low concentrations of ferrocene (e.g., 2%) and a pyrolysis temperature of 900°C. These CNBas were formed due to a low content of ferrocene at low pyrolysis temperature, as reported by Liu et al. (2002). The xylene might have assisted in the formation of these CNBas since low catalyst content at high temperature is known to result in CNTs with a high content of amorphous impurities (Afolabi et al., 2009).

4.2 Diamond films

The production of diamond films followed the experimental procedure in section 3.4. At an acetylene flow rate of 248 ml min^{-1}, hydrogen flow rate of 181 ml min^{-1} and pyrolysis temperatures between 1000°C and 1100°C, diamond particles as well as glassy carbon were produced alongside CNTs (Afolabi et al., 2009). The diamond films were, however, thermodynamically unstable as they disappeared after a few minutes. The production of these diamond particles (Figure 12) could be attributed to the nucleation by already produced CNTs in the reactor. This can be explained as follows (Hou et al., 2002); first it must be noted that diamond and CNTs - both being allotropes of carbon - have some structural resemblances. Under the conditions that were prevailing in the reactor, especially in the presence of hydrogen, the CNTs were broken into pieces lengthwise forming carbon nano-onions. The *sp*^{3} bonds at the broken ends of the tubes turn to dangling bonds. Furthermore, due to the action of hydrogen and high temperature, the *sp*^{2} bonds at the middle of the tubes between two ends can open and turn into *sp*^{2} dangling bonds, and finally turn into *sp*^{3} bonds again (Meilunas et al., 1991; Gruen et al., 1994). This results in countless heterogeneous nucleation sites, and the improvement of nucleation and growth for diamond.

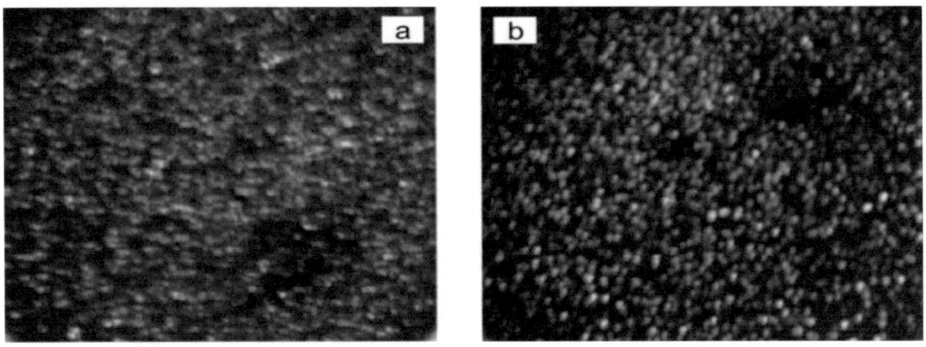

Fig. 12. Photographic images (EuroTEK 10 Megapixels) of diamond films obtained at (a) 10% ferrocene, 1000°C and a hydrogen flow rate of 118 ml min^{-1}, and (b) 10% ferrocene, 1100°C and a hydrogen flow rate of 181 ml min^{-1} (Afolabi et al., 2009).

In the case of seeding with the low quality CNT mixture, the percentage of CNTs is very low (Hou et al., 2002). The separation action of hydrogen on CNT is weak. There are only few nano anion ends. So it is difficult for diamond to nucleate and grow while graphite film can grow and develop easily.

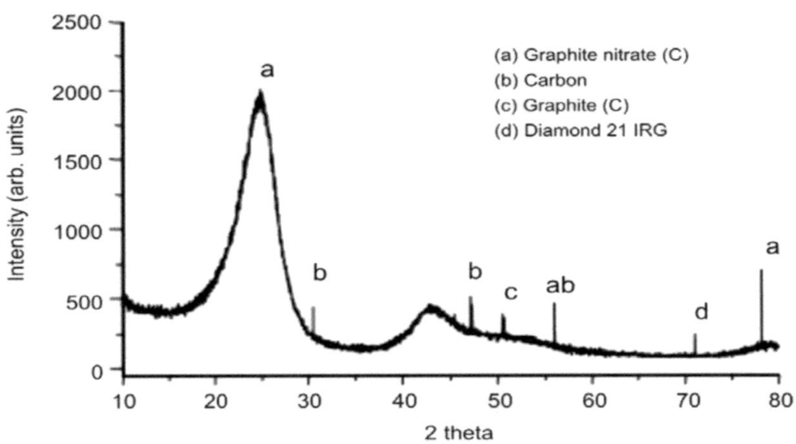

Fig. 13. X-ray diffraction analysis of a sample produced at 10% ferrocene, 1100°C, acetylene flow rate of 248 ml min^{-1}, and a hydrogen flow rate of 181 ml min^{-1}, confirming traces of diamond films (Afolabi et al., 2009).

Figure 13 shows the XRD analysis of the samples obtained at a temperature of 1100°C, an acetylene flow rate of 248 ml min^{-1}, hydrogen flow rate of 181 ml min^{-1}, and 10wt% concentrations of ferrocene, respectively (Afolabi et al., 2009). The analysis confirms the presence of diamond flakes and glassy carbon identified as graphite nitrate; the nitrogen in the nitrate may have been from the traces contained in the argon and hydrogen gases from the supplier, which were 4 and 2 volume per million (vpm), respectively. It was observed that the rate of production of diamond films increased with increases in the concentration of ferrocene, hydrogen flow rate and pyrolysis temperature.

5. Optimisation of the production of carbon nanomaterials

For practical commercial applications of CNTs, one would need quantities in the kilogram range (Subramoney, 1999). As a result, there is still a need to develop and optimise processing routes for the production of uniform and well defined carbon nanostructures in large quantities (Yu et al., 2005). We have seen in this Chapter that the production of carbon nanomaterials is variable, depending on interrelated controllable factors. Therefore, growth and optimisation of carbon nanomaterials need a targeted approach whereby parameters are tuned based on the type of nanomaterials required.

The huge number of parameters required in the growth process is a big challenge in the optimisation of as-produced carbon nanomaterials. However, we have seen that temperature and carbon source quantity are the important determinants in the production of carbon nanomaterials.

Figure 14 shows the effect of decomposition temperature and acetylene flow rate on the quantity of nanoparticles produced. The figure reveals that the quantity of nanoparticles produced increased with temperature and the flow rate of acetylene at constant flow rate of argon carrier gas of 181 ml/min. This indicates that at higher temperature, there is enough heat to effect the pyrolysis of acetylene leading to the formation of CNTs. The highest quantity of nanoparticles produced was obtained at temperature of 1050°C and

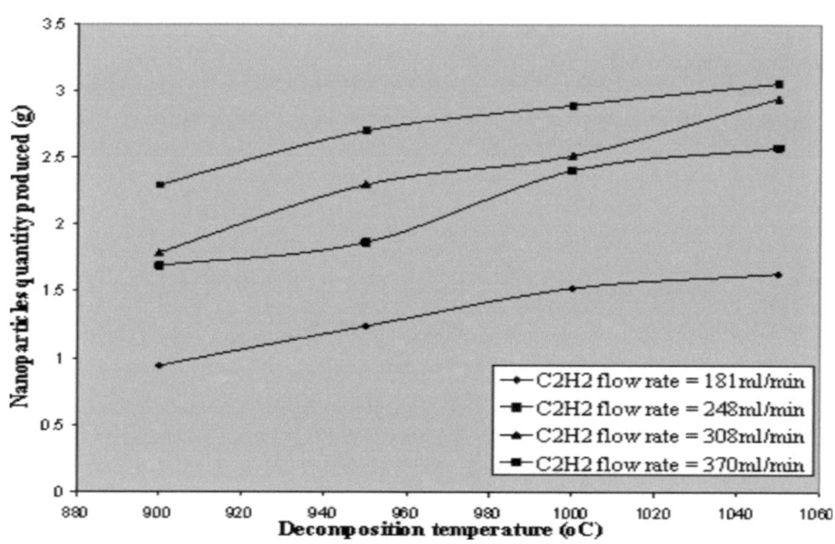

Fig. 14. Decomposition temperature versus quantity of nanoparticles produced at different flow rates of acetylene and constant hydrogen flow rate of 181 ml min^{-1}(Afolabi et al., 2007).

acetylene flow rate of 370 ml/min. Even in the case of CNBa, the rate of production increased with increase in the feed stock flow rate at all pyrolysis temperatures (Mhlanga et al., 2010). However, at very high temperatures (results not shown here) there was a decrease in production rate. Furthermore, low values of amorphous carbon were observed as the temperature increased as a result of amorphous carbon burning off at high temperatures (Iyuke et al., 2007). In general, a higher temperature results in a modest increase in yield.
Raman spectroscopy results presented in Figure 15 at different temperatures also shows that the crystallinity of CNTs produced depends on the pyrolysis temperature. The higher the temperature, the better the crystal structure of the CNTs.

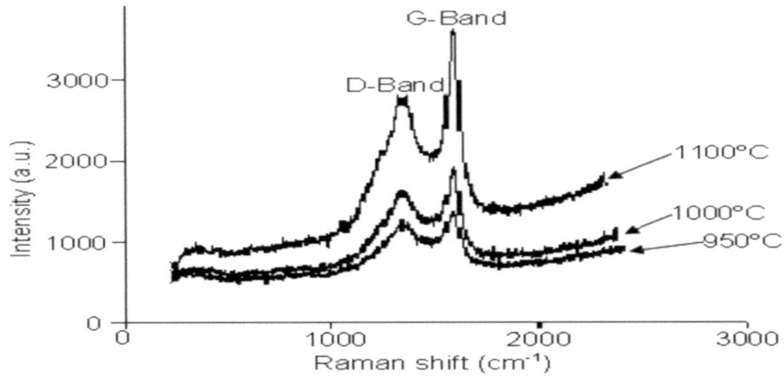

Fig. 15. Raman spectrometra of the CNTs sample produced (Abdulkareem et al., 2007; Iyuke et al., 2007).

Results obtained on the effect of temperatures on the rate of production of CNTs at different flow ratios of acetylene and hydrogen is shown in Figure 16 (Iyuke et al., 2006; Abdulkareem et al., 2007). It can be observed from the figure that the rate of production of CNTs at different flow ratios of acetylene and hydrogen is influenced significantly by the temperature. As the temperature increases, the rate of production of CNTs also increases as well as the quantity produced.

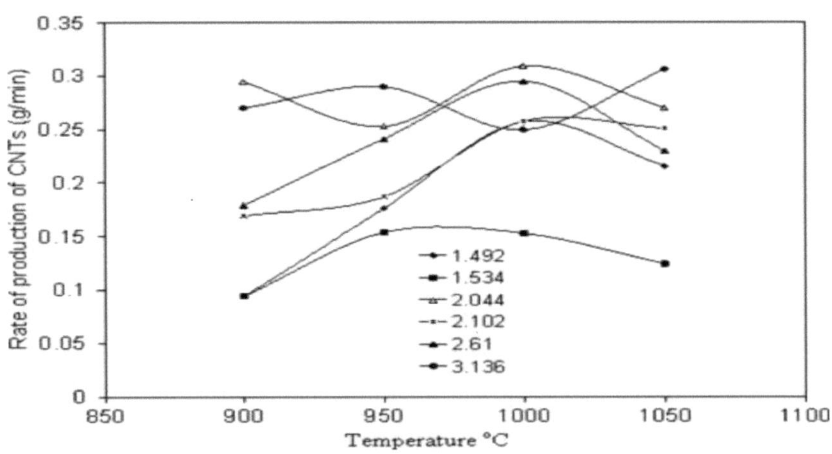

Fig. 16. Rate of production of CNTs at different temperatures and, flow ratios of acetylene to hydrogen (Iyuke et al., 2006; Abdulkareem et al., 2007).

Figure 16 also shows the effect of different flow ratios of acetylene to hydrogen at different temperatures. The figure shows that the rate of production initially increases with increase in C_2H_2/H_2 ratio at all temperatures. Different CNTs production rate peaks were observed for different C_2H_2/H_2 flow ratios at different temperatures. However, the highest CNTs production rate was observed at 1050°C and this corresponds to C_2H_2/H_2 flow ratio of approximately 3 while the minimum CNTs production rate was obtained at flow ratio of 1.5 at 900°C. High quality and long CNTs were also obtained at C_2H_2/H_2 ratio of approximately equal to 3 and CNT production rate of 0.31 g/min (Iyuke et al., 2006; Abdulkareem et al., 2007). This shows the important role of the ratio of carbon source to hydrogen for good and high quality production of CNTs.

The effect of hydrogen flow rate on the rate of production of CNTs was also evaluated. At low flow rates (e.g., 118 - 181 ml/min), increases in flow rates increased the rate of CNT production (Abdulkareem et al., 2007; Iyuke et al., 2007). Here, hydrogen is not only required to create velocity profile in the reactor, but also takes part in the pyrolysis of acetylene. According to Kuwana et al. (2005), the three stages of reaction during pyrolysis of acetylene to give CNTs confirm the major role of hydrogen as shown in the equations below:

$$C_2H_2 \rightarrow 2C + H_2 \tag{12}$$

$$C_2H_2 + 3H_2 \rightarrow 2CH_4 \tag{13}$$

$$2CH_4 \rightarrow 2C + 4H_2 \tag{14}$$

It can be observed from Equation 14 that the methane produced in Equation 13 decomposed at reactor temperature to produce CNTs and hydrogen. This reveals the positive influence of hydrogen in the production of CNTs.

However, it was seen that an increase in hydrogen flow rate above 181ml/min decreased the production rate monotonically from 4.2mg/s to zero (Abdulkareem et al., 2007; Iyuke et al., 2007). Reduction in rate of production of CNTs at higher hydrogen flow rate could be attributed to low residence time of acetylene in the reactor as a result of high velocity profile created by high hydrogen flow rate thereby suppressing the formation of CNTs. In other words, at high flow rate of hydrogen, the residence time of acetylene in the reactor becomes low because the hydrogen flow creates high velocity profile and push off the acetylene at a faster rate from the reaction zone. Singh et al. (2003) also made similar observations about the effect of hydrogen concentration on the production of CNTs using CVD technique. Apart from creating a velocity profile in the reactor, hydrogen may have also reduced ferrocene to atomic iron clusters or nanoparticles (Abdulkareem et al., 2007), thus making ferrocene inactive.

6. Conclusions, challenges and future prospects

CNTs (and other carbon nanomaterials) have reached the forefront of many industrial research projects since their rediscovery. However, there is still a great deal of work to be done for the full potential of CNTs to be fully realised. We have seen in this Chapter that various carbon nanomaterials can be prepared in an SFCCVD by careful controlling of parameters such as temperature and flow rates of carbon source and carrier gas. More importantly, the studies in this Chapter have clearly demonstrated that the SFCCVD method could be scaled up for continuous or semi-continuous production of carbon nanomaterials.

Though we have managed to produce CNTs and various other carbon nanomaterials in larger quantities, there are several challenges which need to be overcome before they can be widely used. They need to be synthesized in longer lengths, and improved techniques are required to align and evenly distribute them. Commercially available nanotubes are usually 0.5–5 µm long. In the design of conventional composites, for example, it is well known that the fiber length has a major influence on strengthening and stiffening of the matrix. For effective load transfer, the fiber length has to exceed a certain critical length, l_c, given by the following equation (Esawi & Faragi, 2007):

$$l_c = \sigma_f d / 2\tau_c \tag{15}$$

Where σ_f is the ultimate or tensile strength of the fiber, d is the fiber diameter and τ_c is the fiber–matrix bond strength. If the fiber length is less than l_c, the matrix cannot effectively grip the fibers and as a result they will slip.

The difficulty of dispersing the CNTs due to the fact that they tend to stick together and the difficulty of aligning the tubes are also major challenges. Although such issues have not yet been completely resolved, extensive efforts are underway to overcome them.

Researchers also agree that one of the major obstacles to using CNT is cost (Breuer & Sundararaj, 2004). Generally, the synthesis techniques used for making nanotubes are expensive. High quality/high purity CNTs could cost $800/g and even ones with defects and impurities (metal catalyst and amorphous carbon) could cost $5–35/g (Esawi & Farag, 2007).

Due to the great importance of CNTs, it is clear that novel technologies will emerge in future. It is also expected that more applications of CNTs will be found as some of their unknown unique properties are discovered.

7. References

Abdulkareem, A. S., Afolabi, A. S., Iyuke, S. E. & Piennar, C. H. vZ. (2007). Synthesis of carbon nanotubes by swirled floating catalyst chemical vapour deposition method, *Journal of Nanoscience and Nanotechnology* 7(10): 1-6.

Afolabi, A. S., Abdulkareem, A. S. & Iyuke, S. E. (2007). Synthesis of carbon nanotubes and nanoballs by swirled floating catalyst chemical vapour deposition method, *Journal of Experimental Nanoscience* 2(4): 269-277.

Afolabi, A. S., Abdulkareem, A. S. & Iyuke, S. E. (2009). Continuous production of carbon nanotubes and diamond films by swirled floating catalyst chemical vapour deposition method, *South African Journal of Science* 105: July/August.

Agboola, A. E., Pike, R. W., Hertwig, T. A. & Lou, H. H. (2007). Conceptual design of carbon nanotube processes, *Clean Technologies and Environmental Policy* 9: 289-311.

Ajayan, P.M. (2000). Carbon nanotubes, *In*: Nalwa, H.S. (ed.), *Handbook of nanostructured materials and nanotechnology*, Academic Press, Inc., New York, Vol. 5, pp 375 – 403.

Amama, P. B., Pint, C. L., McJilton, L., Kim, S. M., Stach, E. A., Murray, P. T., Hauge, R. H. & Maruyama, B. (2009). Role of Water in super growth of single-walled carbon nanotube carpets, *Nano Letters* 9: 44–49.

Bower, C., Zhou ,O., Zhu, W., Werder, D.J. & Jin ,S. (2000). Nucleation and growth of carbon nanotubes by microwave plasma chemical vapor deposition, *Applied Physics Letters* 77(17):2767–2769.

Breuer, O. & Sundararaj, U. (2004). Big returns from small fibers: a review of polymer/carbon nanotube composites, *Polymer Composites* 25(6): 630–645.

Caldero-Moreno, J. M., Labarta, A., Batlle, X., Pradell, T., Crespo, D & Thien Binh, V. (2005). Magnetic properties of dense carbon nanospheres prepared by chemical vapor deposition, *Chemical Physics Letters* 447: 295–299.

Ci, L., Wei, J., Wei, B., Liang, J., Xu, C. & Wu, D. (2001). Carbon nanofibers and single-walled carbon nanotubes prepared by the floating catalyst method, *Carbon* 39(3): 329-335.

Coleman, K. S. (2008). Nanotubes, *Annual Reports on the Progress of Chemistry, Section A: Inorganic Chemistry* 104: 379-393.

Dai, H. (2001). Nanotube growth and characterisation, *In*: Dresselhaus, M. S., Dresselhaus, G., Avouris, P. (eds.), *Carbon Nanotubes: Synthesis, Structure, Properties and Applications*, Springer, New York, pp 29-51.

Dai, H. (2002). Carbon nanotubes: opportunities and challenges, *Surface Science* 500: 218-241.

Deshmukh, A. A., Mhlanga, S. D. & Coville, N. J. (2010). Carbon spheres, *Materials Science and Engineering Review* 70 (1-2): 1–28.

de Jong, K. P. & Geus, J. W. (2000). Carbon nanofibers: catalytic synthesis and applications, *Catalysis Review* 42(4):481-510.

Dresselhaus, M. S., Dresselhaus, G., Avouris, P. (eds.)(2001). *Carbon Nanotubes: Synthesis, Structure, Properties and Applications*, Springer, New York.

Ducati, C., Alexandrou, I., Chhowalla, M., Robertson, J. & Amaratunga, G.A.J.(2004). The role of the catalytic particle in the growth of carbon nanotubes by plasma enhanced chemical vapor deposition, *Journal of Applied Physics* 95(11):6387-6391.

Endo, M. (1988). Grow carbon-fibers in the vapor-phase, *Chemtech* 18(9): 568–576.

Endo, M., Kim, Y. A., Hayashi, T., Nishimura, K. & Matusita, T. (2001). Vapour grown carbon fibres (VGCFs): basic properties and their battery applications, *Carbon* 39: 1287-1297.

Eres, G., Kinkhabwala, A. A., Cui, H., Geohegan, D. B., Puretzky, A. A. & Lowndes, D. H. (2005). Molecular beam-controlled nucleation and growth of vertically aligned single-wall carbon nanotube arrays, *Journal of Physical Chemistry* B 109: 16684–16694.

Esawi, A. M. K. & Farag, M. M. (2007). Carbon nanotube reinforced composites: Potential and current challenges, *Materials and Design* 28: 2394–2401.

Fogler, H. C. (2006). *Elements of Chemical Reaction Engineering*, 4th Edition, Pearson Education, New Jersey.

Gohier, A., Ewels, C.P., Minea, T.M. & Djouadi, M.A. (2008). Carbon nanotube growth mechanism switches from tip- to base-growth with decreasing catalyst particle size, *Carbon* 46: 1331-1338.

Gommes, C., Blacher, S., Bossuot, Ch., Marchot, P., Nagy, J. B. & Pirard, J-P. (2004). Influence of the operating conditions on the production rate of multi-walled carbon nanotubes in a CVD reactor, *Carbon* 42: 1472-1482.

Gruen, D. M., Liu, S., Krauss, A. R., Pan, X. (1994). Buckyball microwave plasma fragmentation and diamond-film growth, *Journal of Applied Physics* 75(3): 1758-1763

Hofmann, S., Csányi, G., Ferrari, A.C., Payne, M.C. & Robertson, J. (2005). Surface diffusion: the low activation energy path for nanotube growth, *Physical Review Letters* 95 (036101):1–4.

Hou, Y. Q., Zhuang, D. M., Zhang, G., Wu, M. S. & Liu, J. J. (2002). Preparation of diamond films by hot filament chemical vapour deposition and nucleation by carbon nanotubes, *Applied Surface Science* 185: 303–308.

Iijima, S. (1991). Helical microtubules of graphitic carbon, *Nature* 354: 56-58.

Iyuke, S.E., Piennar, C. H. vZ., Abdulkareem, A. S. & Afolabi, A. S. (2006). Optimising carbon nanotubes continuous production in a swirled fluid chemical vapour deposition reactor, In: *SACE* 2006, Available from http://uir.unisa.ac.za/bitstream/10500/3261/1/abdulkareem.pdf

Iyuke, S. E. (2007). A process for production of carbon nanotubes, PCT Application WO2007/026213 (Published 2007/03/08) priority data ZA 2005/03438 2005/08/29.

Iyuke, S.E., Abdulkareem, A. S., Afolabi, A. S. & Piennar, C. H. vZ. (2007). Catalytic production of carbon nanotubes in a swirled fluid chemical vapour deposition reactor, *International Journal of Chemical Reactor Engineering* 5(Note S5):1-9.

Iyuke, S.E., Mamvura,T. A., Liu, K., Sibanda, V., Meyyappan, M. & Varadan V. K. (2009). Process synthesis and optimization for the production of carbon nanostructures, *Nanotechnology* 20:375602-375611.

Jacques, D., Villain, S., Rao, A.M., Andrews, R., Derbyshire, F., Dickey, E.C. & Qian, D. (2000). Synthesis of multiwalled carbon nanotubes, *Materials Research Society* 593: 15-20.

Journet, C. & Bernier, P. (1998). Production of carbon nanotubes, *Applied Physics* A 67: 1-9.

Journet, C., Maser, W. K., Bernier, P., Loiseau, A., Lamy de la Chapelle, M., Lefrant, S., Denlard, P., Lee, R. & Fischer, J. E. (1997). Large-scale production of single-walled carbon nanotubes by the electric arc discharge, *Nature* 388: 756-758.

Kim, K., Kim, K. J., Jung, W.S., Bae, S.Y., Park, J., Choi, J. & Choo, J. (2005). Chemical Investigation on the Temperature Re-dependent Growth rate of Carbon Nanotubes using Chemical Vapour Deposition of Ferrocene and Acetylene, *Chemical Physics letters* 40:459-469.

Kunadian, I., Andrews, R., Mengüç, M. P. & Qian, D. (2008). Multiwalled carbon nanotube deposition profiles within a CVD reactor: An experimental study, *Chemical Engineering Science* 64(7): 1503-1510.

Kumar, M. & Ando, Y. (2005). Controlling the diameter of carbon nanotubes grown from camphor on a zeolite support, *Carbon* 43: 533-540.

Kuwana, K., Endo, H., Saito, K., Qian, D., Andrews, R. & Grulke, E. A. (2005). Catalyst deactivation in CVD synthesis of carbon nanotubes, *Carbon* 43: 253 - 260.

Kuwana, K. & Saito, K. (2007). Modeling ferrocene reactions and iron nanoparticle formation: application to CVD synthesis of carbon nanotubes, *Proceedings of the Combustion Institute* 31(2): 1857-1864.

Lee, S. J., Baik, H. K., Yoo, J. & Han, J. H. (2002). Large scale synthesis of carbon nanotubes by plasma rotating arc discharge technique, *Diamond and Related Materials* 11: 914-917.

Leonhardt, A., Hampel, S., Müller, C., Mönch, I., Koseva, R., Ritschel, M., Elefant, D., Biedermann, K. & Büchner, B. (2006). Synthesis, Properties, and Applications of Ferromagnetic-Filled Carbon Nanotubes, *Chemical Vapor Deposition* 12(6): 380-387.

Levenspiel, O. (1999). *Chemical Reaction Engineering*, 3rd Edition, John Wiley and Sons, New York.

Lewis, K. E. & Smith, G. P. (1984). Bond dissociation energies in ferrocene, *Journal of the American Chemical Society* 106: 4650–4651.

Liu, Y., Hu, W., Wang, X., Long, C., Zhang, J., Zhu, D., Tang, D. & Xie S. (2000). Carbon nanorods, *Chemical Physics Letters* 331: 31–34.

Liu, X.Y., Huang, B.C. & Coville, N. J. (2002). The $Fe(CO)_5$ catalyzed pyrolysis of pentane: carbon nanotubes and carbon nanoballs formation, *Carbon* 40: 2791–2799.

Liu, R. (2008). The functionalisation of carbon nanotubes, PhD Thesis, University of New South Wales, Australia.

Maphutha, S. (2009).Thermodynamic study of carbon nanotube production from greenhouse gases during syngas synthesis, MSc Dissertation, University of the Witwatersrand.

Meilunas, R., Chang, R.P.H., Liu, S.Z. & Kappes, M. (1991). Nucleation of diamond films on surfaces using carbon clusters, *Applied Physics Letters* 59(26): 3461-3463.

Mhlanga, S. D., Coville, N. J., Iyuke, S. E., Afolabi, A. S., Abdulkareem, A. S. & Kunjuzwa, N (2010). Controlled syntheses of carbon spheres in a swirled floating catalytic chemical vapour deposition vertical reactor, *Journal of Experimental Nanoscience* 5(1): 40–51.

Merkoci, A. (2006). Carbon nanotubes in analytical sciences, *Microchimica Acta* 152(3-4):157-174.

Meyyappan, M. (2004). *Growth: CVD and PECVD. Carbon nanotubes: science and applications,* CRC Press, Boca Raton

Moisala, A., Nasibulin, A. G. & Kauppinen, E. I. (2003). The role of metal nanoparticles in the catalytic production of single-walled carbon nanotubes - a review, *Journal of Physics: Condensed Matter* 15: S3011-S3035.

Moisala, A., Nasibulin, A. G., Brown, D. P. & Jiang, H. (2006). Single-walled carbon nanotube synthesis using ferrocene and iron pentacarbonyl in a laminar flow reactor, *Chemical Engineering Science* 61: 4393-4402

Monthioux, M. & Kuznetsov, V. L. (2006). Who should be given the credit for the discovery of carbon nanotubes?, *Carbon* 44: 1621-1623

Moothi, K. (2009). Carbon nanotube production from greenhouse gases during syngas synthesis, MSc Dissertation, University of the Witwatersrand.

Nolan, P. E., Schabel, M. J., Lynch, D. C. & Cutler, A. H. (1995). Hydrogen control of carbon deposit morphology, *Carbon* 33: 79-85.

Paradise, M. & Goswami, T. (2007). Carbon nanotubes – Production and industrial applications, *Materials and Design* 28:1477-1489.

Qian, W., Wei, L., Cao, F., Chen, Q. & Qian, W. (2006). Low temperature synthesis of carbon nanospheres by reducing supercritical carbon dioxide with bimetallic lithium and potassium, *Carbon* 44(7): 1298-1352

Robertson, J. 2004. Realistic applications of CNTs, *Materials Today* 7(10): 46-52

Saito, Y. (1995). Nanoparticles and filled nanocapsules, *Carbon* 33:979-988.

See, C .H. & Harris, A. T. (2007). A review of carbon nanotube synthesis via fluidized-bed chemical vapour deposition, *Industrial and Engineering Chemistry Research* 46(4): 997-1012.

Simate, G. S., Iyuke, S. E., Ndlovu, S., Yah, C. S. & Walubita, L. F. (2010). The production of carbon nanotubes from carbon dioxide: challenges and opportunities, *Journal of Natural Gas Chemistry* 19(5): 453-460.

Singh, C., Shaffer, M. S. P. & Windle, A. H. (2003). Production of controlled architectures of aligned carbon nanotubes by an injection chemical vapour deposition method, *Carbon* 41(2):359-368.

Sinnott, S.B., Andrews, R., Qian, D., Rao, A. M., Mao, Z., Dickey, E. C. & Derbyshire, F. (1999). Model of carbon nanotube growth through chemical vapour deposition, *Chemical Physics Letters* 315: 25-30.

Snoeck, J.W., Froment, G. F. & Fowles, M. (1997). Filamentous carbon formation and gasification: Thermodynamics, driving force, nucleation, and steady-state growth, *Journal of Catalysis* 169: 240-249.

Soneda, Y., Duclaux, L. & Béguin, F. (2002). Synthesis of high quality multi-walled carbon nanotubes from the decomposition of acetylene on iron-group metal catalysts supported on MgO, *Carbon* 40(6): 965–969.

Song, I.K., Cho, Y,S., Choi, G.S., Park, J.B. & Kim ,D.J. (2004). The growth mode change in carbon nanotube synthesis in plasma-enhanced chemical vapor deposition, *Diamond and Related Materials* 13(4-8):1210–1213.

Subramoney, S. (1999). Carbon nanotubes - a status report, *Interface* 8: 34-37.

Tsoufis, T., Xidas, P., Jankovic, L., Gournis, D., Saranti, A., Bakas, T. & Karakassides, M. A. (2007). Catalytic production of carbon nanotubes over Fe-Ni bimetallic catalysts supported on MgO, *Diamond and Related Materials* 16(1): 155 -160.

Vivekchand, S.R.C., Cele, L.M., Deepak, F.L., Raju, A.R. & Govindaraj, A. (2004). Carbon nanotubes by nebulized spray pyrolysis, *Chemical Physics Lett*ers 386: 313-318.

Winterbottom, J. M. & King, M. B. (eds.) (1999). *Reactor Design for Chemical Engineers,* Stanley Thornes (Publishers) Ltd, Gloucestershire.

Wood, R. F., Pannala, S., Wells, J. C., Puretzky, A. A. & Geohegan, D. B. (2007). Simple model of the interrelation between single- and multiwall carbon nanotube growth rates for the CVD process, *Physical Review B* 75: 235446-235459.

Xiang, R., Einarsson, E., Okawa, J., Miyauchi, Y. & Maruyama, S. (2009). Acetylene-accelerated alcohol catalytic chemical vapor deposition growth of vertically aligned single-walled carbon nanotubes, *Journal of Physical Chemistry C* 113: 7511–7515.

Yah, C. S., Iyuke, S.E., Simate, G. S., Unuabonah, E. I., Bathgate, G., Mattews, G. & Cluett, J.D. (2011). Continuous synthesis of multiwalled carbon nanotubes from xylene using the swirled floating catalyst chemical vapour deposition technique, *Journal of Materials Research* 26(5): 640-644.

Yamada, T., Maigne, A., Yudasaka, M., Mizuno, K., Futaba,D. N., Yumura, M., Iijima, S. & Hata, K. (2008). Revealing the secret of water-assisted carbon nanotube synthesis by microscopic observation of the interaction of water on the catalysts, *Nano Letters* 8: 4288–4292.

Yang, K. L. & Yang, R. T. (1986). The accelerating and retarding effects of hydrogen on carbon deposition on metal surfaces, *Carbon* 24 (6): 687-693.

Yoon, Y.J. & Baik, H. K. (2001). Catalytic growth mechanism of carbon nanofibers through chemical vapour deposition, *Diamond and Related Materials* 10: 1214–1217.

Young, J. L. & DeSimone, J. M. (2000). Frontiers in green chemistry utilising carbon dioxide for polymer synthesis and applications, *Pure and Applied Chemistry* 72(7): 1357-1363.

Yu, Z., Chen, D., Tøtdal, B., Zhao, T., Dai, Y., Yuan, W. & Holmen, A. (2005). Catalytic engineering of carbon nanotube production, *Applied Catalysis* A: General 279: 223-233.

Zhang, F., Shen, J., Sun, J. & McCartney, D. G. (2006). Direct synthesis of diamond from low purity carbon nanotubes, *Carbon* 44(14): 3136–3138.

Zhong, Z., Chen, H., Tang, S., Ding, J., Lin, J. & Tan, K. L. (2000). Catalytic growth of carbon nanoballs with and without cobalt encapsulation, *Chemical Physics Letters* 330: 41–47.

Zhong, G., Hofmann, S., Yan, F., Telg, H., Warner, J. H., Eder, D., Thomsen, C., Milne, W. I. & Robertson, J. (2009). Acetylene: A key growth precursor for single-walled Carbon nanotube forests, *Journal of Physical Chemistry* C 113: 17321–17325.

3

Synthesis of Carbon Nanotubes Using Metal-Modified Nanoporous Silicas

Pezhman Zarabadi-Poor and Alireza Badiei
School of Chemistry, College of Science, University of Tehran, Tehran, Iran

1. Introduction

Some years ago a scientist, Sumio Iijima, who was working in NEC laboratories in Japan as an electron microscopist has discovered a new class of carbon allotropes which attracted a lot of interests among material science investigators. Briefly, after some investigation on the produced soot on walls of arc discharge reactor he examined the precipitated soot on cathode and has found really interesting graphitic structures. The taken transmission electron microscopy (TEM) images revealed a tubular graphitic structure that has fascinated the scientist due its extraordinary physical and chemical properties (Iijima, 1991).

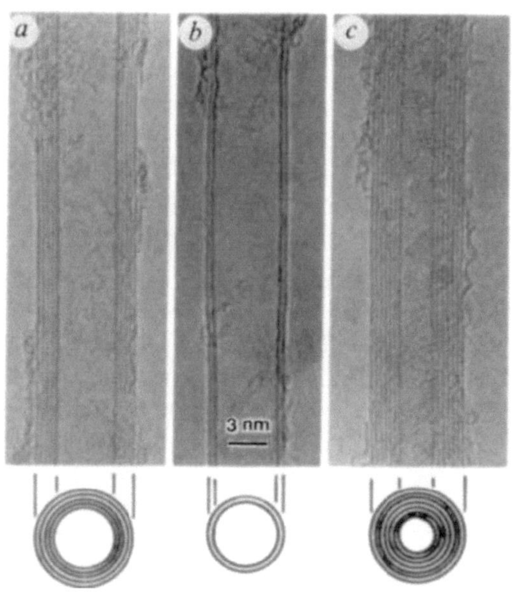

Fig. 1. TEM micrographs of carbon nanotubes consisting (a) five, (b) two, (c) seven graphitic sheets (Iijima, 1991)

After 1991, considerable attention of scientists to carbon nanotubes (CNTs) caused remarkable increase of publication around carbon nanotube related subjects of interests (Fig. 2). Increasing the number of publications from 77 in 1995 to 5619 in 2010 and total number of 37112 publications until the end of 2010, all shows that there are some interesting points in carbon nanotubes which led to this enormous attention of researchers. Basically, unique electronic, optoelectronic, and mechanical properties of carbon nanotubes (Saito & Dresselhaus, 1998) are among the most important reasons for this huge attention to them. These properties make CNTs suitable for a lot of applications such as electronic devices (Fuhrer et al., 2000), chemical sensors (Kong et al., 2000), hydrogen storage (Dillon et al., 1997), mechanical devices (Frank et al., 1998), field emission tips (Fan et al., 1999), nanotweezers (Kim & Lieber, 1999), and so on.

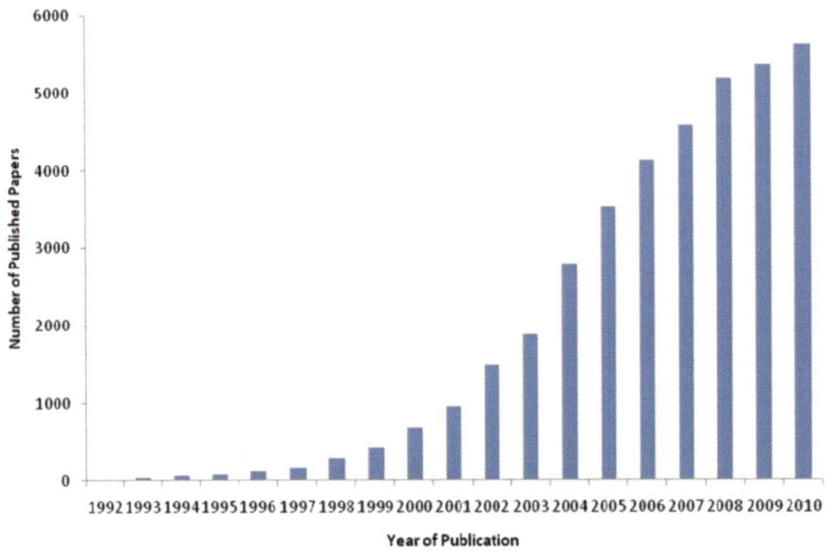

Fig. 2. The annual number of published papers on carbon nanotubes since their discovery until the end of 2010. Data were taken from www.scopus.com using carbon nanotubes* as keyword for article title.

Accordingly, application of carbon nanotubes in aforementioned and other devices needs production of CNTs. Up to know, several synthetic approaches such as arc discharge (Iijima, 1991; Iijima & Ichihashi, 1993; Tans et al., 1997), laser ablation (Guo et al., 1995; Thess et al., 1996), chemical vapor deposition (CVD) (Jose-Yacaman et al., 1993), were introduced to produce carbon nanotubes. Based on the related scientific literature, each of these methods has its charasteristics, advantageous and disadvantageous which define the applicability of mentioned methods. Among these methods, CVD has been shown to be of more industrial interest due to its lower reaction temperature in comparison with arc discharge and laser ablation (Morgan & Mokaya, 2008) and also possibility for industrial scale up (Geng et al., 2002; Somanathan et al., 2006). Indeed, catalytic CVD synthesis of CNTs can be accomplished via two general approaches, fluidized bed method (García-García et al., 2008; Wei et al., 2008) or fixed solid supported metal catalysts (Zarabadi-Poor et al., 2010). Therefore, these procedures need supporting of metals on solid supports to catalyze the carbon nanotubes preparation. Alumina (Zarabadi-Poor et al., 2010), nanoporous silica (Somanathan et al., 2006), MgO (Steplewska & Borowiak-Palen, 2010), and silicon (Mizuno et al., 2005) are among the most studied substrates.

In this chapter, the application of metal-modified nanoporous silicas for CNT growth in a CVD reactor will be reviewed. Herein, different nanoporous silicas which were used for production of carbon nanotubes considering different metals and catalyst preparation methods are the main subjects of following discussions.

2. Nanoporous silicas

In 1992, researchers of Mobil introduced a family of ordered nanoporous materials named M41S (Beck et al., 1992; Kresge et al., 1992). Then attraction of other scientists to this materials caused discovery of other types of nanoporous materials such as SBA-15 (Zhao et al., 1998), LUS-1 (Bonneviot et al., 2003),FSM-16 (Inagaki et al., 1993), KIT-1 (Ryoo et al., 1996), MSU (Bagshaw et al., 1995), and HSM (Tanev & Pinnavaia, 1995)which were synthesized within highly acidic to strongly basic pH range using cationic, anionic, neutral and nonionic structure directing agents. Among these materials, MCM-41, LUS-1 and SBA-15 are most studied ones in our previous researches such as studies on metal modification (Badiei & Bonneviot, 1998; Béland et al., 1998), pre-concentration of metals (Ganjali et al., 2006; Javanbakht et al., 2009b; Javanbakht et al., 2010), modification with 8-hydroxyquinoline (Badiei et al., 2011b; Badiei et al., 2011a), carbon paste electrodes (Ganjali et al., 2010; Javanbakht et al., 2007a; Javanbakht et al., 2007b; Javanbakht et al., 2009a), solid phase micro extraction (Hashemi et al., 2009) and hydroxylation catalysts (Arab et al.). Therefore, these materials are the major subject in the coming discussions about catalytic synthesis of CNTs.

One of most important application of solid supports such as nanoporous silicas is related to the heterogenization of transition metals (Cornils & Herrmann, 1996). Briefly, several methods have been used in this manner such as adsorption or encapsulation of homogenous catalyst (Coronado et al., 2000; Goettmann et al., 2006; Kuil et al., 2006; Zhu et al., 2004), covalent immobilization (Pagar et al., 2006), modification of supported metal catalysts (Marchetti et al., 2004) and etc. To the best of our knowledge, iron, cobalt and nickel are the most studied metals in catalytic production of carbon nanotubes and therefore they will be reviewed in following sections.

3. Synthesis of CNTs using Fe-modified nanoporous silicas

Several reports have been published on the synthesis of CNTs using iron-modified nanoporous silicas up to now. Duxiao et al. (Duxiao et al., 2001) investigated the CVD synthesis of carbon nanotubes on hexagonal nanoporous silica (HMS). They used acetylene as carbon precursor and showed that in the case of HMS the formation of CNTs started at the internal surface of Fe-HSM. In 2005, Lu et al. (Lu et al., 2005) addressed a rare report on the synthesis of carbon nanotubes. It was remarkable because it is related to the synthesis of amorphous carbon nanotubes via CVD procedure and using SBA-15 as the solid support for the iron based catalysts. They have employed impregnation method for introduction of iron species using $Fe(acac)_3$ as iron precursor into the structure of SBA-15 nanoporous silica. The main purpose of using SBA-15 as the solid support in the as mentioned published letter is the porous structure of SBA-15. They believe that it can lead the formation of metal catalysts with well-controlled particle size. They concluded from the formation of carbon nanotubes on the outer side (Fig. 3) of channels that particle migration happened during the CVD process and proposed "top growth mechanism" herein. Also, they mentioned that the outer diameter of CNTs (about 20nm) is much higher than pore diameter of SBA-15 (about 7nm) which caused by the thickness of nanotubes.

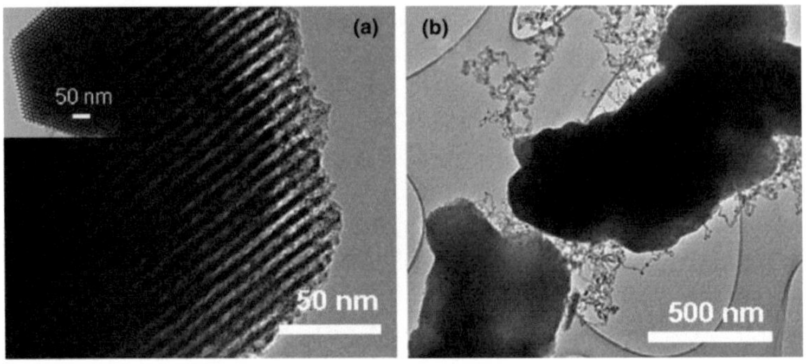

Fig. 3. TEM micrographs of a) Fe-SBA-15 and b) grown CNTs on the outer surface of catalyst (Lu et al., 2005)

Wang and his co-workers presented the synthesis of CNTs using Fe-SBA-15 catalysts (Wang et al., 2005a; Wang et al., 2005b). They, (Wang et al., 2005a), have synthesized Fe-SBA-15 via direct synthesis approach. Iron species were introduced into the structure through addition of iron nitrate nonahydrate during the synthesis of SBA-15 besides the increasing of pH to 7. According to their research, synthesis of multi-walled carbon nanotubes (MWNTs) by CVD decomposition of acetylene on the Fe-SBA-15 produced straight, uniform, open and highly graphitized CNTs. Increasing the amount of iron in the catalysts caused obtaining CNTs with larger diameters and as a conclusion the best CNTs can be obtained using catalysts with 3.5% iron content. A typical high resolution transmission electron microscopy (HRTEM) image of this sample is given in Fig. 4.

Fig. 4. HRTEM image of carbon nanotube prepared using catalyst with 3.5% iron (Wang et al., 2005a)

The same research group (Wang et al., 2005b), have tried to obtain Fe-SBA-15 with higher thermal stability. The synthesis procedure for the catalysts is similar to their previous works (Wang et al.,2005a). Their efforts to obtain more thermal stable catalysts have been accomplished by calcination of catalysts in the temperature ranging from 550 to 950 °C. Synthesis of carbon nanotubes were again done via similar procedure to the first article. They have concluded that the obtained CNTs are more uniformed with outer and inner diameters about 35nm and 15nm, respectively. In the other research Barreca et al. (Barreca et al., 2007) used the advantages of direct synthesis approach for preparation of catalysts. They have used P85 triblock copolymer to synthesize hexagonal nanoporous silica under acidic condition. The iron phthalocyanine was used as iron precursor to incorporate the iron nanoparticles into the nanoporous structure. The acid treatment was applied to eliminate the Fe_2O_3 particles from the outer surface of nanoporous silicas. In the acid washed catalysts, MWNTs with narrow diameter distribution (5 and 15 nm) were obtained. On the other hand, unwashed catalysts produced CNTs on the outer surface with a broad diameter distribution ranging from 20 to 90nm. Also, higher metal content in both catalysts caused higher CNT density. The most recent report on the synthesis of carbon nanotubes using iron-modified nanoporous silica has been published by Gokulakrishnan and co-workers (Somanathan et al., 2011). KIT-6 type nanoporous silica was synthesized following typical routes through using P123 structure directing agent under acidic condition. Iron species were incorporated into the structure by impregnation of KIT-6 with iron nitrate solution. They have mentioned that 1-3% iron was loaded into KIT-6 which caused decreasing the specific area from 745 $m^2.g^{-1}$ (for KIT-6) to 653, 551, and 377 $m^2.g^{-1}$for 1%Fe, 2%Fe and 3%Fe, respectively. The CNT synthesis was carried using acetylene as carbon source at 800 °C. The best yield was obtained by 2% Fe loaded catalyst (91%). The 1% Fe catalysts did not show significant activity through the synthesis of CNTs. It was expected to obtain higher production yield when using catalyst with higher iron contents than 2% but it has not been observed. The observed yield for 3% Fe catalyst was reported 79% which is attributed to the agglomeration of iron particles. It has been concluded from TEM micrographs which 10 nm graphene layers consists of 29 graphene sheets. As a main conclusion the Fe-modified KIT-6 has been introduced as a potential for large scale production of multi-walled carbon nanotubes.

4. Synthesis of CNTs using Co-modified nanoporous silicas

To the best of our knowledge, MCM-41 is the major solid support in the most of the researches on the synthesis of CNTs using Co metal as the active site.

Lim et al. (Lim et al., 2003) utilized direct synthesis method to incorporate Co species into the structure of MCM-41. Quaternary ammonium surfactants, $C_nH_{2n+1}(CH_3)_3NBr$, with n= 10, 12, 14, 16, and 18 were used as structure directing agents in this research. Investigation of Co atoms status inside the Co-MCM-41 materials revealed that about 30-40 atoms uniformly exist in each pore as well as they present as a mixture of tetrahedral and distorted tetrahedral structures like Co_3O_4. The crystallization time (20-160 hr) and temperature (100 and 150 °C) were monitored in order to obtain optimums for both MCM-41 and Co-MCM-41. It has been mentioned that for MCM-41 4 days and 100 °C are optimum synthetic conditions, however, for Co-MCM-41 6 days and 100 °C produce better products.

In the next step, C16 Co-MCM-41 has been selected for further catalytic activity evaluation on the synthesis of single-walled carbon nanotubes (SWNTs). It exhibited good results in this manner with 90% selectivity over SWNT and 4 wt% over carbon deposition. TEM micrograph and corresponding Raman spectra (Fig. 5) shows the successful synthesis of CNTs. Gary L. Haller's research group continued the aforementioned research by following works. They have investigated the effect of catalyst pre-reduction, nanotube growth temperatures (Chen, 2004b), CO pressure, and reaction time (Chen, 2004a) using Co-MCM-41 catalysts. Chen et al. used C16 Co-MCM-41 with 1 wt% cobalt as catalyst (Lim et al., 2003). As pointed out in (Chen, 2004b), single-walled carbon nanotubes were synthesized by

CO disproportionation under following conditions: reaction time, 60 min; reaction temperature, 650-900 °C; and CO pressure, 6 atm. Basically, the influence of catalyst pre-reduction condition and growth temperature on the selectivity and uniformity of SWNT were observed. In order to investigate the effect of pretreatment on the catalyst, two techniques were applied: *in-situ* X-ray Absorption Near-Edge Structure (XANES) for investigation of cobalt state during pretreatment, and *ex-situ* Extended X-ray Absorption Fine Structure (EXAFS) to monitor the formation of cobalt clusters during SWNT growth. The aim of this research, obtaining SWNTs with narrow diameter distribution have been accomplished by applying optimized condition; pre-reducing of catalysts at 500-600 °C and growing SWNTs at 750 -800 °C. Figure 6 indicates the correlation between quality and yield of synthesized carbon nanotubes which shows the optimum synthesis condition. The I_D/I_G ratio in Figure 6a indicates the ratio between two major peaks in Raman spectrum of CNTs. D-band appears around 1340 cm^{-1} and is related to disordered graphite. High frequency G-band appears around 1580 cm^{-1} and is attributed to splitting of the E_{2g} stretching mode of graphite.

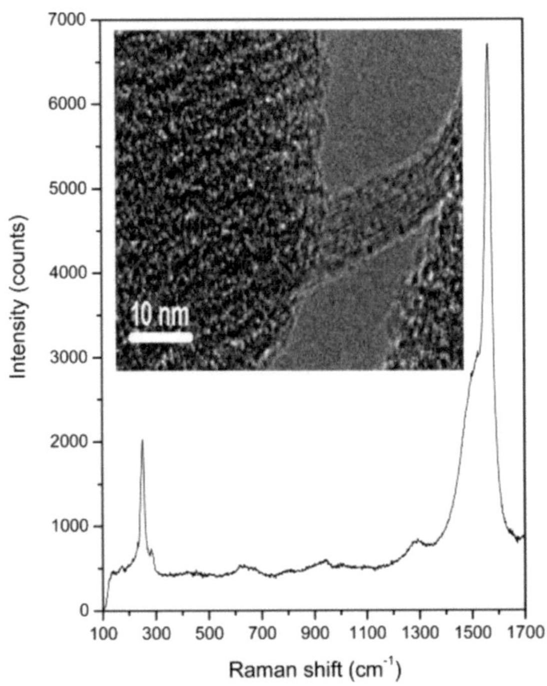

Fig. 5. Raman spectrum and TEM image of CNT synthesized using C16 Co-MCM-41 (Lim et al., 2003)

In the second article (Chen, 2004a), Chen et al. applied the optimized pretreatment and reaction temperature (500 and 800 °C, respectively) to monitor the effect of CO pressure and reaction time on disproportional decomposition of CO in SWNT synthesis. They have chosen 2-6 atm. range for CO partial pressure and 5-120 min. for reaction time. It has been concluded that following processes affects both selectivity and diameter uniformity of SWNTs under applied synthetic conditions: 1) reduction of cobalt, 2) nucleation of the reduced cobalt atoms into clusters and 3) initiation and growth of the SWNTs.

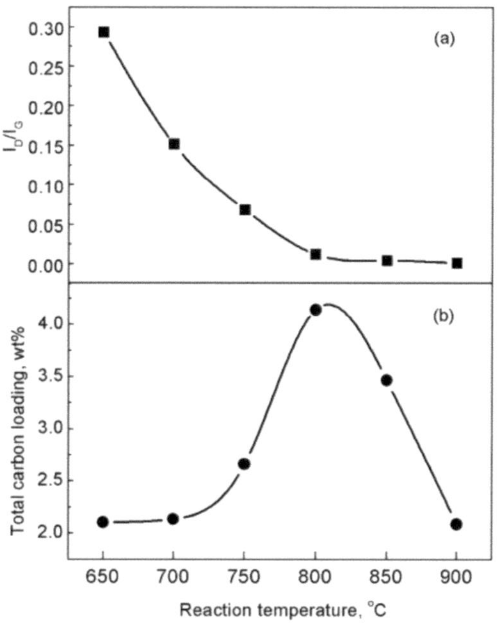

Fig. 6. Correlation diagram between synthesis temperatures (°C), quality index (I_D/I_G), and total carbon loading (%) (Chen, 2004b)

TEM images of synthesized CNTs under different CO pressures are given in Fig. 7. According to these images, it is obvious that higher CO pressures causes better uniformity of synthesized CNTs. Therefore, synthesis of SWNTs under different CO partial pressures revealed two remarkable results. Applying low CO pressures in the synthesis provides SWNTs with broad diameter distribution. It has been explained that in such circumstances reduction of cobalt species and growth of SWNTs are slow (Chen, 2004a). Therefore, cobalt clusters, which are the sites of SWNT formation, are allowed to grow into larger sizes. On the other hand, synthesis of SWNTs under high pressure of CO provides SWNTs with uniform and narrow diameter distribution. Considering the aforementioned discussion about low pressure of CO, it is reasonable.

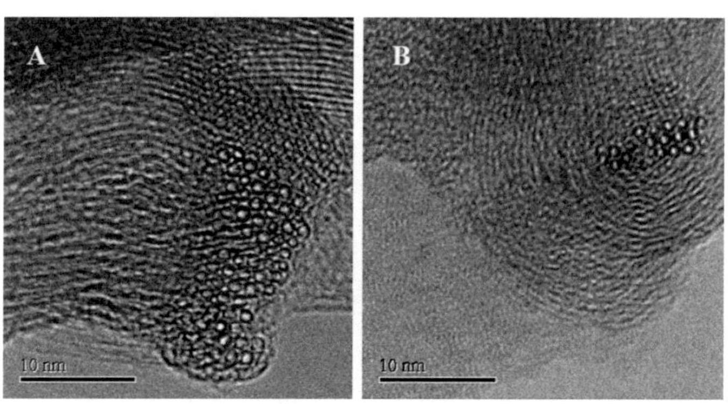

Fig. 7. TEM micrographs synthesized CNTs under CO pressure of a) 2 and b) 6 atm (Chen, 2004a)

After above remarkable researches (Yang et al., 2004), Garry Haller and co-workers have carried out a comprehensive statistical research to recognize how different variable parameters affect the synthesis of Co-MCM-41 catalysts in order to obtain aligned SWNTs. The independent variables which have been considered in this article are: *alkyl chain length, initial cobalt concentration, surfactant silica ratio, TMA silica ratio, water silica ratio, pore diameter, metal composition and structural order.* Considering these variables, several material synthesis have been designed, ID: Co01-Co28, which has produced a scatter plot matrix for analyzing the effect of main synthesis variables. It has been raised to a normalized model which is helpful in quantitative measurement of interaction between different variables considering corresponding correlation coefficients. The most important correlation coefficients are given in Table 1.

Correlation coefficient	Considered variables
0.9974	Metal composition in Co-MCM-41 and initial Co concentration in the synthesis solution
0.9261	Pore size and alkyl chain length of surfactant
0.5756	Structural order and alkyl chain length of surfactant

Table 1. Correlation coefficients between different variables in Yang et al. paper (Yang et al., 2004)

It has been concluded that the presented model can be used to produce samples with varying pore size and constant cobalt content while having a high degree of structural order.

As it has mentioned above regarding Chen et al.'s articles (Chen, 2004b; a), the size of cobalt clusters has a strong effect on the properties of resulting carbon nanotubes. Ciuparu et al.'s paper (Ciuparu et al., 2004) presents a deeper investigation on the mechanism of cobalt cluster size control. The interesting result of this work is the effect of hydrogen pretreatment of catalyst which facilitates the synthesis of SWNTs with small diameters.

Another statistical analysis following Haller research group works addresses investigation of C10 Co-MCM-41 catalyst in order to obtain SWNTs with smaller diameters (Yang et al., 2005). This work is similar to their previous published article (Yang et al., 2004). Herein, Yang et al. defined experimental parameters as follows: *surfactant to silica ratio, pH, TMASi to Si ratio, Co concentration, water to Si ratio, slope of isotherm capillary condensation step and ratio of peak intensity of (110) to (100) of X-ray diffraction pattern.* Considering the first five variables (independent variables) synthesis condition for 27 samples have been designed and carried out. It has been concluded that the most significant effects on the structural order of C10 Co-MCM-41 samples rise from the first three variables. The other independent variables may negatively affect the structural order of mentioned catalyst. Raman spectra of synthesized SWNTs using C10 Co-MCM-41 and C16 Co-MCM-41 are shown in Fig. 8. Appearing the small peak at 301 cm^{-1} reveals the successful synthesis of SWNTs with C10 Co-MCM-41 (3 wt% cobalt) with 0.77 nm diameter.

The effect of cobalt content on the synthesis of SWNTs via CO disproportion on Co-MCM-41 catalysis was presented in L. D. Pfefferle co-workers paper (Chen et al., 2006). Chen et al. synthesized Co-MCM-41 samples with Co concentration ranging from 0.5 to 4 %. The lowest concentration of Co leads to obtain few Co clusters which are large enough to initiate the growth of single-walled carbon nanotubes. While using Co concentration higher than 3 % causes the formation of large cobalt clusters and consequently production of undesired products.

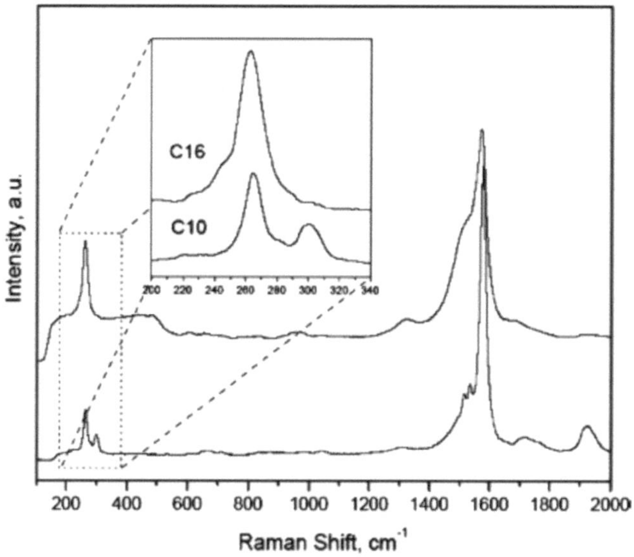

Fig. 8. Raman spectra of synthesized SWNTs using C10 Co-MCM-41 and C16 Co-MCM-41 (Yang et al., 2005)

The last article which has been reviewed here is a report on the application of Co-MCM-41 materials for the synthesis of SWNTs via CVD method. Somanathan et al. (Somanathan et al., 2006) incorporated Co species into the structure of MCM-41 through direct synthesis approach with different Si/Co ratios ranging from 25 to 100. SWNTs were synthesized at 750 °C using acetylene as carbon source with 40 ml.min^{-1} flow rate for 10 min. They have concluded that Co-MCM-41 is very stable under severe reaction condition as well as the obtaining high selectivity toward SWNT using Co-MCM-41 with Si/Co =100 (Fig. 8). The proposed growth mechanism is mentioned top-growth considering different evidences such as TEM image (Fig. 10).

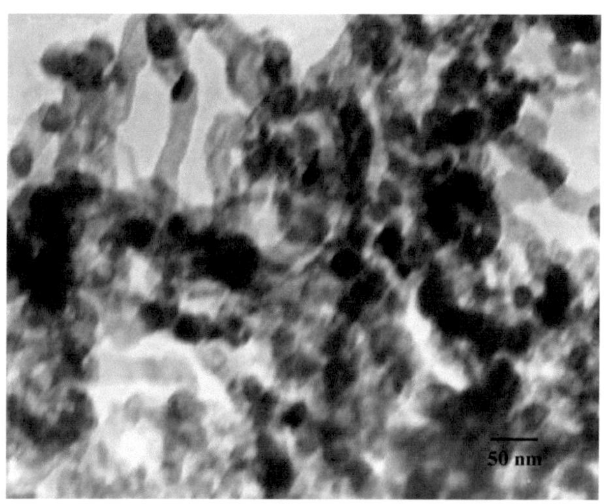

Fig. 9. TEM image of CNTs synthesized using Co-MCM-41 (100) (Somanathan et al., 2006)

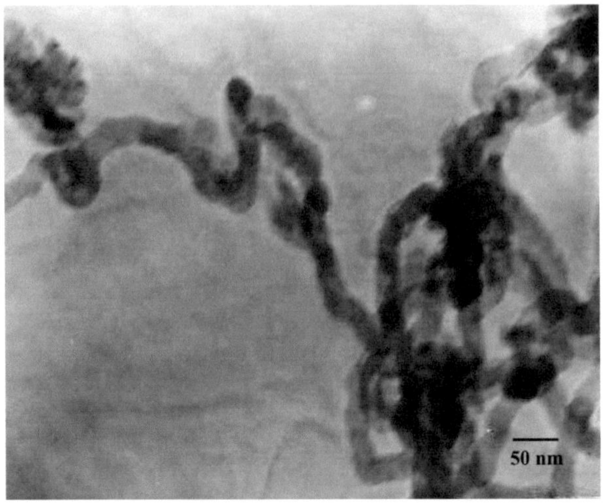

Fig. 10. TEM micrographs with cobalt particles at the tip of nanotubes (Somanathan et al., 2006)

5. Synthesis of CNTs using Ni-modified nanoporous silicas

Herein, some other articles which address application of Ni-MCM-41 in the synthesis of carbon nanotubes are presented.

Chen et al. besides their researches on the Co-MCM-41 catalysts also have carried out research on the Ni-MCM-41 materials. In some of their works (Chen et al., 2005), Ni (1 wt%) have been incorporated into the structure of MCM-41 based on their previous report (Lim et al., 2003). The SWNT synthesis approach is also similar. It has been concluded that using Ni-MCM-41 in comparison with Co-MCM-41 produces SWNTs with broader diameter distribution and lower carbon deposition. This has been related to the tendency of Ni to CO which causes faster reduction of nickel particles and migration of nickel clusters. Chen et al also proposed, Ni-MCM-41 catalyst, while using other carbon sources can improve the properties of SWNTs. Then, Chen et al. tried to investigate the effect of various carbon sources on the synthesis of SWNTs using Ni-MCM-41 (Chen et al., 2007). Synthesis of SWNTs using Ni-MCM-41 using CO, ethanol and methane under similar conditions revealed that CO produced SWNTs with narrower diameter distribution than ethanol. On the other hand, using methane as the carbon precursor was not a successful experiment under same conditions to produce CNTs .

Somanathan et al. also followed their previous works on Co-MCM-41 catalysts by reporting synthesis of SWNTs over Ni-MCM-41 nanoporous materials (Somanathan & Pandurangan, 2006). Herein, Ni-MCM-41 materials were synthesized through similar methods mentioned in their work on Co-MCM-41 (Somanathan et al., 2006). Performing experiments using catalysts with different Si/Ni ratios and reaction conditions provided optimum synthesis circumstances. It has been mentioned that using Ni-MCM-41 with Si/Ni=100 at 750 °C for 10 min at N_2 and acetylene flow rates of 140 and 60 ml.min^{-1} gives the highest carbon yield (71.01%). Presence of metal particles at the tip of grown carbon nanotubes brought authors to the "tip-growth" mechanism occurring in these experiments.

6. Concluding remarks and future directions

Basically, reviewing mentioned articles here has shown that using metal-modified nanoporous silica as catalysts for the synthesis of CNTs was interesting for scientists to improve the synthetic approaches in order to obtain CNTs with desired properties and conditions. Although there are a lot of published works in this manner, taking a look at these researches reveals that still there are a lot of points which need to being studied for further development on the synthesis procedures of CNTs.

Briefly, industries need different class of carbon nanotubes such as single-walled, multiwalled, with junction and so on. Therefore, future studies should be done in order to improve the current methods. These researches should consider two major points, 1) obtaining CNTs with desired properties, 2) decreasing the cost of production. In this manner, studying the effect of different methods for introducing metal species into the structure of nanoporous silicas, effect of calcination of the metal-modified nanoporous silicas, influence of different carbon precursors, combination of two metals and etc. were subject of some few previous reports and can be the interesting subject of research for upcoming works.

7. Acknowledgment

The authors thank the University of Tehran Research Council for support of this work.

8. References

Arab, P., Badiei, A., Koolivand, A. & Mohammadi Ziarani, G. Direct Hydroxylation of Benzene to Phenol over Fe3O4 Supported on Nanoporous Carbon, *Chin. J. Catal.* 32:258-263.

Badiei, A. & Bonneviot, L. (1998). Modification of Mesoporous Silica by Direct Template Ion Exchange Using Cobalt Complexes, *Inorg. Chem.* 37:4142-4145.

Badiei, A., Goldooz, H. & Ziarani, G. M. (2011a). A novel method for preparation of 8-hydroxyquinoline functionalized mesoporous silica: Aluminum complexes and photoluminescence studies, *Appl. Surf. Sci.* 257:4912-4918.

Badiei, A., Goldooz, H., Ziarani, G. M. & Abbasi, A. (2011b). One pot synthesis of functionalized SBA-15 by using an 8-hydroxyquinoline-5-sulfonamide-modified organosilane as precursor, *J. Colloid Interface Sci.* 357:63-69.

Bagshaw, S. A., Prouzet, E. & Pinnavaia, T. J. (1995). Templating of Mesoporous Molecular Sieves by Nonionic Polyethylene Oxide Surfactants, *Science* 269:1242-1244.

Barreca, D., Blau, W. J., Croke, G. M., Deeney, F. A., Dillon, F. C., Holmes, J. D., Kufazvinei, C., Morris, M. A., Spalding, T. R. & Tondello, E. (2007). Iron oxide nanoparticle impregnated mesoporous silicas as platforms for the growth of carbon nanotubes, *Microporous Mesoporous Mater.* 103:142-149.

Beck, J. S., Vartuli, J. C., Roth, W. J., Leonowicz, M. E., Kresge, C. T., Schmitt, K. D., Chu, C. T. W., Olson, D. H. & Sheppard, E. W. (1992). A new family of mesoporous molecular sieves prepared with liquid crystal templates, *J. Am. Chem. Soc.* 114:10834-10843.

Béland, F., Echchahed, B., Badiei, A. & Bonneviot, L. 1998. EXAFS multiple scattering characterization of metal modified mesoporous sieves. Pages 567-574 in: Stud. Surf. Sci. Catal. L. Bonneviot, F. Beland, C. Danumah, S. Giasson and S. Kaliaguine, eds. Elsevier.

Bonneviot, L., Morin, M. & Badiei, A. (2003). Mesostructured metal or non-metal oxides and method for making same, US 2003/0133868.

Chen, Y. (2004a). Synthesis of uniform diameter single wall carbon nanotubes in Co-MCM-41: effects of CO pressure and reaction time, *J. Catal.* 226:351-362.

Chen, Y. (2004b). Synthesis of uniform diameter single-wall carbon nanotubes in Co-MCM-41: effects of the catalyst prereduction and nanotube growth temperatures, *J. Catal.* 225:453-465.

Chen, Y., Ciuparu, D., Yang, Y., Lim, S., Wang, C., Haller, G. L. & Pfefferle, L. D. (2005). Single-wall carbon nanotube synthesis by CO disproportionation on nickel-incorporated MCM-41, *Nanotechnology* 16:S476-S483.

Chen, Y., Ciuparu, D., Lim, S., Haller, G. & Pfefferle, L. (2006). The effect of the cobalt loading on the growth of single wall carbon nanotubes by CO disproportionation on Co-MCM-41 catalysts, *Carbon* 44:67-78.

Chen, Y., Wang, B., Li, L., Yang, Y., Ciuparu, D., Lim, S., Haller, G. & Pfefferle, L. (2007). Effect of different carbon sources on the growth of single-walled carbon nanotube from MCM-41 containing nickel, *Carbon* 45:2217-2228.

Ciuparu, D., Chen, Y., Lim, S., Yang, Y., Haller, G. L. & Pfefferle, L. (2004). Mechanism of Cobalt Cluster Size Control in Co-MCM-41 during Single-Wall Carbon Nanotubes Synthesis by CO Disproportionation, *J. Phys. Chem. B* 108:15565-15571.

Cornils, B. & Herrmann, W. A. (1996). *Applied Homogeneous Catalysis with Organometallic Compounds*, VCH, New York.

Coronado, J. M., Coloma, F. & Anderson, J. A. (2000). Styrene hydroformylation over modified Rh/SiO2·Al2O3 catalysts, *J. Mol. Catal. A: Chem.* 154:143-154.

Dillon, A. C., Jones, K. M., Bekkedahl, T. A., Kiang, C. H., Bethune, D. S. & Heben, M. J. (1997). Storage of hydrogen in single-walled carbon nanotubes, *Nature* 386:377-379.

Duxiao, J., Nongyue, H., Yuanying, Z., Chunxiang, X., Chunwei, Y. & Zuhong, L. (2001). Catalytic growth of carbon nanotubes from the internal surface of Fe-loading mesoporous molecular sieves materials, *Mater. Chem. Phys.* 69:246-251.

Fan, S., Chapline, M. G., Franklin, N. R., Tombler, T. W., Cassell, A. M. & Dai, H. (1999). Self-Oriented Regular Arrays of Carbon Nanotubes and Their Field Emission Properties, *Science* 283:512-514.

Frank, S., Poncharal, P., Wang, Z. L. & Heer, W. A. d. (1998). Carbon Nanotube Quantum Resistors, *Science* 280:1744-1746.

Fuhrer, M. S., Nygård, J., Shih, L., Forero, M., Yoon, Y.-G., Mazzoni, M. S. C., Choi, H. J., Ihm, J., Louie, S. G., Zettl, A. & McEuen, P. L. (2000). Crossed Nanotube Junctions, *Science* 288:494-497.

Ganjali, M., Daftari, A., Hajiagha-Babaei, L., Badiei, A., Saberyan, K., Ziarani, G. & Moghimi, A. (2006). Pico Level Monitoring of Silver with Modified Hexagonal Mesoporous Compound (MCM-41) and Inductively Coupled Plasma Atomic Emission Spectrometry, *Water, Air, Soil Pollut.* 173:71-80.

Ganjali, M., Asgari, M., Faridbod, F., Norouzi, P., Badiei, A. & Gholami, J. (2010). Thiomorpholine-functionalized nanoporous mesopore as a sensing material for Cd²⁺ carbon paste electrode, *J. Solid State Electrochem.* 14:1359-1366.

García-García, F. R., Pérez-Cabero, M., Nevskaia, D. M., Rodríguez-Ramos, I. & Guerrero-Ruiz, A. (2008). Improving the synthesis of high purity carbon nanotubes in a catalytic fluidized bed reactor and their comparative test for hydrogen adsorption capacity, *Catal. Today* 133-135:815-821.

Geng, J., Singh, C., Shephard, D. S., Shaffer, M. S. P., Johnson, B. F. G. & Windle, A. H. (2002). Synthesis of high purity single-walled carbon nanotubes in high yield, *Chem. Commun.*:2666-2667.

Goettmann, F., Le Floch, P. & Sanchez, C. (2006). Highly regioselective terminal alkynes hydroformylation and Pauson-Khand reaction catalysed by mesoporous organised zirconium oxide based powders, *Chem. Commun.*:180-182.

Guo, T., Nikolaev, P., Thess, A., Colbert, D. T. & Smalley, R. E. (1995). Catalytic growth of single-walled manotubes by laser vaporization, *Chem. Phys. Lett.* 243:49-54.

Hashemi, P., Shamizadeh, M., Badiei, A., Zarabadi-Poor, P., Ghiasvand, A. R. & Yarahmadi, A. (2009). Amino ethyl-functionalized nanoporous silica as a novel fiber coating for solid-phase microextraction, *Anal. Chim. Acta* 646:1-5.

Iijima, S. (1991). Helical microtubules of graphitic carbon, *Nature* 354:56-58.

Iijima, S. & Ichihashi, T. (1993). Single-shell carbon nanotubes of 1-nm diameter, *Nature* 363:603-605.

Inagaki, S., Fukushima, Y. & Kuroda, K. (1993). Synthesis of highly ordered mesoporous materials from a layered polysilicate, *J. Chem. Soc., Chem. Commun.*:680-682.

Javanbakht, M., Badiei, A., Ganjali, M. R., Norouzi, P., Hasheminasab, A. & Abdouss, M. (2007a). Use of organofunctionalized nanoporous silica gel to improve the lifetime of carbon paste electrode for determination of copper(II) ions, *Anal. Chim. Acta* 601:172-182.

Javanbakht, M., Ganjali, M. R., Norouzi, P., Badiei, A., Hasheminasab, A. & Abdouss, M. (2007b). Carbon Paste Electrode Modified with Functionalized Nanoporous Silica Gel as a New Sensor for Determination of Silver Ion, *Electroanalysis* 19:1307-1314.

Javanbakht, M., Divsar, F., Badiei, A., Fatollahi, F., Khaniani, Y., Ganjali, M. R., Norouzi, P., Chaloosi, M. & Ziarani, G. M. (2009a). Determination of picomolar silver concentrations by differential pulse anodic stripping voltammetry at a carbon paste electrode modified with phenylthiourea-functionalized high ordered nanoporous silica gel, *Electrochim. Acta* 54:5381-5386.

Javanbakht, M., Khoshsafar, H., Ganjali, M. R., Badiei, A., Norouzi, P. & Hasheminasab, A. (2009b). Determination of Nanomolar Mercury(II) Concentration by Anodic-Stripping Voltammetry at a Carbon Paste Electrode Modified with Functionalized Nanoporous Silica Gel, *Curr. Anal. Chem.* 5:35-41.

Javanbakht, M., Rudbaraki, H., Sohrabi, M. R., Attaran, A. M. & Badiei, A. (2010). Separation, pre-concentration and determination of trace amounts of lead(II) ions in environmental samples using two functionalised nanoporous silica gels containing a dipyridyl sub-unit, *International Journal of Environmental Analytical Chemistry* 90:1014 - 1024.

Jose-Yacaman, M., Miki-Yoshida, M., Rendon, L. & Santiesteban, J. G. (1993). Catalytic growth of carbon microtubules with fullerene structure, *Appl. Phys. Lett.* 62:202-204.

Kim, P. & Lieber, C. M. (1999). Nanotube Nanotweezers, *Science* 286:2148-2150.

Kong, J., Franklin, N. R., Zhou, C., Chapline, M. G., Peng, S., Cho, K. & Dai, H. (2000). Nanotube Molecular Wires as Chemical Sensors, *Science* 287:622-625.

Kresge, C. T., Leonowicz, M. E., Roth, W. J., Vartuli, J. C. & Beck, J. S. (1992). Ordered mesoporous molecular sieves synthesized by a liquid-crystal template mechanism, *Nature* 359:710-712.

Kuil, M., Soltner, T., van Leeuwen, P. W. N. M. & Reek, J. N. H. (2006). High-Precision Catalysts: Regioselective Hydroformylation of Internal Alkenes by Encapsulated Rhodium Complexes, *J. Am. Chem. Soc.* 128:11344-11345.

Lim, S., Ciuparu, D., Pak, C., Dobek, F., Chen, Y., Harding, D., Pfefferle, L. & Haller, G. (2003). Synthesis and Characterization of Highly Ordered Co−MCM-41 for Production of Aligned Single Walled Carbon Nanotubes (SWNT), *J. Phys. Chem. B* 107:11048-11056.

Lu, A.-H., Schmidt, W., Tatar, S.-D., Spliethoff, B., Popp, J., Kiefer, W. & Schüth, F. (2005). Formation of amorphous carbon nanotubes on ordered mesoporous silica support, *Carbon* 43:1811-1814.

Marchetti, M., Paganelli, S. & Viel, E. (2004). Hydroformylation of functionalized olefins catalyzed by SiO2-tethered rhodium complexes, *J. Mol. Catal. A: Chem.* 222:143-151.

Mizuno, K., Hata, K., Saito, T., Ohshima, S., Yumura, M. & Iijima, S. (2005). Selective Matching of Catalyst Element and Carbon Source in Single-Walled Carbon Nanotube Synthesis on Silicon Substrates, *J. Phys. Chem. B* 109:2632-2637.

Morgan, S. & Mokaya, R. (2008). Aligned Bundles of Carbon Nanotubes Are Easily Grown on As-Synthesized Mesoporous Silicate Substrates, *J. Phys. Chem. C* 112:15157-15162.

Pagar, N., Deshpande, R. & Chaudhari, R. (2006). Hydroformylation of olefins using dispersed molecular catalysts on solid supports, *Catal. Lett.* 110:129-133.

Ryoo, R., Kim, J. M., Ko, C. H. & Shin, C. H. (1996). Disordered Molecular Sieve with Branched Mesoporous Channel Network, *J. Phys. Chem.* 100:17718-17721.

Saito, M. S. D. R. & Dresselhaus, G. (1998). *Physical Properties of Carbon Nanotubes*, World Scientific Publishing, Singapore.

Somanathan, T. & Pandurangan, A. (2006). Effective Synthesis of Single-Walled Carbon Nanotubes Using Ni−MCM-41 Catalytic Template through Chemical Vapor Deposition Method, *Ind. Eng. Chem. Res.* 45:8926-8931.

Somanathan, T., Pandurangan, A. & Sathiyamoorthy, D. (2006). Catalytic influence of mesoporous Co-MCM-41 molecular sieves for the synthesis of SWNTs via CVD method, *J. Mol. Catal. A: Chem.* 256:193-199.

Somanathan, T., Gokulakrishnan, N., Chandrasekar, G. & Pandurangan, A. (2011). Fabrication of multiwalled carbon nanotubes in the channels of iron loaded three dimensional mesoporous material by catalytic chemical vapour deposition technique, *Appl. Surf. Sci.* 257:2940-2943.

Steplewska, A. & Borowiak-Palen, E. (2011). Study on the effect of the metal–support (Fe-MgO and Pt-MgO) interaction in alcohol-CVD synthesis of carbon nanotubes, *J. Nanopart. Res.* 13:1987-1994.

Tanev, P. T. & Pinnavaia, T. J. (1995). A Neutral Templating Route to Mesoporous Molecular Sieves, *Science* 267:865-867.

Tans, S. J., Devoret, M. H., Dai, H., Thess, A., Smalley, R. E., Geerligs, L. J. & Dekker, C. (1997). Individual single-wall carbon nanotubes as quantum wires, *Nature* 386:474-477.

Thess, A., Lee, R., Nikolaev, P., Dai, H., Petit, P., Robert, J., Xu, C., Lee, Y. H., Kim, S. G., Rinzler, A. G., Colbert, D. T., Scuseria, G. E., Tománek, D., Fischer, J. E. & Smalley, R. E. (1996). Crystalline Ropes of Metallic Carbon Nanotubes, *Science* 273:483-487.

Wang, X., Wang, M., Jin, H., Li, Z. & He, P. (2005a). Preparation of carbon nanotubes at the surface of Fe/SBA-15 mesoporous molecular sieve, *Appl. Surf. Sci.* 243:151-157.

Wang, X. Q., Ge, H. L., Jin, H. X. & Cui, Y. J. (2005b). Influence of Fe on the thermal stability and catalysis of SBA-15 mesoporous molecular sieves, *Microporous Mesoporous Mater.* 86:335-340.

Wei, F., Zhang, Q., Qian, W.-Z., Yu, H., Wang, Y., Luo, G.-H., Xu, G.-H. & Wang, D.-Z. (2008). The mass production of carbon nanotubes using a nano-agglomerate fluidized bed reactor: A multiscale space–time analysis, *Powder Technol.* 183: 10-20.

Yang, Y., Lim, S., Wang, C., Du, G. & Haller, G. L. (2004). Statistical analysis of synthesis of Co-MCM-41 catalysts for production of aligned single walled carbon nanotubes (SWNT), *Microporous Mesoporous Mater.* 74:133-141.

Yang, Y., York, J. D., Xu, J., Lim, S., Chen, Y. & Haller, G. L. (2005). Statistical design of C10-Co-MCM-41 catalytic template for synthesizing smaller-diameter single-wall carbon nanotubes, *Microporous Mesoporous Mater.* 86:303-313.

Zarabadi-Poor, P., Badiei, A., Yousefi, A. A., Fahlman, B. D. & Abbasi, A. (2010). Catalytic chemical vapour deposition of carbon nanotubes using Fe-doped alumina catalysts, *Catal. Today* 150:100-106.

Zhao, D., Huo, Q., Feng, J., Chmelka, B. F. & Stucky, G. D. (1998). Nonionic Triblock and Star Diblock Copolymer and Oligomeric Surfactant Syntheses of Highly Ordered, Hydrothermally Stable, Mesoporous Silica Structures, *J. Am. Chem. Soc.* 120:6024-6036.

Zhu, H. J., Ding, Y. J., Yan, L., Xiong, J. M., Lu, Y. & Lin, L. W. (2004). The PPh3 ligand modified Rh/SiO2 catalyst for hydroformylation of olefins, *Catal. Today* 93-95:389-393.

4

Dispersions Based on Carbon Nanotubes – Biomolecules Conjugates

Ignác Capek
Slovak Academy of Sciences, Polymer Institute, Institute of Measurement Science, Dúbravská cesta, Bratislava, Trenčín University, Fac. Ind. Technol., Púchov, Slovakia

1. Introduction

The discovery of carbon nanotubes (CNTs) [1] followed by their large-scale production have paved the way to wide CNT integration into modern nanotechnology by taking advantage of their excellent mechanical properties and high electrical and thermal conductivities [2]. Carbon nanotubes have emerged as new class nanomaterials that are receiving considerable interest because of their unique structure, high chemical stability and high surface-to-volume ratio. Composite nanomaterials based on integration of CNTs and some other materials to possess properties of the individual components with a synergistic effect have gained growing interest [3].

The use of carbon nanotubes as "building blocks" in nano-/microelectronic devices could revolutionize the electronic industry in the same way that the microchips have revolutionized the computer industry. However, it has been a long-standing big challenge to efficiently integrate the carbon nanotube "building blocks" into multicomponent/ multifunctional structures or devices. It has been shown that carbon nanotubes could promote electron transfer with various redox active proteins, including glucose oxidase, cytochrome c, and horseradish peroxidase. Li and co-workers have demonstrated that carbon nanotubes can promote electron transfer with certain proteins and enzymes, and the electrochemical behavior with cytochrome c [4].

The CNT is nearly inert and therefore its functionalization is needed to increase its reactivity and to form with other components conjugates or hybrids. Chemical functionalization of CNTs usually destroys the sp^2 structure of CNTs, therefore, damages the intrinsic properties of them. Thus, non-covalent modification of CNTs is of great significance. Surfactants can disperse CNTs in water, however, which needs relatively higher amount and cannot be used for possible biological and chemical application [5].

Pioneering studies have reported on the use of single-walled carbon nanotubes (SWCNTs) as atomic force microscopy (AFM) imaging tips of biomacromolecules, such as antibodies, DNA, proteins, viruses... [6]

The integration of biomaterials (e.g., DNA, proteins/enzymes, or antigens/antibodies) with CNTs provides new hybrid systems that combine the conductive or semiconductive properties of CNTs with the recognition or catalytic properties of the biomaterials. This may yield new bioelectronic systems (biosensors, field-effect-transistors,...) or templated nanocircuitry. Functionalized CNTs are able to cross cell membranes and accumulate in the cytoplasm, and even reach the nucleus, without being cytotoxic (in concentrations up to 10 mM) [7]. CNTs could act as carriers that transport and deliver other bioactive components into cells. In fact, the effective delivery of biomolecules into cells has been used for their immunization and enhanced generation of antibodies [8].

New materials for the intracellular transport of biological cargos such as DNA, proteins, and drug molecules have been actively sought to effectively breach the cell-membrane barriers for delivery and enabling functionality of extracellular agents. Single-walled carbon nanotubes (SWNT) have been recently shown to shuttle various molecular cargos inside living cells including proteins, short peptides, and nucleic acids [9, 10]. The internalized nanotubes were found to be biocompatible and nontoxic at the cellular level. The utilization of the intrinsic physical properties of SWNTs allows the realization of a new class of biotransporters and opens up new possibilities in drug delivery and near infrared (NIR) radiation therapy.

Carbon nanotubes as well as other nanotube structures, such as self-assembled lipid microtubes or peptide nanotubes, have been explored for possible applications in nanobiotechnology [11]. Also, biomedical applications of biomaterial-functionalized CNTs are envisaged. At the present time, several fundamental issues remain to be addressed for the use of carbon nanotubes as potential biological transporters. One such issue is the entry mechanism that regulates the cellular internalization of SWNTs and their cargos.

The attractive properties of CNT make them promising candidates for DNA hybridization detection [12]. Wang et al. [13] demonstrated the use of CNT loaded alkaline phosphatase through cross linking for dramatically amplifying enzyme-based bioaffinity electrical sensing of proteins and DNA. Khairoutdinov et al. [14] and Panhuis and co-workers [15] reported the approach to covalent attachment of ruthenium complex to carboxylated single-wall carbon-nanotubes (SWNT) and to amino functionalized multiwall carbon nanotubes, respectively.

A brief summary of the most recent research development in the field of carbon nanotube/biomolecules is reported. Even this brief account has revealed the versatility of carbon nanotubes for making nanoconjugates with excellent physical and chemical properties. Within the present chapter we focus on decoration of CNTs with DNA to produce unique and smart nanocomposites. Continued research and development in this field favours the possibility of producing biosensors and/or smart nanostructures based on carbon nanotube/DNA conjugates.

2. Carbon nanotubes

Carbon nanotubes (CNTs) are anisotropic structures with sp^2 bonding properties. A carbon nanotube consists of either one cylindrical graphite sheet (single-walled carbon nanotube, SWCNT) or several nested cylinders (multi-walled carbon nanotube, MWCNT). Carbon nanotubes are macromolecules with radii of as small as a few nanometers, which can be grown up to 20 cm in length. Multi-walled nanotubes can reach diameters of up to 100 nm. Single-walled nanotubes possess the simplest geometry, and have been observed with diameters ranging from 0.4 to 3 nm [16]. The rolling-up of the hexagonal lattice can be performed in different ways. The sheet can be rolled-up along one of the symmetry axes, producing, either a zigzag or an armchair tube.

Single-walled carbon nanotubes (SWCNTs) exhibit excellent optical properties such as Raman, fluorescence and absorption spectra. Due to their small size and sensitive optical characteristics, SWCNTs have been shown to be ideal candidates for optical nano-biomarkers and/or nano-biosensors [17]. SWCNTs individually dispersed in aqueous medium show strong fluorescence in the near infrared region and sharp absorption peak distributions caused by van Hove singularities [18]. Absorption spectrum gives a variety of information including electronic structures such as metallic, semiconducting and chiralities [19].

It is well known that individually dispersed carbon nanotubes exhibit a series of sharp absorption peaks in the visible and near infrared regions due to quasi one-dimensionality. The absorption peaks become broad due to the disturbance and superposition of electronic structures by the aggregation of nanotubes. Some materials such as surfactants (sodium dodecyl sulfate (SDS), cetyltrimethylammonium bromide (CTAB),..), DNA, and proteins are known to individually disperse SWCNTs without disturbing the electronic properties of nanotubes [20]. There can also be a slight shift in the wavelength for the different dispersing medium. This could be due to the hypsochromic shift which was also observed in Ref. [20]. The change in dielectric environments around nanotubes, caused by the different surface coverage of wrapping agents, can lead to a shift in wavelength.

It has been stated that nonionic surfactants cause suspension of nanotubes by coating them [20]. The presence of nonionic surfactant could lead to complex micelle formation with CNT in both aqueous and nonaqueous systems. It was shown that the high-molecular weight Pluronic surfactants enhance dispersal of SWCNTs. It is possible that such surfactant-nanotube interactions alter the nanotubes' surface properties, thus modulating their interaction with other additives. The cylindrical micelles were formed in aqueous dispersions of sodium dodecyl sulfate (SDS)-SWNT [21]. For all samples, the UV-vis spectra exhibited the sharp van Hove transitions anticipated from individualized nanotubes. For the dispersion of SWNTs the SDS molecules were considered to form noninteracting core-shell cylindrical micelles with a single nanotube acting as the core. It was reported that the behavior of a surfactant in dispersing the carbon nanotubes is similar to that in the case of solid particles [22]. Since surfactant effects depend strongly on the medium's chemistry, aqueous and organic polymeric systems of carbon nanotubes should, in principle, obey different colloidal processes. However, a common factor is that surfactants having long tail groups and more unsaturated carbon-carbon bonds greatly contribute to stabilizing the carbon nanotube dispersions and reducing the size of CNT agglomerates. In a water-soluble polymer, e.g. poly(ethylene glycol), cationic surfactants show some advantage, owing to their preferential attraction to negatively charged CNT surfaces.

To covalently bond (bio)molecules to the CNTs, it first requires the formation of functional groups on the CNTs. The carboxylic group is often the best choice because it can undergo a variety of reactions and is easily formed on CNTs via oxidizing treatments. It is reported that the presence of carboxylic group at the nanotube ends and at defects on the sidewalls has advantages to perform acid base chemistry and to introduce on the nanotube amide, ester linkage, and so on [23]. The control of reactants and/or reaction conditions may control the locations and density of the functional groups on the CNTs, which can be used to control the locations and density of the attached biomolecules. For example, concentrated acids are known to introduce acidic groups to the sidewalls and ends of CNTs [24].

The SEM images of as-synthesized MWCNTs showed that MWCNTs are held together into bundles via Van der Waals forces [25]. The f-MWCNTs are discrete and shorter than that of as-synthesized MWCNTs due to the acid treatments. The TEM image of as-synthesized

MWCNTs illustrated that the nanotubes are entangled and randomly oriented. The outer surface of MWCNTs is smooth. The diameter and length of MWCNTs are ~ 30-80 nm and 10-20 μm, respectively. After the acid treatment, the MWCNTs are dispersed and most of the nanotubes are shortened (length of MWCNTs ~ 1-5 μm). The TEM image of amino f-MWCNTs indicated that the nanotubes surface is rough compared with the nanotubes without functionalization treatment.

Carbon nanotubes can be functionalized with various biomolecules without their covalent coupling [26]. Open-ended carbon nanotubes provide internal cavities (1-2 nm in diameter) that are capable of accommodating organic molecules and biomolecules of respective sizes. Functionalized carbon nanotubes are able to cross cell membranes and accumulate in the cytoplasm, and even reach the nucleus, without being cytotoxic (in concentrations up to 10 mM) [27].

Thus, carbon nanotubes could act as carriers that transport and deliver other bioactive components into cells. In fact, the effective delivery of biomolecules into cells has been used for their immunization and enhanced generation of antibodies [28]. Pioneering studies have reported on the use of single-walled carbon nanotubes as atomic force microscopy (AFM) imaging tips of biomacromolecules, such as antibodies, DNA… [29].

Carbon nanotubes (CNTs) possess a hollow core and large specific surface area suitable for storing guest molecules [30]. It was demonstrated that CNTs could promote electron transfer reactions with enzymes [31] and enhance the electrochemical activities of many biomolecules, which could allow them to be used as mediators in biosensor systems [32].

The attractive structural, electrical and mechanical properties of CNT make them promising candidates for electrochemical biosensors [33]. Most CNT-sensing work has focused on the ability of surface-confined CNT to accelerate the electron-transfer reactions in connection to amperometric enzyme electrodes. CNT have been recently used as transducers for enhanced electrical detection of DNA hybridization [34].

As the leading nanodevice candidate, SWNTs have shown great potential applications ranging from molecular electronics to ultrasensitive biosensors [35]. Single-stranded DNA (ssDNA) has recently been demonstrated to interact noncovalently with SWNTs, and forms stable complexes with individual SWNTs by wrapping around them by means of π-π stacking between nucleotide bases and SWNT sidewalls [36]. Double-stranded DNA (dsDNA) has also been proposed to interact with SWNTs, but its affinity is significantly weaker than that of ssDNA [37]. Also, scatter examples of noncovalent interactions of SWNTs with organic dyes or dye-labeled biomolecules have now been reported [38] and SWNTs can act collectively as fluorescence quenchers for dyes [39].

3. CNTs/DNA nanoconjugates

Deoxyribonucleic acid (DNA) is a naturally occurring polymer that plays a central role in biology, and now it has gained increasing attention in various biotechnology fields such as biosensor, bioimplant, and so forth [40]. It has been shown that single-stranded DNA (ssDNA) exhibits sequence-dependent effects of non-covalent binding to the surface of SWCNTs through π-stacking whereas double-stranded DNA (dsDNA) can hardly do [36]. Such novel properties enable SWCNTs to be widely applied in nanobiosystems. SWCNTs dispersed by ssDNA were used as molecular tags for Southern blotting assay [17]. They were also utilized in electrochemical analysis of dopamine [41] and cellular uptake observations [9].

Arrayed carbon nanotubes (CNTs) represent an ideal scaffold for the generation of ordered nanostructures featuring biomolecular components. These structures, exhibiting high electrical conductivity [42], also constitute a useful base material for nanoscale biosensors [43]. Most of the hybrid CNT-DNA structures reported to date used unordered CNTs that were functionalized by nonspecific and random adsorption of biomolecular components [44]. Taft et al. have described a rational strategy that permits discrete regions of arrayed CNTs to be functionalized simultaneously and specifically with DNA oligonucleotides [45]. These authors have exploited the different chemical properties of two regions on single CNTs and orthogonal chemical coupling strategies to derivatize CNTs within highly ordered arrays with multiple DNA sequences.

The bifunctional chemical structure of CNTs was suggested to facilitate the selective attachment of multiple DNA sequences using two distinct DNA-CNT linking strategies [45]. In one strategy, by accessing the free carboxyl groups of CNTs, single-stranded, amine-terminated DNA oligonucleotides are attached to the CNT array using amide-coupling chemistry in aqueous/organic solvent mixtures. A second and orthogonal modification strategy involves the attachment of oligonucleotides to the sidewalls of the CNTs through hydrophobic (pyrene unit) interactions. This dual functionalization then allowed to use differential hybridization to deliver two gold nanoparticles with distinct dimensions to discrete regions of an individual CNT. Single-stranded DNA attachment was performed in series, while the hybridization of complementary nanoparticle-labeled strands was performed in parallel.

The results clearly show that the specificity of DNA duplex formation permits each Au-DNA conjugate to be selectively directed toward a target site on the CNT. These experiments illustrate that individual CNTs can be functionalized with special selectivity and can be used to differentiate between two DNA sequences. In addition, they represent an augur for using DNA to controllably produce assemblies of hybrid nanostructures. Delivering different payloads to specific areas of functionalized nanotubes may facilitate the production of new nanomaterials.

The ability of DNA-CNTs conjugates to hybridize reversibly with high specificity to complementary DNA sequences [46] suggests that such conjugates are very promising genetherapy, conducting and semiconducting substrates medical apparatus and so on. However, these devices have inherent limitations in terms of precision, specificity, and interconnection, which obstruct their integration in large scale devices and complex circuits. Among the solutions to solve the problem, self-assembly is the most promising alternative.

A variety of techniques have been developed for DNA hybridization detection, including fluorescence imaging [47], electrochemical [48], micro-gravimetrical [49], bioluminescence [50], chemiluminescence [51] and electrogenerated chemiluminescence (ECL) techniques [52]. ECL technique has many distinct advantages over fluorescence technique because it does not involve a light source and avoids the attendant problems of scattered light and impurities luminescent.

Single-walled carbon nanotubes (SWNTs) self-assembly have been a rapidly evolving research area targeted at integrating nanoscale building blocks into functioning devices. To this end, various researchers have shown that DNA molecules can serve especially well to create highly definable supramolecular networks that can be used to advantage for the programmed self-assembly of objects with nanometer precision [53]. It has been demonstrated that DNA hybridization can induce the self-assembly of gold nanoparticles [54]. So far, DNA has been used to guide the assembly of gold nanoparticles into discrete structures with defined numbers of particles [55] as well as one-dimensional (1D) [56] or two-dimensional (2D) [53] arrays. To enrich the family of objects that can be used for DNA-directed self-assembly, researchers have tried to covalently conjugate DNA to CNTs and have attempted to use DNA hybridization to drive the self-assembly of CNTs [43].

4. CNT/DNA-based nanosensors

Nanostructures, such as nanowires, nanotubes and nanoparticles, offer new and sometimes unique opportunities that can be exploited for sensing [57, 58]. Specifically, the modification of transducers with carbon nanotubes has recently attracted considerable attention in the field of electro-analytical chemistry. The high surface area and the useful mechanical properties of CNTs combined with their electronic conductivity and ability to promote electron transfer reactions provide new exciting nanoelectrodes for the catalysis of biomolecules and inorganic compounds [59]. For the design of a genosensor the crucial step is the immobilization of single stranded DNA probes onto the electrode surface with sufficient stability, activity and well controlled packing density [60].

Carbon nanotubes [61, 62] show great potential for use as highly sensitive electronic biosensors. Single-walled carbon nanotubes arguably are the ultimate biosensor in this class for a number of reasons: SWNTs have the smallest diameter (~ 1 nm), directly comparable to the size of single biomolecules and to the electrostatic screening length in physiological solutions [63]. Furthermore, the low charge carrier density of SWNTs [64] is directly comparable to the surface charge density of proteins, which intuitively makes SWNTs well suited for electronic detection that relies on electrostatic interactions with analyte (bio)molecules. Finally, in SWNTs all the atoms are in direct contact with the environment, allowing optimal interaction with nearby biomolecules. Although an appreciable amount of biosensing studies has been conducted using carbon nanotube transistors, the physical mechanism that underlies sensing is still under debate [61]. Some mechanisms of biosensing are electrostatic gating [65], changes in gate coupling [66], carrier mobility changes [63] and Schottky barrier effects [67].

Tama et al. have developed a DNA sensor based on multi-walled carbon nanotubes (MWCNTs) [68]. The functionalized MWCNTs act as linkers to immobilize the probe DNA strands on the sensor surface for direct and label-free detection of influenza virus. The DNA – based sensor, a member of the biosensor family, is considered a promising tool in pre-diagnostics, and in the prevention and control of infectious diseases in real-time and on site analysis [69]. These sensors have numerous potential applications including the diagnosis of genetic diseases, the detection of infectious agents, and identification in forensic and environmental cases [70]. There are various types of DNA sensors which have developed over the years. Methods used for DNA sequence detection in those sensors have been reported to be based on radiochemical, enzymatic, fluorescent, electrochemical, optical, and acoustic wave techniques [71]. Some disadvantages of the optical sensors, however, include the requirement of a separate labeling process and an equipment to stimulate the transducer; they are also highly complex, and thus, entail higher cost in order to conduct an analysis [72].

The DNA sequence attachment on the surface of the sensor is a key to high sensitivity, long life-span, and short response time. In the immobilization technique, it is necessary that the binding chemistry is stable during subsequent assay steps; the sequence of the DNA probe should not change the chemical structure, and the bio-recognition molecules have to be attached with an appropriate orientation. Nowadays, various methods are used to immobilize the DNA strands on the sensor surface, such as the covalent bonds to the functionalized support [73], electrochemical [74], physical absorption [75] and monolayer self-assembling [76]. Among these methods, the covalent bond induced immobilization provides advantages over other methods in terms of simplicity, efficiency, ordered binding, and low cost. In this method, various mediators can be used to attach the DNA sequences on the sensor surface such as carbon nanotubes (CNTs), aminopropyltriethoxy Silane (APTES), alkanethiols and so on. The covalent immobilization of the carbon nanotubes (CNTs) is usually performed by reacting amino-terminated DNA with the carboxylic acid groups of the CNTs, or directly reacting with the amino group of the oxidized CNTs. Several groups have reported using the covalent binding of the CNTs to immobilize the DNA sequences. Krishna et al. reported the synthesis of functionally engineered single - walled carbon nanotubes (SWNTs) - peptide nucleic acid (PNA) conjugates especially for nanoelectronic applications [77]. Jung et al. demonstrated that the DNA strands can be covalently attached to immobilized SWNT multilayer films [78]. They showed that the SWCNTs multilayer films were constructed via consecutive condensation reactions creating stacks of functionalized SWCNTs layers linked together by dianiline derivatives.

The developer, a multistep route to the formation of covalently linked adducts of single-wall carbon nanotubes (SWNT) and deoxyribonucleic acid (DNA) sequence was reported by Baker et al. [46]. In their report, the DNA molecules covalently linked to SWNTs are accessible to hybridization and strongly favored hybridization with molecules having complementary sequences compared with non-complementary sequences [78]. Recently, Zhang et al. [79] synthesized a type of compound, MWCNTs— CONH—$(CH_2)_2$—SH, via carboxylation, and investigated a thickness-tunable multilayer film DNA biosensor built layer-by-layer (LBL) covalent attachment of gold nanoparticles (AuNPs) and multi-walled carbon nanotubes on an Au electrode [46].

Nanoparticles-based materials offer excellent prospects for DNA detection because of its unique physical and chemical properties. It is another efficient way to improve the sensitivity of DNA-probe. Many new protocols are based on colloidal gold tags, semiconductor quantum dot tracers and polymeric carrier. The power and scope of such nanopatricles can be greatly enhanced by coupling them with biological recognition reactions and electrical process [80]. Magnetite nanoparticles have both the properties of nanoparticles and magnetism. It can collect DNA by magnetic field and trigger DNA detection easily [81]. Palecek and Fojta [82] reported that non-specific adsorption, which is the important error in DNA detection could be remarkably suppressed by hybridization and transduction at the surface of magnetic beads.

5. Electrochemical nanosensors

Electrochemical methods of hybridization detection present a good alternative in comparison with well-developed fluorescent detection. Over the past decade, a significant progress has been made towards the development of the electrochemical DNA sensors. Considerable advantages have been ascribed to these devices owing to their potential for obtaining specific information in a faster, simpler, and less expensive way. These sensors rely on the conventional hybridization signal of the DNA sequences into useful electrical signal.

The modification of electrochemical sensors with carbon nanotubes (CNTs) has attracted considerable attention in the field of DNA sensing technology due to its attractive electronic, chemical, and mechanical performances. Thus, many different schemes for electrochemical DNA sensing based on CNTs have been reported [83]. In the Niu et al.'s work, carboxyl-functionalized multi-walled carbon nanotubes (MWCNTs-COOH) and redox intercalators were utilized in the fabrication of DNA electrochemical biosensor [84]. The presence of carboxyl groups on carbon nanotubes is necessary for the covalently bonding of the oligonucleotides. Oligonucleotide probes with an amino group at the 5′ -phosphate end can form covalent bonds with the carboxyl groups in MWCNTs–COOH with the aid of 1-ethyl-3-(3-dimethylaminopropyl) carbodiimide (EDC). A complex of rutin (R) $C_{54}H_{58}MnO_{32}$ (abbreviated by MnR_2) was synthesized and used as the indicator for the detection of hybridization between the probe DNA and the target sequence. Interaction between MnR_2 and double-stranded salmon sperm DNA was studied using voltammetry and fluorescence spectroscopy. Using MnR_2 as a novel electroactive indicator, ssDNA fragment could be selectively detected on the new electrochemical DNA biosensor with a detection limit of 3.81×10^{-11}M and a linear range from 1.60×10^{-9} M to 4.80×10^{-8} M [84].

In recent years, microfabricated interdigitated array microelectrodes have received great attention in the areas of biosensing [85]. The use of interdigitated microelectrodes is perhaps the most successful of all the recently introduced simple and rapid methods in biosensing for detection of various biological species [86]. Special configurations of the interdigitated microelectrodes have also been developed for improvement of their sensing performance [87]. The development of electrochemical DNA sensors with high potential for miniaturization and integration has become a subject of intense research, with the hope to make sophisticated and challenging molecular diagnostics available for low-cost routine clinical practice. Newly reported research on SWNT - Field-Effect Transistors (FET) based protein [88], and DNA [89] sensors has indicated that the sensing mechanism differs significantly when applied to different analyte molecules despite the commonality among the devices themselves.

The development of sequence-selective DNA sensors for diagnosis of genetic or pathogenic disease has attracted increasing interest. Most DNA detection methods rely on optical, piezoelectric, or electrochemical transductions. However, these sensors may have significant device-to-device variations and their fabrication requires high production costs. Recently, field-effect transistors based on single-walled carbon nanotube networks have been fabricated [90] and their electrical properties depend on the percolation paths of SWNTs in conduction channels, where device variations are expected to be small.

Label-free electrical detection of DNA and biomolecules using SWNT network FETs (SNFETs) has been successfully achieved [91], with typical detection limits on the order of ca. 1 nM of DNA. Dong et al. have reported that the detection sensitivity of SNFETs for DNA can be further improved to ca. 100 fM by using a "nanoparticle enhancement" approach, in which the target DNAs are hybridized with probe DNAs on the device, and reporter DNAs labeled with Au nanoparticles (AuNPs) flank a segment of the target DNA sequence [92]. It was noted that the enhancement of DNA detection by incorporating nanoparticles (e.g., CdS and Au) has been reported by using electrochemical approaches [93]. On the other hand, enhancing the sensitivity of SNFETs from 1 nM to 1 pM by adding a bivalent salt ($MgCl_2$) during the hybridization process has also been reported [89]. In some of these approaches, adsorption of target DNA on a SiO_2 surface via divalent coordination (DNA–Mg^{2+}–SiO_2) [94] rather than specific binding cannot be ruled out and may confound the sensing results. Dong et al. blocked the SiO_2 surface by octyltrichlorosilane (OTS) treatment to reduce possible non-specific binding (NSB) of DNA to SiO_2. In addition, blocking of vacant SWNT surfaces by using polyethylene glycol (PEG; molecular weight 400 kg mol^{-1}) has also been performed to reduce the NSB of DNA to SWNTs.

Li et al. have demonstrated the electrochemical properties of DNA–SWNTs sensor and the interaction between DNA and Riboflavin (VB_2) (Figure 1) [95]. First, the self-assembled monolayers of SWNTs on an amide platinum electrode were made by covalent linking of CO-NH bonds. Then, DNA was attached to SWNTs film via carboxyl-amine coupling.

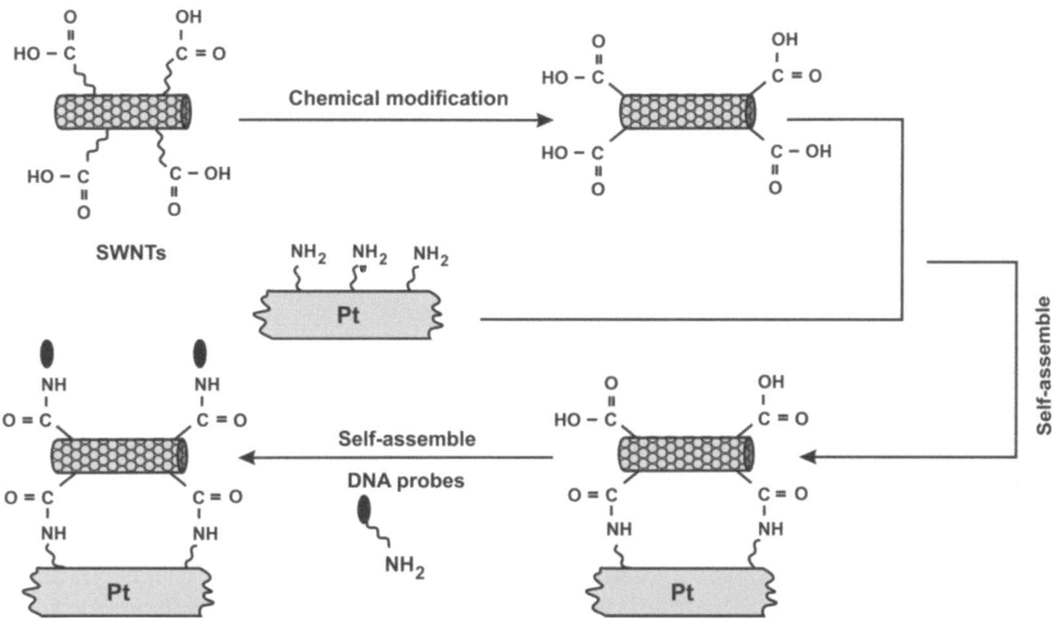

Fig. 1. Schematic diagram of constructing DNA sensor via covalent self-assembly.

Electrochemical experiments demonstrated that DNA still retained the bioactivity and the ability to interact with other biomolecules. Furthermore, the self-assembled biosensor could easily detect VB_2 in a wide range of linearity and exhibited low detection limit for VB_2. The new biosensor design was based on the incorporation of the biomolecule into amino modified Pt electrode, and exhibited improved analytical performance with respect to previous methods. The well-known capabilities of Pt electrode could retain the biological activities of DNA upon adsorption, and the electrocatalytic abilities of carbon nanotubes allow the electrooxidation of biochemical molecules with interest. The coupling of Pt electrode and carbon nanotubes led to robust biosensors with enhanced analytical characteristics that are useful for many applications [95].
Tang et al. have developed fully electronic DNA sensors based on carbon nanotube field effect devices, which are readily scalable to high density sensor arrays and amenable to integration with "lab-on-a-chip" microanalysis systems [96]. The generality of the sensors was demonstrated with synthetic oligonucleotides consisting of random generated sequences and also two different oligo lengths (15mer and 30mer). The random sequenced 15mer thiolated ssDNA probe (p15), its completely complementary target ssDNA (CM15), and its randomly generated mismatched target ssDNA (MM15) were used.
SWNT serves as the transducer which translates and amplifies DNA hybridization on Au into a directly detectable electrical signal. Compared to optical and other electrochemical methods, the essentially two-terminal SWNT DNA sensors involve much simpler chemistry and easier setup. It is highly desirable to fully utilize the surface and electrical properties of SWNT for biosensing in general, where chemical schemes for SWNT-biomolecule conjugation are in critical need that (1) preserve pristine nanotube property, (2) maintain biomolecule functionality, and (3) facilitate efficient SWNT-biomolecule charge transfer.
The sensing mechanism suggested by Star et al. attributes the electrical conductance change to the electron doping by DNA hybridization on the SWNT sidewall [89]. It is well accepted

that SWNT-FETs operate as unconventional Schottky barrier (SB) transistors, in which switching occurs primarily by modulation of the contact resistance rather than the channel conductance [97]. Results suggest that the strong binding between the directly absorbed ssDNA molecules and the sidewalls of SWNTs largely inhibits further hybridization.

It is proposed that the modulation of the Schottky barrier at the metal-tube contact by efficient hybridization on Au electrodes is the dominate sensing mechanism [96]. Furthermore, the DNA hybridization kinetics observed real-time in these sensing experiments, consistent with that on gold surface, also strongly suggests that the electrical signal originates from hybridization events on the gold contact. Quartz crystal microbalance and XPS data conclude that the ssDNA probes wrapped on SWNTs played little role in hybridization; instead they blocked the NSB of analyte ssDNA oligos complementary or mismatched alike. The slight response to mismatched target DNAs is believed to be a combined result of the sequence-dependent DNA-SWNT affinity and the disruption of probe packing due to photoresist residues on the sensor surface. Nonspecific interactions of oligonucleotides with carbon nanotubes could enhance the polymerase chain reaction (PCR), due to the local increase in the reaction components on the surface of CNTs [98]. DNA could also enter into the carbon nanotube cavities [99].

Tama et al. have developed a DNA sensor based on multi-walled carbon nanotubes (MWCNTs) [68]. The functionalized MWCNTs act as linkers to immobilize the probe DNA strands on the sensor surface for direct and label-free detection of influenza virus. These developed sensors comprise a highly sensitive, low-cost, and rapid method and therefore, they have potential application in controlling this disease. To detect influenza virus DNA strands, the DNA sensor was soaked into a solution containing the target DNA sequences. Upon hybridization, double-stranded DNA molecules were formed on the sensor surface. In the case of a perfect match between the target DNA and the immobilized DNA sequences, a concentration-dependent change in surface conductance was detected (Fig. 2 [68]). The output signal of our sensor was linearly proportional to the target DNA concentration in a range between 1 and 10 nM. The conductance remained unchanged when non-matching DNA strands were used (Fig. 2 [68]).

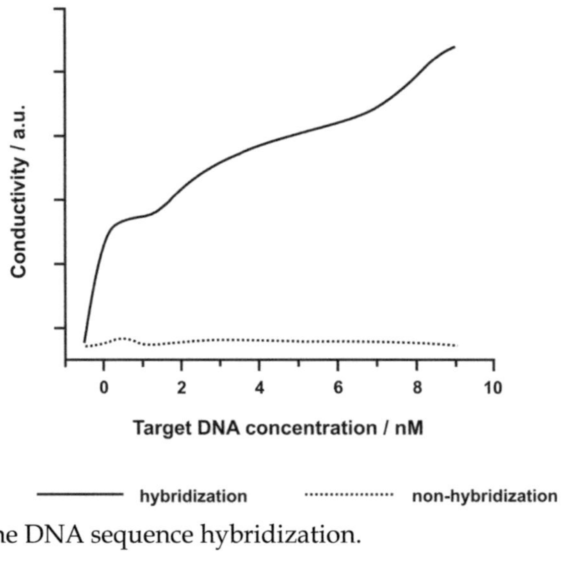

Fig. 2. The curve of the DNA sequence hybridization.

Here, the detection limit of the sensor was about 0.5 nM concentration of the influenza virus sample. This is lower than that of the electrochemical transducer using square wave voltammetry and fluorescence of Vincent Noel (25 nM) [100], or the one using the electrochemical impedance spectroscopy of Hui Peng (0.98 nM) [101]. The sensitivity of the DNA sensor was 0.06 mV/nM. The conductance modulation of these sensors can be explained based on the mechanism which has been well studied in literature [65]. The mechanism for electrical detection of the DNA hybridization in semiconducting carbon nanotube network devices in this work was likely due to the modification of junction barrier energy, whereas the conductance change forming the metal-nanotube's contact by efficient hybridization on the Pt electrodes was the dominant sensing mechanism.

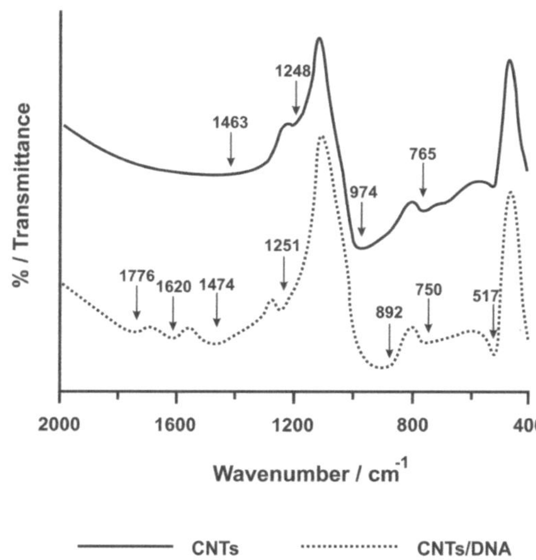

Fig. 3. The FTIR spectra of the MWCNTs and the DNA probe sequences-MWCNTs bonds.

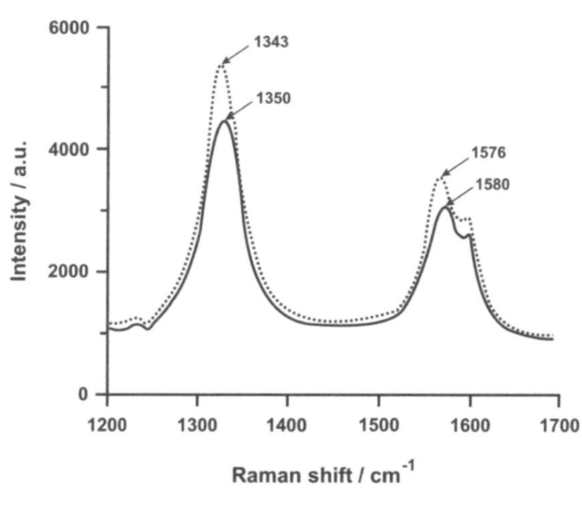

Fig. 4. The Raman spectra of the MWCNTs and the MWCNTs/DNA film.

As observed in the FTIR spectra (Fig.3) for MWCNTs, there were peaks at 1463 cm^{-1}, which correspond to the CH_2 of the MWCNTs [79]. During interaction with the DNA strands, a shift in the CH_2 stretch could be seen centered around 1474 cm^{-1}, and also at 1248 cm^{-1} corresponding to the C—O vibration [102]. Fig. 3 also shows the existence of DNA/CNTs interaction at around 750 cm^{-1}, 892 cm^{-1}, corresponding to the asymmetric stretching mode of the phosphate group and the DNA backbone of the DNA sequence, respectively [79]. The change in the wave number of mode for CH2—O—P—O was indicated at 1251 cm^{-1} [102].
Fig. 4 shows the Raman spectra of the MWCNTs and the MWCNTs/DNA film. The main features of these samples in the Raman spectra are the G band at 1580 cm^{-1} and the D band at 1350 cm^{-1}. A downshift of the tangential G band and D band were observed in the Raman spectra of the DNA-MWCNTs, which corresponded to peaks at 1576 cm^{-1} and 1343 cm^{-1}, respectively. This is attributed to the results of the charge transfer between the oxygen groups on the CNT surface and the DNA matrix [103].

6. CNTs-DNA hybrids

The formation of ssDNA-SWCNT hybrids and their purification is schematically illustrated in Fig. 5 [104].

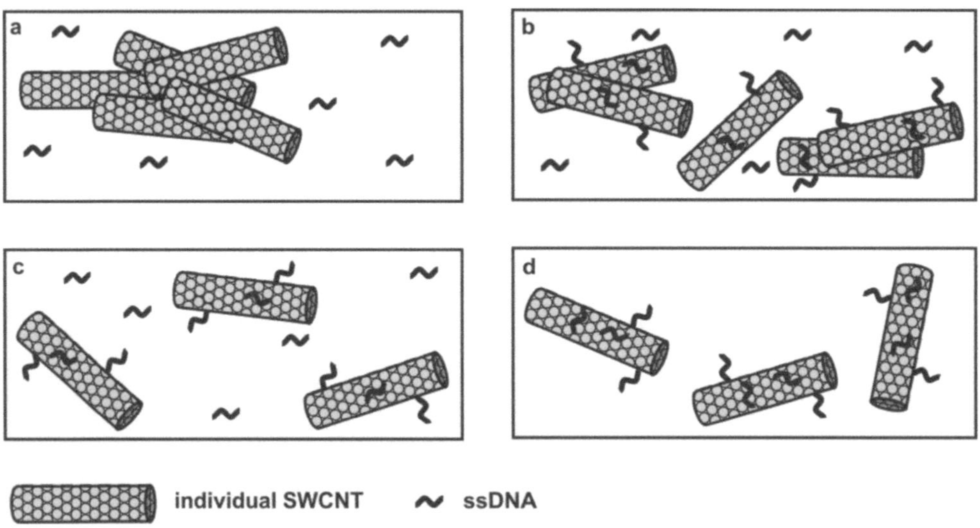

Fig. 5. Schematic of preparing ssDNA-SWCNT hybrids: (a) SWCNT powder in DNA solution; (b) after sonication; (c) after centrifugation, (d) after dialysis.

The ssDNA-SWCNT hybrids show well-resolved absorption spectra (Fig. 6 [104]). The reaction was carried out with the complementary DNA (cDNA) and noncomplementary DNA (ncDNA), respectively. The absorption spectra exhibit sharp peaks composed of the first van Hove transition of metallic SWCNTs ($M_{11} \approx$ 400–600 nm), the first ($S_{11} \approx$ 800–1400 nm) and second ($S_{22} \approx$ 550–800 nm) van Hove transition of semiconducting SWCNTs. A red shift was observed after hybridization with cDNA whereas there was no shift in the absorption spectra after the reaction with ncDNA. Also, semiconducting species showed clear red shift whereas the change in the spectra was negligible for metallic nanotubes. The reaction with ncDNA did not result in any shift in the fluorescence spectra [105]. However, a blue shift was observed after hybridization with cDNA. Jeng et al. also observed a blue shift in the fluorescence spectra after hybridization with cDNA [106]. As shown by AFM, the average height was increased from 2.02 nm to 2.81 nm after the reaction, indicating that DNA hybridization was achieved between ssDNA and cDNA.

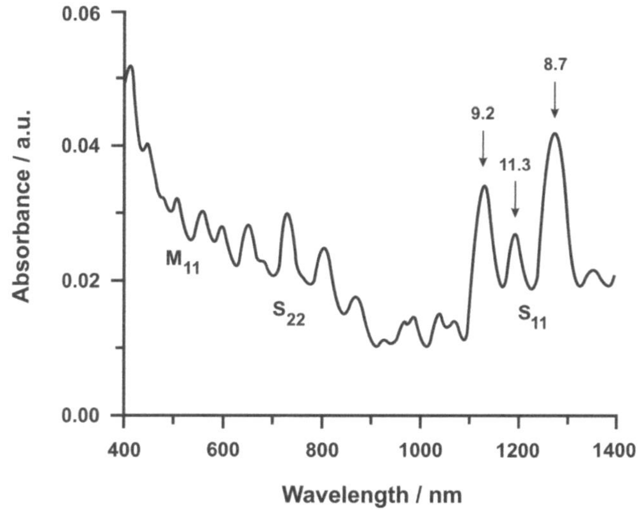

Fig. 6. Absorption spectrum of ssDNA-SWCNT hybrids.

7. CNTs-enzyme based nanoconjugates

CNTs are attractive materials for application to biosensors due to the low-potential detection of hydrogen peroxide and β-Nicotinamide adenine dinucleotide and the minimal surface passivation during the electrochemical oxidation of NADH [107]. The electrocatalytic behavior of CNTs can be attributed to the ends of CNTs [108]. These characteristics of CNTs are very useful for the development of novel enzyme-based sensors where rapid electron transfer at the electrode is required. Many enzymes, over 100 oxidases and 200 dehydrogenases, catalyze specifically the reactions of important analytes to generate the electrochemicaly detectable products hydrogen peroxide and NADH. The importance of electrochemical biosensors lies in the fact that they combine the specificity of biological systems with the advantages of electrochemical transduction. CNTs have been widely used for application in enzyme-based biosensors.

The electrocatalytic oxidation of NAD(P)H [NAD(P)H=1,4-dihydronicotinamide adenine dinucleotide (phosphate)] cofactors and the reduction/oxidation of H_2O_2 stimulated by CNTs are particularly important, since these electrocatalytic reactions may be easily coupled to enzymatic transformations [109]. For example, two kinds of amperometric biosensors have been prepared for analyzing ethanol and glucose by the encapsulation of alcohol dehydrogenase or glucose oxidase, respectively, in a carbon nanotube/Teflon composite material [109]. These sensors operate in diffusion mode, yielding NADH or H_2O_2 in the presence of NAD^+ and ethanol or O_2 and glucose (GOx), respectively. The biocatalytically generated NADH or H_2O_2 were then detected electrochemically by the catalytic electrodes. Enzymes generating H_2O_2 were covalently linked to carbon nanotubes or encapsulated into polymer coatings associated with carbon nanotubes. For example, GOx was covalently coupled to carboxylic groups at the ends of short CNTs using carbodiimide coupling [110] or cross-linked with a CNT/Pt-nanoparticle hybridlayer [111] yielding amperometric glucose sensors based on H_2O_2 detection [26]. Coaxial nanowires consisting of a concentric layer of polypyrrole uniformly coated onto aligned carbon nanotubes have provided a template for making glucose sensors with a large amount of electrochemically entrapped GOx in the ultrathin polypyrrole film. Similarly, amperometric detection of organophosphorus pesticides and nerve agents was performed using a screen-printed biosensor based on co-immobilized acetylcholine esterase, choline oxidase, and CNTs [112].

8. CNTs/protein conjugates

Proteins can be non-specifically bound to the external sides of the carbon nanotube walls [113].

Proteins adsorb individually, strongly, and noncovalently along the nanotubes. For example, it has been shown that open single-walled carbon nanotubes can accommodate small proteins such as cytochrome c inside their internal cavities, leading to the stable immobilization of the proteins in bioactive conformations [114].

Ortiz-Acevedo et al. used a novel class of cyclic peptides containing alternating L- and D-amino acids (AAs), called reversible cyclic peptides (RCPs), for the diameter-selective solubilization of HiPco single-walled carbon nanotubes [115] .

In L-/D-peptides, all side chains reside on one face of the backbone, encouraging a ringlike conformation with the side chains on the ring exterior. In addition, present cyclic peptides have N- and C-termini that are derivatized to contain thiol groups, allowing reversible peptide cyclization through a disulfide bond. These results suggest that peptides with different N-to-C-terminal lengths wrap around SWNTs having sufficiently small diameters, promoting selective enrichment of small-diameter CNTs dispersed in solution. By controlling the length of the reversible cyclic peptides, the authors demonstrated limited diameter-selective solubilization of single-walled carbon nanotubes, which may prove useful in SWNT purification. In addition, RCPs covalently closed around SWNTs do not dissociate from the SWNTs unless the disulfide bond is reduced. RCPs thus provide a platform to which other functional groups could be attached without disturbing the covalent structure of SWNTs.

The attachment of BSA protein and DNA to the amino f-MWCNTs was verified by comparing the FTIR spectrum of as-prepared amino f-MWCNTs and amino f-MWCNTs-BSA/DNA samples [25]. The biomolecules contain both amine and carboxylic groups. In the present experiment, the carboxylic groups of biomoelcules (such as BSA protein and DNA) react with the free amine groups of the amino f-MWCNTs. As a result, the carboxylic bonds in biomolecules have been converted into amide bonds (-NH-C=O). The interaction between amino f-MWCNTs and biomolecules (protein and DNA) is noticed by the shift of the amide bond (C=O) peak (1 650 to 1 642 and 1 650 to 1 645 cm^{-1} for amino f-MWCNTs-BSA and amino f-MWCNTs-DNA samples, respectively) in the FTIR spectrum. Furthermore, TEM studies confirmed the success of the attachment of BSA protein molecules and DNA to amino f-MWCNTs. The BSA protein molecules densely decorate the side walls of the MWCNTs. The location of the BSA protein is representative of where the amine groups were present.

Kam et al. have presented the investigation of the cellular uptake mechanism and pathway for carbon nanotubes [116]. The authors showed that intracellular transportation of proteins and DNA by SWNTs is general, thus further confirming the transporter ability of these materials. They presented evidence that shows clathrin-dependent endocytosis as the pathway for the uptake of various SWNT conjugates with proteins and DNA. They also discussed the differences between the nanotube materials and the experimental procedures used in our work and by Pantarotto et al. who suggested an energy-independent nonendocytotic uptake of nanotubes.

9. Conclusion

One of the most exciting classes of nanomaterials is represented by the carbon nanotubes. Carbon nanotubes have emerged as new class nanomaterials that are receiving considerable interest because of their unique structure, high chemical stability and high surface-to-volume ratio. Composite nanomaterials based on integration of CNTs and some other materials to possess properties of the individual components with a synergistic effect have gained growing interest. The use of carbon nanotubes as "building blocks" in nano-/microelectronic devices could revolutionize the electronic industry in the same way that the microchips have revolutionized the computer industry. Significant enhancement of optical, mechanical, electrical, structural, and magnetic properties are commonly found through the use of novel nanomaterials. Composite materials based on integration of carbon nanotubes and some organic and bioorganic materials to possess properties of the individual components with a synergistic effect have gained growing interest. The integration of carbon nanotubes (CNTs) with these materials has led to the development of new hybrid nanomaterials and sensors. The integration of carbon nanotubes with biological systems to form functional hybrid assemblies is, however, a new and relatively unexplored area. Carbon nanotubes have been explored for possible applications in nanobiotechnology. At the present time, several fundamental issues remain to be addressed for the use of carbon nanotubes as potential biological transporters.

Single-walled carbon nanotubes (SWCNTs) exhibit excellent optical properties such as Raman, fluorescence and absorption spectra. Due to their small size and sensitive optical characteristics, SWCNTs have been shown to be ideal candidates for optical nano-biomarkers and/or nano-biosensors. SWCNTs individually dispersed in aqueous medium show strong fluorescence in the near infrared region and sharp absorption peak distributions caused by van Hove singularities. Absorption spectrum gives a variety of information including electronic structures such as metallic, semiconducting and chiralities.

Ionic and nonionic surfactants and some biomolecules such as DNA and protein cause suspension of nanotubes by coating them. These materials, individually disperse SWCNTs without disturbing the electronic properties of nanotubes. The presence of nonionic surfactant leads to complex micelle formation with CNT in both aqueous and nonaqueous systems. It is possible that such surfactant-nanotube interactions alter the nanotubes' surface properties, thus modulating their interaction with other additives. The cylindrical micelles were formed in aqueous dispersions of sodium dodecyl sulfate (SDS)-SWNT. For all dispersions, the UV-vis spectra exhibit the sharp van Hove transitions anticipated from individualized nanotubes. For the dispersion of SWNTs the SDS molecules were considered to form noninteracting core-shell cylindrical micelles with a single nanotube acting as the core.

To covalently bond molecules to the CNTs, it first requires the formation of functional groups on the CNTs. The control of reactants and/or reaction conditions may control the locations and density of the functional groups on the CNTs, which can be used to control the locations and density of the attached biomolecules. The edges of carbon nanotubes are more reactive than their sidewalls, thus allowing the attachment of functional groups to the nanotube ends. Concentrated acids are known to introduce acidic groups to the sidewalls and ends of CNTs. The FTIR spectra of as-synthesized MWCNTs and after functionalization showed the presence of carboxylic and amino groups. Arrayed carbon nanotubes (CNTs) represent an ideal scaffold for the generation of ordered nanostructures featuring biomolecular components. These structures, exhibiting high electrical conductivity, also constitute a useful base material for nanoscale biosensors. Most of the hybrid CNT-DNA structures reported to date used unordered CNTs that were functionalized by nonspecific and random adsorption of biomolecular components.

It has been shown that single-stranded DNA (ssDNA) exhibits sequence-dependent effects of non-covalent binding to the surface of SWCNTs through π-stacking whereas double-stranded DNA (dsDNA) can hardly do. Such novel properties enable SWCNTs to be widely applied in nanobiosystems. SWCNTs dispersed by ssDNA were used as molecular tags for Southern blotting assay. Single-stranded DNA (ssDNA) has recently been demonstrated to interact noncovalently with SWNTs, and forms stable complexes with individual SWNTs by wrapping around them by means of π-π stacking between nucleotide bases and SWNT sidewalls. Double-stranded DNA (dsDNA) has also been proposed to interact with SWNTs, but its affinity is significantly weaker than that of ssDNA. The ssDNA–SWCNT hybrids show well-resolved absorption spectra The absorption spectra exhibit sharp peaks composed of the first van Hove transition of metallic SWCNTs ($M_{11} \approx$ 400–600 nm), the first ($S_{11} \approx$ 800–1400 nm) and second ($S_{22} \approx$ 550–800 nm) van Hove transition of semiconducting SWCNTs. A red shift was observed after hybridization with cDNA whereas there was no shift in the absorption spectra after the reaction with ncDNA.

Hybrid nanoscale materials are well established in various processes such as nucleic acid detachment, protein separation, and immobilization of enzymes. Those nanostructures can be used as the building blocks for electronics and nanodevices because uniform bioorganic coatings with the small and monodisperse domain sizes are crucial to optimize conductivity and absorption and to detect changes in conductivity and absorption induced by analyte adsorption on the CNT surfaces. DNA, protein and enzymes are useful as an engineering material for the construction of smart objects at the nanometer scale because of its ability to selforganize into desired structures via the specific selfassembling or hybridization of complementary sequences. Particularly, color changes induced by the association of nanometer-sized metal nanoparticles with CNTs conjugates provide a basis of a simple yet highly selective method for detecting specific biological reactions between anchored ligand molecules and receptor molecules in the milieu. Arrayed carbon nanotubes/DNA represent an ideal scaffold for the generation of ordered nanostructures. These structures, exhibiting high electrical conductivity, also constitute a useful base material for nanoscale biosensors

The ability of DNA-CNTs conjugates to hybridize reversibly with high specificity to complementary DNA sequences suggests that such conjugates are very promising genetherapy, conducting and semiconducting substrates medical apparatus and so on. A variety of techniques have been developed for DNA hybridization detection, including fluorescence imaging, electrochemical, micro-gravimetrical, bioluminescence, chemiluminescence and electrogenerated chemiluminescence (ECL) techniques. Among these methods, the covalent attachment immobilization provides advantages in terms of simplicity, efficiency, ordered binding, and low cost.

The DNA molecules covalently linked to SWNTs are accessible to hybridization and strongly favored hybridization with molecules having complementary sequences compared with non-complementary sequences. Nanoparticles-based materials offer excellent prospects for DNA detection because of its unique physical and chemical properties. It is another efficient way to improve the sensitivity of DNA-probe. Many new protocols are based on colloidal gold tags, semiconductor quantum dot tracers and polymeric carrier. The power and scope of such nanopatricles can be greatly enhanced by coupling them with biological recognition reactions and electrical process. Magnetite nanoparticles have both the properties of nanoparticles and magnetism. It can collect DNA by magnetic field and trigger DNA detection easily. It was reported that non-specific adsorption, which is the important error in DNA detection could be remarkably suppressed by hybridization and transduction at the surface of magnetic beads.

During the past decade, considerable progress has been made towards the development of the electrochemical DNA sensors. Considerable advantages have been ascribed to these devices owing to their potential for obtaining specific information in a faster, simpler, and less expensive way. These sensors rely on the conventional hybridization signal of the DNA sequences into useful electrical signal. The modification of electrochemical sensors with carbon nanotubes (CNTs) has attracted considerable attention in the field of DNA sensing technology due to its attractive electronic, chemical, and mechanical performances. Thus, many different schemes for electrochemical DNA sensing based on CNTs have been reported. Field-effect transistors based on single-walled carbon nanotube networks have been fabricated and their electrical properties depend on the percolation paths of SWNTs in conduction channels, where device variations are expected to be small. Label-free electrical detection of DNA and biomolecules using SWNT network FETs (SNFETs) has been successfully achieved, with typical detection limits on the order of ca. 1 nM of DNA. This sensing mechanism attributes the electrical conductance change to the electron doping by DNA hybridization on the SWNT sidewall. It is well accepted that SWNT-FETs operate as unconventional Schottky barrier (SB) transistors, in which switching occurs primarily by modulation of the contact resistance rather than the channel conductance.

Proteins can be bound to the external sides of the carbon nanotube walls. Proteins adsorb individually, strongly, and noncovalently along the nanotubes. For example, it has been shown that open single-walled carbon nanotubes can accommodate small proteins inside their internal cavities, leading to the stable immobilization of the proteins in bioactive conformations.

The CNTs are very useful for the development of novel enzyme-based sensors where rapid electron transfer at the electrode is required. Many enzymes, over 100 oxidases and 200 dehydrogenases, catalyze specifically the reactions of important analytes to generate the electrochemicaly detectable products hydrogen peroxide and NADH. The importance of electrochemical biosensors lies in the fact that they combine the specificity of biological systems with the advantages of electrochemical transduction. CNTs have been widely used for application in enzyme-based biosensors.

10. Acknowledgements

This research is supported by the VEGA project No. 2/0037/10 and 2/0160/10.

11. Nomenclature

1D	one-dimensional
2D	two-dimensional
AFM	atomic force microscopy
APTES	aminopropyltriethoxy Silane
AuNPs	Au nanoparticles
AuNPs	gold nanoparticles
BSA	bovine serum albumin
cDNA	complementary DNA
CNT	carbon nanotube
CTAB	cetyltrimethylammonium bromide
DNA	deoxyribonucleic acid
dsDNA	double-stranded DNA
ECL	electrogenerated chemiluminescence
EDC	1-(3-dimethylaminopropyl)-3-ethylcarbodiimide hydrochloride
FET	field effect transistor
LBL	layer-by-layer
MWCNT	multi-walled carbon nanotube
NAD	β-nicotinamide adenine dinucleotide
NIR	near IR
NSB	nonspecific binding
OTS	octyltrichlorosilane
PCR	polymerase chain reaction
PEG	polyethylene glycol
PNA	peptide nucleic acid
R	complex of rutin
RCPs	reversible cyclic peptides
SB	schottky barrier
SDS	sodium dodecyl sulfate
SNFETs	biomolecules using SWNT network FETs
ssDNA	single-stranded DNA
SWCNT	single-walled carbon nanotube
SWNT	single-wall carbon nanotubes
TEM	transmission electron microscopy

12. References

[1] S. Iijima, Nature 354 (1991) 56.
[2] M. Fujii, X. Zhang, H. Xie, H. Ago, K. Takahashi, T. Ikuta, H. Abe, T. Shimizu, Phys. Rev. Lett. 95 (2005) 065502.
[3] S.H. Lim, J. Wei, J. Lin, Q. Li, J.K. You, Biosens. Bioelectron. 20 (2005) 2341.
[4] J.X. Wang, M.X. Li, Z.J. Shi, N.Q. Li, Z.N. Gu, Anal. Chem. 74 (2002) 1993.
[5] M.F. Islam, E. Rojas, D.M. Bergey, A.T. Johnson, A.G. Yodh, Nano Lett. 3 (2003) 269.
[6] M.C. Daniel, D. Astruc, Chem. Rev. 104 (2004) 293.
[7] X.R. Ye, Y.H. Lin, C.M. Wang, M.H. Engelhard, Y. Wang, C.M. Wai, J. Mater. Chem. 14 (2004) 908.

[8] B. Kim, W.M. Sigmund, Langmuir 20 (2004) 8239.
[9] N.W.S. Kam, M. O'Connell, J.A. Wisdom, H. Dai, Proc. Natl. Acad. Sci. USA 102 (2005) 11600.
[10] N.W.S. Kam, Z. Liu, H. Dai, J. Am. Chem. Soc. 127 (2005) 12492.
[11] Y. Lin, S. Taylor, H. Li, K.A.S. Fernando, L. Qu, W. Wang, L. Gu, B. Zhou, Y.P. Sun, J. Mater. Chem. 14 (2004) 527.
[12] Q. Zhao, Z.H. Gan, Q.K. Zhuang, Electroanalysis 14 (2002) 1609.
[13] J. Wang, G. Liu, M.R. Jan, J. Am. Chem. Soc. 126 (2004) 3010.
[14] R.F. Khairoutdinov, L.V. Doubova, R.C. Haddon, L. Saraf, J. Phys. Chem. B 108 (2004) 19976.
[15] F. Frehill, J.G. Vos, S. Benrezzak, A.A. Koos, Z. Konya, M.G. Ruther,W.J. Blau, A. Fonseca, J.B. Nagy, L.P. Biro, A.I. Minett, M. in het Panhuis, J. Am. Chem. Soc. 124 (2002) 13694.
[16] J.M. Bonard, H. Kind, T. Stöckli, L.O. Nilsson, Solid-State Electronics 45 (6) (2001) 893.
[17] E.S. Hwang, C. Cao, S. Hong, H.J. Jung, C.Y. Cha, J.B. Choi, Y.J. Kim, S. Baik, Nanotechnology 17 (2006) 3442.
[18] M.J. O'Connell, S.M. Bachilo, C.B. Huffman, V.C. Moore, M.S. Strano, E.H. Haroz, K.L. Rialon, P.J. Boul, W.H. Noon, C. Kittrell, J. Ma, R.H. Hauge, R.B.Weisman, R.E. Smalley, Science 297 (2002) 593.
[19] S.M. Bachilo, M.S. Strano, C. Kittrell, R.H. Hauge, R.E. Smalley, R.B. Weisman, Science 298 (2002) 2361.
[20] V.C. Moore, M.S. Strano, E.H. Haroz, R.H. Hauge, R.E. Smalley, Nano Lett. 3 (2003) 1379.
[21] K. Yurekli, C.A. Mitchell, R. Krishnamoorti, J. Am. Chem. Soc. 126 (2004) 9902.
[22] L. Vaisman, G. Marom, H.D. Wagner, Adv. Funct. Mater. 16 (2006) 357.
[23] K. Jiang, L.S. Schadler, R.W. Siegel, X. Zhang, H. Zhang, M. Terrones, J. Mater. Chem. 14 (2004) 37.
[24] L. Liu, Y. Qin, Z.X. Guo, D. Zhu, Carbon 41 (2003) 331.
[25] K. Awasthi, D.P. Singh, S.K. Singh, D. Dash, O.N. Srivastava, New Carbon Materials, 24(4) (2009) 301.
[26] E. Katz, I. Willner, Chem. Phys. Chem. 5 (2004) 1084.
[27] D. Pantarotto, J.P. Briand, M. Prato, A. Bianco, Chem. Commun. 1 (2004) 16.
[28] D. Pantarotto, C.D. Partidos, J. Hoebeke, F. Brown, E. Kramer, J.-P. Briand, S. Muller, M. Prato, A. Bianco1, Chem. Biol. 10 (2003) 961.
[29] J.H. Hafner, C.L. Cheung, A.T. Woolley, C.M. Lieber, Prog. Biophys. Mol. Biol. 77 (2001) 73.
[30] W. Joseph, Electroanalysis 17 (2005) 1341.
[31] Y. Zang, J. Li, Y. Shen, M. Wang, J. Li, J. Phys. Chem. B 108 (2004) 1543.
[32] J.P. Kim, B.Y. Lee, S. Hong, S.J. Sim, Anal. Biochem. 381 (2008) 193.
[33] J. Wang, M. Musameh, Y. Lin, J. Am. Chem. Soc. 12 (2003) 2408.
[34] J. Li, H. Ng, A. Cassell, W. Fan, H. Chen, Q. Ye, J. Koehne, J. Han, M. Meyyappan, Nano Lett. 5 (2003) 597.
[35] M. Valcarcel, S. Cardenas, B.M. Simonet, Anal. Chem. 79 (2007) 4788.
[36] M. Zheng, A. Jagota, E.D. Semke, B.A. Diner, R.S. McLean, S.R. Lustig, R.E. Richardson, N.G. Tassi, Nat. Mater. 2 (2003) 338.
[37] M. Franchini, D. Veneri, Ann. Hematol. 84 (2005) 347.

[38] P.J. Boul, D.G. Cho, G.M. Rahman, M. Marquez, Z. Ou, K.M. Kadish, D.M. Guldi, J.L. Sessler, J. Am. Chem. Soc. 129 (2007) 5683.
[39] Q. Lu, K.O. Freedman, R. Rao, G. Huang, J. Lee, L.L. Larcom, A.M. Rao, P.C. Ke, J. Appl. Phys. 96 (2004) 6772.
[40] J.J.J.P. van den Beucken, M.R.J. Vos, P.C. Thune, T. Hayakawa, T. Fukushima, Y. Okahata, X.F. Walboomers, N.A.J.M. Sommerdijk, R.J.M. Nolte, J.A. Jansen, Biomaterials 27 (2006) 691.
[41] C. Hu, Y. Zhang, G. Bao, Y. Zhang, M. Liu, Z. Wang, J. Phys. Chem. B 109 (2005) 20072.
[42] M.S. Dresselhaus, G. Dresselhaus, P. Avouris, Carbon Nanotubes: Synthesis, Structure, Properties, and Applications; Springer: Berlin (2001).
[43] K.A. Williams, P.T.M. Veenhuizen, B.G. de la Torre, R. Eritja, C. Dekker, Nature 420 (2002) 761.
[44] M. Hazani, R. Naaman, F. Hennrich, M.M. Kappes, Nano Lett. 3 (2003) 153.
[45] B.J. Taft, A.D. Lazareck, G.D. Withey, A. Yin, J.M. Xu, S.O. Kelley, J. Am. Chem. Soc. 126 (2004) 12750.
[46] S.E. Baker, C. Wei, T.L. Lasseter, K.P. Weidkamp, R.J. Hamers, Nano Lett. 2 (2002) 1413.
[47] Y. Dharmadi, R. Gonzales, Biotechnol. Prog. 20 (2004) 1309.
[48] E.L.S. Wong, F.J. Mearns, J.J. Gooding, Sens. Actuators B 111–112 (2005) 515.
[49] X. Su, R. Robelek, Y.J. Wu, G.Y. Wang, W. Knoll, Anal. Chem. 76 (2004) 489.
[50] L.J. Kricka, Clin. Chem. 45 (1999) 453.
[51] B.J. Cheek, A.B. Steel, M.P. Torres, Y.-Y. Yu, H. Yang, Anal. Chem. 73 (2001) 5777.
[52] W.J. Miao, A.J. Bard, Anal. Chem. 75 (2003) 5825.
[53] J.D. Le, Y. Pinto, N.C. Seeman, K. Musier-Forsyth, T.A. Taton, R.A. Kiehl, Nano Lett. 2004, 4, 2343.
[54] T.A. Taton, C.A. Mirkin, R.L. Letsinger, Science 289 (2000) 1757.
[55] A.H. Fu, C.M. Micheel, J. Cha, H. Chang, H. Yang, A.P. Alivisatos, J. Am. Chem. Soc. 126 (2004) 10832.
[56] Z.X. Deng, Y. Tian, S.H. Lee, C.D. Mao, Angew. Chem. 117 (2005) 3648.
[57] A. Erdem, H. Karadeniz, A. Caliskan, Electroanalysis 21 (2009) 464.
[58] F. Patolsky, C.M. Lieber, Mater. Today 8 (2005) 20.
[59] S.N. Kim, J.F. Rusling, F. Papadimitrakopoulos, Adv. Mater. 19 (2007) 3214.
[60] M.I. Pividori, S. Alegret, Immob. DNA on Chips I 260 (2005) 1.
[61] B.L. Allen, P.D. Kichambare, A. Star, AdV. Mater. 19 (2007) 1439.
[62] E.D. Minot, A.M. Janssens, I. Heller, H.A. Heering, C. Dekker, S.G. Lemay, Appl. Phys. Lett. 91 (2007) 093507.
[63] A. Maroto, K. Balasubramanian, M. Burghard, K. Kern, Chem. Phys. Chem. 8 (2007) 220.
[64] I. Heller, J. Kong, K.A. Williams, C. Dekker, S.G. Lemay, J. Am. Chem. Soc. 128 (2006) 7353.
[65] E.L. Gui, L.J. Li, K. Zhang, Y. Xu, X. Dong, X. Ho, P.S. Lee, J. Kasim, Z.X. Shen, J.A. Rogers, S.G. Mhaisalkar, J. Am. Chem. Soc. 129 (2007) 14427.
[66] K. Besteman, J.O. Lee, F.G.M. Wiertz, H.A. Heering, C. Dekker, Nano Lett. 3 (2003) 727.
[67] H.R. Byon, H.C. Choi, J. Am. Chem. Soc. 128 (2006) 2188.
[68] P.D. Tama, N.V. Hieu, N.D. Chien, A.T. Le, M.A. Tuan, J. Immunological Methods 350 (2009) 118.
[69] T.G. Drummond, M.G. Hil, J.K. Barton, Nature Biotech. 21 (2003) 1192.

[70] B.D. Malhotra, R. Singhal, A. Chaubey, S.K. Sharma, A. Kumar, Current Appl. Phys. 5 (2005) 92.
[71] P. Kara, B. Meric, Anal. Chim. Acta 518 (2004) 69.
[72] E. Pearson, A. Gill, P. Vadgama, Ann. Clin. Biochem. 37 (2000) 119.
[73] R. Gabl, H.D. Feucht, H. Zeininger, G. Eckstein, M. Schreiter, R. Primig, D. Pitzer, W. Wersing, Biosens. Bioelectron. 19 (2004) 615.
[74] N. Zammatteo, L. Jeanmart, S. Hamels, S. Courtois, P. Louette, L. Hevesi, J. Remacle, Anal. Biochem. 280 (2000) 143.
[75] E. Komarova, M. Aldissi, A. Bogomolova, Bioelectron. 21 (2005) 182.
[76] D. Peelen, L.M. Smith, Langmuir 21 (2005) 266.
[77] K.V. Singh, R.R. Pandey, X. Wang, R. Lake, C.S. Ozkan, K. Wang, M. Ozkan, Carbon 44 (2006) 1730.
[78] D.H. Jung, B.H. Kim, Y.K. Ko, M.S. Jung, S. Jung, S.Y. Lee, H.T. Jung, Langmuir 20 (2004) 8886.
[79] Y. Zhang, H. Ma, K. Zhang, S. Zhang, J. Wang, Electrochim. Acta 54 (2009) 2385.
[80] J. Wang, G.D. Liu, A. Merkoci, Anal. Chim. Acta 482 (2003) 149.
[81] J. Wang, W. Danke, K. Abdel-Nasser, P. Ronen, J. Am. Chem. Soc. 124 (2002) 4208.
[82] E. Palecek, M. Fojta, Bioelectrochemistry 59 (2002) 85.
[83] A. Erdem, P. Papakonstantinou, H. Murphy, Anal. Chem. 78 (2006) 6656.
[84] S. Niu, M. Zhao, L. Hu, S. Zhang, Sensors and Actuators B 135 (2008) 200.
[85] L. Yang, R. Bashir, Biotechnol. Adv. 26 (2008) 135.
[86] Y.S. Kim, J.H. Niazi, M.B. Gu, Anal. Chim. Acta 634 (2009) 250.
[87] D.D. Venuto, S. Carrara, B. Ricco, Microelectron. J. 40 (2009) 1358.
[88] R.J. Chen, H.C. Choi, S. Bangsaruntip, E. Yenilmez, X. Tang, Q. Wang, Y. Chang, H.J. Dai, J. Am. Chem. Soc. 126 (2004) 1563.
[89] A. Star, E. Tu, J. Niemann, J.P. Gabriel, C.S. Joiner, C. Valcke, Proc. Natl. Acad. Sci. USA 104 (2006) 921.
[90] W. Lee, K. Zhang, H. Tantang, A. Lohani, T. Nagahiro, K. Tamada, Y. Chen, S.G. Mhaisalkar, L.J. Li, Appl. Phys. Lett. 91 (2007) 103515.
[91] E.L. Gui, L.J. Li, P.S. Lee, A. Lohani, S.G. Mhaisalkar, Q. Cao, J. Kang, J.A. Rogers, N.C. Tansil, Z. Gao, Appl. Phys. Lett. 89 (2006) 232104.
[92] X. Dong, C.M. Lau, A. Lohani, S.G. Mhaisalkar, J. Kasim, Z. Shen, X. Ho, J.A. Rogers, L.J. Li, Adv. Mater. 20 (2008) 2389.
[93] J. Zhang, S. Song, L. Zhang, L. Wang, H. Wu, D. Pan, C. Fan, J. Am. Chem. Soc. 128 (2006) 8575.
[94] S.I. Tanaka, M. Taniguchi, T. Kawai, Japan. J. Appl. Phys. 43 (2004) 7346.
[95] J. Li, Y. Zhang, T. Yang, H. Zhang, Y. Yang, P. Xiao, Materials Science and Engineering C 29 (2009) 2360.
[96] X. Tang, S. Bansaruntip, N. Nakayama, E. Yenilmez, Y.L. Chang, Q. Wang, Nano Lett. 6 (2006) 1632.
[97] S. Heinze, J. Tersoff, R. Martel, V. Derycke, J. Appenzeller, Ph. Avouris, Phys. Rev. Lett. 89 (2002) 106801.
[98] D. Cui, F. Tian, Y. Kong, I. Titushikin, H. Gao, Nanotechnology 15 (2004) 154.
[99] T. Ito, L. Sun, R.M. Crooks, Chem. Commun. (2003) 1482.
[100] Z. Li, Z. Wu, K. Li, Anal. Biochem. 387 (2009) 267.
[101] S. Reisberg, B. Piro, V. Noel, M.C. Pham. Bioelectrochemistry 69 (2006) 172.

[102] S.R. Bhattarai, S. Aryal, K.C.R. Bahadur, N. Bhattarai, P.H. Hwang, H.K. Yi, H.Y. Kim, Mater. Sci. Eng. C 28 (2008) 64.
[103] M. Gong, T. Han, C. Cai, T. Lu, J. Du, J. Electroanal. Chem. 623 (2008) 8.
[104] C. Cao, J.H. Kim, D. Yoon, E.S. Hwang, Y.J. Kim, S. Baik, Mater. Chem. Phys. 112 (2008) 738.
[105] R.A. Graff, J.P. Swanson, P.W. Barone, S. Baik, D.A. Heller, M.S. Strano, Adv. Mater. 17 (2005) 980.
[106] E.S. Jeng, A.E. Moll, A.C. Roy, J.B. Gastala, M.S. Strano, Nano Lett. 6 (2006) 371.
[107] Y.C. Tsai, S.C. Li, S.W. Liao, Biosens. Bioelectron. 22 (2006) 495.
[108] C.E. Banks, R.R. Moore, T.J. Davies, R.G. Compton, Chem. Commun. 16 (2004) 1804.
[109] J. Wang, M. Musameh, Anal. Chem. 75 (2003) 2075.
[110] Y. Lin, F. Lu, Y. Tu, Z. Ren, Nano Lett. 4 (2004) 191.
[111] K. Wu, X. Ji, J. Fei, S. Hu, Nanotechnology 15 (2004) 287.
[112] Y. Lin, F. Lu, J. Wang, Electroanalysis 16 (2004) 145.
[113] Z. Guo, P.J. Sadler, S.C. Tsang, Adv. Mater. 10 (1998) 701.
[114] J.J. Davis, M.L.H. Green, H.A.O. Hill, Y.C. Leung, J. Sloan, S.C. Tsang, Inorg. Chim. Acta 272 (1998) 261.
[115] A. Ortiz-Acevedo, H. Xie, V. Zorbas, W.M. Sampson, A.B. Dalton, R.H. Baughman, R.K. Draper, I.H. Musselman, G.R. Dieckmann, J. Am. Chem. Soc. 127 (2005) 9512.
[116] N.W.S. Kam, Z. Liu, H. Dai, Angew. Chem. Int. Ed. 45 (2006) 577.

5

Defected and Substitutionally Doped Nanotubes: Applications in Biosystems, Sensors, Nanoelectronics, and Catalysis

Charles See Yeung, Ya Kun Chen and Yan Alexander Wang
Department of Chemistry, University of British Columbia, Vancouver, Canada

1. Introduction

Carbon nanotubes (CNTs) have been the subject of intensive research since their discovery by Iijima in the early 1990s (Iijima 1991). Single-walled carbon nanotubes (SWCNTs, Iijima & Ichihashi 1993; Bethune et al. 1993) are of particular interest because the electronic properties of these nanomaterials can vary from semiconducting to metallic depending on its molecular structure. This contrasts multi-walled carbon nanotubes (MWCNTs), which are metallic and exhibit a zero bandgap. Fundamental understanding of these supramolecular carbon allotropes (Tasis et al. 2006; Ajayan 1999) is essential to the development of new nanomaterials for applications in biosystems, sensors (Wang & Yeow 2009; Hu & Hu 2009; Li et al. 2008), optics (Avouris 2008), nanomechanics (Li et al. 2008; Park 2004), nanoelectronics (Park 2004; Fuhrer 2003; Tsukagoshi 2002), and catalysis (Serp 2003, Tian et al. 2006, Yeung et al. 2011).

The molecular structure of SWCNTs can be described as a cylindrical roll of an infinite graphene sheet and is characterized by a chiral circumferential vector **AB** = $ma + nb$, a linear combination of two unit lattice vectors a and b where m and n are integers (Figure 1-1, Ajayan 1999; Moniruzzaman & Winey 2006). The pair of indices (m,n) for any given nanotube determines its diameter, chirality, and electronic character. For all $n = m$, the nanotube is termed armchair and is metallic, exhibiting a zero bandgap. For $n \neq m$ and neither n and m are zero, the nanotube exhibits chirality and supramolecular helicity, having important implications in optical properties. For $n = 0$ or $m = 0$, the nanotube is termed zigzag. For $n - m = 3p$, where p is a non-zero integer, the nanotube is semimetallic with a band gap on the order of meV. For $n - m \neq 3p$, where p is a non-zero integer, the nanotube is semiconducting with a band gap on the order of 1 eV; as a general rule of thumb, the observed band gaps are roughly proportional to the reciprocal of the tube radius.

Each individual C atom in the sidewall of a CNT exhibits pyramidalization and partial sp^3 hybridization as a result of sidewall curvature. This phenomenon leads to a weakening of the overall π-conjugation of the SWCNT and slight misalignment of π-orbitals between adjacent atoms. Curved π-conjugation can be quantified using Haddon's π-orbital axis vector (POAV) method (Figure 1-2, Haddon & Scott 1986). In this analysis, the pyramidalization angle $\theta_p = \theta_{\sigma\pi} - 90°$, where $\theta_{\sigma\pi}$ is the angle between the π-orbital and the σ-bond of the C atom of interest. In contrast to planar (i.e., $\theta_p = 0°$) and pyramidal (i.e.,

θ_p = 19.5°) geometries, nanotubes exhibit varying degrees of pyramidalization depending on the tube diameter. In general, SWCNTs are more inert than corresponding fullerenes with similar diameters. Alternatively, Li introduced the concept of bond curvature K to better describe the supramolecular curvature of CNTs and offer more accurate predictions for chemical reactivity based on the local curvature of the independent C–C bonds (Li et al. 2007). K is characterized by directional curvature K_D (corresponding to the arc curvature of C–C bonds) and its mean K_M (corresponding to arc curvature of the vertex carbon atoms).

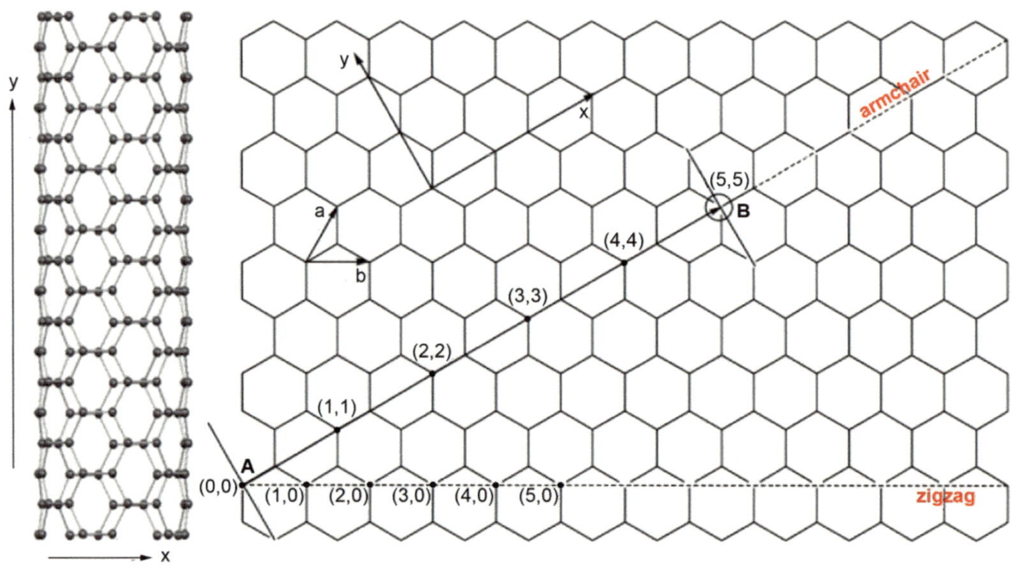

Fig. 1.1. A schematic drawing of graphene wrapping to form SWCNTs. The (5,5) SWCNT on the left is drawn highlighting wrapping vector x and translation vector y (Tian et al. 2006).

Carbon nanotubes are characterized by a very high aspect ratio (i.e., length per width) and display a very large surface area. Attractive interactions by van der Waals or weak long-range covalent bonding on the order of 0.50 to 0.95 eV per nanometer of tube-to-tube contact have been observed. Aggregation of CNTs often results in the generation of hexagonally packed bundles or ropes. SWCNTs exhibit extensive levels of electron delocalization and strong confinement effects that result in a pronounced one-dimensionality and weak interaction between electrons and phonons. These nanomaterials are ballistic conductors and exhibit two units of quantum conductance (i.e., $4e^2/h$) and mean free paths on the order of μm (Javey 2003; White 1998).

Natural carbon nanotubes can exhibit defects formed either during their synthesis or as a result of environmental stress that have significant effects on the geometric and electronic properties of carbon nanomaterials (Collins 2010; Charlier 2002). The most typical type of defects are point vacancies in which an atom is missing in the crystalline lattice of the nanotube backbone and occur as a result of external radiation (e.g., high energy electron, ion, or neutron radiation) in which a carbon atom is either dislodged or fully removed. A single vacancy of this type generates three dangling bonds (DBs) and immediately undergoes structural reorganization, for example, yielding the 5-1DB defect in which one five-membered carbocycle is formed, leaving a single C atom with a reactive DB (Berber &

Oshiyama 2006; Ajayan 1998). Different from point vacancies, other structural defects arising from geometric rearrangements may also be present in nanotube sidewalls. The simplest of these rearrangements is the pentagon-heptagon defect in which a single C-C bond positioned between two adjacent hexagon rings undergoes a simple bond rotation, resulting in the formation of two non-hexagonal rings. Pentagon-heptagon defects have minimal effect on the local environment of the nanotube but induce a global buckling of the macromolecule, leading to significant changes in conductivity and aggregation (Chico 2010; Sonia & Niranjan 2010). Of note, two adjacent pentagon-heptagon defects can form a pentagonal-heptagonal-heptagonal-pentagonal (5775) structure known also as the Stone-Wales (SW) defect (Stone & Wales 1986).

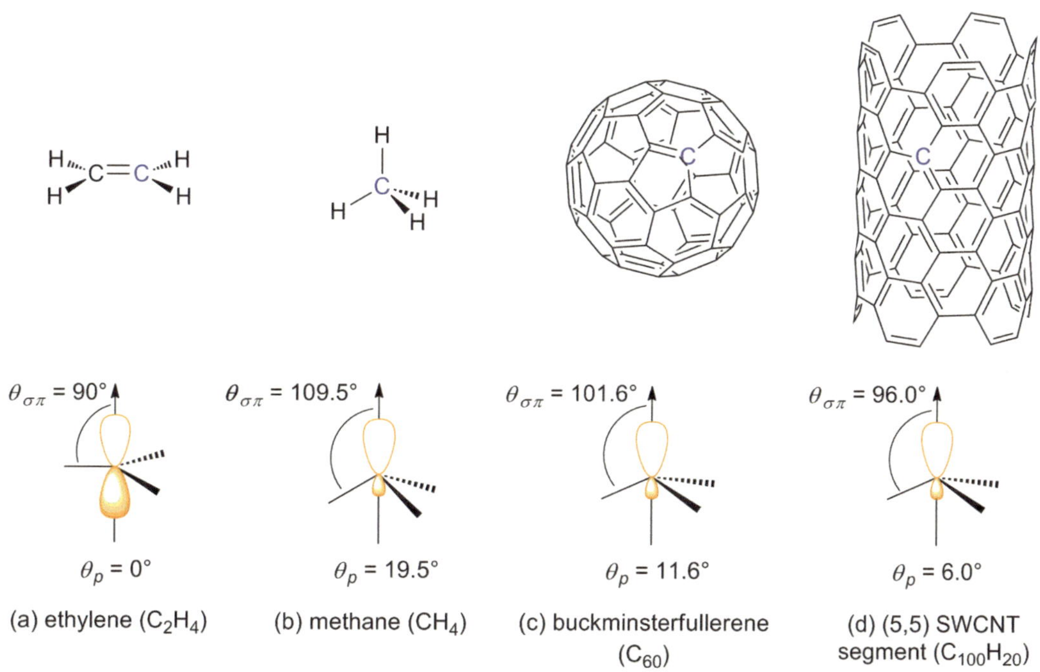

Fig. 1.2. π-orbital axis vector (POAV) analysis of curved π-conjugation in (a) ethylene, (b) methane, (c) buckminterfullerene, and (d) a (5,5) SWCNT segment (Yeung et al. 2011).

Chemical functionalization provides scientists an opportunity for fine-tuning the electrical properties of carbon nanotubes for the desired application. CNTs are known to exhibit interstitial defects, or extra atoms not present in a lattice site, as a result of atmospheric oxidation (Collins 2010). Covalent modification by treatment of these carbon macromolecules with chemical reagents has been accomplished (Karousis et al. 2010; Tasis et al. 2006). The chemical reactivity of nanotube sidewalls bears some resemblance to that of olefins (i.e., C=C double bonds). As such, strategies including ozonolysis, Diels-Alder cycloaddition, osmylation, hydroboration, carbene addition, nitrene addition, dipolar cycloaddition of azomethine ylides, and vinylcarbonylation via zwitterionic intermediates are possible. Other methods include fluorination followed by nucleophilic substitution, dissolving metal reduction (i.e., Billups reaction), electrophilic oxidations, and free radical reactions including alkylation, perfluoroalkylation, and arylation.

Direct substitutional replacement of a sidewall C atom with a heteroatom is another alternative for electronic tuning (Ewels et al. 2010; Glerup et al. 2010). Substitutional doping of a SWCNT produces a hetero-SWCNT (HSWCNT) that exhibits geometric and electronic properties distinct from pristine carbon nanotubes depending on the doping concentration. Replacement of a single C atom with N or B is of particular interest because of the generation of quasibound states consisting of *p*-orbitals above and below the Fermi energy, respectively, in analogy to *p*- and *n*-type semiconductors. This is because boron has one less electron than carbon, while nitrogen has one more electron. Both N- and B-doped SWCNTs have been synthesized by thermal treatment, chemical vapor deposition, laser ablation, and arc methods, but generally with poor control of doping position and concentration. Srivastava et al. predicted that a free gas-phase neutral N atom, if brought into close vicinity of a backbone vacancy defect, could induce a selective substitution to occur (Srivastava et al. 2004). Alternatively, our group proposed that nitrogen monoxide (NO) could be used as a nitrogen source by interaction with the 5-1DB defects (Liu et al. 2006). Substitutional doping, such as N-doping, typically causes minimal geometric changes in comparison to their undoped congeners.

Although physisorption of transition metals onto CNTs is a known process, substitutional doping in which the transition metal atom is embedded into the sidewall remains an unsolved challenge in nanoscience, although some researchers have proposed that this transformation may be assisted by the presence of vacancy defects (Yang et al. 2006; Zhuang et al. 2008). The corresponding transition metal substitutionally doped fullerenes, however, have been prepared experimentally and characterized (Branz et al. 1998; Poblet et al. 1999; Kong et al. 2003). For example, *d*-block transition metals Pt-, Ir-, Rh-, Sc-, and Y-, and *f*-block La- and Sm-doped fullerenes have been synthesized by laser ablation or ionization and can be detected by mass spectrometry. Although the mechanistic details concerning substitution are sparse, it is believed that a fullerene-transition metal adsorbate complex is formed as a result of strong interactions between the metal *d*-orbitals and extended π-framework of the fullerene, followed by direct insertion into the carbon framework (Changgeng et al. 2001). In some cases, the initial generation of vacancy defects may induce the observed structural rearrangement. The synthesis of second- and third-row transition metal-doped fullerenes has been predicted by the chemical reaction of transition metal chloride (MCl_n) with buckminsterfullerene (C_{60}) under an atmosphere of chlorine gas (Cl_2, Sparta et al. 2006).

Theoretical studies of transition metal-doped fullerenes suggest a decrease in the energy gap between the highest occupied molecular orbital (HOMO) and the lowest unoccupied molecular orbital (LUMO), leading to both higher conductivity and reactivity (Changgeng et al. 2001). As such, transition metal-doped SWCNTs may display enhanced electronic properties in comparison to their undoped analogs. The introduction of transition metals is an attractive prospect because of the inherently rich chemistry of transition metal surfaces and organometallic complexes. By adding *d*-orbitals to the extended π-framework of the C atoms, it may be possible to enhance the ability of nanotubes to interact and bind small adsorbates such as NO, a molecule critical to biological systems. Since coordination of transition metals affect the electronic nature of the metal center, it is possible that a HSWCNT containing a transition metal atom embedded in the sidewall of the nanotube may exhibit changes in conductance in the presence of gases such as carbon monoxide (CO), water (H_2O), and ammonia (NH_3). By monitoring current, the development of new nanosensors may be possible (Peng & Cho 2003; Sinha et al. 2006; Li et al. 2008; Rouxinol et al. 2010; Yeung et al. 2010). This prospect also has important implications in efforts toward new nanoelectronic devices. Controlled storage, capture, activation, and release of other relevant molecules, including hydrogen (H_2), is another potential application of transition metal-doped SWCNTs based on the known reactivity of transition metals. We have also conducted preliminary investigations highlighting the similarity between transition metal-doped SWCNTs and simple alkyl complexes, suggesting that nanocatalysis (e.g., Lewis acid catalysis) may be optimal using sidewall substitution as the key strategy for immobilizing these metals (Yeung et al. 2007; Yeung & Wang 2011). Indeed, transition metal-doped SWCNTs could sit at the interface between heterogeneous and homogeneous catalysts and provide complementary reactivity to traditional systems.

Our work in SWCNTs and their substitutionally doped analogs (i.e., HSWCNTs) is driven by a fundamental interest in the unique chemical reactivity afforded by these macromolecules. We have focused our investigations primarily on defected and substitutionally doped nanotubes because of their potential applications in biological systems, sensory technology, nanoelectronics, and catalysis. In this account, we highlight our theoretical studies on the geometric and electronic structure of defected and doped SWCNTs, their chemical reactivity, and proposed applications of these nanomaterials. We also describe the related defected and doped boron-nitride nanotubes (BNNTs) and their behavior.

2. Models and computational details

We chose the (5,5) armchair metallic SWCNT and (5,0) zigzag SWCNT as models for all computations with suitable capping groups (i.e., either a fullerene hemisphere or H atoms). All calculations were done with Gaussian 03 (Frisch et al. 2003). Natural bond orbital (NBO) analysis was conducted with Gaussian NBO Version 3.1 (Reed et al. 1988) to obtain the frontier molecular orbitals (FMOs, Fukui 1982): the HOMO and the LUMO. Density of states (DOS) and local DOS (LDOS) studies were performed using PyMOlyze Version 1.1 (Tenderholt 2005). The Hessian was calculated to verify the nature of stationary points on the potential energy surface (PES, either a local minimum or transition state). Optimizations using spin-restricted and spin-unrestricted methods provided the same results.

To study the substitutional doping of SWCNTs, we envisioned a chemical reaction between a single vacancy defect on the backbone of the nanotube and a suitable atom transfer reagent. To this end, we evaluated the use of NO and ozone (O_3) for N- and O-substitutional doping, respectively. For N-doped SWCNT synthesis with NO (Liu et al. 2006), a single C atom was removed from the middle of a (5,5) segment capped with H atoms ($C_{199}H_{20}$), yielding a metastable conformation with three dangling bonds. Geometry optimization with semiempirical MNDO-PM3 method (Stewart 1989) and hybrid Hartree-Fock/density functional theory (DFT) method B3LYP (Becke 1988; Lee et al. 1988) using Pople's 6-31G basis set (Ditchfield et al. 1971) afforded the 5-1DB defect. To study the chemical reactivity, we employed a two-layer ONIOM (our own N-layered integrated molecular orbital and molecular mechanics) model (Dapprich et al. 1999), with the nine-membered ring surrounding the defect modeled with C_9H_8 as the higher layer at B3LYP/6-31G(d), while all other atoms were the lower layer treated with universal force field (UFF, Rappe et al. 1992). For O-doped SWCNT synthesis with O_3 (Liu et al. 2006b), an analogous system was used ($C_{119}H_{20}$). Geometry optimization was achieved with semiempirical AM1 (Dewar et al. 1985) and B3LYP/6-31G. Static quantum mechanical calculations were validated by the atom-centered density matrix propagation (ADMP) based *ab initio* molecular dynamics (AIMD) simulation (Schlegel et al. 2001)

To study transition metal-doped SWCNTs, we initiated our studies using one-dimensional periodic boundary conditions with 100 carbon atoms in the primitive cell (Chen et al. 2011). Replacement of a single C atom in the middle of the cell with a transition metal was performed, followed by geometry optimization with PBEPBE (Perdew et al. 1996) using LANL2DZ (Dunning Jr. & Hay 1976), Hay and Wadt's relativistic 18-electron Los Alamos National Laboratory effective core pseudopotential (ECP, Hay & Wadt 1985). We then focused our investigations on Pt due to its prevalence in well-defined organometallic complexes and heterogeneous catalysis. Two (5,5) models were chosen in which a C atom was substituted with Pt: 1) $C_{69}H_{20}Pt$ with H caps (Yeung et al. 2007; Yeung et al. 2010), and 2) C_{170} with D_{5h} symmetry and hemispherical caps (Tian et al. 2006). For $C_{69}H_{20}Pt$, DFT calculations were performed with PBEPBE using LANL2MB, followed by LANLDZ. For $C_{169}Pt$, we used BPW91 (Becke 1988; Perdew et al. 1996a). We also chose a (5,0) model containing of formula $C_{49}H_{10}Pt$ (Yeung & Wang 2011), optimized with B3LYP/LANL2DZ. NMR chemical shifts (δ) were calculated using the gauge-independent atomic orbital (GIAO) method (Wolinski et al. 1990). We further examined model Pt complexes. Alkylplatinum species $PtMe_3^+$ and its C̲O-adsorbates (i.e., with a C-end coordinated CO) was evaluated with PBEPBE. $PtMe_3^+$, $PtPh_3^+$, Pt-doped phenaline, Pt-doped sumanene, Pt-doped corannulene, and Pt-doped C_{24} fullerene (with Pt substitution at the junction between three pentagons being PPP and between a hexagon and two pentagons being HPP) were also considered with B3LYP/LANL2MB and LANL2DZ.

To study the importance of the carbon backbone, the isoelectronic transition metal-doped single-walled boron nitride nanotube (BNNT) and its undoped analog were investigated (Chen et al. 2010). Defected BNNT models containing SW defects (Stone & Wales 1986) had formula $B_{45}N_{45}H_{20}$, while Pt substitutional doping gave two isomers, $B_{44}N_{45}H_{20}Pt$ and $B_{45}N_{44}H_{20}Pt$. DFT calculations were performed using B3LYP/LANL2MB, followed by LANL2DZ.

3. Defected SWCNTs

3.1 Defected SWCNT rods

We investigated vacancy defects of SWCNTs by removing a single atom from the middle of model (5,5) SWCNT rod capped with H atoms ($C_{200}H_{20}$), yielding a defected SWCNT ($C_{199}H_{20}$) that undergoes subsequent structural rearrangement to the 5-1DB defect (Figure 3-1, Liu et al. 2006). In our model system, a single C atom protrudes to the exterior of the sidewall and is the active carbon (C1). Other than the nine C atoms surrounding the vacancy, the remainder C atoms in the hexagonal lattice remain relatively unaffected. Because of coordinative unsaturation and less steric hindrance, a higher chemical reactivity is expected. The electronic structure was affected by introduction of the defect, as determined by a decrease in HOMO−LUMO gap from 1.38 to 0.84 eV. The HOMO of the defected SWCNT contains a large contribution from the lone pair on the active carbon atom and the π-bonds of the other atoms surrounding the vacancy. We also determined that a smaller model system (C_9H_8) with fixed geometry can be used to represent most of the chemical properties of the defect by qualitative comparison of the FMOs. Introduction of the vacancy defect also resulted in several small changes in the DOS (Figure 3-2). We have also studied the effect of multiple vacancies and have observed related structural rearrangements (Liu et al. 2009).

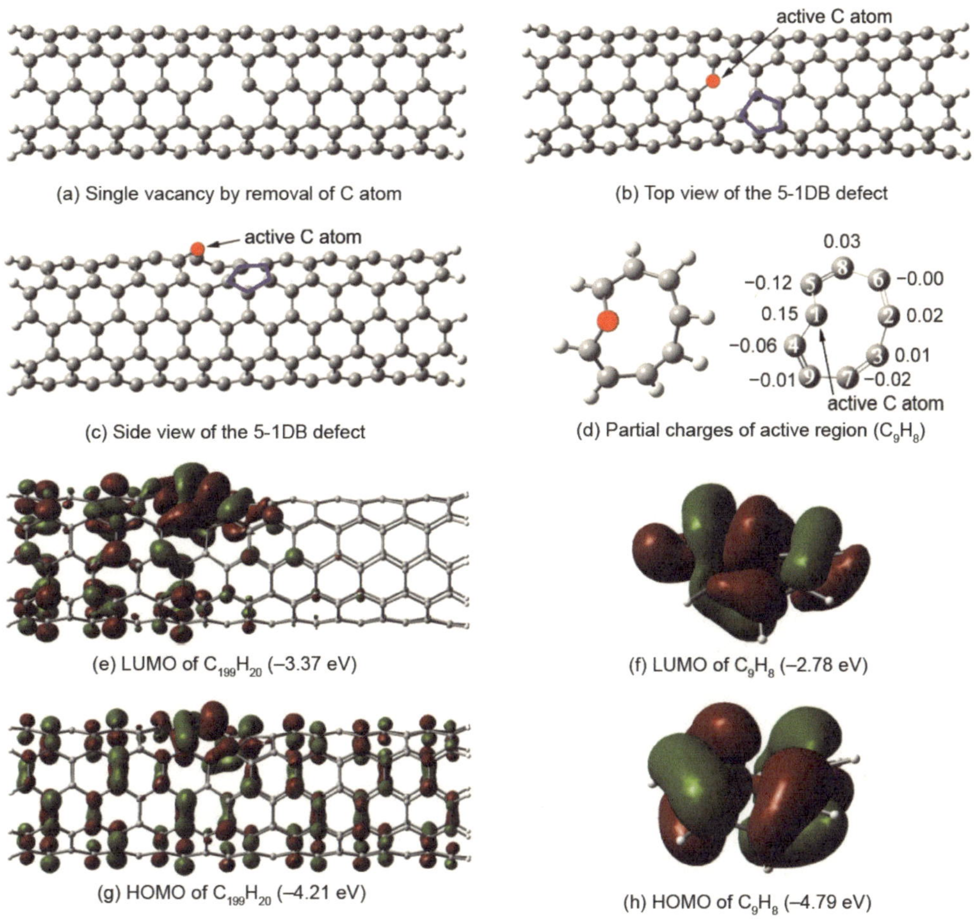

Fig. 3.1. Optimized geometry and FMOs of a 5-1DB defected SWCNT and higher-layer model C_9H_8. Orbital energies are in parentheses (Liu et al. 2006). Legend: dark grey = C, light grey = H.

3.3 Substitutional N-doping of SWCNTs via reaction with NO

Based on FMO analysis, the active carbon (C1) of the 5-1DB defect, expected to be reactive toward small molecules, contains an sp^2-orbital with 1.51 electrons, a p-orbital with 0.51 electrons, and an overall charge of 0.149 (Liu et al. 2006). As the first example, we chose nitrogen monoxide (NO) as the representative small-molecule reagent.

Nitrogen monoxide is of particular interest due to its biological relevance (Miller & Megson 2007). In fact, while NO has a reputation for destroying O_3, causing cancer, and fostering acid rain, it is a critical signaling molecule in biological systems, important for learning and memory, blood pressure regulation, skin homeostasis, inflammation, and carcinoma metastasis. As such, NO was named "Molecule of the Year" in 1992 (Culotta & Koshland Jr. 1992). Physiological NO is biosynthesized via a five-electron oxidation of the amino acid arginine (Figure 3-3) and can impact biosystems based on its ability to interact with molecular oxygen to yield reactive oxygen species (ROS) including hydroxyl radicals (HO•), nitrogen dioxide (N_2O_4), and peroxynitrite ($ONOO^-$). Free NO exhibits polarization where the O atom bears a charge of −0.181, thus making chemical reactivity likely.

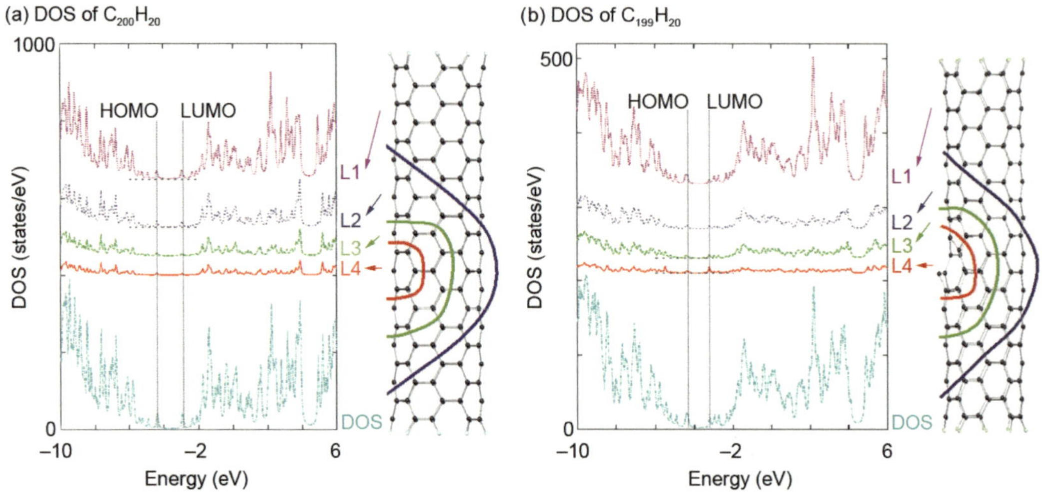

Fig. 3.2. Optimized geometries, DOS, and LDOS of (a) $C_{200}H_{20}$ (ε_{HOMO} = −4.35 eV, ε_{LUMO} = −2.97 eV) and (b) $C_{199}H_{20}$ (ε_{HOMO} = −4.21 eV, ε_{LUMO} = −3.37 eV). L1–L4 are the LDOS for the specified layer of atoms (Liu et al. 2006).

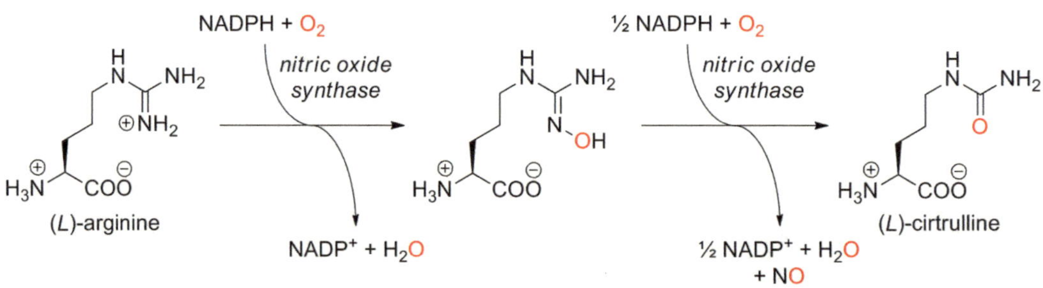

Fig. 3.3. Biosynthesis of NO from (*L*)-arginine (Griffith & Stuehr 1995).

The reaction of NO with the defected SWCNT proceeds in two distinct stages. First, an electrostatic interaction between the defected SWCNT and NO takes place, involving an O-end attack of NO onto the active C atom via an orbital interaction between the singly-occupied molecular orbital (SOMO) of NO and the HOMO of the SWCNT (Figure 3-4), producing heterocycle **INT1** containing an N–O bond. This rate-limiting step takes place rapidly with a low activation barrier (8.6 kcal/mol). Next, the N atom undergoes insertion into the nearby C2–C3 bond of the adjacent pentagon, expanding the pentagon via a three-membered transition state to liberate **INT2** containing a six-membered N-containing heterocycle. C1–N bond formation then occurs with concomitant cleavage of the C1–O bond to give **INT3**, formally an N-doped SWCNT that has undergone pyrimidalizing oxygenation where the N atom has filled the single vacancy where the C atom was removed (i.e., substitutional doping). The N–O bond is relatively weak, with a bond length of 1.41 Å and a partial charge of −0.574 residing on the O atom. A second equivalent of NO initiates cleavage of this N−O bond via a four-membered concerted and asynchronous transition state, liberating one molecule of nitrogen dioxide (NO_2) and **Product**, the N-doped SWCNT ($C_{199}H_{20}N$), a thermodynamically feasible process (ΔE = 78.3 kcal/mol).

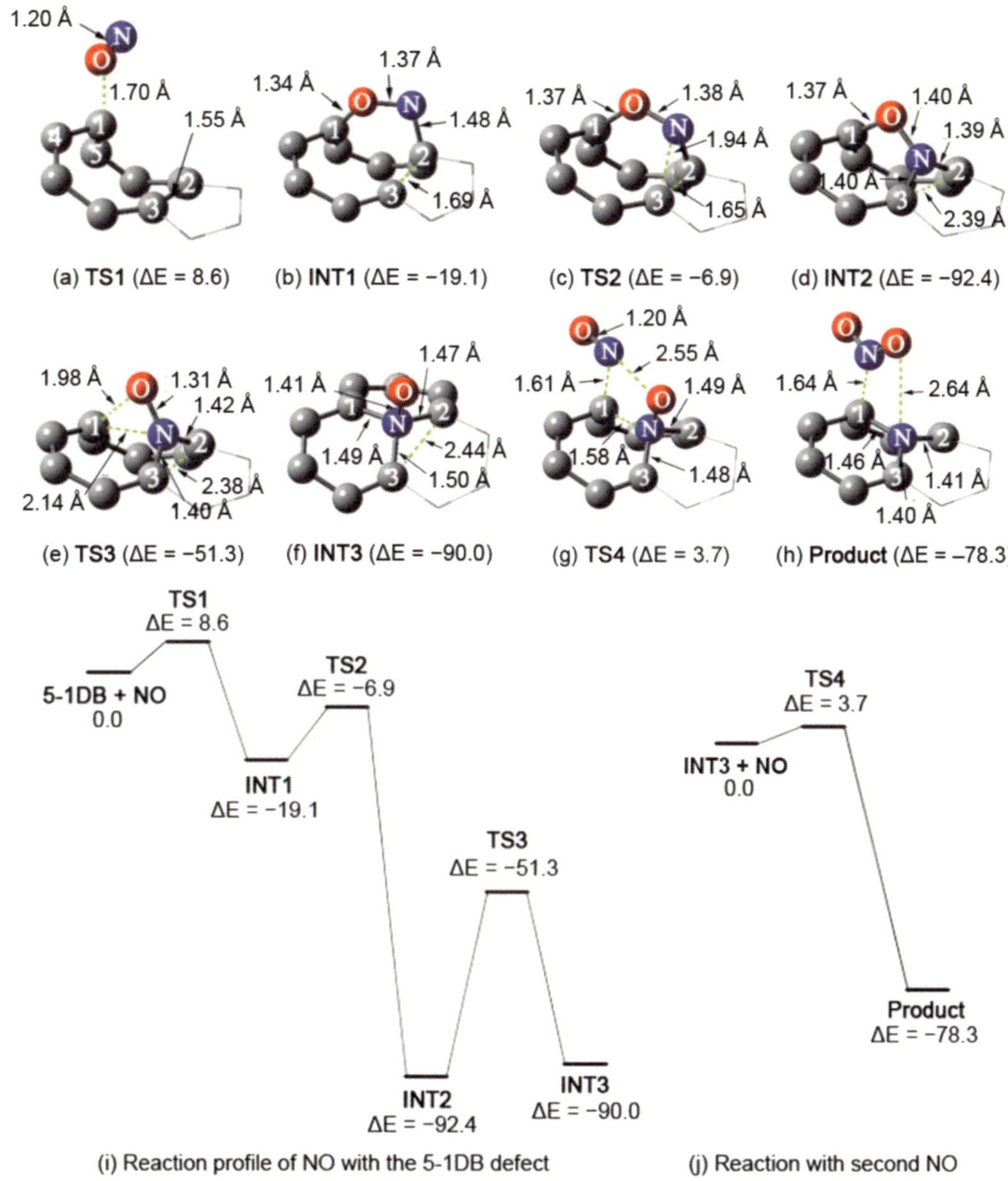

Fig. 3.4. Optimized geometries and energy profiles (with energies in kcal/mol) for the reaction of the defected SWCNT with NO (Liu et al. 2006). Legend: dark grey = C, navy blue = N, red = O.

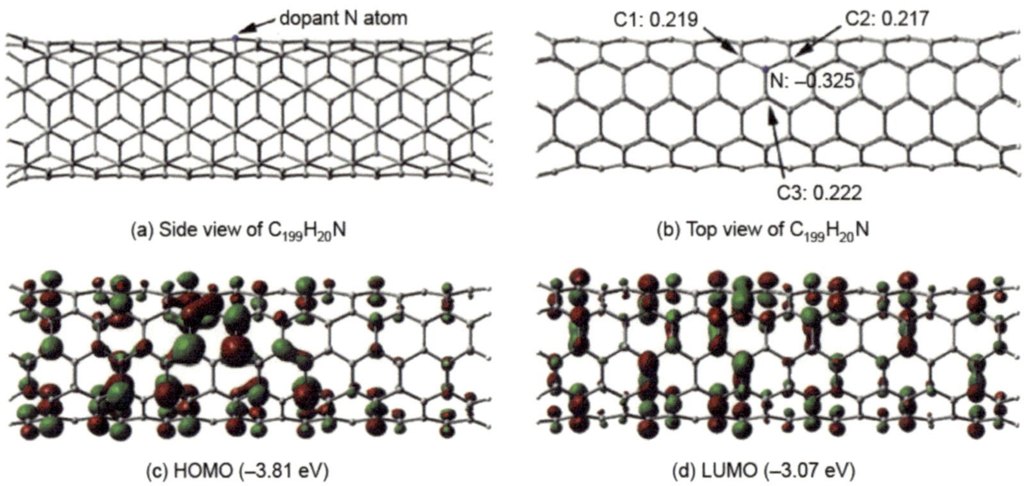

Fig. 3.5. Optimized geometry and FMOs of $C_{199}H_{20}N$. Orbital energies are in parentheses (Liu et al. 2006).

In accord with literature data (Nevidomskyy et al. 2003), the geometry of N-doped HSWCNT model ($C_{199}H_{20}N$) strongly resembles the undoped, defect-free SWCNT ($C_{200}H_{20}$, Figure 3-5). Substitutional doping leads to a decreased HOMO−LUMO gap (0.74 eV) and slight polarization of the nanotube, with the N atom bearing a charge of −0.325.

3.4 Biological applications of defected SWCNTs and reactions with NO: Therapeutics and biosensing

Maintaining homeostasis of physiological NO is critical to biological systems (Miller & Megson 2007). As such, if future nanodevices are implanted into living organisms, it may interact with the biocycles of NO and strongly impact the amount of NO present. Given our studies (Liu et al. 2006), we envision the use of vacancy defective nanotubes in reducing the local concentration of NO by the irreversible chemical reaction depicted in Figure 3-4. For every single vacancy, two molecules of NO are consumed, producing a N-doped SWCNT and one molecule of NO_2 as a byproduct. This process can be controlled in both location and defect concentration by inducing the selective formation of single vacancies by applying external stimuli to the area of interest. We can imagine, for example, the employment of ultraviolet light as a form of medical treatment. This strategy would provide a platform for physicians to carefully control the amount of NO present in any given biological system. Manipulating the local concentration of NO would be achieved by a hybrid approach involving chemically designed pharmaceutical agents such as glyceryl trinitrate and sodium nitroprusside (Figure 3-6), as well as new technologies involving materials such as zeolites (Wheatley et al. 2006) and Pt nanoparticles (Caruso et al. 2007). Recent work on the antithrombotic activity of NO drugs has confirmed that controlled release is essential to these types of treatments and we expect CNTs to play a critical role in the development of new therapeutic techniques in the future.

We can also imagine a scenario in which nanotubes are used as a preventative measure to control the amount of ROS present in the body; in other words, CNTs could act as an artificial antioxidant. Because it is known that the introduction of N atoms into nanotube sidewalls increases clearance rates (Singh et al. 2006), their removal from the biological system may be possible following irreversible NO capture. This suggests that the antioxidant properties of these carbon macromolecules may reverse the damage caused by high-energy processes, including harmful UV-A and UV-B rays from the sun. If defect-free, pristine SWCNTs can be implanted into the human body, environmental stresses may trigger the formation of defects, leading to a decrease in the amount of free biological NO.

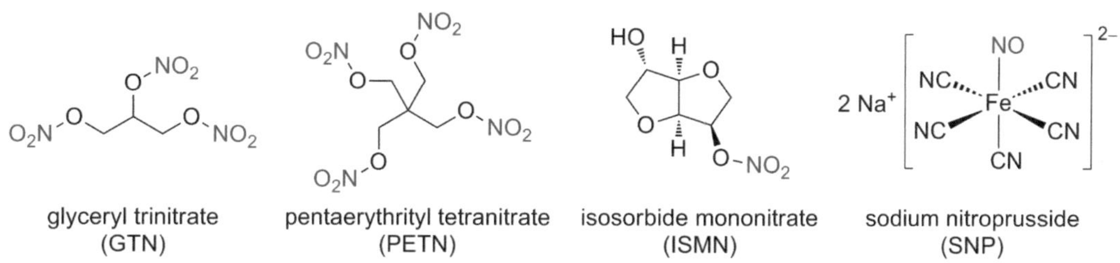

Fig. 3.6. Clinically used NO donor drugs (Miller & Megson 2007).

Defected SWCNTs, if embedded in physiological systems, can serve as a sensor for biological NO. Unlike conventional methods, which involve breath analysis (Menzel et al. 2001), the use of these nanomaterials offers a direct evaluation of NO concentrations at any given location based on the known ability of environmental stresses to induce vacancy defect formation. Quantitative analysis may be achieved via isolation of excreted N-doped SWCNTs in comparison to pristine SWCNTs assuming equal clearance rates. Additionally, we expect that substitutional N-doping results in increased conductivity, since our studies have revealed a decreased HOMO−LUMO gap from 1.38 to 0.74 eV (Liu et al. 2006). If we can monitor electrical current passing through the nanotube, we would have an *in vivo* technique of determining physiological NO. Although the specific details are yet to be determined, it may be possible to orient a nanotube across a lipid bilayer (such as a cell membrane) by capping the ends with hydrophilic ends and subsequently measure its resistance. Alternatively, voltage-sensitive or potentiometric dyes (which change spectral properties in response to voltage changes) may offer another method for non-intrusive analysis (Grinvald & Hildesheim 2004).

As promising as they are, therapeutic applications of SWCNTs necessitate an understanding of nanotoxicology (Oberdorster et al. 2007; Buzea et al. 2007; Lam et al. 2006; Karakoti et al. 2006; Nel et al. 2006). Because macromolecules resist metabolism by macrophages, nanotubes are postulated to persist indefinitely in biological systems, a cause for concern due to bioaccumulation. These effects are further complicated by subtle differences between methods of preparation and the types of defects present that strongly affect intertube interactions and bundle formation. Functionalization is also critical. The introduction of carbonyl (C=O), carboxyl (COOH), and hydroxyl (OH) groups, for example, has resulted in increased cytotoxicity either after contact with cell membranes or following internalization. On the other hand, functionalization with ammonium cations has demonstrated enhanced clearance rates of these nanomaterials (Singh et al. 2006). Nanotoxicity is proposed to be a result of oxidative stress caused by the generation of ROS in which light-induced formation of electron-hole pairs and unpaired electrons triggers oxidation of molecular oxygen (O_2) to superoxide (O_2^-), promoted by the presence of reactive groups and defects (Karakoti et al. 2006; Nel et al. 2006). Although nature counteracts the effect of ROS, such as by superoxide dismutase or glutathione, accumulation of these reactive species remains problematic. CNTs exhibit a known cytotoxicity in human skin fibroblasts, macrophages, and developing zebrafish embryos, and the particulate nature of these nanomaterials can cause interstitial fibrosis and pulmonary toxicity. Clinically relevant data, however, remains sparse, and recent research has suggested that nanotubes can persist without causing toxicity (Schipper et al. 2008) or any immunological or inflammation reactions (Chiaretti et al. 2008). It is clear that further work is necessary before any clear conclusions can be made regarding nanotoxicology and its impact on human health.

3.5 Substitutional O-doping of SWCNTs via reaction with O_3

In analogy to NO, 5-1DB defects react exothermically with O_3 (Liu et al. 2006a). In traditional organic chemistry, O_3 is a strong oxidizing agent that is capable of cleaving C=C double bonds, generating two C=O bonds via the Criegee mechanism (Kuczkowski 1992). Because the sidewalls of CNTs exhibit π-frameworks of C atoms, it is reasonable to presume that ozonization can take place. We considered a $C_{120}H_{20}$ segment of the (5,5) armchair SWCNT and its defected analog $C_{119}H_{20}$ by removal of a single C atom in the middle of the nanotube. C1 was confirmed as the active C atom within the 5-1DB defect, analogous to our earlier work (Liu et al. 2006).

Only the nine C atoms surrounding the vacancy defect were evaluated in this SWCNT ozonization reaction. Although these C atoms exhibit partial C=C double bond character within the backbone of the defect, the pathway that is most energetically favored involves the direct interaction between ozone and C1 (Figure 3-7). The first step is physisorption of O_3 onto the surface of the SWCNT, forming **INT**, resulting in a transfer of 0.19 electrons from the carbon backbone to O_3. A concerted oxygenation and loss of singlet dioxygen liberates **Product** ($C_{119}H_{20}O$) via a four-membered transition state **TS**. This step is barrierless (−0.3 kcal/mol) and produces a substitutionally O-doped SWCNT with the release of 88.7 kcal/mol. Subsequent relaxation of singlet to triplet O_2 results in further stabilization. We also performed ADMP-AIMD dynamics simulations to confirm our static analysis of the 5-1DB defect ozonization and found that at 300 K, a spontaneous reaction occurs between O_3 and C1 in less than 50 fs. The structure of this O-doped SWCNT differs from our N-doped SWCNT derived from NO (*vide supra*). This is because the bond between C1 and O1 is very strong and exhibits significant double bond character. Hence, it may be more accurate to describe this reaction as a net oxidation of C1 to a ketone. No significant interaction could be observed between C5 and O1.

Because of the ease of oxygenation by ozonization of 5-1DB defects, it has been proposed that such single vacancies are often short-lived species that may rapidly travel through the backbone and annihilate each other (Collins 2010). It would be interesting to see whether other weaker oxygen sources including O_2 and H_2O may be able to oxidize a 5-1DB defect. By studying these processes, we would come to a better understanding of the fate of defect-free and defected SWCNTs within biological systems.

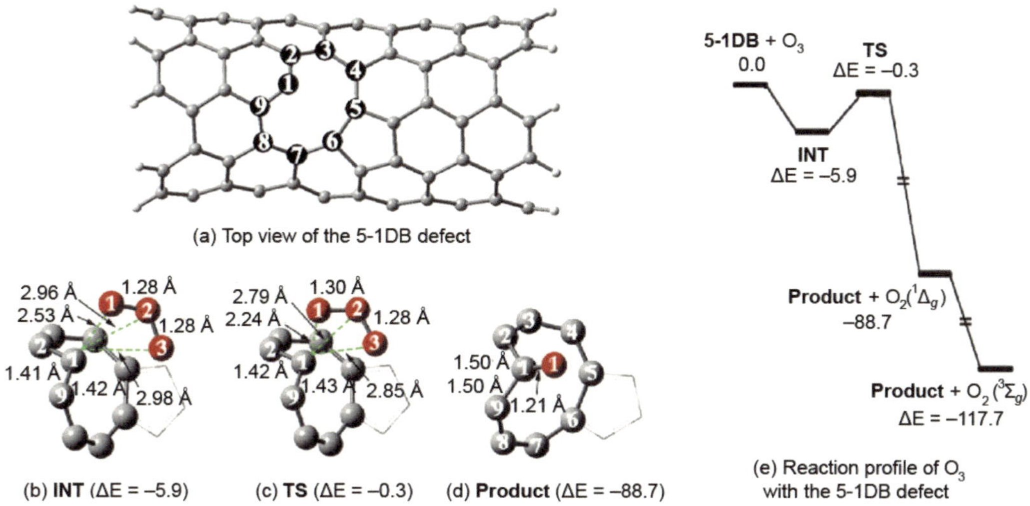

Fig. 3.7. Optimized geometries and energy profiles (with energies in kcal/mol) for the reaction of the defected SWCNT with O_3 (Liu et al. 2006a). Legend: dark grey = C, light grey = H, red = O.

4. Substitutionally doped SWCNTs

4.1 Transition metal-doped SWCNT rods with hydrogen caps

Substitutional doping of SWCNTs with transition metals is yet to be experimentally realized. However, we became particularly interested in this material from theoretical perspective, particularly because of its tremendous potential in nanosensory technology and nanocatalysis based on the rich chemical reactivity of transition metals.

Our studies with N- and O-doped SWCNTs suggest that heteroatoms typically prefer to be situated to the exterior of the nanotube superstructure in which the dopant forms a tripodal structure with the three adjacent C atoms (Liu et al. 2006; Liu et al. 2006a). To confirm this assessment, we used a truncated (5,5) SWCNT capped with H atoms and substituted a single C atom in the middle of the segment with a transition metal considering both the possibility of external doping and internal doping (Chen et al. 2011). Our calculations are consistent with the proposal that *exo* substitution is energetically more favored than the corresponding *endo* HSWCNTs by approximately 2 eV due to an unfavorable interaction between the dopant and the nanotube sidewall, in addition to a slight geometric deformation (Figure 4-1). Overall, the introduction of these heteroatoms does not significantly elongate the nanotube along the longitudinal direction. The net charge distribution in a transition metal-doped SWCNT is mostly localized within the adjacent layers of C atoms surrounding the substitutional dopant, but spin density in open-shell systems can be significantly delocalized. From left to right across the transition metal series, the HOMO (or SOMO for open-shell systems) energy decreases. Thus, transition metal-doped SWCNTs become less reactive toward incoming electrophiles. Within group 10 metals (i.e., Ni, Pd, Pt), the difference in energies between *exo* and *endo* substitutions are larger, which is not unexpected since larger atomic radii will induce more structural strain.

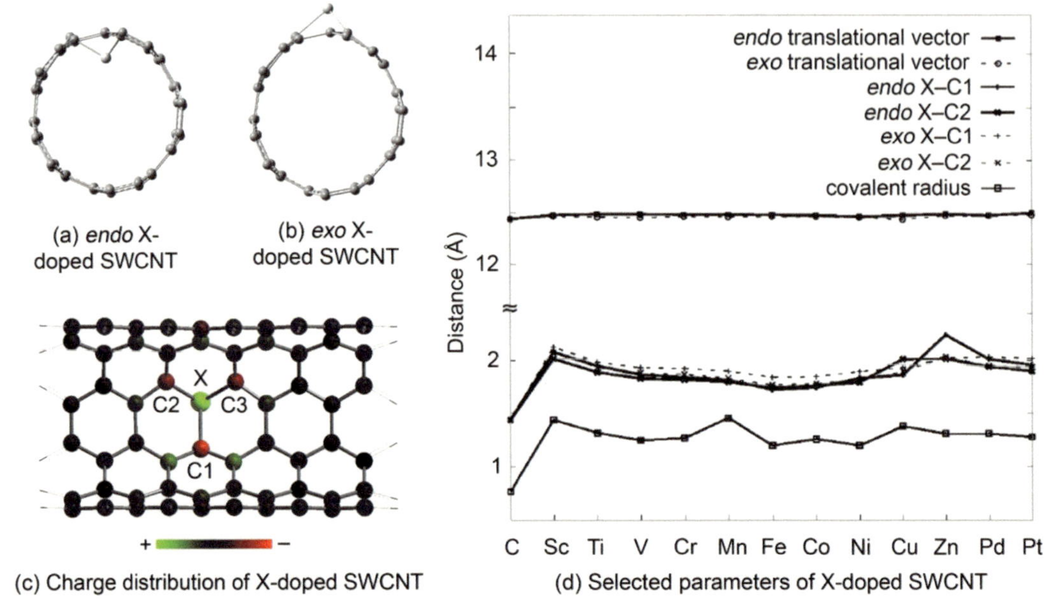

Fig. 4.1. *Endo* and *exo*-substituted transition metal-doped SWCNTs.

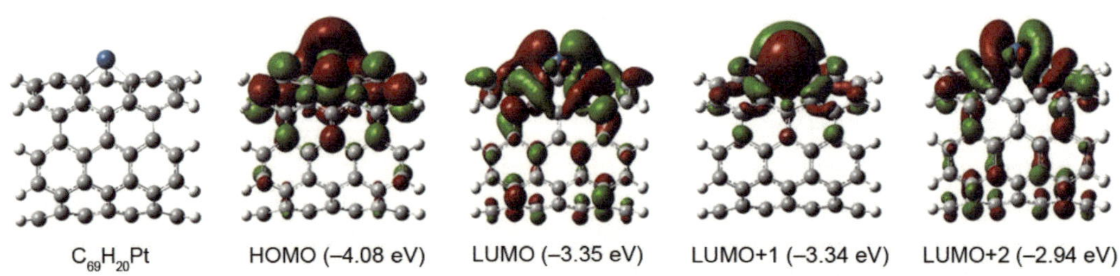

Fig. 4.2. Optimized geometry of $C_{69}H_{20}Pt$ and relevant FMOs. Orbital energies are in parentheses (Yeung et al. 2007). Legend: dark grey = C, light grey = H, blue = Pt.

4.2 Pt-doped SWCNT rods with hydrogen caps

A smaller segment of the (5,5) SWCNT with H caps revealed results consistent with the structure and electronic properties of the larger model described above (Figure 4-2, Yeung et al. 2007). A brief comparison of the FMOs of both models confirms that the shorter $C_{69}H_{20}Pt$ segment suitably describes sidewall doping with the transition metal without the need to use periodic boundary conditions. The dopant Pt atom prefers to adopt a structural arrangement in which it protrudes to the exterior of the sidewall of the SWCNT, due to the larger atomic radius of the Pt atom and subtle interactions between metal *d*-orbitals and the C delocalized π-orbitals. The HOMO–LUMO gap, however, increased to 0.74 eV. In this model, a positive charge of 0.83 resides on the Pt atom in the optimized structure.

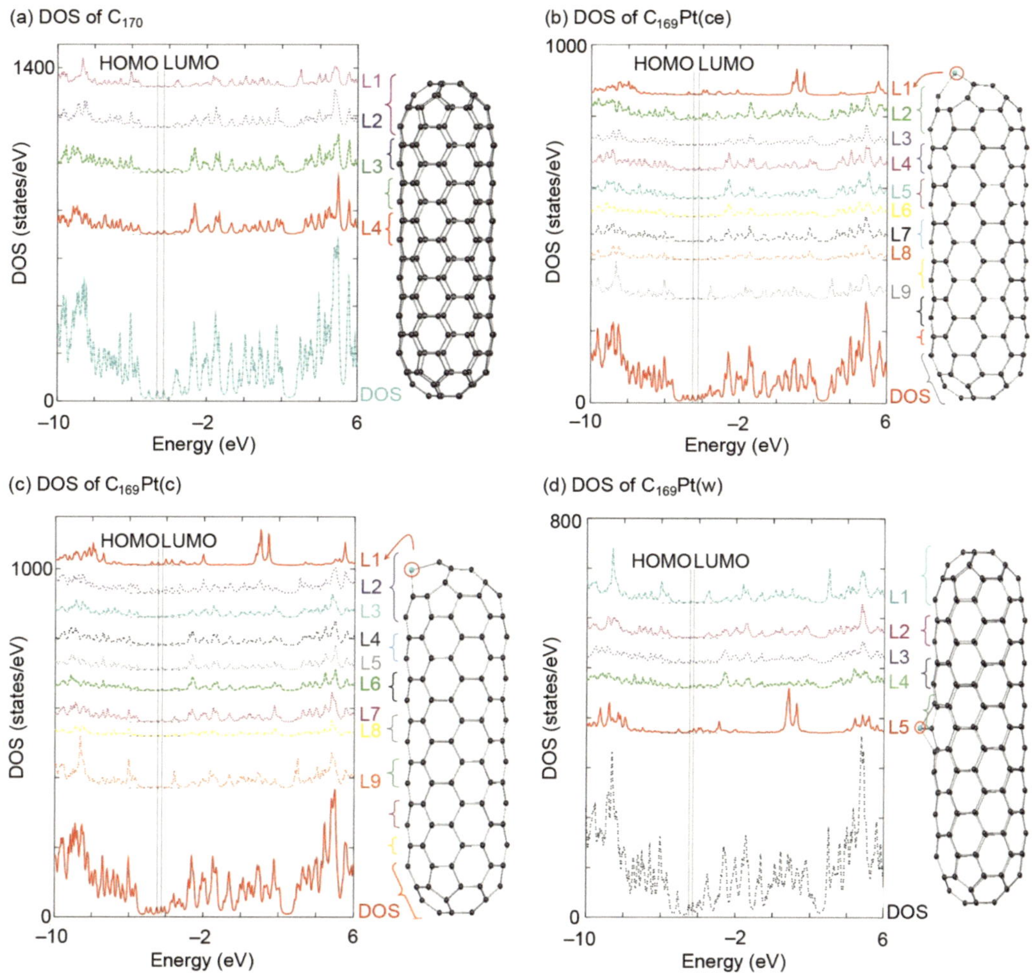

Fig. 4.3. Optimized geometries, DOS, and LDOS of (a) C_{170} with D_{5h} symmetry (ε_{HOMO} = −4.61 eV, ε_{LUMO} = −4.26 eV), (b) C_{169}Pt(ce) with C_s symmetry (ε_{HOMO} = −4.51 eV, ε_{LUMO} = −4.21 eV), (c) C_{169}Pt(c) with C_s symmetry (ε_{HOMO} = −4.48 eV, ε_{LUMO} = −4.25 eV), and (d) C_{169}Pt(w) with C_s symmetry (ε_{HOMO} = −4.46 eV, ε_{LUMO} = −4.26 eV). L1−L9 are the LDOS for each specified layer of atoms as marked on the diagram (Tian et al. 2006).

4.3 Pt-doped SWCNT rods with fullerene caps

Next, we constructed a (5,5) SWCNT rod capped with fullerene hemispheres (C_{170}, Figure 4-3a) and substituted a single C atom with a Pt atom at the cap-end (C_{169}Pt(ce), Figure 4-3b), at the cap (C_{169}Pt(c), Figure 4-3c), and along the sidewall (C_{169}Pt(w), Figure 4-3d, Tian et al. 2006). These nanorods exhibit slightly decreased HOMO−LUMO gaps in comparison to the pristine all-carbon compound.

Pt substitution at the cap-end is the most stable type of doping, with cap-doping and wall-doping sitting 0.8 and 17.9 kcal/mol higher in energy. In our analysis, the triplet spin state was higher in energy than the singlet state. By calculating the FMOs of these model systems (Figure 4-4), we observed a strong interaction between the Pt 5*d* atomic orbitals and the extended π-framework of the SWCNT backbone, with the greatest effect observed in $C_{169}Pt(w)$. This is because of the better overlap of orbitals where the Pt substitution occurs with roughly mirror-plane symmetry, a feature that is not true for either the cap-end or cap Pt-doped macromolecule.

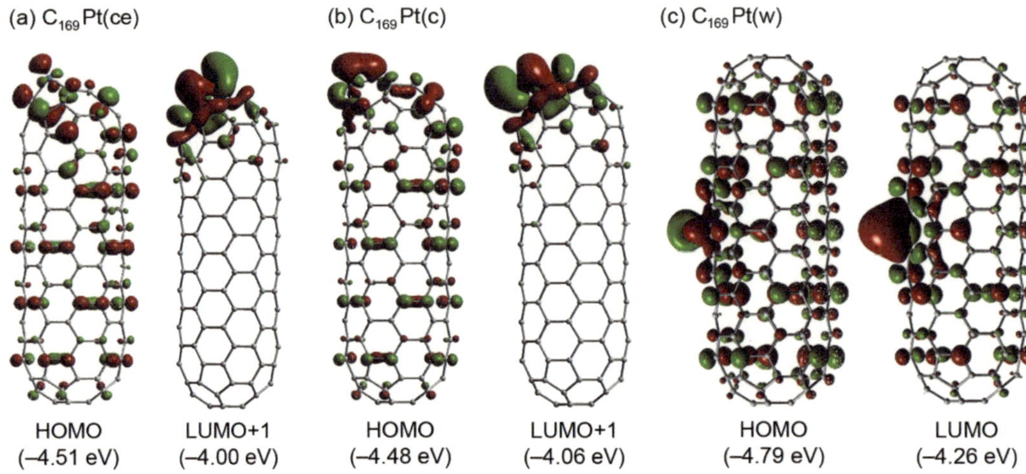

Fig. 4.4. Relevant FMOs for $C_{169}Pt(ce)$, $C_{169}Pt(c)$, and $C_{169}Pt(w)$. Orbital energies are in parentheses (Tian et al. 2006).

4.4 Pt-doped BNNT rods with hydrogen caps

Boron nitride nanotubes are isoelectronic with carbon nanotubes that exhibit different structural, electronic, and chemical features (Wang et al. 2010, Golberg et al. 2010). Indeed, the B–N couple is isoelectronic with the C–C bond but locally exhibits charge separation due to the difference in electronegativity. While B has an empty p-orbital and is a coordinatively unsaturated trigonal plane, N has a full octet with an non-bonding pair of electrons in a filled p-orbital. Molecular borazine ($B_3N_3H_6$), for example, is an aromatic compound that is isoelectronic and isostructural to benzene (C_6H_6) but is more susceptible to nucleophilic and electrophilic reactions than the all-carbon analog. Hexagonal boron nitride (hBN) is analogous to graphite but displays mechanical hardness and chemical inertness, while graphite is used as a lubricant. In nanotubes, zigzag BNNTs exhibit a non-zero dipole moment because of alternating layers of B and N atoms with band gaps and electronic structures relatively invariant to changes in nanotube diameter and chirality. Importantly, because BNNTs are incombustible at high temperatures and chemically inert, these macromolecules are promising in nanoelectronics.

Natural defects in boron nanotubes can be induced from mechanical fractures, with SW defects consisting of a 5775 structure being more stable than corresponding quadrilateral-octagonal-octagonal-quadrilateral (4884) structure (Karousis 2010). Using a truncated (5,5) BNNT model capped with H atoms ($B_{45}N_{45}H_{20}$, Figure 4-5, Chen et al. 2010), we considered two SW defective structures resulting from a slant B–N bond rotation (SW1) or a vertical B–N bond rotation (SW2). In our defect-free BNNT model, the HOMO–LUMO gap is 6.02 eV, rendering this molecule a good insulator. By introducing SW defects, a decrease in the HOMO–LUMO gap was observed (4.98 and 5.19 eV for SW1 and SW2, respectively).

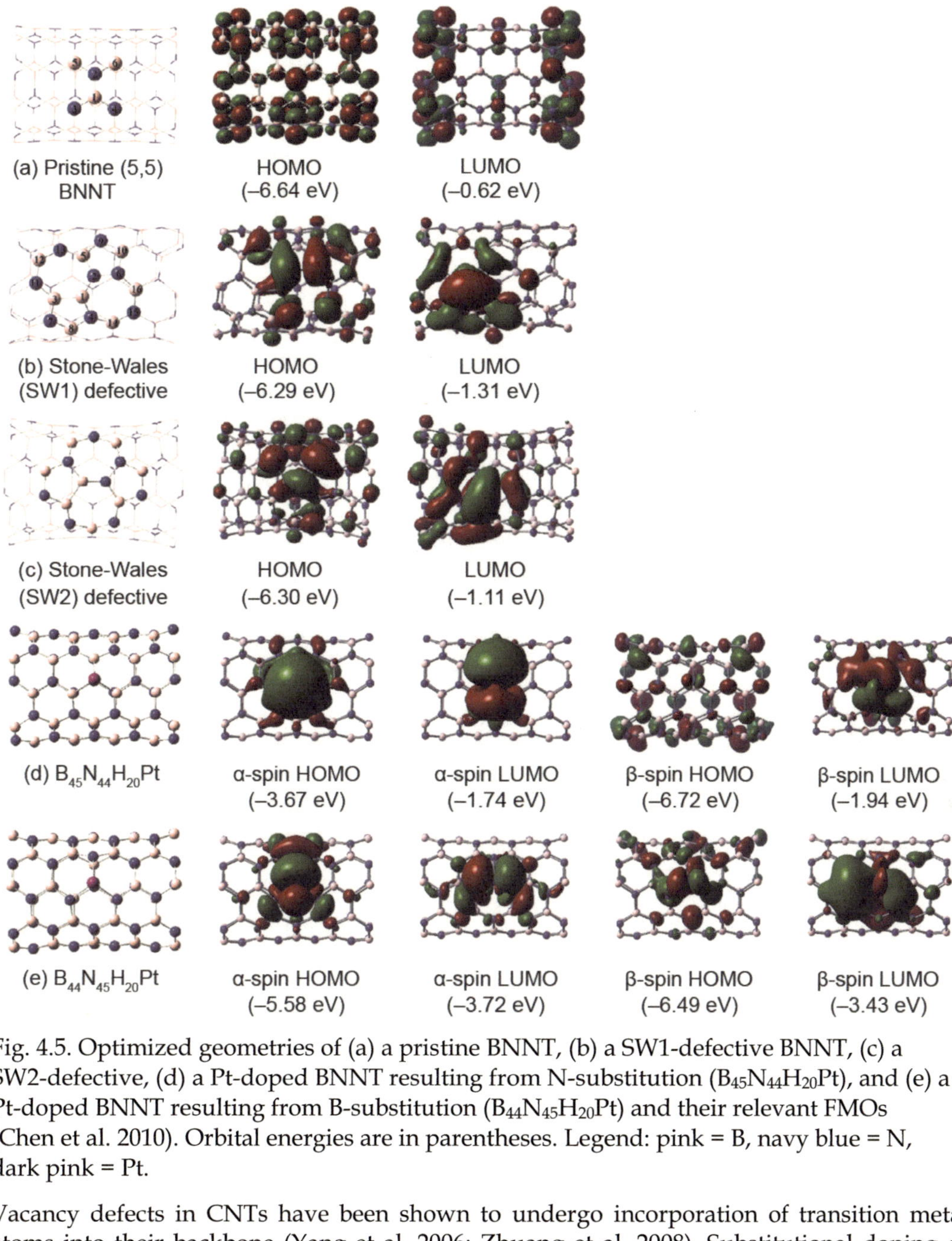

Fig. 4.5. Optimized geometries of (a) a pristine BNNT, (b) a SW1-defective BNNT, (c) a SW2-defective, (d) a Pt-doped BNNT resulting from N-substitution ($B_{45}N_{44}H_{20}Pt$), and (e) a Pt-doped BNNT resulting from B-substitution ($B_{44}N_{45}H_{20}Pt$) and their relevant FMOs (Chen et al. 2010). Orbital energies are in parentheses. Legend: pink = B, navy blue = N, dark pink = Pt.

Vacancy defects in CNTs have been shown to undergo incorporation of transition metal atoms into their backbone (Yang et al. 2006; Zhuang et al. 2008). Substitutional doping of BNNTs is a logical extension, although replacement of either a single N or B atom yields two different materials (Figure 4-5). Although the mechanism of Pt doping is unclear, our studies confirm the existence of doublet ground states and a decreased band gap in these nanomaterials (Chen et al. 2010). Pt-doped BNNTs, unlike pristine and SW-defective BNNTs, are semiconducting.

4.5 Adsorption of gases onto Pt-doped SWCNTs

Discrete organometallic Pt complexes and metal surfaces alike can undergo coordination or adsorption to gas molecules. Hence, we expect similar reactivity with Pt atoms embedded within the sidewall of a carbon nanotube (Yeung et al. 2010).

Carbon monoxide is an important ligand for organometallic complexes and has demonstrated the ability to undergo physisorption onto metal surfaces (Orita et al. 2004). Although the C≡O triple bond of CO is polarized with a reactive lone-pair of electrons on the C, we imagined that adsorption would be possible via either C-end adsorption (a C̲O-adsorbed HSWCNT) or O-end adsorption (an O̲C-adsorbed HSWCNT, Figure 4-6). Both processes were found to be exothermic, with C-end adsorption liberating 41.2 kcal/mol while O-end adsorption only produced 7.0 kcal/mol (Table 4-1). This coordination is a coordinate dative bond, in which the C atom donates its pair of electrons into a metal *d*-orbital, causing a backdonation of electrons to the π*-orbital of the adsorbate. The result is a lengthening the C≡O triple bond. Indeed, we observe a slight elongation in the adsorbed CO molecule for C̲O-adsorption, from 1.18 Å in the free gas to 1.19 Å, and a net charge of −0.12 on the adsorbate. The bond length between the Pt atom and the CO molecule was 1.97 Å. In contrast, O̲C-adsorption did not display the same type of bond stretching and was a weaker interaction overall.

Since Pt complexes commonly hosts six ligands in its coordination sphere, we explored the possibility of multiple adsorption. We successfully identified a structure in which two CO molecules underwent coordination to the Pt center through C-end adsorption (Figure 4-6). The $(\underline{C}O)_2$-adsorbed HSWCNT was formed with a further release of energy (36.8 kcal/mol per CO molecule) and displayed almost identical orientation of both adsorbates. The second adsorption triggered an increase in distance between the Pt atom and the adsorbate to 2.00 Å, while the net charges on each CO molecule decreased to −0.05. Attempts to probe the possibility of triple adsorption did not afford structures in which the CO molecules were bound strongly to the Pt atom via C-end adsorption; instead, a third additional molecule of CO displayed a weak interaction with the nanotube sidewall. This process is uphill in energy from the doubly-adsorbed intermediate. Clearly, the steric bulk of the nanotube precludes the possibility of triple adsorption.

Nitrogen monoxide is not commonly encountered in coordination chemistry, unlike its related charged cousins NO^+ and NO^-, but are known to interact with Pt surfaces (Backus et al. 2004). In analogy to CO-adsorption, coordination between the Pt atom and the adsorbate can take place via N-end adsorption or O-end adsorption. Our calculations revealed that N̲O-adsorption is energetically preferential to O̲N-adsorption (Figure 4-6). In all geometries examined, the quartet spin state exhibited higher energy than doublet states (which can arise from either singlet or triplet Pt-doped SWCNT starting materials). The adsorption of NO onto the Pt atom embedded in the nanotube results in a complex bearing a charge of 0.99 on the Pt center and an increased HOMO−LUMO gap to 0.84 eV. Multiple adsorption was observed to occur in an unsymmetrical fashion in which one of the O atoms of the adsorbate molecules points away from the surface of the nanotube sidewall. This process is less exothermic than single adsorption, and consequently, simultaneous coordination of three molecules of NO was not considered.

Ammonia is a common N-containing ligand for Pt complexes, including the famous chemotherapy drug cisplatin, $PtCl_2(NH_3)_2$ (Boulikas & Vougiouka 2003). Heterogeneous Pt surfaces are critical to NH_3 chemistry, such as in the industrial oxidation of NH_3 to NO in nitric acid production. Unlike CO and NO, however, NH_3 does not exhibit a low-lying π*-orbital to allow backbonding to occur. In our studies, the adsorption energy between NH_3

and the Pt-doped SWCNT was -31.8 kcal/mol and the resulting <u>N</u>H_3-adsorbed HSWCNT exhibited a HOMO−LUMO gap of 0.73 eV and a Pt–N bond distance of 2.25 Å (Figure 4-6, Yeung et al. 2008). Biadsorption of NH_3 occurred with 57.1 kcal/mol release of energy (28.5 kcal/mol per NH_3 molecule).

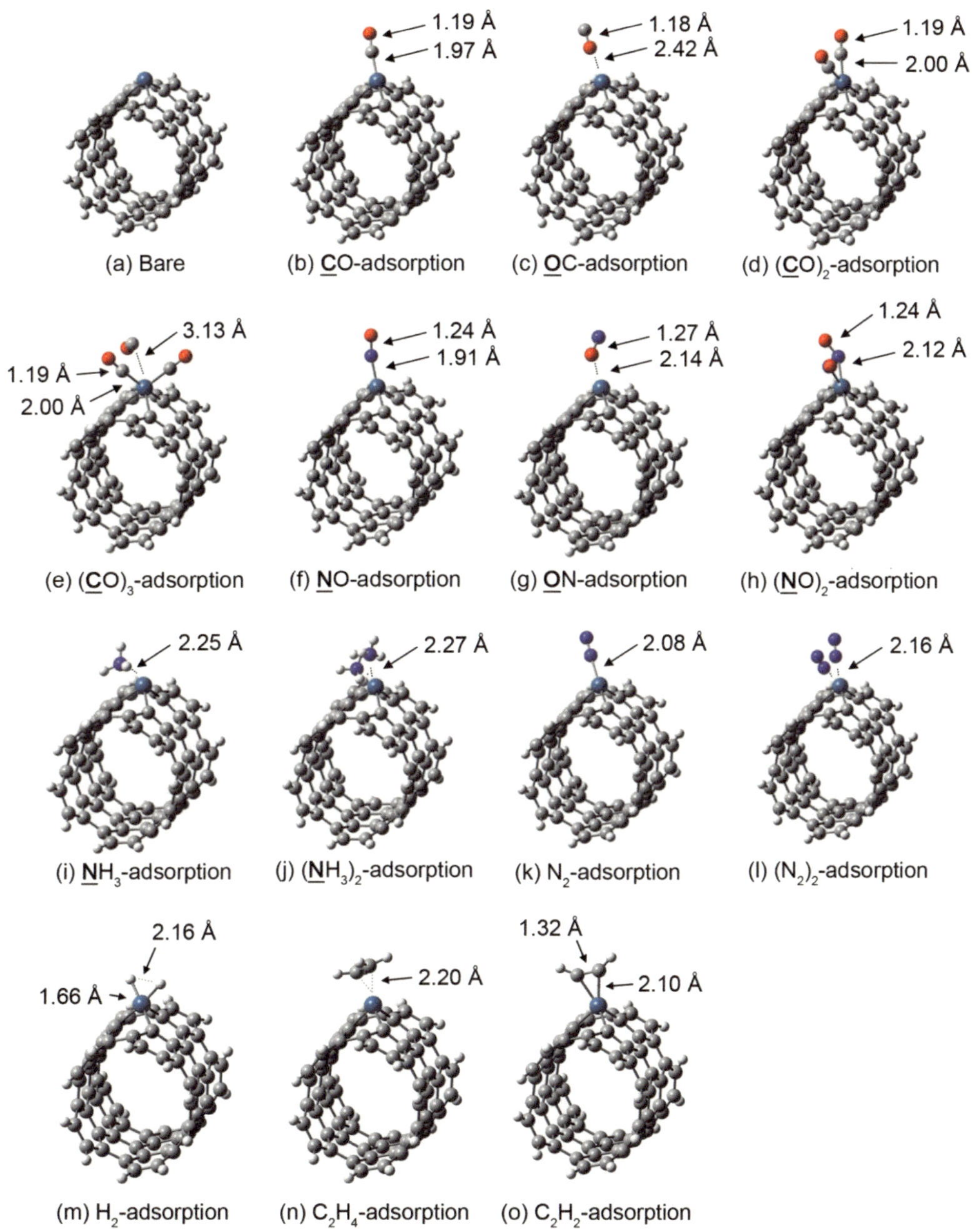

Fig. 4.6. Optimized geometries for gas-adsorbed HSWCNTs (Yeung et al. 2010). Legend: dark grey = C, light grey = H, navy blue = N, red = O, blue = Pt.

Adsorbate	Spin state	ΔE[a]	*d*(PtX)[b]	*d*(XY)[c]	*q*(Pt)[d]	*q*(XY)[e]	*q*(C)[f]
None	Singlet	0	N/A	N/A	0.82	N/A	−0.41
	Triplet	0	N/A	N/A	0.84	N/A	−0.49
$\underline{C}O$[g]	Singlet	−41.2	1.97	1.19	0.85	−0.12	−0.43
	Triplet	−33.1	1.96	1.20	0.83	−0.15	−0.46
$\underline{O}C$[g]	Singlet	−7.0	2.42	1.18	0.87	−0.02	−0.41
	Triplet	−6.8	2.35	1.19	0.88	−0.06	−0.45
$(\underline{C}O)_2$[g]	Singlet	−73.7	2.00, 2.00	1.19, 1.19	0.83	−0.05, −0.05	−0.49
	Triplet	−59.0	2.00, 2.00	1.19, 1.19	0.78	−0.02, −0.02	−0.49
$(\underline{C}O)_3$[g]	Singlet	−71.4	1.98, 2.00, 3.13	1.19, 1.19, 1.19	0.77	0.01, −0.01, −0.06	−0.37
	Triplet	−73.7	2.03, 2.03, 2.41	1.18, 1.18, 1.19	0.76	0.05, 0.05, −0.13	−0.46
$\underline{N}O$[g]	Doublet	−47.0 (−56.5)[h]	1.91	1.24	0.99	−0.39	−0.39
	Quartet	−46.2	2.00	1.25	0.95	−0.38	−0.34
$\underline{O}N$[g]	Doublet	−23.2 (−32.7)[h]	2.14	1.27	0.98	−0.37	−0.34
	Quartet	−30.0	2.12	1.29	1.00	−0.44	−0.33
$(\underline{N}O)_2$[g]	Singlet	−69.9	2.10, 2.12	1.24, 1.24	0.99	−0.24, −0.28	−0.36
	Triplet	−77.8	2.12, 2.12	1.24, 1.24	0.99	−0.25, −0.25	−0.35
$\underline{N}H_3$[i]	Singlet	−31.8	2.25	N/A	0.88	0.18	−0.44
	Triplet	−31.3	2.23	N/A	0.87	0.18	−0.50
$(\underline{N}H_3)_2$[i]	Singlet	−57.1	2.27, 2.27	N/A	0.92	0.18, 0.18	−0.48
	Triplet	−55.3	2.26, 2.26	N/A	0.86	0.19, 0.19	−0.49
N_2	Singlet	−24.7	2.08	1.16	0.91	−0.14	−0.41
	Triplet	−25.9	2.05	1.16	0.91	−0.18	−0.44
$(N_2)_2$[i]	Singlet	−43.9	2.16, 2.16	1.16, 1.16	0.93	−0.09, −0.09	−0.42
	Triplet	−41.7	2.16, 2.16	1.16, 1.16	0.89	−0.09, −0.09	−0.41
H_2[j]	Singlet	−11.3	1.66, 1.68[k]	2.16	0.77	−0.34	−0.33
	Triplet	−6.0	1.66, 1.66[k]	2.19	0.74	−0.34	−0.43
C_2H_4[j]	Singlet	−26.4	2.31, 2.20[k]	1.43	0.95	−0.25	−0.35
	Triplet	−24.7	2.31, 2.30[k]	1.42	0.92	−0.16	−0.37
C_2H_2[j]	Singlet	−31.4	2.13, 2.10[k]	1.32	0.98	−0.38	−0.34
	Triplet	−25.3	2.17, 2.15[k]	1.30	0.92	−0.27	−0.31

[a]Total stabilization energy (in kcal/mol).
[b]Distance (in Å) between Pt and X of XY.
[c]Distance (in Å) between X and Y of XY.
[d]Partial charge on Pt.
[e]Net partial charge on XY.
[f]Net partial charge on the C atoms of the SWCNT adjacent to Pt.
[g]$(\underline{X}Y)_n$ refers to an *n* X-end adsorbed HSWCNT fragment.
[h]Total stabilization energy (in kcal/mol) from triplet ground state bare Pt-doped SWCNT in parentheses.
[i]End-on adsorption.
[j]X = Y. These adsorbates are bound in a side-on fashion.
[k]Distance (in Å) between Pt and both X in X_2.

Table 4.1. Binding energy and geometrical data for nanotube-adsorbate complexes $C_{69}H_{20}Pt(XY)$ (Yeung et al. 2007; Yeung et al. 2010).

Nitrogen gas is distinct from the aforementioned adsorbates because of its lack of inherent polarity. If our Pt-doped nanotubes will be applied as nanosensors, the ability to detect simple and unpolarized small molecules is critical. The interaction between N_2 and Pt surfaces is important to the industrial fixation of nitrogen via the Haber-Bosch process. In transition metal complexes, N_2 can coordinate with either an end-on orientation (with the N atom acting as a σ-donor) or a side-on orientation (with the N≡N triple bond acting as a π-donor). We found that Pt-doped SWCNTs prefer to bind N_2 with end-on geometry, as both end-on and side-on binding converged to give the same structure (Figure 4-6). The energy released was significantly less than CO adsorption (24.7 kcal/mol), because the electrostatic interactions between an unpolarized adsorbate such as N_2 and the HSWCNT is weak. The observed N≡N bond distance was lengthened slightly from 1.15 Å in the free gas to 1.16 Å in this complex and the net charge on the adsorbate was −0.14, signifying a net donation of electrons from the Pt atom to the ligand. The calculated HOMO−LUMO gap was 0.81 eV. We also identified a doubly adsorbed complex in which two molecules of N_2 bound to the Pt atom via end-on coordination. The energy released per molecule was 22.0 kcal/mol, close to that of single adsorption, possibly because the weaker interaction between the Pt and the adsorbate minimizes steric repulsion between the two coordinated N_2 molecules.

Hydrogen gas is the lightest diatomic molecule. Activation of H_2 with Pt has found tremendous application in organic synthesis in the catalytic hydrogenation of alkenes and alkynes. Since our Pt-doped SWCNT has the capability of binding linear diatomic molecules, we reasoned that either H_2 physisorption or chemisorption is possible. With our (5,5) $C_{69}H_{20}Pt$ nanotube model, we obtained a geometry in which H–H bond cleavage occurred. In fact, a distance of 2.16 Å was observed between the two H atoms. The energetic preference for this reaction was −11.32 kcal/mol. In the optimized structure, the HOMO−LUMO gap was 0.70 eV, and a charge of 0.77 was present on the Pt atom, while 0.34 had transferred to the chemisorbed H_2. (For other studies of H_2 adsorption, see section 4.7.)

Carbon-carbon multiple bonds are excellent substrates for coordination to transition metals. In particular, the adsorption of C=C and C≡C bonds to Pt surfaces is an important process to catalytic hydrogenation reactions. To this end, we considered the adsorption of the simplest two-carbon alkene ethylene (C_2H_4) onto the (5,5) Pt-doped SWCNT model and found that 26.4 kcal/mol were released upon physisorption. The HOMO−LUMO gap increased to 0.88 eV in this case and the observed C=C bond distance was 1.43 Å. A net negative charge of -0.25 on the adsorbate was determined. When we optimized structures involving the simplest two-carbon alkyne acetylene (C_2H_2), we observed a strong interaction between the adsorbate and the HSWCNT. Alkynes are known to interact with transition metals through weak donation of π-electrons of the adsorbate to the metal and in some circumstances can result in oxidative addition, generating a metallocyclopropene complex. In our studies, C_2H_2 coordination liberated 31.4 kcal/mol and resulted in a significant lengthening of the C≡C bond distance from 1.23 Å in the free gas to 1.32 Å. Additionally, the terminal C-H bonds are no longer co-linear with the C≡C bond but instead are bent. The HOMO−LUMO gap roughly paralleled that of C_2H_4 adsorption (0.88 eV), and a net charge of −0.38 was detected on the alkyne. The Pt center was found to bear a charge of 0.98 and the separation between the Pt atom and either C atom of C_2H_2 was approximately 2.15 Å. Hence, the interaction between C_2H_2 and CNTs substitutionally doped with Pt is indeed chemisorptive in nature and yields a metallocyclopropene. In other words, C_2H_2 adsorption induces a rehybridization of the C atoms from *sp* to sp^2.

4.6 Applications of Pt-doped SWCNTs: Chemical sensors and nanoelectronics

Because of the ability for Pt-doped SWCNTs to adsorb different gases, it is reasonable to consider using these nanomaterials as chemical sensors. Developing a nanodevice capable of sensing the presence of various analytes depends primarily upon the ability to monitor

minor changes in electronic structure, such as by measuring conductance across a nanowire. Designing new nanosensory technology for molecules including CO, NO, NH_3, and H_2 would be worthwhile (Rouxinol et al. 2010; Wang & Yeow 2009).
The first CNT gas sensor was reported by Kong, in which a semiconducting SWCNT acted as a channel to conduct the source-drain current in a field-effect transistor (FET, Kong et al. 2000). In a typical FET, the SWCNT connects the source and drain, and current is passed by applying an external voltage. The surface of the SWCNT is exposed to the atmosphere, hence allowing detection to take place (Rouxinol et al. 2010; Wang & Yeow 2009). Current is measured as a function of bias voltage (V_B) and gate voltage (V_G).
A variant of this design involves interdigitated electrodes (IDE). In a system developed by Li (Li et al. 2003), a solution casting process was used to prepare a device containing a network or mesh of SWCNTs on a Ti and Au IDE on a layer of silicon dioxide (SiO_2). Drop-deposition of the nanotubes was the method of choice for delivery. This nanosensor combines the advantages of single nanotube transistors and film-based sensors since there is a large density of nanotubes present on the IDE and the array configuration provides excellent electrical contact between the SWCNTs and the electrodes while maintaining good accessibility for gas adsorption.
A complementary strategy known as the resonant-circuit sensor (RCS) method was described by Chopra (Figure 4-7, Chopra et al. 2002). In this scenario, carbon nanotubes are adsorbed onto a Cu resonator and irradiated by a microwave signal from a radio frequency transmitter. As a result of interaction with the sensor resonator, a strong signal is generated at its resonant frequency. Detection of analytes hence depends on measuring the shift of the frequency between the received signal and the transmitted signal.

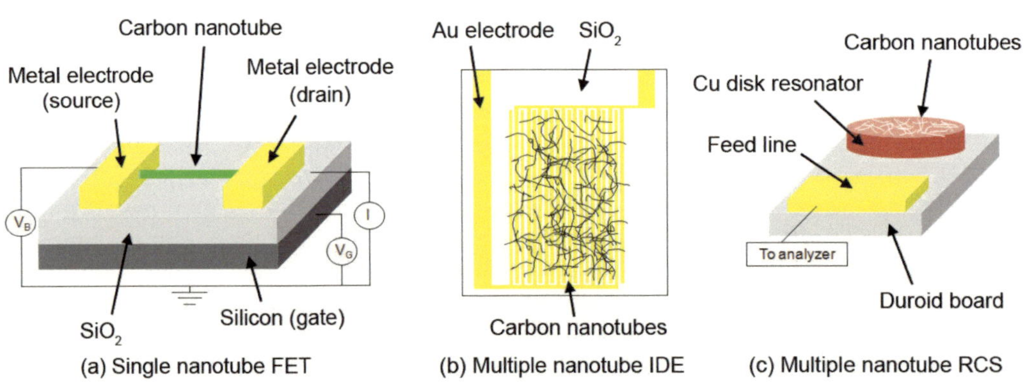

Fig. 4.7. Schematic diagrams for nanotube sensors. (a) A single nanotube field-effect transistor (Rouxinol et al. 2010; Wang & Yeow 2009). V_B = bias voltage, V_G = gate voltage, I = current. (b) A multiple nanotube interdigitated electrode device (Li et al. 2003). (c) A multiple nanotube resonant-circuit sensor (Chopra et al. 2002).

In devices in which measurements of conductivity form the basis of quantitative analysis, minor fluctuations in electronic structure can be evaluated by considering the amount of charge transfer that takes place between the SWCNT and the small molecule adsorbate. Analysis of the HOMO–LUMO gap may also provide some insight. Peng and co-workers have described that charge transfer resulting from adsorption can change conductance along the axis as follows:

$$\Delta Q = C_g \cdot \Delta V_g = \delta\theta \frac{\pi dl}{\sigma} \tag{13}$$

where ΔQ is the charge transfer, C_g is the capacitance, and ΔV_g is the observed voltage change (Peng et al. 2003). The nanotube length and diameter are l and d, respectively, and σ and θ are the molecular cross-section area and coverage, respectively. For the majority of analytes considered (Table 4-2), adsorption results in a net charge transfer from the Pt-doped SWCNT to the gas and an increased HOMO−LUMO gap. This suggests a net decrease in conductivity and a net increase in resistivity.

In our investigations (Yeung et al. 2010), the magnitude of charge transfer between adsorbate and the Pt atom is relatively small. However, it has been demonstrated a change of 0.28 is sufficient for being experimentally measurable. We predict that extending our model Pt-doped SWCNTs in either direction *ad infinitum* should enhance sensitivity enough such that a practically useful device can be prepared using these macromolecules. Cooperativity and synergy between nearby transition metals embedded in the nanotube sidewall is another possible feature that warrants further investigation. Other strategies, such as using polymer matrices to encapsulate our HSWCNTs, may help. In fact, the inclusion of CNTs in insulating polymer matrices can produce a hybrid material that is conductive above a certain filler concentration known as the percolation threshold (Moniruzzaman & Winey 2006). Yodh and co-workers have confirmed that the percolation threshold can be as low as 0.005 vol% in SWCNT/epoxy composites (Bryning et al. 2005). Should an amplification of signal be observed in the CNT/polymer hybrid, selective chemical sensing of individual gas molecules may become possible. The ideal nanosensor would be a small handheld device with short response times, high sensitivities, and the ability to differentiate between analytes. Using Pt-doped SWCNTs is potentially advantageous to undoped CNTs and may revolutionize biomedical, automotive, food, agriculture, and manufacturing industries. Detection techniques of relevance to environmental pollutants and national security will also benefit from these materials.

Analyte	Charge transfer[a]	G_{HL} (eV)	ΔG_{HL}[f] (eV)
None	N/A	0.74	0.00
CO[b]	0.05	0.78	0.04
NO[c]	0.28	0.74	0.00
NH_3[d]	−0.18	0.75	0.01
N_2[e]	0.09	0.81	0.07
H_2	0.34	0.70	−0.04
C_2H_4	0.25	0.88	0.14
C_2H_2	0.38	0.88	0.14

[a]Represents charge donated from the Pt-doped SWCNT to the adsorbate.
[b]Represents ($\underline{C}O)_2$-adsorbed HSWCNT.
[c]Represents ($\underline{N}O)_2$-adsorbed HSWCNT.
[d]Represents ($\underline{N}H_3)_2$-adsorbed HSWCNT.
[e]Represents $(N_2)_2$-adsorbed HSWCNT.
[f]Represents the change in the HOMO−LUMO gap upon adsorption.

Table 4.2. Average charge transfer and HOMO−LUMO gap (G_{HL}) data for nanotube-adsorbate complexes with side-on binding motif (Yeung et al. 2010).

Other than chemical sensors, Pt-doped SWCNTs may be applied to nanoelectronics. There have already been significant studies involving the utilization of CNTs as FETs, optoelectronic devices, and nanoelectromechanical systems (NEMS, Park 2004; Srivastava et al. 2003; Fuhrer 2003; Tsukagoshi et al. 2002). Technology using CNTs in FETs has already been commercialized (e.g., Infineon Technologies). In this specific FET, which features a back-gate, the CNT is situated at the top of the electrodes, separated from each other by a dielectric spacer to the back-gate. The conductive nanochannel hence provides an electrical connection between the source and drain electrodes and acts as a molecular wire. Applying these devices to more complex components such as nonvolatile molecular memory elements is a rapidly advancing field.

Because Pt-doped SWCNTs undergo conductivity changes when exposed to different types of gases, we envision the design of a new element of control in nanoelectronics in which varying the gas atmosphere will allow control of electrical current. For instance, if the conductance of a CNT nanowire is too high, introducing an atmosphere of CO should provide an immediate increase in resistivity and a decreased current. This effect would be compounded if the nanotube backbone is substitutionally doped at multiple sites. The key advantage to this type of electronic manipulation is the ability to control conductivity and resistivity *in a reversible fashion*. By simply purging the atmosphere with an inert gas, such as argon, any adsorbed molecules that were previously used to modulate electrical current can be removed. Because of the high diffusion rates of gases, extremely fast circuit breaks are possible. In the future, more complicated devices may contain substitutionally doped SWCNTs of different elements. Each CNT nanowire would be affected to a different extent depending on the adsorbate. Such careful control of nanocircuits would provide engineers with new tools for the development of new nanodevices.

4.7 Adsorption of H_2 onto Pt-doped SWCNTs and BNNTs

The incorporation of Pt into the sidewall of CNTs provides a unique chemical reactivity, particularly with small molecules, that is otherwise not possible. In our studies of H_2 adsorption (Yeung et al. 2010), our (5,5) $C_{69}H_{20}Pt$ model revealed a strong interaction between hydrogen and the Pt atom, resulting in a chemisorptive process in which the two H atoms became split. We also pursued investigations of Pt-doped nanorods capped with fullerenes and their ability to perform H–H bond cleavage (Tian et al. 2008). To our surprise, with the larger $C_{169}Pt$ nanorods, both cap-doped HSWCNTs underwent physisorption with H_2, releasing only 2.0 kcal/mol of energy. We were pleased to find, however, that wall-doped nanorods performed a similar chemisorption as predicted by our H-capped model.

The isoelectronic Pt-doped BNNTs was also considered in H_2 adsorption (Li et al. 2009). In their work, up to three equivalents of H_2 can undergo physisorption to the Pt center. This is in stark contrast to our Pt-doped SWCNTs which can perform chemisorption. Clearly, the electronic nature of the carbon backbone plays a role in determining the chemical reactivity of the Pt atom.

4.8 Applications of H_2 adsorption onto Pt-doped SWCNTs and BNNTs: Hydrogen storage and nanocatalytic hydrogenations

The use of CNTs in hydrogen storage remains an active area of research (Yao 2010). For instance, defective CNTs and their ability to undergo adsorption of H_2 has also been investigated (Gayathri et al. 2007). If Pt-doped SWCNTs and Pt-doped BNNTs are to be used as hydrogen storage materials, then their ability to adsorb H_2 reversibly becomes important. As we have described above, Pt-doped SWCNTs (especially wall-doped) tend to undergo chemisorption with H_2 and release of hydrogen may require external input of energy (Yeung et al. 2010; Tian et al. 2008). On the other hand, a reversible interaction between Pt-doped BNNTs and hydrogen appears to be much more favorable. Careful tuning of the electronic properties of the Pt atom will be necessary to achieve the right balance to provide facile and reversible adsorption and desorption. In an infinitely long SWCNT with numerous Pt dopants, we expect that efficiency will be improved.

Perhaps more interesting than H_2 storage is the ability for Pt to activate H_2. While many transition metals are known to cleave hydrogen homolytically, forming a formal metal dihydride, the substitutional doping strategy that we have described is of particular interest because of potential use of these nanomaterials as catalysts for catalytic hydrogenation. We have already discovered that C=C double bonds (in C_2H_4) bind reversibly to the Pt atom of Pt-doped SWCNTs, while H_2 undergoes complete dissociation. At an appropriate doping concentration and geometry, we envision two Pt atoms working together to perform hydrogenation of alkenes such as ethylene. On the first Pt atom, H_2 would undergo H–H bond breaking via oxidative addition. Simultaneously, on the second Pt atom, one molecule of alkene can bind transiently. Because of the weak interaction, it is reasonable to presume that the alkene can insert into the Pt–H bond and form a new Pt–C bond. Finally, reductive elimination will furnish the reduced alkane product. Should this transformation be possible, it would be a modern variant of Pt metal catalyzed hydrogenation. The only reason that catalytic hydrogenation would become possible using nanotubes is because of the innate chemical reactivity of Pt. Hence, substitutional doping may open the door for examining other types of nanocatalysis, one of which is described below.

4.9 Applications of Pt-doped SWCNTs: Lewis acidity and nanocatalysis

Pt-doped SWCNTs can undergo adsorption with CO (Yeung et al. 2007). Based on this reactivity, we investigated model alkylplatinum complexes through DFT, considering all possible trimethylplatinum species with formulae $PtMe_3(CO)_x^{n+}$ (where $x = 0$ to 3 and $n = -3$ to 1, Figure 4-8). Based on the observed geometry and partial charge analysis, we chose $PtMe_3(CO)_x^+$ for comparison purposes. By examining the FMOs of $PtMe_3(CO)_x^+$ and comparing it with the model (5,5) Pt-doped SWCNT, we found remarkable similarities, hence suggesting that the reactivity of the Pt atom as a substitutional dopant should roughly parallel that of $PtMe_3^+$. Interestingly, we observed facile triple adsorption in the case of the monomeric $PtMe_3^+$ complex, while this process was not possible in the HSWCNT. This dichotomy in reactivity can be attributed to the steric bulk of the nanotube sidewall and a decreased electron density in the reactive *d*-orbital of the Pt dopant atom.

$PtMe_3^+$ is a known Lewis acid catalyst (Hsieh et al. 2007; Procelewska et al. 2005). Hence, we considered the use of Pt-doped SWCNTs as nanocatalysts (Yeung et al. 2011). To evaluate Lewis acidity, we first performed vibrational analyses on CO-adsorbed Pt complexes, since the C≡O stretch is an indicator for the level of backdonation from the metal center to the ligand. Next, the nuclear magnetic resonance (NMR) chemical shift (δ) of the olefinic H in the β position of an α,β-unsaturated carbonyl compound, specifically maleic anhydride (MA), was evaluated upon coordination to the Pt atom, a technique first described by Child and co-workers (Childs et al. 1982; Laszlo & Teston 1990). Compared to the uncomplexed MA, Lewis acid coordination should generate a positive downfield shift of the proton resonance at the β position due to a deshielding effect and consequently, chemical shift difference ($\Delta\delta$) is a direct measure of Lewis acidity. Finally, to truly test our hypothesis of Pt-doped SWCNTs as Lewis acids in nanocatalysis, we envisioned the use of these macromolecules as catalysts for the Diels-Alder reaction between MA and cyclopentadiene (CPD), a [4+2] cycloaddition. This reaction generates two possible products, the *exo* and *endo* adducts. If substitutional doping of Pt into a nanotube sidewall bears some resemblance to $PtMe_3^+$, then catalysis is likely. We were particularly interested to evaluate both rate enhancement and effects on stereoselectivities in this transformation.

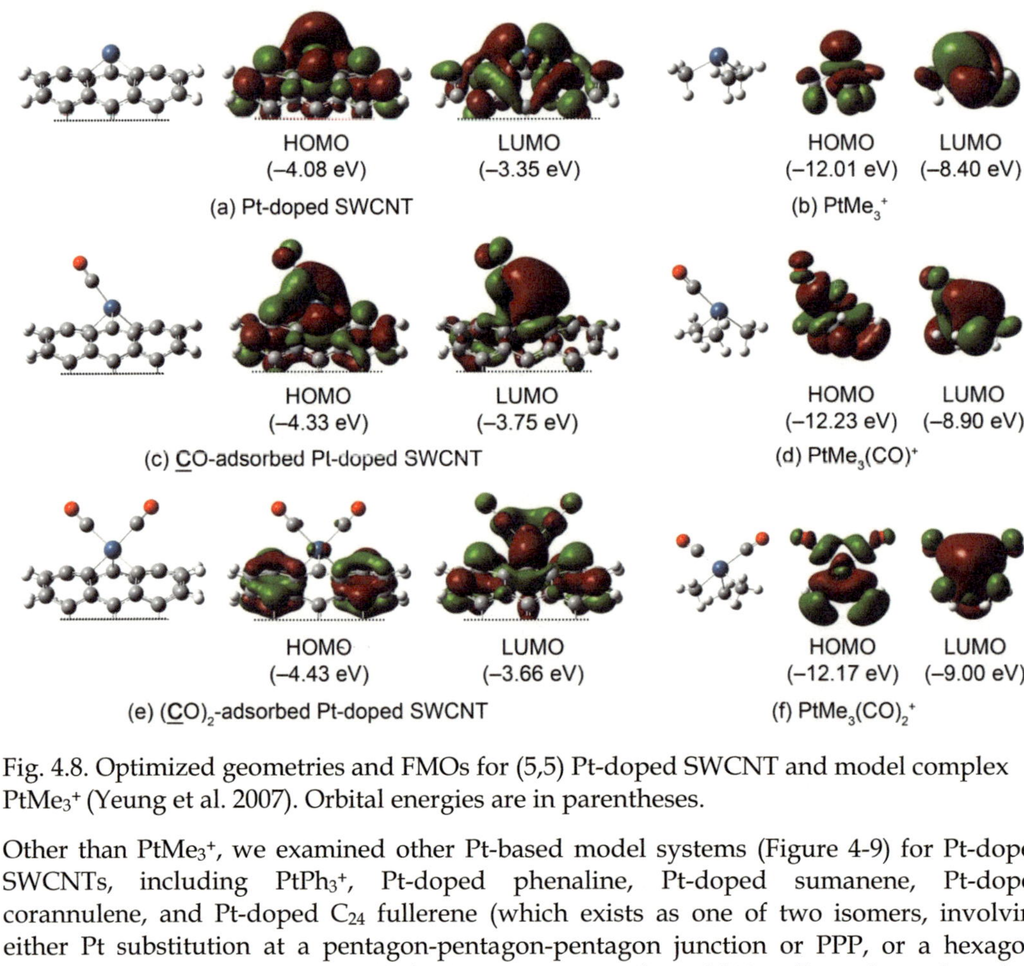

Fig. 4.8. Optimized geometries and FMOs for (5,5) Pt-doped SWCNT and model complex $PtMe_3^+$ (Yeung et al. 2007). Orbital energies are in parentheses.

Other than $PtMe_3^+$, we examined other Pt-based model systems (Figure 4-9) for Pt-doped SWCNTs, including $PtPh_3^+$, Pt-doped phenaline, Pt-doped sumanene, Pt-doped corannulene, and Pt-doped C_{24} fullerene (which exists as one of two isomers, involving either Pt substitution at a pentagon-pentagon-pentagon junction or PPP, or a hexagon-pentagon-pentagon junction or HPP). We were intrigued to additionally evaluate the role of the delocalized π-electrons, particularly as it pertained to curvature (Lu & Chen 2005). Because of computational limitations, we chose a (5,0) Pt-doped SWCNT segment capped with ten H atoms of molecular formula $C_{49}H_{10}Pt$ and calculated a geometry in which the Pt atom protrudes to the exterior of the nanotube sidewall. Of the model complexes considered, only $PtMe_3^+$ and $PtPh_3^+$ displayed less pyramidalization at the Pt atom than the HSWCNT ($\sum$ = 297.5° and 315.4°, respectively, versus $\sum$ = 282.1°). Pt-doped sumanene, corannulene, and C_{24} are all more highly pyramidalized ($\sum < 270°$).

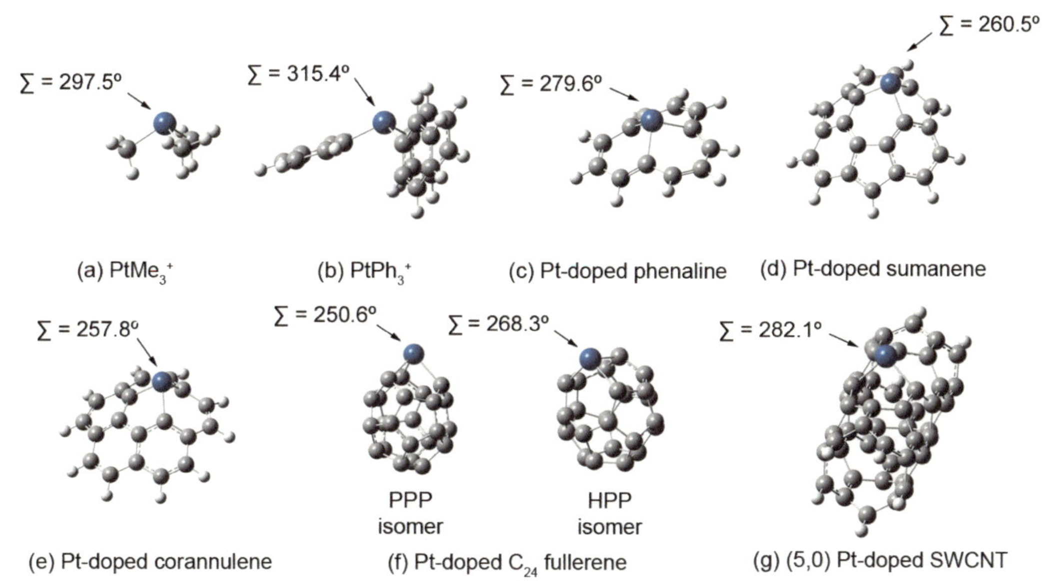

Fig. 4.9. Optimized geometries of model alkyl- and arylplatinum complexes (Yeung et al. 2011).

By means of C≡O vibrational frequency analysis (Table 4-3), we found that Pt-doped (5,5) SWCNTs demonstrated net backdonation to CO (Yeung et al. 2010), while the corresponding (5,0) SWCNT system underwent adsorption with net donation of electron density from the adsorbate (Yeung et al. 2011). In contrast, all other model complexes evaluated have a tendency to deplete electron density from CO. Based on similar stretching frequencies of C̲O-adsorbed Pt-doped C_{24} and corannulene, higher levels of π-conjugation appear to have a positive impact on the ability to backbond to incoming ligands, leading to a tighter Pt–CO interaction (~2.00 Å). The most acidic complexes, however, are $PtPh_3^+$ and $PtMe_3^+$, which display weaker Pt–CO interactions (~2.11 Å). Based on this investigation, the predicted trend of Lewis acidity is as follows:

> (5,0) Pt-doped SWCNT ~ Pt-doped corannulene < Pt-doped C_{24} PPP isomer < Pt-doped C_{24} HPP isomer < Pt-doped sumanene < Pt-doped phenaline < $PtPh_3^+$ < $PtMe_3^+$.

By means of NMR analysis (Table 4-4), we were surprised to observe decreased chemical shifts of the β-H of coordinated MA in both (5,0) Pt-doped SWCNT and Pt-doped corannulene, implying that the charge-separated resonance structure is less important than in uncoordinated MA. This effect can be attributed to larger steric hindrance toward the productive coordination of the C=O group of MA and the Pt atom. All other complexes are in fact strong Lewis acids ($\Delta\delta$ ~ 0.5 ppm). Lewis acid/Lewis base pair formation decreases the C1–C2 bond length and increases C1=O and C2=C3 bond lengths and suggests that reactivity at C2 and C3 towards electrophiles and nucleophiles, respectively, should be enhanced. Complexation of the model organoplatinum complexes resulted in larger geometric deviations than from the simpler $PtMe_3^+$ and $PtPh_3^+$ systems and suggests that higher levels of π-curvature and increased ability to favor delocalization of the electron density may facilitate this productive interaction. The trend of Lewis acidity suggested by chemical shift analysis is nonetheless identical to the conclusion derived from the above C≡O vibrational frequency analysis.

Pt complex	Σ[a]	q(Pt)[b]	E_{elec}[c]	ΔG[d]	d(Pt–C)[e]	d(C≡O)[e]	$\nu_{C≡O}$[f]	q(Pt/CO)[g]	q(CO)[h]
$PtMe_3^+$	297.5°	0.77	−24.0	−8.25	2.116	1.156	2089.3	0.54	0.25
$PtPh_3^+$	315.4°	0.77	−16.8	−0.46	2.113	1.160	2055.1	0.54	0.22
Pt-doped phenaline	279.6°	0.74	−28.6	−13.07	2.057	1.161	2050.3	0.40	0.23
Pt-doped sumanene	260.5°	0.77	−35.2	−19.67	2.025	1.163	2043.4	0.39	0.21
Pt-doped corannulene	257.8°	0.59	−34.3	−18.22	1.983	1.179	1949.4	0.26	0.08
Pt-doped C_{24} PPP	250.6°	0.69	−31.6	−16.74	2.027	1.167	2017.3	0.32	0.17
Pt-doped C_{24} HPP	268.3°	0.66	−32.7	−17.59	1.995	1.166	2026.4	0.30	0.22
(5,0) Pt-doped SWCNT	292.2°	0.53	−36.0	−20.31	1.971	1.179	1949.8	0.12	0.12

[a]Sum of angles around Pt center (without adsorbate).
[b]Partial charge on Pt atom (without adsorbate).
[c]Electronic energy of stabilization (in kcal/mol) upon CO adsorption.
[d]Free energy of stabilization (in kcal/mol) upon CO adsorption at 298 K.
[e]Bond length (in Å).
[f]Vibration frequency (in cm^{-1}) of bound C≡O molecule (*cf.* 2028.8 cm^{-1} for unbound free CO).
[g]Partial charge of Pt atom in CO-adsorbed Pt complex.
[h]Partial charge on bound CO molecule.

Table 4.3. Electronic structural data for alkyl- and arylplatinum complexes and their CO-adsorbates (Yeung et al. 2011).

LA/MA complex	d(C1=O)[a]	d(C1–C2)[a]	d(C2=C3)[a]	q(O)[b]	q(C3)[c]	δ(H)[d]	Δδ(H)[e]
Uncoordinated MA	1.223	1.499	1.351	−0.496	−0.250	6.25	0.00
$PtMe_3^+$/MA	1.248	1.488	1.354	−0.583	−0.215	7.02	0.77
$PtPh_3^+$/MA	1.243	1.488	1.354	−0.557	−0.221	6.94	0.69
Pt-doped phenaline/MA	1.249	1.486	1.356	−0.584	−0.229	6.80	0.55
Pt-doped sumanene/MA	1.249	1.484	1.356	−0.577	−0.221	6.71	0.46
Pt-doped corannulene/MA	1.278	1.440	1.384	−0.643	−0.311	5.53	−0.72
Pt-doped C_{24} PPP/MA	1.254	1.472	1.363	−0.595	−0.266	6.30	0.05
Pt-doped C_{24} HPP/MA	1.245	1.479	1.359	−0.552	−0.250	6.39	0.14
(5,0) Pt-doped SWCNT/MA	1.276	1.441	1.385	−0.636	−0.317	5.62	−0.63

[a]Bond length (in Å).
[b]Partial charge on O atom.
[c]Partial charge on C3 atom.
[d]Chemical shift (in ppm) on H atom (referenced to tetramethylsilane).
[e]Difference in chemical shift (in ppm) on H atom relative to uncoordinated maleic anhydride (referenced to tetramethylsilane).

Table 4.4. Electronic structural data for Lewis acid/maleic anhydride complexes (Yeung et al. 2011)

By means of enhancing the reactivity of chemical reactions such as the Diels-Alder cycloaddition, we observed a surprising *decrease* in reaction efficiency (Table 4-5) when employing the (5,0) Pt-doped SWCNT as a catalyst (Figure 4-10). While our vibrational frequency and NMR analyses both suggested rate enhancements, this was not the case for our nanotube-based catalyst. This suggests that although the LUMO energy of the dienophile (MA) was decreased by binding, the steric demand for the productive [4+2] cycloaddition made catalysis less likely. In agreement with the literature (Hsieh et al. 2007), we determined that $PtMe_3^+$ is a viable catalyst for the Diels-Alder reaction in addition to the other model Pt complexes. The Lewis acid catalyzed cycloaddition of MA and CPD varied inversely in efficiency with catalyst bulk. Of note, while Pt-doped fullerene gave an approximate threefold increase in rate, other systems yielded greater enhancements of up to 10^7-fold! Pyramidalization of the carbon framework has relatively little impact on the effectiveness of the proposed catalysis.

The cyclohexene product formed in the Diels-Alder cycloaddition can exist as one of two diastereomeric compounds (Rulisek et al. 2005; Dewar & Pierini 1984). The so-called *endo* product is *kinetically* favored and involves *secondary orbital interactions* between the two unreactive C atoms on the diene (CPD) and the π-orbitals of the neighboring C=O groups on the alkene (MA). The *exo* product is accessible through thermodynamic control. In our studies, if we assume kinetic control of the Diels-Alder reaction, Pt complexes with higher levels of delocalization of electron density in the catalyst tend to give a lower preference for the *endo* product. To our delight, while Pt-doped SWCNTs were not excellent catalysts for this transformation, the *exo* product became favored over the *endo* product (Yeung et al. 2011). This result suggests that nanocatalysts have the potential to effect *complementary* selectivities to other traditional catalysts. By controlling externally applied voltages, the Lewis acidity of transition metal dopants can be manipulated, hence providing an opportunity for tuning the efficiency and selectivity of nanocatalyzed reactions. The advantage of using nanomaterials in catalysis is that they sit uniquely at the boundary between homogeneous and heterogeneous catalysis. This provides organic chemists an opportunity to establish the connection between reaction mechanisms of these two catalytic systems. In addition, substitutional doping of CNTs with Pt atoms may be a useful and effective strategy for developing heterogeneous-like Lewis acid catalysts. Immobilization of reactive Lewis acids is expected to ease catalyst recovery and reuse of precious metals, as well as afford an enhanced stability of these reactive species. Moreover, we can possess greater controllability of the reaction process, e.g., to pause the reaction by simply separating the catalyst from the solution. Other transformations may be catalyzed by our HSWCNTs, such as Mukaiyama aldol reactions, and should be explored both theoretically and experimentally.

More complex, novel nanosystems may even become possible. Because nanotubes are channels, we may one day be able to achieve nanoscale microfluidics. We can imagine a transition metal-doped SWCNT embedded within a much larger SWCNT. If this is possible, the nanochannels sitting between the two carbon backbones may provide enhanced chemical reactivity due to higher local concentrations of the reactive partners. In the ideal world, it may be possible that reactants be introduced at one end of a nanochannel and products come out of the other end. Cooperative catalysis between different transition metals may allow for unique types of catalysis that are not possible in bulk solutions, on flat surfaces, or at pointy small clusters. Needless to say, nanocatalysis remains an emerging field that will continue to develop in the years to come.

Catalyst	$\Delta G_{exo}^{\ddagger a}$	ΔG_{exo}^{b}	$\Delta G_{endo}^{\ddagger c}$	ΔG_{endo}^{d}	$\Delta\Delta G^{\ddagger e}$	k_{endo}/k_{exo}^{f}	k_{cat}/k_{uncat}^{g}
none	27.3	−3.8	26.7	−3.0	0.6	2.78	1.00
$PtMe_3^+$	15.9	−8.1	14.4	−9.5	1.4	11.43	1.27×10^3
$PtPh_3^+$	17.4	−8.2	16.8	−8.5	0.6	2.79	1.58×10^3
Pt-doped phenaline	17.7	−8.2	16.7	−7.7	1.0	5.60	1.47×10^3
Pt-doped sumanene	16.8	−8.6	16.2	−8.4	0.6	2.63	3.01×10^3
Pt-doped corannulene	16.0	−13.3	15.3	−13.3	0.7	3.43	8.22×10^6
Pt-doped C_{24} PPP	23.5	−4.0	22.6	−3.5	0.9	4.26	1.42
Pt-doped C_{24} HPP	23.5	−4.5	22.5	−4.2	1.0	5.44	2.99
(5,0) Pt-doped SWCNT	31.0	6.4	31.9	6.5	-0.9	0.21	1.6×10^{-3}

[a]Free energy required to reach transition state leading to *exo* product at 298 K.
[b]Free energy change to *exo* product at 298 K.
[c]Free energy required to reach transition state leading to *endo* product at 298 K.
[d]Free energy change to *endo* product at 298 K.
[e]Difference in free energy changes required to reach transition states leading to *endo* and *exo* products at 298 K.
[f]Relative rate leading to *endo* product versus *exo* product.
[g]Relative rate of catalyzed Diels-Alder reaction versus uncatalyzed Diels-Alder reaction.

Table 4.5. Free energies (in kcal/mol) and rates of the Diels-Alder reaction of maleic anhydride and cyclopentadiene as catalyzed by alkyl- and arylplatinum complexes (Yeung et al. 2011)

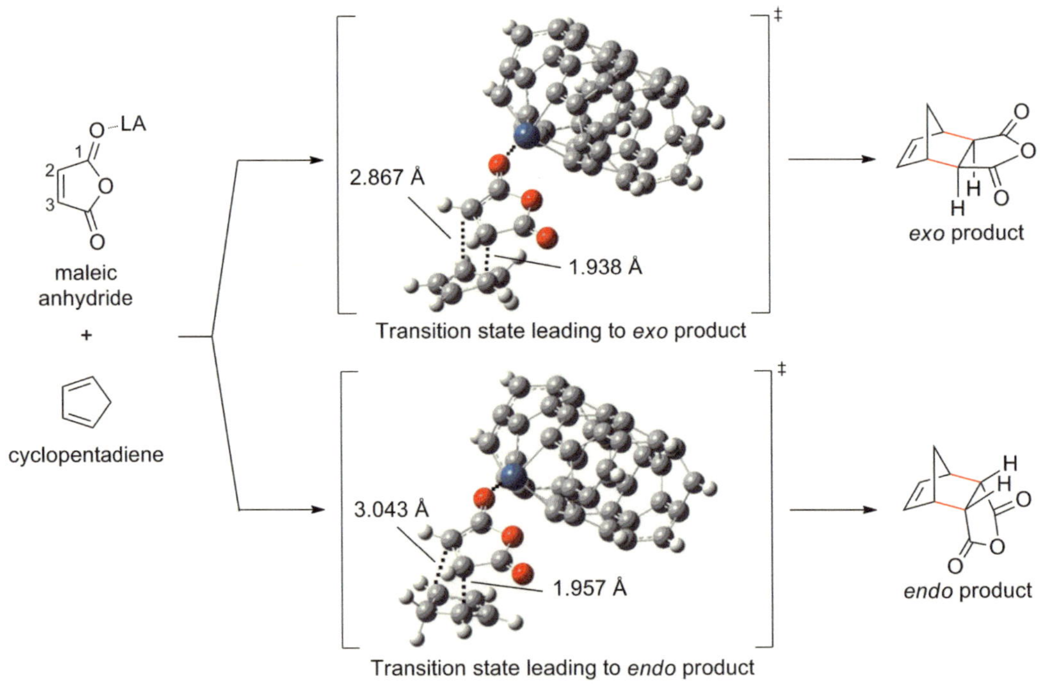

Fig. 4.10. Transition states of the Diels-Alder reaction of maleic anhydride and cyclopentadiene (Yeung et al. 2011). Legend: dark grey = C, light grey = H, red = O, blue = Pt.

5. Concluding remarks

Within DFT, the rich chemistry of substitutionally doped SWCNTs has been investigated. We summarize our findings below:

1. A mono vacancy formed by removing a single C atom from the sidewall of a SWCNT rearranges to form a five-membered carbocycle and a single C atom with a dangling bond (i.e., the 5-1DB defect). 5-1DB defects are reactive with NO, liberating NO_2 and N-doped SWCNTs, and with O_3, generating O_2 and O-doped SWCNTs.
2. Substitutional doping of SWCNTs with transition metals results in buckling of the nanotube sidewall with the larger atom protruding to the exterior.
3. Adsorption of gases onto Pt-doped SWCNTs is analogous to the coordination chemistry of monomeric Pt complexes. Because the nanotube sidewall provides much larger steric bulk, multiple adsorption is thermodynamically unfavorable. The ability for Pt atoms to perform physisorption and chemisorption depends strongly on the doping location. Adsorption results in a change in the electronic structure of Pt-doped SWCNTs.
4. Pt-doped SWCNTs exhibit Lewis acidity similar to that of simple alkyl- and arylplatinum complexes. Evaluation of Lewis acid strength can be done by measuring the stretching frequency of a bound molecule of CO or by ^{1}H NMR studies of the β-H of Lewis acid/Lewis base complexes formed with α,β-unsaturated carbonyl compounds. However, because the nanotube provides larger steric bulk than simple alkyl- and arylplatinum complexes, the ability to activate a substrate for conventional catalysis may be impeded, but alternative catalytic activities might be released instead.

Transition metal-doped SWCNTs have yet to be prepared experimentally. In this Chapter, we have provided a comprehensive overview of the potential of these nanomaterials and their applications in nanosensing, nanoelectronics, and nanocatalysis. Our initial investigations here should provide inspiration and guidance to experimentalists in their own endeavors.

6. Acknowledgements

We thank the Natural Sciences and Engineering Research Council (NSERC) of Canada for financial support. WestGrid and C-HORSE have provided the necessary computational resources. C.S.Y. gratefully acknowledges NSERC for an Undergraduate Student Research Award.

7. References

Ajayan, P. M., Ravikumar, V. & Charlier, J.-C. (1998) Surface reconstructions and dimensional changes in single-walled carbon nanotubes. *Phys. Rev. Lett.*, *81*, 1437-1440.

Ajayan, P. M. (1999) Nanotubes from carbon. *Chem. Rev.*, *99*, 1787-1800.

Avouris, P., Freitag, M. & Perebeinos, V. (2008) Carbon-nanotube photonics and optoelectronics. *Nature Photonics*, *2*, 341-350.

Backus, E. H. G., Eichler, A., Grecea, M. L., Kleyn, A. W. & Bonn, M. (2004) Adsorption and dissociation of NO on stepped Pt (533). *J. Chem. Phys.*, *121*, 7946-7954.

Becke, A. D. (1988) Density-functional exchange-energy approximation with correct asymptotic-behavior. *Phys. Rev. A*, *38*, 3098-3100.

Berber, S. & Oshiyama, A. (2006) Reconstruction of mono-vacancies in carbon nanotubes: atomic relaxation vs. spin polarization. *Physica B, 376-377*, 272-275.

Bethune, D. S., Kiang, C. H., de Vries, M. S., Gorman, G., Savoy, R., Vazquez, J. & Beyers, R. (1993) Cobalt-catalyzed growth of carbon nanotubes with single-atomic-layer walls. *Nature, 363*, 605-607.

Boulikas, T. & Vougiouka, M. (2003) Cisplatin and platinum drugs at the molecular level. *Oncol. Rep., 10*, 1663-1682.

Branz, W., Billas, I. M. L., Malinowski, N., Tast, F., Heinebrodt, M. & Martin, T. P. (1998) Cage substitution in metal-fullerene clusters. *J. Chem. Phys., 109*, 3425-3430.

Bryning, M. B., Islam, M. F., Kikkawa, J. M. & Yodh, A. G. (2005) Very low conductivity threshold in bulk isotropic single-walled carbon nanotube-epoxy composites. *Adv. Mater., 17*, 1186-1191.

Buzea, C., Blandino, I. I. P. & Robbie, K. (2007) Nanomaterials and nanoparticles: sources and toxicity. *Biointerphases, 2*, MR17-MR172.

Caruso, E. B., Petralia, S., Conoci, S., Giuffrida, S. & Sortino, S. (2007) Photodelivery of nitric oxide from water-soluble platinum nanoparticles. *J. Am. Chem. Soc., 129*, 480-481.

Changgeng, D., Jinlong, Y., Rongsheng, H. & Kelin, W. (2001) Formation mechanism and structural and electronic properties of metal-substituted fullerenes $C_{69}M$ (M = Co, Rh, and Ir). *Phys. Rev. A., 64*, 043201.

Chen, Y. K., Liu, L. V. & Wang, Y. A. (2010) Density functional study of interaction of atomic Pt with pristine and Stone-Wales-defective single-walled boron nitride nanotubes. *J. Phys. Chem. C, 114*, 12382-12388.

Chen, Y. K.; Liu, L. V., Tian, W. Q. & Wang, Y. A. (2011) Theoretical studies of transition-metal-doped single-walled carbon nanotubes. *J. Phys. Chem. C, 115*, 9306-9311.

Chiaretti, M., Mazzanti, G., Bosco, S., Bellucci, S., Cucina, A., le Foche, F., Carru, G. A., Mastrangelo, S., di Sotto, A., Masciangelo, R., Chiaretti, A. M., Balasubramanian, C., de Bellis, G., Micciulla, F., Porta, N., Deriu, G. & Tiberia, A. (2008) Carbon nanotubes toxicology and effects on metabolism and immunological modification in vitro and in vivo. *J. Phys.: Condens. Matter, 20*, 474203.

Chico, L., Santos, H., Ayuela, A., Jaskolski, W., Pelc, M. & Brey, L. (2010) Unzipped and defective nanotubes: rolling up graphene and unrolling tubes. *Acta Phys. Polonica A, 118*, 433-441.

Childs, R. F., Mulholland, D. L. & Nixon, A. (1982) The Lewis acid complexes of α,β-unsaturated carbonyl and nitrile compounds. A nuclear magnetic resonance study. *Can. J. Chem., 60*, 801-808.

Chopra, S., Pham, A., Gaillard, J., Parker, A. & Rao, A. M. (2002) Carbon-nanotube-based resonant-circuit sensor for ammonia. *Appl. Phys. Lett., 80*, 4632-4634.

Collins, P. G., Bradley, K., Ishigami, M. & Zetti, A. (2000) Extreme oxygen sensitivity of electronic properties of carbon nanotubes. *Science*, 287, 1801-1804.

Collins, P. G. (2010) Defects and disorder in carbon nanotubes, *in* The Oxford Handbook of Nanoscience and Nanotechnology, Oxford University Press: New York, NY, Volume 2, Chapter 2, pp. 31-93.

Culotta, E. & Koshland Jr., D. E. (1992) NO news is good news. *Science, 258*, 1862-1865.

Dapprich, S., Komaromi, I., Byun, K. S., Morokuma, K. & Frisch, M. J. (1999) A new ONIOM implementation in Gaussian 98. 1. The calculation of energies, gradients and vibrational frequencies and electric field derivatives. *J. Mol. Struc. (Theochem), 462*, 1-21.

Dewar, M. J. S. & Pierini, A. B. (1984) Mechanism of the Diels-Alder reaction. Studies of the addition of maleic anhydride to furan and methylfurans. *J. Am. Chem. Soc., 106*, 203-208.

Dewar, M. J. S., Zoebisch, E. G. & Healy, E. F. (1985) AM1: A new general purpose quantum mechanical molecular model. *J. Am. Chem. Soc.*, *107*, 3902-3909.

Ditchfield, R., Hehre, W. J. & Pople, J. A. (1971) Self-consistent molecular orbital methods. 9. Extended Gaussian-type basis for molecular-orbital studies of organic molecules. *J. Chem. Phys.*, *54*, 724.

Dunning Jr., T. H. & Hay, P. J. (1976) Gaussian basis sets for molecular calculations, *in* Modern Theoretical Chemistry, Plenum: New York, NY, Volume 3, pp. 1-28.

Ewels, C., Glerup, M. & Krstic, V. (2010) Nitrogen and boron doping in carbon nanotubes, *in* Doped Nanomaterials and Nanodevices, American Scientific Publishers: Valencia, CA, Volume 3.

Frisch, M. J., Trucks, G. W., Schlegel, H. B., Scuseria, G. E., Robb, M. A., Cheeseman, J. R., Montgomery, J. A., Vreven, T., Kudin, K. N., Burant, J. C., Milliam, J. M. Iyengar, S. S., Tomasi, J., Barone, V., Mennucci, B., Cossi, M. , Scalmani, G., Rega, N., Petersson, G. A., Nakatsuji, H., Hada, M. Ehara, M., Toyota, K., Fukuda, R., Hasegawa, J., Ishida, M., Nakajima, T., Honda, Y., Kitao, O., Nakai, H., Li, X., Knox, J. E., Hratchian, H. P., Cross, J. B., Bakken, V., Adamo, C., Jaramillo, J., Gomperts, R., Stratmann, R. E., Yazyev, O., Austin, A. J., Cammi, R., Pomelli, C., Ochterski, J. W., Ayala, P. Y., Morokuma, K., Voth, G. A., Salvador, P., Dannenberg, J. J. Zakrzewski, V. G., Dapprich, S., Daniels, A. D., Farkas, O., Malick, D. K., Rabuck, A. D., Raghavachari, K., Foresman, J. B., Ortiz, J. V., Cui, Q., Baboul, A. G., Clifford, S., Coislowski, J., Stefanov, B. B., Liu, G., Liashenko, A., Piskorz, P., Komaromi, I., Martin, R. L., Fox, D. J., Keith, T., Al-Laham, M. A., Peng, C. Y., Nanayakkara, A., Challacombe, M., Gill, P. M. W., Johnson, B., Chen, W., Wong, M. W., Gonzalez, C. & Pople, J. A. (2003) Gaussian 03, Revision B.05, Gaussian, Inc.: Wallingford CT.

Fuhrer, M. S. (2003) Single-walled carbon nanotubes for nanoelectronics, *in* Advanced Semiconductor and Organic Nano-Techniques (Part II), Academic Press: London, Chapter 6, pp. 293-343.

Fukui, K. (1982) Role of frontier orbitals in chemical reactions. *Science*, *218*, 747-754.

Gayathri, V. & Geetha, R. (2007) Hydrogen adsorption in defected carbon nanotubes. *Adsorption*, *13*, 53-59.

Golberg, D., Bando, Y., Huang, Y., Terao, T., Mitome, M., Tang, C. & Zhi, C. (2010) Boron nitride nanotubes and nanosheets. *ACS Nano*, *4*, 2979-2993.

Glerup, M., Krstic, V., Ewels, C., Holzinger, M. & Lier, G. V. (2010) Doping of carbon nanotubes, *in* Doped Nanoamterials and Nanodevices, American Scientific Publishers: Valencia, Volume 3.

Griffith, O. W. & Stuehr, D. J. (1995) Nitric oxide synthases: properties and catalytic mechanism. *Ann. Rev. Physiol.*, *57*, 707-736.

Grinvald, A. & Hildesheim, R. (2004) VSDI: a new era in functional imaging of cortical dynamics. *Nature Rev. Neurosci.*, *5*, 874-885.

Haddon, R. C. & Scott, L. T. (1986) π-Orbital conjugation and rehybridization in bridged annulenes and deformed molecules in general: π-orbital axis vector analysis. *Pure Appl. Chem.*, *58*, 137-142.

Hay, P. J. & Wadt, W. R. (1985) Ab initio effective core potentials for molecular calculations – potentials for K to Au including the outermost core orbitals. *J. Chem. Phys.*, *82*, 299-310.

Hsieh, V., de Crisci, A. G., Lough, A. J. & Fekl, U. (2007) Lewis-acidity of trimethylplatinum(IV) with labile oxygen-donor ligands. *Organometallics*, *26*, 938-944.

Hu, C. & Hu, S. (2009) Carbon nanotube-based electrochemical sensors: principles and applications in biomedical systems. *J. Sensors*, *2009*, 187615.

Iijima, S. (1991) Helical microtubules of graphitic carbon. *Nature, 354*, 56-58.
Iijima, S. & Ichihashi, T. (1993) Single-shell carbon nanotubes of 1-nm diameter. *Nature, 363*, 603-605.
Javey, A., Guo, J., Wang, Q., Lundstrom, M. & Dai, H. (2003) Ballistic carbon nanotube field-effect transistors. *Nature, 424*, 654-657.
Karakoti, A. S., Hench, L. L. & Seal, S. (2006) The potential toxicity of nanomaterials - the role of surfaces. *JOM, 58*, 77-82.
Karousis, N., Tagmatarchis, N. & Tasis, D. (2010) Current progress on the chemical modification of carobn nanotubes. *Chem. Rev., 110*, 5366-5397.
Kong, J., Franklin, N. R., Zhou, C., Chapline, M. G., Peng, S., Cho, K. & Dai, H. (2000) Nanotube molecular wires as chemical sensors. *Science, 287*, 622-625.
Kong, J., Chapline, M. G. & Dai, H. (2001) Functionalized carbon nanotubes for molecular hydrogen sensors. *Adv. Mater., 13*, 1384-1386.
Kong, Q., Zhuang, J., Xu, J., Shen, Y., Li, Y., Zhao, L. & Cai, R. (2003) Contrasting behaviors of metal fullerides $C_{60}Rh_x$ and $C_{60}M_x$ (M = La, Y) in the metallofullerrene formation by laser ablation. *J. Phys. Chem. A, 107*, 3670-3677.
Kuczkowski, R. L. (1992) The structure and mechanism of formation of ozonides. *Chem. Soc. Rev., 21*, 79-83.
Lam, C.-W., James, J. T., McCluskey, R., Arepalli, S. & Hunter, R. L. (2006) A review of carbon nanotube toxicity and assessment of potential occupational and environmental health risks. *Crit. Rev. Toxicol., 36*, 189-217.
Laszlo, P. & Teston, M. (1990) Determination of the acidity of Lewis acids. *J. Am. Chem. Soc., 112*, 8750-8754.
Lee, C., Yang, W. & Parr, R. G. (1988) Development of the Colle-Salvetti correlation-energy formula into a functional of the electron density. *Phys. Rev. B, 37*, 785-789.
Li, C., Thostenson, E. T. & Chou, T.-W. (2008) Sensors and actuators based on carbon nanotubes and their composites: a review. *Compos. Sci. Technol., 68*, 1227-1249.
Li, J., Lu, Y., Ye, Q., Cinke, M., Han, J. & Meyyappan, M. (2003) Carbon nanotube sensors for gas and organic vapor detection. *Nano Lett., 3*, 929-933.
Li, J., Jia, G. & Zhang, Y. (2007) Chemical anisotropies of carbon nanotubes and fullerenes caused by the curvature directivity. *Chem. Eur. J., 132*, 6430-6436.
Li, X. M., Tian, W. Q., Huang, X.-R., Sun, C.-S. & Jiang, L. (2009) Adsorption of hydrogen on novel Pt-doped BN nanotubes. *J. Molec. Struc.: THEOCHEM, 901*, 103-109.
Liu, L. V., Tian, W. Q. & Wang, Y. A. (2006) Chemical reaction of nitric oxides with the 5-1DB defect of the single-walled carbon nanotube. *J. Phys. Chem. B, 110*, 1999-2005.
Liu, L. V., Tian, W. Q. & Wang, Y. A. (2006a) Ozonization at the vacancy defect site of the single-walled carbon nanotube. *J. Phys. Chem. B, 110*, 13037-13044.
Liu, L. V., Tian, W. Q. & Wang, Y. A. (2009) Ab initio studies of vacancy-defected fullerenes and single-walled carbon nanotubes. *Int. J. Quantum Chem., 109*, 3441-3456.
Lu, X. & Chen, Z. (2005) Curved pi-conjugation, aromaticity, and the related chemistry of small fullerenes ($<C_{60}$) and single-walled carbon nanotubes. *Chem. Rev., 105*, 3643-3696.
Menzel, L., Kosterev, A. A., Curl, R. F., Tittel, F. K., Gmachi, C., Capasso, F., Sivco, D. L., Baillargeon, J. N., Hutchinson, A. L, Cho, A. Y. & Urban, W. (2001) Spectroscopic determination of biological NO with a quantum cascade laser. *Appl. Phys. B, 72*, 859-863.
Miller, M. R. & Megson, I. L. (2007) Recent developments in nitric oxide donor drugs. *Br. J. Pharmacol., 151*, 305-321.

Moniruzzaman, M. & Winey, K. I. (2006) Polymer nanocomposites containing carbon nanotubes. *Macromolecules, 39,* 5194-5205.

Nel, A., Xia, T., Maedler, L. & Li, N. (2006) Toxic potential of materials at the nanolevel. *Science, 311,* 622-627.

Nevidomskyy, A. H., Csanyi, G. & Payne, M. C. (2003) Chemically active substitutional nitrogen impurity in carbon nanotubes. *Phys. Rev. Lett., 91,* 105502.

Oberdorster, G., Stone, V. & Donaldson, K. (2007) Toxicology of nanoparticles: a historical perspective. *Nanotoxicology, 1,* 2-25.

Orita, H., Itoh, N. & Inada, Y. (2004) All electron scalar relativistic calculations on adsorption of CO on Pt(111) with full-geometry optimization: a correct estimation for CO site-preference. *Chem. Phys. Lett., 384,* 271-276.

Park, J.-Y. (2004) Nanoelectronics/mechanics with carbon nanotubes. http://nanohub.org/resources/157.

Peng, S. & Cho, K. (2003) Ab initio study of doped carbon nanotube sensors. *Nano Lett., 3,* 513-517.

Perdew, J. P., Burke, K. & Ernzerhof, M. (1996) Generalized gradient approximation made simple. *Phys. Rev. Lett., 77,* 3865-3868.

Perdew, J. P., Burke, K. & Wang, Y. (1996a) Generalized gradient approximation for the exchange-correlation hole of a many-electron system. *Phys. Rev. B, 54,* 16533-16539.

Poblet, J. M., Munoz, J., Winkler, K., Cancilla, M., Hayashi, A., Lebrilla, C. B., Balch, A. L. & Winkler, K. (1999) Geometric and electronic structure of metal-caged fullerenes, $C_{59}M$ (M = Pt, Ir) obtained by laser ablation of electrochemically deposited films. *Chem. Commun.,* 493-494.

Procelewska, J., Zahl, A., Liehr, G., van Eldik, R., Smythe, N. A., Williams, B. S. & Goldberg, K. I. (2005) Mechanistic information on the reductive elimination from cationic trimethylplatinum(IV) complexes to form carbon-carbon bonds. *Inorg. Chem., 44,* 7732-7742.

Rappe, A. K., Casewit, C. J., Colwell, K. S., Goddard III, W. A. & Skiff, W. M. (1992) UFF, a full periodic-table force-field for molecular mechanics and molecular-dynamics simulations. *J. Am. Chem. Soc., 114,* 10024-10035.

Reed, A. E., Curtiss, L. A. & Weinhold, F. (1988) Intermolecular interactions from a natural bond orbital, donor-acceptor viewpoint. *Chem. Rev., 88,* 899-926.

Rouxinol, F. P., Gelamo, R. V. & Moshkalev, S. A. (2010) Gas sensors based on decorated carbon nanotubes, *in* Carbon Nanotubes. InTech, Chapter 19, pp. 357-374.

Rulisek, L., Sebek, P., Havlas, Z., Hrabal, R., Capek, P. & Svatos, A. (2005) An experimental and theoretical study of stereoselectivity of furan-maleic anhydride and furan-maleimide Diels-Alder reactions. *J. Org. Chem., 70,* 6295-6302.

Schipper, M. L., Nakayama-Ratchford, N., Davis, C. R., Kam, N. W. S., Chu, P., Liu, Z., Sun, X., Dai, H. & Gambhir, S. S. (2008) A pilot toxicology study of single-walled carbon nanotubes in a small sample of mice. *Nature Nanotech., 3,* 216-221.

Schlegel, H. B., Millam, J. M., Iyengar, S. S., Voth, G. A., Scuseria, G. E., Daniels, A. D. & Frisch, M. J. (2001) Ab initio molecular dynamics: propagating the density matrix with Gaussian orbitals. *J. Chem. Phys., 114,* 9758-9763.

Serp, P., Corrias, M., & Kalck, P. (2003) Carbon nanotubes and nanofibers in catalysis. *Appl. Catal. A, 253,* 337-358.

Singh, R., Pantarotto, D., Lacerda, L., Pastorin, G., Klumpp, C., Prato, M., Bianco, A. & Kostarelos, K. (2006) Tissue biodistribution and blood clearance rates of intravenously administered carbon nanotube radiotracers. *Proc. Natl. Acad. Sci. U.S.A., 103,* 3357-3362.

Sinha, N., Ma, J. & Yeow, J. T. W. (2006) Carbon nanotube-based sensors. *J. Nanosci. Nanotech., 6*, 573-590.

Sonia, K. & Niranjan, M. (2010) Quantum conductance of defected carbon nanotubes. *AIP Conf. Proc., 1324*, 415-418.

Sparta, M., Borve, K. J. & Jensen, V. R. (2006) Structure and stability of networked metallofullerenes of the transition metals. *J. Phys. Chem. A, 110*, 11711-11716.

Srivastava, D., Chenyu, W. & Cho, K. (2003) Nanomechanics of carbon nanotubes and composites. *Appl. Mech. Rev., 56*, 215-230.

Srivastava, D., Menon, M., Daraio, C., Jin, S., Sadanadan, B. & Rao, A. M. (2004) Vacancy-mediated mechanism of nitrogen substitution in carbon nanotubes. *Phys. Rev. B, 69*, 153414.

Stewart, J. J. P. (1989) Optimization of parameters for semiempirical methods. I. Method. *J. Comput. Chem., 10*, 209-220.

Stone, A. J. & Wales, D. J. (1986) Theoretical studies of icosahedral footballene sixty-carbon-atom molecules and some related species. *Chem. Phys. Lett., 128*, 501-503.

Tasis, D., Tagmatarchis, N., Bianco, A. & Prato, M. (2006) Chemistry of carbon nanotubes. *Chem. Rev., 106*, 1105-1136.

Tenderholt, A. (2005) *PyMOlyze*, Version 1.1.

Tian, W. Q., Liu, L. V. & Wang, Y. A. (2006) Electronic properties and reactivity of Pt-doped carbon nanotubes. *Phys. Chem. Chem. Phys., 8*, 3528-3539.

Tsukagoshi, K., Yoneya, N., Uryu, S., Aoyagi, Y., Kanda, A., Ootuka, Y. & Alphenaar, B. W. (2002) Carbon nanotube devices for nanoelectronics. *Physica B, 323*, 107-114.

Wang, J., Lee, C. H. & Yap, Y. K. (2010) Recent advances in boron nitride nanotubes. *Nanoscale, 2*, 2028-2034.

Wang, Y. & Yeow, J. T. W. (2009) A review of carbon nanotubes-based gas sensors. *J. Sensors*, 493904.

Wheatley, P. S., Butler, A. R., Crane, M. S., Fox, S., Xiao, B., Rossi, A. G., Megson, I. L. & Morris, R. E. (2006) NO-releasing zeolites and their antithrombotic properties. *J. Am. Chem. Soc., 128*, 502-509.

White, C. T. & Todorov, T. N. (1998) Carbon nanotubes as long ballistic conductors. *Nature, 393*, 240-242.

Wolinski, K., Hilton, J. F. & Pulay, P. (1990) Efficient implementation of the gauge-independent atomic orbital method for NMR chemical shift calculations. *J. Am. Chem. Soc., 112*, 8251-8260.

Yang, S. H., Shin, W. H., Lee, J. W., Kim, S. Y., Woo, S. I. & Kang, J. K. (2006) Interaction of a transition metal atom with intrinsic defects in single-walled carbon nanotubes. *J. Phys. Chem. B, 110*, 13941-13946.

Yao, Y. (2010) Hydrogen storage using carbon nanotubes *in* Carbon Nanotubes, InTech, Chapter 28, pp. 543-562.

Yeung, C. S., Liu, L. V. & Wang, Y. A. (2007) Novel nanotube-coordinated platinum complexes. *J. Comput. Theor. Nanosci., 4*, 1108-1119.

Yeung, C. S., Liu, L. V. & Wang, Y. A. (2008) Adsorption of small gas molecules onto Pt-doped single-walled carbon nanotubes. *J. Phys. Chem. C, 112*, 7401-7411.

Yeung, C. S. & Wang, Y. A. (2011) Lewis acidity of Pt-doped buckybowls, fullerenes, and single-walled nanotubes. *J. Phys. Chem. C, 115*, 7153-7163.

Zhuang, H. L., Zheng, G. P. & Soh, A. K. (2008) Interactions between transition metals and defective carbon nanotubes. *Comput. Mater. Sci., 43*, 823-828.

Part 2

CNTs in Biological and Medical Applications

6

Carbon Nanotubes in Biomedicine and Biosensing

Yingyue Zhu, Libing Wang and Chuanlai Xu
Jiangnan University,
China

1. Introduction

Since carbon nanotubes (CNTs) were discovered by Iijima in 1991(Iijima, 1991), they have become the subject of many studies because of their unique electrical, optical, thermal, and mechanical properties(Ouyang, Huang, & Lieber, 2002; Thostenson, Ren, & Chou, 2001; Troiani, Miki-Yoshida, Camacho-Bragado, Marques, Rubio, Ascencio, et al., 2003; Wan, Dong, & Xing, 1998). Carbon nanotubes (CNTs) can be visualized as a sheet of carbon atoms rolled up into a tube with a diameter of around tens of nanometers. There are two main types of CNTs, Single-walled (SWCNTs) and multi-walled carbonnanotubes (MWCNTs), the latter being formed by several concentric layers of rolled graphite (Figure 1). In particular, SWCNTs are characterized by a high aspect ratio. Moreover, Their versatile physicochemical features enable the covalent and noncovalent introduction of several biomedicine and biosensing application relevant entities. Thus exploitation of their unique electrical, optical, thermal, and spectroscopic properties in a biological context is hoped to yield great advances in the therapy of disease and detection biomolecules such as DNA, antigen–antibody, cells, and other biomolecules.

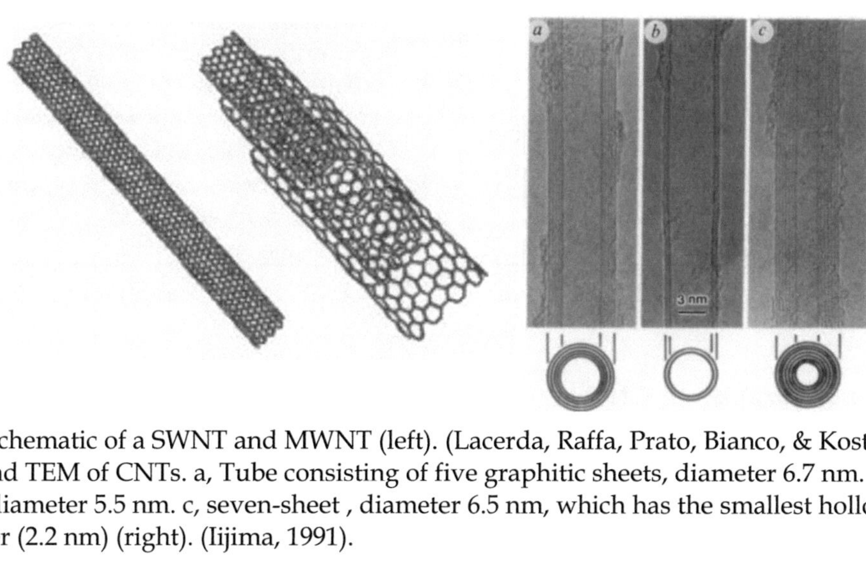

Fig. 1. Schematic of a SWNT and MWNT (left). (Lacerda, Raffa, Prato, Bianco, & Kostarelos, 2007) and TEM of CNTs. a, Tube consisting of five graphitic sheets, diameter 6.7 nm. b, two-sheet , diameter 5.5 nm. c, seven-sheet , diameter 6.5 nm, which has the smallest hollow diameter (2.2 nm) (right). (Iijima, 1991).

Part A. Carbon nanotubes in biomedicine

CNTs have been used as efficient electrochemical and optical sensors, substrates for directed cell growth, supporting materials for the adhesion of liposaccharides to mimic the cell membrane(Bianco & Prato, 2003; Lin, Taylor, Li, Fernando, Qu, Wang, et al., 2004), transfection(Kam, Jessop, Wender, & Dai, 2004; Pantarotto, Briand, Prato, & Bianco, 2004; Pantarotto, Singh, McCarthy, Erhardt, Briand, Prato, et al., 2004), and controlled drug release(Kam, Kim, & Dai, 2004; Kam, Liu, & Dai, 2005; Kam, O'Connell, Wisdom, & Dai, 2005; Lacerda, Raffa, Prato, Bianco, & Kostarelos, 2007). Some researches have shown the ability of single-walled carbon nanotubes (SWNTs) to cross cell membranes and to enhance deliver peptides, proteins, and nucleic acids into cells because of their unique structural properties(Kam & Dai, 2005; Pantarotto, Briand, Prato, & Bianco, 2004; Prato, Kostarelos, & Bianco, 2008). For this reason, carbon nanotubes could serve as an excellent vehicle to administer therapeutic agent providing effective utilization of drug and less elimination by the macrophage.

One key advantage of carbon nanotubes is their ability to translocate through plasma membranes, allowing their use for the delivery of therapeutically active molecules in a manner that resembles cell-penetrating peptides. Moreover, utilization of their unique electrical, optical, thermal, and spectroscopic properties in a biological context is hoped to yield great advances in the detection, monitoring, and therapy of disease.

Advantage	Disadvantage
• Unique mechanical properties offer in vivo stability • Extremely large aspect ratio, offers template for development of multimodal devices • Capacity to readily cross biological barriers; novel delivery systems • Unique electrical and semiconducting properties; constitute advanced components for in vivo devices • Hollow, fibrous, light structure with different flow dynamics properties; advantageous in vivo transport kinetics • Mass production - low cost; attractive for drug development	• Nonbiodegradable • Large available surface area for protein opsonization • As-produced material insoluble in most solvents; need to surface treat preferably by covalent functionalization chemistries to confer aqueous solubility (i.e. biocompatibility) • Bundling; large structures with less than optimum biological behavior • Healthy tissue tolerance and accumulation; unknown parameters that require toxicological profiling of material • Great variety of CNT types; makes standardization and toxicological evaluation cumbersome

Table 1. Advantage and Disadvantage of using CNTs for biomedical applications. (Lacerda, Raffa, Prato, Bianco, & Kostarelos, 2007).

1. Functionalization of CNTs

For biological applications, the improvement of solubility of CNTs in aqueous or organic solvents is a major task. Great efforts have devoted to search cost-effective approaches to functionalize CNTs for attachment of biomolecules as recognition elements. Generally, this procedure can be performed by noncovalent and covalent functionalization strategy.

2. Noncovalent interaction

The noncovalent approach via electrostatic interaction, Van der Waals force, or π-π stacking is a feasible immobilization method for biomolecules. Particularly, it is promising for improving the dispersion proteins of CNTs without destructing of the nanotube structure. Generally, this route can be performed by physical adsorption or entrapment.(Arnold, Guler, Hersam, & Stupp, 2005; Richard, Balavoine, Schultz, Ebbesen, & Mioskowski, 2003)

2.1 Physical adsorption

A variety of proteins can strongly bind to the CNTs exterior surface via physical adsorption. When the ends of the CNTs are open as a result of oxidation treatment, smaller proteins can be inserted into the tubular channel (~5-10 nm in diameter).

The combined treatment of strong acids and cationic polyelectrolytes is known to reduce the CNTs length and enhance the solubility under physiological. After this treatment, cationic polyelectrolytes molecules adsorb on the surface of the nanotubes by van der Waals force to produce the distribution of positive charges, which prevents the aggregation of CNTs.

2.2 Entrapment

Another method for immobilizing biomolecules on CNTs is to entrap them in biocompatible polymer hydrogen and sol-gel. Single strand DNA (ssDNA) can wrap around SWCNTs through aromatic interaction to form a soluble DNA-SWCNT complex, which has been used for construction of effective delivery for gene therapy.

Sol-gel chemistry has paved a versatile path for the immobilization of biomolecules with acceptable stability and good activity retention capacity.

3. Covalent interaction

Since the as-produced CNT contain variable amounts of impurities, such as amorphous carbon and metallic nanoparticles, the initial efforts in their purification focused on the selective oxidation of the impurities with respect to the less reactive CNT. The combined treatment of strong acids and sonication is known to purify the CNTs and generate anionic groups (mainly carboxylate) along the sidewalls and ends of the nanotubes (see figure 2). Also, dangling bonds can react similarly, generating other functions at the sidewalls.(Bahr, Yang, Kosynkin, Bronikowski, Smalley, & Tour, 2001; Williams, Veenhuizen, de la Torre, Eritja, & Dekker, 2002)

4. CNTs for biomedical applications

4.1 CNTs for protein delivery

Various low molecular weight proteins can adsorb spontaneously on the sidewalls of acid-oxidized single-walled carbon nanotubes(Kam & Dai, 2005). The proteins are found to be readily transported inside mammalian cells with nanotubes acting as the transporter via the endocytosis pathway. This research was reported by Dai group. The results shown streptavidin (SA) and cytochrome c (Cyt-c) could easily transport into the cytoplasm of cells by the CNTs and take effect of their physiological action in the cell. Carbon nanotubes could become new class of protein transporters for various in vitro and in vivo delivery applications.

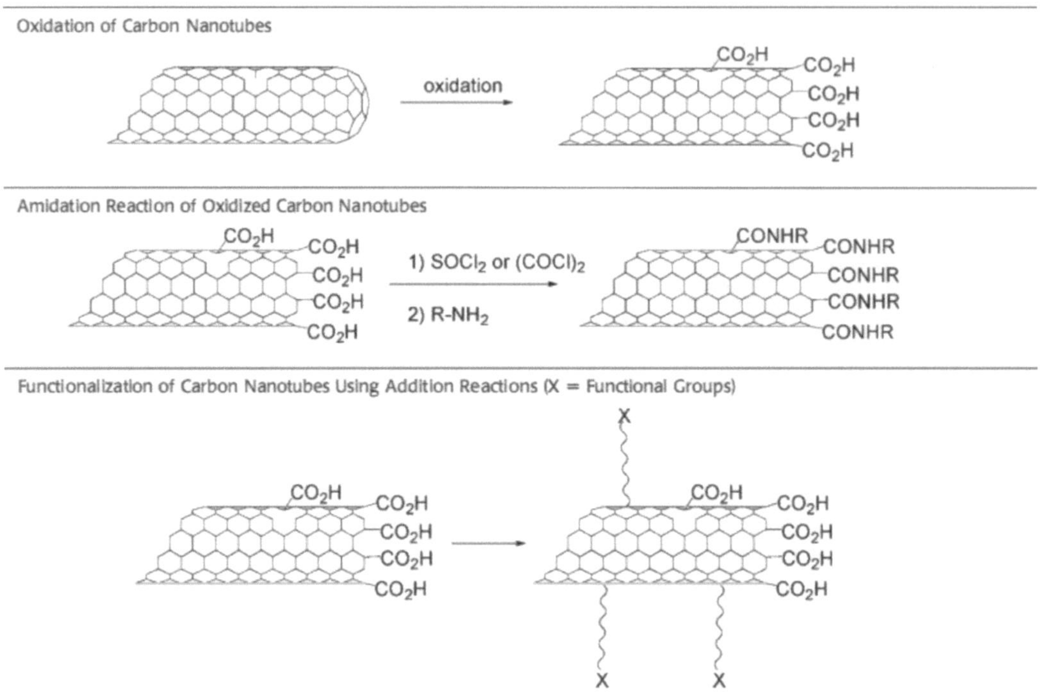

Fig. 2. Oxidation and Functionalization of CarbonNanotubes (Prato, Kostarelos, & Bianco, 2008).

4.2 CNTs for gene delivery

One of the most promising concepts to correct genetic defects or exogenously alter the cellular genetic makeup is gene therapy. Some challenges have existed in gene therapy. Primary concerns are the stability of molecules, the amount of intracellular uptake, their susceptibility to enzyme degradation, and the high impermeability of cell membranes to foreign substances. To overcome this problem, the CNTs are used as vector able to associate with DNA, RNA, or another type of nucleic acid by self-assembly and assist its intracellular translocation. These systems offer several advantages, including easy upscaling, flexibility in terms of the size of nucleic acid to be delivered, and reduced immunogenicity compared with viruses.

The Kostas group reported CNT-mediated gene delivery and expression leading to the production of marker proteins encoded in double-stranded pDNA(Y. Liu, Wu, Zhang, Jiang, He, Chung, et al., 2005; Pantarotto, et al., 2004). The delivery of pDNA and expression of β-galactosidase (marker gene) in Chinese hamster ovary (CHO) cells is five to ten times higher than naked pDNA alone.

The concept of gene delivery systems based on CNTs has also been reported by Liu group(Y. Liu, et al., 2005). They report a noncovalent association of pDNA with PEI–CNTs by electrostatic interaction. They have tested CNT–PEI:pDNA complexes at different charge ratios in different cell lines. The levels of expression of luciferase (marker gene) are much higher for the complexes incorporating CNTs than pDNA alone and about three times higher than PEI alone.

4.3 CNTs for chemical delivery

Recently, Dai group reported that using supramolecular π–π stacking to load a cancer chemotherapy agent doxorubicin (DOX) onto branched polyethylene glycol (PEG) functionalized SWNTs for in vivo drug delivery applications(Z. Liu, Fan, Rakhra, Sherlock, Goodwin, Chen, et al., 2009). It has been found that the surface of PEGylated SWNTs could be efficiently loaded with DOX by supramolecularπ–π stacking. These methods offer several advantages for cancer therapy, including enhanced therapeutic efficacy and a marked reduction in toxicity compared with free DOX.

4.4 CNTs for cancer therapy

More interestingly, CNTs can be used as platforms for multiple derivatization by loading their surface with therapeutic agents (treatment), fluorescent, magnetic, or radionuclide probes (tracking), and active recognition moieties (targeting).

We present a strategy for using SWNTs as intracellular vectors for delivery of ASODNs modified with gold nanoparticles (figure 3). This strategy allows intracellular delivery and localization to enhance the therapeutic efficiency of the ASODNs by the conjugations of SWNTs and GNPs compared with the naked ASODNs in this experiment.

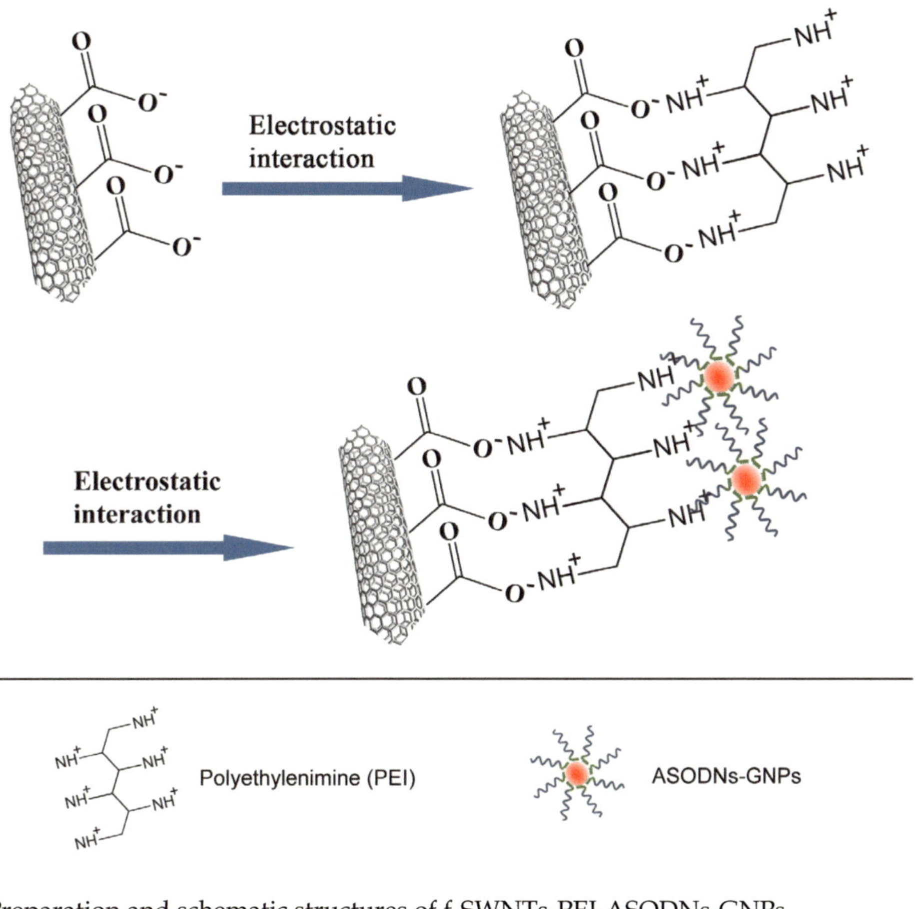

Fig. 3. Preparation and schematic structures of f-SWNTs-PEI-ASODNs-GNPs.

Recently, Jia et al, have explored a novel double functionalization of a carbon nanotube delivery system containing antisense oligodeoxynucleotides (ASODNs) as a therapeutic gene and CdTe quantum dots as fluorescent labeling probes via electrostatically layer-by-layer assembling(N. Jia, Lian, Shen, Wang, Li, & Yang, 2007). With this novel functionalization, it has demonstrated efficient intracellular transporting, strong cell nucleus localization and high delivery efficiency of ASODNs by the PEI -MWNTs carriers. Furthermore, the ASODNs bound to PEI-MWNTs show their effective anticancer activity.

Another strategy to achieve this is used CNTs covalently bound to Pt (IV) to deliver a lethal dose of an anticancer drug and to a noncovalently bound (via a lipid coating of the CNTs) fluorescein to track the system(figure 4)(Feazell, Nakayama-Ratchford, Dai, & Lippard, 2007). Here the toxic effect of the anticancer drug is dependent upon its release and reduction inside the cell, only possible at lower pH environments such as endocytic vesicles, which was exemplified using a testicular carcinoma cell line NTera-2.

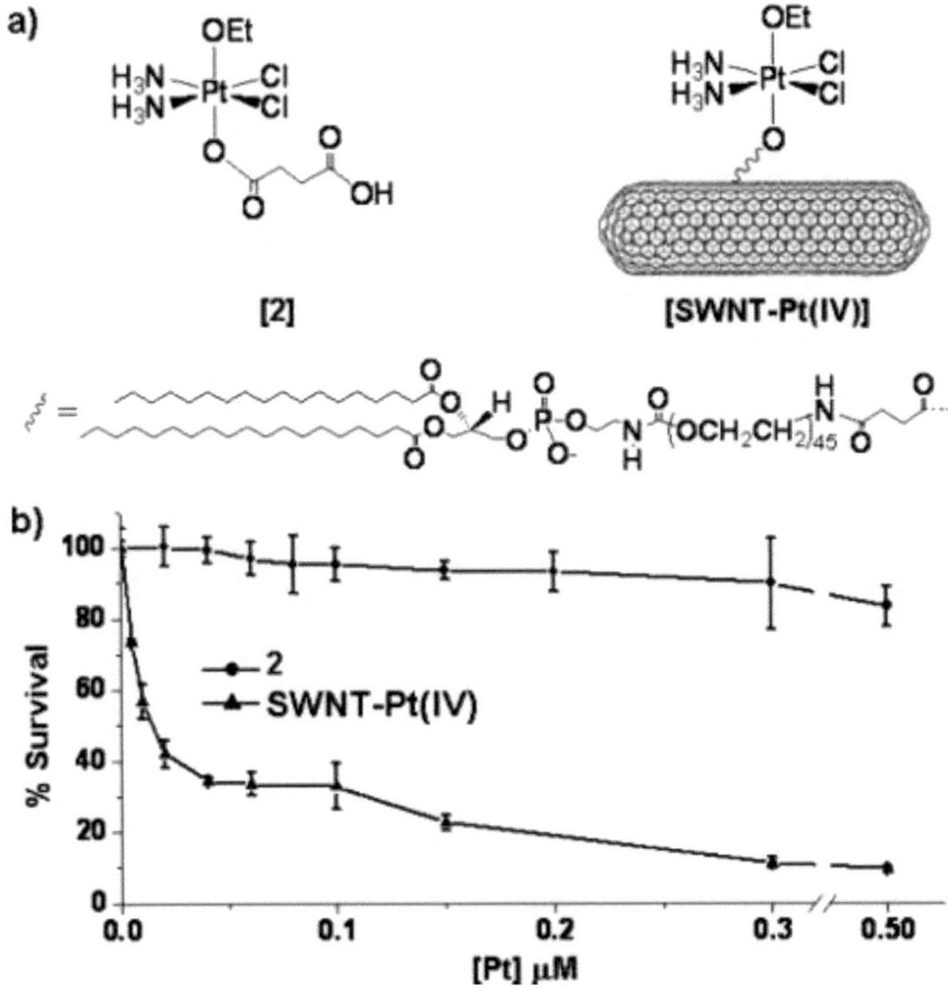

Fig. 4. (a) Preparation and schematic structures of Longboat Delivery Systems. (b) Cytotoxicity of free Pt(IV) and SWNT-tethered Pt(IV) inNTera-2cells.

4.5 CNTs for HIV/AIDS therapy

Recently, Liu et al.(Z. Liu, Winters, Holodniy, & Dai, 2007) have shown the delivery of siRNA molecules conjugated to CNT to human T cells and primary cells. The results Show that nanotubes are capable of siRNA delivery to afford efficient RNAi of CXCR4 and CD4 receptors on human T cells and peripheral blood mononuclear cells (PBMCs).The siRNA sequences used in these studies are able to silence the expression of the cell-surface receptors CD4 and coreceptors CXCR4 necessary for HIV entry and infection of T cells. This work demonstrates that siRNA linked through cleavable disulfide bonds to lipid molecules coating CNTs can be efficiently delivered, leading to knockdown (about 60%) of the CD4 and CXCR4 expression. Furthermore, the siRNA-S-S-lipid coated CNT conjugates greatly improve the silencing in T cells compared with Lipofectamine 2000 and other liposome-based transfection agents. Even though preliminary at this stage, these results indicate the potential use of CNTs for the treatment of HIV.

Part B. Nanotubes in biosensing

Carbon nanotubes (CNTs) have recently emerged as novel electronic and optical biosensing materials for the detection of biomolecules such as DNA, antigen–antibody, cells, and other biomolecules.(W. Cheng, L. Ding, S. J. Ding, Y. B. Yin, & H. X. Ju, 2009; Drouvalakis, Bangsaruntip, Hueber, Kozar, Utz, & Dai, 2008; Hu, Huang, Li, Ling, Liu, Fei, et al., 2008) Among widespread nanoscale building blocks, such as organic or inorganic nanowires and nanodots, CNTs are considered as one of the most versatile because of their superior mechanical and electrical properties and geometrical perfection. DNA analysis plays an ever-increasing role in a number of areas related to human health including diagnosis of infectious diseases, genetic mutations, drug discovery, food security, and warning against biowarfare agents. etc. And thus make electrical DNA hybridization biosensors has attracted considerable research efforts due to their high sensitivity, inherent simplicity and miniaturization, and low cost and power requirements.

1. Optical DNA sensors

Alternatively, an effective sensing platform has been presented via the noncovalent assembly of SWCNTs and dye-labeled ssDNA.(Yang, Tang, Yan, Kang, Kim, Zhu, et al., 2008) Figure 1(a) shows the signaling scheme. When the SWCNTs are added to the dye-labeled ssDNA solution, the ssDNA/SWCNT hybrid structure can be formed, in which the dye molecule is in close proximity to the nanotube, thus quenching the fluorescence of dye molecule.(Nakayama-Ratchford, Bangsaruntip, Sun, Welsher, & Dai, 2007)
The dye-labeled ssDNA can restore the fluorescence signal to an initial state in the presence of the target. Figure 1(b) illustrates no significant variation in the fluorescence intensity of fluorescein derivative (FAM)-labeled oligonucleotides (P1) in the absence of CNTs. In the presence of SWCNT, a dramatic increase of the fluorescence intensity at 528 nm can be observed in the DNA concentration range of 5.0–600 nM, suggesting that the SWCNT/DNA assembly approach is effective in biosensing target DNA (Yang, et al., 2008)Yang, et al., 2008).
Furthermore, a visual sensor has been designed to detect DNA hybridization by measuring the light scattering signal with DNA modified MWCNT as recognition element as shown in figure 2. (Hu, et al., 2008) This sensor can be reused for at least 17 times and is stable for more than 6 months.

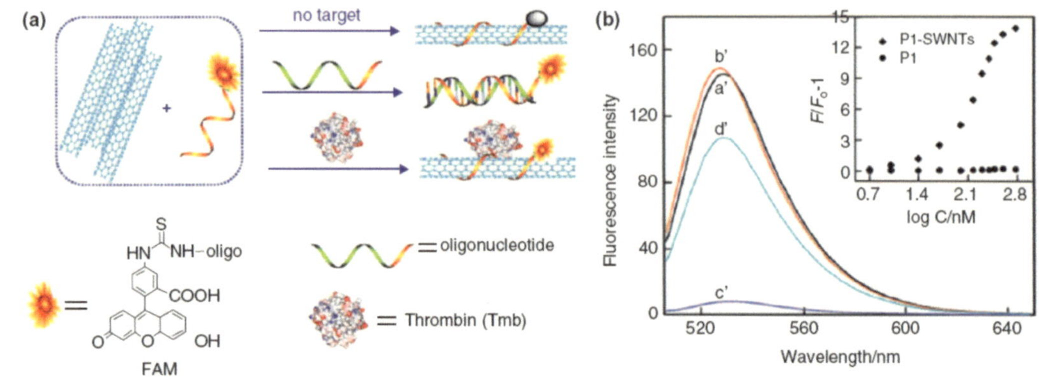

Fig. 1. (a) Scheme for signaling biomolecular interaction by the assembly of single-walled carbon nanotubes (SWCNT) and dye-labeled single strand DNA. (b) Fluorescence emission spectra of 50 nM FAM-labeled oligonucleotides (P1) in (a) phosphate buffer (PBS), (b)300 nM perfect cDNA (T1), (c) SWCNT, and (d_) SWCNT + 300 nM T1. Inset: fluorescence intensity ratio of P1 and P1–SWCNT with *F/F0* plotted against the logarithm of the concentration of T1. Excitation was at 480 nm, and emission was monitored at 528 nm. (Yang, et al., 2008).

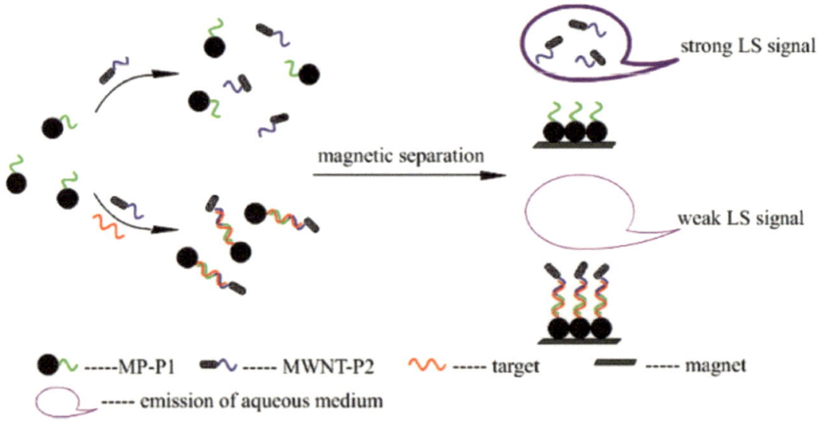

Fig. 2. The DNA Detection with Magnetic Particles-Carbon Nanotubes Coupled Sandwich Probe.

2. Antigen–antibody immunoreactions

There are two different types of detection patterns for CNT-based immunosensors: label-free immunosensors and immunosensors that employ labels and mediators. The label-free immunosensor shows a convenient fabricating and detection procedure. Several label-free peptide-coated CNTs based immunosensors has been proposed for the direct assay of human serum sample using square wave stripping voltammetry (Ly & Cho, 2009), quartz crystal microbalance measurements, and differential pulse voltammetry (DPV) (Okuno, Maehashi, Kerman, Takamura, Matsumoto, & Tamiya, 2007). Based on CNT-FET, a label-free protein biosensor was also prepared for monitoring of a prostate cancer marker (Kim, Lee, Lee, Hong, & Sim, 2009). As one of the most popular tracer labels, enzymes, including ALP (Aziz, Park, Jon, & Yang, 2007), HRP (Wang, Liu, & Jan, 2004), and GOD(Lai, Yan, & Ju, 2009) have been immobilized on CNTs for enhancing the enzymatic signal. Typically, a novel immunosensor array was constructed by coating layer-by-layer colloidal Prussian blue (PB), gold nanoparticles (AuNPs), and capturing antibodies on screen-printed carbon electrodes (Figure 3) and coupling with a new tracer nanoparticle probe labeled antibody (Ab2) that was prepared by one-pot assembly of GOD and the antibodies on AuNPs attached CNTs (Lai, Yan, & Ju, 2009).

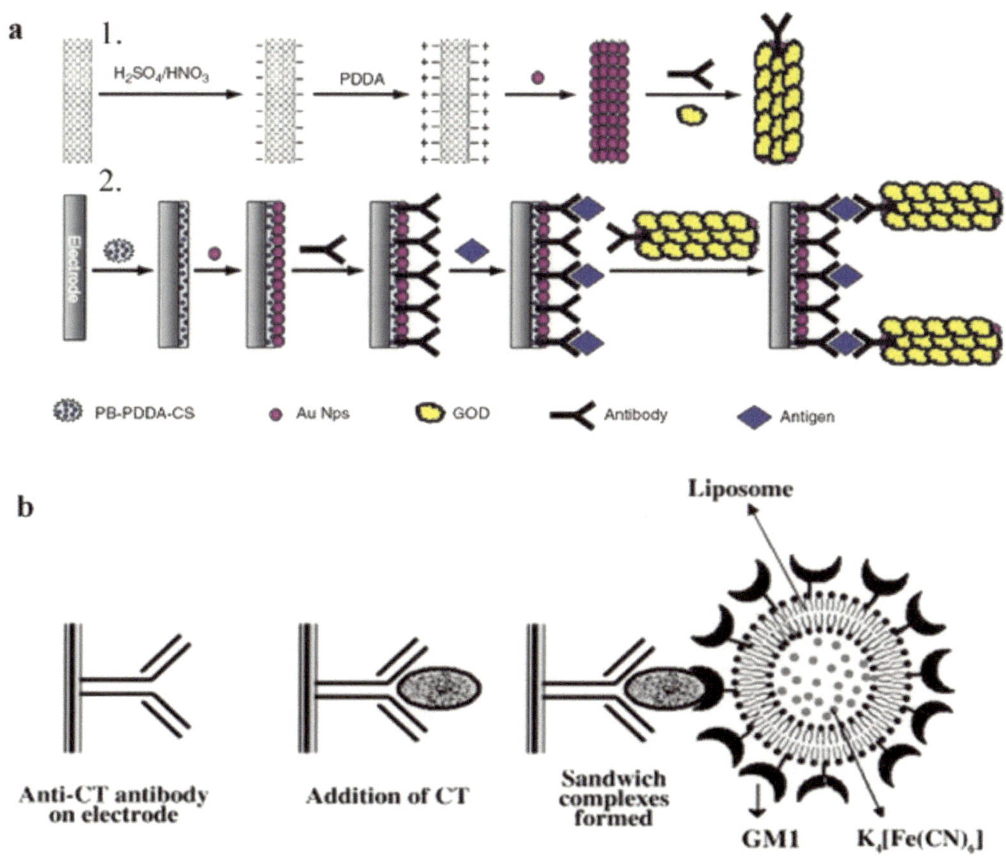

Fig. 3. Schematic representation of (a1) preparation procedure of glucose oxidase (GOD)–Au Nps/carbon nanotubes (CNTs)-Ab2 tracer and (a2) preparation of immunosensors and sandwich-type electrochemical immunoassay (Lai, Yan, & Ju, 2009). c. Schematic outlines of Immunosensor for Cholera Toxin.

The immobilized PB could not only eliminate the electrochemical cross talk but also avoid the interference of dissolved oxygen. Using carcinoembryonic antigen and α-fetoprotein as model analytes, the simultaneous multiplexed immunoassay method showed the linear ranges of three orders of magnitude with the detection limits down to 1.4 and 2.2 pg mL−1, respectively (Lai, Yan, & Ju, 2009). This assay approach showed a great potential in clinical applications and detection of low-abundant proteins. In addition, a sensitive method for the detection of cholera toxin (CT) using an electrochemical immunosensor with liposomic magnification has been proposed as shown in Figure 3c (Viswanathan, Wu, Huang, & Ho, 2006). The sensing interface consists of monoclonal antibody against the B subunit of CT that is linked to poly (3, 4-ethylenedioxythiophene) coated on Nafion-supported MWCNT caste film on a glassy carbon electrode. The sandwich assay provides the amplification route for the detection of CT ranging from 10^{-14} to 10^{-7} g mL^{-1} with a detection limit of 10^{-15} g mL^{-1}. In the same group, a disposable electrochemical immunosensor for carcinoembryonic antigen using ferrocene liposome and MWCNT modified screen-printed carbon electrode was also developed (Viswanathan, Rani, Vijay Anand, & Ho, 2009).

3. Sensing of cells

To achieve biocompatible interactions between CNTs and living cells, a strategy to functionalize CNTs with biomolecules such as peptide as shown in Figure 4 (W. Cheng, L. Ding, S. Ding, Y. Yin, & H. Ju, 2009) (Cheng, Ding, Lei, Ding, & Ju, 2008)and monosaccharides was presented (Sudibya, Ma, Dong, Ng, Li, Liu, et al., 2009). A novel

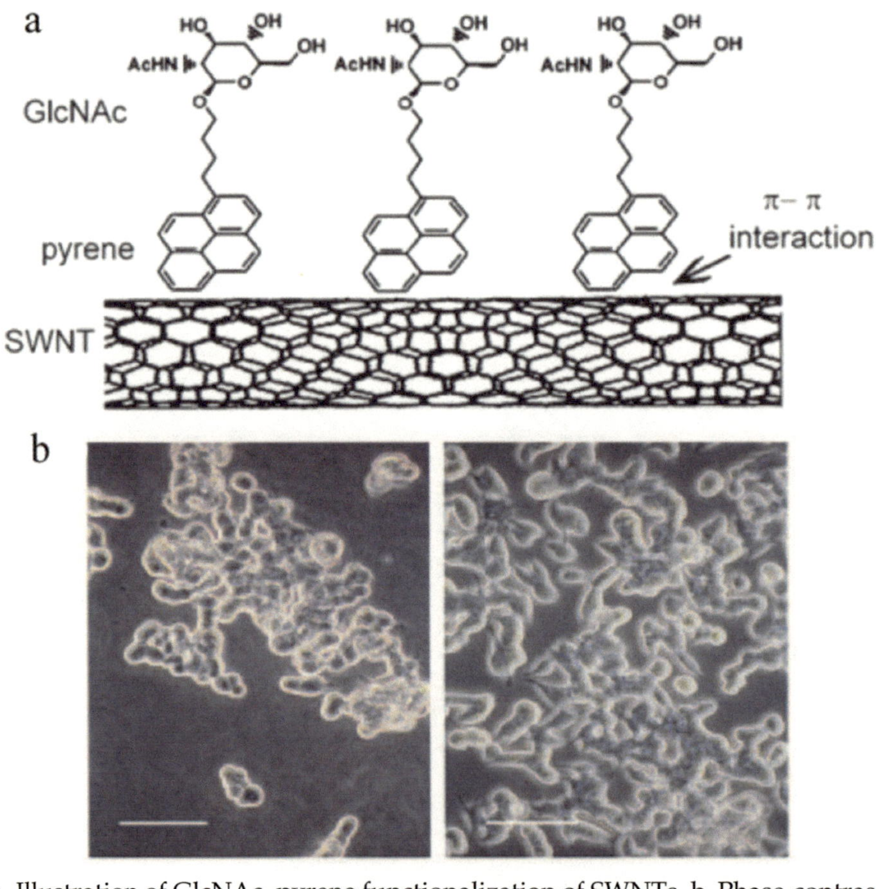

Fig. 4. a. Illustration of GlcNAc–pyrene functionalization of SWNTs. b. Phase-contrast images of PC12 cells cultured on a bare SWNT-net (left) and GlcNAc–SWNTs (right).

electrochemical cytosensing strategy was designed based on the specific recognition of integrin receptors on cell surface to arginine-glycine-aspartic acid-serine (RGDS)-functionalized SWCNT (Figure 4) (Cheng, Ding, Lei, Ding, & Ju, 2008). The conjugated RGDS showed a predominant ability to capture cells on the electrode surface by the specific combination of RGD domains with integrin receptors. On the basis of the dual signal amplification of SWCNT and enzymatic catalysis, the cytosensor could respond down to 620 cells mL^{-1} of BGC cells with a linear calibration range from 1.0×10^3 to 1.0×10^7 cells mL^{-1}. Furthermore, the mannosyl group on a single living intact BGC cell was evaluated to be 5.3×10^7 molecules of mannose. The same group further prepared a cytosensor array for multiplex evaluation of both the glycan expression profile on an intact cell surface and the dynamic changes in the glycome during drug treatment.(W. Cheng, L. Ding, S. Ding, Y. Yin, & H. Ju, 2009) The further functionalization of the metal-cluster-decoration CNTs with Tween 20 could suppress non-specific binding and enabled label-free and selective detection of *A. anophagefferens (Ishikawa, Stauffer, Caron, & Zhou, 2009).*

4. Detection of other biomolecules

4.1 NADH

The electrochemical oxidation of NADH at the electrode surface has received considerable interest due to the need to develop amperometric biosensors for substrates of NAD^+ dependent dehydrogenases. Dihydronicotinamide adenine dinucleotide (NADH) and its oxidized form, nicotinamide adenine dinucleotide (NAD^+), are the key central charge carriers in living cells. However, the oxidation of NADH at a conventional solid electrode surface is highly irreversible with considerable overpotentials, which limits the selectivity of the determination in a real sample. CNTs have been devoted to decreasing the high overpotential for NADH oxidation on carbon paste electrodes (Blackburn,R.S. & Burkinshaw,S.M 2007) and microelectrodes (Wang, Deo, Poulin, & Mangey, 2003). By integrating the hydrophilic ion-conducting matrix of CHITn with electron mediator toluidine blue O and CNTs, the produced NADH sensor shows very low oxidation overpotential and good analytical performance (Zhang & Gorski, 2005).

4.2 Glucose

The detection of glucose in blood is one of the most frequent performances for human healthy, since some diseases are related to the blood glucose concentration. However, the direct electron transfer for oxidation of $FADH_2$ or reduction of FAD(Shan, Yang, Song, Han, Ivaska, & Niu, 2009) is hard to realize at conventional electrodes, because the FAD is deeply seated in a cavity and not easily accessible for conduction of electrons from the electrode surface. Thus, many CNTsbased nanohybrids, such as MWCNT/AuNPs/ionic liquid (F. Jia, Shan, Li, & Niu, 2008), SWCNT/GOD/Nafion (Lyons & Keeley, 2008), polyaniline (PANI)-coated Fe_3O_4 nanoparticle/MWCNT (Zhun Liu, Wang, Xie, & Chen, 2008), and palladium/SWCNT (Meng, Jin, Yang, Lu, Zhang, & Cai, 2009), have been explored to immobilize GOD for glucose biosensing. More interestingly, Willner's group demonstrated that aligned reconstituted GOD on the edge of SWCNT as conductive nanoneedles can be linked to an electrode surface for fast glucose response (G. Liu & Lin, 2006).

4.3 Organophosphate pesticides

The rapid detection of these toxic agents in the environment and public places has become increasingly important for homeland security and health protection. The flow injection amperometric biosensor for OPs has been developed by assembling AChE on CNTs-modified GCE. Under optimal conditions, the biosensor has been used to measure paraoxon as low as 0.4 pM with a 6-min inhibition time (G. Liu & Lin, 2006).

4.4 H_2O_2

H_2O_2 is a product of the enzymatic reactions between most oxidases and their substrates. This detection is very interesting for the development of biosensors for oxidase substrates. The earlier work on the electrocatalytic action of CNTs toward H_2O_2 was reported at an apparently decreased overvoltage using the CNTs/Nafion-coated electrode. With the introduction of MWCNT, the polyaniline-PB/MWCNT hybrid system showed the synergy between the PANI-PB and MWCNT, which amplified the sensitivity greatly (Zou, Sun, & Xu, 2007).

5. Near-IR fluorescent based CNTs biosensor

Generally, the change modes of SWCNT NIR can be modulated to uniquely fingerprint agents by either the emission band intensity or wavelength. CNTs have been found to be useful optical materials with high photostability and efficiency for sensing applications because of their NIR fluorescence properties from 900 to 1600 nm. Other than optical detection, SWCNTs as sensing elements have a particular advantage due to the fact that all atoms are surface atoms causing the nanotube to be especially sensitive to surface adsorption events (Strano & Jin, 2008) (Barone, Parker, & Strano, 2005).

5.1 Sensing with change of emission intensity

Quenching of SWCNT fluorescence by means of oxidative charge transfer reactions with small redox-active organic dye molecules has been demonstrated by suspending in SDS solution and biotin–avidin test system. The NIR optical properties of SWCNT have attracted particular attention for nanobiosensors based on the redox chemistry. At the most sensitive band of 1270 nm, the detection limit for H_2O_2 is found to be 8.8, 0.86, and 0.28 ppm by three different methods based on the concentrationdependent rate constant, pectral intensity change, and signal-to-noise ratio (Tu, Pehrsson, & Zhao, 2007). Another NIR optical protein assay based on aptamers wrapped on the sidewall of SWCNT was designed. After the target protein (thrombin) was added into the SWCNT–aptamer solution, the NIR absorption at 1142 nm decreased linearly upon the increasing concentration from 0.2 to 6.3 nM. This signal provides a label-free and separation-free optical method for aptamer-based protein assays (H. Chen, Yu, Jiang, Zhang, Liu, & Kong, 2009).

5.2 Sensing with shift of emission wavelength

The shift of emission wavelength has also been a usefulway to make sensing in addition to emission intensity-based sensing, When cations adsorb onto the negatively charged backbone of DNA, DNA oligonucleotides transform from the native, righthanded B form to the left-handed Z form, which modulates the dielectric environment of SWCNT and decreases their NIR emission energy up to 15 meV. The change of the emission wavelength results in an effective ion sensor, especially formercuric ions. These NIR ion sensors can operate in strongly scattering or absorbing mediator to detect mercuric ions in whole blood, black ink, and living mammalian cells and tissues (D. A. Heller, Jeng, Yeung, Martinez, Moll, Gastala, et al., 2006).

5.3 Single-molecule detection

Nanoscale sensing elements offer promise for single molecule detection through NIR fluorescence in physically or biologically constrained environments. A single-molecule detection of H_2O_2 has been demonstrated by stepwise NIR photoluminescence quenching of surface-tethered DNA-SWCNT complexes (Figure 5(a)).

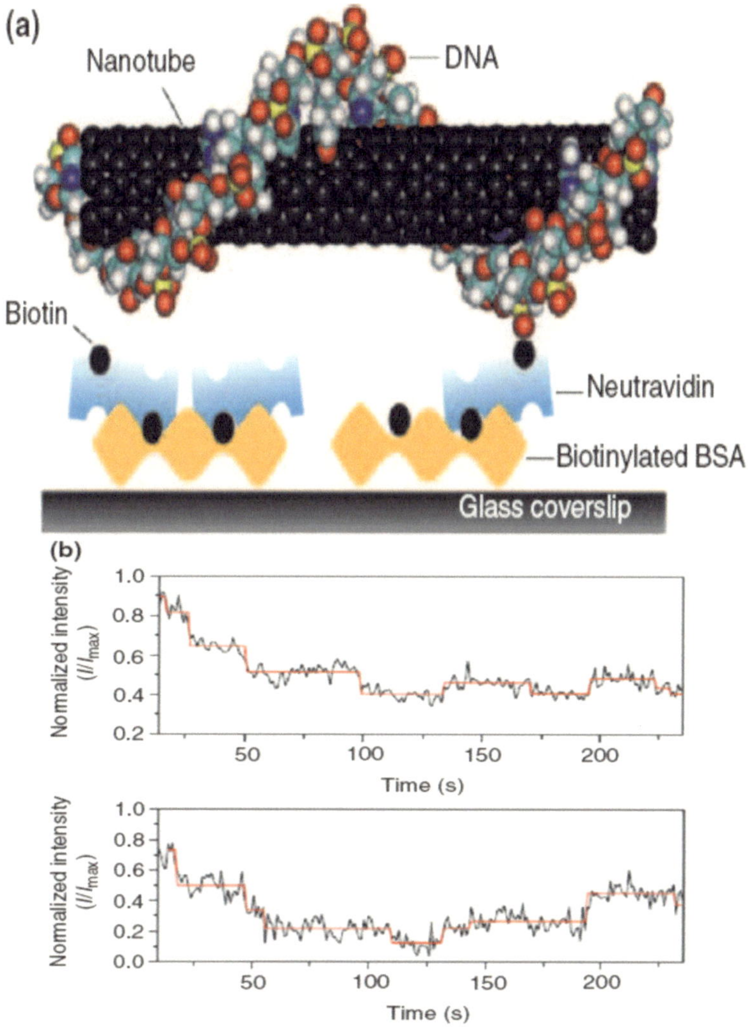

Fig. 5. Single-molecule H_2O_2 detection: (a) Schematic of biotinylated DNA–single-walled carbon nanotubes (SWCNT) binding to a glass surface with bovine serum albumin–biotin and Neutravidin. (b) Fitted traces from a movie showing single-step SWCNT emission quenching upon perfusion of H_2O_2. ((D. A. Heller, Jin, Martinez, Patel, Miller, Yeung, et al., 2009)).

The time trace of SWCNT quenching was obtained by measuring the intensity of four-pixel spots in movies recorded at one frame per second (Figure 5(b)), resulting in multiple traces that exhibited single-step attenuation upon perfusion of H_2O_2. These measurements demonstrated singlemolecule detection of H_2O_2 and provided promise for new classes of biosensors with the single-molecular level of sensitivity.

6. SWCNT-based field-effect biosensor

Currently, four possible mechanisms have been proposed to account for the observed changes in the SWCNT conductance: electrostatic gating (I. Heller, Männik, Lemay, & Dekker, 2008) (Gui, Li, Zhang, Xu, Dong, Ho, et al., 2007), Schottky barrier effect (R. J. Chen, Choi, Bangsaruntip, Yenilmez, Tang, Wang, et al., 2004), change in gate coupling (Besteman, Lee, Wiertz, Heering, & Dekker, 2003), and carrier mobility change (Hecht, Ramirez, Briman, Artukovic, Chichak, Stoddart, et al., 2006), among which the electrostatic gating and Schottky barrier effect are dominant in the SWCNTbased FET biosensing device (I. Heller, Janssens, Mannik, Minot, Lemay, & Dekker, 2007). The label-free CNTs-based filed-effect sensor offers a new approach for a new generation of DNA biosensing. For example, a synthetic polymer is well adsorbed to the walls of CNTs and carries activated succinimidyl ester groups to fix the NH_2-ssDNA probes for constructing a large array of CNTs-FETs. Furthermore, a simple and generic protocol for label-free detection of DNA hybridization is demonstrated by random sequencing of 15 and 30 mer oligonucleotides. DNA hybridization on gold lectrodes, instead of on SWCNT sidewalls, is mainly responsible for the acute electrical conductance change due to the modulation of energy level alignment between SWCNT and gold contact. Aptamer is artificial oligonucleotides (DNA or RNA) that can bind to a wide variety of entities with high selectivity, specificity, and affinity, equal to or often superior to those of antibodies. The first SWCNT-FET-based biosensor comprising aptamer was proposed by Lee's group (So, Won, Kim, Kim, Ryu, Na, et al., 2005). Briefly, aptamer immobilization was performed by modifying the side wall of the CNTs with carbodiimidazole-activated Tween 20 through hydrophobic interaction, and covalently attaching the 3-amine group of the thrombin aptamer (Figure 6(a)). The conductance dropped sharply upon addition of 1.5 μmol thrombin. The sensitivity became saturated around protein concentration of 300 nM, where the linear response regime of the sensor was expected to occur within the 0–100 nM range (Figure 6(b)). The addition of elastase did not affect the conductance of the thrombin aptamer functionalized SWCNT-FET. Again, adding thrombin to the thrombin aptamer functionalized SWCNT-FET surface caused a sharp decrease in conductance (Figure 6(c)), thereby demonstrating the selectivity of the immobilized thrombin aptamers.

The aptamer modified SWCNT-FETs are another promising sensor for the development of label-free protein detection. SWCNT-FET are also promising devices for the specific recognition of proteins. The first biosensor based on an individual SWCNT was reported by Dekker's group (Besteman, Lee, Wiertz, Heering, & Dekker, 2003). GOD was attached to the sidewalls of a semiconductive CNT by a bifunctional reagent with a pyrene group. GOD-coated semiconducting SWCNTs acted as sensitive pH sensors due to the strong pH-dependent conductance of GOD immobilized SWCNT. Moreover, change of conductance of GOD coated semiconducting SWCNT upon addition of glucose indicated that an enzyme-activity sensor could be constructed at the single-molecule level of an individual SWCNT. In the presence of redox mediators such as K3Fe(CN)6/K4Fe(CN)6 and K2IrCl6/K3IrCl6, the SWCNT-FETs were shown to linearly detect the enzyme activity of the blue copper oxidase, laccase, varied over two orders of magnitude of enzyme concentration in the picomolar range (Boussaad, Diner, & Fan, 2008).

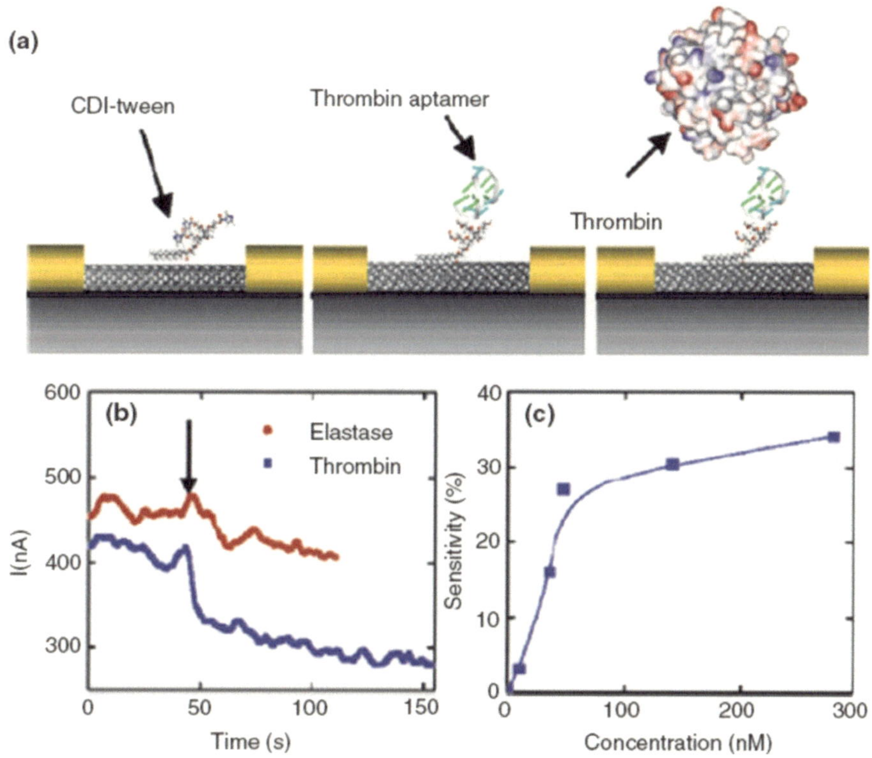

Fig. 6. (a) Binding of thrombin on an single-walled carbon nanotubes-field-effect transistor (SWCNT-FET)-based aptamer sensor. (b) The sensitivity of SWCNT-FET aptamer sensor as a function of thrombin concentration. (c) The sensitivity of SWCNT-FET aptamer sensor as a function of thrombin concentration (So, et al., 2005).

7. Electrochemical sensors

Electrochemical DNA sensors can convert the hybridization event into an electrochemical signal. DNA sensing approaches include the intrinsic electroactivity of DNA, electrochemistry of DNA-specific redox indicators, electrochemistry of enzymes, and conducting polymers. The direct electrochemical oxidation of guanine or adenine residues of ssDNA leads to an indicator-free DNA biosensor. For example, Wang's group used CNTs for dramatically amplifying alkaline phosphatase (ALP) enzyme-based bioaffinity electrical sensing of DNA with a remarkably low detection limit of around 1 fg mL^{-1} (54 aM).(Wang, Liu, & Jan, 2004) Professor Kotov and collaborator (Professor Xu) had demonstrated that CNT/cotton threads can be used to detect albumin, the key protein of blood, with high sensitivity and selectivity(Shim, Chen, Doty, Xu, & Kotov, 2008). In this method, cotton yarn has been coated with CNTs and polyelectrolytes. This method provides a fast, simple, robust, low-cost, and readily scalable process for making e-textiles, reminiscent of layer-by-layer assembly processes used before. The resulting CNT/cotton yarns showed high electrical conductivities as well as some functionality due to biological modification of inter-nanotube tunneling junctions. When the CNT/cotton yarn incorporated anti-albumin, it became an e-textile biosensor that quantitatively and selectively detected albumin, the essential protein in blood. The same sensing approach can easily be extended to many other proteins and biomolecules. Single-walled and multi-walled carbon nanotubes (SWNTs, MWNTs) were dispersed in dilute Nafion™-ethanol or poly(sodium 4-styrene sulfonate) (PSS)-water solutions. A general commodity cotton thread (1.5 mm in diameter) was dipped in the prepared CNT dispersions and dried (Figure 7ab). After several repetitive dips, reminiscent of the layer-by-layer assembly process, the cotton thread became conductive with a resistivity as low as 20 Ω/cm. As a demonstration of the conductivity, we easily powered an LED device connected to a battery by the prepared threads (Figure 7c).

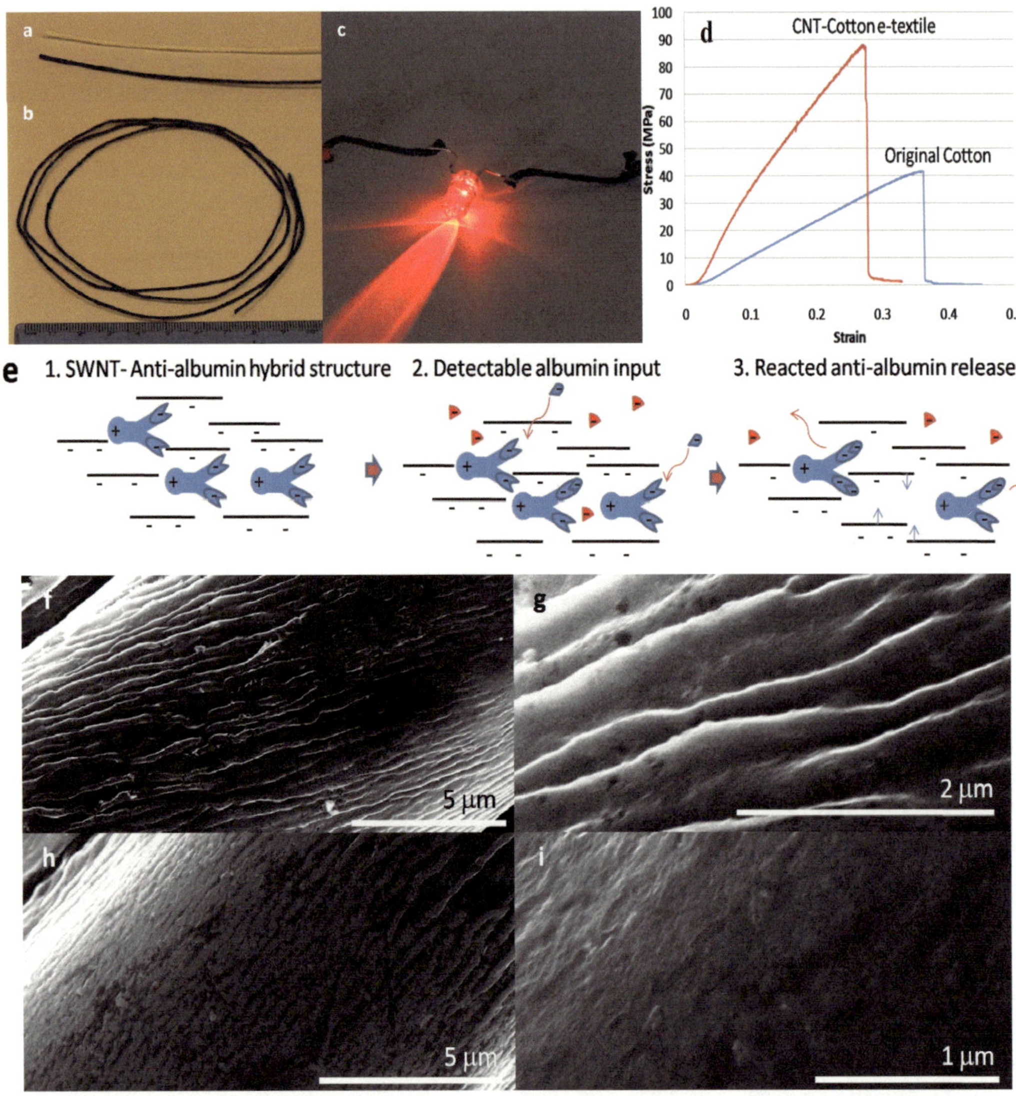

Fig. 7. Photographs of SWNT-cotton yarn. (a) Comparison of the original and surface modified yarn. (b) 1 μ m-long piece as made. (c) Demonstration of LED emission with the current passing through the yarn. (d) Stress-strain curves for the CNT-cotton yarn and the original cotton thread. (e) Suggested detection mechanism of antibody-antigen reaction. SEM images before (f,g) and after (h,i) the antibody/antigen reaction.

The incorporation of CNTs into the cotton yarn was much more efficient than their adsorption into carbon fibers, which was tried elsewhere. This could be a result of the efficient interaction of polyelectrolytes with cotton and other natural polysaccharide- and cellulose-based materials, such as paper, which is well known in industry. Additionally, the flexibility of the CNTs allowed them to conform to the surface of the cotton fibers. Both SWNTs and MWNTs stabilized in Nafion™ seamlessly cover. In comparison with other electronic textiles, fabrics, and threads, the resistivity of the yarn in Figure 7ab is two orders of magnitude lower than the resistivity of comparable CNT-dyed textiles (7.8 kΩ/cm). Furthermore, the reported resistivity of 1 cm-long yarn drawn from CNT forests is at best, if converted to the scale used here, in the range of a few kΩ/cm.
The strength of the CNT/cotton yarn is more than 2 times higher than that of the original cotton thread due to a reduction of the overall diameter, densification and stronger adhesion of the fibers to each other by the polymer material. Even though the cotton yarn became slightly harder after being coated with SWNTs, it is still very flexible and soft, both of which are important for the wearability of electronic fabric. Single exposure of the produced yarn to different solvents imitating washing did not appreciably affect the electrical properties.
The low electrical resistance of CNT/cotton yarn allows for convenient sensing applications which may not require any additional electronics or converters. It also reduces the power necessary for sensing. PSS is more hydrophilic than Nafion™, and, thus, CNT-Nafion™ is more advantageous for dry-state sensing while CNT-PSS will be more advantageous in humid conditions. For intelligent fabric demonstrations, the CNT-Nafion™ yarn was tested as a humidity sensor in a dry state while CNT-PSS yarn served as a wet-state bio-sensor platform. As the humidity was raised, the resistance increased. This is most likely a result of reversible hygroscopic swelling of both Nafion™ and cotton, which readily disrupts the electron transport between CNTs. The change in the resistance was almost instantaneous, and the signal was strong even in the very dry conditions of 20% humidity. Sensitivity to humidity changes also gives a good indication of the so-called "breathability" of the material, which is also an important parameter for smart fabrics.
Another example of an integrated, functional biosensor was demonstrated using SWNT-PSS yarn. the choose of the antigen/antibody reaction between human serum albumin (HSA) and its respective immunoglobulin (IgG) anti-HSA for the model system that can be generalized to many other relevant antigen/antibody systems of interest. PSS is known as an excellent stabilizer of proteins and can be used to form a layer-by-layer film with IgG antibodies. After the adding of anti-HSA directly to the SWNT-PSS solution and coated the cotton yarn as before, the CNT-IgG/cotton yarn was frozen and then dried under vacuum in order to minimize antibody denaturation. This cycle was repeated three times before use. For sensing experiments, two different albumin proteins were used; human serum albumin (HSA, 67 kDa) and bovine serum albumin (BSA, 66 kDa). Each experiment involved the measurements of conductivity of yarns being in contact with a 500 μl aqueous volume of water. 50 μl aliquots of bovine and human albumins at different concentration were added to this starting volume. Detection of the antigen with CNT-IgG/cotton yarn was very sensitive and selective. The presence of analyte around the CNT-IgG/cotton yarn was indicated by an increase in conductivity (Figure 8ab). The detectable concentration of HSA was as low as 119 nM (Figure 8a, ×100), producing a signal drop of 2980 Ω, which is a 2.5% change from the baseline. As a reference, the HSA concentration in our blood ranges from 446 μM to 746 μM. The presence of analyte around the CNT-IgG/cotton yarn was indicated

by an increase in conductivity. The detectable concentration of HSA was as low as 119 nM. As a reference, the HSA concentration in our blood ranges from 446 nM to 746nM. The high sensitivity obtained in these experiments is comparable or exceeds that of sensing devices based on surface modified cantilevers similar to those used in AFM. At the same time, the selectivity of the SWNT-cotton yarn sensor was also high.

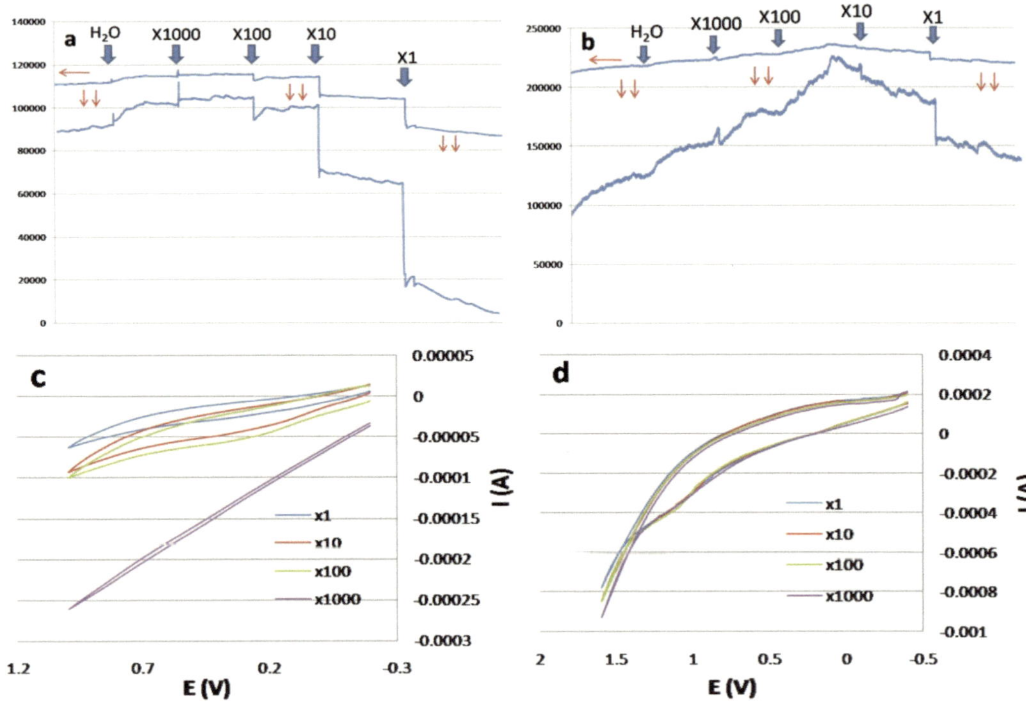

Fig. 8. Demonstration of the biosensing functionality of SWNT-modified yarn using a generic antibody-antigen reaction. (a) Effect of the concentration of HSA (11.9μM at ×1 dilution) and (b) BSA (30 μM at ×1 dilution) on conductivity of a CNT-PSS-anti-HSA coated yarn. (c, d) Cyclic voltammetry measurements of HSA (11.9 μM at ×1 dilution) on (c) a CNT-PSS-anti-HSA coated yarn and (d) a CNT-PSS yarn.

The signal transduction mechanism is believed to involve the release or significant rearrangement of IgGs from the CNT/cotton yarn. Negatively-charged HSA reacts with anti-albumin, which is followed by the process of expulsion from the SWNT-cotton matrix by the negatively charged polyelectrolyte, such as PSS. As a result, more extensive SWNT contacts are formed producing a more conductive network, resulting in the drop of the resistance. Because the contact resistance between SWNTs is affected by changes in the tunneling junction as small as a few angstroms, the removal or rearrangement of large protein macromolecules with diameters of a few nanometer results in a very substantial change in resistivity as one can see above from exceptional sensitivity obtained. SEM observations and cyclic voltammetry (CV) measurements corroborate the suggested signal transduction mechanism. SEM images show substantial restructuring upon exposure to the target protein. Before the biosensing reaction, the SWNT-PSS-anti-HSA coating displays a wavy morphology (Figure 7fg), which likely originates from the drying of frozen SWNT-

PSS-anti-HSA yarn under vacuum. After HSA detection, the wavy structures have disappeared; flat coatings with clearly visible SWNT networks can be seen. It is evident that after reaction with HSA, the SWNTs formed a more compact phase and, thus, more efficient percolation routes. These observations were further validated by CV measurements in which the anti-HSA coated smart yarn was set as a working electrode. CV data indicate a clear increase of conductivity of the smart fabric upon the less diluted antigen proteins (Figure 8c) in solution confirming the partial removal of the insulating spacing between the SWNTs. This effect is clearly absent when no antibody was incorporated between the nanotubes (Figure 8d). This finding also correlates well with the general sensing scheme outlined above. The suggested signal transduction mechanism implies one-time sensing upon complete removal of the antibodies, or cumulative sensing of the protein until it has been completely removed. From a fundamental standpoint, it would be interesting to engineer a coating with reversible sensing functionality. From a practical standpoint, however, which must consider (1) the limited life-time of antibodies and (2) the actual circumstances that can result in the appearance of blood, the multiple use of this sensor is unlikely. So, the reversible sensor to HAS might be interesting from academic point of view but its practicality is questionable.

Based on previously reported the SWNT coated cotton yarns to detect proteins in solutions, it would be fundamentally interesting as well as practically important to establish whether the similar method of analysis can be applied to the environmental needs and food safety. With this idea in mind, we have prepared and characterized the SWNT coated paper as the sensor for MC-LR toxin in the water. We attribute it to greater flexibility of SWNTs and their stronger adherence to paper originating in strong non-covalent cooperative interactions between the polyelectrolytes and cellulose. It is also probably relevant to mention that, even under high electrical current, no detachment of CNTs from the SWNT-modified paper electrode was observed.

Regular filter paper strips were dip-coated with the SWNT and the dip-dry cycles were repeated until the desirable electrical parameters of the sensor were obtained. The number of the cycles is treated as the number of SWNT layers deposited. The deposition of SWNTs can be observed by the change in color from white to black (Figure 9b). SEM images of the SWNT-coated paper indeed indicate the typical paper morphology, presence of the finely integrated nanotubes, and excellent physical integrity of the material. (Figure 9cd) As expected the conductivity of the produced material increases with increasing the SWNT contents and the number of layers of SWNT/PSS dispersion deposited (Figure 9b). The gradual increase of conductivity is quite important because in perspective the conductivity of the paper electrode needs to be within a specific range of values depending on the parameters of electrical circuit being used in order to get the best noise-to-signal ratio and the detection linearity for sensing in aqueous environments.

For sensing, we employed the standard three-electrode electrochemical station to measure changes in electrical properties of the SWNT-paper strips, which were used as work electrodes. Pt wire and the saturated Hg_2Cl_2 were used as a counter and referenced electrodes, respectively. The standard electrochemical set-up gives more accurate results than a simple clamping of the SWNT-paper material between two electrodes due to interfacial potential drops at electrode-SWNT interfaces of different nature including the Schottky barrier. Different concentrations of the MC-LR were obtained by dilution of a stock solution of 0.156 nmol/L, 0.313 nmol/L, 0.625 nmol/L, 1.25 nmol/L, 2.5 nmol/L, 5 nmol/L, 10 nmol/L, 20 nmol/L, 40 nmol/L, 100 nmol/L. Aliquots of this solution were added into

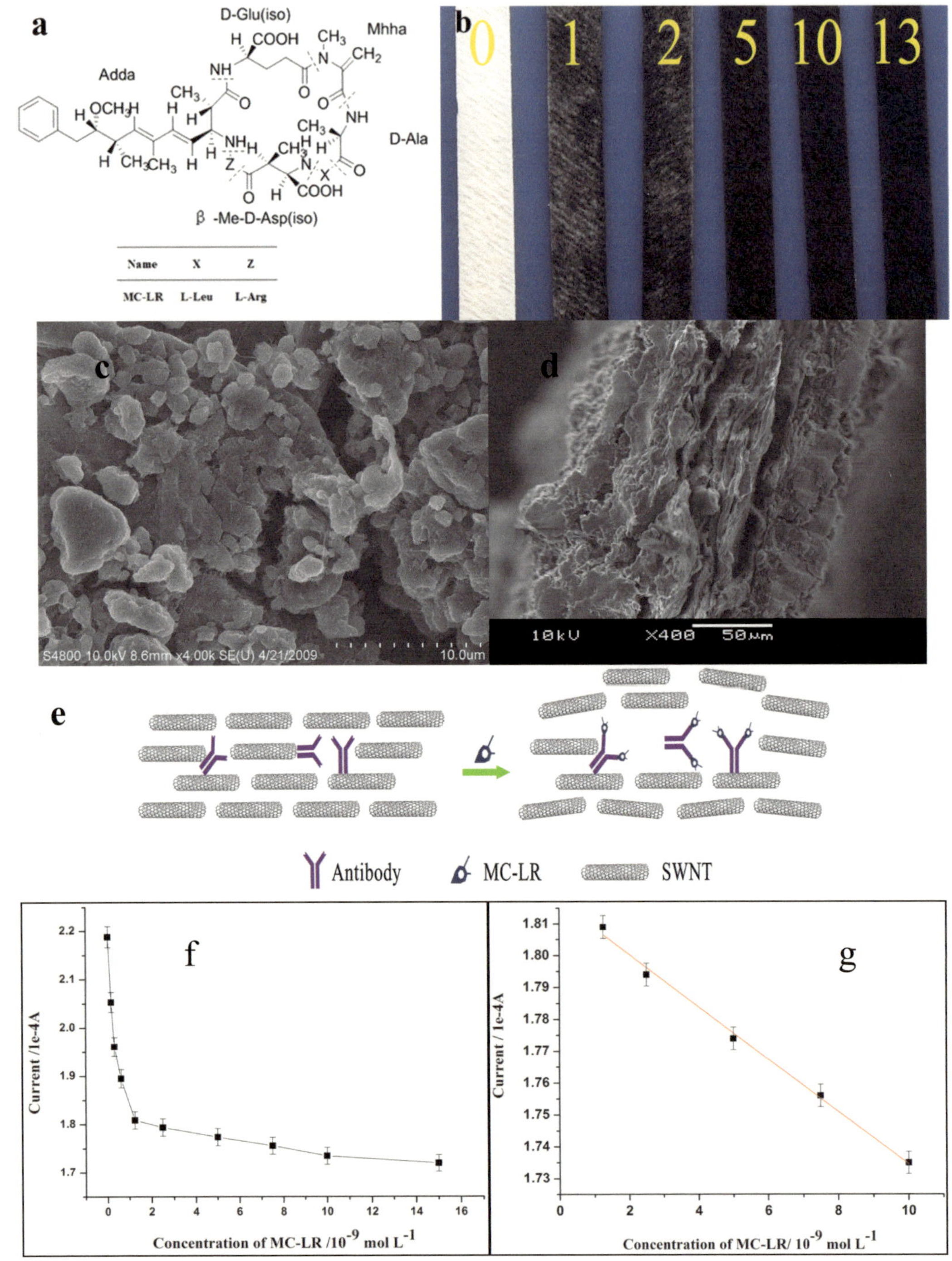

Fig. 9. (a) Chemical Structure of MC-LR. (b) Optical photographs of the SWNT-impregnated filter paper with a different number of the deposition cycles. The SEM images of (c) the surface morphology and (d) the edge of the paper electrode. (e) The sense mechanism of the developed method. The calibration curve of the determination for MC-LR (f) and (g).

the reaction cell one by one at specific time points to obtain the trend from low to high concentrations. After each addition at least *300 s* was allowed to pass to make sure that immunoreaction has enough time to proceed before the corresponding *i-t* curve was recorded. The reaction time 300 s adopted here was based on the optimization result with the best signal intensity. The current values after the start of detection where the *i-t* is transient plateaus (i.e. in the "flat" portion of the curve in the Supporting Information) were used as the analytical signal to be correlated to the concentration of MC-LR. (Figure 9e) As indicated in Figure 9e, the presence of the target analyte, i.e. MC-LR in this case reduces the current through the electrode. This corresponds to the reduction of the conductivity of SWNT-paper composite, which is quite different than the observations made for SWNT and anti-albumin Ab on cotton, where the resistivity decreased when antigen was present in solution. It was explained by the removal of Ab from the SWNT layers, resulting in shrinking of nanotubes-nanotube gaps and improvement of charge transport. In the case of electrodes described here, a different mechanism is apparently at play. Antigen penetrates through the SWNT polymer layer on the surface of paper fibers and forms the immunocomplex with Ab. This spreads apart the nanotubes, increases the nanotubes-nanotube contact resistance and hence, reduces the current passing through the material.

The SWNT-paper electrode can sense even the minor change of the MC-LR in the solution with limit of detection (LOD) of 0.6 nmol/L, which is correspond to 0.6 ng/mL and the sensitivity to the detection of MC-LR is 0.6 ng/mL (Figure 9g). It is also highly specific. The control sample of ochratoxin, which belongs to the family of micotoxints and is also a carcinogen, produced only slight variations in the current probably due to manipulations with the solution but no systematic correlation with the concentration of the control sample was observed. The calibration curve for MC-LR on SWNT-paper electrodes in the range of 0.125 to 40 nmol/L has a prevalent L-shape (Figure 9f). In the most important range of 1.25 to 10 μmol/L the calibration curve displayed excellent linearity with R^2 of 0.99426. Such behavior is indicative of the saturation phenomena when most of the antibodies in the SWNT-paper electrode formed complexes with MC-LR. According the requirements of the WHO, the content of the MC-LR in the daily water should be less than 1ng/mL, which corresponds to 1 nmol/L. The SWNT-paper based sensor could be used to monitor the quality of the drinking water for safety control. Comparing with the traditional ELISA method, the newly developed method has the similar detection range, LOD, and sensitivity with the ELISA, but in much shorter detection time. It is also much easier to operate. The time necessary for the analysis by ELISA usually exceeds 2 hours. In cases of our method, the entire analysis takes no longer than 30 min. This is much more suitable for the task of everyday monitoring of water supply. The water from Tai lake was spiked with MC-LR, the technique affords excellent recoveries of the spiked samples and acceptable relative standard deviation (n=3). Overall, excellent correlation between the MC-LR concentration values obtained by ELISA and SWNT-paper method was observed.

8. Conclusion

Different types of CNT delivery has been explored in various biomedical applications. The mechanisms of the cellular uptake of CNTs are primarily dependent on the cell type and the chemical nature and characteristics of the molecules used to functionalize the nanotube surface. Consideration of all possible mechanisms leading to CNT uptake by cells is

essential to transform one of the most promising types of novel nonmaterial into a useful and clinically relevant biotechnological and biomedical tool.
The introduction of a probe biomolecule on the surface of the CNTs as recognition element results in highly specific recognition and detection of the biomolecules from the biological samples. Meanwhile, CNT is in direct contact with the environment, which permits them to act as chemical and biological sensors in single-molecular detection of biomolecules.
To meet the urgent demand of monitoring different analytes, the carbon nanotubes based sensor may provide a very simple, rapid, sensitive, and inexpensive electrical sensor. The detection limit, sensitivity, specificity and the repeatability of the developed sesnor can be compared to that of other analytic methods while the sensor is much easier to use. It is believed that the carbon nantube based sensor could be a potential and powerful method for the monitoring of targets. Importantly, future researches on CNTs-based biosensing have attractive interest in vivo detection with less cytotoxicity, high sensitivity, and long-term stability for reliable point-of-care diagnostics under physiological conditions.

9. References

Part A.

Arnold, M. S., Guler, M. O., Hersam, M. C., & Stupp, S. I. (2005). Encapsulation of carbon nanotubes by self-assembling peptide amphiphiles. *Langmuir, 21*(10), 4705-4709.

Bahr, J. L., Yang, J. P., Kosynkin, D. V., Bronikowski, M. J., Smalley, R. E., & Tour, J. M. (2001). Functionalization of carbon nanotubes by electrochemical reduction of aryl diazonium salts: A bucky paper electrode. *Journal of the American Chemical Society, 123*(27), 6536-6542.

Bianco, A., & Prato, M. (2003). Can carbon nanotubes be considered useful tools for biological applications? *Advanced Materials, 15*(20), 1765-1768.

Feazell, R. P., Nakayama-Ratchford, N., Dai, H., & Lippard, S. J. (2007). Soluble single-walled carbon nanotubes as longboat delivery systems for Platinum(IV) anticancer drug design. *Journal of the American Chemical Society, 129*(27), 8438-+.

Iijima, S. (1991). Helical Microtubules of Graphitic Carbon. *Nature, 354*(6348), 56-58.

Jia, N., Lian, Q., Shen, H., Wang, C., Li, X., & Yang, Z. (2007). Intracellular Delivery of Quantum Dots Tagged Antisense Oligodeoxynucleotides by Functionalized Multiwalled Carbon Nanotubes. *Nano Letters, 7*(10), 2976-2980.

Kam, N. W. S., & Dai, H. J. (2005). Carbon nanotubes as intracellular protein transporters: Generality and biological functionality. *Journal of the American Chemical Society, 127*(16), 6021-6026.

Kam, N. W. S., Jessop, T. C., Wender, P. A., & Dai, H. J. (2004). Nanotube molecular transporters: Internalization of carbon nanotube-protein conjugates into mammalian cells. *Journal of the American Chemical Society, 126*(22), 6850-6851.

Kam, N. W. S., Kim, W., & Dai, H. J. (2004). Phospholipids-functionalized carbon nanotubes for chemical, biological and electronic applications. *Abstracts of Papers of the American Chemical Society, 227*, U508-U508.

Kam, N. W. S., Liu, Z., & Dai, H. J. (2005). Functionalization of carbon nanotubes via cleavable disulfide bonds for efficient intracellular delivery of siRNA and potent gene silencing. *Journal of the American Chemical Society, 127*(36), 12492-12493.

Kam, N. W. S., O'Connell, M., Wisdom, J. A., & Dai, H. J. (2005). Carbon nanotubes as multifunctional biological transporters and near-infrared agents for selective cancer

cell destruction. *Proceedings of the National Academy of Sciences of the United States of America, 102*(33), 11600-11605.
Lacerda, L., Raffa, S., Prato, M., Bianco, A., & Kostarelos, K. (2007). Cell-penetrating CNTs for delivery of therapeutics. *Nano Today, 2*(6), 38-43.
Lin, Y., Taylor, S., Li, H. P., Fernando, K. A. S., Qu, L. W., Wang, W., Gu, L. R., Zhou, B., & Sun, Y. P. (2004). Advances toward bioapplications of carbon nanotubes. *Journal of Materials Chemistry, 14*(4), 527-541.
Liu, Y., Wu, D. C., Zhang, W. D., Jiang, X., He, C. B., Chung, T. S., Goh, S. H., & Leong, K. W. (2005). Polyethylenimine-grafted multiwalled carbon nanotubes for secure noncovalent immobilization and efficient delivery of DNA. *Angewandte Chemie-International Edition, 44*(30), 4782-4785.
Liu, Z., Fan, A. C., Rakhra, K., Sherlock, S., Goodwin, A., Chen, X. Y., Yang, Q. W., Felsher, D. W., & Dai, H. J. (2009). Supramolecular Stacking of Doxorubicin on Carbon Nanotubes for In Vivo Cancer Therapy. *Angewandte Chemie-International Edition, 48*(41), 7668-7672.
Liu, Z., Winters, M., Holodniy, M., & Dai, H. J. (2007). siRNA delivery into human T cells and primary cells with carbon-nanotube transporters. *Angewandte Chemie-International Edition, 46*(12), 2023-2027.
Ouyang, M., Huang, J. L., & Lieber, C. M. (2002). One-dimensional energy dispersion of single-walled carbon nanotubes by resonant electron scattering. *Physical Review Letters, 88*(6), -.
Pantarotto, D., Briand, J. P., Prato, M., & Bianco, A. (2004). Translocation of bioactive peptides across cell membranes by carbon nanotubes. *Chemical Communications*(1), 16-17.
Pantarotto, D., Singh, R., McCarthy, D., Erhardt, M., Briand, J. P., Prato, M., Kostarelos, K., & Bianco, A. (2004). Functionalized carbon nanotubes for plasmid DNA gene delivery. *Angewandte Chemie-International Edition, 43*(39), 5242-5246.
Prato, M., Kostarelos, K., & Bianco, A. (2008). Functionalized carbon nanotubes in drug design and discovery. *Accounts of Chemical Research, 41*(1), 60-68.
Richard, C., Balavoine, F., Schultz, P., Ebbesen, T. W., & Mioskowski, C. (2003). Supramolecular self-assembly of lipid derivatives on carbon nanotubes. *Science, 300*(5620), 775-778.
Thostenson, E. T., Ren, Z. F., & Chou, T. W. (2001). Advances in the science and technology of carbon nanotubes and their composites: a review. *Composites Science and Technology, 61*(13), 1899-1912.
Troiani, H. E., Miki-Yoshida, M., Camacho-Bragado, G. A., Marques, M. A. L., Rubio, A., Ascencio, J. A., & Jose-Yacaman, M. (2003). Direct observation of the mechanical properties of single-walled carbon nanotubes and their junctions at the atomic level. *Nano Letters, 3*(6), 751-755.
Wan, X. G., Dong, J. M., & Xing, D. Y. (1998). Optical properties of carbon nanotubes. *Physical Review B, 58*(11), 6756-6759.
Williams, K. A., Veenhuizen, P. T. M., de la Torre, B. G., Eritja, R., & Dekker, C. (2002). Nanotechnology - Carbon nanotubes with DNA recognition. *Nature, 420*(6917), 761-761.

Part B.

Arnold, M. S., Guler, M. O., Hersam, M. C., & Stupp, S. I. (2005). Encapsulation of carbon nanotubes by self-assembling peptide amphiphiles. *Langmuir, 21*(10), 4705-4709.

Aziz, M. A., Park, S., Jon, S., & Yang, H. (2007). Amperometric immunosensing using an indium tin oxide electrode modified with multi-walled carbon nanotube and poly(ethylene glycol)-silane copolymer. *Chemical Communications*(25), 2610-2612.

Bahr, J. L., Yang, J. P., Kosynkin, D. V., Bronikowski, M. J., Smalley, R. E., & Tour, J. M. (2001). Functionalization of carbon nanotubes by electrochemical reduction of aryl diazonium salts: A bucky paper electrode. *Journal of the American Chemical Society, 123*(27), 6536-6542.

Barone, P. W., Parker, R. S., & Strano, M. S. (2005). In Vivo Fluorescence Detection of Glucose Using a Single-Walled Carbon Nanotube Optical Sensor: Design, Fluorophore Properties, Advantages, and Disadvantages. *Analytical Chemistry, 77*(23), 7556-7562.

Besteman, K., Lee, J.-O., Wiertz, F. G. M., Heering, H. A., & Dekker, C. (2003). Enzyme-Coated Carbon Nanotubes as Single-Molecule Biosensors. *Nano Letters, 3*(6), 727-730.

Bianco, A., & Prato, M. (2003). Can carbon nanotubes be considered useful tools for biological applications? *Advanced Materials, 15*(20), 1765-1768.

Boussaad, S., Diner, B. A., & Fan, J. (2008). Influence of Redox Molecules on the Electronic Conductance of Single-Walled Carbon Nanotube Field-Effect Transistors: Application to Chemical and Biological Sensing. *Journal of the American Chemical Society, 130*(12), 3780-3787.

Chen, H., Yu, C., Jiang, C., Zhang, S., Liu, B., & Kong, J. (2009). A novel near-infrared protein assay based on the dissolution and aggregation of aptamer-wrapped single-walled carbon nanotubes. *Chemical Communications*(33), 5006-5008.

Chen, R. J., Choi, H. C., Bangsaruntip, S., Yenilmez, E., Tang, X., Wang, Q., Chang, Y. L., & Dai, H. (2004). An investigation of the mechanisms of electronic sensing of protein adsorption on carbon nanotube devices. *Journal of the American Chemical Society, 126*(5), 1563-1568.

Cheng, W., Ding, L., Ding, S., Yin, Y., & Ju, H. (2009). A Simple Electrochemical Cytosensor Array for Dynamic Analysis of Carcinoma Cell Surface Glycans. *Angewandte Chemie International Edition, 48*(35), 6465-6468.

Cheng, W., Ding, L., Ding, S. J., Yin, Y. B., & Ju, H. X. (2009). A Simple Electrochemical Cytosensor Array for Dynamic Analysis of Carcinoma Cell Surface Glycans. *Angewandte Chemie-International Edition, 48*(35), 6465-6468.

Cheng, W., Ding, L., Lei, J., Ding, S., & Ju, H. (2008). Effective Cell Capture with Tetrapeptide-Functionalized Carbon Nanotubes and Dual Signal Amplification for Cytosensing and Evaluation of Cell Surface Carbohydrate. *Analytical Chemistry, 80*(10), 3867-3872.

Drouvalakis, K. A., Bangsaruntip, S., Hueber, W., Kozar, L. G., Utz, P. J., & Dai, H. J. (2008). Peptide-coated nanotube-based biosensor for the detection of disease-specific autoantibodies in human serum. *Biosensors & Bioelectronics, 23*(10), 1413-1421.

Feazell, R. P., Nakayama-Ratchford, N., Dai, H., & Lippard, S. J. (2007). Soluble single-walled carbon nanotubes as longboat delivery systems for Platinum(IV) anticancer drug design. *Journal of the American Chemical Society, 129*(27), 8438-+.

Gui, E. L., Li, L.-J., Zhang, K., Xu, Y., Dong, X., Ho, X., Lee, P. S., Kasim, J., Shen, Z. X., Rogers, J. A., & Mhaisalkar. (2007). DNA Sensing by Field-Effect Transistors Based

on Networks of Carbon Nanotubes. *Journal of the American Chemical Society, 129*(46), 14427-14432.
Hecht, D. S., Ramirez, R. J. A., Briman, M., Artukovic, E., Chichak, K. S., Stoddart, J. F., & Grüner, G. (2006). Bioinspired Detection of Light Using a Porphyrin-Sensitized Single-Wall Nanotube Field Effect Transistor. *Nano Letters, 6*(9), 2031-2036.
Heller, D. A., Jeng, E. S., Yeung, T.-K., Martinez, B. M., Moll, A. E., Gastala, J. B., & Strano, M. S. (2006). Optical Detection of DNA Conformational Polymorphism on Single-Walled Carbon Nanotubes. *Science, 311*(5760), 508-511.
Heller, D. A., Jin, H., Martinez, B. M., Patel, D., Miller, B. M., Yeung, T.-K., Jena, P. V., Hobartner, C., Ha, T., Silverman, S. K., & Strano, M. S. (2009). Multimodal optical sensing and analyte specificity using single-walled carbon nanotubes. *Nat Nano, 4*(2), 114-120.
Heller, I., Janssens, A. M., Mannik, J., Minot, E. D., Lemay, S. G., & Dekker, C. (2007). Identifying the Mechanism of Biosensing with Carbon Nanotube Transistors. *Nano Letters, 8*(2), 591-595.
Heller, I., Männik, J., Lemay, S. G., & Dekker, C. (2008). Optimizing the Signal-to-Noise Ratio for Biosensing with Carbon Nanotube Transistors. *Nano Letters, 9*(1), 377-382.
Hu, P., Huang, C. Z., Li, Y. F., Ling, J., Liu, Y. L., Fei, L. R., & Xie, J. P. (2008). Magnetic particle-based sandwich sensor with DNA-modified carbon nanotubes as recognition elements for detection of DNA hybridization. *Analytical Chemistry, 80*(5), 1819-1823.
Iijima, S. (1991). Helical Microtubules of Graphitic Carbon. *Nature, 354*(6348), 56-58.
Ishikawa, F. N., Stauffer, B., Caron, D. A., & Zhou, C. (2009). Rapid and label-free cell detection by metal-cluster-decorated carbon nanotube biosensors. *Biosensors and Bioelectronics, 24*(10), 2967-2972.
Jia, F., Shan, C., Li, F., & Niu, L. (2008). Carbon nanotube/gold nanoparticles/polyethylenimine-functionalized ionic liquid thin film composites for glucose biosensing. *Biosensors and Bioelectronics, 24*(4), 945-950.
Jia, N., Lian, Q., Shen, H., Wang, C., Li, X., & Yang, Z. (2007). Intracellular Delivery of Quantum Dots Tagged Antisense Oligodeoxynucleotides by Functionalized Multiwalled Carbon Nanotubes. *Nano Letters, 7*(10), 2976-2980.
Kam, N. W. S., & Dai, H. J. (2005). Carbon nanotubes as intracellular protein transporters: Generality and biological functionality. *Journal of the American Chemical Society, 127*(16), 6021-6026.
Kam, N. W. S., Jessop, T. C., Wender, P. A., & Dai, H. J. (2004). Nanotube molecular transporters: Internalization of carbon nanotube-protein conjugates into mammalian cells. *Journal of the American Chemical Society, 126*(22), 6850-6851.
Kam, N. W. S., Kim, W., & Dai, H. J. (2004). Phospholipids-functionalized carbon nanotubes for chemical, biological and electronic applications. *Abstracts of Papers of the American Chemical Society, 227*, U508-U508.
Kam, N. W. S., Liu, Z., & Dai, H. J. (2005). Functionalization of carbon nanotubes via cleavable disulfide bonds for efficient intracellular delivery of siRNA and potent gene silencing. *Journal of the American Chemical Society, 127*(36), 12492-12493.
Kam, N. W. S., O'Connell, M., Wisdom, J. A., & Dai, H. J. (2005). Carbon nanotubes as multifunctional biological transporters and near-infrared agents for selective cancer

cell destruction. *Proceedings of the National Academy of Sciences of the United States of America, 102*(33), 11600-11605.

Kim, J. P., Lee, B. Y., Lee, J., Hong, S., & Sim, S. J. (2009). Enhancement of sensitivity and specificity by surface modification of carbon nanotubes in diagnosis of prostate cancer based on carbon nanotube field effect transistors. *Biosensors and Bioelectronics, 24*(11), 3372-3378.

Lacerda, L., Raffa, S., Prato, M., Bianco, A., & Kostarelos, K. (2007). Cell-penetrating CNTs for delivery of therapeutics. *Nano Today, 2*(6), 38-43.

Lai, G., Yan, F., & Ju, H. (2009). Dual Signal Amplification of Glucose Oxidase-Functionalized Nanocomposites as a Trace Label for Ultrasensitive Simultaneous Multiplexed Electrochemical Detection of Tumor Markers. *Analytical Chemistry, 81*(23), 9730-9736.

Lin, Y., Taylor, S., Li, H. P., Fernando, K. A. S., Qu, L. W., Wang, W., Gu, L. R., Zhou, B., & Sun, Y. P. (2004). Advances toward bioapplications of carbon nanotubes. *Journal of Materials Chemistry, 14*(4), 527-541.

Liu, G., & Lin, Y. (2006). Biosensor Based on Self-Assembling Acetylcholinesterase on Carbon Nanotubes for Flow Injection/Amperometric Detection of Organophosphate Pesticides and Nerve Agents. *Analytical Chemistry, 78*(3), 835-843.

Liu, Y., Wu, D. C., Zhang, W. D., Jiang, X., He, C. B., Chung, T. S., Goh, S. H., & Leong, K. W. (2005). Polyethylenimine-grafted multiwalled carbon nanotubes for secure noncovalent immobilization and efficient delivery of DNA. *Angewandte Chemie-International Edition, 44*(30), 4782-4785.

Liu, Z., Fan, A. C., Rakhra, K., Sherlock, S., Goodwin, A., Chen, X. Y., Yang, Q. W., Felsher, D. W., & Dai, H. J. (2009). Supramolecular Stacking of Doxorubicin on Carbon Nanotubes for In Vivo Cancer Therapy. *Angewandte Chemie-International Edition, 48*(41), 7668-7672.

Liu, Z., Wang, J., Xie, D., & Chen, G. (2008). Polyaniline-Coated Fe3O4 Nanoparticle–Carbon-Nanotube Composite and its Application in Electrochemical Biosensing. *Small, 4*(4), 462-466.

Liu, Z., Winters, M., Holodniy, M., & Dai, H. J. (2007). siRNA delivery into human T cells and primary cells with carbon-nanotube transporters. *Angewandte Chemie-International Edition, 46*(12), 2023-2027.

Ly, S. Y., & Cho, N. S. (2009). Diagnosis of human hepatitis B virus in non-treated blood by the bovine IgG DNA-linked carbon nanotube biosensor. *Journal of Clinical Virology, 44*(1), 43-47.

Lyons, M. E. G., & Keeley, G. P. (2008). Immobilized enzyme-single-wall carbon nanotube composites for amperometric glucose detection at a very low applied potential. *Chemical Communications*(22), 2529-2531.

Meng, L., Jin, J., Yang, G., Lu, T., Zhang, H., & Cai, C. (2009). Nonenzymatic Electrochemical Detection of Glucose Based on Palladium−Single-Walled Carbon Nanotube Hybrid Nanostructures. *Analytical Chemistry, 81*(17), 7271-7280.

Nakayama-Ratchford, N., Bangsaruntip, S., Sun, X., Welsher, K., & Dai, H. (2007). Noncovalent Functionalization of Carbon Nanotubes by Fluorescein−Polyethylene Glycol: Supramolecular Conjugates with pH-Dependent Absorbance and Fluorescence. *Journal of the American Chemical Society, 129*(9), 2448-2449.

Okuno, J., Maehashi, K., Kerman, K., Takamura, Y., Matsumoto, K., & Tamiya, E. (2007). Label-free immunosensor for prostate-specific antigen based on single-walled carbon nanotube array-modified microelectrodes. *Biosensors and Bioelectronics, 22*(9-10), 2377-2381.

Ouyang, M., Huang, J. L., & Lieber, C. M. (2002). One-dimensional energy dispersion of single-walled carbon nanotubes by resonant electron scattering. *Physical Review Letters, 88*(6), -.

Pantarotto, D., Briand, J. P., Prato, M., & Bianco, A. (2004). Translocation of bioactive peptides across cell membranes by carbon nanotubes. *Chemical Communications*(1), 16-17.

Pantarotto, D., Singh, R., McCarthy, D., Erhardt, M., Briand, J. P., Prato, M., Kostarelos, K., & Bianco, A. (2004). Functionalized carbon nanotubes for plasmid DNA gene delivery. *Angewandte Chemie-International Edition, 43*(39), 5242-5246.

Prato, M., Kostarelos, K., & Bianco, A. (2008). Functionalized carbon nanotubes in drug design and discovery. *Accounts of Chemical Research, 41*(1), 60-68.

Richard, C., Balavoine, F., Schultz, P., Ebbesen, T. W., & Mioskowski, C. (2003). Supramolecular self-assembly of lipid derivatives on carbon nanotubes. *Science, 300*(5620), 775-778.

Shan, C., Yang, H., Song, J., Han, D., Ivaska, A., & Niu, L. (2009). Direct Electrochemistry of Glucose Oxidase and Biosensing for Glucose Based on Graphene. *Analytical Chemistry, 81*(6), 2378-2382.

Shim, B. S., Chen, W., Doty, C., Xu, C. L., & Kotov, N. A. (2008). Smart Electronic Yarns and Wearable Fabrics for Human Biomonitoring made by Carbon Nanotube Coating with Polyelectrolytes. *Nano Letters, 8*(12), 4151-4157.

So, H.-M., Won, K., Kim, Y. H., Kim, B.-K., Ryu, B. H., Na, P. S., Kim, H., & Lee, J.-O. (2005). Single-Walled Carbon Nanotube Biosensors Using Aptamers as Molecular Recognition Elements. *Journal of the American Chemical Society, 127*(34), 11906-11907.

Strano, M. S., & Jin, H. (2008). Where is it Heading? Single-Particle Tracking of Single-Walled Carbon Nanotubes. *Acs Nano, 2*(9), 1749-1752.

Sudibya, H. G., Ma, J., Dong, X., Ng, S., Li, L.-J., Liu, X.-W., & Chen, P. (2009). Interfacing Glycosylated Carbon-Nanotube-Network Devices with Living Cells to Detect Dynamic Secretion of Biomolecules. *Angewandte Chemie International Edition, 48*(15), 2723-2726.

Thostenson, E. T., Ren, Z. F., & Chou, T. W. (2001). Advances in the science and technology of carbon nanotubes and their composites: a review. *Composites Science and Technology, 61*(13), 1899-1912.

Troiani, H. E., Miki-Yoshida, M., Camacho-Bragado, G. A., Marques, M. A. L., Rubio, A., Ascencio, J. A., & Jose-Yacaman, M. (2003). Direct observation of the mechanical properties of single-walled carbon nanotubes and their junctions at the atomic level. *Nano Letters, 3*(6), 751-755.

Tu, X., Pehrsson, P. E., & Zhao, W. (2007). Redox Reaction of DNA-Encased HiPco Carbon Nanotubes with Hydrogen Peroxide: A Near Infrared Optical Sensitivity and Kinetics Study. *The Journal of Physical Chemistry C, 111*(46), 17227-17231.

Viswanathan, S., Rani, C., Vijay Anand, A., & Ho, J.-a. A. (2009). Disposable electrochemical immunosensor for carcinoembryonic antigen using ferrocene liposomes and MWCNT screen-printed electrode. *Biosensors and Bioelectronics, 24*(7), 1984-1989.

Viswanathan, S., Wu, L.-c., Huang, M.-R., & Ho, J.-a. A. (2006). Electrochemical Immunosensor for Cholera Toxin Using Liposomes and Poly(3,4-ethylenedioxythiophene)-Coated Carbon Nanotubes. *Analytical Chemistry, 78*(4), 1115-1121.

Wan, X. G., Dong, J. M., & Xing, D. Y. (1998). Optical properties of carbon nanotubes. *Physical Review B, 58*(11), 6756-6759.

Wang, J., Deo, R. P., Poulin, P., & Mangey, M. (2003). Carbon Nanotube Fiber Microelectrodes. *Journal of the American Chemical Society, 125*(48), 14706-14707.

Wang, J., Liu, G., & Jan, M. R. (2004). Ultrasensitive Electrical Biosensing of Proteins and DNA: Carbon-Nanotube Derived Amplification of the Recognition and Transduction Events. *Journal of the American Chemical Society, 126*(10), 3010-3011.

Williams, K. A., Veenhuizen, P. T. M., de la Torre, B. G., Eritja, R., & Dekker, C. (2002). Nanotechnology - Carbon nanotubes with DNA recognition. *Nature, 420*(6917), 761-761.

Yang, R., Tang, Z., Yan, J., Kang, H., Kim, Y., Zhu, Z., & Tan, W. (2008). Noncovalent Assembly of Carbon Nanotubes and Single-Stranded DNA: An Effective Sensing Platform for Probing Biomolecular Interactions. *Analytical Chemistry, 80*(19), 7408-7413.

Zhang, M., & Gorski, W. (2005). Electrochemical Sensing Platform Based on the Carbon Nanotubes/Redox Mediators-Biopolymer System. *Journal of the American Chemical Society, 127*(7), 2058-2059.

Zou, Y., Sun, L.-X., & Xu, F. (2007). Biosensor based on polyaniline-Prussian Blue/multi-walled carbon nanotubes hybrid composites. *Biosensors and Bioelectronics, 22*(11), 2669-2674.

7

Carbon Nanotubes - A Potential Material for Affinity Biosensors

Vepa K. Rao, S. Suresh, Mukesh K. Sharma, Ajay Gupta and R. Vijayaraghavan
Defence Research & Development Establishment, Gwalior, India

1. Introduction

The detection of bacteria, virus and toxins in food, clinical and environmental samples is an important area of research. Normally, the identification of pathogens is based on immunological and DNA based methods. In case of immunological methods, the interaction between an antigen and antibody was exploited. Techniques such as agglutination, ELISA are based on this principle. Another method is based on attraction between complementary sequences of DNA. PCR method is a very popular method used to detect DNA. All these techniques either have poor sensitivity or need qualified man power. During the past two decades, efforts were made to develop biosensors which are faster and more sensitive than traditional techniques. The development of biosensors need materials which should be biocompatible and has good electrochemical properties. Nanomaterials such as metal oxides, carbon nanotubes and quantum dots were found to be promising materials. Carbon nanotubes have several important properties such as biocompatibility, good electrochemical and electrical properties. They are amenable for immobilization of biomolecules. This chapter highlights on the recent developments and improvements in the field of affinity based biosensors by using CNTs.

2. Biosensor: An introduction

Biosensor is an analytical device consisting of a sensing element which is in close contact with a transducer and an electronic circuit for display of results. The sensing element may be a biological material (enzyme, antibody), biologically derived material (recombinant antibodies, proteins, and aptamers) or biomimic (synthetic catalysts, imprinted polymer, conducting polymer). The transducer will convert any physicochemical changes taking place in its proximity into electrical signals. Various types of transducers have been used in realizing biosensor. These are based on electrochemical, optical, mass-sensitive, thermal and electronics principles. Based on the biological recognition elements, biosensor can be catalytic or an affinity based device. The catalytic biosensors are made by using enzymes which utilize their catalytic efficiency towards an analyte. While affinity based biosensors utilize the complexing forming ability of antigen and antibody. And also they utilize the attraction between complimentary sequences in DNA. The carbon nanotubes are mainly used in electrochemical biosensors and Field effect transistor (FET) based biosensors.

2.1 Electrochemical biosensor

The electrochemical biosensors can be further subdivided into amperometric, potentiometric, impedometric and conductometric sensors. The electrochemical biosensors are cheap and require simple instrumentation. An amperometric biosensor contains a three electrode system consisting of sensing electrodes, reference electrode and auxiliary electrode. All these electrodes are to be immersed in a suitable electrolyte. And a constant potential on the sensing electrode is applied and the resulting current is related to the concentration of analyte. Some electrochemical sensors utilize voltammetric techniques such as cyclic voltammetry, differential pulse voltammetry or square wave voltammetry. In potentiometric experiments the potential developed between two electrodes is measured by a high impedance voltmeter. The biological element is attached to the sensing electrode and the other electrode serves as reference electrode. In the impedometric sensors, a three electrode system is used and the impedance plots were made in presence of a redox compound such as ferrocyanide. In a conductometric sensor the conductivity of the sensor is measured. CNTs are mainly used in amperometric and impedometric sensor and voltammetry based biosensors.

2.2 FET based biosensor

Typically a field effect transistor device (FET) will have a source and a drain. A current passes from source to drain. The FET also contains a gate, whose properties will be able to control the current passing between the source and drain. The gate material will generate an electrical field and controls the current flow. Ion sensitive field effect transistors (ISFET) are found to be suitable for pH sensing. Another form of FET utilizes a nano wire between two conducting materials. The nanowire has its atoms concentrated on its surface. Thus, any small changes in the charges present on the nanowire will cause a change in the flow of current. The electrical properties of one dimensional material such as Silicon nanowire, conducting polymer based nanowires, metal oxide nanowires and carbon nanotubes are sensitive to the recognizing element attached to them. This is because the high surface to volume ratio associated the one dimensional materials. The single walled carbon nanotubes (SWCNT) have a band gap varying from 0 eV to 2 eVs. Hence, SWCNT can behave as metallic or as semiconductor. Hence, they are suitable for FET devices.

3. Nanotechnology

Nanotechnology refers to materials and structures having dimensions less than 100 nm. These nanomaterials and nanostructures offer newer methods of sensing. The nanomaterials possess special properties due to quantum confinement effect, high surface area and high aspect ratio. In addition, the electrochemical devices have high sensitivity with transducers using nano materials because of edge diffusion phenomena. Nanomaterials such as precious metal nanoparticles, metal oxides, nano-structured conducting polymers and carbon nanotubes have recently attracted much interest owing to their application in nano-scaled devices, sensors and detectors. Nanotechnology also offer new devices such as interdigitated electrodes, nano gap electrodes, cantilevers and FETs. These devices have high sensitivity.

3.1 Properties of CNTs

The development of biosensors need novel material with suitable properties. Depending on the type of transducer, the material property requirements varies. For example, electrochemical immunosensors require materials having good electrochemical and good electrical conducting properties. While the electrical sensors such as FETs require nano

wires having semiconductor properties. Besides these properties, the materials should have good biocompatibility and amenable for immobilization of biomolecules.
The carbon nanotubes have several desirable properties such as good electronic, mechanical and optical properties. Their remarkable mechanical strength stems from covalent sp^2 bonds between individual carbon atoms [Bockrath et al., 1997]. Thus they have good thermal stability. Depending on their structure, carbon nanotubes can be metallic or semiconducting in nature and hence some nanotubes have conductivities higher than that of copper, while others behave more like silicon. And this characteristic opens the way for carbon nanotubes applications in electronic devices including field effect transistor, single-electron transistors and rectifying diodes [Popov et al., 2004]. Carbon nanotubes have large surface area due to their nanometer diameter range. In case of semiconducting nanotubes, transport measurement depicts that a nanotube is connected to two metal electrodes and hence characteristic of field effect transistor. By applying potential to a gate electrode, nanotubes can be switched from a conducting to an insulating state. This type of system was shown to be operative even at room temperatures thus meeting the requirement for potential practical application [Robertson et al., 1992]. The FET operation was explained by semiclassical band-bending model. Carbon nanotubes (CNTs) have been widely discussed as materials with enormous potential for various applications. Due to superior electronic and mechanical properties along with nanoscale dimensions, a lot of attention has been drawn toward CNTs for bio-applications ranging from drug delivery to highly sensitive biosensors. CNT-based immunosensors and applications are still in the nascent stage, and there are many challenges to be overcome for the successful commercialization of the concepts. This chapter highlights on the recent developments and the improvement in the sensitivity of immunosensors by using CNTs.

	CNTs	SiNWs	CPNWs
Materials	Carbon	Silicon	Metal alloy/oxide, conduct. polymers
Deposition technique	Arch discharged Laser assisted CVD	Laser assisted Super critical fluid solution VLS method	Electrochemical methods
Manufacturability	Difficult	Difficult	Easy
Surface modification	Limited	Well known	Well known
Functionality	Single process	Single process	Multiple process
	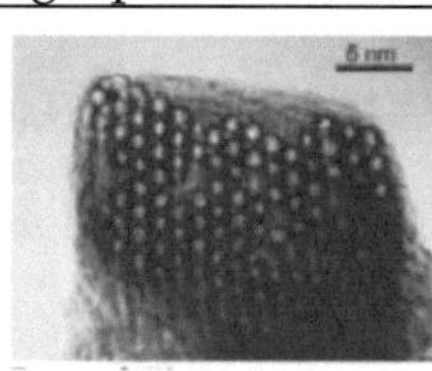	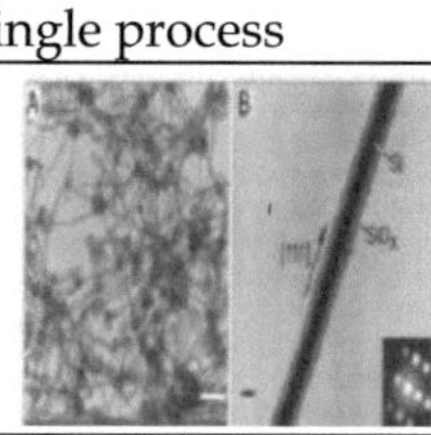	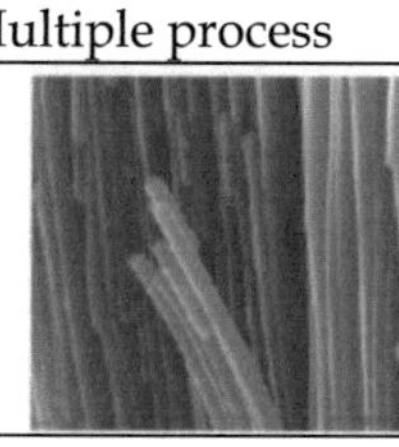

Table. 1. State-of-the Art Nanosensor Materials

4. Methods of immobilization of biomolecules on CNTs

For realizing a biosensor, it is essential to fix biomolecules on to the transducer. Carbon nanotubes can be a part of transducer. In general, several classes of immobilization methods have been used, such as physical adsorption, microencapsulation, entrapment, covalent attachment and cross-linking. The CNTs offer newer opportunities and special abilities than the existing materials. Physical adsorption is the simplest method of immobilization in which biomolecules are mechanically attached on to the surface with the help of van der Waals forces. And in this method no conformational change occurs. This method also has a disadvantage that biomolecules may leak from the surface during experiments due to weak binding force between biomolecules and the surface. It is very easy to adopt physical adsorption methods because it involves dropping of a buffer solution containing the biological molecule on to the electrode. Carbon nanotubes are in the range of nanometer size, so it has large surface area. Also they are hydrophobic in nature. Hence, the biomolecules can get adsorbed easily. The CNTs were used in various forms for development of immunosensors. They are used in making or modifying screen printed electrodes. They are also used in modifying the glassy carbon electrode. Normally, a binder such as nafion or chitosan has been used for this purpose [Tsai et al., 2005, 2007; Liaw et al., 2006]. The CNT paste electrodes were prepared by mixing CNTs with mineral oil. It was found that CNT paste has lot of crevices and voids, which help in holding the antibodies. Below is the SEM Image of CNTs which shows the crevices and voids on the surface of a MWCNT paste electrode [Fig. 1]. By spectroscopic experiments it was proved that the MWCNT paste can adsorb more quantity of antibody than graphite paste electrode [Suresh et al., 2010]. Similar property was also reported for nano porous zinc oxide.

Fig. 1. SEM image of MWCNT. Reprinted with permission from Talanta, 81, 703 (2010).

In microencapsulation, biomolecules are trapped between membranes. In the entrapment method, biomolecule is trapped in a matrix of a gel, paste or polymer and it is the very popular method. In covalent attachment, there is a formation of covalent chemical bonds between biomolecules and transducer. The CNTs are a part of transducer. The best stability, accessibility and selectivity can be achieved through covalent bonding. Covalent bonding has the capability to control the location of the biomolecules; therefore, it improves the stability, accessibility and selectivity. For the covalent bonding of molecules to the nanotubes, it is essential to form functional groups on the carbon nanotubes [Scheme 1]. The carboxylic acid group is often the best choice because it can undergo a variety of reactions and is easily formed on carbon nanotubes via oxidizing treatments, e.g. sonication in sulphuric and nitric acid, refluxing in nitric acid and air oxidization. The control of reactants and/or reaction conditions may control the locations and density of the carboxylic groups on the nanotubes which can be used for controlled attachment of biomolecules. One of the universal methods for connecting biomolecules to other materials is diimide-activated amidation, by direct coupling of carboxylic acid to proteins using N-ethyl-N′-(3-dimethylaminopropyl) carbodiimide hydrochloride (EDAC) or N,N′ dicyclohexyl carbodiimide (DCC) as a coupling agent. However, this process leads to undesirable side reactions of intermolecular conjugation of proteins, because most proteins are rich in both amine groups and carboxylic acid groups on their surface. This intermolecular connection can be avoided by using a two-step process: carboxylic acid groups are first converted to active esters via diimide-activation, and then the active esters are reacted with the amine groups on proteins without the presence of diimide. Thus, the process can guarantee homogenous attachment of proteins onto carbon nanotubes.

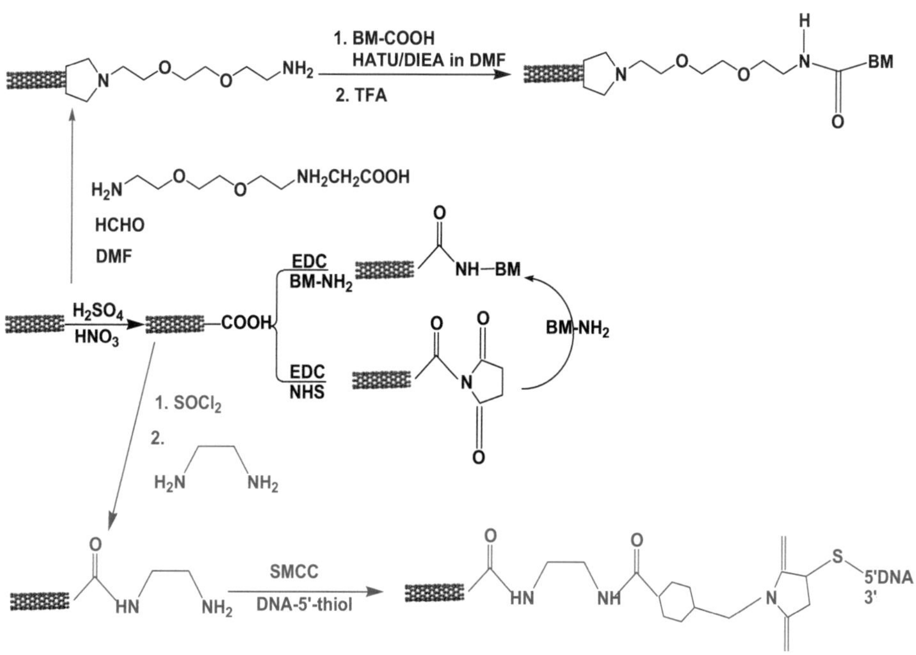

Scheme 1. Covalent functionalization with biomolecules

Kuiyang et al. reported the covalent immobilization of the protein molecules on carbon nanotubes via a two-step process of diimide-activated amidation. Here, ferritin and bovine serum albumin (BSA) proteins are chemically bonded to nitrogen-doped multi-walled carbon nanotubes (CNx MWNTs) through a two-step process of diimide-activated amidation. This two-step process avoids the intermolecular conjugation of proteins, and guarantees the uniform attachment of proteins on carbon nanotubes. This approach provides an efficient method to attach biomolecules to carbon nanotubes at ambient conditions (Kuiyang et al., 2004).

Biofunctionalization and manipulating of carbon nanotubes (CNTs) is important for biomedical research and application. Cy5 labeled goat anti-rabbit IgG (anti-IgG-Cy5) is chemically bonded to CNTs via a two-step process of diimide-activated amidation. This process can avoid the intermolecular connection of proteins (Xu et al., 2008).
The phage display to identify peptides with selective affinity for CNTs has been explored (Wang et al., 2003). Binding specificity has been confirmed by demonstrating direct attachment of nanotubes to phage and free peptides immobilized on microspheres. Consensus binding sequences show a motif rich in histidine and tryptophan at specific locations. The analysis of peptide conformations shows that the binding sequence is flexible and folds into a structure matching the geometry of carbon nanotubes. The hydrophobic structure of the peptide chains suggests that they act as symmetric detergents. An IgG monoclonal antibody against the fullerene C_{60} (Braden et al., 2000) was also studied to show binding to CNTs with some selectivity (Erlanger et al., 2001). The combination of multiwalled carbon nanotubes (MWCNT) as a transducer with polysulfone (PS) polymer enables easy incorporation of biological moieties (hormones or antibodies). It provides three dimensional composites with high electrochemical response to corresponding analytes. For biomedical purposes, human chorionic gonadotropin (hCG) hormone was tested by competitive immunoassay. The detection limit was determined to be 14.6 mIU/mL with a linear range up to 600 mIU/mL (Sanchez et al. 2008).
The nanotechnology applied in biosensing must have relevant methods to immobilize biomolecules, whose activity may be preserved for long periods of time. The most used methods for this purpose are the electrostatic layer-by-layer (LbL) (Decher et al.,1992) and the Langmuir–Blodgett techniques (LB technique)(Blodgett, 1934), which are complementary to each other in terms of the types of material that can be employed. The LbL method utilizes alternating layers of positive and negatively charged materials soluble in water, which is suitable for proteins. The LbL films are obtained via transfer of insoluble films from the air/water interface onto solid supports. Traditional materials forming stable monolayers are fatty acids, phospholipids, sterols, and substances with a long alkyl chain and a hydrophilic moiety. Soluble substances with affinity to the air–water interface (proteins and nucleic acids) can also be incorporated into the monolayers by adsorption from the aqueous subphase. Therefore, a variety of materials may be immobilized on solid matrices through the LB method, opening the way to fabricate hybrid systems. In particular, LbL technique is promising for silicon-based sensors as this method allows a control of film architecture and thickness, in addition to the synergy between properties of distinct materials, including carbon nanotubes [Lee et al., 2009], proteins [Lvov et al.,1995], antigen-antibody pairs [Zucolotto et al., 2007], DNA [Elbakry et al., 2009] and nanoparticles [Crespilho et al., 2006].
The most widely used advanced technique is patterning of biological macromolecules onto solid surfaces in the form of microarrays and/or chips. The target-capture process is performed on the substrates (e.g., silicon wafer, glass slide) via biological recognition. A signal probe (fluorescent dye molecules are used usually) is utilized to signal such biological interactions. Sensitivity is a central factor for bioanalytical technique. To achieve a high sensitivity, a large amount of research has focused on signal amplification by utilizing various nanomaterials (e.g., quantum dots, metal nanoparticles) as strong and photostable signal probes [Cao et al.,, 2002; Maxwel et al., 2002; Li et al., 2002]. Although these approaches have made condsiderable progress in biomolecular detection, they still have several drawbacks: (1) these techniques involve a complex procedure for immobilization of

the biomolecules on the flat substrate; (2) since the target-catching procedure is carried out on the flat surface of microarray or titer plate, such heterogeneous procedure increases assay time and decreases the sensitivity due to the slow target-binding kinetics; (3) some nanomaterials (e.g., nanoparticles of Ag, CdS, CdSe) used for signal amplification are sensitive to air, which causes reduced reproducibility. In cross-linking method, a bifunctional agent is used to bond chemically the transducer to the biomolecules. There are a number of advantage of immobilizing biomolecules to the surface- (i) single batch of biomolecules can be used multiple times (ii) reaction can be stop rapidly by removing the biomolecule from the reaction solution (iii) there is less chance of contamination of product with biomolecules (iv) immobilization provide long life to the biomolecules.

5. CNTs and metal nanoparticles composites

There are many reports on the application of CNTs with metal nanoparticle such as gold, platinum, copper, silver etc. [Lin et al., 2009; Wen et al., 2009; Valentini et al., 2007; Lin et al., 2009] . Below are the SEM images of MWCNTs/SPE, nano-gold/SPE and MWCNTs/nano gold/SPE (screen printed electrode).

(a) (b) (c)

Fig. 2. (a) MWCNTs/SPE, (b) nano-Au/SPE, (c) MWCNTs/nano-Au/SPE

In these figures, figure 2(a) shows the nanotube like structure on the electrode surface, figure 2(b) shows the nanoparticle like structure of gold which are electro-generated on the surface of bare screen printed electrode and figure 2(c) shows the nanohybride structure of MWCNTs and gold [Sharma et al., 2008]. The electrocatalytic efficiency of SPE modified electrode towards oxidation of naphthol was found to be very high. The 1-naphthol is produced by hydrolysis of 1-naphthyl phosphate by the enzyme alkaline phosphatase. Hence, these electrodes were suited in the detection of malaria by sandwich ELISA system, in which alkaline phosphatase is used as an enzyme tagged to revealing antibody.

Preparation of nanocomposite materials from carbon nanotubes (CNTs) and metal or metal oxide nanoparticles has important implications to the development of advanced catalytic and sensory materials. Molecularly mediated assembly of monolayer-capped nanoparticles on multiwalled CNTs via a combination of hydrophobic and hydrogen-bonding interactions between the capping/mediating shell and the CNT surface have advantage that it does not require tedious surface modification of CNTs. It shows simplicity and effectiveness for assembling alkanethiolate-capped gold nanoparticles of 2-5 nm core sizes onto CNTs with controllable coverage and spatially isolated character. The loading and distribution of the nanoparticles on CNTs depend on the relative concentrations of gold nanoparticles, CNTs, and mediating or linking agents. The composite nanomaterials can be dispersed in organic solvent, and the capping/linking shells can be removed by thermal treatment to produce controllable nanocrystals on the CNT surfaces.

6. CNTs and affinity biosensors

There are two types of affinity based biosensors. These are called immunosensors and DNA sensors. The immunosensors exploit the property of complex formation between an antigen and its antibody. And the other utilizes attraction between the complimentary sequences of the DNA strands. Mainly CNTs were utilized in developing electrochemical, optical and electronic sensors.

6.1 Immunosensors

The immunosensors utilizes the affinity between the antigen and its antibody to form a complex. The ELISA methods were well established techniques and have variety of formats. Since, the formation of the complex can not be determined directly by earlier methods, an enzyme tagged antibody is used to reveal the formation of the complex between the antigen and antibody [Suresh et al., 2010; Sharma et al., 2008]. The enzyme tagged antibody will provide an optical or electrochemical signal based on the substrate used. Carbon nanotubes are used both in label less methods and labeled methods. The enzymes used for conjugation with antibodies are HRP or alkaline phosphatase. The CNTs are mainly used because of their good electrochemical properties and binding properties towards biomolecules. In the case of electrochemical immunosensors CNTs were used in various forms. They are mixed with a binder to modify glassy carbon electrodes or screen printed electrodes (SPE). The electrochemical properties of these are enhanced due to modifications. The main reason for using SPEs is disposability and cost effectiveness. The binder materials normally used are nafion, chitosan or sodium alginate. These three are known to be biocompatible. In some publications the MWCNTs modified electrodes were further modified by electrodeposition of gold nano particles [Sharma et al., 2008]. In some reports the CNTs were initially modified with gold nano particles and then used to modify the GC electrodes using a binder. Various electrochemical techniques such as amperometry, square wave voltammetry and impedance were used in these immunosensors. Even pure carbon nanotubes were used. Highly aligned multiwalled carbon nanotubes were grown on a Fe/Al2O3/SiO2/Si substrate by chemical evaporation. An elaborate procedure is used to expose the ends on the CNTs . These are oxidized at their tips electrochemically to functionalize. The resulting carboxylic acid is attached to the antibodies. This is a label less sensor working on the principle of impedance.

Carcinoembryonic (CEA) antigen is present in serum samples and other biofluids in patients suffering from cancer. Viswanathan et al. developed a disposable electrochemical immunosensor by attaching monoclonal anti carcinoembryonic antibodies (α-CEA) covalently on polyethyleneimine wrapped multi walled carbon nanotubes modified SPE by sandwich immunoassay. A sandwich immunoassay with CEA & α-CEA was tagged to ferrocene carboxylic acid encapsulated liposomes. Square wave voltammentric technique (SWV) was used to detect the antigen in the range of $5X10^{-12}$ to $5X10^{-7} gmL^{-1}$ and the detection limit of $1x10^{-12}$ gmL^{-1} [Viswanathan et al., 2009].

Alpha Fetoprotein (AFP) is known as an important tumor marker and has an average concentration of about 25 ng/mL in healthy human serum [Zhu et al., 2000]. An elevated level of AFP concentration in serum can be determined and the disease cancer can be diaginised earlier. Jiang et al. also reported on α-fetoprotein sensor based on MWCNTs / Prussian blue / nanogold modified GC electrode and found the linear range 0.01-300ng/mL with detection limit of 3pg/mL [Jiang et al.,2010]. In this study Jiang et al. deposited Prussian blue (PB) and gold nanoparticles electrochemically on MWCNT modified GC electrode. The leakage of PB was prevented due to sequential deposition of PB and Gold on the modified electrode. This GNP film also helps to prevent shedding of MWCNT/PB composite film to electrode surface. The electron transfer between the PB and electrode gets affected due to antigen-antibody interaction. It results in decrement in cyclic voltammetric response. This decrement is related to antigen concentration. This is a lablel less immunosensor. Lin et al. also reported amperometric immunosensor for detection of alpha fetoprotein by using GNP/CNT/Chitosan modified GC electrode, formed by one step synthesis through direct redox reaction. For this immuno detection they used sample AFP, immobilized AFP and alkaline phosphatase tagged antibodies. The linear range of AFP was found from 1-55ng/mL with a detection limit of 0.6ng/mL [Lin et al, 2009].

Tang et al. reported enzyme free electrochemical immunoassay for AFP detection by using carbon nanotube enriched Au nanoparticles as a nanocatalyst on anti-AFP/glutaraldehyde/thionine-modified GCEs with a wide linear range of 8.0×10^{-7}–2.0×10^{2} ng/mL and detection limit (LOD) of 0.8 fg/mL of AFP which was lower 6 orders than that of commercially available ELISA. This high sensitivity was possible due to redox cycling of p-aminophenol and p-quinone imine and that was resulted in continuously increasing of signaling. In this process, initially the p-nitrophenol molecules were reduced to p-aminophenol by the catalysis of the AuNP labels on the CNT-AuNPs with the help of $NaBH_4$, then the generated p-aminophenol molecules were oxidized to p-quinone imine by an electron mediator of thionine, and then the oxidized QI molecules were reduced back to APs by $NaBH_4$ [Tang et al., 2011].

Zhao et al. reported on the disposable Shigella flexneri immunosensor based on MWCNT/sodium alignate (SA) composite SPE [Zhao et al., 2011]. They used HRP labeled antibodies to S. flexneri and immoblised these antibodies on MWCNT/sodium alignate composite screen printed electrode by physical adsorption. Shigella is major causes of human infectious diseases and is responsible for millions of cases of diarrhea [Li et al., 2006]. Alginate is a copolymer of β-D-mannuronic acid and α-L-guluronic acid linked together by 1–4 linkages [Rehm et al., 1997]. It is a natural and biocompatible polymer that can provide microenvironments to improve the biomolecules stability and maintain their bioactivity [Liu et al., 2009]. Here sodium alignate biocomposite acts as a matrix to adsorb and immobilize antibodies. Linear range was found to be 10^{4}cfu/mL to 10^{11} cfu/mL and detection limit was $3.1\text{x}10^{3}$ cfu/mL. There is noncovalent association of MWCNTs with SA chains and has proved to be preferable for electrochemical immunosensing due to the unique structural and electronic properties of MWCNTs [Zhao et al., 2009]. In this system, MWCNTs provide a signal transduction, whereas SA works as a biocompatible and chemically modifiable scaffold for biomolecule immobilization. SA can effectively improve the solubility of MWCNTs, providing a useful way for preparing MWCNT–binder composite modified electrode for a wide range of sensing applications [Zhao et al., 2009].

Zhu et al. studied the amperometric immunosensor for the detection of neomycin by using GNP decorated electrode. Neomycin performs activity against both gram negative bacteria as well as gram positive bacteria and this behaviour of neomycin is similar to other aminoglycosides. In this study Zhu et al., used poly-[2,5-di-(2-thienyl)-1H-pyrrole-1-(p-benzoic acid)] (pDPB). It is a conducting polymer and behaves as a sensor probe to detect neomycin in sandwitch manner in which secondary antibody was attached to GNP decorated MWCNTs labeled with hydrazine (Hyd–MWCNT(AuNP)–Ab2). Hydrazine works as catalyst for the reduction of H_2O_2. The linear range was found to be 10ng/mL-250ng/mL with a detection limit of 6.76±0.17ng/mL [Zhu et al., 2010]. In this study Zhu et al., immobilized neomycin antibody on conducting polymer covalently and this type of immobilization of biomolecules on conduction polymer has attracted wide attention due to great compatibility of polymer with biomolecule in neutral solution [Lee and Shim, 2001; Abdelwahab et al., 2010; García-Aljaro et al.,2010].

Sanchez et al., reported immunosensor for rabbit IgG (RIgG) by using MWCNTs/polysulfone/RIgG modified SPE and based on competitive immunoassay between free and labelled anti-RIgG. Labeling of antibody was done by horseradish peroxidase (HRP) and hydroquinone was used as mediator. This resulting immunosensor have high sensitivity as well as double roughness in comparison to graphite/polysulfone/RIgG immunosensor. Incorporation of RIgG in MWCNTs/polysulfone was resulted from phase inversion. MWCNTs/polysulfone works both as reservoir of immunological material and transducer while offering high surface area, high toughness and mechanical flexibility. Linear range of anti RIgG was found to be 2-5μg/mL with Detection limit of 1.66μg/mL and C_{50} value at 3.56μg/mL. The polysulfone have high resistance in extreme pH conditions, good adhesion and susceptibility to incorporate biological molecules and good thermal stability, while MWCNT has good conductivity. Hence the composite made by using polysulfone and MWCNT will be ideal for use in immunosensors. [Sanchez et al., 2007]. In order to immobilize antibodies in polysulfone matrix by phase inversion technique Mulder et al. modify the chemical nature of the polysulfone matrix [Mulder et al., 2000]. There has been little research also done on the basis of adsorption of antibody for example Yu et al. and O'Connor et al. reported immunosensor by immobilizing antibody on SWCNTs, oriented perpendicularly [Yu et al.,2004; O'Connor et al.,2005].

Recently, He et al. reported on label free electrochemical immunosensor for rapid determination of clenbuterol by using CNTs. In this study, Clenbuterol and to MWCNTs was linked covalently in a two-step process using 1-(3-(dimethylamino)-propyl)-3-ethylcarbodiimide and N-hydroxysulfo-succinimide as crosslinkers after that casted on GC electrode and found the detection limit of 0.32ng/mL [He et al., 2009]. There are two strategies which were applied to immobilize immunological molecules in literatures (1) antibody immobilised on the electrode was directly applied to determine the antigen in solution by observing the response change of ferrocyanide redox marker on the electrode [Lei et al., 2003; Zhang et al., 2006; Zhou et al., 2005]. (2) Antigen immobilised on the electrode was used to compete with free antigen in solution for a specific antibody [Chen et al., 2005]. But the above method suited for protein, virus, and cell factor detection mostly and also this type of direct cross-linking of immunological molecules might lose their activities, and then regeneration of the sensor required complete removal of all immobilized materials from the electrode surface.

Recently, Cao et al. studied electrochemical immunosensor for casein based on GNP/Poly (L-arginine)/MWCNTs modified GC electrode and observed linear range was $1x10^{-7}$ - $1x10^{-5}$

gmL^{-1} with detection limit of $5x10^{-8}$ [Cao et al., 2011]. Cui et al. reported on human IgG detection based GNPs/CNTs hybrid modified GC electrode. Linear range was observed between 0.125 and 80 ng/mL with a detection limit of 40pg/mL. This high sensitivity was possible due to using of bioconjugates featuring HRP labels and secondary antibodies linked to GNPs at high HRP/Ab2 molar ratio [Cui et al., 2008]. Bourigua et al. studied label free impedemetric biosensor for the detection of deep venous thrombosis biomarker by using SWCNT-COOH modified gold microelectrodes and found the linear range in between of 0.1pg/mL-2μg/mL with detection limit of 0.1pg/mL and response time of 10min [Bourigua et al., 2010]. Deep venous thrombosis is named due to formation of blood clot in deep vein [Firkin and Nandurkar, 2009]. Panini et al 2008 reported that Integrated microfluidic systems with an immunosensor for detection of prostate specific antigen (PSA) in human serum samples by using MWCNTs modified GC electrodes with observed detection limit of 0.08 μgL^{-1} which was very high than the observed detection limit by ELISA. Hetero-geneous enzyme immunoassays, coupled with flow injection system and electrochemical detection, represent a powerful analytical tool for the determination of low levels of many analytes such as antibodies, hormones, drugs, tumour markers, and viruses [Gubitz and Shellum, 1993; Wu et al., 2006]. Okuno et al., 2007, studied Label-free immunosensor for prostate-specific antigen based on single-walled carbon nanotube array-modified microelectrodes and found the detection of 0.25ng/mL. Li et al., 2010 reported reagentless amperometric immunosensor for the detection of cancer antigen 15-3 based on enzyme-mediated direct electrochemistry and found the linear range in between of 0.1-160U/mL with detection limit of 0.04U/mL. Recently, Piao et al., 2011 reported sensitive and high-fidelity electrochemical immunoassay for the detection of human IgG by using CNT coated with enzymes and magnetic nanoparticles. In this report, Piao et al. used tyrosinase as enzyme. In this immunosensor firstly enzyme and magnetic nanoparticle were immobilized covalently on the surface of CNT then cross-linked via glutaraldehyde to form multilayered cross-linked tyrosinase-magnetic nanoparticles composite that was further conjugated with primary antibody against human IgG. Secondary antibody was conjugated with alkaline phosphatase and used in sandwich immunoassay pattern. Detection limit was found to be 0.19ng/mL. Zhao et al., 2011, reported on determination of lutiolin in peanut hulls by using MWCNT modified GC. Linear range was found to be $2x10^{-10}$ - $3x10^{-9}$ with detection limit of $6x10^{-11}$. Luteolin is member of flavonoid and it is actually 3′,4′, 5, 7-tetrahydroxyflavone and has a wide range of biochemical and pharmacological effects and also have anti-oxidation, anti-bacteria, anti-virus, anti-inflammatory, anti-carcinogenic and other beneficial properties [Jian & Xiao, 1986; Merken & Beecher, 2000; Robards & Antolovich, 1997]. Highly precise and sensitive electrochemical immunosensor for the detection of carcinoembryonic antigen (CEA) in saliva and serum was reported [Vishwanathan et al., 2009]. Monoclonal anti-CEA antibodies (anti-CEA) were covalently immobilized on polyethyleneimine wrapped multiwalled carbon nanotubes screen-printed electrode. A sandwich immunoassay was performed with CEA and anti-CEA tagged ferrocene carboxylic acid encapsulated liposomes (anti-CEA-FCL). The squarewave voltammetry (SWV) was employed to analyze faradic redox responses of the released ferrocene carboxylic acid from the immunoconjugated liposomes on the electrode surface. The calibration curve for CEA concentration was in the range of 5×10^{-12} to 5×10^{-7} gmL^{-1} with a detection limit of 1×10^{-12} gmL^{-1} (S/N = 3). Sharma et al. have reported sensitive immunosensor for the detection of *Plasmodium falciparum* histidine- rich protein 2 (PfHRP-2) in the sera of humans with *P. falciparum* malaria based on modified SPEs. For this purpose, disposable SPEs were modified with multiwall carbon nanotubes (MWCNTs) and Au nanoparticles. The immunosensing experiments were performed on bare SPEs, MWCNT-modified SPEs, and Au nanoparticle and MWCNT-modified SPEs (Nano-Au/MWCNT/SPEs) for the amperometric detection of PfHRP-2. Nano-Au/MWCNT/SPEs yielded the highest-level immunosensing performance among the electrodes, with a detection limit of 8ng/ml. The analytical results of immunosensing experiments with human serum samples were compared with the results of a commercial Paracheck Pf test, as well as the results of microscopy. The Paracheck Pf kit exhibited a sensitivity of 79% and a specificity of 81%, whereas the amperometric immunosensor showed a sensitivity of 96% and a specificity of 94%.

Carboxylated multiwalled carbon nanotubes (MWCNT-COOH) were used to modify the working electrode surface of different SPEs. The effect of this modification on the electrodic characteristics (double layer capacitance, electroactive area and heterogeneous rate constants for the electron transfer) was evaluated and optimized for the cyclic voltammetric determination of *p*-aminophenol. The enzymatic hydrolysis of *p*-aminophenylphosphate was employed for the quantification of alkaline phosphatase, one of the most important label enzymes in immunoassays. Finally, ELISA assays were carried out to quantify pneumolysin using this enzymatic system. Results obtained indicated that low superficial densities of MWCNT-COOH (0.03–0.06 $\mu g\ mm^{-2}$) yielded the same electrodic improvements but with better analytical properties [Lamas et al., 2008].

Vishwanathan et al described a sensitive method for the detection of cholera toxin using an electrochemical immunosensor with liposomic magnification followed by adsorptive square-wave stripping voltammetry is described [Vishwanathan et al., 2006]. In this immunodetection potassium ferrocyanide-encapsulated and ganglioside (GM1)-functionalized liposomes act as highly specific recognition labels for the amplified detection of cholera toxin. The sensing interface consists of monoclonal antibody against the B subunit of CT that is linked to poly (3,4-ethylenedioxythiophene) coated on nafion-supported MWCNTs caste film on a GC electrode. The CT is detected by a "sandwich-type" assay on the electronic transducers, where the toxin is first bound to the anti-CT antibody and then to the GM1-functionalized liposome. The potassium ferrocyanide molecules are released from the bounded liposomes on the electrode by lyses with methanolic solution of Triton X-100. The released electroactive marker is measured by adsorptive square-wave stripping voltammetry. The sandwich assay provides the amplification route for the detection of the CT present in ultratrace levels. The calibration curve for CT had a linear range of 10^{-14}–10^{-7}g mL^{-1}. The detection limit of this immunosensor was 10^{-16} g of cholera toxin (equivalent to 100 μL of 10^{-15} g mL^{-1}).

Typically, a field-effect transistor (FET) device will have a nanowire in contact with a source and drain along with a gate. Typically a nanowire such as silicon, metal oxide or carbon nanotubes was used. In case of single walled carbon nanotube, all the carbons will be on its surface. The conductivity of this can be greatly influenced by any molecules attached to it. Since, various methods are available for immobilization of biomolecules on CNTs. The CNTs possesses good electrical properties and are ideal for FET based Biosensors also. Dekker's group reported the first carbon nanotube-FET (CNTFET) [Tans et al., 1997]. CNTFETs have been proved suitable system for biosensing applications due to its properties to combining the principles of molecular recognition through the recognition layer with the transduction capabilities of the carbon nanotubes. CNTFETs have been used to detect DNA [Gui et al., 2007; Star et al., 2006; So et al, 2006,], proteins like thrombin and IgE and IgG by means of antibodies and aptamers as molecular receptors [Cid et al., 2008; Maehashi et al., 2007; So et al., 2005], viruses by means of peptide nucleic acids [Dastagir et al., 2007] and bacteria [Villamizar et al., 2008; So et al., 2008].

Recently, Villamizar et al., 2009 reported CNTFET based detection of candida albicans. CNTFET have a network of SWCNTs which works as the conductor channel. This CNTFET based biosensor was able to detect atleast 50 cfu/mL within 1h and remained stable for more than 10 days. This CNTFET was also tested with potential competing yeasts like Cryptococcus albidus and Saccharomyces cerevisiae for C. albicans to evaluate the selectivity of FET devices. Kim et al., 2008, reported a method for ultrasensitive CNT-FET based biosensors by using antibody binding fragment. In this study, CNTFETs were functionalized with antibody binding fragment as a receptor, and the binding event of target IgG onto the fragments was detected by observing gating effect caused by charges of target IgG. And it was observed that CNTFET biosensors based on small antibody fragments (Fab fragments) have very high sensitivity (detection limit 1pg/mL) in comparison to those CNTFET biosensors based on whole antibody segment (detection limit 1000 ng/mL).

6.2 DNA sensors

DNA sensor is an affinity based biosensor and considered a promising tool in pre-diagnostics as well as in the prevention and control of infectious diseases in real-time and on site analysis [Drummond et al., 2003]. DNA sensors have several potential applications including the diagnosis of genetic diseases, detection of infectious agents and environmental cases [Malhotra et al., 2005; Wang et al., 2002]. Methods used for DNA sequence detection reported are based on radiochemical, enzymatic, fluorescent, electrochemical, optical, and acoustic wave techniques [Kara and Meric, 2004]. Optical DNA sensors gave promising results but some disadvantages are also there including the requirement of a separate labeling process and an equipment to stimulate the transducer of high cost [Pearson et al., 2000]. Electrochemical methods for hybridization detection present a good alternative in comparison fluorescent detection due to having considerable advantages ascribed to their potential for obtaining specific information in a faster, simpler, and less expensive way and huge progress has been made towards the development of the electrochemical DNA sensors. These sensors rely on the conventional hybridization signal of the DNA sequences into useful electrical signal.

DNA is potentially a building block for the assembly of nanoscale electronic devices and has all the basic properties necessary for it with their application in decentralized clinical testing, food safety and environmental monitoring. The DNA nanoscale device construction will be a predominant technique of new molectron with attention paying benefits like high efficiency, low power requirement, miniaturization and low heat generation [Guo et al., 2004].

DNA immobilization considered as a fundamental methodology for construction of DNA biosensor, requires intimate connection between nucleic acid and electronic transducer. High sensitivity, long life-span as well as short response time depends mainly on attachment of DNA sequence on the surface of the sensor. So it is necessary that the binding chemistry should be stable during subsequent assay steps.There are various methods like chemical adsorption, covalent binding, electrostatic attraction, co-polymerization and avidin-biotin affinity system for DNA immobilization [Cai et al., 2003]. Besides DNA biosensor, some other approaches have been studied like DNA mineralization, the use of DNA to nanopaticles/nanotubes assembly, and larger colloidal particles. There are various methods to the DNA strands immobilization on the sensor surface and several groups have reported using the covalent binding of the CNTs to immobilize the DNA sequences. Such as the covalent attachment on the functionalized support [Gabl et al., 2004], physical adsorption [Komarova et al., 2005] and monolayer self-assembling [Saoudi et al., 1997; Huang et al., 2001; Peelen & Smith, 2005]. The most ideal approach for DNA immobilization in CNTs is covalent binding on a solid surface via a single point attachment. Among these methods, immobilization by means of covalent attachment has advantages such as simplicity, efficiency, ordered binding, and low cost. In covalent attachment, various mediators can be used to attach the DNA sequences on the sensor surface such as carbon nanotubes (CNTs), Aminopropyltriethoxy silane (APTES), Alkanethiols etc.. Covalent immobilization of the CNTs is usually performed by amino-terminated DNA reaction with the carboxylic acid groups of the CNTs, or directly reacting with the amino group of the oxidized CNTs. Jung et al., 2004 demonstrated that the DNA strands can be covalently attached to immobilized SWNT multilayer films. These multilayer films were constructed via consecutive condensation reactions and resulted into stacks of functionalized SWCNTs layers linked together by 4,4-oxydianiline. Singh et al., 2006 reported that aminated- or carboxylated- DNA strands were covalently immobilized to the carboxylated or aminated SWNT multilayer films respectively through formation of amide bond by help of using 1-ethyl-3-(3-dimethylaminopropyl) carbodiimide hydrochloride. Baker et al., 2002 reported a multistep route for the formation of covalently linked adducts of SWNT and DNA sequence. Most of the applications of immobilized oligonucleotide are based on the hybridization between immobilized oligonucleotide and its complementary DNA sequence in the sample [Lee et al., 2003]. Singh et al. studied the synthesis of functionally engineered SWCNTs–peptide nucleic acid conjugates for nanoelectronic applications [Singh et al., 2006].Wang et al., 2004 reported that the strong accumulation of phenolic products of alkaline phosphatase onto CNT-modified electrodes allow the detection of extremely low levels of the target DNA. Enhanced voltammetric response of phenolic compounds at CNT modified GC electrode was measured in connection with enzyme-based electrochemical detection of DNA hybridization. The detailed mechanism behind the dissolution of nanotubes in DNA is not clear at present. Nakashima et al., 2003 suggested that the π-π interactions between the nanotube sidewalls and the nucleic acid bases may be responsible. There may also be some weak interaction between the major and minor grooves of the DNA and the nanotubes. Cai et al.,2003 reported the application of CNTs to the fabrication of an electrochemical DNA biosensor for the specific DNA sequence detection. Moghaddam et al., 2004 firstly reported the azide photolysis for the functionalization of CNTs. They used the azide-photochemistry to functionalize the sidewalls and tips of CNTs in a solid-state reaction and the subsequent synthesis of a DNA oligonucleotide from the reactive group on each photo-adduct. Lu et al., 2005 investigated a system consisting of B-DNA and an array of carbon nanotubes periodically arranged to fit into major grooves of the DNA. They discussed in detail about the system used as an electronic switch or as a sensor device for ultra fast DNA sequencing. A novel sensitive electrochemical biosensor based on magnetite nanoparticle for monitoring DNA hybridization by using MWCNT – COOH/ppy – modified GC electrode was described by Cheng et al., 2005. CNTs can amplify DNA, protein recognition and transduction events. This property was used for the ultrasensitive method for the electrical biosensing of DNA or proteins. An effective DNA sensing system to detect specific nucleic acid sequences is playing an important role in many areas such as clinical diagnosis, medicine, epidermic prevention, environmental protection and bioengineering.

The delivery of gold nanoparticles to CNTs using the self assembly properties of DNA represents an advance towards building higher order nanostructures with rational control. The controlled self-assembly of CNTs was recently achieved by interphasing the CNTs with biomolecules. This approach has considerable potentials for driving self-assembly of CNT-based devices in the light of the immense richness of biological recognition. The approach involves two steps. In the first step, a self-assembled nanolayer of single stranded DNA is adsorbed onto Au contacts by reaction with thiol terminated oligonucleotides and in the next step, the oxidized CNTs modified with oligonucleotides of the complementary sequence are allowed to hybridize with the DNA located on the Au contacts. Hazani et al., 2004 used a long DNA molecule featuring Rec A proteins as a scaffold, onto which streptavidin-functionalized SWCNTs were assembled utilizing anti-ReeA primary antibodies and biotinylated secondary antibodies. Wang et al., 2004 reported a novel DNA immobilization strategy, in which the DNA probes are adsorbed on self-assembled multiwalled nanotubes. Their results showed that this immobilization strategy based on self assembled MWCNTs yields higher hybridization efficiency than that adsorbed on random MWCNTs. Li et al., 2005 demonstrated that a wide range of multicomponent structures of CNTs can be constructed by DNA directed self-assembling of CNTs and gold nanoparticles. Wang et al., 2004 developed DNA biosensors based on the self-assembly of CNTs. They assayed the hybridization by the changes in the voltammetric peak of the indicator methylene blue and the results showed that the DNA biosensors based on the self-assembled MWCNTs have higher hybridization efficiency than that based on random MWCNTs.

Recently, Tam et al., 2009, reported label free DNA sensor for detection of Influenza virus (type A) by using CNTs and found the detection limit of 0.5M with a response time of 4 min. DNA was attached to sensor surface By means of covalent bonding between amine and phosphate group of DNA sequence.

Recently, Weber et al., 2011 reported on electrochemical impedance based DNA sensor for detection of Salmonella enterica serovar *Typhimurium* by using SWCNTs modified electrode. In this study, Weber et al., reported on the use of SWCNTs with a diameter range of 20–30 nm for DNA attachment, and its subsequent hybridization. SWCNTs covalently bonded via N-Ethyl-N'-(3-dimethylaminopropyl) carbodiimide hydrochloride (EDAC) to oligonucleotide probes having amino groups on the 3' end. After achieving hybridization, the capacitance change and charge transfer resistance of the electrode to the redox-active compound $Fe(CN)6^{-3/-4}$ were measured by electrochemical impedance spectroscopy (EIS). In this study, Weber et al. used Salmonella specific probes. This method of genosensing is a quick, facile approach to detecting DNA without the use of additional labels. By this impedance method, detection was found to be at 1x10-9 $molL^{-1}$. Wang et al., 2011 reported on label free DNA biosensor for detection of short DNA species by using MWCNTs/Chitosan nanocomposite modified GC electrode with glutaraldehyde as an arm linker and found the linear range between $1x10^{-13}$ - $5x10^{-10}$ M with detection limit of $8.5x10^{-14}$.

Carboxylic group-functionalized SWNTs were assembled vertically on the electrode using ethylenediamine as linking agent to fabricate an aligned electrode (SWNTE). Single-stranded DNA (ssDNA) wrapped around the SWNTs to form ssDNA-wrapped SWNTE structures based on the interaction between ssDNA and SWNT. A sensitive differential pulse voltammetric (DPV) response was obtained at the ssDNA-wrapped SWNTE owing to the electro-oxidation of guanine bases. Double-stranded DNA (dsDNA) was formed when ssDNA on the ssDNA-wrapped SWNTE was hybridized with complementary ssDNA (cDNA). The dsDNA was removed from the SWNTs by undergoing a process of preconditioning at −0.6 V. Consequentially, the DPV response of guanine bases decreased. The used SWNTE could be renewed easily via ultrasonically rinsing. On the basis of this mechanism, a label-free and readily reusable electrochemical DNA hybridization biosensor was designed by directly monitoring the current change of guanine bases. Under optimum conditions, the plot of the measurement signal of guanine bases versus the cDNA concentrations was a good straight line in the range of 40−110 nM with a detection limit of 20 nM (3s). The biosensor can be switched to detect different target DNAs easily.

CNTs modified electrodes have been used for biochemical detection [Wang et al., 2004] and recently have been used as transducers for more sensitive DNA hybridization detection electrically [Wang et al., 2003]. CNTs can play an important role in selective and sensitive recognition of DNA in electrical DNA biosensing due to its ability to amplify DNA recognition and transduction events. For this selective and sensitive recognition, single strands DNA can be grafted chemically on to aligned CNT-electrodes. Successful integration of CNTs in sensors needed controlled deposition at well defined location. Interphasing of CNTs with biological molecule provide controlled self assembly of CNTs and this approach has considerable potential for self assembly of CNTs based devices with intense recognition of biological molecule. Due to charge transfer characteristics of CNTs while approaching the size of biomolecules, CNTs utilities will be beneficial in electrochemical biosensing. CNTs are helpful in amplifying enzyme based bioaffinity and hence can be used in electrical sensing of DNA [Wang et al., 2004].

Recently, DNA attracted several attentions in connection with CNTs and through sequence-specific pairing interaction DNA chains have been used to create various functional structures and devices. Also the principle based on DNA based biomolecular recognition has been applied to construct CNT-DNA electrochemical sensors [Daniel et al., 2007]. Guo et al., 2004 described the electrochemical characteristics of DNA functionalization of CNTs holds interesting prospects in various fields including solubilization in aqueous media, nucleic acid sensing, gene-therapy and controlled deposition on conducting or semiconducting substrates. DNA molecules can increase the CNT solubility and also be used to distinguish metallic CNT from semiconducting CNTs. The DNA attachment occurs predominantly at or near the nanotube ends. The rare attachment of DNA to other regions of CNTs indicates that it is the result of sequence specific polynucleic acid–DNA base pairing rather than nonspecific interactions [William et al., 2002]. The individual CNTs can be functionalized with special selectivity and can be used to differentiate between the two DNA sequences. The concept of using DNA to direct the assembly of nanotubes into nanoscale devices is attracting attention because of its potential to assemble a multicomponent system in one step by using different base sequence for each component. The reactive sites on the CNTs were created by the acid treatment to introduce the carboxyl groups on their tips. DNA molecules with functional linkers are then coupled to the carboxyl groups on the CNTs. Chen et al., 2005 developed a multistep method to covalently functionalize multiwall carbon nanotubes with DNA and oligonucleotides. Thus, the bioconjugates of carbon nanomaterials and DNA will have potential uses in many areas due to the combination of unusual structure of carbon nanomaterials and bioactivity of DNA [Yan et al., 2005]. Hazani et al., 2003 reported the confocal fluorescence imaging of SWCNT–DNA adducts obtained by carbodiimide assisted coupling of amine functionalized oligonucleotides to oxidized SWCNTs. CNTs have been recently used as transducers for enhanced electrical detection of DNA hybridization . The DNA sensing application requires high sensitivity through amplified transduction of the oligonucleotide interaction. The wrapping of CNTs in DNA results in some interesting effects. The DNA nanotube species are highly soluble in water removing the requirement for surfactants. Also the negative charges on the phosphate group of DNA results in the charging of DNA nanotube species. The bifunctional chemical structure of CNTs would facilitate the selective attachment of multiple DNA sequences using two distinct DNA–CNT linking strategies. In one strategy, by accessing the free carboxyl groups of CNTs, single stranded, amine terminated DNA oligonucleotides are attached to the CNT array using amide coupling chemistry in aqueous/organic solvent mixtures. The second strategy involves the attachment of oligonucleotides to the sidewalls of the CNTs through hydrophobic interactions. Taft et al., 2004 reported the immobilization of DNA through the interaction of the hydrophobic pyrene group with the graphite-based sidewalls of the CNTs, which was highly specific and DNA-dependent process.

The unique property of the specific molecular recognition of DNA coupling with SWCNTs and hybridizing these macromolecular wires will provide a versatile means of incorporating SWCNTs into larger electronic devices by recognition-based assembly and using SWCNTs as probes in biological systems by sequence-specific attachment [William et al., 2002]. Buzaneva et al., 2002 developed the DNA nanotechnology for the formation of the multifunctional CNT cells using theoretical predictions of chemical activity changing of SWCNT under its localization into the biopolymer surrounding and electronic/optical properties of these cells that are determined by SWCNT–biopolymer heterostructure properties.

7. Conclusions

Applications of nanomaterials in the field of affinity biosensors have become advanced greatly. For example, nanomaterials-based biosensors, which represent the integration of material science, molecular engineering, chemistry and biotechnology, can markedly improve the sensitivity and specificity of biomolecule detection, hold the capability of detecting or manipulating atoms and molecules, and have great potential in applications such as biomolecular recognition, pathogenic diagnosis and environment monitoring.

CNTs have unique properties. Unique properties lead to fabrication of different devices. Improvements of current synthesis of CNTs needed to make available commercial products. The totally new nanoelectronic architecture may be constructed on CNTs. There is a little knowledge about growth mechanism, structural defects and their influence on practical properties of CNTs. There are many attractive phenomena hidden within the tiny, mysterious world that exists inside the CNTs.

CNTs have several interesting properties suitable for developing affinity based biosensors. They have good electrical properties, electrochemical properties and amenable for immobilisation of biomolecules by various methods. They are stable and biocompatible. They are already exploited for detection of several pathogens and biomarkers for various diseases. Their properties can be enhanced by incorporation of nano gold and platinum. Since they are having one dimensional structure and semiconducting properties they are being studied as potential materials for FET based biosensor. Since they are nano materials, it is possible to make array of sensors on a single platform so that it is possible to develop multianalyte systems. The results of affinity based sensors using CNTs indicate high sensitivity. The practical use of sensors will depend on the shelf life and ease of use. Research work in this direction will be beneficial.

CNT based nanobiosensors may also be used to detect DNA sequences in the body. These instruments detect a very specific piece of DNA that may be related to a particular disease. Therefore, these sensors can possibly diagnose patients as having specific sequences related to a cancer gene. The use of CNT-based sensors will avoid problems associated with the current much-larger implantable sensors, which can cause inflammation, and eliminate the need to draw and test blood samples.

In time, nanodiagnostics may become very cost-effective, as is currently the case with nanotubes arrays. This should allow better clinical diagnostic services. These nanodiagnostics can also be applied to point-of-care testing and lab-on-a-chip technologies. Whether nanodiagnostics will replace current diagnostic methods remains to be seen. Many aspects of these nanodiagnostics techniques need to be evaluated further, especially the safety issues. However, the advantages that these new technologies offer are too good to ignore.

8. References

Abdelwahab, A.A., Won, M.-S., Shim, Y.-B. 2010, Direct electrochemistry of cholesterol oxidase immobilized on a conducting polymer: Application for a cholesterol biosensor. Electroanalysis, 22, 21–25.

Baker, S.E., Cai, W., Lasseter, T.L., Weidkamp, K.P., Hamers, R.J. 2002, Covalently bonded adducts of deoxyribonucleic acid (DNA) oligonucleotides with single-wall carbon nanotubes: synthesis and hybridization. Nano Lett., 2, 1413-1417.

Blodgett, K.B. 1934, Monomolecular films of fatty acids on glass. J. Am. Chem. Soc., 56, 495.

Bourigua, S., Hnaien, M., Bessueille, F., Lagarde, F., Dzyadevych, S., Maaref, A., Bausells, J, Errachid, A., Jaffrejic Renault, N. 2010, Impedimetric immunosensor based on SWCNT-COOH modified gold microelectrodes for label-free detection of deep venous thrombosis biomarker. Biosens, and Bioelectron., 26, 1278-1282.

Braden, B.C., Galbanum, F.A., Chen, B.X., Kirschner, A.N., Wilson, S.R.,Erlanger, B.F. 2000, X-ray crystal structure of an anti-Buckminsterfullerene antibody Fab fragment: Biomolecular recognition of C_{60}. Proc. Natl. Acad. Sci. USA, 97, 12193-12197

Bremer, K., Micus, S., Bremer, G., 1995. CEA, CA 15-3 and MCA: comparative clinical relevance in breast cancer. Eur. J. Cancer, 31, s262.

Buzaneva, E., Karlash, A., Yakovkin, K., Shtogun, Y., Putselyk, S., Zherebetskiy, D., Gorchinskiy, A., Popova, G., Prilutska, S., Matyshevska, O., Prilutskyg, Y., Lytvyn, P., Scharff, P., Eklund, P. 2002, DNA nanotechnology of carbon nanotube cells: Physio-chemical model of self organization and properties. Mater. Sci. Eng. C, 19, 41–45

Cai, H., Cao, X., Jiang, Y., He, P., Fang, Y. 2003, Carbon nanotube-enhanced electrochemical DNA biosensor for DNA hybridization detection. Anal. Bioanal. Chem., 375, 287–293.

Cao, Q., Zhao, H., Yang, Y., He, Y., Ding, N., Wang, J., Wu, Z., Xiang, K., Wang, G. 2011, Electrochemical immunosensor for casein based on gold nanoparticles and poly(l-Arginine)/multi-walled carbon nanotubes composite film functionalized interface. Biosens. and Bioelectron., doi: 10.1016/j.bios.2011.01.027.

Cao, Y.W.C., Jin, R.C., Mirkin, C.A., 2002 Nanoparticles with Raman spectroscopic fingerprints for DNA and RNA detection. Science, 297, 1536–1540.

Cid, C., Riu, J., Maroto, A., Rius, F.X. 2008, Carbon nanotube field effect transistors for the fast and selective detection of human immunoglobulin G. Analyst, 133, 1005–1008.

Chen, W.W., Tzang, C.H., Tang, J.X., Yang, M.S., Lee, S.T. 2005, Covalently linked deoxy ribonucleic acid with multiwall carbon nanotubes–synthesis and characterization. Appl. Phys. Lett., 86, 3114 (Art No. 103114).

Chen, J., Yan, F., Dai, Z., Ju, H. 2005, Reagentless amperometric immunosensor for human chorionic gonadotrophin based on direct electrochemistry of horseradish peroxidase. Biosensors & Bioelectronics, 20, 330–336.

Cheng, G.F., Zhao, J., Tu, Y.H., He, P.A., Fang, Y.H. 2005, A sensitive DNA electrochemical biosensor based on magnetite with a glassy-carbon electrode modified by multiwalled carbon nanotubes in polypyrrole. Anal. Chim. Acta., 533, 11–13.

Cui, R., Huang, H., Yin, Z., Gao, D., Zhu, J-J. 2008, Horseradish peroxidase-functionalized gold nanoparticle label for amplified immunoanalysis based on gold nanoparticles/carbon nanotubes hybrids modified biosensor. Biosens. And Bioelectron., 23, 1666-1673.

Crespilho, F.N., Nart, F.C., Oliveira Jr., O.N., Brett, C.M.A., 2007, Oxygen reduction and diffusion in electroactive nanostructured membranes (ENM) using a layer-by-layer dendrimer-gold nanoparticle approach. Electrochim. Acta, 52, 4649–4653

Daniel, S., Rao, T.P., Rao, K.S., Rani, S. U., Naidu, G.R.K., Li, H-Y., Kawai, T. 2007, A review of DNA functionalized/grafted carbon nanotubes and their characterization. Sens. and Actuator B, 122, 672-682.

Dastagir, T., Forzani, E.S., Zhang, R., Amlani, I., Nagahara, L.A., Tsui, R., Tao, N. 2007, Electrical detection of hepatitis C virus RNA on single wall carbon nanotube-field effect transistors. Analyst, 132, 738–740.

Decher, G., Hong, J. D., Schmitt, J. 1992 Build-up of ultrathin multilayer films by a self-assembly process: III. Consecutively alternating adsorption of anionic and cationic polyelectrolytes on charged surfaces, Thin Solid Films, 210-211, 831-835.

Duffy, M.J., van Dalen, A., Haglund, C., Hansson, L., Klapdor, R., Lamerz, R., Nilsson, O., Sturgeon, C., Topolcan, O., 2003. Clinical utility of biochemical markers in colorectal cancer: European Group on Tumour Markers (EGTM) guidelines. Eur. J. Cancer, 39, 718–727.

Drummond, T.G., Hil, M.G., Barton, J.K. 2003. Electrochemical DNA sensors. Nature Biotech., 21, 1192-1199.

Elbakry, A., Zaky, A., Liebkl, R., Rachel, R., Goepferich, A., Breunig, M., Layer-by-layer assembled gold nanoparticles for siRNA delivery, 2009 Nano Lett., 9, 2059–2064.

Engelen, M.J.A., de Bruijn, H.W.A., Hollema, H., ten Koor, K.A., Willemse, P.H.B., Aalders, J.G., van der Zee, A.G.J., 2000. Serum CA 125, carcinoembryonic antigen, and CA 19-9 as tumor markers in borderline ovarian tumors. Gynecol. Oncol., 78, 16–20.

Erlanger, B. F., Chen, B.X., Zhu, M., Brus L. 2001 Binding of an anti-fullerene IgG monoclonal antibody to single wall carbon nanotubes, Nano Lett., 1, 465-467

Firkin, F., Nandurkar, H. 2009. Australian Prescriber, 32, 148–150.

Gabl, R., Feucht, H.D., Zeininger, H., Eckstein, G., Schreiter, M., Primig, R., Pitzer, D., Wersing, W. 2004, First results on label-free detection of DNA and protein molecule using a novel integrated sensor technology based on gravimetric detection principles, Biosens. Bioelectron., 19, 615-620.

García-Aljaro, C., Bangar, M.A., Baldrich, E., Munoz, F.J., Mulchandani, A. 2010, Conducting polymer nanowire-based chemiresistive biosensor for the detection of bacterial spores, Biosens. Bioelectron., 25, 2309–2312.

Gold, P., Freedman, S.O., 1965. Demonstration of tumor-specific antigens inhuman colonic carcinomata by immunological tolerance and absorption techniques. J. Exp. Med., 121, 439–462.

Gubitz, G., Shellum, C. 1993, Flow-injection immunoassay. Anal. Chim. Acta, 283, 421–428.

Gui, E.L., Li, L.J., Zhang, K., Xu, Y., Dong, X., Ho, X., Lee, P.S., Kasim, J., Shen, Z.X., Rogers, J.A., Mhaisalkar, S.G. 2007, DNA sensing by field-effect transistors based on networks of carbon nanotubes. JACS, 129 14427–14432.

Guo, M., Chen, Liu, J., D., Nie, L., Yao, S. 2004, Electrochemical characteristics of the immobilization of calf thymus DNA molecules on multi-walled carbon nanotubes. Bioelectrochemistry, 62, 29–35.

Hazani, M., Hennrich, F., Kappes, M., Naaman, R., Peled, D., Sidorov, V., Shvarts D. 2004, DNA-mediated self-assembly of carbon nanotube-based electronic devices. Chem. Phys. Lett., 391, 389–392.

Hazani, M., Naaman, R., Hennrich, F., Kappes, M.M. 2003, Confocal fluorescence imaging of DNA-functionalized carbon nanotubes. Nano Lett., 3, 153–155.

Hernandez, L., Espasa, A., Fernandez, C., Candela, A., Martin, C., Romero, S., 2002. CEA and CA 549 in serum and pleural fluid of patients with pleural effusion. Lung Cancer, 36, 83–89.

He, P., Wang, Z., Zhang, L., Yang, W. 2009, Development of a label-free electrochemical immunosensor based on carbon nanotube for rapid determination of clenbuterol. Food Chemistry, 112, 707–714.

Huang, E., Satjapipat, M., Han, S., Zhou, F. 2001, Surface structure and coverage of an oligonucleotide probe tethered onto a gold substrate and its hybridization efficiency for a polynucleotide target. Langmuir, 17, 1215-1224.

Jian, J. W., Xiao, Q. X. 1986, Phytomedicine effective component book. Beijing: People's Hygiene Press.

Jezersek, B., Cervek, J., Rudolf, Z., Novakovic, S., 1996. Clinical evaluation of potential usefulness of CEA, CA 15-3, and MCA in follow-up of breast cancer patients. Cancer Lett., 110, 137–144.

Jiang, Wen., Yuan, R., Chai, Y-Q., Yin, B. 2010, Amperometric immunosensor based on multiwalled carbon nanotubes / Prussian blue / nanogold-modified electrode for determination of α-fetoprotein. Anal. Biochem., 407, 65-71.

Jung, D.-H., Kim, B.H., Ko, Y.K., Jung, M.S., Jung, S., Lee, S.Y., Jung, H.-T. 2004, Covalent attachment and hybridization of DNA oligonucleotides on patterned single-walled carbon nanotube films. Langmuir, 20, 8886-8891

Kara, P., Meric, B. 2004, Electrochemical DNA biosensor for the detection and discrimination of herpes simplex Type I and Type II viruses from PCR amplified real samples. Anal. Chim. Acta, 518, 69-76

Kim, J.P., Lee, B.Y., Hong, S., Sim, S.J. 2008, Ultrasensitive carbon nanotube-based biosensors using antibody-binding fragments. Anal. Biochem., 381, 193-198.

Kleisbauer, J.P., Castelnau, O., Thomas, P., Ramirez, J., Lanteaume, A., Roux, F., 1996. Prognostic value of carcinoembryonic antigen in lung carcinoma before treatment. Lung Cancer, 15, 148.

Komarova, E., Aldissi, M., Bogomolova, A. 2005, Direct electrochemical sensor for fast reagent-free DNA detection. Biosens. Bioelectron., 21, 182-189.

Kuiyang J., Linda, S. Schadler, R., Siegel, W., Xinjie Z., Haifeng, Z., Mauricio, T. 2004 Protein immobilization on carbon nanotubes via a two-step process of diimide-activated amidation, J. Mater. Chem., 14, 37–39.

Lamas- Ardisana, P.,A., Queipo, P., Fanjul-Bolado, P., Costa-garcia, A. 2008, Multiwalled carbon nanotube modified screen printed electrodes for the detection of p-amino phenol: Optimization and application in alkaline phosphatase- based assays. Anal. Chem., 615, 30-38

Lee, S.W., Kim, B.S., Chen, S., Shao Horn, Y., Hammond, P.T., 2009 Layer-by-layer assembly of all carbon nanotube ultrathin films for electrochemical applications. J. Am. Chem. Soc., 131, 671–679.

Lee, T.-Y., Shim, Y.-B. 2001, Hybridization biosensor based on the covalent immobilization of probe DNA on chitosan - mutiwalled carbon nanotubes nanocomposite by using glutaraldehyde as an arm linker. Anal. Chem., 73, 5629–5632.

Lee, G., Chen, C., Wang, T., Lee, W. 2003, Affinity chromatography of DNA on nonporous copolymerized particles of styrene and glycidyl methacrylate with immobilized polynucleotide, Anal. Biochem., 312, 235 - 241.

Lei, C., Gong, F., Shen, G., Yu, R. 2003, Amperometric immunosensor for Schistosoma japonicum antigen using antibodies loaded on a nano-Au monolayer modified chitosan-entrapped carbon paste electrode. Sens. and Actuators B, 96, 582–588.

Li, S., He, P., Dong, J., Guo, Z., Dai, L. 2005, DNA-directed self-assembling of carbon nanotubes. J. Am. Chem. Soc., 127, 14–15.

Li, W., Yuan, R., Chai, Y., S. Chen. 2010, Reagentless amperometric cancer antigen 15-3 immunosensor based onenzyme-mediated direct electrochemistry. Biosens. and Bioelectron., 25, 2548-2552.

Li, Y., Liu, D., Cao, B., Han, W., Liu, Y., Liu, F., Guo, X., Bastin, D.A., Feng, L., Wang, L. 2006, Development of a serotype-specific DNA microarray for identification of some Shigella and pathogenic Escherichia coli strains. J. Clin. Microbiol., 44, 4376–4383.

Li, Z., Jin, R.C., Mirkin, C.A., Letsinger, R.L., 2002 Multiple thiol- anchor capped DNA-gold nanoparticle conjugates, Nucleic Acids Res., 30, 1558–1562.

Liaw.H.W., Chen.J.M., Tsai. Y.C. 2006, Development of an amperometric ethanol biosensor based on a multiwalled carbon nanotube-Nafion-alcohol dehydrogenase nanobiocomposite. J. Nanosci. Nanotechnol., 6, 2396–2402

Lin. J., He. C., Zhao. C., Zhang. S. 2009, One step synthesis of silver nanoparticles/carbon nanotubes/ chitosan films and its application in glucose biosensor. Sens. and Actuator B, 137, 768-773.

Lin,J., He, C., Zhang,L., Zhang S. 2009, Sensitive amperometric immunosensor for a-fetoprotein based on carbon nanotube/gold nanoparticle doped chitosan film. Anal. Biochem., 384, 130-135.

Liu, C.H., Guo, X.L., Cui, H.T., Yuan, R. 2009, An amperometric biosensor fabricated from electro-co-deposition of sodium alginate and horseradish peroxidase. J. Mol. Catal. B, 60, 151–156

Liu. Y., Wang. M., Zhao. F., Xu. Z., Dong. S., 2005, The direct electron transfer of glucose oxidase and glucose biosensor based on carbon nanotubes/chitosan matrix. Biosens. Bioelectron., 21, 984–988

Lu, G., Maragakis, P., Kaxiras E. 2005, Carbon nanotube interaction with DNA. Nano Lett., 5, 897–900.

Lvov, K., Ariga, I., Ichinose, Kunitake, T., 1995 Assembly of Multicomponent Protein Films by Means of Electrostatic Layer-by-Layer Adsorption, J. Am. Chem. Soc., 117, 6117–6123.

Maehashi, K., Katsura, T., Kerman, K., Takamura, Y., Matsumoto, K., Tamiya, E. 2007, Label-free protein biosensor based on aptamer-modified carbon nanotube field-effect transistors. Anal. Chem., 79, 782–787.

Malhotra, B.D., Singhal, R., Chaubey, A., Sharma, S.K., Kumar, A. 2005, Recent trends in biosensors. Current Appl. Phys., 5, 92-97.

Martel, R., Schmidt, T., Shea, H.R., Hertel, T., Avouris, Ph. 1998, Single- and multi-wall carbon nanotube field-effect transistors. Appl. Phys. Lett., 73 2447.

Maxwel, D.J., Taylor, J.R., Me, S.M., 2002 Self-assembled nanoparticles probes for recognition and detection of biomolecules, J. Am. Chem. Soc., 124, 9606–9612.

Merken, H. M., Beecher, G. R. 2000, Measurement of food flavonoids by highperformance liquid chromatography: A review. Journal of Agricultural and Food Chemistry, 48, 3, 577–599.

Moghaddam, M.J., Taylor, S., Gao, M., Huang, S., Dai, L., McCall, J.M. 2004, Highly efficient binding of DNA on the sidewalls and tips of carbon nanotubes using photochemistry. Nano Lett., 4, 89–93.

Mulder, M. 2000, Basic principles of membrane technology. Kluwer Academic Publishers, Dordrecht, The Netherlands.

Nakashima. N., Okuzono. S., Marakami. H., Nakai. T., Yoshikawa. K. 2003, DNA dissolves single-walled carbon nanotubes in water. Chem. Lett., 32, 456–457.

O'Connor, M., Kim, S.N., Killard, A.J., Forster, R.J., Smyth, M.R., Papadimi-trakopoulos, Rusling, J.F. 2004, Mediated amperometric immunosensing using single walled carbon nanotube forests. Analyst, 129, 1176–1180.

Okuno, J., Maehashi. K., Kerman, K., Takamura, Y., Matsumoto, K., Tamiya, E. 2007, Label-free immunosensor for prostate-specific antigen based on single-walled carbon nanotube array-modified microelectrodes. Biosens. and Bioelectron., 22, 2377-2381.

Panini, N. V., Messina, G. A., Salinas, E., Fernandez. H., Raba, Julio. 2008, Integrated microfluidic systems with an immunosensor modified with carbon nanotubes for detection of prostate specific antigen (PSA) in human serum samples.Biosens. and Bioelectron., 23, 1145-1151.

Pearson, E., Gill, A., Vadgama, P. 2000, Analytical aspect of biosensors, Ann. Clin. Biochem., 37, 119-145.

Peelen, D., Smith, L.M. 2005, Immobilization of amine-modified oligonucleotides on aldehyde-terminated alkanethiol monolayers on gold. Langmuir, 21, 266-271.

Phuong Dinh Tama, Nguyen Van Hieu , Nguyen Duc Chien , Anh-Tuan Le a, Mai Anh Tuan DNA sensor development based on multi-wall carbon nanotubes forlabel-free influenza virus (type A) detection Journal of Immunolological methods, 350 (2009) 118–124.

Piao, Y., Zin, Z., Lee, D., Lee, H-J., Na, H-B., Hyeon, T., Oh, M-K., Kim, J., Kim, H-S. 2011, Sensitive and high-fidelity electrochemical immunoassay using carbon nanotubes coated with enzymes and magnetic nanoparticles. Biosens. and Bioelectron., 26, 3192-3199

Popov, V.N. 2004, Carbon nanotubes: Properties and application. Material Science and Engineering R., 43, 61-102.

Rehm, B.H.A., Valla, S. 1997, Bacterial alginates: biosynthesis and applications. Appl. Microbiol. Biotechnol., 48, 281–288.

Robards, K., Antolovich, M. 1997, Analytical chemistry of fruit bioflavonoids: A review. Analyst, 122, 11R–34R.

Robertson, D.H., Brenner, D.W., Mintmire, J.W. 1992, Energetics of nanoscale graphitic tubules, Phys. Rev. B, 45, 12592.

Tsai. Y.C., Li. S.C., Chen. J.M. 2005, Cast thin film biosensor design based on a Nafion backbone, a multiwalled carbon nanotube conduit, and a glucose oxidase function. Langmuir, 21, 3653–3658.

Tsai. Y.C., Chen. S.Y., Liaw. H.W. 2007, Immobilization of lactate dehydrogenase within multiwalled carbon nanotube-chitosan nanocomposite for application to lactate biosensors. Sens. Actuators B: Chem., 125, 474–481

Sahin, B., Paydak, V., Paydas, S., 1996. Hepatosteatosis and alterations of CA15-3 and CEA in patients with breast cancer receiving tamoxifen. Eur. J. Cancer, 32, s24.

Sanchez, S., Monika, R., Perez, S., Fabregas, E., 2008 Toward a Fast, Easy, and Versatile Immobilization of Biomolecules into Carbon Nanotube/Polysulfone-Based Biosensors for the Detection of hCG Hormone, Anal. Chem., 80, 6508–6514.

Sanchez, S., Pumera, M., Fabregas, E. 2007, Carbon nanotube/polysulfone screen-printedelectrochemical immunosensor. Biosens. and Bioelectron., 23, 332-340.

Saoudi, B., Jammul, N., Abel, M.-L., Chehimi, M.M., Dodin, G. 1997, DNA adsorption onto conducting polypyrrole. Synthetic Metals, 87, 97-103.

Schlageter, M.H., Larghero, J., Cassinat, B., Toubert, M.E., Borschneck, C., Rain, J.D., 1998. Serum carcinoembryonic antigen, cancer antigen 125, cancer antigen 15-3, squamous cell carcinoma, and tumor-associated trypsin inhibitor concentrations during healthy pregnancy. Clin. Chem., 44, 1995–1998.

Sharma, M., K., Rao, V., K., Agarwal, G., S., Rai, G., P., Gopalan, N., Prakash, S., Sharma, S., K., Vijayaraghvan, R. 2008, Highly Sensitive Amperometric Immunosensor for Detection of *Plasmodium falciparum* Histidine-Rich Protein 2 in Serum of Humans with Malaria: Comparison with a Commercial Kit, 46, J.Clin. Microbiology., 3759-3765

Singh, K.V., Pandey, R.R., Wang, X., Lake, R., Ozkan, C.S., Wang, K., Ozkan, M. 2006, Covalent functionalization of single walled carbon nanotubes with peptide nucleic acid: nanocomponents for molecular level electronics. Carbon, 44, 1730–1739

Skládal, P.1997, Advances in electrochemical immunosensors. Electroanal., 9, 737–745.

So, H.M., Park, D.W., Jeon, E.K., Kim, Y.H., Kim, S.K., Lee, C.K., Choi, S.Y., Kim, S.C., Chang, H., Lee, J.O. 2008, Detection and titer estimation of Escherichia coli using aptamer- functionalized single-walled carbon-nanotube field-effect transistors. Small, 4, 197–201

So, H.M., Won, K., Kim, Y.H., Kim, B.K., Ryu, B.H., Na, P.S., Kim, H., Lee, O.J. 2005, Single-walled carbon nanotube biosensors using aptamers as molecular recognition elements. JACS, 127, 11906–11907.

So, H.M., Won, K., Kim, Y.H., Kim, B.K., Ryu, B.H., Na, P.S., Kim, H., Star, A., Tu, E., Niemann, E.J., Gabriel, J.C.P., Joiner, C.S., Valcke, C. 2006, Label-free detection of DNA hybridization using carbon nanotube network field-effect transistors. PNAS, 103, 921–926.

Suresh, S., Gupta, A.K., Rao, V.K., Om Kumar., Vijayaraghavan, R. 2010, Amperometric immunosensor for ricin by using on graphite and carbon Nanotube paste electrodes. Talanta, 81, 703–708.

Star, A., Tu, E., Niemann, J., Gabriel, J.C.P., Joiner, C.S., Valcke, C. 2006, Label-free detection of DNA hybridization using carbon nanotube network field-effect transistors. PNAS 103, 921–926.

Taft, B.J., Lazareck, A.D., Withey, G.D., Yin, A., Xu, J.M., Kelley, S.O. 2004, Site-specific assembly of DNA and appended cargo on arrayed carbon nanotubes. J. Am. Chem. Soc., 126, 12750–12751

Tam, P.D., Hiew, N.V., Chein, N.D., Le, A-T., Tuan, A-N. 2009, DNA sensor development based on multi-wall carbon nanotubes for label-free influenza virus (type A) detection. J. Immunological methods, 350, 118-124

Tang, J., Tang, D., Su, B., Huang, J., Qiu, B., Chen, G. 2011, Enzyme-free electrochemical immunoassay with catalytic reduction of p-nitrophenol and recycling of p-aminophenol using gold nanoparticles-coated carbon nanotubes as nanocatalysts. Biosens. and Bioelectron., 26, 3219-3226.

Tans, S.J., Verschueren, A.R., Dekker, C. 1998, Room-temperature transistor based on a single carbon nanotube. Nature, 393, 49–52

Valentini. F., Biagiotti. V., Lete. C., Palleschi. G., Wang. J. 2007, The electrochemical detection of ammonia in drinking water based on multiwalled carbon nanotube/copper nanoparticle composite paste electrode. Sens. and Actuators B, 128, 326-333.

Villamizar, R.A., Maroto, A., Rius, F.X., Inza, I., Figueras, M.J. 2008, Fast detection of Salmonella Infantis with carbon nanotube field effect transistors. Biosens. Bioelectron., 24, 279–283

Villamizar, R. A., Maroto, A., Rius, F. X. 2009, Improved detection of Candida albicans with carbon nanotube field-effect transistors. Sens. And Actuator B, 136, 451-457.

Viswanathan, S., Rani. C., Vijay, A.A., Annie, H.J. 2009, Disposable electrochemical immunosensor for carcinoembryonic antigen using ferrocine liposomes and MWCNT screen printed electrode. Biosens. Bioelectron., 24, 1984-1989.

Vishwanathan. S., Wu. L-c., Huang. M-R., Ho. J-a. A. 2006, Electrochemical immunosensor for cholera toxin using liposomes and poly (3,4-ethylene deoxy thiophene)-coated carbon nanotubes. Anal. Chem., 78, 1115-1121.

Wang, J. 2002, Electrochemical nucleic acid biosensors. Anal. Chim. Acta, 469, 63-71.

Wang, J., Kawde, A., Jan, M.R. 2004, Carbon-nanotube-modified electrodes for amplified enzyme-based electrical detection of DNA hybridization. Biosens. Bioelectr., 20, 995–1000.

Wang, J., Liu, G., Jan, M.R., Zhu, Q. 2003, Electrochemical detection of DNA hybridization based on carbon-nanotubes loaded with CdS tags. Electrochem. Commun. 5, 1000–1004.

Wang, J., Liu, G., Jan, M.R. 2004, Ultrasensitive electrical biosensing of proteins and DNA: carbon-nanotube derived amplification of the recognition and transduction events,. J. Am. Chem. Soc., 126, 3010–3011.

Wang, Q., Zhang, B., Lin, X., Weng, W. 2011, Hybridization biosensor based on the covalent immobilization of probe DNA on chitosan - mutiwalled carbon nanotubes nanocomposite by using glutaraldehyde as an arm linker. Sens. and actuator B, doi:10.1016/j.snb.2011.02.004

Wang, S.G., Wang, R., Sellin, P.J., Zhang, Q. 2004, DNA biosensors based on self-assembled carbon nanotubes. Biochem. Biophys. Res. Commun., 325, 1433–1437

Wang, S., Humphreys, E.S., Chung, S.Y., Delduco, D.F., Lustig, S,R., Wang, H., Parker, K.N., Subramoney, S., Chiang, Y.M., Jagota, A, 2003 Peptides with selective affinity for carbon nanotubes Nat. Mater. 2, 196–200.

Wen. Z., Ci. S., Li. J. 2009, Pt nanoparticle inserting in carbon nanotube arrays: Nanocomposites for glucose biosensors. J. Phys. Chem., 113, 13482-13487.

Williams, K.A., Peter, T.M., Veenhuizen Torre, B.G., Eritja, R., Dekker, C. 2002, Nanotechnology: carbon nanotubes with DNA recognition. Nature, 420, 761–762.

Weber, J.E., Pillai, S., Ram, M.K., Kumar, A., Singh, S.R. 2011, Electrochemical impedance-based DNA sensor using a modified single walled carbon nanotube electrode. Material science and Engineering C, 10.1016/j.msec.2010.12.009.

Wu, J. Tang. J., Dai. Z., Yan. F., Ju. H., Murr. N. E. 2006, A disposable electrochemical immunosensor for flow injection immunoassay of carcinoembryonic antigen. Biosens. Bioelectron., 22, 102-108.

Xu, Z., Hu, P., Wang, S., Wang, X., 2008 Biological functionalization and fluorescent imaging of carbon nanotubes, Appl. Sur. Sci., 254, 1915-1918.

Yan. W., Pang. D.W., Wang. S.F., Zho. Y.D. 2005, Carbon nanomaterials–DNA bioconjugates and their applications. Fullerenes, nanotubes, and carbon nanostructures, 13, 309–318.

Yu, H., Yan, F., Dai, Z., Ju, H. 2004, A disposable amperometric immunosensor for α-1-fetoprotein based on enzyme-labeled antibody/chitosan-membrane-modified screen-printed carbon electrode. Anal. Biochem., 331, 98-105.

Yu, X., Kim, S.N., Papadimitrakopoulos, F., Rusling, J.F. 2005, Protein immunosensor using single-wall carbon nanotube forests with electrochemical detection of enzyme labels. Mol. BioSyst., 1, 70–78.

Zhang, X.,Wang, S., Hu, M., Xiao, Y. 2006, An immunosensor for ferritin based on agarose hydrogel. Biosens. Bioelectron., 21, 2180–2183.

Zhao, G., Zhan, X., Dou, W. 2011, A disposable immunosensor for Shigella flexneri based on multiwalled carbon nanotube/sodium alginate composite electrode. Anal. Biochem., 408, 53-58.

Zhao, H.Y., Zheng, W., Meng, Z.X., Zhou, H.M., Xu, X.X., Li, Z., Zheng, Y.F. 2009, Bioelectrochemistry of hemoglobin immobilized on a sodium alginate-multiwall carbon nanotubes composite film. Biosens. Bioelectron., 24, 2352–2357.

Zhao, D., Zhang, X., Feng, L., Qi, Q., Wang, S. 2011, Sensitive electrochemical determination of luteolin in peanut hulls using multi-walled carbon nanotubes modified electrode. Food Chemistry, 126, 694-698.

Zhou, Y., Yuan, R., Chai, Y., Tang, D., Zhang, Y., Wang, N., Li, X., Zhu, Q. 2005, A reagentless amperometric immunosensor based on gold nanoparticles/ thionine/Nafion-membrane-modified gold electrode for determination of a-1-fetoprotein. Electrochemistry Communications, 7, 355–360.

Zhu, Y., Son, J.I., Shim, Y-B. 2010, Amplification strategy based on gold nanoparticle-decorated carbon nanotubes for neomycin immunosensors. Biosens. Bioelectron., 26, 1002-1008.

Zhu. Z.Z. 2000, Modern Medical Labelled immunology (in Chinese), People's Military Doctor Press 56.

Zucolotto, V., Daghastanli, K.R.P., Hayasaka, C.O., Riul Jr., A., Ciancaglini, Oliveira Jr., O.N. 2007, Using capacitance measurements as the detection method in antigen-containing layer-by-layer films for biosensing. Anal. Chem., 79, 2163–2167.

8

Imaging and Biomedical Application of Magnetic Carbon Nanotubes

O. Vittorio[1,2], S. L. Duce[3], V. Raffa[1] and A. Cuschieri[1,4]
[1]*Medical Science lab, Scuola Superiore Sant'Anna, Pisa*
[2]*Department of Oncology, Transplantation and Advanced Technologies in Medicine, University of Pisa*
[3]*Division of Biological Chemistry and Drug Discovery, College of Life Sciences, University of Dundee*
[4]*Institute for Medical Science and Technology, University of Dundee*
[1,2]*Italy*
[3,4]*UK*

1. Introduction

The concepts of 'nano-technology' was introduced by the physicist Richard Feynman in his talk "There's Plenty of Room at the Bottom" at an American Physical Society meeting at Caltech on December 29, 1959. Feynman described a process by which individual atoms and molecules might be manipulate, using a set of precise tools to build and operate another smaller scale set of tools, scaling down to the nano-scale. In the course of this, he noted, scaling issues would arise from the changing magnitude of various physical phenomena: gravity would become less important, surface tension and Van Der Waals attraction would become increasingly more significant, etc. The term "nanotechnology" was defined by Professor Norio Taniguchi of Tokyo Science University in a 1974 paper in which he states "'Nano-technology' mainly consists of the processing of, separation, consolidation, and deformation of materials by one atom or by one molecule." Almost four decades later, nanotechnology has had an impact on all sectors of human life including electronics, computers and mobile phones; food and agriculture industries; composite materials, textiles, paints and cosmetics; and, of course, healthcare. In today's market there are thousands of products based on nanotechnology, produced by hundreds of companies worldwide. Essentially, nanotechnology entails the manufacturing and manipulation of matter at a scale ranging from a single atom to micron-sized objects. In biology, nanomaterials are a comparable size to many biological functional molecules such as proteins and are often small enough to fit inside a cell. Being at the same microscopic scale as biological functions allow nanoparticles to interact with many biological processes, this potentially can have an impact in many aspects of the healthcare. The new field of 'Nanomedicine' (Allhoff, 2009) will not meets its immense potential until the safety of nanomaterials is fully demonstrated. It is prudent to investigate any potential adverse effects on health or the environment of nanomaterials. However with suitable safe guards and internationally agreed standards emanating from these nano-safety and toxicity studies, there seems little reason to doubt the overriding benefit that nanomedicines will provide in the 21st Century (Jain et al., 2008).

Nanomedicine has been defined as the application of nanotechnology in healthcare. Their size and shape confers them with unique electrical, thermal, optical and magnetic properties (Emerich et al., 2007). They have large surface area to volume ratio and if chemically modified increases their application. In principle, it is possible to fabricate nanoparticles that can be used in the early detection and prevention of diseases, to improve diagnosis, treatment and follow-up. It has already enabled miniaturization of many current devices resulting in faster operation or enhanced integration of several operations. There are many biomedical nanoparticles on the market, with different physical and chemical properties. In the following discussion we will be investigating the potential use of carbon nanotubes in biomedical applications. This will include safety issue, and an overview of the current and the future perspectives of the exploitation of their magnetic properties for imaging and therapies.

2. Carbon nanotubes

Carbon nanotubes are either single-wall (SWCNTs) consisting of a single graphite lattice rolled into a perfect cylinder or multi-wall (MWCNTs) made up of several concentric cylindrical graphite shells (Russian doll configuration). CNTs are usually produced by catalytic chemical vapour deposition and contain metals, mainly Fe at their closed ends (Kim et al., 2005). For this reason they are paramagnetic - a valuable property for certain biomedical applications. CNTs vary in diameter (from a few nm to 100 nm) and widely in length (up to several mm). Their molecular structure accounts for their unique properties: high tensile strength, high electrical conductivity, heat resistance and efficient thermal conduction and relative chemical inactivity (constituent atoms not easily displaced). The exact structure of CNT especially their n - m chirality determines their electric properties. When n-m is a multiple of 3 (armchair type), the CNTs have static dielectric properties, i.e. exhibit a metallic longitudinal and an insulator transverse response. In practice all multiwalled CNTs behave in this way. By virtue of their nano-scale, electron transport in CNTs occurs through quantum effects and thus only propagates uni-dimensionally along the axis of the tube. One of the problems which have impeded the use of CNTs for biomedical applications which has since been resolved is their insolubility in aqueous solution. This is an essential requirement for biological interactions and biocompatibility. Coating their surface with covalent and non-covalent polymer improves their solubility, enabling in-vitro cell viability assays and in-vivo studies on biocompatibility (Kagan et al., 2010; Dutta et al., 2007). Carbon nanotubes will interact with cells and their 'needle-like' shape helps them cross cell membranes. By virtue of this characteristic, they can be used as carriers for drug and DNA delivery. Several groups have demonstrated that both SWCNTs and MWCNTs can be internalized by a variety of cell types with no external agent required facilitating the delivery of therapeutic and diagnostic small molecules (Kostarelos et al., 2007; Shi Kam et al., 2005).

2.1 Open issue: CNTs toxicity

In recent years, conflicting data have been reported concerning safety and biocompatibility of these nanotubes. In a study on the cytotoxicity of unrefined SWCNT to immortalized human epidermal keratinocytes (HaCaT), Shvedova et al showed accelerated oxidative stress, loss in cell viability and morphological alterations of cellular structures. They concluded that those effects were the result of high concentration of residual iron catalyst

(30%) present in the unrefined SWCNTs (Shvedova et al, 2003). Other groups confirmed these toxic effects, e.g., induction of intracellular reactive oxygen species (ROS) (Pulskamp et al., 2007), DNA damage (Zhu et al., 2007) and cell apoptosis (Bottini et al., 2007). Recently, Poland et al demonstrated that exposure of the mesothelial lining of the peritoneal cavity of mice to long (10-15mm) multi-walled carbon nanotubes results in asbestos-like, length-dependent, pathological inflammation and the formation of giant cell granulomas (Poland et al., 2008). In sharp contrast, other reports have demonstrated that CNTs do not induce toxic effects on cells. Huczko et al showed that CNTs exhibited negligible risk of skin irritation and allergy (Huczko et al., 2001). Cherukuri et al observed that macrophages phagocytosed SWCNTs at the rate of approximately one SWCNT per second without any apparent cytotoxicity (Cherukuri et al, 2004). Kam et al reported no cytotoxicity for the pristine SWCNTs and Pantarotto et al concluded that CNTs coated DNA provided a useful vector for safe gene delivery (Shi Kam et al., 2004; Pantarotto et al., 2004).

The exact fate of carbon nanotubes inside the cells and in animals remains controversial. In vitro studies showed that SWCNTs can be degraded by living cells which cause their complete biodegradation (Kagan et al., 2010); however the degradation of MWCNTs has not been demonstrated. Only a few studies have been published on the bio-distribution of carbon nanotubes in vivo. Singh et al. studied CNT bio-distribution following intraperitoneal administration in mice and they observed rapid blood clearance from systemic blood circulation through renal excretion. Moreover, urinary excretion studies using both f-SWNT and functionalized MWCNT followed by electron microscopy analysis of urine samples revealed that nanotubes were excreted as intact nanotubes after three hours (Singh et al., 2006). More work is needed to understand the clearance of nanotubes and more fully characterise what influences their biodistribution, degradation and excretation. A major source of toxicity is the presence of impurities, so it is important that CNT should be fully characterised reporting their length, purity, metal content and carbon soot. However, the research in this field is not conclusive for two reasons: 1) the absence of detailed data about the characterization of the nanomaterials confuses the interpretation of toxicity data; 2) the reaction that CNTs have with certain reagents in the in vitro cell viability studies invalidates the results of these assays.. Wörle-Knirsch et al. demonstrates interferences of carbon nanotubes with the MTT proliferation assay (Wörle-Knirsch et al., 2006). For this reason, the use of different and independent assays to study toxicological effects of nanoparticles has been proposed and generally accepted. In order to define valid guidelines for toxicological studies on nanomaterials, the EU has recently established the Nano-safety Cluster - a network of researchers and toxicologists. They will consider the somewhat conflicting results and develop appropriate guidelines. In addition, there is an urgent need for the definition of 'medical grade' CNT in terms of length, purity and metal content.

2.2 CNTs properties and their biomedical applications: The future of nanomedicine

In the last ten years, several groups have shown that SWCNTs and MWCNTs can be used as excellent intracellular transporters to deliver therapeutic and diagnostic small molecules and macromolecules to cells. Different uptake mechanisms (phagocytosis, diffusion and endocytosis) have been reported in the literature. Some physico-chemical characteristics of the carbon nanotubes (e.g., the nanotube dispersion, the formation of supramolecular complexes, and the nanotube length) drive the uptake pathway (Raffa et al., 2010). Both SWCNTs and MWCNTs can be internalized by a variety of cell types and thus used to deliver therapeutic and diagnostic molecules. Conjugation of CNTs with different molecules can be used for new vaccine production, novel therapies against retrovirus infection and tumor cell proliferation. Pantarotto et al reported that VP-1 protein of the foot-and-mouth disease virus (FMDV) covalently linked to SWCNT induced a specific anti-body response in vivo without any cross reactivity (Pantarotto et al., 2003). Liu et al transfected human T cells and peripheral blood mononuclear cells with siRNA molecules conjugated to CNTs to abrogate the expression of cell-surface receptors CD4 and co-receptors CXCR4 necessary for HIV entry and T cells infection (Liu et al., 2007). Additionally McDevitt et al constructed a specific CNT conjugated antibody to target the CD20 epitope on Human Burkitt lymphoma cells and simultaneously deliver a radionuclide (McDevitt et al 2007). The physical and chemical properties of MWCNTs (e.g., high strength, electrical conductivity, flexibility, functionalization with biomolecules) make them attractive as nano-vectors for enhanced cell and tissue growth on scaffolds in vitro (Balani et al., 2007) and for the development of complex neural prosthetic implants (Khabashesku et al 2005).

CNTs also possess intriguing magnetic properties which derive from the metal catalyst impurities entrapped at the CNT extremities during their manufacture, these magnetic fields will interact with external magnetic fields. This property has been utilised by Cai et al. to develop a physical technique for in-vitro and ex-vivo gene transfer known as 'nanotube spearing', capable of effective cell transfection with plasmid DNA (Cai et al., 2005). Similarly, we have recently demonstrated that MWCNTs are able to interact with cells and, when exposed to a magnetic field, induce their migration towards the magnetic source (Vittorio et al 2010). This control of cell movement has important medical applications both in cell therapy for regeneration, cell transplantation and in cancer therapy (anti-metastasis). Moreover, we recently have demonstrated that MWCNTs with low metal impurities (2.57 % iron) can be used as MRI (magnetic resonance imaging) contrast agents even at concentrations of tens of μg/ml, as the MWCNTs have a significant effect on the observed ^{1}H transverse ($1/T_2$) relaxation rate of water. Consequently, cells labelled with MWCNTs exhibit a reduced image intensity in T_2-weighted MR images compared to cells without internalized MWCNTs. The 3D MRI cellular study suggests that it should be possible to track stem cells injected in vivo by labeling cells with these low Fe MWCNTs. It is possible that MRI could be used in the studies of biodistribution and the fate of carbon nanotubes *in vivo* (Vittorio et al 2011). Another important property of CNTs is their strong absorbance of near infra red (NIR) light and subsequent release of this heat that can be utilised in the destruction of cells – nanohyperthermic ablation of tumours. Shi Kam and colleagues achieved selective cancer cell destruction in vitro by using folate functionalized nanotubes heating and continuous NIR radiation (Shi Kam et al 2005 a). Additionally CNTs acquire and release heat on application of radio-frequency waves (58). Gannon et al. induced efficient heating of aqueous suspensions of SWNTs by applying RF waves. In particular they produced a selective and SWNT concentration dependent thermal destruction in vitro of human cancer cells that contained internalized SWNTs. Moreover they observed that intratumoral in vivo injection of SWNTs in the liver followed by immediate exposure to RF waves induced necrosis in the tumor mass with no apparent adverse effects on the healthy tissues (Gannon et al., 2007). A recent discovery (and patented by the NINIVE consortium) is the ability of CNTs to act as a dipole antenna and acquire a charge on exposure to electromagnetic radiation in the microwave range. This is the basis of an exciting new technology for remote wireless low voltage electro-chemotherapy and gene transfection and novel forms of electro-stimulation therapies for neurodegenerative disease (advanced

Parkinson's disease unresponsive to medication), electrostimulation for skeletal and visceral muscle palsies and for cardiac arrhythmias.

Several groups have investigated the potential of using nanoparticles in neurological applications. As carbon nanotubes/fibers have excellent electrical conductivity, strong mechanical properties and similar nano-scale dimensions to neurites, researcher have been explored their ability to guide axonal regeneration and to improve neural activity by acting as biomimetic scaffolds at sites of nerve injury. Mattson et al were the first group to demonstrate neurons grow on MWCNTs (Mattson et al., 2000). Moreover they reported an increase of over 200% in total neurite length and approximately 300% increase in the number of branches and neurites on MWCNTs coated with 4-hydroxynonenal compared to uncoated MWCNTs. Hu et al reported that positively charged MWCNTs significantly increased the number of growth cones and neurite branches compared to negatively charged MWCNTs (Hu et al., 2004). Gheith et al investigated the biocompatibility of a free-standing positively charged single wall carbon nanotubes (SWCNT)/polymer thin-film membrane prepared by layer-by-layer assembly (Gheith et al., 2005). They observed a 94-98% viability of neurons on the SWCNT/polymer films after 10 days of incubation and this induced neuronal cell differentiation, guided neuron extension and directed more elaborate branches than controls. Moreover, Lovat et al demonstrated that purified MWCNTs have the potential to boost electrical signal transfer of neuronal networks (Lovat et al., 2005). Recently Cellot et al investigated the nature of CNT-neuron interactions and proposes a mechanism in which carbon nanotubes (CNTs) can boost neuronal activity by providing a shortcut for electrical coupling between somatic and dendritic neuronal compartments (Cellot et al., 2009).

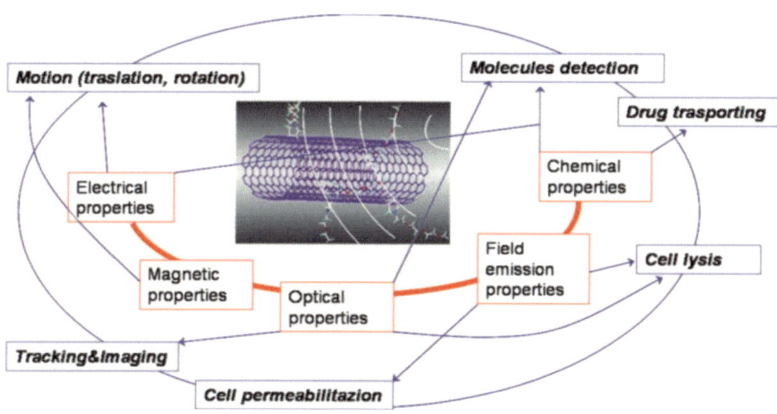

Fig. 1. CNT properties and their application in the biomedical field (Ciofani, G., Raffa, V., 2009).

2.3 Bio-medical imaging of CNTs

One of the goals of nano-medicinal research is to produce bespoke multifunctional "nano-devices" that can act as diagnostic devices, tissue-specific drug delivery systems or vehicles for gene therapy, ideally combining this with multimodal imaging capabilities (Hong et al., 2009; Kostarelos et al,. 2009; Lui et al., 2009; Lu et al., 2009; Liu et al., 2010). Carbon nanotubes have many properties that make them particularly useful as biomedical devices (Hong et al., 2009; Kostarelos et al,. 2009; Lui et al., 2009; Lu et al., 2009; Liu et al., 2010).

Their microscopic morphology confers them with unique electrical, thermal, optical and magnetic properties. Coating the surface of CNT improves their solubility and their 'needle-like' shape allows internalized by cells. By functionalizing the surface with biological active compounds, it is possible to use CNTs as 'nano-vehicles' delivering compounds into cells such as small molecule drugs, proteins, vaccines or in the case of gene therapy delivering siRNA. Thus CNTs have proved to be very effective carriers for gene delivery, with the advantage of avoiding the viral vectors used by other methods (Kostarelos et al., 2009). In addition, there are a number of CNTs strategies being developed to target CNTs to a region of interest, either by exploiting their magnetic properties or attaching targeting-ligands, such as antibodies or peptides, to their surface. With the exciting research published in this area, one should not forget that this technology is still in a developmental stage. Much work is still required in for example testing: (i) the effectiveness of CNT atfor targeting disease tissues; (ii) the stability efficacy of using CNTs to carry drugs to a region of interest and the profile of drug release; (iii) the precise mechanism of cell internalization as well as , and observing the sub-cellular compartments in which the CNTs accumulate; and also (iv) efficiency of using CNT as a heating methods for hyperthermia treatments. Undoubtedly there is still a lot to be done assessing the efficacy of the different CNT therapies as well as understanding the safety issues associated with CNTs.

Bio-medical imaging has a vital role to play in testing both the efficacy of these new therapies and investigating their biodistribution. To date, the majority of imaging of CNTs has involved optical methods, such as fluorescence and bioluminescence microscopy (Hang et al., 2009). Optical microscopy has high spatial resolution, however it requires optically transparent samples. This is less of an issue in near-infrared photoluminescence imaging, Raman imaging, or photo acoustic tomography. THowever, there is a reduction in spatial resolution and sensitivity with these methods, and they still have tissue depth limitations. Attaching radionuclide labels onto CNTs allows three dimensional (3D) whole body, in vivo molecular imaging such aseg position emission tomography (PET) and single photon emission computed tomography (SPECT). These techniques have high sensitivity, moderate resolution and have the advantage of being well established medical techniques, but they do involve ionising radiation.

Magnetic resonance imaging (MRI) is an extremely versatile imaging modality, and particularly suitable for testing novel medical therapies as it is available as pre-clinical and clinical platforms, allowing a seamless transition from the laboratory into the clinic. MRI produces high resolution 3D images, non-invasively, from within optically opaque samples. The MRI signal originates from protons in water and lipid, and as a result the images contain impressive anatomical and pathological information. Furthermore, magnetic resonance (MR) can be used to interrogate tissue morphology, physiology, function, vascularity and metabolism. The versatility of MRI originates from the fact that signal intensity is dependent upon a number of physical parameters including the density of protons in the liquid state, 1H longitudinal (T1) and transverse (T2) relaxation times, and mass transport processes such as diffusion or flow (Modo et al 2007). In a simple 90o-180o-acquire spin echo imaging sequence, the intensity of the MR magnetisation (Mt) observed in the spin echo can be expressed as:

$$Mt = Mo \cdot \exp(-TE/T2) \cdot [1 - \exp(-TR/T1)] \quad (1)$$

if diffusion and flow are negligible, and when Mo is magnetisation at time zero, TE is echo time between excitation and detection of magnetisation, and TR is repetition time between successive imaging pulse sequences.

A limitation of MRI is that it is an inherently insensitive technique; this is due to its quantum physics and the Boltzmann's distribution in particular. Its low sensitivity and low signal-to-noise ratios are the reason most MR images map the location of water and lipids, as these molecules are found in high concentrations. It is also the reason that the intrinsic MRI spatial resolution is rarely below the order of 10's of microns. Consequently, that it is not possible to detect a MRI signal directly from solid CNTs with low proton concentrations. Second, even if CNTs did produce a MRI detectable signal, they are so small that it would not be possible to see the shape of an an individual nanoparticle in an MRI image due to insufficient spatial resolution. However it is possible to use MRI to observe the location of CNTs indirectly, because magnetic CNTs can act asare MRI contrast agents and this can modify the appearance of an image. There are two classes of MRI contrast agents (CAs): T1 CAs reduces the 1H T1 relaxation times of the water molecules in their vicinity, whilst T2 CAs reduces 1H T2 relaxation times. T1 MRI CAs produces hyper-intensity in T1-weighted images thus appears white. They typically contain lanthanide ions; gadolinium is the most potent in this class with seven unpaired f-electrons. There are several Gd-chelated contrast agents with FDA approval including Omniscan, Multihance and Magnevist. T2 MRI contrast agents produce hypo-intensity in T2-weighted images thus appears black; this can be difficult to interpret if there are other black features in the image. Super-paramagnetic iron oxide nanoparticles are particularly potent T2 MRI CAs and formulations such as Feridex and Resovist have FDA approval for clinical use. The main purpose of contrast agents in the clinic is to improve disease detection and increase diagnostic confidence.

The effectiveness of MRI contrast agents are evaluated by determining their longitudinal or transverse relaxivity, r1 or r2 respectively. Relaxivity is defined as the change in the relaxation rate of water protons per molar concentration of the contrast agent with units of s-1mM-1 and is expressed as:

$$ri\,[CA] = 1/Ti - 1/Tio \qquad (2)$$

where ri is relaxivity; i=1 or 2; [CA] is the concentration of contrast agent; 1/Ti is the longitudinal or transverse relaxation rate in the presence of contrast agent [CA]; and 1/Tio is the relaxation rate of the medium in the absence of contrast agent. To assist comparison of MRI data from different laboratories, it would be useful if new CNTs were tested against a set of 'standard'standardized conditions such as using 1% agarose gels as a phantom and also using similar cell types.

Choi et al attached superparamagnetic iron oxide nanoparticles onto SWCNTs to demonstrate the potential of CNTs as T2 MRI contrast agent. They imaged murine machrophage cells that had been incubated with these CNTs. The image contrast is generated by localised magnetic inhomogeneities induced by the magnetic CNTs, which reduces the T2 relaxation time of nearby water protons. Mesenchymal stem cells containing MWCNTs produced regions of hypo-intensity compared to untreated cells in T2-weighted RARE images (Vittorio et al. 2011) (Figure). Ananta et al measured the r2 relaxivity of three different types of SWCNTs. Interestingly, the r2 of pristine, low Fe content CNTs was twice as large as the relaxivity of raw, unpurified SWCNT; even though pristine SWCNT had three times less iron. Similarly, the relaxivity r2 of MWCNT with low iron content (2.57%) measured by Vittorio et al was 564 ± 41 s-1mM-1, (Vittorio et al. 2011), which was over three times higher than the relaxivity of commercial contrast agent Feridex measured under similar conditions (Chen et al. 2010). They attributed the relaxivity r2 to the presence of iron oxide impurities in the CNTs, which originate from the fabrication process, and to the carbon MWCNT structure itself. Although the equation shown in Eq 2 is conventionally used to determine relaxivity, in this situation to measure relaxivity relative to iron concentration alone is slightly misrepresentative. A study using magnetic iron oxide nanoworms neatly demonstrated that the r2 relaxivity was dependent on the intrinsic shape of the nanoparticles as well as to the presence of iron oxide impurities (Park et al., 2008). These results suggest that these nanoparticles are members of a new class of T2 relaxation agents where shape of the particle also contributes to relaxivity r2, this warrants further investigations.

CNT T1 contrast agents have been produced by loading Gd3+ ions onto carbon nanotubes;. Sitharaman et al used ultra-short SWCNTs (Sitharaman et al., 2005), whilst Richard et al used MWCNTs (Richard et al., 2008). Superparamagnetic ultra-short single walled carbon nanotubes called 'gadonanotubes' have been produced (Sitharaman et al., 2010) with r1 and r2 relaxivity of 170 s-1mM-1 and 578 s-1mM-1 respectively. They are substantially more potent than the paramagnetic Gd chelates currently in clinical use. These contrast agents have the advantage that they can be used as either T1 positive CAs or T2 negative CAs. An in vivo study with rats using gadonanotubes generated negative hypo-intensity (Sitharaman et al., 2010). There are investigations into using these gadonanotubes as pH-probes since the T1 relaxation is sensitive to pH (Hartman et al., 2008).

In vivo MRI has been utilised in detecting the bio-distribution and potential impact of CNTs and potential impact. CNTs are extremely stable and not easily bio-degraded and v. Very little is known of about how their long term they behaviour in vivo or their pharmacological profile. Faraj et al have used MRI to evaluated the bio-distribution of SWCNT in animals. A Their longitudinal, time course, in vivo MRI study observed the accumulation of carbon nanotubes in the spleen and liver of rats (Al Faraj et al., 2011) after intravenous injection, although no acute toxicological effect on the liver's metabolic profile was observed. They also assessed the effects on the lungs of rats after inhaling SWCNTs using hyperpolarized 3-He MRI (Al Faraj et al., 2009). The results from a 3 month follow-up study showed that granulomatous and inflammatory reactions were produced over time in a dose dependent manner (Al Faraj et al., 2010).

In summary, the objective of the majority of MRI studies involving CNTs, to-date, have been evaluating the potential of CNTs as MRI contrast agents. The development of these bio-nanomaterials is still in its early stage,P and consequently the fabrication, functionalisation and characterisation of novel CNTs is an important and fruitful area of research. Producing CNTs that are potent and stable MRI contrast agents is particularly essential when using this technique to track their movement in vivo the behaviour of CNTs in vivo. Obvious applications of this would be the tracking of stem cells and guiding their delivery. MRI also has a crucial role to play in assessing the behaviour of CNTs in vivo. Multinuclear MRI studies have been published following CNTs bio-distribution in animal models after intravenous injection and inhalation. The non-invasive nature of MRI makes it particularly suitable for such longitudinal time course, follow-up studies. However, it is our belief that the most important role of MRI in carbon nanotubes research will be evaluating the effectiveness of the various bio-medical applications and therapies associated with CNTs both in animal models and in the clinic. In hospitals, MRI is a work-horse imaging modality for the diagnosis of a broad range of diseases. It contributes to almost every aspect of disease management from diagnosis and staging a disease, to selecting and assessing therapies, and in follow-up and to detecting recurrence. It is theits ability of MRI to produce high resolution images with excellent anatomical and pathological contrast in animal models of diseases as well as in the clinic, which will be needed when evaluating these new nanomaterials in targeted chemotherapy, focussed electrotherapy and gene therapy etc.

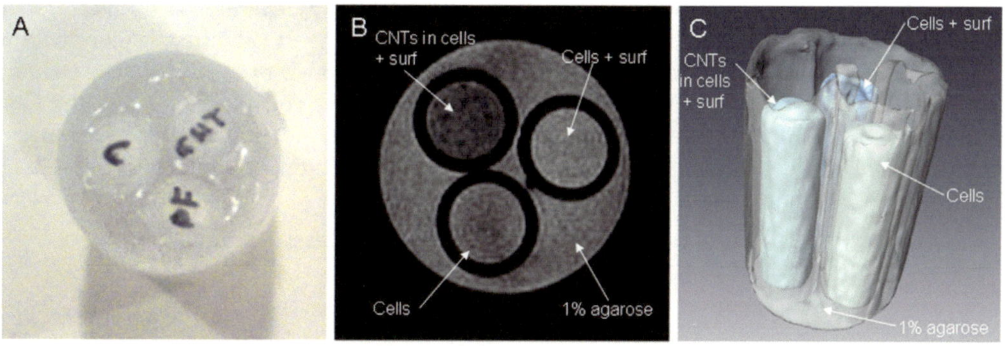

Fig. 2. MWCNTs MRI study of mesenchymal stem cells (MSCs). Three vials were embedded in container of 1 % agarose: vial 1 (test) contains 97% MWCNTs in MSCs treated with surfactants held in agarose gels (cnt); vial 2 (control i) contains MSCs treated with surfactants held in agarose gels (pf), and vial 3 (control ii) contains MSCs cells held in agarose gels (c). (A) Photo of the dorsal view of the sample; (B) T2-weighted RARE-4 axial image of sample from the 128 by 128 by 128 RARE-4 (TR/TE=250/40 ms) data set; (C) 3D surface reconstruction of the sample from the same MRI data set as (B). MRI measurements were completed at 7.1 T, at 19°C, with field of view is 30 mm, and voxel spatial resolution of 0.234 mm/pixel. (Vittorio et al., 2011 b)

2.4 Magnetic properties of MWCNTs: Application for cell displacement *in vitro*

The magnetic properties of MWCNTs can be exploited for many biomedical applications. An interesting application is represented by the possibility of labelling the cells with the magnetic nanoparticels and guide their localization by an external magnetic field. It is easy to image the exploitation of this methodology to move and collect metastatic cells, or to guide the cell transpalatation towards the target organ. We recently used the same approach to guide and improve the nerve regeneration after the axon treatment with magnetic nanoparticles (www.marvene-project.org). The magnetic interactions of MWCNTs can be attributed to the metal particles encapsulated in the graphene sheet (Zhang et al., 2001; Glenis et al., 2004). After dispersion treatment, the tubes become water dispersible and as a result are shorter in length. These effects are propitious for the interaction between cells and CNTs: the reduced length and the reduced degree of agglomeration should limit toxicity response and at the same time could favour passive endocytosis (Wick et al., 2007). We have experience in the study of interactions between nanomaterials and different cell lines. We spent much efforts in finding the best conditions to label cells with MWCNTs and to move them by magnetic field without compromising their physiological conditions.

Under the effect of a permanent dipole magnet, cells have been seeded for three days with the CNT-modified medium: a progressive displacement of cells toward the more intensive magnetic fields was visible in the dishes where the CNTs have been added in culture, while in the control dishes (without CNTs) there was negligible translations of cells during the same period (Pensabene et al.,2008). Our results suggest that cell displacement occurs during cell duplications. However, the displacement mechanism is still not clear, but the present result is a starting point to further investigation of the interaction between CNTs and cells and the controlled displacement of cells for selective cancer therapy. Concerning these issues, there are currently contrasting opinions and the mechanisms of interaction are still not clear. Two mechanisms can be supposed: the uptaking of the tubes (by endocytosis or pinocytosis) or their attachment to the cell membrane (Monch et al., 2005). The evidence of Figure 3 leads us to affirm that the binding of MWCNTs to cells is strong and the CNTs are able to drag along the cells, following the traction force generated by the permanent magnet. This effect could be used also to study how a mechanical stimulation of cells by CNTs can influence the cell activity (Cartmell et al., 2002).

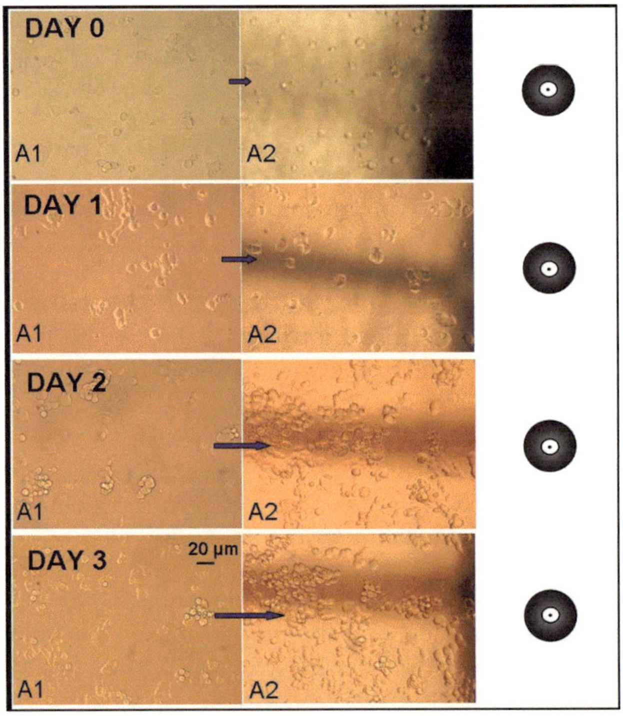

Fig. 3. Neuroblastoma cells displacement after three days in culture with MWNTs-modified medium. Control sample not showed (with Nikon TE2000U inverted optical microscope, magnifications 20X). (Pensabene et al., 2008)

We also studied the interaction of magnetic carbon nanotubes with PC12 cells which represent a valid in vitro model to study the effect of nanoparticles in neuronal cells and the effect on differentiation in neurons (Vittorio et al 2011c). We demonstrate that when PC12 cells were cultured in a MWCNT-containing medium, the nanotubes interact with the cells without compromising the cell's viability and their ability to differentiate into neurons following exposure to NGF. As a result of this interaction, CNTs in the presence of a magnetic field are able to translocate cells, in response to a permanent magnetic field. In contrast, no displacement was observed in control samples when cells were cultured with a CNT-free cell culture medium. These results confirm the ability of MWCNTs to shepherd PC-12 cells in response to the force generated by a permanent magnet. We believe that recent advances in nanoscience and nanotechnology may provide new therapeutic options as alternative/supplemental therapies to established surgical repair techniques for the treatment of peripheral nerve injuries. We studied the interaction between neuronal cell line and magnetic carbon nanotubes, because this is an essential step to investigate the exploitation of magnetic nanoparticles to enhance/accelerate nerve regeneration and to provide guidance for the regenerating axons.

2.5 Magnetic properties of MWCNTs: Application for stem cell displacement *in vitro* and to drive the homing of mesenchymal stem cells after their transplantation *in vivo*

Next, we decided to try to move stem cells after their incubation with MWCNTs. With the aim of finding the best protocol to guide the cell homing towards the target organs and improve stem cell transplantation strategy. Many experts in stem cells transplatation report the lack of efficient technology to inject stem cells in vivo in target sites and avoiding their dispersion.

The experimental data obtained in our study confirm that magnetic fields can be used to control the movement and location of MSCs cultured with carbon nanotubes (Vittorio et al., 2011 a). The cell distribution in control cultures MSCs in the Petri dish was homogeneous (350-400 cells/mm^2), whereas the cell density was more heterogeneous in the test culture. Specifically, in the test culture the MSCs density ranged from 200 cells/ mm^2 to 800 cells/mm^2, correlating respectively with distances from the magnetic pole of 9-14 mm and 5 mm (Figure 4) . This cell streaming behaviour was closely related to the intensity of the flux density within the Petri dish which ranged from a very low value (<0.1 T) when the distance from the magnetic pole was >9 mm to about 0.5 T at a distance of 5 mm. Mathematically, an MSC interacting with CNTs is subjected to a translational force in the presence of a gradient field of:

$$F_m = \frac{1}{\mu_0} \cdot \chi_{rp} \cdot V \cdot B \cdot \frac{dB}{dr} \tag{3}$$

where μ_0 is the magnetic permeability of free space, X_{rp} and V are respectively the magnetic susceptibility and the total volume of magnetic nanotubes attached to the cell. Eq. 3 is a usefull mathematical tool to design the manipulation of MSCs cultured with carbon nanotubes by a magnetic flux gradient. The migration dynamics of a single cell shows that the cell moves towards the magnetic source with a speed of approximating 30 μm/h. Cell proliferation assays were performed to identify if there was any adverse effect of the nanotubes and/or the magnetic field on the viability, proliferation and functionality of MSCs. The results showed that the PF127 surfactant marginally decreased the cell viability but the cell viability reduction was negligible with PF127-CNTs. This can be explained by the action of the surfactant which wraps the nanotubes surface and reduces the amount of free PF127 in the medium (Vittorio et al 2009). Additionally, the data on the cell growth assays showed that the MSCs doubling time was not significantly influenced by the presence of either the nanotubes or the magnetic field and no apoptotic cells were observed in any of the samples studied.

The in vitro studies also confirmed that cells treated with CNTs for up to 5 days maintain their ability to differentiate under specific conditions just as well as the untreated cells. The lack of adverse effects on cell viability and proliferation by MWCNTs observed in the present study needs to be compared with the reported findings by Mooney et al. In this study, hMSCs were treated for 24 h with both COOH-functionalised SWCNTs and OH-functionalised MWCNTs, at various concentrations. The authors demonstrated that COOH-functionalised SWCNT, at

concentration up to 32 μg/ml, did not affect cell viability, proliferation, differentiation and metabolic activity; but at higher concentrations, SWCNTs exerted detrimental effects on the cells. In contrast, OH-functionalised MWCNTs were found toxic at all concentrations (Mooney et al., 2008). In the present study, the MWCNTs were not functionalised but simply dispersed in an anionic surfactant and used at concentration of 10 μg/ml. Finally, the staining of the actin filaments suggested clearly that the treatment with the nanotubes and the exposure to the magnetic field did not alter cell morphology.
The results of the *in vivo* experiments provide a proof of concept that MSCs cultured with CNTs can be shepherded by means of an external magnetic source towards a specific organ; specifically, we demonstrated that the application of an external magnetic field alters the biodistribution of CNT-labelled MSCs after intravenous injection into rats, increasing the accumulation of cells into the target organ. This observation assumes importance in view of the widespread interest in MSC-based cell therapy for tissue engineering, regeneration/ repair of damaged organs and allogeneic transplantation. As MSCs also exhibit immuno-modulatory and anti-inflammatory effects they may play a role in the in vivo induction of tolerance. It was observed that MSCs reduce the incidence and severity of Graft-Versus-Host-Disease (GVHD) (Le Blanc et al., 2007), as well as prolong skin graft survival.
The clinical applications of MSCs require the administration of cells by the intravenous route, but their subsequent dispersion in many tissue and organs (Allers et al., 2004) reduces the number of cells which colonize the intended target organ. There is therefore a need to increase targeted stem cell localization and homing to the diseased site. The approach proposed in this paper has the potential to achieve the objective of site-specific localization of the CNT-labelled MSCs, by reducing their colonisation of other sites with consequent adverse effects resulting from their proliferation and differentiation in ectopic sites. In the animal model studied we were able to achieve a 3-fold increment of MSC localization in the target organ (liver), and a corresponding decrease of MSC localization in the lung and kidney, which represent a natural filter for stem cells (Figure 5). In a recent work, Gao demonstrated that intravenous infusion of sodium nitroprusside, a vasodilator, administered prior to the cells infusion, reduced by 15% the number of cells present in the lungs and increased by 10% the cells in the liver (Gao et al., 2001). Based on this result we are currently investigating the combined action of a vasodilator and the magnetic guidance to enhance the targeted homing of cells. Additionally, we are optimising the protocol of CNT-labelled MSC transplantation and we are designing a magnetic applicator device which would allow a more accurate and controlled configuration of the magnetic field gradient applied to the target site. Further improvements could be the design of the magnetic field by finite element modelling (FEM), the development of a wearable device and the use of electromagnets for switching the magnetic field.
In summary, we have demonstrated that when mesenchymal stem cells are cultured in a CNT- containing medium, the nanotubes interact with the cells without compromising the cell's viability, proliferation rate, cell phenotype and cytoskeletal conformation. Moreover we confirmed that magnetically labelled cells maintained the ability to differentiate in adipocytes and osteocytes. As result of the cells interaction with CNTs, the application of a magnetic field, enables shepherding of MSCs to the desired location in vitro. Moreover, in the experimental model used we were able to increase significantly the localization of mesenchymal stem cells within the liver, with a reduction of their migration to other organs. This paves the way for the development of a new methodology for shepherding cells in a target tissue/organ. The application of such technology would significantly improve both

the range and efficacy of therapies based on transplanted cells, included totipotent, pluripotent multipotent stem cells, reprogrammable adult cells, induced pluripotent stem cells (iPSC) and embryonic stem cells. Compared to the existing methodologies, CNT-labelled MSCs maintain their magnetization for a period over 24 hours and up to 3 weeks. The long-time magnetization should allow for an efficient cell manipulation via magnetic fields and a reliable cell tracking via magnetic resonance imaging. The application of such technology could significantly improve the range and efficacy of the current and future cell therapies. Our findings pave the way for the exploitation of the magnetic properties of biocompatible carbon nanotubes to localize the stem cells in a target organ, after their transplantation. A controlled and localized stem cell transplantation represent the future of regenerative medicine.

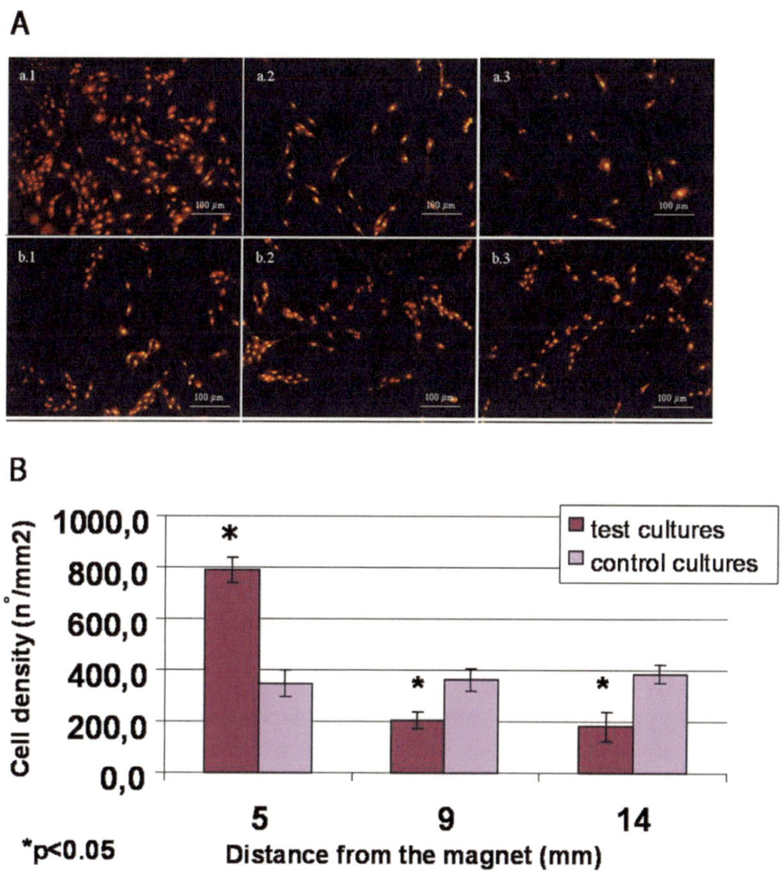

Fig. 4. Imaging of MSCs shepherding: (A). MSC-CNT-cultures (a) MSC-control (b). Cultures 72 h after magnet placement. Images taken at different distances from the magnet pole (1, 2 and 3 correspond respectively to 5, 9 mm and 14 mm from the magnetic pole). Cell is stained with Syto 82 fluorescent dye. (B). MSC density for test and control cultures after 72 h from the magnet placement at different distances from the magnet pole. Assays performed in quadruple and results are the mean±S.E.M. (vertical bars) (Vittorio et al 2011 a)

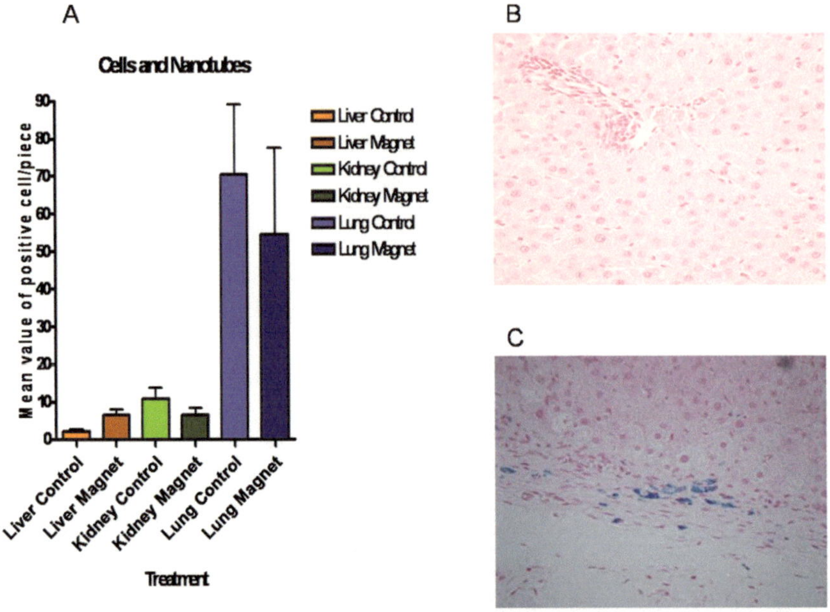

Fig. 5. Histochemical analysis: (A) Mean and S.E.M. values of positive cells for Perls staining in each type of organ;(B) Liver's section of control sample; (C) Liver's section with Perls labelled mesenchymal stem cells with magnetic nanotubes (Vittorio et al., 2011 a)

2.6 Exploitation of magnetic CNTs in gene therapy

The magnetic properties of CNTs derive from metal catalyst impurities entrapped at their extremities during their manufacture or from filling the nanotubes with tailored materials in which the active content is encapsulated by a protecting carbon shell (Weissker et al., 2010).

One medical application of ferromagnetic material is the so-called "magnetic fluid hyperthermia" (MFH). MFH is based on a controled transfer of power to magnetic nanoparticles by an alternating magnetic field which results in local generation of heat. Depending on the equilibrium temperature of the tumour tissue, this heat may either destroy the tumor cells directly (thermoablation) or result in a synergic reinforcement of radiation efficacy (hyperthermia) (Latorre et al., 2009). Much of the current research focuses on iron oxide nanoparticles which have proven their feasibility in animal experiments (Johannsen et al., 2005; Matsuoka et al., 2004)] and are now under clinical trials (Johannsen et al., 2007). The protective carbon shell of the magnetic CNT may avoid the toxicity problem associated of iron oxidation. Degradation of the filling materials is avoided and potential toxicity and adverse effects suppressed, so that CNT become effective carrier system on the nanometer scale. Magnetic studies on the feasibility of iron-containing carbon nanotubes for magnetic hyperthermia have been shown (Krupskaya et al., 2009). The authors presented a detailed magnetic study of iron containing carbon nanotubes (Fe-CNT), and highlighted their potential for contactless magnetic heating in hyperthermia cancer treatment. DC magnetization studies showed a different magnetic response of Fe-CNT powder compared to Fe-CNT dispersed in aqueous solution. The ferromagnetic Fe-CNT in powder did not show any hysteresis when dispersed in a liquid.

Such magnetic behaviour implies the rotation of the magnetic moments following the magnetic field. The fact that in the frozen dispersion they observed a ferromagnetic-like hysteresis, clearly indicates that the rotation of the magnetic moments in liquid is associated with the motion of the whole nanotube, but not of the moments within each particle. AC susceptibility measurements showed an increase of the susceptibility of Fe-CNT in a liquid dispersion compared to the powder. Again, this can be explained by the motion of Fe-CNT in the AC magnetic field. AC inductive heating experiments showed a substantial temperature increase of Fe-CNT dispersions in AC magnetic fields, which was dependent on the magnetic field. In summary, these results showed the feasibility of Fe-containing CNT in magnetic hyperthermia. Another intriguing application of magnetic CNTs is the so-called magnetofection.

Magnetic nanoparticle-based transfection methods are based on the principles developed in the late 1970s by Widder and others for magnetically targeted drug delivery. Transfection using of magnetic microparticles was first demonstrated in 2000 by Cathryn Mah, Barry Byrne and others at the University of Florida, both in vitro with C12S cells and in vivo in mice using an adenoassociated virus (AAV) linked to magnetic microspheres via heparin (Ma et al., 2002). Since these initial studies, the efficiency of this technique has been demonstrated in a variety of cells (Dobson, 2006). The technique is based on the coupling of genetic material to magnetic nanoparticles. In the case of in vitro magnetic nanoparticle based transfection, the particle/DNA complex (normally in suspension) is introduced into the cell culture where the field gradient produced by rare earth magnets (or electromagnets) placed below the cell culture. They increase sedimentation of the complex and increase the speed of transfection. In vivo applications apply magnetic fields focused over the target site have the potential to not only enhance transfection but also target the therapeutic gene to a specific organ or site within the body. Generally, particles carrying the therapeutic gene are injected intravenously and strong, magnets gradients are applied to capture the particles as they flow through the blood-stream. Once captured by the field, the particles are held at the target, where they are taken up by the tissue. The therapeutic genes can be released either via enzymatic cleavage of the cross-linking molecules, charge interactions, or degradation of the polymer matrix. Alternatively, if the DNA is embedded within the matrix, such as with hydrogels, alternating fields may be applied to heat the particles and release the genes from the magnetic carrier. An application of magnetically-driven drug-carrying CNTs in cells was suggested (Cai et al., 2005).

Cai et al have designed an alternative physical method of *in vitro* and *ex vivo* gene transfer, called nanotube "spearing": CNTs grown from plasma-enhanced chemical vapour deposition contain nickel particle catalysts entrapped into their tips, allowing them to respond to a magnetic field. The tubes were covalently functionalized with a DNA strain containing the sequence coding for the enhanced green fluorescent protein (pEGFP-c1). Dividing and non dividing cells like Bal17, B-lymphoma, *ex vivo* splenic B cells and primary neurons were grown on a substrate and incubated with magnetic pDNA/CNT. First a rotating magnetic field drives the nanotubes to mechanically spear the cells. In a subsequent step, a static magnetic field pulled the tubes into the cells. The spearing set-up and procedure are illustrated in Figure 6. The cells were efficiently transfected as confirmed by fluorescent microscopy measurements. They demonstrated that both spearing steps are necessary for efficient transduction. The efficiency was equal to viral approaches even for non-dividing cells, such as primary B cells and neurons which are generally more difficult to transfect.

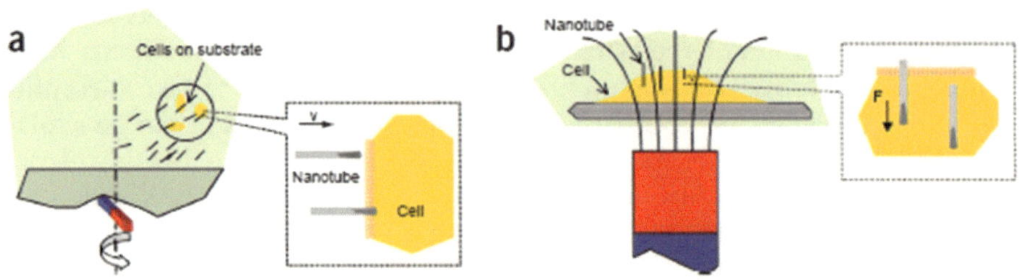

Fig. 6. Two step procedure for nanotube spearing: a) a rotating magnetic field drives nanotubes to spear the cells on a substratum (surface) and b) a static field pulls nanotubes into the cells. (Cai et al.,2005).

The same researchers investigated the compatibility of the nanospearing with primary *ex vivo* cultures of B lymphocytes (Cai et al., 2007). They reported that, by applying the original 2-step procedure of the nanospearing, they noted by phase contrast microscopy B-cell blasts and cellular aggregation, suggesting non-specific B-cell activation (i.e., increased cell size, activation of signal transduction pathways, increased protein and RNA content, normally associated with stimulation by extrinsic growth factors). They conducted a comprehensive characterization of the potential effects of exposing *ex vivo* primary B cells to CNTs with respect to cellular activation, survival, and signal transduction. The original multistep nanospearing protocol was modifield to a single step process (cells were speared by positioning the cover slips directly over a permanent magnet) and by using non-covalent DNA immobilization strategy. The results indicate that CNT-mediated nanospearing of primary B cells does not result in non-specific activation of naıve B lymphocytes but facilitate efficient delivery of nucleic acids into B cells (Figure 7).

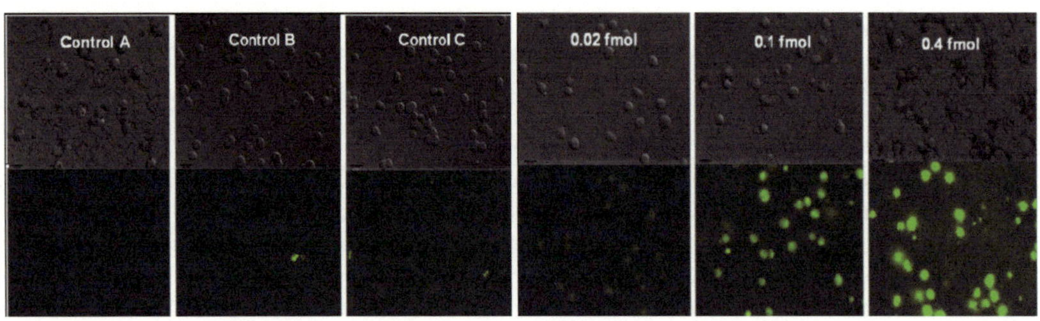

Fig. 7. CNT were covalently functionalised with poly(L-lysine) (PLL-CNT) and incubated with FITC-oligos in order to achieve FITC-oligo–PLL–CNTs. The pictures show the delivery of FITC-oligo–PLL–CNTs into B lymphocytes following nanospearing. Bal17 cells were speared with 0.4 fmol non-FITC-oligo–PLL–CNT complexes (Control A) or with 0.02, 0.1, or 0.4 fmol FITC-oligo PLL–CNTs. More control results are shown as 'Control B' (FITC-oligo alone) and 'Control C' (FITC-oligo-PLL) with the corresponding reagent treatments to cell following the spearing procedures (Cai et al., 2007).

Recently, Cai and colleagues showed that CNT–cell complexes form in the presence of a magnetic field (Cai et al., 2008). The complexes were analyzed by flow cytometry as a quantitative method for monitoring the physical interactions between CNTs and cells. They observed an increase in side scattering signals, where the amplitude was proportional to the amount of CNTs that are associated with cells. Even after the formation of CNT–cell complexes, cell viability was not significantly decreased. The association between CNTs and cells was strong enough to be used for manipulating the complexes and thereby conducting cell separation with magnetic force.

2.7 Summary and concluding

The emergence of CNTs in the biomedical field have raising great hopes. CNTs have been proposed as components for DNA and protein biosensors, ion channel blockers and as bioseparators and biocatalysts. Their use is becoming relevant in many fields such as neuroscience research and tissue engineering. CNTs have been developed as scaffolds for neuronal and ligamentous tissue growth for regenerative interventions of the central nervous system (e.g. brain, spinal cord) and orthopaedic sites They have also been used as new platforms to detect antibodies associated with human autoimmune diseases with high specificity. This findings pave the way to the development of CNT-based diagnostic devices for the discrimination and identification of different proteins from serum samples and in the fabrication of microarray devices for proteomic analyses. Plus CNTs covalently modified with DNA and PNA (peptide nucleic acid) have led to innovative systems for hybridization of complementary DNA strands allowing ultrasensitive DNA detection. Furthermore, CNTs have also emerged as a new efficient, alternative tool for transporting and translocating therapeutic molecules. The development of new and efficient drug delivery systems is of fundamental importance in improving the pharmacological profiles of many classes of therapeutic molecules. CNTs can be functionalised with bioactive peptides, proteins, nucleic acids and drugs, and used to deliver their cargos to cells and organs.

With the prospect of gene therapy, cancer treatments, and innovative therapies, the science of nanomedicine has become an fast-growing field that has an incredible ability to bypass barriers previously thought unavoidable. The properties and characteristics of CNTs are still being explored and scientists have barely begun to mine the potential of these structures. CNTs have already proven to serve as safer and more effective alternatives to previous drug delivery methods. They can serve as nano-vehicles, carrying therapeutic drugs, vaccines, and nucleic acids deep into the cell to targets previously unreachable in response to static and dynamic energetic fields.

Carbon nanotubes have the potential of revolutionizing bio-medical research. They frequently show superior performance over other nanoparti-cles. The advantage lies in their unique combination of electrical, magnetic, optical and chemical properties which can be exploited in the development of new classes of CNT-based drugs and therapy.

In this chapter we showed the extraordinary physical properties of carbon nanotubes and how such properties can play an important role in their biomedical application. In particular we discussed to the magnetic properties of CNTs and considered how they can improve the imaging, the cell labelling, the guided homing of stem cell and the selected delivery of drug towards tumour cells.

The magnetic properties of nanotubes can be also exploited to study in vivo biodistribution by using magnetic resonance. Al Faraj and colleagues performed a non-invasive follow-up in vivo study to evaluate the biodistribution of CNTs and effect of nanotube deposition after exposure (Al Faraj et al., 2009). They used both helium-3 and proton magnetic resonance (MRI) to evaluate the biodistribution and biological impact of raw single-wall CNTs (raw- SWNTs) and superpurified SWNTs (SP-SWNTs) in a rat model. Superpurified SWNTs, thanks to low content of metal residuals, represent the best samples in terms of biocompatibility, but it is not possible with only MRI to fully understand their biodistribution in vivo.

In a series of other experiments, the effectiveness of cellular imaging as T1 and T2 contrast agent was investigated.

Cai et al. have designed an alternative physical method of in vitro and ex vivo gene transfer, called nanotube "spearing". The tubes were functionalized with DNA plasmids immobilized onto the CNTs and subsequently speared into target cells via external magnetic fields.

Moreover, Cai and colleagues showed that CNT–cell complexes are formed in the presence of a magnetic field (Cai et al., 2008). Even after the formation of CNT–cell complexes, cell viability was not significantly decreased. The association between CNTs and cells was strong enough to be used for manipulating the complexes and thereby con-ducting cell separation with magnetic force.

We demonstrated that, by combining the magnetic response with the ability to interact with cells (Vittorio et al., 2011a), CNTs can also be used for cell manipulation. Our data showed that cells treated with nanotubes with -3% iron content are able to internalize this nanoparticle and to move towards a magnetic source.

In cancer therapy, there is potential to prevent cell migration in metastasis. For example, CNTs could be functionalized so they bind selectively to cancer cells, by applying external magnetic field in order to constrain cancer cell migration. In addition, cancer cells treated with CNTs could act as intracellular transporters of chemotherapies or as heater probes. Moreover magnetic fields could be used to control movement and location of MSCs cultured with carbon nanotubes to shepherd MSCs towards the magnetic source, increasing the accumulation of cells into the target organ (liver). Undoubtedly, MWCNTs hold a distinct potential for use as nano-devices to improve therapeutic protocols for transplantation and homing of stem cells in vivo.

We believe Nanomedicine will revolutionize the future of medical approaches.

3. Acknowledgment

Authors thank the Caripi Foundation for the economic support.

4. References

Al Faraj, A, Cieslar, K; Lacroix, G et al. (2009) In Vivo Imaging of Carbon Nanotube Biodistribution Using Magnetic Resonance Imaging. *Nano Lett.* 9 (3):1023-1027.

Al Faraj, A, Fauvelle, F, Luciani, N et al. (2011) In vivo biodistribution and biological impact of injected carbon nanotubes using magnetic resonance techniques. *Inter. J. of Nanomed.* 6 351–361.

Al Faraj, A; Bessaad, A; Cieslar, K et al. (2010) Long-term follow-up of lung biodistribution and effect of instilled SWCNTs using multiscale imaging techniques. *Nanotechnology* 21 (17):175103

Allers, C, Sierralta, WD, Neubauer, S et al. (2004) Dynamic of distribution of human bone marrow-derived mesenchymal stem cells after transplantation into adult unconditioned mice. *Transplantation.* 78;503-508.

Allhoff, F. (2009). The coming era of nano medicine. *Am J Bioethics.* 9:3–11.

Ananta, JS, Matson, ML, Tang, AM et al. (2009) Single-Walled Carbon Nanotube materials as T-2-Weighted MRI contrast agents. *J. of Physical Chem.* 113 (45):19369-19372.

Balani, K, Andersonb, R, Lahaa, T et al. (2007) Plasma-sprayed carbon nanotube reinforced hydroxyapatite coatings and their interaction with human osteoblasts in vitro. *Biomaterials.* 28:618-624.

Bottini, M, Bruckner, S, Nika, K, et al. (2006) Multi-walled carbon nanotubes induce T lymphocyte apoptosis. *Toxicol. Lett.* 160:121–126.

Cai, D, Blair, D, Dufort, FJ et al. (2008) Interaction between carbon nanotubes and mammalian cells: characterization by flow cytometry and application. *Nanotechnology* 19:345102:10.

Cai, D, Doughty, CA, Potocky, TB et al. (2007) Carbon nanotube-mediated delivery of nucleic acids does not result in non-specific activation of B lymphocytes. *Nanotechnology* 18:365101:10.

Cai, D, Mataraza, JM, Qin, ZH et al. (2005) Highly efficient molecular delivery into mammalian cells using carbon nanotube spearing. *Nature Methods* 2: 449–454.

Cartmell, SH, Dobson, J, Verschueren ,SB, El Haj, AJ. (2002) Development of magnetic particle techniques for long-term culture of bone cells with intermittent mechanical activation. *IEEE Trans. Nanobiosci.* 1;2;92–97.

Cellot, G, Cilia, E, Cipollone, S et al. (2009) Carbon nanotubes might improve neuronal performance by favouring electrical shortcuts. *Nat. Nanotechnol.* 4; 126-133.

Chen, S, Wang, L, Duce, SL, et al. (2010) Engineered biocompatible nanoparticles for in vivo imaging applications. *J. Am. Chem. Soc.* 132(42): 15022–15029.

Cherukuri, P, Bachilo, SM, Litovsky, SH (2004). Weisman RB. Near-infrared fluorescence microscopy of single-walled carbon nanotubes in phagocytic cells. *J. Am. Chem. Soc.* 126:15638-15639.

Choi, JH, Nguyen, FT, Barone, PW et al. (2007) Multimodal biomedical imaging with asymmetric single-walled carbon nanotube/iron oxide nanoparticle complexes. *Nano Lett.* 7 (4):861-867.

Ciofani, G, Raffa, V. (2009) Chemically functionalized carbon nanotubes: emerging vectors for cell therapy. *Mini-Rev Med. Chem.* 9, 000.

Dobson, J. (2006) Gene therapy progress and prospects: magnetic nanoparticle-based gene delivery. Gene Therapy. 13:283–287 *Toxicol.*

Dutta, D, Sundaram, SK, Teeguarden, JG et al. (2007) Adsorbed proteins influence the biological activity and molecular targeting of nanomaterials. Sci. 100:303-15.

Emerich, DF, Halberstadt, C, Thanos, C. (2007) Role of Nanobiotechnology in Cell-Based Nanomedicine: A Concise Review. *J Biomed Nanotech.* 3:235–44.

Gannon, CJ, Cherukuri, P, Yakobson, BI et al. (2007) Carbon Nanotube-enhanced Thermal Destruction of Cancer Cells in a Noninvasive Radiofrequency Field. *Cancer.* 110: 2654-5.

Gao, J, Dennis, JE, Muzic, RF, et al. (2001) The dynamic in vivo distribution of bone marrow-derived mesenchymal stem cells after infusion. *Cells Tissues Organs.* 169(1);12–20.

Gheith, MK, Sinani, VA, Wicksted, JP et al. (2005) Single-Walled Carbon Nanotube polyelectrolyte multilayer and freestanding films as a biocompatible platform for neuroprosthetic implants. *Adv. Mater.* 17(22);2663-2670.

Glenis, S, Likodimos, V, Guskos, N, Lin, CL. (2004) Magnetic properties of multiwall nanotubes. *J. Magn. Magn. Mater.* 272–276.

Hartman, KB, Laus, S, Bolskar, RD et al. (2008) Gadonanotubes as ultrasensitive pH-smart probes for magnetic resonance imaging. *Nano Lett.* 8(2):415-419.

Hong, H, Gao, T, Cai, W. (2009) Molecular imaging with single-walled carbon nanotubes. *Nano Today* 4(3):252-261.

Hu, H, Ni, Y, Montana, V et al. (2004) Chemically functionalized carbon nanotubes as substrates for neuronal growth. *Nano Letters.* 4;507-511.

Huczko, A, & Lange, H. (2001). Carbon nanotubes: experimental evidence for a null risk of skin irritation and allergy. *Fullerene Sci. Technol.* 9:247–250.

Jain, KK. (2008). Nanomedicine: Application of Nanobiotechnology in Medical Practice. *Med Princ Pract.* 17:89–101

Johannsen, M, Gneveckow, U, Taymoorian, K et al. (2007) Morbidity and quality of life during thermotherapy using magnetic nanoparticles in locally recurrent prostate cancer: Results of a prospective phase I trial. *International Journal of Hyperthermia.* 23:315–323.

Johannsen, M, Thiesen, B, Jordan, A et al. (2005) Magnetic fluid hyperthermia (MFH) reduces prostate cancer growth in the orthotopic Dunning R3327 rat model. *The Prostate* 64, 283–292.

Kagan, VE, Konduru, VN, Feng, W, et al. (2010) Carbon nanotubes degraded by neutrophil myeloperoxidase induce less pulmonary inflammation. *Nature Nanotech.* 5:354–359.

Khabashesku, VN, Margrave, JL, Barrera, EV. (2005) Functionalized carbon nanotubes and nanodiamonds for engineering and biomedical applications. *Diamond Related Materials.* 14:859-866

Kim, DY, Yang, CM, PaKr, YS, et al. (2005). Characterization of thin multi-walled carbon nanotubes synthesized by catalytic chemical vapor deposition. *Chem Phys Lett.* 413:135-41.

Kostarelos, K, Lacerda, L, Pastorin, G et al.(2007) Cellular uptake of functionalized carbon nanotubes is independent of functional group and cell type. *Nat. Nanotech.* 2:108-113.

Kostarelos, K; Bianco, A; Prato, M. (2009) Promises, facts and challenges for carbon nanotubes in imaging and therapeutics. *Nat. Nanotech.* 4 (10):627-633.

Krupskaya, Y, Mahn, C, Parameswaran, A, et al. (2009) Magnetic study of iron-containing carbon nanotubes: Feasibility for magnetic hyperthermia. *J. Magn. Magn. Mater.* 321:4067–4071.

Latorre, M, Rinaldi, C. (2009) Applications of magnetic nanoparticles in medicine: magnetic fluid hyperthermia. *PR Health Sci J.* 28(3):227-38.

Le Blanc, K, & Ringdén, O. (2007) Immunomodulation by mesenchymal stem cells and clinical experience. *J Intern Med.* 262;509-25

Liang, F, & Chen, B. (2010) A Review on Biomedical Applications of Single-Walled Carbon Nanotubes. *Curr Med. Chem.* 17(1):10-24.

Liu, Z, Winters, M, Holodniy, M, Dai, H. (2007) siRNA delivery into human T cells and primary cells with carbon-nanotube transporters. *Angew Chem Int Ed Engl.* 46:2023-2027

Liu, ZA; Yang, K; Lee, ST. (2011) Single-walled carbon nanotubes in biomedical imaging. *J of Mat. Chem.* 21(3):586-598.

Lovat, V, Panzarotto, D, Lagostena, L et al. (2005) Carbon Nanotube Substrates Boost Neuronal Electrical Signaling. *Nano Letters.* 5(6);1107-1110.

Mah, C, Fraites, TJ, Zolotukhin, I et al. (2002) Improved method of recombinant AAV2 delivery for systemic targeted gene therapy. *Molecular Therapy.* 6:1:106-112.

Matsuoka, F, Shinkai, M, Honda, H et al. (2004) Hyperthermia using magnetite cationic liposomes for hamster osteosarcoma. *Biomagnetic Research and Technology.* 2:3::1–6.

Mattson, MP, Haddon, RC, Rao, AM. (2000) Molecular functionalization of carbon nanotubes and use as substrates for neuronal growth. *J. Mol. Neurosci.* 14;175-182.

McDevitt, MR, Chattopadhyay, D, Kappel, BJ et al. (2007) Tumor targeting with antibody-functionalized, radiolabeled carbon nanotubes. *J. Nucl. Med.* 48:1180-1189

Modo, MMJ, & Bulte, JWM. (2007) Molecular and cellular MR Imaging. *Eds CRC Press* 421 pages.

Monch, I, Meye, A, Leonhardt, A et al. (2005) Ferromagnetic filled carbon nanotubes and nanoparticles: Synthesis and lipid-mediated delivery into human tumor cells. *J. Magn. Magn. Mater.* 290–291;1;276–278.

Mooney, E, Dockery, P, Greiser, U et al. (2008) Carbon Nanotubes and Mesenchymal Stem Cells: Biocompatibility, Proliferation and Differentiation. *Nano Lett.* 8;2137–2143.

Pantarotto, D, Hoebeke, J, Graff, R, et al. (2003). Synthesis, structural characterization, and immunological properties of carbon nanotubes functionalized with peptides. *J. Am. Chem. Soc.* 125:6160-6164.

Pantarotto, D, Singh, R, McCarthy, D, et al. (2004). Functionalized carbon nanotubes for plasmid DNA gene delivery. *Angew. Chem. Int. Ed. Engl.* 43:5242–5246

Park, JH, Von Maltzahn, G, Zhang, L et al. (2008) Magnetic iron oxide nanoworms for tumor targeting and imaging. *Adv. Mater.* 20 1630–1635.

Pensabene, V, Vittorio, O, Raffa, V et al. (2008) Neuroblastoma cells displacement by magnetic carbon nanotubes. *IEEE T. Nanobiosci.* 7(2), 105-110.

Poland, CA, Duffin, R, Kinloch, I et al. (2008). Carbon nanotubes introduced into the abdominal cavity of mice show asbestos-like pathogenicity in a pilot study. *Nat. Nanotechnol.* 3:423-428.

Pulskamp, K, Diabat´e, S, Krug, HF. (2007). Carbon nanotubes show no sign of acute toxicity but induce intracellular reactive oxygen species in dependence on contaminants. *Toxicology Letters.* 168:58–74.

Raffa, V, Ciofani, G, Vittorio, O et al. (2010) Physicochemical properties affecting cellular uptake of carbon nanotubes. *Nanomedicine.* 5:89-97.

Richard, C, Doan, BT, Beloeil, JC et al. (2008) Noncovalent functionalization of carbon nanotubes with amphiphilic Gd3+ chelates: Toward powerful T-1 and T-2 MRI contrast agents. *Nano Lett.* 8 (1):232-236.

Shi Kam, W, Jessop, TC, Wender, PA, Dai, H. (2004). Nanotube molecular transporters: internalization of carbon nanotube-protein conjugates into Mammalian cells. *J. Am. Chem. Soc.* 126:6850-6851.

Shi Kam, W & Dai, H (2005). Carbon Nanotubes as Intracellular Protein Transporters: Generality and Biological Functionality. J. Am. Chem. Soc. 127:6021–6026.

Shi Kam, W, O'Connell, M, Wisdom, JA, Dai H. (2005a) Carbon nanotubes as multifunctional biological transporters and near-infrared agents for selective cancer cell destruction. Proc. Natl. Acad. Sci. USA. 102;11600-11605.

Shvedova, AA, Castranova, V, Kisin, ER et al. (2003) Exposure to carbon nanotube material: assessment of nanotube cytotoxicity using human keratinocyte cells. J. Toxicol. Environ. Health A 66:1909-26.

Singh, R, Pantarotto, D, Lacerda, L et al. (2006) Tissue biodistribution and blood clearance rates of intravenously administered carbon nanotube radiotracers. PNAS. 103-9:3357–62.

Sitharaman, B, Kissell, KR, Hartman, KB et al. (2005) Superparamagnetic gadonanotubes are high-performance MRI contrast agents. Chem. Commun. (31):3915-3917.

Sitharaman, B, Van Der Zande, M, Ananta, JS et al. (2010) Magnetic resonance imaging studies on gadonanotube-reinforced biodegradable polymer nanocomposites. J. of Biomed Mater Res Part A. 93A (4):1454-1462.

Vittorio, O, Quaranta, P, Raffa, V, et al. (2011a) Magnetic carbon nanotubes: a new tool for shepherding mesenchymal stem cells by magnetic fields. Nanomedicine. 6:1:43-54

Vittorio, O, Duce, SL, Pietrabissa, A, Cuschieri, A. (2011b) Multiwall carbon nanotubes as MRI contrast agents for tracking stem cells. Nanotechnology 4;22(9).

Vittorio, O, Raffa, V, Cuschieri, A. (2009) Biocompatibility of multi-wall carbon nanotubes on human neuroblastoma cell line and effects of metal impurity and surface oxidation. Nanomedicine NBM. 5(4):424-31.

Vittorio, O, Raffa, V, Riggio, C et al. (2011c) PC-12 interaction with magnetic carbon nanotubes: effects on viability and influence of cell differentiation on cell translocation induced by a permanent magnetic field. Current Nanoscience 7(3):337-344.

Weissker, U, Hampel , S, Leonhardt, A, Büchner, B. (2010) Carbon Nanotubes filled with ferromagnetic materials. Materials. 3(8):4387-4427.

Wick, P, Manser, P, Limbach, LK et al.(2007) The degree and kind of agglomeration affect carbon nanotube cytotoxicity. Toxicol Lett. 168:121–131.

Wörle-Knirsch, JM, Pulskamp, K, Krug, HF. (2006) Oops they did it again! Carbon nanotubes hoax scientists in viability assays. Nano Lett. 6:1261-8.

Zhang, X, Wen, GH, Huang, S et al. (2001) Magnetic properties of Fe nanoparticles trapped at the tips of the aligned carbon nanotubes. J. Magn. Magn. Mater. 231;L9–L12.

Zhu, L, Chang, DW, Dai, L, Hong Y. (2007). DNA Damage Induced by Multiwalled Carbon Nanotubes in Mouse Embryonic Stem Cells. Nano letters. 7:3592-3597.

9

Organically Structured Carbon Nanotubes for Fluorescence

Jianguo Tang and Qingsong Xu
[1]*Institute of Hybrid Materials,*
[2]*Laboratory of New Fiber Materials and Modern Textile,*
The Growing Base for State Key Laboratory,
[3]*Department of Materials Science and Engineering, College of Chemistry,*
Chemical and Environmental Engineering, Qingdao University, Qingdao
P.R. China

1. Introduction

Carbon nanotubes (CNTs), an one-dimensional (1D) nanostructural material, despite its inactivity, has the advantages of chemical flexibility and sensitivity arising from the susceptibility of their surfaces to interacting species. The researches emerged recently on functionalizing CNTs with functional organic macromolecules and oligomers have achieved novel light-emitting, light-electric or electric-light converting materials. In this chapter, we aim to capture recent advances and present our research achievements of rational design and chemical functionalization of CNTs for the purpose to obtain enhanced fluorescence property in the variety of organically modified structures, lanthanide-existed hybrid structures, polymer-embedded composites as well as the wide variety of applications for novel organic light-emitting diodes (OLEDs), laser resource, optical signal amplification, solar cells and biosensors. To provide a deeper understanding of the fluorescence property, this review will also survey the proposed mechanisms. As demonstrated by remarkable examples, the relationship between the structures of modified CNTs and the fluorescence property helps to offer attractive new prospects for constructing CNT-based molecular optoelectronic and photon devices with desired functionalities.

2. Organic modification chemistry of CNTs

Within the scope of this chapter, we shall focus on the fluorescence property of CNTs. Two main approaches are now considered for the surface modifications of CNTs. The first one is noncovalent attachment of functional molecules (NAFM) on to the walls of CNTs that is based mainly on van der Waals forces, controlled by thermodynamic parameters. NAFM can change the nature of CNTs surface and make it more compatible with different matrixes. The advantage of NAFM is that the perfect structure of the nanotube remains intact, and its mechanical properties also retain unchanged. However, its main disadvantage is that the binding forces between modification molecules and CNT surface might be too weak. The second approach is covalent attachment of functional molecules (CAFM). The functional molecules or oligomers can be attached on CNT surface via covalent bonds. It improves the chemical interface bonding between modifiers and CNT surface and provides stable surface properties of modified CNTs. However, the modification methods will be complicated, compared with NAFM, and introduce defects on the walls of CNTs.

2.1 Noncovalent attachment of functional molecules (NAFM)

The endeavors to improve the poor surface properties have been started as early as CNTs' discovery in 1991[1]. So far there are lots of publications [2-4] focusing on the modifications of CNTs for their compatibility with solvents in solutions and with matrixes in solid materials. For functionalized fluorescent CNTs, a successful surface modification strategy on CNTs pursues not only the compatibility but also the structures to be extended for luminescent groups.

Although the solubility can be improved by covalent modification of the SWNTs [5,6], these methods can disturb the natural properties of SWNTs. Therefore, NAFM has its merits to obtain the non-damaged structures. The early reports of R. E. Smalley [6] and S. M. F. Islam [7] indicated the efficient solubilization of SWNTs can be achieved by using the noncovalent wrapping, adsorption, and encapsulation. In this review, we will give the comments on the related developments. Because SWNTs structured by NAFM organically offer a unique combination of electrical, mechanical, thermal, and optical properties [2,3], they make them highly promising materials for a huge number of applications [4] ranging from nanocomposites, solid-state nanoelectronics, sensors, biomedical devices, and cellular delivery [8].

2.1.1 Wrapping by oligomers

Molecular engineering (cutting, solubilization, chemical functionalization, purification, manipulation, and assembly) of single walled carbon nanotubes (SWNTs) will play a vital role in exploring and developing their applications. Noncovalent wrapping of carbon nanotubes, as shown in Figure 1, is of particular interest, because it enables one to tailor their properties while still preserving nearly all of the nanotube intrinsic properties. SWNTs have become solubilized in organic solvents and water after wrapped by oligomers.

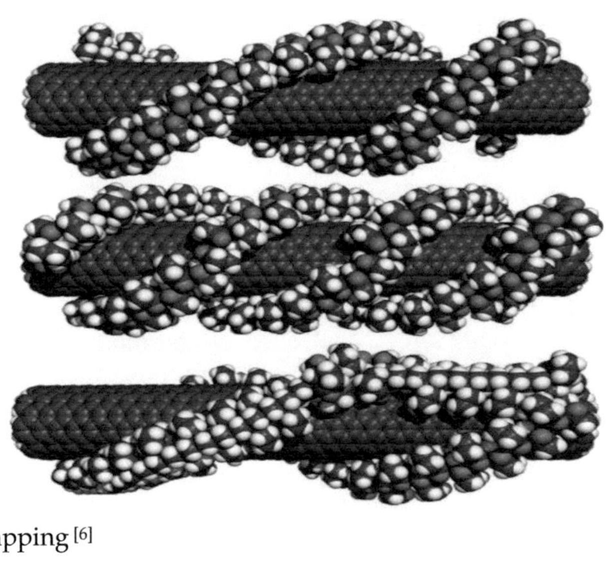

Fig. 1. Polymer wrapping [6]

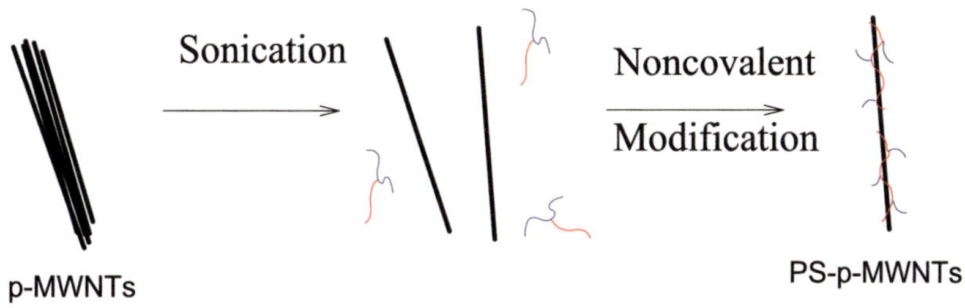

Fig. 2. The noncovalent surface modification of pristine-MWNTs with PS-g-(GMA-co-St) [10]

A simple, nondestructive method to modify MWNTs with a graft polymer (PS-g-(GMA-co-St)) noncovalently has been shown in Figure 2. This strategy is based on the affinity of the poly(styrene)(PS) main chains to the surface of pristine MWNTs (p-MWNTs) [11, 12], and the modified MWNTs can be solubilized in a wide variety of polar and nonpolar organic solvents at the same time.

Recently, Gupta and his coworkers found a simple method to functionalize the CNTs with fluorescent ink noncovalently [13]. 1 mg of CNTs was stirred with 10 ml of dilute ink solution in water for 8 to 10 h and subjected to sonication afterwards for two hours. The extra insoluble CNTs were then removed by centrifuging at 1600g (3500 rpm) for one hour. The collected solution was found to be stable for a few months. One to two drops of the solution of functionalized CNTs was dispensed onto a carbon-coated copper grid (300 mesh) for a few seconds, and extra solution was soaked off with a filter paper. The composite shows spectroscopic features of the fluorescent ink indicating noncovalent bonding between the CNTs and the ink molecules. The results shown here throw light upon the feasibility of designing efficient nanocomposite materials via attaching well known optical materials to CNTs.

T. D. Krauss et al. reported significant increases in the fluorescence efficiency of individual DNA-wrapped SWNTs upon addition of reducing agents, including dithiothreitol, trolox, and β-mercaptoethanol [14]. Brightening was reversible upon removal of the reducing molecules, suggesting that a transient reduction of defect sites on the SWNT sidewall caused the effect. These results implied that SWNTs were intrinsically bright emitters and that their poor emission aroused from defective nanotubes.

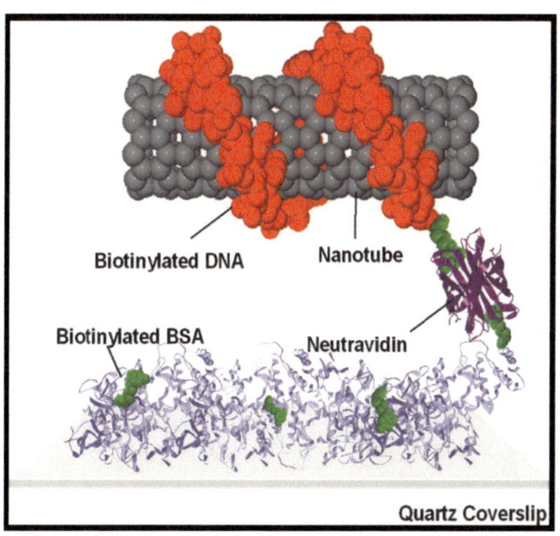

Fig. 3. Sample configuration for fluorescence measurements of individual SWNTs[14].

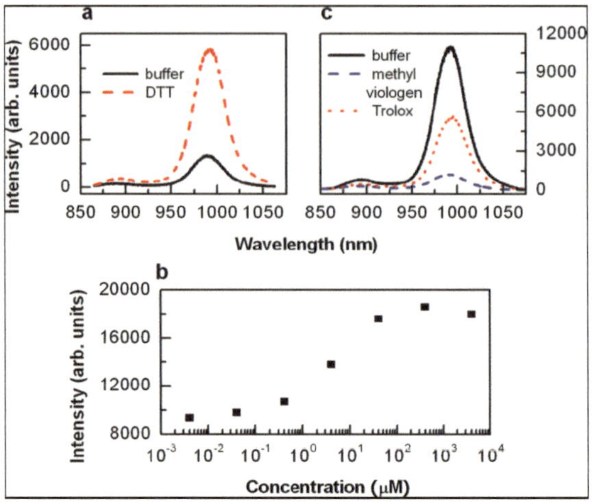

Fig. 4. Enhancement and quenching of fluorescence intensity for SWNT ensembled by reductants and oxidants. (a)Fluorescence spectra for an ensemble of SWNTs, displaying the enhancement of the fluorescence intensity upon addition of DTT. (b) Saturation curve of SWNT fluorescence intensity for increasing concentrations of added Trolox. (c) Fluorescence spectra for an ensemble of SWNTs measured in buffer, with addition of methyl viologen and with addition of Trolox[14].

Short, rigid conjugated polymers, poly (aryleneethynylene) s (PPE) (Figure 5), are used to solubilize SWNTs [15]. In contrast to previous work [10], the rigid backbone of PPE cannot wrap around the SWNTs. The major interaction between polymer backbone and nanotube surface is most likely π -stacking. This approach allows to control over the distance between functional groups on the carbon nanotube surface, through variation of the polymer backbone and side chains. This approach represents the carbon nanotube solubilization via π-stacking without polymer wrapping and enables the introduction of various neutral and

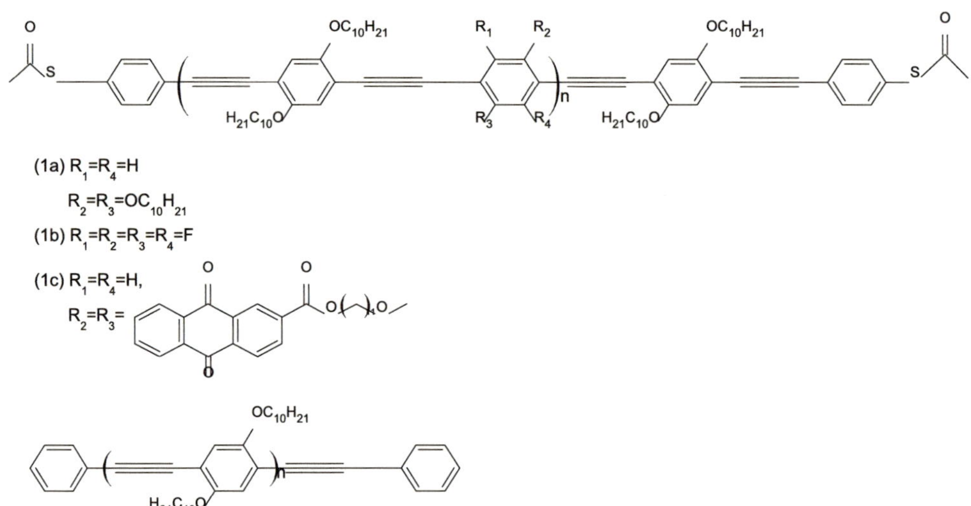

2

Fig. 5. Molecular structures of poly (aryleneethynylene) s (PPE) [15]

ionic functional groups onto the carbon nanotube surface. The optical spectroscopy supports a significant π-π interaction between the polymer and the nanotube (Figure 6). The strong fluorescence of 1a is efficiently quenched in 1a-SWNTs$_{HiPCO}$. The quenching likely arises from efficient energy transfer between 1a and SWNTs, rather than the disruption of π-conjugation caused by a conformational change.

PPE ($n_{average}$)	avg length (nm)	SWNT$_{Hipco}$ Solubility (mg/mL)
1a(19.5)	27.9	~2
1a(13)	19.6	~2.2
2(16)	12.0	~2
2(10)	7.9	~1.5

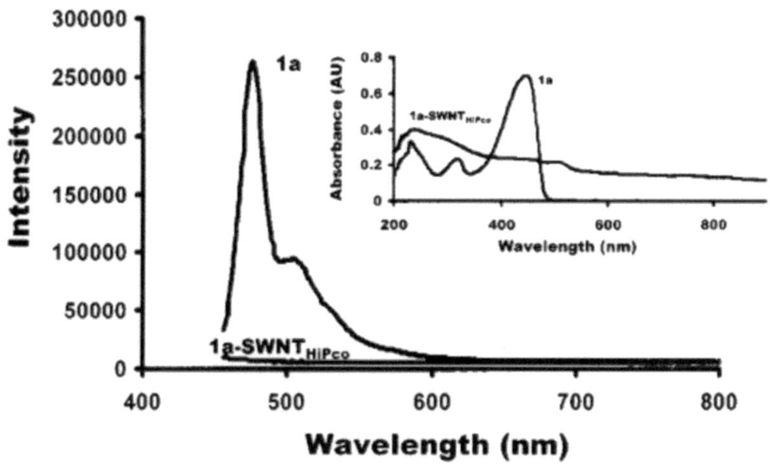

Fig. 6. Room temperature solution-phase ($CHCl_3$) fluorescence spectra (excitation wavelength: 400nm) and UV-visible spectra (inset) of 1a and 1a-SWNTs$_{HiPco}$ complex [15].

B. X. Li and coworkers had developed a new multicolor fluorescent sensing system to detect multiple analytes in one pot [16]. This design was built on the noncovalent assembly of dye-labeled aptamer with SWNTs by π-stacking between the nucleotide bases and the SWNTs sidewalls. That is to say, they combine the highly specific binding ability of aptamers with the ultrahigh quenching ability of SWNTs to develop a multicolor fluorescent sensing system. This multicolor fluorescent system is used to simultaneously detect thrombin and ATP in a single solution.

C. Fantini et al. carried on an insightful research on the influence of the nanotube and surfactant concentrations on the absorption and emission of light by individualized CNTs [17]. SWNTs dispersed in different surfactant solutions and at different concentrations were investigated by optical absorption and photoluminescence, aiming to investigate how higher photoluminescence efficiency (emission/absorption ratio) can be obtained for SWNT dispersion by choosing the type of surfactant and controlling the SWNT and surfactant concentrations. The result showed that the concentrations whose best efficiency of PL measurements was obtained correspond to the dispersions with higher ratio between individually dispersed nanotubes and bundles.

2.1.2 Adsorption of semiconductor nanoparticles

Because of their unique size-tunable chemical and physical properties, semiconductor nanoparticles have attracted much attention. Several semiconductor nanoparticles such as Ag_2S, CdS, have already been bound to the surfaces of CNTs. Metal sulfides, as one kind of important semiconductors, have been used in many new application areas, such as laser communication and light-emitting diodes. Metal sulfides nanoparticles such as Ag_2S and CdS with size of less than 30 nm are coated onto MWNTs successfully by a simple and effective in situ synthetic method without severely affecting the energy states of the MWNTs (Figure 7) [9]. This method could be extended to other transition metal compound nanostructures. This new type of hybrid carbon nanotubes with coated metal sulfides nanoparticles on the sidewall may have more interesting potential applications in field emitters, nanometer-scale optoelectronic devices and other related sides.

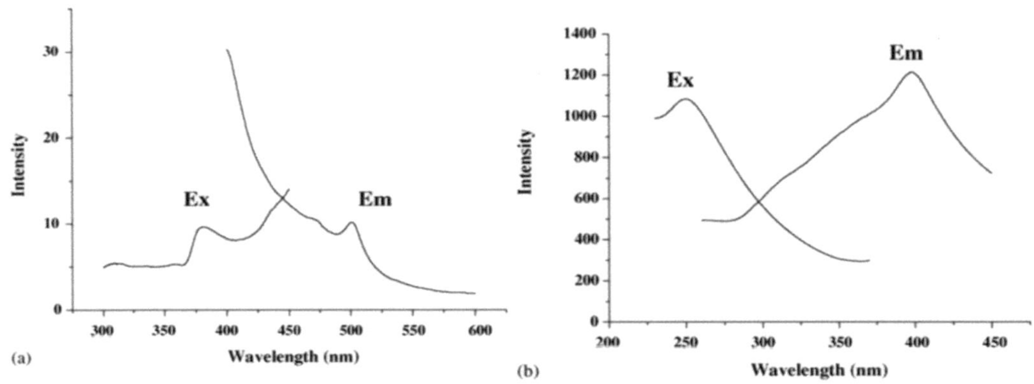

Fig. 7. Emission and excitation spectra of (a) CdS/MWNTs, (b) Ag_2S/MWNTs [9]

2.1.3 Noncovalent encapsulation

Filling its interior cavity with other molecules is another novel means to modify the properties of a SWNT. For example, SWNTs filled with 1-D chains of C_{60}, can be manufactured via a vapor phase or surface diffusion mechanism. The presence of interior C_{60} could decrease the SWNTs compressibility and increase its elastic modulus, which has been shown by molecular dynamics simulation. The method of encapsulation has been extended to other related molecules such as metallofullerenes La_2-C_{80} and Gd-C_{82}. The case of La_2-C_{80}-SWNT is regarded as the definitive proof that a non-intrinsic molecule could be inserted in bulk into SWNTs [18].

2.2 Covalent Attachment of functional molecules (CAFM) or oligomers (CAFO)

We can find in recent reports on the chemical compatibility and dissolution properties of CNTs that most researchers put special emphasis on developing modifications or functionalizations of their surface. When tailoring the properties of these materials and the engineering of nanotube devices, the modification chemistry of the open ends, the exterior walls, and the interior cavity of the CNTs is expected to play a vital role. Until now, several methods have been applied to graft or assemble synthetic oligomers, polymers or biomacromolecules onto the exterior surface of CNTs using covalent bonds.

Chemical modification of the SWNTs, as well as the MWNTs, has been carried out with a mixture of sulfuric acid and nitric acid, which is used to form carboxyl acid groups on the surface. The resultant carboxylic acid groups are formed along the nanotube walls and the ends. Then, we could attach the desired groups to the exterior walls by the reaction between the groups and the carboxyl acid groups.

2.2.1 Reactions with carboxyl groups on CNTs

The surface modification of MWNTs using highly branched molecules covalently attached onto the surface of MWNTs has been developed, which proves to be a representative example for the modification of CNTs by reacting with the carboxyl groups on the sidewall. The general strategy for modification is described in Figure 8. During this process, the volume of the MWNTs expands several times, perhaps because of exfoliation of the MWNTs bundles to give individual nanotubes.

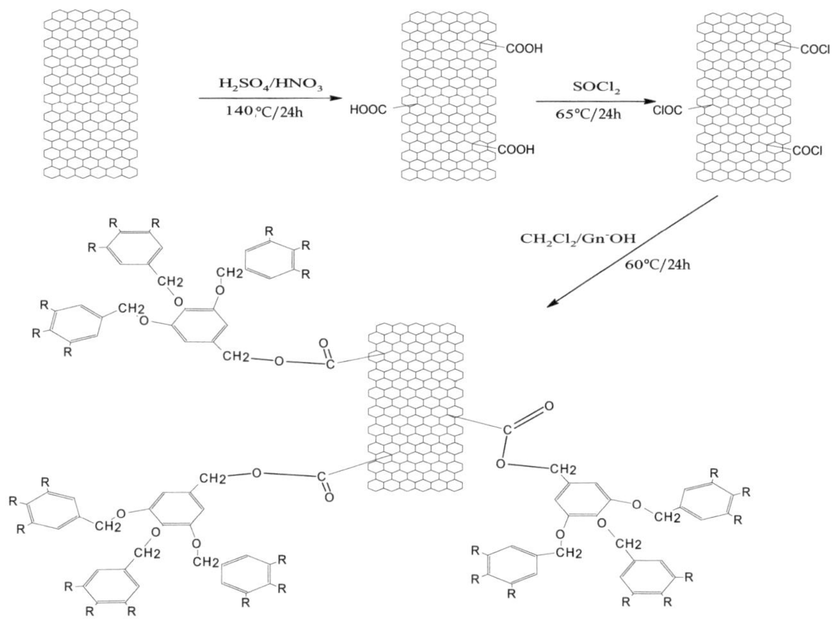

Fig. 8. A method of surface modification of multiwalled carbon nanoparticles (MWNTs) using highly branched molecules covalently attached onto the surface of MWNTs [19].

The MWNTs with carbonyl chloride groups (MWNT-COCl) were prepared via reaction of thionyl chloride with carboxyl-contained MWNT (MWNT-COOH) previously made by oxidation of the raw MWNT [20]. After centrifugation, the brown-black supernatant was decanted and the remaining solid was washed with anhydrous tetrahydrofuran (THF). After centrifugation, the pale yellow-colored solution was decanted. The remaining solid was then dried at room temperature in vacuum. An anhydrous dichloromethane (CH_2Cl_2) mixture of MWNT-COCl and 0.5 g poly(benzyl ether) highly branched molecules was heated at 60°C for 24 h. After cooling to room temperature, the excess highly branched molecules are removed by washing with ethanol for four times (5 to 10 min sonication at 40 kHz). The remaining solid is dried at room temperature under vacuum. The yield of resultant product is usually >60% (based on raw MWNTs). In summary, MWNTs can be successfully modified with highly branched molecules by reaction of carbonyl chloride groups functionalized MWNT (MWNT-COCl) and highly branched molecules that have hydroxyl groups at the focal point. The

modification of CNTs may provide valuable properties in many application areas such as optoelectronics, information storage, and catalysis [19].
The first study on functionalization of SWNTs with enzymes has already been achieved by the initial acylation of SWNTs followed by amidation with the desired amine or enzyme [21]. The two-step chemical method needs mild conditions and results in tethering of the organic functionality through a covalent bond [Figure 9]. It is a simple, but practical and highly effective process. The two enzymes tethered to SWNTs are porcine pancrease lipase (PPL) and amino lipase (AK). The same method is also used to functionalize SWNTs with various amines, which include three primary amines (cis-myrtanylamine, 2, 4-dinitroaniline, 2, 6-dinitroaniline) and two secondary amines N-decyl-2, 4, 6-trinitroaniline and N-(3-morpholinopropyl)-2, 4, 6-trinitroaniline. Linkage of chiral molecules and enzymes to SWNTs further makes the applications of CNTs possible in areas such as medicinal and biological fields, biosensor or chemically modulated nanoelectronic devices. Tethering to nitrated molecules is also hoped for the use of SWNTs in nanoscale energetic systems.

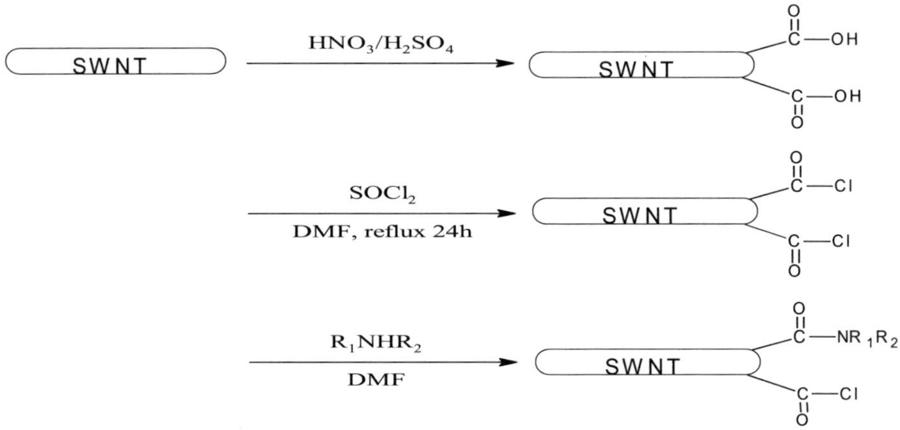

Fig. 9. CNTs modification with amines and enzymes [21]

The exact functional groups of R_2 for different functionalizations are those compounds, such as 2, 4-dinitroaniline and N-decyl-2, 4, 6-trinitroaniline, while the R_1 is mainly hydrogen.
Both MWNTs and SWNTs have been considered as attractive candidates for fabricating novel materials with desirable properties, owing to their tubular nanostructures, unique and promising mechanical properties. However, compared with MWNTs, SWNTs exhibit simpler structures and are easier to control as regards diameter during fabrication, so most previous academic researches on CNTs are focused on SWNTs. However, as noted by Dalton et al. [22], the high cost of SWNTs severely hindered its commercialization for most applications severely. This problem can be released by using MWNTs, which have been scaled up recently in the industrial scale, resulting in a relatively lower price. Most of the excellent properties and merits of MWNTs are comparable with those of SWNTs. Therefore, it is desirable to pay more attention to MWNTs, particularly as regards functionalization, in the future.
Chemical (covalent) functionalization has been achieved through ultrasonication with organic materials [23], diimide-activated amidation, and 1, 3-dipolar cyclo additions, which we are going to introduce later [24]. End-to-end and end-to-side SWNT interconnects are formed by reacting chloride terminated SWNT with aliphatic diamine [25].

European scientists first reported their approach to covalently attach an inorganic metal complex to MWNTs [26] [Figure 10]

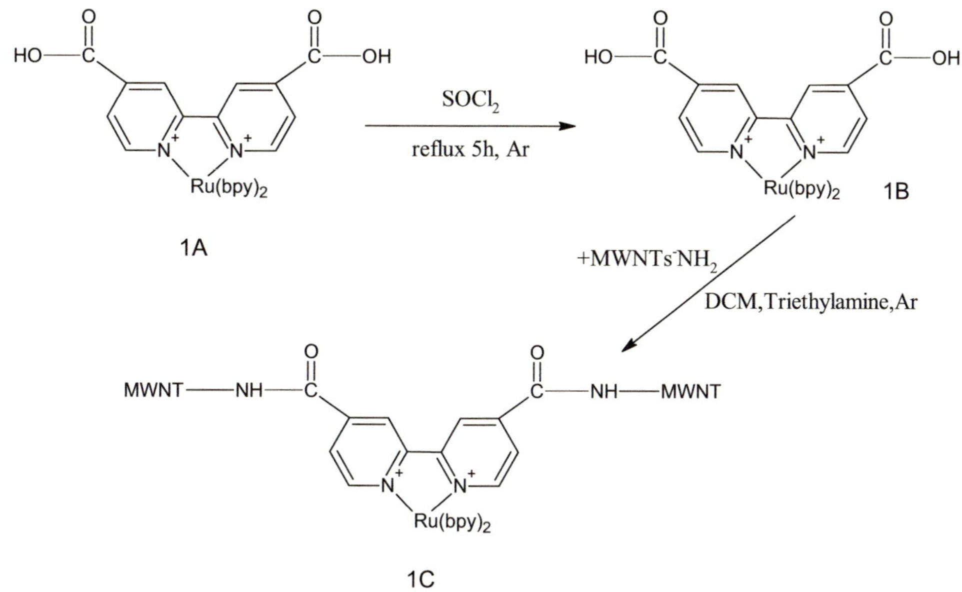

Fig. 10. Connecting CNTs with an Inorganic Metal Complex [26]

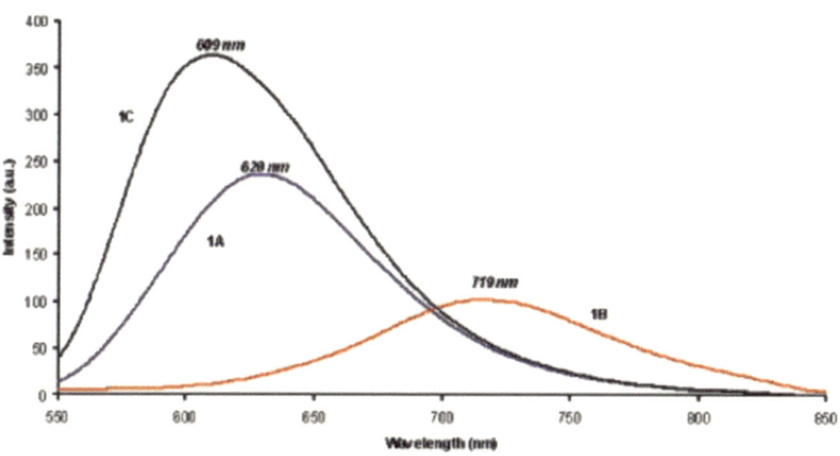

Fig. 11. Emission spectroscopy (recorded in dichloromethane) on Ruthenium complex (1A), chlorinated product (1B), and Ruthenium nanotube complex (1C) [26].

The experimental procedure is shown in Figure 10. Ten milligrams (0.011 mmol) of $[Ru(dcbpy)(bpy)_2](PF_6)_2$ (1A) was dissolved in 15 ml of thionyl chloride. The reaction mixture (1B) was refluxed under argon for 5h. The thionyl chloride was then removed by vacuum distillation. The remaining solid was partially dissolved in dichloromethane (DCM). Two milligrams of the MWNT functionalized with NH_2 (MWNT-NH_2) was

sonicated in 5 ml of dichloromethane for 2 min and then were added to the refluxed mixture. Ten milliliters of triethylamine were added. The solution mixture was stirred at room temperature under argon for 72 h. Then, the product would be filtered to remove the solvents and washed with DCM. The product was placed in 20 ml of DCM and sonicated for 2 min. At last, the solution was allowed to settle for 24 h. Excess MWNT-NH_2 settled at the bottom, and the "functionalized ruthenium MWNT" (1C) product was in suspension in solution. A color change from dark red-orange (1B) to dark brown-green (1C) would be observed after the completion of the reaction scheme [Figure 11] [26].
Strong evidences by emission spectroscopy had been presented, which confirmed the successful creation of MWNTs interconnection through amide linkage with a ruthenium complex [26] [Figure 11].

2.2.2 Reactions with other groups on CNTs

MWNTs can be functionalized and solubilized via attaching aminopolymers to the CNTs [27], making it soluble in certain solvents. The acylation-amidation method and the direct heating method are both effective for this kind of functionalization, just as shown in Figure 12. The aqueous solubility of these functionalized samples as a result of the hydrophilicity of the aminopolymers may find applications in introducing CNTs into biologically significant systems.

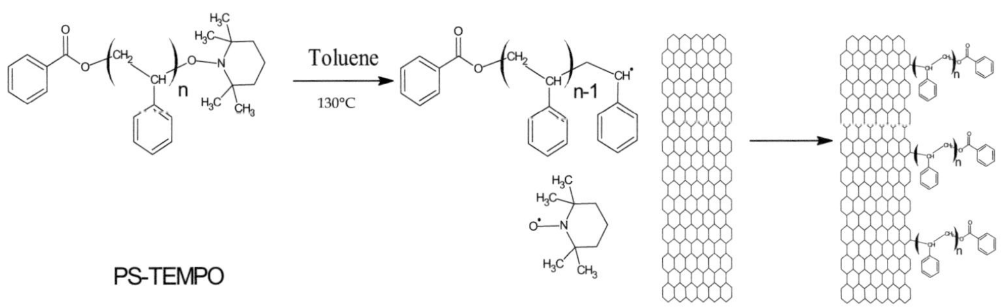

Fig. 12. A strategy to graft alkoxyamine end-capped (co)polymers onto MWNTs [28]

Additionally, MWNTs have been modified by PS, PCL, and PCL-b-PS as results of the addition reaction of the parent polymeric radicals. Grafted MWNTs can easily be dispersed in solvents such as toluene and THF with the help of these grafted polymers [28].
A general route for the 1, 3-dipolar cycloaddition of azomethine ylides functionalization of nanotubes is described in Figure 13. Derivatives on the substituents of either the aminoacid or the aldehyde may lead to numerous structurally different functionalized CNT materials which are potentially useful in diverse applications on nanotechnology.
Modified SWNTs are bonded like bundles or ropes with the diameter of about 50-100 nm and the length of several micrometers. Every rope consists of small bundles of two or three nanotubes, or even an individual isolated nanotube. In contrast, pristine SWNTs are usually bonded in bundles with an average diameter of 10 nm.
The development on architectural superstructures in the level of nanometer has become possible with the help of spontaneous self-assembly of structurally different fulleropyrrolidines. Numerous novel materials such as the organic functionalized nanotube derivatives or nanocomposites mentioned above can be synthesized, with a wide variety of properties resulted from the attached functional group.

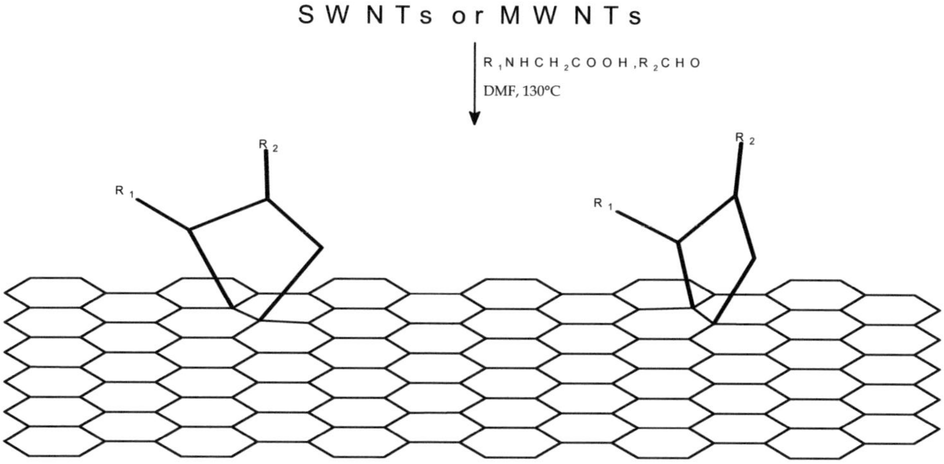

Fig. 13. Modification of nanotubes via 1, 3-dipolar cycloaddition of azomethine ylides [29].

Japanese researchers presented a method for simultaneously solubilizing and labeling CNTs by using a detergent covalently coupled with a variety of fluorophores commonly used in biology [30]. Because of their stability under physiological conditions and their varying fluorescence properties, fluorescently labeled nanotubes can be easily utilized in combination with biomolecules such as proteins.

2.2.3 New approach to the modification of CNTs

Just as mentioned earlier, functionalization of CNTs is advantageous to prevent the aggregation of nanotubes and to favor their solubilization in organic solvents at the same time [31, 32]. The attached functional groups can be used as precursors for the subsequent attachment of a wide variety of other functional groups [33]. To obtain functionalized nanotubes, direct fluorination by F_2 gas is a method widely used, which is known to be very corrosive, making the reaction difficult to handle. Besides, it is shown that fluorine can attack the nanotubes over 50°C, and at higher temperatures some undesirable reactions may take place [33, 34]. The average length of nanotubes after the preparation step is usually much higher than their diameter, which makes them unsuitable to be used as a nanometer-scale material [35]. The nanotubes whose length/diameter ratio is too high to be used as electron emitters are also difficult to disperse in polymer matrices. Boiling in oxidative media or grinding in a ball mill can effectively reduce the length of CNTs [35-37]. Compared to other shortening methods such as ultrasound power or STM voltage, by ball milling we can obtain shortened nanotubes in large quantities. Ko'nya et al. performed the functionalization of multi-walled carbon nanotubes by ball milling in reactive atmospheres (H_2S, NH_3, Cl_2 etc.) and proved that this method was appropriate for large scale production of short functionalized nanotubes [38]. Recent research has carried out the functionalization of SWNTs in an agate ball mill by using trifluoromethane (TFM), trichloromethane (TCM), tetrachloroethylene (TCE), hexafluoropropene (HFP) and chlorine (Cl_2), which had demonstrated that ball milling of single-walled carbon nanotubes in reactive atmospheres was an effective method in large-scale production of functionalized short SWNTs.

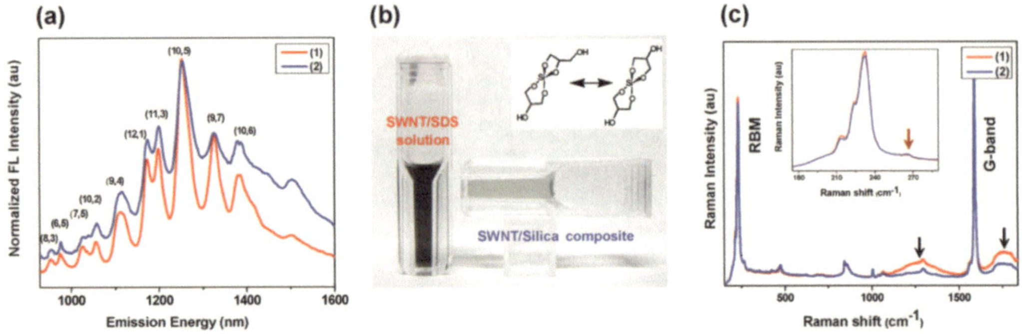

Fig. 14. (a) Normalized fluorescence spectra recorded on (1) SWNT/SDS dispersion in D_2O and (2)SWNT/silica monolith prepared using the SWNT/SDS solution shown in (1); (b) Photograph showing a SWNT/SDS aqueous dispersion (upright cuvette) and a SWNT/silica monolith (tilted cuvette). Inset showed the possible molecular structures of the silica precursor DGS; (c) Raman spectra of (1) a SWNT/SDS dispersion in D_2O and (2) a SWNT/silica monolith. Inset showed the radial breathing mode (RBM) spectra of the samples [40].

It was shown that alkyl-halides were suitable for substitution of very corrosive fluorine and chlorine gases in the process of functionalization of carbon nanotubes [39]. A. M. Dattelbaum and his coworkers had developed a new approach for the preparation of SWNT/silica composite materials, which were fluorescently active [40]. This approach made use of diglyceryl silane, a sugar alcohol based silica precursor molecule, which would condense under conditions that did not promote significant aggregation of the surfactant-nanotube assemblies, as was shown by fluorescence and Raman spectroscopy [Figure 14].

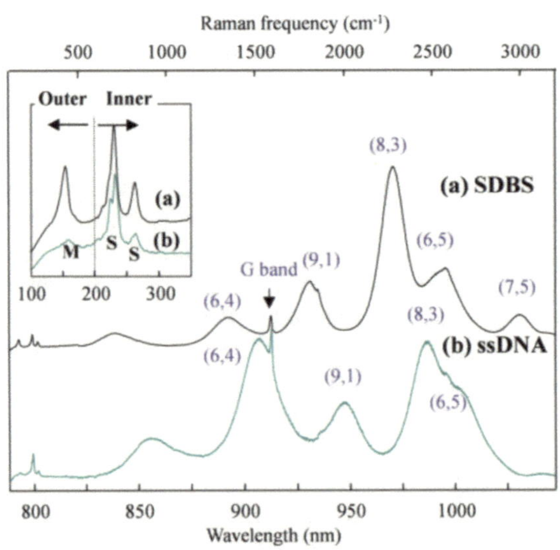

Fig. 15. Comparative Raman/fluorescence spectra taken with laser excitation of 785 nm for SDBS and DNA-dispersed DWNT supernatants. The inset showed the magnified low frequency Ramen spectra, where M indicated metallic and S indicated semiconducting tubes. [42]

Double-walled carbon nanotubes (DWNTs), have attracted a great deal of attention due to their intrinsic coaxial structures make them mechanically, thermally, and structurally more stable than SWNTs[41]. Kim et al. reported for the first time, detailed Raman/luminescence spectroscopic studies on single-standard DNA-dispersed DWNT solutions at different dispersion states [42], in comparison with an SDBS-dispersed DWNT solution using three different laser lines, in order to understand the interactions between DNA and the outer tubes, and the effect of these different DWNT environments on the vibrational and luminescence behaviors.

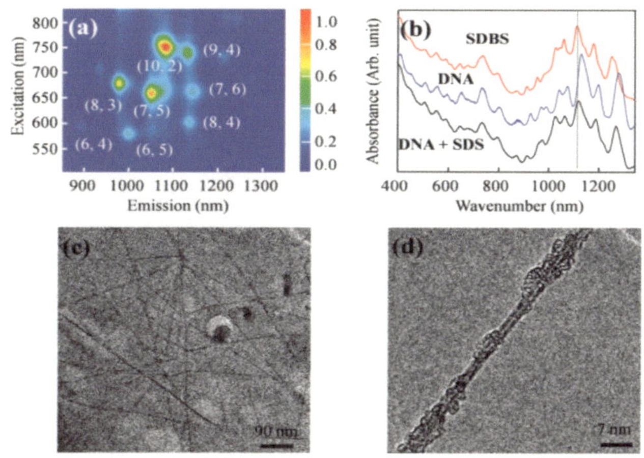

Fig. 16. (a) PL map and (b) UV/visible absorption spectra of DNA-dispersed DWNT solution at pH=8.0,(c,d) their corresponding TEM images. Note that DWNTs are individualized with the help of help of helically wrapped DNA. Their color represented the PL intensity on a linear scale. [42]

3. Fluorescent CNTs grafted by fluorescent groups

CNTs are now attracting more and more attentions because of their high potentials in exploring novel nanoscale electronic and optical devices. For example, room temperature single electron transport devices have been developed which are expected to be the building block of integrated circuits in the future. Artificial atoms have also demonstrated their use in quantum computing. It has also been shown that semiconducting SWNTs can emit light from visible to infrared spectral region and light-emission based on current injection has been demonstrated, all of which arise the study on optical properties of SWNTs. It has been found that the exciton binding energy in a CNT is as large as 0.4 eV, several tens of that of a traditional semiconductor. Large nonlinearity has also been found in SWNTs [43]. With rising temperature, suspended SWNTs exhibit discontinuous changes in their emission energy which is much different compared to a traditional semiconductor. These results show that more efforts are still needed to discover the novel physical and optical properties of CNTs.

As mentioned earlier, the insolubility of nanotubes in most solvents has hindered quantitative investigations. A feasible way to solubilize carbon nanotubes is to covalently attach them to highly soluble linear polymers. An interesting finding has been reported that the polymer-bound CNTs in homogeneous organic and aqueous solutions are luminescent or even strongly luminescent.

Shortened MWNT (S-MWNT), shortened SWNT (S-SWNT) and SWNT samples were treated with HCl solution to fully recover the carboxylic acid groups on the surface of CNTs, and refluxed in $SOCl_2$ for 24 h to convert the carboxylic acids into acryl chlorides. These functionalized CNTs were then mixed with poly-(propionylethylenimine-*co*-ethylenimine) (PPEI-EI, M_W≈200000, EI mole fraction≈15%) and reacted at 165°C for 20 minutes. The reaction mixtures were repeatedly extracted with chloroform to obtain the soluble fractions, which were then purified via repeated precipitations. It appeared that the polymer attachment is at the end of nanotube, as illustrated in Figure 17.

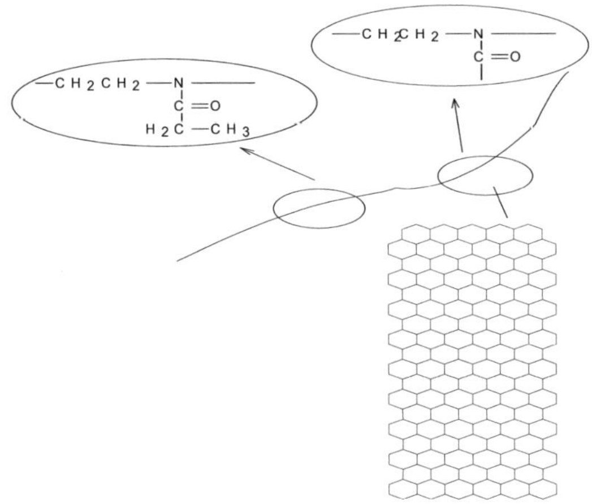

Fig. 17. An illustration of the PPEI-EI polymer-bond carbon nanotubes [44]

sample	solvent	λ_{EX}(nm)	Φ^a	τ_1(ns)	τ_2(ns)	a_1/a_2
S-MWNT–	$CHCl_3$	632		2.2	5.6	1.2
PPEI-EI		400	0.11	2.3	8.0	2.9
S-SWNT–	Water	400	>0.03	2.3	8.9	4
PPEI-EI	$CHCl_3$	632		1.9	5.8	1.2
SWNT–PPEI-	$CHCl_3$	440	0.06			
EI		400	>0.003	1.5	7.3	4.2
S-MWNT–	Water	365		2.1	9.8	2.3
PVA-VA	$CHCl_3$	365	>0.03	1.6	6.0	7.3

Table 1. Luminescence Parameters of the Polymer-Bond Carbon Nanotubes in Solution [44]

The same reaction conditions were used to attach S-MWNT to poly (vinyl acetate-*co*-vinyl alcohol) (PVA-VA, MW≈110 000, alcohol mole fraction≈40%) via ester linkages [45, 46]. These samples of polymer-bound CNTs were soluble in both organic solvents and water, forming highly colored homogeneous solutions. After being repeatedly filtered through 0.2 μm

Teflon filters, the chloroform and aqueous solutions were used for spectroscopic measurements. UV/vis absorption spectra of the chloroform solutions at room temperature (~22°C) are compared in Table 1, and spectra of the aqueous solutions are similar. The band-gap of fluorescence of semiconducting SWNTs in the near-IR is very sensitive to surface chemistry and often attenuated upon doping and functionalization, as is already known in the literature [47, 48]. On the other hand, the functionalized CNTs do show significant emission in the visible when excited at a shorter wavelength. The emission intensities and quantum yields can be very high, with yields of 4.5% and 3% for the spectra of PPEI-EI-SWNT and PEG_{1500N}-SWNT, respectively, shown in Figure 18. The luminescence decays are mostly nonexponential but generally fast, with the scale of a few nanoseconds [49, 50]. The decay results suggest inhomogeneity in the emitting species and the excited states responsible for the observed emissions. The excitation wavelength dependence, with the observed emission spectra progressively moving toward the red with longer excitation wavelengths, is consistent with the presence of significant inhomogeneity.

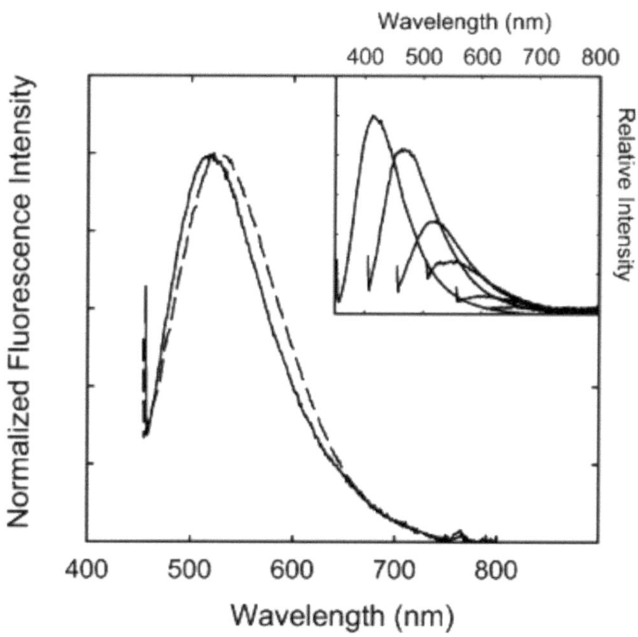

Fig. 18. Luminescence emission spectra (normalized, 450 nm excitation) of PPEI-EI-SWNT (—) and PEG_{1500N}-SWNT (- - -) in aqueous solution [50]. Inset: the spectra of PPEI-EI-SWNT excited at 350, 400, 450, 500, 550, 600 nm (intensities shown in relative quantum yields).

This is demonstrated well in a comparison of the visible luminescence emissions between purified nanotubes dispersed in a stable suspension with the aid of polymers and functionalized nanotubes in solution. While the two samples appear indistinguishable and are of similar absorption spectra and optical densities at the excitation wavelength, the observed luminescence emission intensities are very different. As compared in Figure 19, the solution of functionalized SWNTs is considerably more luminescent, which maybe due to the fact that more SWNTs are dispersed at individual nanotube level in the functionalized sample.

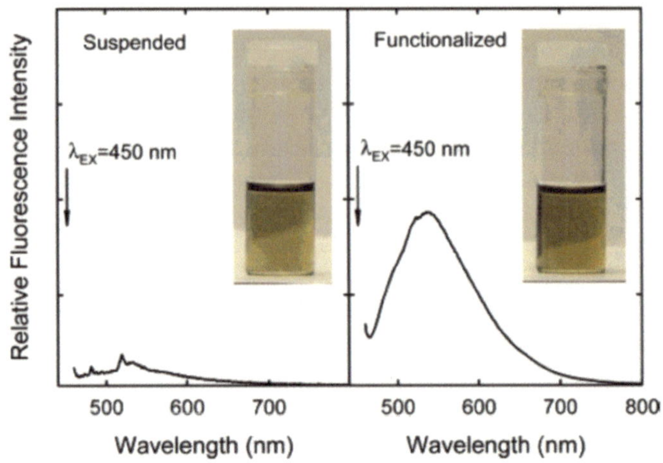

Fig. 19. Luminescence emission spectra from SWNTs dispersed with the aid of polyimide in DMF (left) and the polyimide-functionalized SWNTs (PI-NH_2-SWNT) in DMF solution (right) [50].

That is to say, there are strong luminescence emissions from well-dispersed CNTs in most functionalized samples. The broad luminescence emissions are logically owing to the trapping of excitation energy by defect sites in the nanotube structure, which are passivated upon the appropriate functionalization of the nanotubes. The better the dispersion and functionalization of the nanotubes, the more intense the luminescence emissions.

It is well known that the bleaching of nanotube fluorescence and absorbance spectra from the reaction of surfacted SWNTs with small organic electron-acceptor molecules depends solely on the reduction potential of the organic molecule. Thus, SWNTs can be perceived as behaving in a manner similar to that of other fluorescent organic molecules [51]. Metal ions have been used in emission studies to quench the fluorescence of various organic molecules including pyrenes, anthracenes, flavins, bipyridines, and acridinium ions [52-55]. However, no work before has focused on the interactions between SWNTs and metal ions. In order to further explore this phenomenon and determine the generality of M^{n+} quenching of SWNT fluorescence, A. R. Barron et al. had investigated the charge-transfer reaction between sodium dodecylbenzen-sulfonate(SDBS) surfacted SWNTs with group 2, 12, and 13 metal ions[51]. They found that the larger the ionic radii (lower the charge density) of the ion, the greater the efficiencies of quenching the smaller the SWNT, the greater the quenching effect of a particular M^{2+} ion.

K. J. Ziegler et al. reported a general method for coating SWNTs with polymer using emulsion-like microenvironments surrounding SWNTs [56]. Nylon 6, 10 were chosen as model systems for in situ emulsion polymerization. The reaction was going on at the surface of the nanotube, resulting in a thin polymer coating around individual SWNTs. The nylon-coated SWNTs were easily redispersed in water after freeze-drying. The fluorescence intensity of the nylon-coated SWNTs remains high at both acidic and basic pH conditions.

4. Polymer embedded CNTs with fluorescence emission centers

The use of polymers that are structurally close to the matrix polymer for the functionalization of CNTs is a favorable strategy in the development of polymeric carbon nanocomposites. It ensures compatibility of the functionalized CNTs with the polymer matrix to avoid any potential microscopic phase separation in the nanocomposites. Generally speaking, the species used in the functionalization and solubilization of CNTs become "impurities" in the final nanocomposites. Some units in polymers, such as derivatized styrene units in the polystyrene copolymers, which are structurally closed to the matrix polymers, are also regarded as impurities. Thus, an ideal polymeric carbon nanocomposite may be prepared by using solubilized CNTs which are functionalized with the matrix polymer. One of the polymer systems that can be used for such a purpose is poly (vinyl alcohol) (PVA) [Figure 20]. PVA is an excellent matrix polymer for nanocomposites.

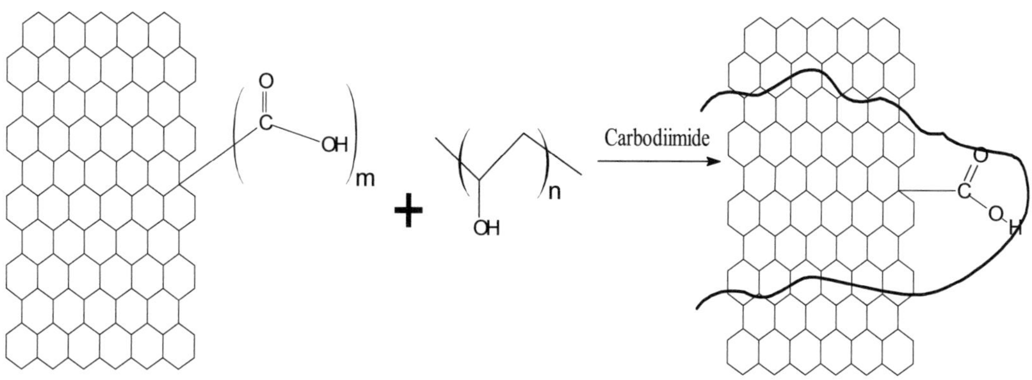

Fig. 20. Functionalization of CNTs with PVA [57]

The brief strategy of this kind of functionalization is shown vividly in Figure 20. Both SWNTs and MWNTs are functionalized with PVA in carbodiimide-activated esterification reactions. [58]

The PVA-CNTs composite films have high optical qualities, without any observable phase separation. The characterization results of the nanocomposite films show that the dispersion of CNTs is as homogeneous as that in solution.

There are studies which described an approach to the use of CNTs to pattern a high molecular weight polymer [Figure 21] [59, 60]. The resulting order of the attached polymer across the tube is surprising and seems to reflect the structural perfection of the tube itself. This templating of crystalline polymer suggests the possibility of constructing uniquely ordered, chemically tailored, and nanostructured materials in bulk from CNTs. The possibilities for such materials are numerous, from simple attachment to a polymer matrix material for enhancing yield strength to the construction of larger polymeric architectures with order over many different length scales. In this approach, MWNT and SWNT nanotubes are used as the beginning or templating nanomaterials [59].

The attachment between CNTs and the polymer results in a highly ordered polymer around the nanotubes, which provides us with a first step toward more complex construction using these nanomaterials. Dissolution of CNTs and the study of their properties in solution have been challenges for chemists. Although some efforts have been made in this direction, most studies by now have involved cutting and chemical functionalization of CNTs, or attachment to polymers with solubilizing features. This approach has two disadvantages. On one hand, tedious chemical derivatization is often required, while on the other hand, these CNT derivatives may have significantly different properties than those of pristine materials. CNTs can be dissolved in aniline without any prior chemical functionalization, and this material is then soluble in a variety of organic solvents, which represents the first observation of significant dissolution of pristine CNTs in standard organic solvents. For future work involving the separation, purification, and chemical functionalization of CNTs, solubilized nanotubes have a distinct advantage. For example, one can envisage CNT-aniline solutions for the formation of nanocomposites or thin films, which would solve some of the practical problems involved in making nanotube-based electronic and field emission display devices [60]. The strong fluorescence emission of CNTs should also be a useful probe to illustration of the physical and biological properties of these materials.

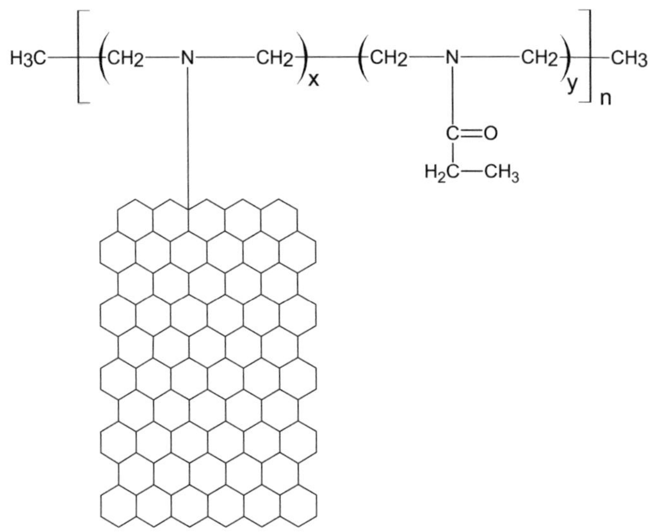

Fig. 21. Idealization of the PPEI-EI attachment [59]

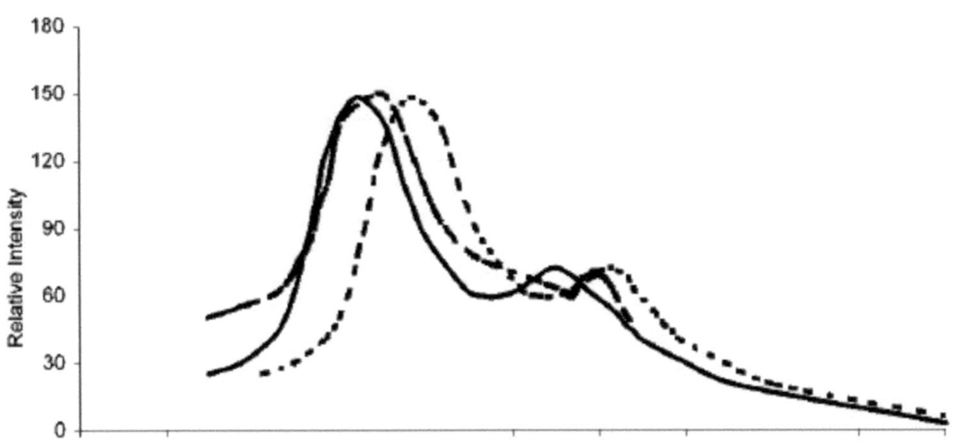

Fig. 22. Emission spectra of aniline dissolved carbon nanotubes in different solvents: in acetone (—), in toluene (-··-), and in methanol (---). All samples were excited at 500 nm [61].

Strong fluorescence can be observed upon exciting the diluted SWNT-aniline solutions at 500 nm. Figure 22 shows the emission spectra of solutions in acetone, toluene, and methanol, respectively. The maximum emission in acetone is observed at 565 nm with a shoulder at 610 nm. In methanol, the emission is redshifted by 20 nm without any change in intensity. The shift of the emission maximum in polar/nonpolar solvents is unanimous with charge separation in the excited state. The quantum yield of fluorescence of SWNT-aniline in acetone is 0.30, considerably higher than that of aromatic molecules. Luminescence has been observed at slightly longer wavelengths for polymer-bound CNTs but it is controversial for the origin of this fluorescence. The quantum yields for CNT-aniline solutions are higher than those for the polymer-bound CNTs as well. Precautions are taken to prevent interference from fluorescence of small aromatic species and other impurities.

Composites of polymers and CNTs have been studied as materials not only for mechanical reinforcement but also for optoelectronic devices such as polymer light-emitting diodes (LEDs) and photovoltaic cells. Curran et al. reported 5 times better thermal stability from the composites LED's of PmPV and MWNTs [62]. Ago et al. fabricated photovoltaic devices using a heterojunction of PPV and MWNTs, and obtained about twice the external quantum efficiency compared to the standard indium-tin oxide (ITO) based devices [63].

The spectral-resolved photoluminescence of pure polymers and composites has been studied [64]. While the photoluminescence intensity of m-PMEH-PPV increases with temperature, composites slightly varies without order.

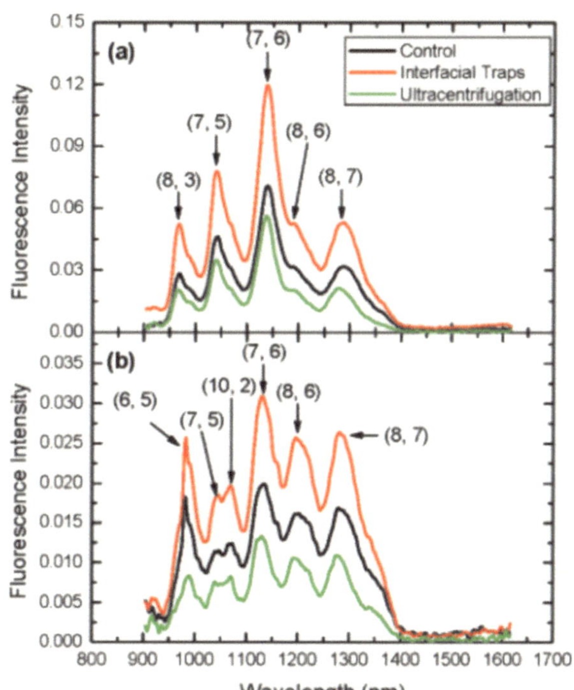

Fig. 23. Fluorescence spectra of Gum Arabic-suspended SWNTs from an initial mass concentration of 0.03 mg/mL of raw material with (a) excitation at 662 nm, and (b) excitation at 784 nm [66]. The control spectra were the samples after homogenization and sonication. This sample was then subjected to either ultracentrifugation or interfacial traps.

The conventional method to disperse individual nanotubes in aqueous solutions is by high-shear homogenization and ultrasonication in various surfactant solutions. While individual nanotubes are coated with a surfactant, some SWNT bundles remain because of large van der Waals attractions [65]. The bundling of nanotubes perturbs the electronic structure quenching the fluorescence of SWNTs. Ultracentrifugation is often used to remove nanotube bundles, but it is limited to analytical scales. Therefore, alternative routes are needed for large-scale removal of SWNT bondles.

R. K. Wang et al. introduced a process to remove nanotube bundles from aqueous suspensions by liquid-liquid interfacial trapping at toluene-water interfaces [66]. The approach is simpler than ultracentrifugation, and its resultant suspensions also have higher fluorescence intensities reflecting a higher concentration of individually suspended nanotubes. Fluorescence spectra of the aqueous phase were recorded with excitation at 662 and 784 nm, shown respectively in Figure 23a and 23b. In order to make a comparison, the spectra after homogenization and ultrasonication was shown as well as the spectra using traditional ultracentrifugation rather than interfacial trapping. The spectra showed that ultracentrifugation results in an obvious decrease in fluorescence intensity, indicating the removal of individual nanotubes.

5. Metal-contained emission systems of modified CNTs

It is an arising hot point that makes the metals cooperate with the CNTs by a certain kind of physical or chemical bonds, froming composites completely new. A large number of metals have been taken into consideration raging from the metals we could see everyday to rare-earth metals as well as the noble ones like Au and Ag. By cooperating with metals, we could obtain the modified CNTs with amazing new functions.

5.1 Lanthanide complexes as emission centers

Rare earth elements are of great importance in magnetic, electronic, and optical materials because of the number of unpaired electrons in their shells [67]. The novel properties of rare earth compounds make them rather appealing for practical applications in, for example, luminescence, catalysis, florescence imaging, and biological fields [68].

At the same time, more and more researchers have focused on the coating of CNTs. Because coating CNTs exhibit better physical and chemical properties and will lead to an even more diverse range of applications. For these reasons, the coating of CNTs with lanthanide related compounds is beginning to emerge.

Chinese researchers reported for the first time that rare earth fluoride EuF_3 and TbF_3 nanoparticles could be in situ bound to MWNTs through a simple and efficient synthetic route without causing a significant modification of the energy states of the MWNTs [69] [Figure 24].

MWNTs with an average outer diameter between 20 and 50 nm and length up to a dozen micrometers were prepared by the thermal catalytic decomposition of hydrocarbon [70] and the purity was over 90%. First, the MWNTs were dispersed in a 1 wt% sodium dodecyl sulfate (SDS) aqueous solution to modify the MWNTs surface by ultrasonication for 4h. After further washing and drying, 100 mg SDS adsorbed MWNTs was sonicated in 20 ml 0.1 mol/L $Ln(NO_3)_3$ (Ln = Eu^{3+}, Tb^{3+}) solution for 5 min, then 20 ml 0.3 mol/L NaF solution were slowly added into the mixture above with vigorous stirring. After 24 h reaction, the final products were washed repeatedly with water and then dried for 12 h at 100°C [69]. EuF_3

and TbF_3 nanoparticles with size of less than 10 nm were coated on MWNTs successfully by a simple and effective in situ synthetic method without severely changing the energy states of the MWNTs. This method could also be extended to other transition metal and rare earth elements compound nanostructures. It should be further noted that this new type of hybrid CNTs with coated rare earth fluoride nanoparticles on the sidewall may have more interesting potential applications in field emitters and nanometer-scale optoelectronic devices.

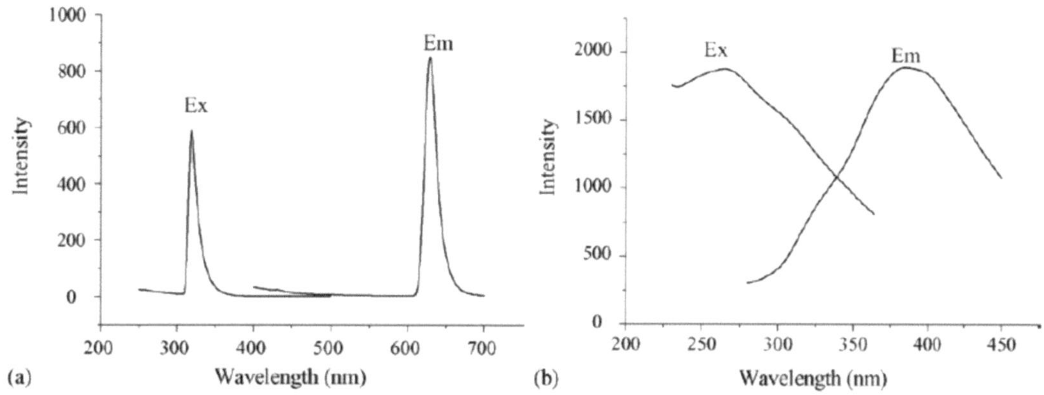

Fig. 24. Excitation and emission spectra of (a) EuF_3/MWNTs and (b) TbF_3/MWNTs [69]

As it is well accepted, the surface of a nanotube is often not ideal for coating; therefore the nanotubes have to be treated before coating. Previous reports have demonstrated that boilling the CNTs with oxidizing acids was an effective way for coating CNTs with CeO_2, while the CNTS are dispersed into nitric acid and heated at 500℃ for 2 h in air, only few CeO_2 particles absorbed on CNTs[71].

CNTs coating with europium oxide by a simple method is first reported by Chinese scientists [72]. Researchers covered the CNTs with a uniform layer of Eu_2O_3. Such kinds of CNTs have the potential applications in the field of emission display materials and luminescent materials. At the same time, europium oxide nanowires may be prepared by using CNTs as removable templates.

Coated CNTs are prepared as follows: CNTs were produced by catalysis and dissociation of hydrocarbon compounds as original material at high temperature. The average diameter of CNTs is found to be about 20 nm by TEM. The CNTs were suspended in a solution of concentrate nitric acid containing europium nitrate and refluxed for 4.5 hours [73]. When the mixture was cooled to room temperature, the ammonia solution with a concentration of 2.5wt% was added dropwise until the pH value reaches 9[74].Then the mixture was filtered and annealed in a stream of N_2, at 700°C for 2 hours. After the Sample was washed with distilled water, the solvent was removed and the samples were dried for 5 hours at 100°C.

Another investigation had been presented by Chinese scientists on the luminescence of MWNTs with carboxylic groups [MWNT (-COOH)] and the Eu (III)/MWNT $(-COOH)_n$ composite [75].

MWNTs were prepared and purified as follows[76].In a pipe face, with reagent based on Ni and ethylene gas, raw MWNTs were eliminated with acid, 20% hydrofluoric acid and hydrochloric respectively. MWNTs which were larger than 98% purity were obtained. Then, the purified MWNTs were suspended in concentrated nitric acid and refluxed for 48 h. The mixture was to be filtered and washed with deionized water repeatedly until the pH value approached 7. After drying, MWNT (-COOH)$_n$ was obtained.

MWNT(-COOH)$_n$ was suspended in ethanol and refluxed for 2 h, then the europium(III) chloride solution in ethanol was slowly dripped into suspension. The pH value of mixture dropped to 3 from 7 gradually in half an hour. Triethylamine was then added to keep the pH value at about 6. The mixture was filtered and washed with ethanol 24 h later until the pH value reached 7. Having been dried, the Eu(III)/MWNTs(-COOH)$_n$ composite was obtained[75].

Both MWNT (-COOH)$_n$ and the Eu^{3+}/MWNT (-COOH)$_n$ composite demonstrated luminescence, while the luminescence from the composite was stronger than that from MWNT (-COOH)$_n$, and the luminescence peak position of the composite slightly red-shifted to a longer wavelength [Figure 25]. The difference could be attributed to the coordination of Eu^{3+} with carboxylic groups on MWNT (-COOH)$_n$. The above results may have potential use in optoelectronic devices.

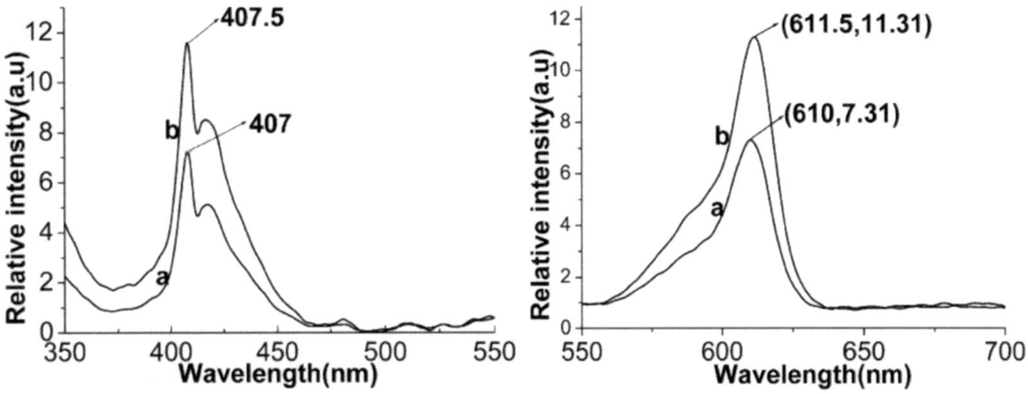

(1) Fluorescence excitation spectra (2) Fluorescence emission spectra at excitation wavelength of 407 nm

Fig. 25. Excitation and fluorescence characteristics of (a) MWNT (-COOH)$_n$ and (b) Eu^{3+}/MWNT (-COOH)$_n$ composite [75]

J. G. Tang and his fellows once worked on the nanoblock building strategy to obtain hybrid fluorescent nanoblocks through grafting ligand-antenna integration oligomer onto nanoscaffolds [77] [Figure 26]. In their research, CNTs were studied as nanoscaffolds to anchor organic oligomers and further to complex with lanthanide (i.e. Tb^{3+}) acceptor to obtain hybrid fluorescent nanoblocks. These built hybrid nanoblocks showed sharp fluorescent emission under ultraviolet excitation [Figure 27]. These results presented an important method to prepare small ligands from nanoscaled scaffolds, and generated the excellent fluorescent hybrid nano blocks, in which CNTs provided stable structural hosts for lanthanide complexes. Through the D-A(Donor-Acceptor) strategy of getting fluorescent nanoblocks, the promising nanohybrid luminescent materials will be emerged.

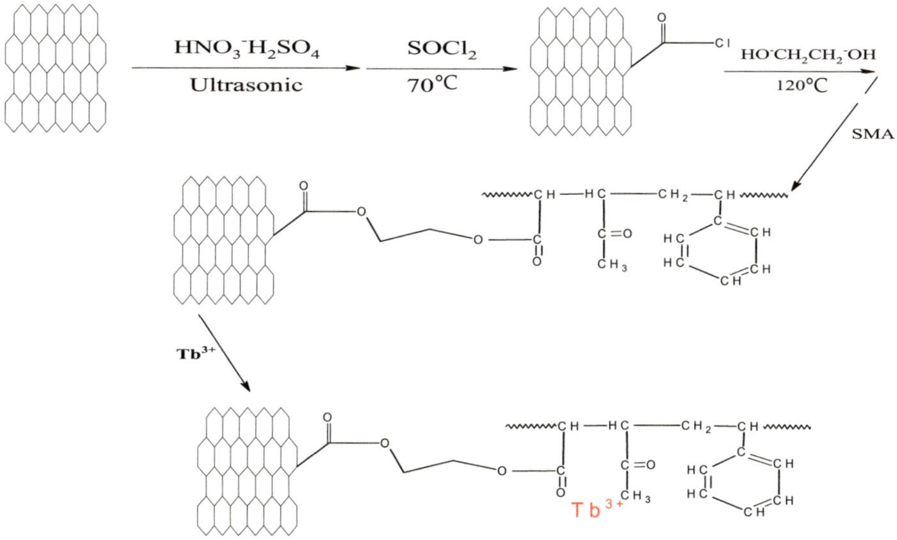

Fig. 26. Steps of preparing fluorescent nanoblock of carbon nanotube [77].

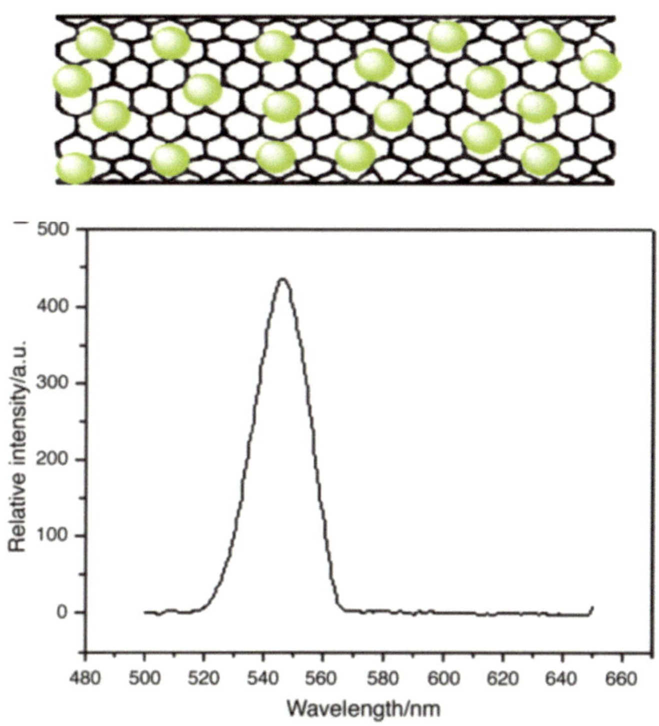

Fig. 27. The sketched images of fluorescent nanoblocks and Emission spectrum of fluorescent nanoblocks of LSMA-CNTs (LSMA-CNTs were excited by ultraviolet radiation: The excitation peak at 270 nm is a symmetric narrow band ranging from about 260–280 nm.)[77].

5.2 Noble metal nanoparticle-existed systems

A novel stragety to attach gold nanoparticles to CNTs selectively has been developed [78]. Just as Figure 28 shows, the MWNTs were firstly chemically modified with an H_2SO_4-HNO_3 treatment, and then were suspended in a concentrated sulfuric acid/nitric acid mixture (3:1 in volume) and sonicated in a sonic bath for 2 h. A CNT mat was obtained after filtration and was then thoroughly washed with a dilute sodium hydroxide aqueous solution.

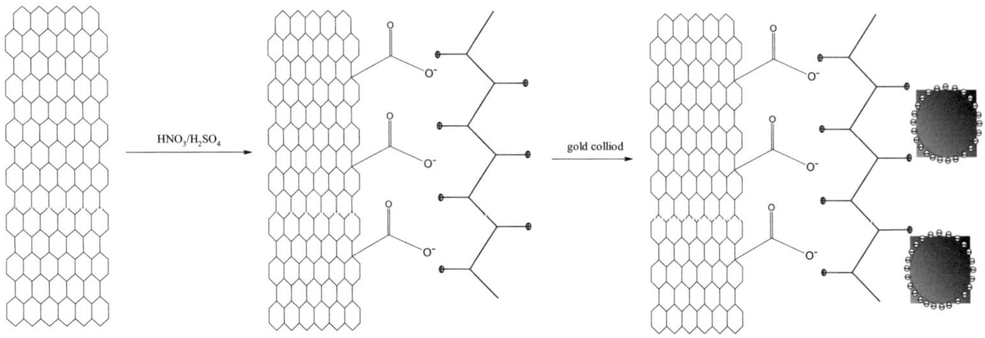

Fig. 28. Attachment of gold nanoparticles onto CNTs [78]

The nanotube suspension was then mixed with a cationic polyelectrolyte, poly (diallyldimethylammonium) chloride (M_w≈100,000-200,000), and NaCl aqueous solution for 30 min. PDADMAC was adsorbed to the surface of the nanotubes because of the electrostatic interaction between the carboxyl groups and the polyelectrolyte. After filtration and thorough washing with a NaCl aqueous solution and deionized water, the nanotubes were dispersed in water again, and mixed with a gold colloid (10 nm) for 30 min. The negatively charged gold nanoparticles were anchored to the surface of the nanotubes through the electrostatic interaction between the polyelectrolyte and the nanoparticles.
By choosing different kinds of polyelectrolytes, the surfaces of CNTs can be tailored to be negatively or positively charged, so many other nanoparticles (such as magnetic nanoparticles, semiconductor nanocrystals) can be selectively attached to the surfaces of nanotubes [Figure 29]. Additionally, this method of decorating nanotubes can be used to identify the location of functional groups. These nanoparticle-decorated nanotube heterostructures could be used in catalytic, electronic, optical, and magnetic applications [79].

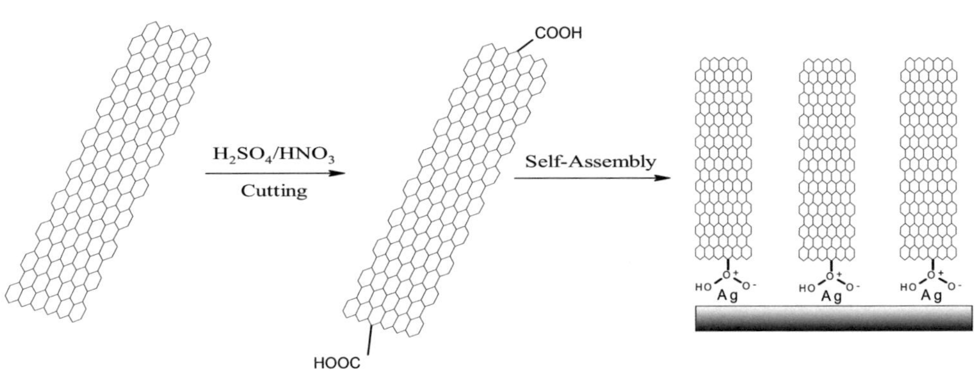

Fig. 29. Chemical alignments of oxidative CNTs on silver surface [79]

Another method has been reported for immobilizing the randomly tangled SWNTs on silver surface, forming an organizing nanotube assembly. The long and randomly tangled SWNTs which have already been prepared are cut into short pipes by chemical oxidation, which produced carboxyl groups at the open end of the tubes [80]. Basing on the fact that spontaneous adsorption of long chain *n*-alkanoic acids can occur via the carboxyl groups on various metal native-oxide surfaces [81, 82, and 83], such oxidatively shortened SWNTs would be immobilized on the silver surface.
H. J. Dai et al. first report on metal enhanced fluorescence of surfactant-coated carbon nanotubes on nanostructured gold substrates [84]. The photoluminescence quantum yield of SWNTs is observed to be enhanced more than 10-fold. It is suggested that the mechanism of SWNT fluorescence enhancement is due to radiative lifetime shortening of the excited state, resulting from resonant coupling of nanotube emission with the scattering and reradiating component of plasmons localized on the surface of the metal substrate.

6. Applications of fluorescent CNT

Due to the unique properties on photology, electromagnetism and chemistry, the modfied CNTs with fluorescent property have already been widely used as production materials of optical devices, electrical equipment and biomedicine kits. The excellent performance of them amazes us greatly. Now we will give a brief introduction to the most important and representative three of all these applications. They are, respectively, OLEDs, solar cells and biosensors.

6.1 OLEDs

CNT thin films now have had amazing applications. Their outstanding optical property as well as electrical behaviors makes them the perfect choice for the OLED materials. There has been a considerable interest to find a new material as the replacement of the traditional materials. On one hand, the traditional indium tin oxide (ITO) anodes are so brittle that they are easy to crack. On the other hand, indium is rare-earth metal, whose supply is lacked. Thus, due to their flexibility and work function (4.7eV-5.2eV) [85, 86], as well as the chemical stability during the wet processing of OLEDs, the CNT thin film-made OLED anodes are now more attractive.
E. C-W Ou and co-workers treat the surface of CNT thin films with nitric acid, PEDOT: PSS and CNT composition (PS^c) and polymer coating, and further study the influence of these treatments on the properties of OLED devices [87]. The result is gratifying-the modification of CNTs will benefit the enhancement of properties of these devices, making people full of confidence to the future of OLEDs.

6.2 Solar cells

As a completely new resource of renewable energy and a potential kind of alternatives of traditional inorganic solar cells, organic photovoltaic solar cells start to attract more and more attentions from researchers all around the world as well as the interest to study on them. The processing of these new organic photovoltaic solar cells is relatively simple. They are made from inexpensive organic materials, making the production on a large scale possible.
There are several reasons for using modified CNTs in this kind of materials. The big surface area of the CNTs sets a good stage for the morphological construction. Also, the high aspect

ratio allows the settlement of percolation at relatively low doping levels, which provides the way to high carrier mobility and efficient charge transfer of certain electrodes. The conducting polymer nanocomposites, such as SWNT-epoxy composites, whose percolation thresholds are extremely low, have already been developed.

Recently, Brazilian scientists [88] announced that the open-circuit voltage was raised to 1.8V by the use of modified CNTs combined with polarized polybithiophene layers, resulting in a conversion of 1.5% [89]. Generally, the CNTs stack into bundles in polymers by the interaction of Van der Waals force with each other. Thus, the thiophene groups located at the edges and defects of the CNTs will improve the dispersion of CNTs in conducting polymer, as well as the compatibility of them. All of these will of course increase the efficiency of polymer/CNTs solar cells.

Maurizio Prato et al. carried their research on the phthalocyanine-pyrene conjugates, ZnPc-Py and H_2Pc-Py, which had emerged as valuable building blocks for assembling electron donor-acceptor hybrids with SWNTs [90]. Owing to the strong ability of pyrene to adhere to SWNT sidewalls by means of π-π interactions, they had exploited this polyaromatic anchor to immobilize metal-free (H_2Pc) as well as zinc (ZnPc) phthalocyanines onto the surface of SWNTs. Encouraged by the charge-transfer features, researchers have utilized ZnPc/SWNT and H_2Pc/SWNT thin films in photoelectrochemical cells to test their solar energy conversion potential. Performances have been realized that are much superior to those of previously reported SWNT conjugates and hybrids [91].

6.3 Biosensors

Due to the weak fluorescence of CNTs [6, 92], a large number of efforts have been invested in developing the fluorescent CNTs by means of covalent or noncovalent modification [93, 94]. Thus, the fluorescent molecules such as pyrene and porphyrin are often used for modification of CNTs. However, most of the organic compounds have a short fluorescence lifetime, which is easy to quench at the same time.

As is known to all of us that the rare-earth compounds are often used to manufacture laser materials, optoelectronic devices and fluorescence probes [95, 96]. Chinese scientists designed and synthesized the SWNTs covalently modified by europium (Eu^{3+}) complex [97] [Figure 30]. The modified SWNTs can emit strong red luminescence upon excitation. Meanwhile, the research on the luminescence change of modified SWNTs after being bonded to DNA was also studied.

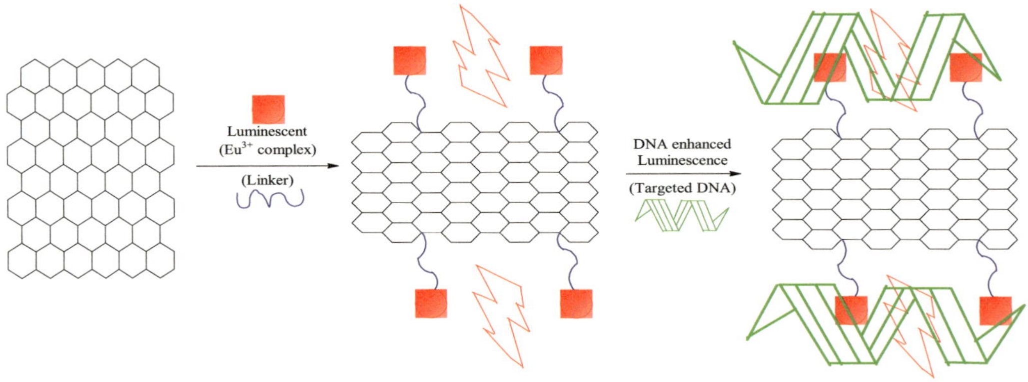

Fig. 30. Schematic representation of the luminescent Eu^{3+}-Complex covalently-modified SWNT and its luminescence enhanced by DNA [97]

Making the use of these unique properties, we can manufacture biosensors and many other devices with these modified CNTs. Another effective and novel self-assembled oligonucleotide/SWNT composite has been reported by Yang and co-workers [98]. This kind of composites are able to judge the sequence of certain DNA by the use of a self-assembled quenched complex of fluorescent Single-stranded DNA(ssDNA) and SWNTs as an efficient molecular beacons which can fluorescently detect single nucleotide differences in DNA in homogeneous solutions. In this way, the application of CNTs in biosensors will be broadened greatly.

7. Conclusions and future remarks

In summary, in this chapter, we focus mainly on the recent advances and achievements on the design and functionalization of CNTs, including noncovalent and covalent modification, polymer-existed system and metal-contained system, for the purpose to obtain enhanced fluorescence property, as well as properties on many other sides. As demonstrated by these remarkable examples, the relationship between the structures of modified CNTs and the fluorescence property helps to offer new attractive prospects for constructing CNT-based molecular optoelectronic and photon devices with desired functionalities. Many of these devices have already been used in practice or even industry, which push forward the development of our society and science, as well as profiting us a lot, both on economy and everyday life. However, there are still so many mechanisms and potentials waiting to be explored. For this reason, the research on the relationship between the structure of CNTs as well as their derivatives and properties of them is still one of the hottest and hopeful fields in nanotechnology.

Now, the production of CNTs has become cheaper and easier than before, which opens the door to more researchers. As the rapid development of nanotechnology, the overwhelming trend of the functionalization of CNTs has turned to the technological application of them. For the achievements we have already scored(only a small portion is listed in this review), there are sufficient reasons to hold an optimistic attitude towards the industrial products of CNTs in the near future, as well as the going-on researches in laboratories.

8. References

[1] S. Iijima, *Nature*, 1991, 354, 56-58.
[2] P.M. Ajayan, *Chem. Rev*, 1999, 99, 1787-1800.
[3] M.S. Dresselhaus, G. Dresselhaus, P.C. Eklund, *Science of Fullerenes and Carbon Nanotubes*, Academic Press, San Diego, CA, 1996.
[4] S.S. Karajanagi, H. Yang, P. Asuri, E. Sellito, J.S. Dordick, E.S. Kane. *Langmuir*, 2006, 22, 1392-1395.
[5] W. Huang, S. Fernando, Y. Lin, B. Zhou, L.F. Allard, Y.P. Sun, *Langmuir*, 2003, 19, 7084-7088.
[6] M.J. O'Connell, S.M. Bachilo, C.B. Huffman, V.C. Moore, M.S. Strano, E.H. Haroz, K.L. Rialon, P.J. Boul, W.H. Noon, C. Kittrell, J. Ma, R.H. Hauge, R. B. Weisman, R.E. Smalley, *Science*, 2002, 297, 593-596.
[7] M.F. Islam, E. Rojas, D.M. Bergey, A.T. Johnson, A.G. Yodh, *Nano Lett*, 2003, 3, 269-273.
[8] A. Hirsch, *Angew.Chem. Int. Ed*, 2002, 41, 1853.
[9] X.W. Wei, X.J. Song, J. Xu, Y.H. Ni, P. Zhang, *Materials Chemistry and Physics*, 2005, 92,159-163.

[10] Y.T. Liu, W. Zhao, Z.Y. Huang, Y.F. Gao, X.M. Xie, X.H. Wang, X.Y. Ye, *Letters to the Editor / Carbon*, 2006, 44, 1581-1616.
[11] K. Liao, S. Li, *Appl Phys Lett*, 2001, 79, 4225-4227.
[12] M. Wong, M. Paramsothy, X.J. Xu, Y. Ren, S. Li, K. Liao, *Polymer*, 2003, 44, 7757-7764.
[13] J. Gupta, C. Vijayan, S. K. Maurya, D. Goswami, *Materials Letters*, 2011, 65, 915-917.
[14] A.J. Lee, X.Y. Wang, L.J. Carlson, J.A. Smyder, B. Loesch, X.M Tu, M. Zheng, T.D. Krauss, *Nano Lett*, 2011, 11, 1636-1640.
[15] J. Chen, H.Y. Liu, W.A. Weimer, M.D. Halls, D.H. Waldeck, G.C. Walker, *J. Am. Chem. Soc*, 2002, 124, 9034-9035.
[16] Y.F. Zhang, B.X. Li, C.G. Yan, L.H. Fu, *Biosensors and Bioelectronics*, 2011, 26, 3505-3510.
[17] C. Fantini, J. Cassimiro, V.S.T. Peressinotto, F. Plentz, A.G. Souza Filho, C.A. Furtado, A.P. Santos, *Chemical Physics Letters*, 2011, 473 , 96-101.
[18] D.E. Luzzi, B.W. Smith, R. Russo, B.C. Satishkumar, F. Stercel, N. Nemes, *AIP Conference Proceedings,* 2001, 591, 622-626.
[19] G.H. Jiang, L. Wang, C. Chen, X.C. Dong, T. Chen, H.J. Yu, *Materials Letters,* 2005, 59, 2085-2089.
[20] H. Kong, C. Gao, D. Yan, *J. Am. Chem. Soc,* 2004, 126, 412-413.
[21] Y.B. Wang, Z. Iqbal, S.V. Malhotra. *Chemical Physics Letters*, 2005,402, 96-101.
[22] A.B. Dalton, S. Collins, J. Razal, E. Munoz, V. Ebron, B.G. Kim, J.N. Coleman, J.P. Ferraris, R.H. Baughm, *J. Mater. Chem,* 2004, 14, 1.
[23] A. Koshio, M. Yudusaka, M. Zhang, S.Iijima, *Nano Lett,* 2001, 1, 361-363.
[24] J.L. Bahr, J. Yang, D.V. Kosynkin, M.J. Bronikowski, R.E. Smalley, J.M. Tour, *J. Am. Chem. Soc,* 2001, 123, 6536-6542.
[25] P.W. Chu, G.S. Duesburg, U. D. Wegiikowska, S. Roth, *Appl. Phys. Lett,* 2002, 80, 3811-3813.
[26] F. Frehill, J.G. Vos, S. Benrezzak, A.A. Koos, Z. Konya, M.G. Ruther, W.J. Blau, A. Fonseca, J.B. Nagy, L.P. Biro, A.I. Minett, M. Panhuis, *J. Am. Chem. Soc,* 2002 , 124, 13694-13695.
[27] Y. Lin, A.M. Rao, B. Sadanadan, E. A. Kenik, Y.P. Sun, *J. Phys. Chem. B,* 2002, 106, 1294-1298.
[28] X.D. Lou, C. Detrembleur, V. Sciannamea, C. Pagnoulle, R. Jerome, *Polymer*, 2004, 45, 6097-6102.
[29] D. Tasis, N. Tagmatarchis, V. Georgakilas, C. Gamboz, M.R. Soranzo, M. Prato, *C. R. Chimie,* 2003, 6, 597-602.
[30] H.S. Yoshimura, S. Khan, H. Maruyama, Y. Nakayama, K. Takeyasu, *Biomacromolecules*, 2011, 12, 1200-1204.
[31] P.R. Marcux, J. Schreiber, P. Batail, S. Lefrant, J. Renouard, G. Jacob, et al, *Phys Chem,* 2002, 4, 2278-2285.
[32] H. Hu, B. Zhao, M.A. Hamon, K. Kamaras, M.E. Itkis, R.C. Haddon, *J. Am. Chem. Soc,* 2003, 125, 14893-14900.
[33] E.T. Mickelson, C.B. Huffman, A.G. Rinzler, R.E. Smalley, R.H. Hauge, J.L. Margrave, *Chem Phys Lett*, 1998, 296,188-194.
[34] H. Touhara, F. Okino, *Carbon,* 2000, 38,241-267.
[35] Z. Jia, Z. Wang, J. Liang, B. Wei, D. Wu, *Carbon*, 1999, 37,903-906.
[36] N. Pierard, A. Fonseca, Z. Konya, I. Willems, G. Van Tendelo, J.B. Nagy, *Chem Phys Lett,* 2001, 335, 1-8.
[37] N. Pierard, A. Fonseca, J.F. Colomer, C. Bossout, J.M. Benoit, G. Van Tendeloo, et al, *Carbon*, 2004, 42,1691-1697.

[38] Z. Konya, I. Vesselenyi, K. Niesz, A. Kukovecz, A. Demortier, A. Fonseca, et al. *Chem Phys Lett*, 2002, 360, 429-435.

[39] R. Barthos, D. Mehn, A. Demortier, N. Pierard, Y. Morciaux, G. Demortier, A. Fonseca, J.B. Nagy, *Carbon*, 2005, 43 , 321-325.

[40] B. C. Satishkumar, S.K. Doorn, G.A. Baker, A.M. Dattelbaum, *ACS Nano*, 2008, 2, 2283-2290

[41] Y.A. Kim, H. Muramatsu, T. Hayashi, M. Endo, M. Terrones, M.S. Dresselhaus, *Chem. Phys. Lett*, 2004, 398, 87-92.

[42] J.H. Kim et al. *ACS Nano*, 2010, 4, 1060-1066.

[43] B.P. Zhang, K. Shimazaki, M. Suzuki, T. Shiokawa, Y. Homma, K. Ishibashi. Fabrication of luminescent carbon nanotubes, *Article in Press*.

[44] J.E. Riggs, Z.X. Guo, D.L. Carroll, Y.P. Sun, *J. Am. Chem. Soc*, 2000, 122, 5879-5880.

[45] G.E. Lawson, Ph.D. Dissertation, Clemson University, Clemson, SC, 1999.

[46] S.S. Wong, E. Joselevich, A.T. Woolley, C.L. Cheung, C.M. Lieber, *Nature*, 1998, 394, 52-55.

[47] S. Lebedkin, F. Hennrich, T. Skipa, M.M. Kappes, *J. Phys. Chem. B*, 2003, *107*, 1949-1956.

[48] S. Qin, D. Qin, W.T. Ford, J.E. Herrera, D.E. Resasco, S.M. Bachilo, R.B. Weisman, *Macromolecules*, 2004, 37, 3965-3967.

[49] D.M. Guldi, M. Holzinger, A. Hirsch, V. Georgakilas, M. Prato, *Chem. Commun*, 2002, 1130-1131.

[50] Y. Lin, B. Zhou, R.B. Martin, K.B. Henbest, B.A. Harruff, J.E. Riggs, Z.X. Guo, L.F. Allard, Y.P. Sun, *J. Phys. Chem. B*, 2005, 109, 14779-14782.

[51] C. Hariharan, V. Vijaysree, A.K. Mishra, *J. Lumen*, 1997, 75, 205.

[52] C. Hariharan, A.K. Mishra, *Radiat. Meas*, 1998, 29, 473.

[53] F. Grieser, R.T. Treml, *J. Am. Chem. Soc*, 1980, 102, 7258-7264.

[54] A.W. Varnes, R.B. Dodson, E.L. Wehry, *J. Am. Chem. Soc*, 1972, 94, 946-950.

[55] A.R. Barron et al, *J. Phys. Chem. C*, 2007, 111, 17804-17806.

[56] W.C. Chen, R.K. Wang, K.J. Ziegler, *Applied Materials and Interfaces*, 2009, 1, 1821-1826.

[57] Y. Lin, B. Zhou, K. A. Shiral Fernando, P. Liu, Lawrence F. Allard, Y. P. Sun, *Macromolecules*, 2003, *36*, 7199-7204.

[58] W. Huang, Y. Lin, S. Taylor, J. Gaillard, A.M. Rao, Y.P. Sun, *Nano Lett*, 2002, 2, 231-234.

[59] R. Czerw, Z.X. Guo, P.M. Ajayan, Y.P. Sun, David L. Carroll, *Nano Lett*, 2001, 1, 423-427.

[60] S. Kazaoui, N. Minami, R. Jacquemin , *Phys. Rev. B*, 1999, 60, 339-342.

[61] Y. Sun, S.R. Wilson, D.I. Schuster, *J. Am. Chem. Soc*, 2001, 123, 5348-5349.

[62] S. A. Curran et al, *Adv. Mater*, 1998, 10, 1091.

[63] H. Ago et al, *Adv. Mater*, 1999, 11, 1281.

[64] S.H. Jhang, S.Y. Kirn, J.H. Park, T. Aim, D.S. Kirn, Y.W. Park, *AIP Conference Proceedings*, 2001, 591, 489-492.

[65] T. Ahn et al., *Macromolecules*, 1999, 32, 3279-3285.

[66] R.K. Wang, R.D. Reeves, K.J. Ziegler, *J. Am. Chem. Soc*, 2007, 129, 15124-15125.

[67] S. Banerjee, S.S. Wong, *Nano Lett*, 2002, 2,195-200.

[68] G.X. Xu, *Rare Earths, 2nd ed.*, Metallurgical Industry Press, Beijing, 1995.

[69] X.W. Wei, J. Xu, X.J. Song, Y.H. Ni, P. Zhang, C.J. Xia, G.C. Zhao, Z.S. Yang, *Materials Research Bulletin*, 2006, 41, 92-98.

[70] B.C. Liu, S.H. Tang, Q. Liang, L.Z. Gao, B.L. Zhang, M.Z. Qu, Z.L. Yu. *Chin. J. Chem.* 2001, 19, 983.

[71] Y.H. Li, J. Ding, J.F. Chen, C.L. Xu, B.Q. Wei, J. Liang, D.H. Wu, *Materials Research Buletin*, 2002, 37, 313.

[72] H.Q. Cao, G.Y. Hong, J.H. Yan, J.L. Zhang, G.X. Liu, *Chinese Chemical Letters,* 2003, 14, 1293-1295.

[73] G.Z. Chen, M.S.P. Shaffer, D. Coleby, G. Dixon, W. Zhou, D.J. Fray, A.H. Windle, *Advanced Materials*，2000，12，7.

[74] P.M. Aiayan, O. Stephan, P. Redlich, C. Colllex, *Nature,* 1995, 375, 564.

[75] W.X. Sun, Z.P. Huang, L. Zhang, J. Zhu, *Letters to the Editor / Carbon,* 2002, 411, 1645-1687.

[76] Z.J. Jia, Z.Y. Wang, J. Liang, *Carbon,* 1999, 37,903-906.

[77] L.B. Kong, J.G. Tang, J.X. Liu, Y. Wang, Li.Y. Wang, F. Cong. *Materials Science and Engineering C,* 2009, 29, 85-91.

[78] K.Y. Jiang, A. Eitan, L.S. Schadler, P.M. Ajayan, R.W. Siegel, *Nano Lett,* 2003, 3, 275-277.

[79] B. Wu, J. Zhang, Z. Wei, S.M. Cai, Z.F. Liu, *J. Phys. Chem. B,* 2001, 105, 5075-5078.

[80] J.Liu, A.G. Rinzler, H. Dai, J.H. Hafner, R.K. Bradley, P.J. Boul, A. Lu, T. Iverson, K. Shelimov, C.B. Huffman, F.R. Macias, Y.S. Shon, T.R. Lee, D.T. Colbert, R.E. Smalley, *Science,* 1998, *280,* 1253-1256.

[81] D.L. Allara, R.G. Nuzzo, *Langmuir,* 1985, 1, 45-52.

[82] D.L. Allara, R.G. Nuzzo, *Langmuir,* 1985, 1, 52-56.

[83] Y.T. Tao, *J. Am. Chem. Soc,* 1993, 115, 4350-4358.

[84] G.S. Hong, Scott M. Tabakman, Kevin Welsher, H.L. Wang, X.R. Wang, H.J. Dai, *J. Am. Chem. Soc,* 2010, 132, 15920-15923.

[85] S. Suzuki, C. Bower, Y. Watanabe, O. Zhou, *Appl. Phys. Lett,* 2000, 76, 4007-4009.

[86] J. Zhao, J. Han, J.P. Lu, *Phys. Rev. B,* 2002, 65, 193401-1-193401-4.

[87] E. Ou, L.B. Hu, G.C.R. Raymond, O.K. Soo, J.S. Pan, Z. Zhang, Y. Park, D. Hecht, G. Irvin, P. Drzaic, G. Gruner, *ACS Nano,* 2009, 3, 2258-2264.

[88] A.F. Nogueira, B.S. Lomba, M. A. Soto-Oviedo, C.R. Duarte Correia, *J. Phys. Chem. C,* 2007, 111, 18431-18438.

[89] R.L. Patyk, B.S. Lomba, A.F. Nogueira, C.A. Furtado, A.P. Santos, R.M.Q. Mello, L. Micaroni, I.A. Hummelgen, *Phys. Status Solidi,* 2007, 1, 43.

[90] J. Bartelmess, B. Ballesteros, Gema de la Torre, D. Kiessling, S. Campidelli, M. Prato, T. Torres, D. M. Guldi, *J. Am. Chem. Soc,* 2010, 132, 16202-16211.

[91] S. Campidelli, B. Ballesteros, A. Filoramo, Diaz Diaz. D, de la Torre.G, T. Torres, G.M.A. Rahman, C. Ehli, D. Kiessling, F. Werner, V. Sgobba, D.M. Guld, C. Cioffi, M. Prato, J.P. Bourgoin, *J. Am. Chem. Soc,* 2008, 130, 11503-11509.

[92] J. Lefebvre, D.G. Austing, J. Bond, P. Finnie, *Nano Lett,* 2006, 6, 1603-1608.

[93] D.M. Guldi, G.M. Rahman, N. Jux, N. Tagmatarchis, M. Prato, *Angew. Chem. Int. Ed,* 2004, 43, 5526-5530.

[94] D. Baskaran, J.W. Mays, X.P. Zhang, M.S. Bratcher, *J. Am. Chem. Soc,* 2005, 127, 6916-6917.

[95] H. Maas, A. Currao, G. Calzaferr, *Angew. Chem. Int. Ed,* 2002, 41, 2495-2497.

[96] K. Maekawa, Y. Kinoshita, I. Nishino, S. Kikuoka, T. Fukunaga, K. Kawamoto, Y. Numata, H. Takemoto, K. Nagata, *Anal. Biochem,* 2006, 351, 229-240.

[97] C. Zhao, Y,J. Song, K.G. Qu, J.S. Ren, X.G. Qu, *Chem. Mater,* 2010, 22, 5718-5724.

[98] R.H. Yang, J.Y. Jin, Y. Chen, N. Shao, H.Z. Kang, Z.Y. Xiao, Z.W. Tang, Y.R. Wu, Z. Zhu, W.H. Tan, *J. Am. Chem. Soc,* 2008, 130, 8351-8358.

10

Simultaneous Detection of Multi-DNAs and Antigens Based on Self-Assembly of Quantum Dots and Carbon Nanotubes

Peng Huang and Daxiang Cui
National Key Laboratory of Nano/Micro Fabrication Technology, Key laboratory for thin film and microfabrication of Ministry of Education, Institute of Micro and Nano Science and Technology, Shanghai Jiao Tong University, Shanghai, P. R. China

1. Introduction

The development of convenient methodologies for simultaneous detection of specific multi-DNAs and antigens in biological and environmental samples has received broad attention because their detection, identification and quantification are very complex, expensive and time consuming (Abu-Salah et al., 2010; Strehlitz et al., 2008; Zhang et al., 2009). Over the past 30 years, biosensors, namely the devices incorporating biological sensing elements either intimately connected to or integrated within transducers, have been designed and fabricated to find effective solutions to these problems, which offer certain operational advantages over other traditional methods, notably with respect to rapidity, ease-of-use, low cost, simplicity, portability, and ease of mass manufacture (Cooper, 2002; Turner, 2000; Turner et al., 1987). The fundamental prerequisite of biosensors depends on specific molecular recognition based on affinity between complementary structures such as enzyme-substrate, antibody-antigen, receptor-hormone, and so forth (Zhang et al., 2009). This specific recognition gives rise to the production of concentration–proportional signals. Up to date, some biosensors have been commercialized for some special applications like blood glucose and lactate measurement or bioprocess control, amongst others. However, they have not still entered the market as much as expected, which is caused by the following reasons: (1) the selectivity and specificity of biosensor highly depends on biological recognition systems (Spichiger-Keller, 1998); (2) the sensitivity detection limit is difficult to achieve trace levels, even the single molecular detection (Sheehan and Whitman, 2005); (3) High throughput assay is desired to simultaneously process multiple samples (Sittampalam et al., 1997); (4) the biological recognition elements of the biosensor (e.g. enzymes, antibodies or cells) are usually instable (So et al., 2005).
With the development of nanoscience and nanotechnology, a series of novel nanomaterials with controlled size and morphologies are being fabricated, their novel properties are being gradually discovered with difference from their corresponding bulk materials, and the applications of nanomaterials in biosensors have also made great advances (Jianrong et al., 2004; Kumar, 2007; Pandey et al., 2008). Nanomaterials can be made from both inorganic and organic materials and are less than 100 nm in length along

at least one dimension (Asefa et al., 2009; Zhong, 2009). This small size scale leads to large surface areas and unique size-related optical properties. For example, the quantum confinement effects that occur in nanometer-sized semiconductors widen their band gap and generate well-defined energy levels at the band edges, causing a blue-shift in the threshold absorption wavelength with decreasing particle size and inducing luminescence that is strictly correlated to particle size (Krishna and Friesner, 1991; Peng et al., 2000). Therefore, the position of the absorption as well as the luminescence peaks can be fine-tuned by controlling the particle size and the size distribution during synthesis, generating a large group of "fluorophores" with diverse optical properties (Nirmal and Brus, 1999; Pradhan et al., 2005). The size- or shape-controllable optical characteristics of nanomaterials facilitate the selection of diverse probes for higher throughput assay (Zhong, 2009). Furthermore, the nanostructure can provide a substrate support for sensing assays with multiple probe molecules attached to each nanostructure, simplifying assay design and increasing the labeling ratio for higher sensitivity (Kumar, 2007). Therefore, nanomaterials have opened up new horizons for biosensors.

Biosensors based on nanomaterials, which represent the integration of material science, molecular engineering, chemistry and biotechnology, can markedly improve the sensitivity, selectivity, specificity and rapidity of bio-molecular detection, offer the promising capability of detecting or manipulating atoms and molecules, and have great potential in the development of the miniaturizebility or portability of analytical system (Zhang et al., 2009). Previous studies have shown that combining the specific molecular recognition ability of biomolecules with the unique structural and photophysical characteristics of inorganic or organic nanomaterials, such as nanocrystals, nanotubes, nanowires, nanomicelles, and nanovesicles, can create new types of analytical tools (Niemeyer, 2001). So far, the nanomaterials are widely used in biosensors mainly including (1) carbon nanomaterials (e.g., fullerenes, carbon nanotubes, carbon nanohorns, graphene, etc.) (Chen et al., 2003; Vamvakaki and Chaniotakis, 2007; Yang et al., 2010b), (2) metallic nanomaterials (e.g., quantum dots, gold nanoparticles, gold nanorods, europium nanoparticles, etc.) (Ao et al., 2006; Liu, 2009; Pan et al., 2005), (3) silica nanomaterials (Slowing et al., 2007), (4) organic polymer nanomaterials (e.g., molecular imprinted polymers) (Hatchett and Josowicz, 2008), or (5) supramolecular aggregates (nanomicelles, nanovesicles) (Kuhn, 1994). Fig. 1 illustrates the correlation between these typical nanomaterials and the major properties exploited for analytical purposes (Valcarcel et al., 2008).

2. Carbon nanotubes as analytical tools

Since their discovery by Iijima in 1991, Carbon nanotubes (CNTs), due to the remarkable structure-dependent electronic, mechanical, optical, and magnetic properties, have triggered intensive studies directed towards numerous applications including nanoelectronics, biomedical engineering, biosensing, and bioanalysis (Dai, 2002; Iijima, 1991). CNTs are rolled up seamless cylinders of graphene sheets. According to the number of graphene layers, CNTs are classified into single-walled carbon nanotubes (SWNTs) and multi-walled carbon nanotubes (MWNTs) (Liu et al., 2009). Their lengths can range from several hundred nanometers to several micrometers, and the diameters from 0.2 to 2 nm for SWNTs and from 2 to 100 nm for coaxial MWNTs (Valcarcel et al., 2005). So far, CNTs has been used as an analytical tool to improving the analytical process (Valcarcel et al., 2007). CNTs-based biosensors have been developed to detect biological species including proteins and DNA.

Fig. 2 illustrates their potential roles in the development of new tools for analytical science, arranged in terms of complexity of design and integration (Valcarcel et al., 2007).

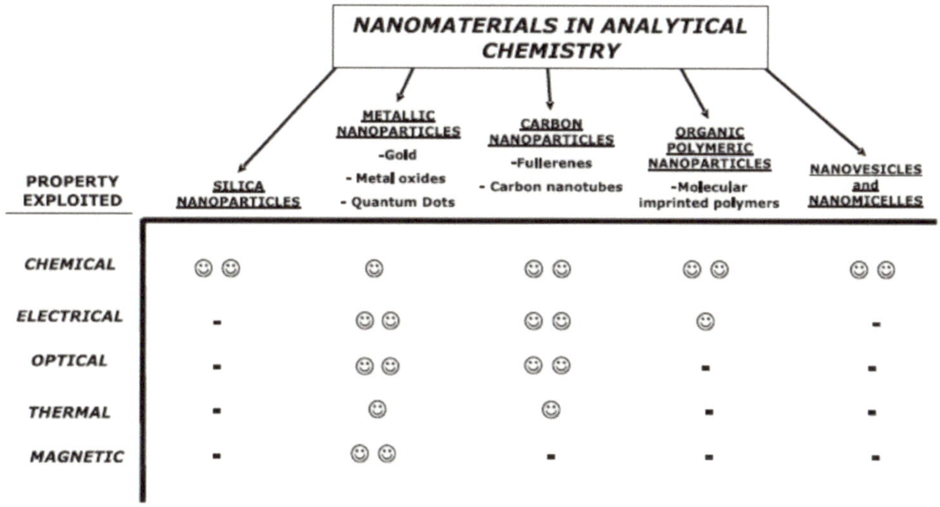

Fig. 1. Correlation between typical nanomaterials and the major property exploited for analytical purposes (Valcarcel et al., 2008). (With permission from Springer)

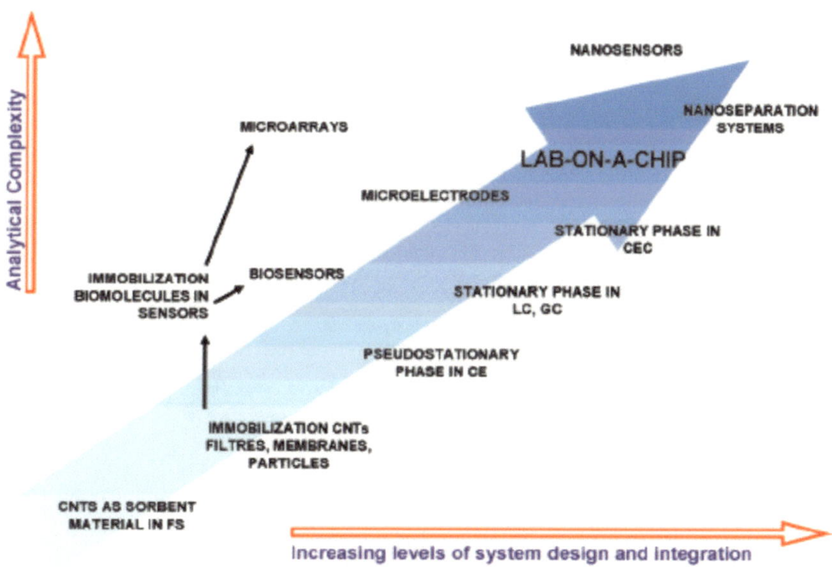

Fig. 2. Classification of analytical tools based on the use of carbon nanotubes according to the analytical complexity and the increasing level of system design and integration (Valcarcel et al., 2007). (With permission from American Chemical Society)

Generally, CNTs are expected to be controllably assembled into designed architectures as integral components of composites or supramolecular structures (Grzelczak et al., 2006). The integration of biomaterials (e.g., proteins/enzymes, antigens/antibodies, or DNA) with CNTs provides new hybrid systems that combine the conductive or semiconductive properties of CNTs with the recognition or catalytic properties of the biomaterials in Fig. 3 (Katz and Willner, 2004). CNTs can provide scaffolds for biomolecules immobilization, allowing subsequent applications in biosensors, utilizing the intrinsic electronic or optical properties of CNTs for signal transduction (Cataldo, 2008). Due to CNTs' the high surface area, semiconducting behavior, band gap fluorescence, and strong Raman scattering spectra, the proximal or adsorbed biomolecules can be measured or detected easily, when their interactions along the CNTs sidewall, at functionalized cap regions (Cheung et al., 1998), and even within the nanotube shell (Lin et al., 2004; Liu et al., 2009). Proximity of reasonably charged or polarized biomolecules yields gating effects on isolated semiconducting CNTs, or net semiconducting networks of CNTs, thus yielding field-effect transistors (FETs) capable of quantifying the degree of specific or non-specific binding of biomolecules (Chen et al., 2003; Liu et al., 2009). Additionally, CNTs were also used as analytical targets. After specific conjugation of targeting ligands to SWNT tags or coupled with sufficient sidewall passivation in order to prevent non-specific binding, the photoluminescent (FL) and Raman scattering properties of SWNTs can be detected as target signals in biosensors (Liu et al., 2009; Valcarcel et al., 2007). Therefore, CNTs are of great interest for the development of highly sensitive and multiplexed biosensors for applications in analytical chemistry.

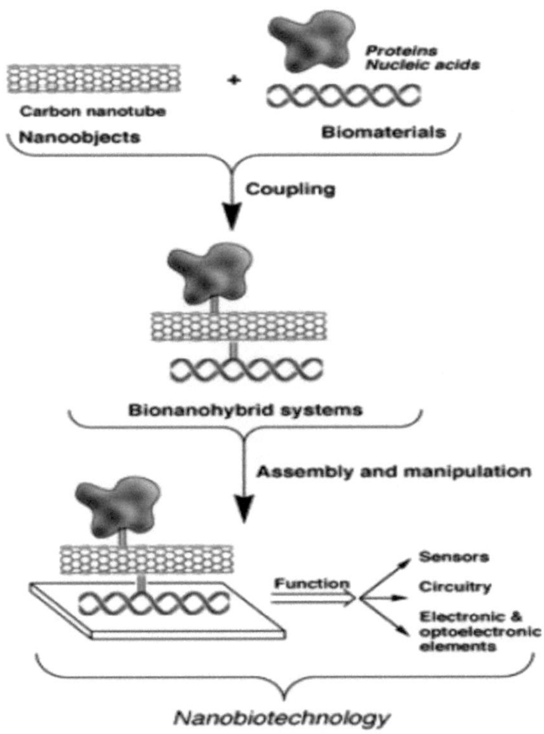

Fig. 3. The conceptual generation of biomolecules-carbon nanotubes conjugates, and their assembly to yield functional devices (Katz and Willner, 2004). (With permission from Wiley-VCH)

3. Quantum dots as analytical tools

Since their discovery in the early 1980s, semiconductor nanomaterials or quantum dots (QDs), have been extensively used as potential luminescence probes due to their high resistance to photobleaching, narrow emission spectra, broad excitation spectra, and longer fluorescence lifetime in the field of biosensing and imaging (Alivisatos, 1996). QDs are usually composed of atoms of elements from groups II to VI (e.g., Cd, Zn, Se, Te) or III–V (e.g., In, P, As) in the periodic table (Chan et al., 2002). As a result of their very small (< 10 nm) dimensions, QDs exhibit quantum confinement effects that are responsible for their wide UV-visible absorption spectra, narrow emission bands, and optical properties, which can be tuned by size, composition, and shape (Alivisatos, 2003; Chan and Nie, 1998). These features come with high flexibility in the selection of excitation wavelength as well as minimal overlap in the emission spectra from multiple QDs, making them excellent labels for high throughput screening (Chan et al., 2002; Han et al., 2001). Additionally, choosing excitation wavelengths far from the emission wavelengths can eliminate background scattering.

Compared with organic fluorophores, QDs have similar quantum yields but extinction coefficients that are 10~50 times larger, and much-reduced photobleaching rates. The overall effect is that QDs have 10~20 times brighter fluorescence and about 100~200 times better photostability (Gao et al., 2005; Zhong, 2009). For applications in biomedical studies, QDs should be water soluble, which can be achieved in two ways: the first is to directly synthesize QDs in aqueous solution; the other is to synthesize QDs in organic solvents and then transfer the hydrophobic QDs into aqueous solution, for example, by ligand exchange or polymer coating (Zhang et al., 2010). So far, several methods have been developed to synthesize water-soluble quantum dots for use in cellular imaging, immunoassays, DNA hybridization and optical bar-coding (Drbohlavova et al., 2009; Yang et al., 2009). Moreover, QDs also have been used to study the interaction between protein molecules or detect the dynamic course of signal transduction in live cells by fluorescence resonance energy transfer (FRET) (Chan and Nie, 1998).

Because QDs are intrinsically fluorescent, they can be widely employed as the reporter molecules for biomolecules detection (Chan and Nie, 1998). For example, QDs-based western blot detection kits can improve the sensitivity detection limit as low as 20 pg protein per lane (Ornberg et al., 2005). The detection limits is around hundreds of picograms of protein per lane in the colorimetric or chemiluminescent detection. So the QDs-based protocol is more sensitive with better image quality in the same measuring time. The test samples can be stored for longer time after staining with minimal loss of signal (Edgar et al., 2006). According to the high resistance of QDs to photobleaching, Genin et al. have reported that the organic dye CrAsH conjugated QDs nanohybrids serve as a probe to bind efficiently and selectively to Cys-tagged proteins and subsequently trace them for more than 150 s, where the fluorescence emission of CrAsH displayed a significant increase after the interaction between CrAsH and cysteine (Genin et al., 2008). While the latter faded rapidly under continuous excitation, emission of the QDs remained unaffected. The persistent fluorescence of the QDs should thus allow extended monitoring of the target protein. In particular, the use of multicolor QDs probes in immunohistochemistry (IHC) is considered one of the most important and clinically relevant applications. Nie et al. have developed antibody-conjugated QDs for multiplexed and quantitative (or semi-quantitative) IHC, and have achieved five-color molecular profiling on formalin-fixed and paraffin-embedded

(FFPE) clinical tissue specimens (Xing et al., 2007). They have also optimized the experimental procedures for QDs bioconjugation (**Fig. 4**), tissue specimen preparation, multi-color staining, image processing and analysis, and biomarker quantification.

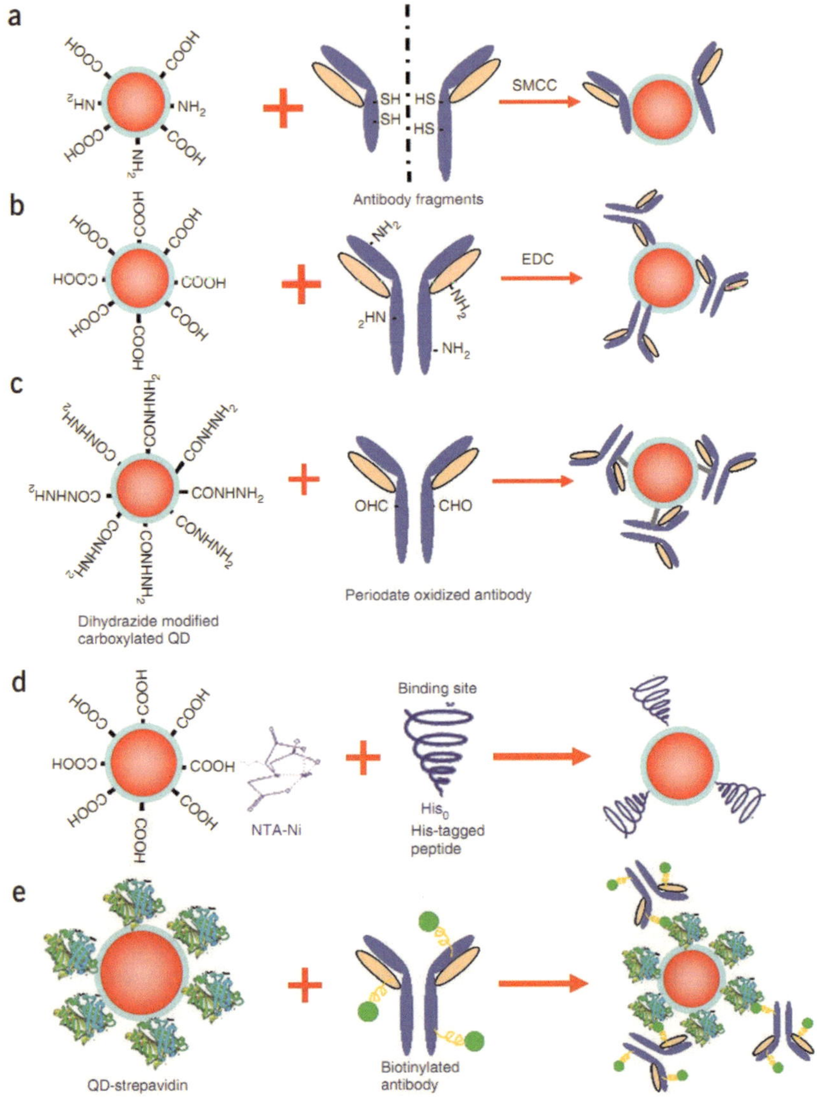

Fig. 4. Schematic diagrams showing various methods for QDs-antibody (QD-Ab) bioconjugation. (a)QDs conjugation to antibody fragments via disulphide reduction and sulfhydryl-amine coupling; (b) covalent coupling between carboxylic acid (-COOH) coated QDs and primary amines (-NH_2) on intact antibodies using EDAC as a catalyst; (c) site-directed conjugation via oxidized carbohydrate groups on the antibody Fc portion and covalent reactions with hydrazide-modified QDs; (d) conjugation of histidine-tagged peptides or antibodies to Ni-NTA modified QDs; and (e) noncovalent conjugation of streptavidin-coated QDs to biotinylated antibodies (Xing et al., 2007). (With permission from Nature)

Recently, our group have reported a quick and parallel analytical method based on QDs for ToRCH-related antibodies including Toxoplasma gondii, Rubella virus, Cytomegalovirus and Herpes simplex virus type 1 (HSV1) and 2 (HSV2) (Yang et al., 2009). We fabricated the microarrays with the five kinds of ToRCH-related antigens and used CdTe QDs to label secondary antibody and then analyzed 100 specimens of randomly selected clinical sera from obstetric outpatients (**Fig. 5**). In comparison with enzyme-linked immunosorbent assay (ELISA) kits, the QDs labeling-based ToRCH microarrays display comparable sensitivity and specificity with ELISA. Besides, the microarrays hold distinct advantages over ELISA test format in detection time, cost, operation and signal stability.

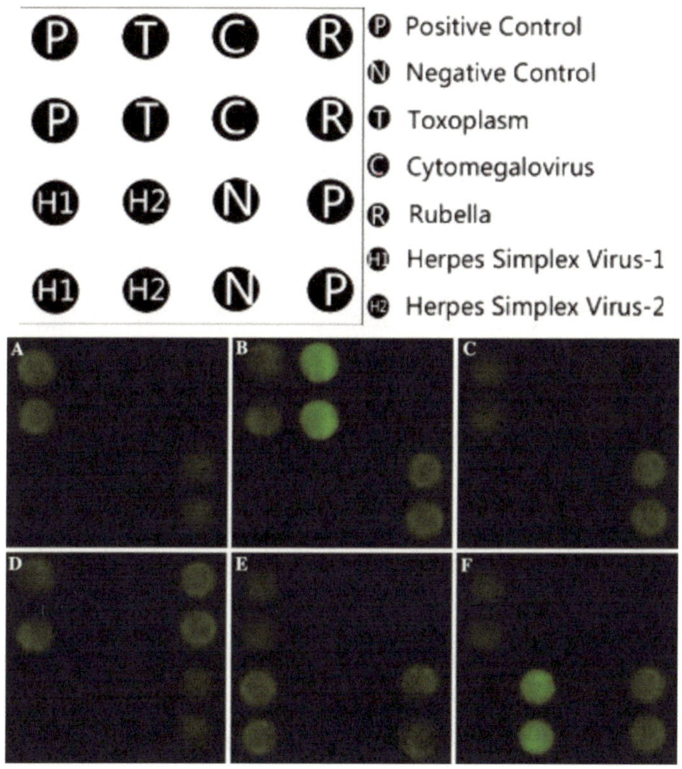

Fig. 5. The microarray results with corresponding control sera (a negative serum; b positive control serum of Toxoplasmosis; c positive control serum of Cytomegalovirus; d positive control serum of Rubella virus; e positive control serum of Herpes simplex virus type 1; f positive control serum of Herpes simplex virus type 2) (Yang et al., 2009). (With permission from Springer)

Additionally, we developed a novel fluorescent POC (Point Of Care) test method to be used for screening for syphilis, combining the rapidness of lateral flow test and sensitiveness of fluorescent method (Yang et al., 2010a). 50 syphilis-positive specimens and 50 healthy specimens conformed by Trep-onema pallidum particle agglutination (TPPA) were tested with QDs-labeled and colloidal gold-labeled lateral flow test strips, respectively. Our results showed that both sensitivity and specificity of the QDs–based method reached up to 100% (95% confidence interval [CI], 91~100%), while those of the colloidal gold-based method were 82% (95% CI, 68–91%) and 100% (95% CI, 91–100%), respectively. We found that the naked-eye detection limit of QDs-based method could achieve 2 ng/mL of anti-TP47 polyclonal antibodies purified by affinity chromatography with TP47 antigen, which was ten-fold higher than that of colloidal gold-based method (**Fig. 6**).

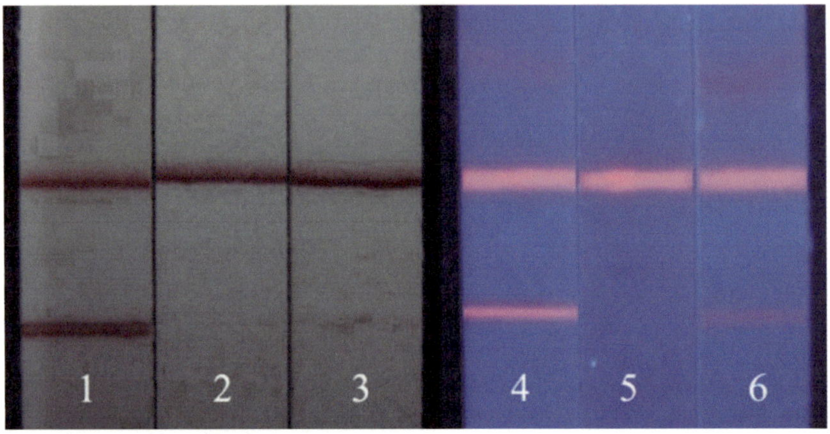

Fig. 6. Result of detection of clinical specimens by colloidal gold and QDs (Yang et al., 2010a). result of detection by colloidal gold: 1 positive; 2 negative; 3 weak positive; result of detection by QDs: 4 positive; 5 negative; 6 weak positive. (With permission from Springer)

Due to their bright intensity and high photostability, QDs also have a wide range of applications in bioimaging (Cui et al., 2009). Our group has successfully prepared the dendrimer-modified QDs with water-soluble, high quantum yield, and good biocompatibility. Our results indicated that arginine-glycine-aspartic acid (RGD) conjugated QDs can specifically target human umbilical vein endothelial cells (HUVEC) and A375 melanoma cells, as well as nude mice loaded with A735 melanoma cells (Li et al., 2010b). Meanwhile, we have successfully synthesized dendrimer-modified QDs nanocomposites, and developed a class of aptamers-conjugated nanoprobes, which could specifically bind with U251 glioblastoma cells and exhibit in vitro molecular imaging (Li et al., 2010a).

4. Carbon nanotubes and quantum dots nanohybrids as analytical tools

Sensitive detection of target analytes present at trace levels in biological samples often requires the labeling of reporter molecules with fluorescent dyes, because fluorescence detection is by far the dominant detection method in the field of sensing technology, due to its simplicity, the convenience of transducing the optical signal, the availability of organic dyes with diverse spectral properties, and the rapid advances made in optical imaging (Zhong, 2009). However, it can be difficult to obtain a low detection limit in fluorescence detection due to the limited extinction coefficients or quantum yields of organic dyes and the low dye-toreporter molecule labeling ratio.

Currently, fluorescence (Förster) resonance energy transfer (FRET) and photoinduced electron transfer (PET) have been widely studied as novel fluorescence detection techniques in biosensors (Zhang et al., 2010). In FRET sensing systems, QDs normally act as donor and transfer excitation energy to a vicinal fluorophore acceptor, leading to a reduced donor PL and a concomitant increased acceptor PL (**Fig. 7**). In PET sensing systems, the excited QDs act as the electron donor and transfer the excited electrons to acceptor molecules. This in turn results in a quenched QD PL and an increased acceptor PL (if possible). In both, FRET and PET sensing systems, the most significant advantage is that the transfer efficiencies can be used as a ratiometric readout without demanding an extra reference. The second advantage is that such sensing systems are more flexible and enable more complicated designs, resulting from the multiple components involved, e.g., donor, acceptor, and spacers.

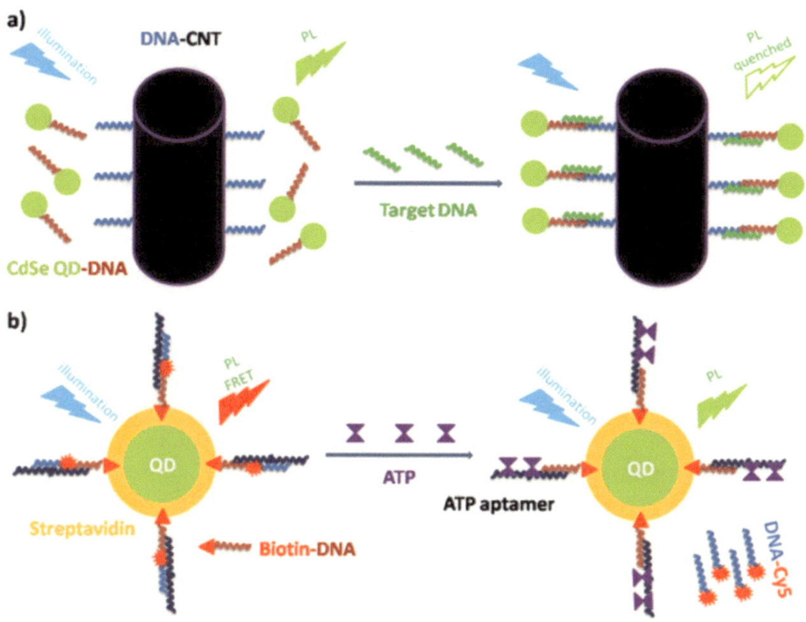

Fig. 7. FRET-based sensing (Zhang et al., 2010). a) CdSe QDs and CNTs are conjugated with two oligonucleotides. The two oligonucleotides can be hybridized upon binding a target DNA fragment. In the absence of target DNA, the QDs have its original PL. In the presence of the target DNA fragment, hybridization brings QDs and CNTs close enough so that the QDs PL is quenched by the CNTs. b) Streptavidin modified commercial QDs 'decorated' with biotinylated DNA fragments. An ATP aptamer binds with one side to the DNA attached to the QDs and with one side to Cy5-labeled DNA. Upon excitation of the QDs donor, FRET reduces the QDs PL and increases the PL of the Cy5 acceptor. In the presence of ATP molecules, the aptamer dehybridizes and complexes the ATP. As soon as the Cy5-labeled DNA becomes detached from the ATP aptamer, the distance to the QDs is increased and there is no longer FRET. (With permission from Springer)

Based on FRET, Wang et al. combined positively charged CdTe QDs capped with cyst-amine with negatively charged gold nanoparticles (GNPs) capped with 11-mercaptoundecanoic acid (MUA) in a FRET system, the PL of donor QDs can be quenched by the close acceptor GNPs, which is due to the high extinction coefficient and broad absorption of GNPs (Wang and Guo, 2009). Firstly, QDs and GNPs assemble into aggregates due to their mutual electrostatic attraction. Then the Pb^{2+} was added, MUA-modified GNPs chelated with it into aggregates, the QDs were released and the restored PL was read out. Finally, the PL changes of the QDs were instead of detecting changes. QD-based FRET sensors have been investigated by Mattoussi (Medintz et al., 2003) and Willner (Patolsky et al., 2003) groups. Our group also designed a unique, sensitive, and highly specific fluoroimmunoassay system for antigen detection using GNPs and QDs nanoparticles (Huang et al., 2010). To demonstrate its analytical capabilities, the CdTe QDs were coated with anti-HBsAg monoclonal antibodies (QDs-MAb1) and GNPs coated with another anti-HBsAg monoclonal antibodies (GNPs-MAb2) which specifically bound with HBsAg could sandwich the HBsAg captured by the immunoreactions. The sandwich-type immunocomplex was formed and the fluorescence intensity of QDs was measured. The results showed that the fluorescence intensity of QDs at 570 nm was negative linear proportional to the HBsAg concentration logarithm, and the limit of detection of the HBsAg was 0.928 ng/mL. This new system can be extended to detect target molecules with matched antibodies and has broad potential applications in immunoassay and disease diagnosis.

CNTs represent one type of unique nanomaterials used in fluorescence-based bioassays. The sensing utilizes the ability of CNTs to quench organic dyes or QDs (Pan et al., 2006b; Pan et al., 2008). Tang group reported that organic dyes could be quenched by CNTs through an energy transfer mechanism (Yang et al., 2008). This feature was employed to develop a non-covalent assembly between CNT and ssDNA for effective sensing of biomolecule interactions (Fig. 8). The strong interaction between CNT and ssDNA quenched the fluorophore conjugated on ssDNA. Hybridization of a complimentary DNA strand or binding of an interactive protein caused ssDNA to be released from the CNT, leading to the restoration of fluorescence signal in increments relative to the fluorescence without a target. The signaling mechanism makes it possible to detect the target by fluorescence spectroscopy.

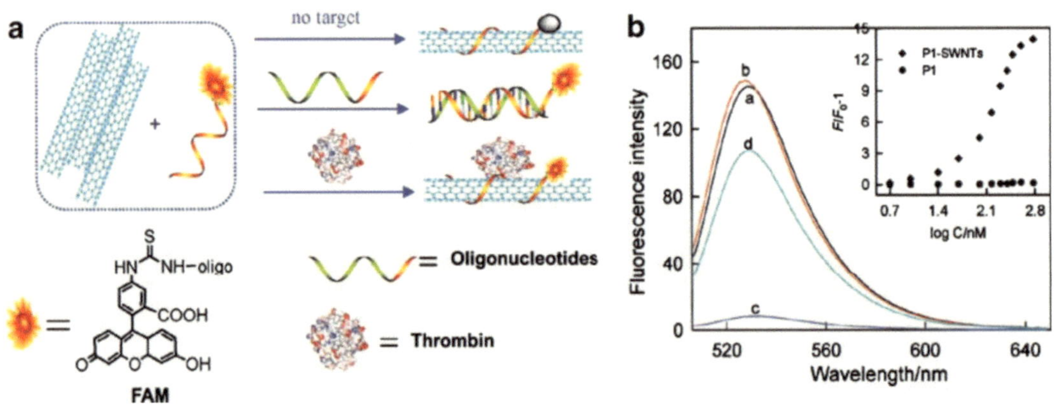

Fig. 8. A Scheme for signaling biomolecular interactions using an assembly between SWNTs and dye-labeled ssDNA. P1 and P2, the FAM-labeled oligonucleotides; P2, the thrombin-binding aptamer; T1 and T2, the perfect cDNA (T1) and one mismatched DNA (T2) of P1. B Fluorescence emission spectra (λ_{ex}=480 nm) of P1 (50 nM) under different conditions: (a) P1 in PBS; (b) P1 + 300 nM T1; (c) P1 + SWNTs; and (d) P1 + SWNTs + 300 nM T1. Inset: fluorescence intensity ratio of P1 (b) and P1-SWNTs with F/F_0–1 plotted against the logarithm of the concentration of T1 (Yang et al., 2008; Zhong, 2009). (With permission from American Chemical Society and Springer)

Recently, nanohybrids containing both semiconductor QDs and CNTs have been the subject of great interest as a consequence of the development of methods for the chemical modification of CNTs and the seeking for novel functional materials in biosensors (Cui, 2007; Pan et al., 2009). When QDs were binding to CNTs, CNTs could promote direct charge transport and efficient charge transfer from the QDs. This system has the potential to significantly increase the efficiency of photovoltaic devices (Kamat, 2007). Besides, CNTs binding with QDs together can provide one kind of novel nanomaterials–luminescent CNTs can afford fluorescent labels and be utilized for real-time detection, molecular imaging and cell sorting in biological applications (Guo et al., 2008; Shi et al., 2006). Our group has synthesized luminescent CNTs as a new functional platform for bioanalytical sciences and biomedical engineering (Cui et al., 2010).

To further understand that how the nanohybrid structure affects the charge transfer and energy transfer behaviors between the QDs and CNTs, the interactions between QDs (such as CdS, CdSe and CdTe) and CNTs have been carefully investigated by several groups (Guldi et al., 2006; Li et al., 2006b; Robel et al., 2005; Sheeney-Haj-Ichia et al., 2005; Si et al., 2009). The charge-transfer efficiencies were evaluated by studying the changes in the photoluminescence (Li et al., 2006b) or photo-electrochemical properties of hybrid materials (Pan et al., 2008). Studies indicated that strong PL quenching by charge-transfer mechanism were found in the CdS/TOAB/CNT (Guldi et al., 2005), CdSe/pyridine/CNT (Li et al., 2006a) and CdSe/pyrene/CNT (Hu et al., 2008) system. Partial emission quenching was observed on nanohybrids consisting of dendron-modified CdS QDs on CNTs (Hwang et al., 2006). In contrast, Marek Grzelczak et al. reported a reproducible procedure based on the combination of both polymer wrapping and LbL self-assembly techniques for the deposition of CdTe nanocrystals onto CNTs, yielding linear colloidal CdTe–CNT composites with a high degree of coverage (Grzelczak et al., 2006). Although quenching of PL from CdTe occurs when the nanocrystals are directly assembled on the CNTs, such quenching can be controlled through the growth of a silica-shell spacer between the CNT surface and the deposited QDs. The main general steps of this method for the deposition of CdTe QDs onto CNTs and onto silica-coated CNTs (CNTs@SiO_2) are summarized in Fig. 9.

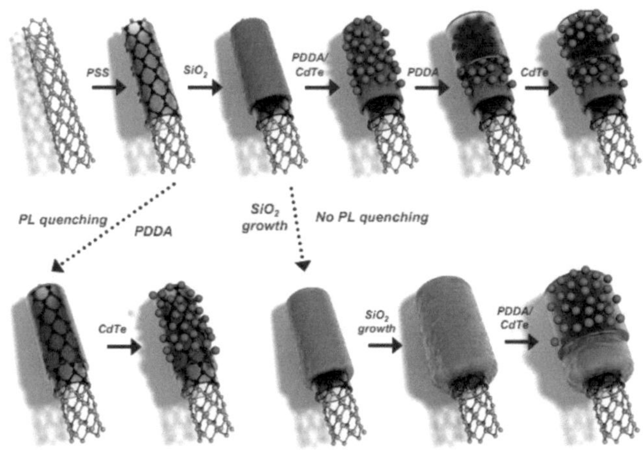

Fig. 9. Various possible routes for the preparation of CdTe-CNT and CdTe–CNT@SiO_2 nanocomposites. PSS: poly (sodium 4-styrene sulfonate); PDDA: poly (diallyldimethylammoniumchloride) (Grzelczak et al., 2006). (With permission from Wiley-VCH)

The previous works have shown that the PL properties of the QD/CNT nanohybrids are strongly dependent on QD-CNT separation, but precise control to the distance between QDs and CNT was difficult to achieve in the available systems. Recently, Hao-Li Zhang et al. reported a facile strategy for attaching CdSe QDs onto CNT surface by electrostatic self-assembly (Fig. 10) (Si et al., 2009). By using different mercaptocarboxylic ligands, the shell thicknesses of the CdSe QDs are well controlled within angstrom-level precision. The efficiency of the PL quenching decreases upon increasing the shell thickness due to the distance-dependent electron transfer efficiency. This work demonstrates that the shell thickness control to the QDs opens up a straightforward methodology for investigating the interaction between fluorescent nanomaterials coupled with CNTs.

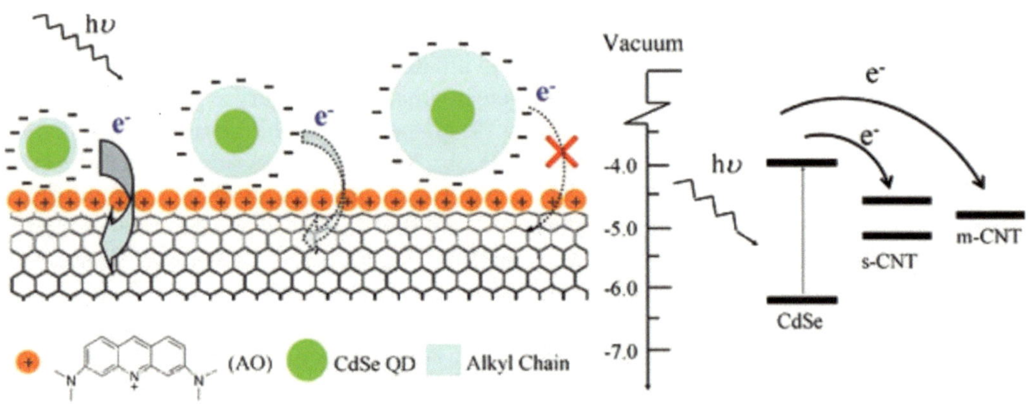

Fig. 10. Schematic illustration to the structures of the CdSe/CNT nanohybrids prepared by electrostatic assembly (left). The photo-induced charge-transfer efficiency within the nanohybrids is controlled by the shell thickness. Energy-level diagram and possible charge-transfer process for the conjugate complex between CdSe QDs and semiconducting CNTs (s-CNT) or metallic CNTs (m-CNT) are illustrated in the right (Si et al., 2009). (With permission from Springer)

An investigation of the effects of MWNTs on the PL properties of CdSe QDs showed that CNTs could suppress the PL of QDs through both dynamic and energy transfer quenching mechanisms (Cui et al., 2008; Pan et al., 2006b; Pan et al., 2008). In order to potentially exploit this feature in bioassays, we reported a novel ultrasensitive DNA or antigen detection strategy based on the CNT-QD assembly (Cui et al., 2008). MWNTs and QDs, their surfaces functionalized with oligonucleotide DNA or antibody (Ab), can be assembled into nanohybrid structure upon the addition of a target complementary oligonucleotide or antigen (Ag). Nanomaterials building blocks that vary in chemical composition, size, or shape are arranged in space on the basis of their interactions with complementary linking oligonucleotides for potential application in biosensors. We show how this oligonucleotide-directed assembly strategy could be used to prepare binary (two-component) assembly materials comprising two differently shaped oligonucleotide-functionalized nanomaterials. Importantly, the proof-of-concept demonstrations reported herein suggest that this strategy could be extended easily to a wide variety of multicomponent systems.

5. Experimental section

5.1 Preparation of water-soluble CdSe QDs

The colloidal CdSe QDs were dissolved in chloroform and reacted with glacial mercaptoacetic acid (1.0 M) for 2 h. Water was added to this reaction mixture at a 1:1 volume ratio. After vigorous shaking and mixing, the chloroform and water layers separated spontaneously. The aqueous layer, which contained mercaptoacetic-coated QDs, was extracted (Scheme 1). Excess mercaptoacetic acid was removed by four or more rounds of centrifugation.

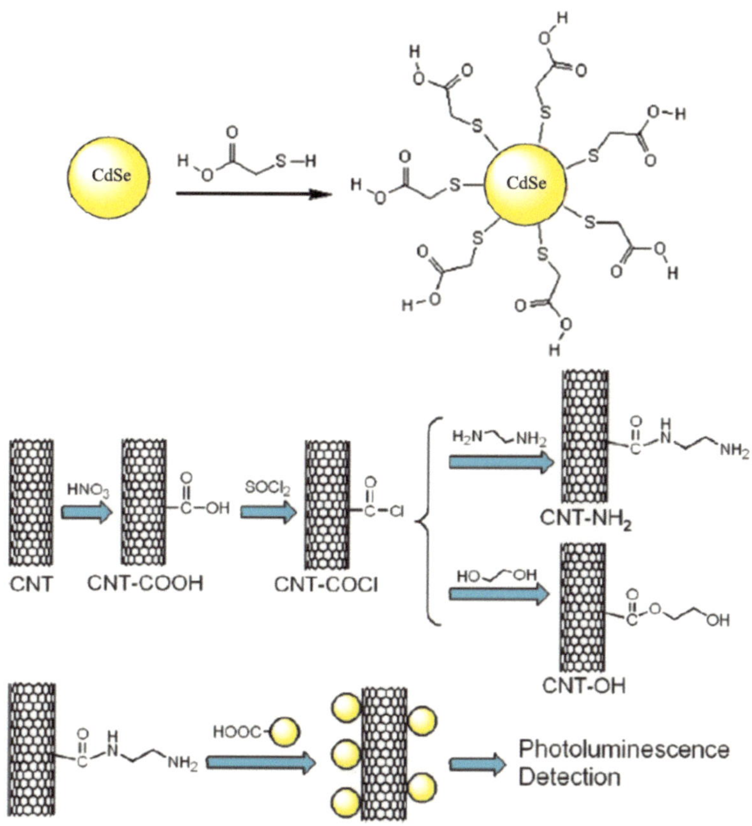

Scheme 1. Water-Soluble CdSe QDs, Preparation of CNT-COOH, CNT-OH, and CNT-NH$_2$, and Electrostatic Interaction between CNT-NH$_2$ and QDs. (With permission from American Chemical Society)

5.2 Preparation of CNT-COOH, CNT-OH, and CNT-NH$_2$

A 100 mg amount of pristine CNTs was added to aqueous HNO_3 (10.0 mL, 60%). The mixture was placed in an ultrasonic bath for 30 min and then stirred for 24 h while being boiled under reflux. The mixture was then in vacuum filtered through a 0.22 μm Millipore polycarbonate membrane and subsequently washed with distilled water until the pH of the filtrate was ca. 7. The filtered solid was dried under vacuum for 12 h at 60 °C. Dried CNT-COOH was suspended in $SOCl_2$ (20 mL) and stirred for 24 h at 65 °C. The solution was filtered, washed with anhydrous THF, and dried under vacuum at room temperature for 24 h, generating CNT-COCl. Dried CNT-COCl was mixed with excess ethylene glycol and stirred for 48 h at 120 °C. The resulting solid was separated by vacuum filtration using a 0.22 µm Millipore polycarbonate membrane filter and subsequently washed with anhydrous THF. After repeated washing and filtration, the resulting solid was dried overnight in a vacuum, generating CNT-OH. On the other hand, CNT-COCl was reacted with ethylenediamine to obtain CNT-NH_2 (Scheme 1).

5.3 Preparation of CNT-DNA probe (Scheme 2)

Binding of the aminoalkyloligonucleotide, in PBS buffer (0.1 M NaCl, 10 mM phosphate buffer, pH 7.0), to the CNT was accomplished by adding 50 µL of 1 µM aminoalkyloligonucleotide to 1.0 mL of 1 mg/L CNT-COCl in phosphate buffer solution (PBS), incubated for 40 h at 25 °C, collected by centrifugation at 4000 rpm for 10 min, and resuspended in 1 mL PBS to form CNT-oligonucleotide composite (probe 1).

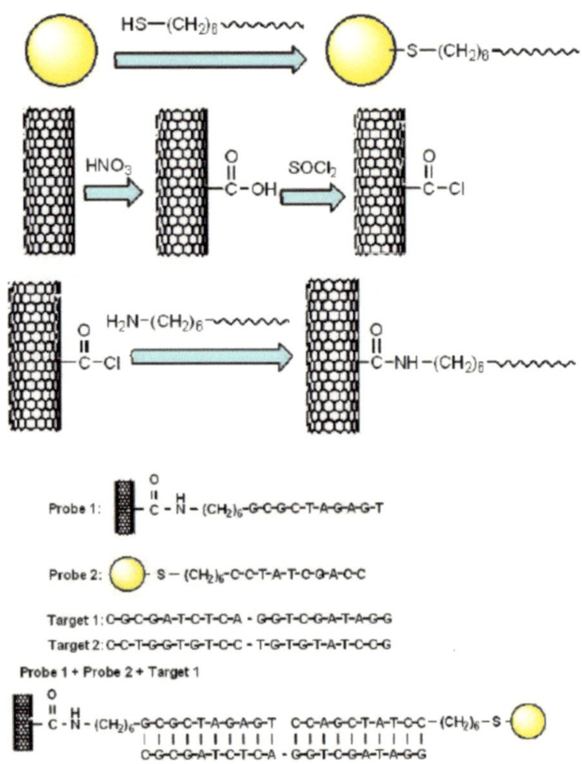

Scheme 2. Surface functionalization with oligonucleotide, and subsequent addition of target oligonucleotide to form CNT-QD assembly. Target 1, complementary DNA; target 2, non-complementary DNA. (With permission from American Chemical Society)

5.4 Preparation of QD-DNA probe

The colloidal CdSe QDs were dissolved in chloroform and were reacted with glacial mercaptoalkyloligonucleotide (1.0 µM) for 2 h. PBS buffer was added to this reaction mixture at a 1:1 volume ratio. After vigorous shaking and mixing, the chloroform and PBS buffer layers separated spontaneously. The PBS buffer layer, containing mercaptoalkyloligonucleotide-coated QDs (probe 2), was extracted from chloroform liquid.

5.5 Preparation of CNT-Ab probe

For the covalent immobilization of antibody on the CNT surface, CNT-NH_2 was exposed to 20 nM anti-BRCAA1 IgG in PBS buffer (pH 7.4) overnight at room temperature, rinsed thoroughly in deionized water for 6 h, and then dried with nitrogen gas.

5.6 Preparation of QD-Ab probe

The QDs-pAb probes were prepared by adding BRCAA1 pAb (40 μg) to an aqueous solution of QDs (5 mL, 2.33 nM) at pH 9.0 for 30 min. Then, the solution was treated with 0.5 mL 10% BSA solution for a night to passivate and stabilize the QDs. The final QDs-pAb probes were re-dispersed in 0.01 M PBS at pH 7.4. CNTs-pAb probes also were prepared by above-mentioned method, and the pAb concentration was 1.5 mg/mL.

5.7 Photoluminescence (PL) measurement of CdSe QDs binding to CNTs

CNTs with concentrations from 0 to 100 mg/L were added to the CdSe aqueous solution. QD-CNT solutions were incubated at room temperature in the dark for 3 h. PL spectra were taken with a Perkin-Elmer LS-55 spectrofluorometer. Samples were thermostated at 25 °C. An excitation wavelength of 460 nm was used. The emission spectra were recorded from 500 to 650 nm. The excitation and emission slit widths were set to 5 and 5 nm, respectively. Samples were contained in 1 cm path length quartz cuvettes and continuously stirred. For control experiments, CNTs were removed by centrifugation, the same buffer solution without nanotubes was added to the CdSe solution, and then the changes of PL were recorded.

5.8 Assembly and characterization of CNTs and QDs through DNA hybridization (Scheme 3, System 1)

The methodology of DNA detection by fluorescence measurement is shown in System 1. Two microliters of target oligonucleotide with different concentrations were added into the mixture of CNT-DNA (50 μL, [CNT]) 1 mg/mL, [DNA]) 0.1 nM) and QD-DNA probes (50 μL, [DNA]) 0.1 nM). This was mixed and incubated at 75 °C for 5 min and then at 25 °C for 20 min. After hybridization with target DNA, CNT-QD assemblies were formed and then removed by centrifugation at 2000 rpm for 5 min. The unbound QD-DNA probes in supernatant PBS buffer were immediately detected by spectrofluorometer. The wavelength λ = 480 nm of the laser source was used for the excitation of the CNT-QD detection system. The fluorescence signal was recorded over a range from λ = 450 nm to λ = 700 nm. NoncDNA target (CGC GAT CTC AGG TCG ATA GG) was used as the control experiment.

5.9 Three-component CNT-QD system with the purpose of detecting three different DNA targets simultaneously (Scheme 3, System 2)

Three QDs with different emission wavelengths at 510, 555, and 600 nm were used to simultaneously detect three target DNA molecules, as shown in System 2, called QD_{510}, QD_{555}, and QD_{600} probes, respectively. There are six probes in this system: three CNT-DNA

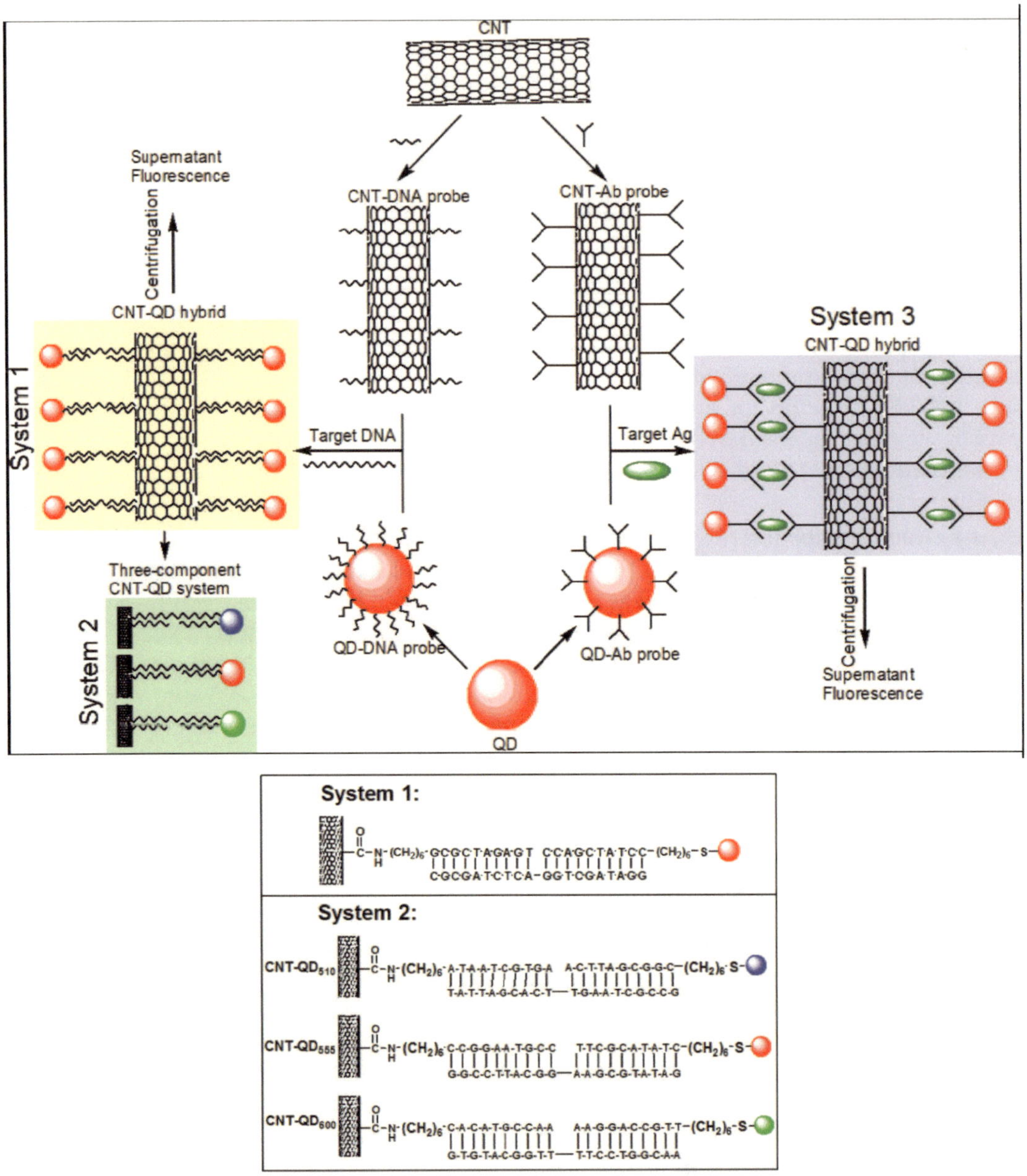

Scheme 3. Surface Functionalization of CNT (or QD) with Oligonucleotide/Antibody (Ab), Forming aCNT-DNA (or -Ab) Probe and QD-DNA (or -Ab) Probe, and Subsequent Addition of Target Oligonucleotide (or Antigen) to Form a CNT-QD Assembly[a] (a The unbound QD probe was obtained by simple centrifugation separation, and the supernatant fluorescence intensity of QDs was monitored by spectrofluorometer. (System 1) Formation of CNT-QD hybrid in the presence of complementary DNA target. (System 2) Three-component CNT-QD system with the purpose to detect three different DNA targets simultaneously. (System 3) CNT-QD protein detection system based on antigen-antibody immunoreaction.) (With permission from American Chemical Society)

probes (50 μL of each, [CNT]) 1 mg/mL and [DNA]) 0.1 nM) and 3 QD-DNA probes (50 μL of each, [DNA]) 0.1 nM) (as shown in Scheme 3, System 2). A mix of the six probes formed a uniform solution. Three cDNA targets were used in this system. Two microliters of each DNA target with different concentration was added into the six-probe mixture, individually. NoncDNA target (CGC GAT CTC AGG TCG ATA GG) was used as the control experiment. In addition, we mixed the three DNA targets (2 μL of each) to form a 6 μL DNA solution containing three different DNA targets, and then incubated the three DNA targets with the six-probe mixture simultaneously. The mixture solution containing probes and targets was incubated at 75 °C for 5 min and then at 25 °C for 20 min. After hybridization with target DNA molecules, CNT-QD assemblies were formed and then removed by centrifugation at 2000 rpm for 5 min. The unbound QD-DNA probes in supernatant PBS buffer were immediately detected by spectrofluorometer.

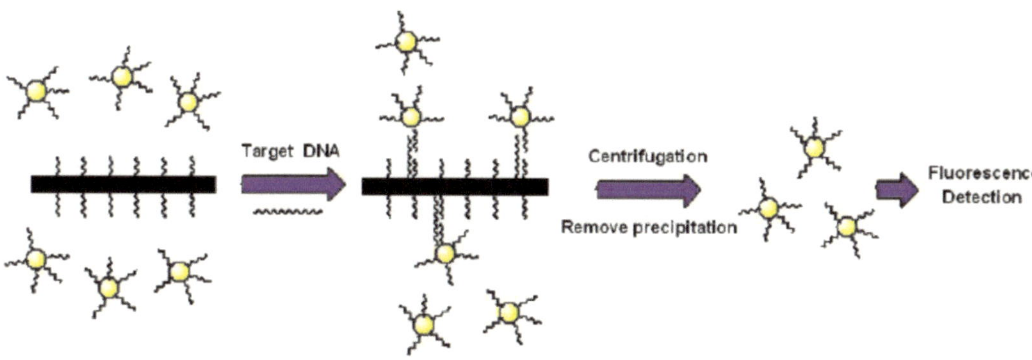

Scheme 4. Formation of CNT-QD assemblies for ultrasensitive DNA detection. (With permission from American Chemical Society)

5.10 CNT-QD system for antigen detection via antigenantibody immunoreaction (Scheme 3, System 3)

To confirm that the CNT-QD system can be used for antigen detection, a novel CNT-QD immunoassay system based on antigen-antibody immunoreaction was established. BRCAA1 protein was chosen as a typical example. CNT-Ab and QD-Ab probes were prepared as shown in the Supporting Information. The immunoreactions procedure is described as follows: 50 μL of CNT-Ab probe ([CNT]) 1.0 mg/mL, [Ab]) 1.0 nM) was reacted with 10 μL of BRCAA1 antigen (concentration from 0 to 1.0 nM) for 30 min, then 50 μL of QD-Ab probe (1.0 nM) was added, and the mixture was incubated for 2 h at room temperature. After the immunoreaction, the sandwich-type immunocomplex was formed on the surface of CNT probes. The unbound QDs-pAb was obtained by simple centrifugation separation and the supernatant fluorescence intensity of QDs was monitored by spectrofluorometer (Scheme 4.).

5.11 Characterization of CNT-QD hybrids

TEM images were taken with a JEM 100-CXII microscope (JEOL, Japan) at 100 kV to demonstrate the formation of CNT-QD assembly. The zeta potentials of CNT-QD aqueous suspension were measured using a MALVERN ZETAZIZER 2000 instrument (U.K.). Fourier transform infrared (FT-IR) spectra were recorded using a PE Paragon 1000 spectrometer. UV-vis absorption spectra were obtained on UNICAM UV 300 spectrometers (Thermo Electronic). Raman measurements were carried out on a Jobin Yvon microRaman system (Ramanor U1000, Instruments SA, USA) using a Spectra-Physics Ar ion laser at an excitation wavelength of 514.5 nm (2.41 eV). All measurements were taken at room temperature, and for each sample the Raman data were collected at different light spots on the sample surface. For every Raman spectrum taken, the position of the peaks was verified by calibrating the spectral positions with respect to the silicon substrate peak seen at 521 cm^{-1}.

6. Results and discussion

6.1 Formation of self-assembled CNT-QD nanocomplex

TEM images of QDs, CNTs, and CNT-QD nanocomplexes are shown in Fig. 11. The size distribution of CdSe nanoparticles is narrow with an average particle size of 3.0 nm in diameter, while the diameters of CNTs are about 20~30 nm. When the colloids of CdSe nanoparticles were mixed with CNT-COOHs or CNT-OHs in aqueous solution, very few CdSe nanoparticles can be bound onto the nanotubes (Fig. 11a, b). On the other hand, the TEM images give a direct view of the CdSe nanoparticles binding to CNT-NH_2 (Fig. 11c); the CNTs with -NH_2 surface groups are densely coated with CdSe nanoparticles. For all three CNT/QD systems no precipitation was observed for at least 4 weeks. For CNT-COOH, most of the CNTs remained suspended for over 1 year after adding CdSe QDs because of the strong electrostatic repulsion between CNTs and QDs, which is consistent with the TEM pictures as shown in Fig. 11.

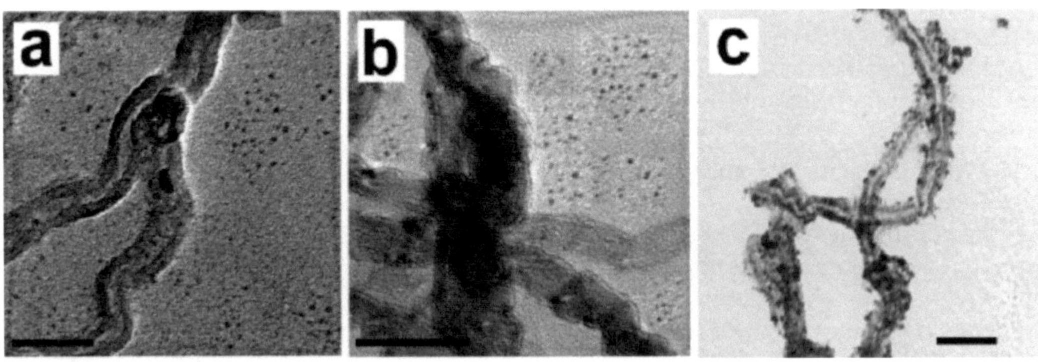

Fig. 11. TEM images of (a) a mixture of CNT-COOH and CdSe, (b) a mixture of CNT-OH and CdSe, and (c) self-assembled CNT-NH_2-CdSe nanocomplexes. Scales bar: 50 nm. (With permission from American Chemical Society)

To further investigate the interaction between CNTs and QDs, the ζ-potential of the CNTs and CdSe samples are summarized in Fig. 12. Positively charged CNT-NH_2, negatively charged CNT-COOH, and negatively charged QD-COOH can be confirmed from ζ-potential analysis. Thus, addition of positively charged –NH_2-modified CNTs to the negatively charged CdSe QDs with a COOH group results in formation of CNT-CdSe complexes by electrostatic interaction as shown in Scheme 1. The interaction forces between QDs and CNTs in aqueous solution are mainly electrostatic forces, which is why they are weakest for negative carboxyl-terminated CNTs. At pH 7.0, CdSe has a negative charge (ζ potential = 36

mV from Fig. 12); therefore, amino-terminated cationic CNTs (ζ potential = +32 mV) have the biggest impact on the QDs.
The UV-vis absorption spectra of CdSe nanocrystals (a) before adding CNT solution and (b) after adding CNT-COOH, (c) CNT-OH, and (d) CNT-NH_2 are shown in Fig. 13A. The spectral feature located at ca. 580 nm has been assigned to the first excitonic transition occurring in CdSe nanocrystals. No obvious change of the band gap (580 nm) of the QDs is observed after adding CNT samples, indicating that no larger QD agglomerates were formed after adding CNT; in particular, for the CNT-NH_2-QD samples, individual QD was separately coated on the CNT surface (Fig. 11c) instead of agglomerate.

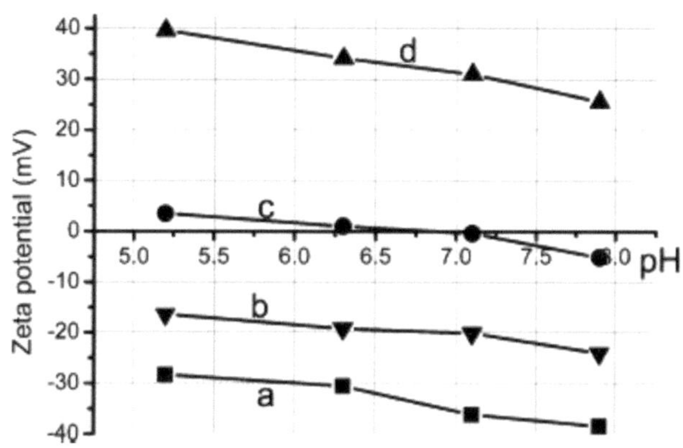

Fig. 12. Zeta potentials of (a) mercaptoacetic acid coated CdSe, (b) CNT-COOH, (c) CNT-OH, and (d) CNT-NH_2 in aqueous solution. (With permission from American Chemical Society)

6.2 Photoluminescence (PL) quenching of CdSe QDs in the presence of CNTs

The PL emission spectra (Fig. 13B-D) show that CdSe QDs have strong PL, and the wavelength of the PL maximum for CdSe was at 552 nm. The CNT concentration dependence of the PL intensity of CdSe QDs is shown in Fig. 13B-D. The decrease in the PL intensity was the most marked change in the PL spectrum observed upon addition of CNTs. For all types of CNTs, their increasing concentrations caused a linear reduction in the PL of CdSe. The effect was strongest for CNT-NH_2 (Fig. 13D), less pronounced for CNT-OH (Fig. 13C), and weakest for CNT-COOH (Fig. 13B).

6.3 Calculation of quenching constants from the Stern-Volmer equation and double-logarithmic equation

The PL quenching data were analyzed by the Stern-Volmer equation (Fan and Jones Jr, 2006; Maurel et al., 2006; Pan et al., 2006c) :

$$\frac{I_0}{I} = 1 + K_{SV}[CNT] \tag{1}$$

where I_0 and I are, respectively, PL intensities in the absence and presence of CNT, K_{SV} is the Stern-Volmer dynamic quenching constant, and [CNT] is the concentration of CNT. The equation assumes a linear plot of I_0/I versus [CNT], and the slope is equal to K_{SV}. The Stern-

Volmer constants express CdSe QD accessibility to the CNT. Fig. 14A shows Stern-Volmer quenching curves describing I_0/I as a function of CNT concentration. The Stern-Volmer constants (K_{SV}) for quenching of CdSe PL intensity by different kinds of CNTs are presented in Table 1. K_{SV} were calculated from the plots shown in Fig. 14A.

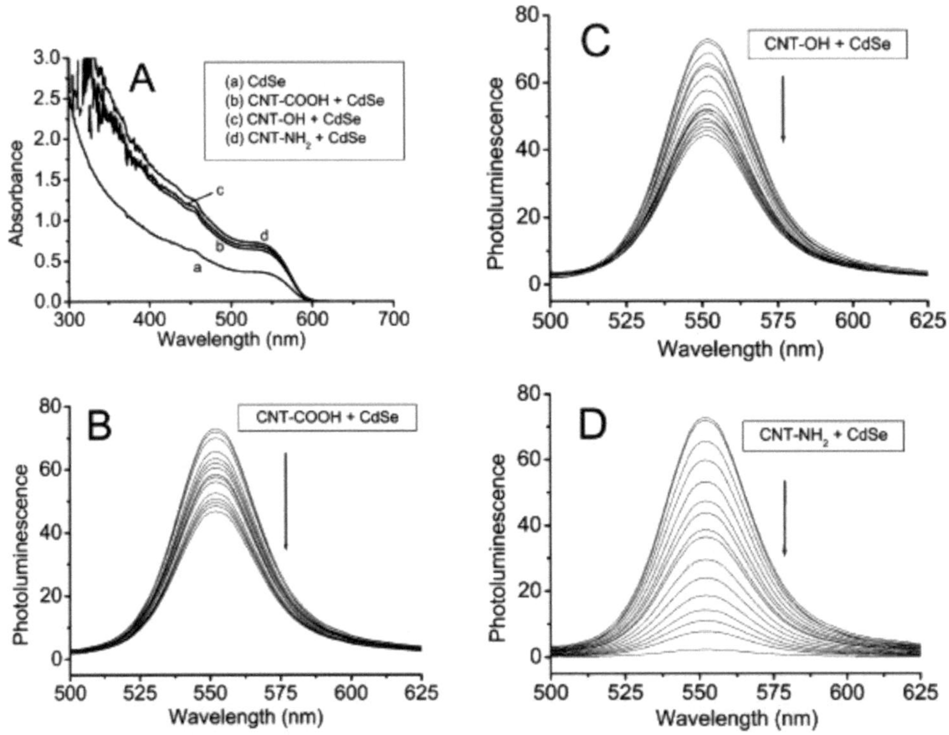

Fig. 13. (A) UV-vis spectra of QDs before and after adding CNTs; (B-D) PL spectra of CdSe QDs by adding CNTs. The concentration of CNT from top to bottom: 0, 0.5, 1.0, 2.0, 3.0, 4.0, 5.0, 6.0, 7.0, 8.0, 9.0, 10.0, 20.0, 30.0, 50.0, and 100.0 (mg/L). (With permission from American Chemical Society)

6.3 Calculation of quenching constants from the Stern-Volmer equation and double-logarithmic equation

The PL quenching data were analyzed by the Stern-Volmer equation (Fan and Jones Jr, 2006; Maurel et al., 2006; Pan et al., 2006c) :

$$\frac{I_0}{I} = 1 + K_{SV}[CNT] \tag{1}$$

where I_0 and I are, respectively, PL intensities in the absence and presence of CNT, KSV is the Stern-Volmer dynamic quenching constant, and [CNT] is the concentration of CNT. The equation assumes a linear plot of I_0/I versus [CNT], and the slope is equal to K_{SV}. The Stern-Volmer constants express CdSe QD accessibility to the CNT. Fig. 14A shows Stern-Volmer quenching curves describing I_0/I as a function of CNT concentration. The Stern-Volmer constants (K_{SV}) for quenching of CdSe PL intensity by different kinds of CNTs are presented in Table 1. K_{SV} were calculated from the plots shown in Fig. 14A.

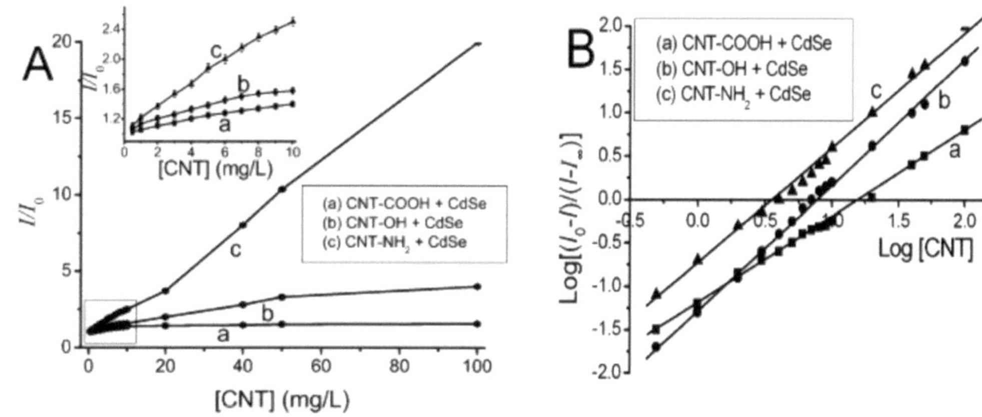

Fig. 14. (A) Stern-Volmer plot and (B) double-logarithmic plot for CdSe PL quenching by CNT with different concentration. (With permission from American Chemical Society)

Alternatively, analyses of the PL data were performed using the double-logarithmic equation (Pan et al., 2006c). According to this approach, the QDs PL intensity scales with the CNT concentration ([CNT]) through the following equation:

$$\frac{I_0 - I}{I - I_s} = \left[\frac{CNT}{K_{diss}}\right]^n \qquad (2)$$

The binding constant K_b is obtained by plotting log[$(I_0 - I)/ (I- Is)$] versus log[CNT], where I_0 and I_s are the PL intensities of the QDs alone and the QDs saturated with the CNTs, respectively. The slope of the double-logarithm plot obtained from the experimental data is the number of n, whereas the value of log[CNT] at log[$(I_0 - I)/ (I - I_s)$]) 0 equals the logarithm of the dissociation constant K_{diss}. The reciprocal of K_{diss} is the quenching constant Kb. The PL intensity values (I) were obtained from the area under the PL spectra. Figure 4B represents the plot of log[$(I_0 - I)/ (I - I_s)$] versus log[CNT] for CNT-COOH, CNT-OH, and CNT-NH_2. Values of K_b and n obtained from Fig.14B are shown in Table 1.

sample	Stern-Volmer equation K_{SV} [L/mg]	Double-logarithmic equation K_b (n) [L/mg]
CNT-COOH	0.060	0.0631(1.000)
CNT-OH	0.106	0.1365(1.434)
CNT-NH_2	0.176	0.269(1.346)

Table 1. Quenching Constants of CdSe by CNTs from the Stern-Volmer Equation and Double-Logarithmic Equation.

PL of CdSe QDs was strongly quenched by amino-terminated CNTs but only poorly quenched by CNT-COOH, as shown by K_{SV} and K_b values. The data were well fitted by a straight line, typical of a simple dynamic quenching mechanism. For [CNT] >10 mg/L, KSV values increase rapidly for CNT-NH_2 + QD solution, but fewer changes can be observed for

CNT-OH + QD and CNT-COOH + QD solution. The data shown in Table 1 indicate that quenching constants of K_{SV} and K_b from the Stern-Volmer equation and double-logarithmic equation are very similar. The quenching constant (K_{SV} and K_b) for CNT-NH_2 is much higher than that of CNT-COOH and CNT-OH, indicating CNT-NH_2 is more strongly bound to the QD compared with CNT-COOH and CNT-OH. These observations support that surface groups and their charges are key factors of the CdSe-CNT interactions, which also confirm that PL properties of CdSe QDs are closely associated with their surface structure and surface charges. The ζ-potential analysis in Fig. 12 showed that both carboxyl and amino surface groups are ionized in water, only hydroxyl-terminated CNTs bear no charge on their surface. Thus, the dynamic process quenched the emission of CdSe QDs because of the electrostatic interactions between the CNT surface and CdSe surface in aqueous solution.
From the results of TEM, zeta-potential analysis, PL quenching experiments, the Stern-Volmer equation, and the doublelogarithmic equation we can make conclusions that the quenching intensity of CdSe caused by CNT highly depends on the surface structure and charge of CNTs and can be ordered as follows: CNT-NH_2 > CNT-OH > CNT-COOH.

6.4 Effect of pH on the quenching constants

Control experiments were carried out by adding the same buffer but without CNTs to QD solution; furthermore, buffer samples with pH ranging from 4.0 to 10.4 were used to investigate the pH effect on QD PL properties. As shown in Fig. 15A, little PL change can be observed for QDs after adding buffer solutions without CNT at pH values from 4.0 to 10.4, which highly suggest that PL quenching was mainly caused by the CNTs in the QD solution.

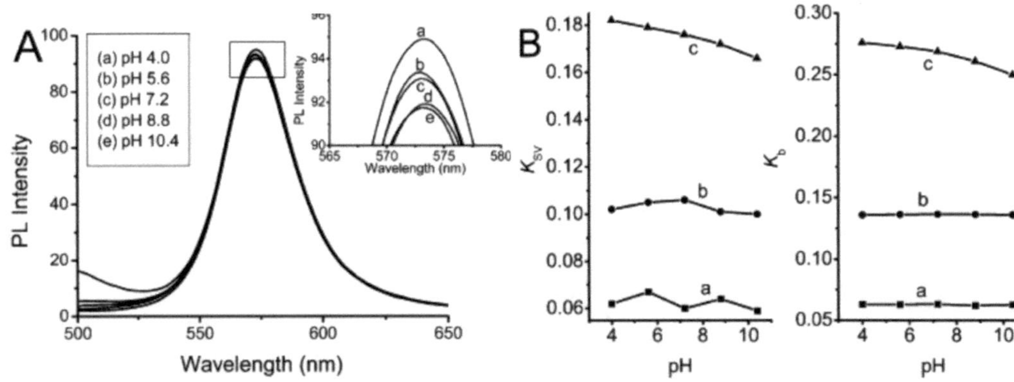

Fig. 15. (A) pH effect of buffer solution on QD PL intensity. (B) Plot of quenching constant versus solution pH: (a) CNT-COOH, (b) CNT-OH, and (c) CNT-NH_2. (With permission from American Chemical Society)

To investigate the pH effect on quenching constant (K_{SV} and K_b), CNTs were added to QD solution at different pH environments, and then K_{SV} and K_b were calculated from the Stern-Volmer equation and double-logarithmic equation. Plots of quenching constant versus pH value are given in Fig. 15B. In the CNT-NH_2 case, the quenching constant (K_{SV} and K_b) values decreased slightly with increasing pH from 4.0 to 10.4. It can be explained as follows. QD is negatively charged by the zeta potential result (Fig. 12). At lower pH, amino groups of CNT-NH_2 are highly protonated, but at higher pH, the surface charge becomes less positive; thus, the electrostatic interactions between amino groups CNT-NH_2

and the QD molecule are weakened, resulting in the decrease of quenching constants. For CNT-COOH and CNT-OH samples, little changes in K_{SV} and K_b can be observed as shown in Fig. 15B.

6.5 FT-IR and raman spectroscopy

As shown in Fig. 16A (a), the FT-IR spectrum of the carboxylic-acid-coated QDs showed a strong asymmetric vibration characteristic of carboxyl groups at 1711 cm^{-1}. The FT-IR spectra of QDs after adding CNT-COOH and CNT-OH are shown as Fig. 16A (b, c), indicating no shift of the -COOH peak at 1711 cm^{-1}. The QD-COOH peak at 1711 cm^{-1} shifts to 1732 cm^{-1} after adding CNT-NH_2, indicating a strong electrostatic interaction between CNT-NH_2 and QDs. Raman spectra of CNT-NH_2 before and after binding with QDs are shown in Fig. 16B. CNT-NH_2 reveals two main peaks at 214 and 1582 cm^{-1} in the Raman spectrum. After attachment of QDs onto the sidewalls of CNT-NH_2, the original G peak at 1582 cm^{-1} of the uncoated CNT-NH_2 was shifted to 1578 cm^{-1} (Fig. 16B). However, adsorption of QD on CNT-NH_2 did not result in any Raman shift at 214 cm^{-1}, indicating that the assembly of QD nanocrystals onto the tubes deformed the sidewall of CNTs (Pan et al., 2006b). For CNT-COOH and CNT-OH, no Raman shift can be observed after addition of QDs (data not shown), indicating CNT-QD complexes cannot be formed by adding QD to the CNT-COOH or CNT-OH.

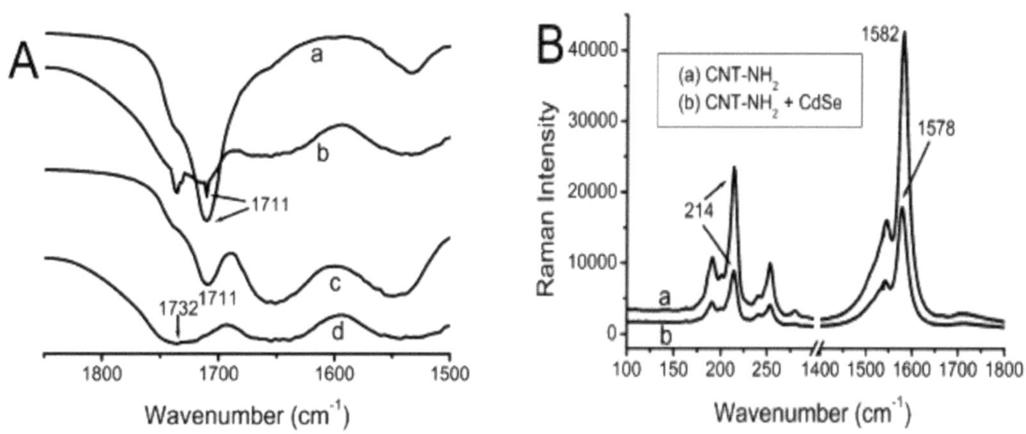

Fig. 16. (A) FT-IR spectra of (a) QD-COOH, (b) CNT-COOH + QD, (c) CNT-OH + QD, and (d) CNT-NH_2 + QD. (B) Raman spectra of CNT-NH_2 and CNT-NH_2 + QDs. (With permission from American Chemical Society)

6.6 The quenching mechanism between CdSe QDs and CNTs

With the linearity, the concentration of CNT can be monitored by measuring the PL intensity of CdSe QDs. PL quenching is a process which decreases the intensity of the QD PL emission. QD PL quenching may occur by several mechanisms (Fan and Jones Jr, 2006; Pan et al., 2006c): (1) dynamic quenching, (2) static quenching, (3) quenching by energy transfer, and (4) charge-transfer reactions. In this paper we can draw the conclusion that QD quenching in the presence of CNTs is mainly caused by dynamic quenching and energy-transfer mechanisms (Clapp et al., 2004; Oh et al., 2005). When quenching occurs by a dynamic mechanism, the quenching is an additional process that deactivates the excited state besides radiative emission (Cui et al., 2004b). The dependence of the QD emission intensity on CNT concentration is given by the Stern-Volmer equation and double-logarithmic equation. The accessibility of CNT to QD is reflected in the quenching constant (including K_{SV} and K_b). For CNT-COOH and CNT-OH, low values of K_{SV} and K_b indicate low exposure of CdSe to CNTs, i.e., the QD did not attach to the CNT surface. In particular, in the case of CNT-COOH, strong repulsion between CNT and QD in aqueous solution can be confirmed by the TEM image and zeta-potential data. The interparticle space between CNT and QD will be the main determinative factor for PL quenching according to the dynamic quenching and energy transfer mechanisms. According to energy-transfer mechanism, energy transfer is proportional to r^{-6} (r = distance between CNT and QD), i.e., PL will be severely quenched when QDs are getting close to the surface of the CNT. It is very difficult for QD-COOH to reach the CNT-COOH surface because of the strong repulsion electrostatic force. However, QD-COOH can easily bind to the CNT-NH_2 surface to form CNT-QD nanocomplex, which can be confirmed by FT-IR and Raman spectroscopy, resulting in strong PL quenching of QDs; thus, the strong quenching of QDs caused by CNT-NH_2 is mainly due to attachment of the QDs onto the CNT surface. These QDs have no shifted emissions at 552 nm (Fig. 13), indicating a relatively unchangeable size. Given that the CNTs quenched the CdSe QDs, the interaction between CNTs and QDs must provide an alternative, nonirradiative decay path (Clapp et al., 2004; Cui et al., 2004b; Oh et al., 2005). It is believed that this nonirradiative decay path occurs because the electron affinity between the CdSe QDs and the CNTs is sufficiently different that it allows electron transfer from the QDs to the CNTs (Clapp et al., 2004; Pan et al., 2006c). In other words, formation of CNT-QD conjugates favors electron transfer from the quantum dots (donor) to the CNTs (acceptor) such that the excited electrons are accepted by the CNTs rather than being emitted as the PL peak. For CNT-COOH and CNT-OH, QDs cannot reach the CNTs surface, so that energy transfer will be very difficult to occur for the CNT-COOH/QD or CNT-OH/QD system; therefore, a dynamic quenching mechanism will be fit for the CNT-COOH/QD or CNT-OH/QD system. For CNT-NH_2/QD system, besides dynamic quenching, electron/energy transfer will be allowed from QDs to CNTs, i.e., an energy-transfer quenching mechanism.

6.7 CNT-QD assembly system for DNA target detection (System 1)

The supernatant fluorescence spectra in Fig. 17 show that the QD-DNA probe has strong fluorescence, and the maximum fluorescence wavelength for CdSe was at ~570 nm. The cDNA target (sequence shown in Scheme 3, system 1) with concentration ranges of 0~200 pM was added to the mixture of CNT-DNA and QD-DNA probes to form CNT-QD nanohybrids. After centrifugation at 2000 rpm for 5 min, the CNT-QD nanohybrids were removed; therefore, only unbounded QD-DNA existed in the supernatant buffer. The decrease in the fluorescence intensity was the most marked change in the fluorescence spectrum observed upon addition of the cDNA target.

Control experiments were carried out by adding cDNA target to the QD-DNA solution in the absence of the CNT-DNA probe (Fig.18A). In addition, buffer solutions with pH ranging from 4.0 to 11.0 were used to investigate the pH effect on the fluorescence properties of the QD-DNA probe (Fig.18B). As shown in Figure 2A and B, little PL change can be observed for QD-DNA probes after adding complementary target DNA solutions without CNT-DNA probe at pH values from 4.0 to 11.0, which highly suggests that the QD fluorescence intensity was

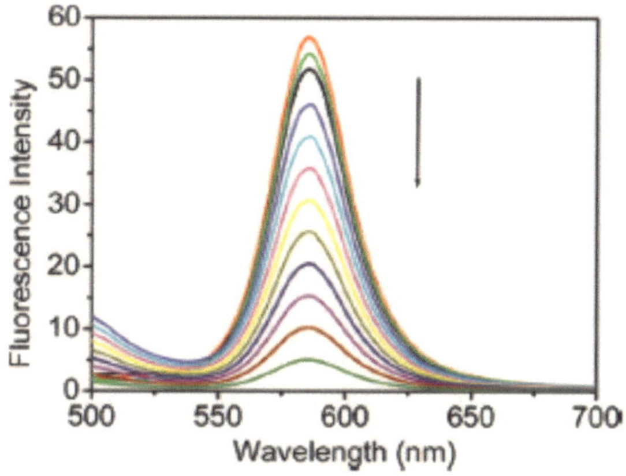

Fig. 17. Fluorescence intensity of supernatant after adding target 1 with different concentrations (from top to bottom): 0, 10, 16, 25, 40, 50, 60, 75, 90, 120, 150, and 200 pM. (With permission from American Chemical Society)

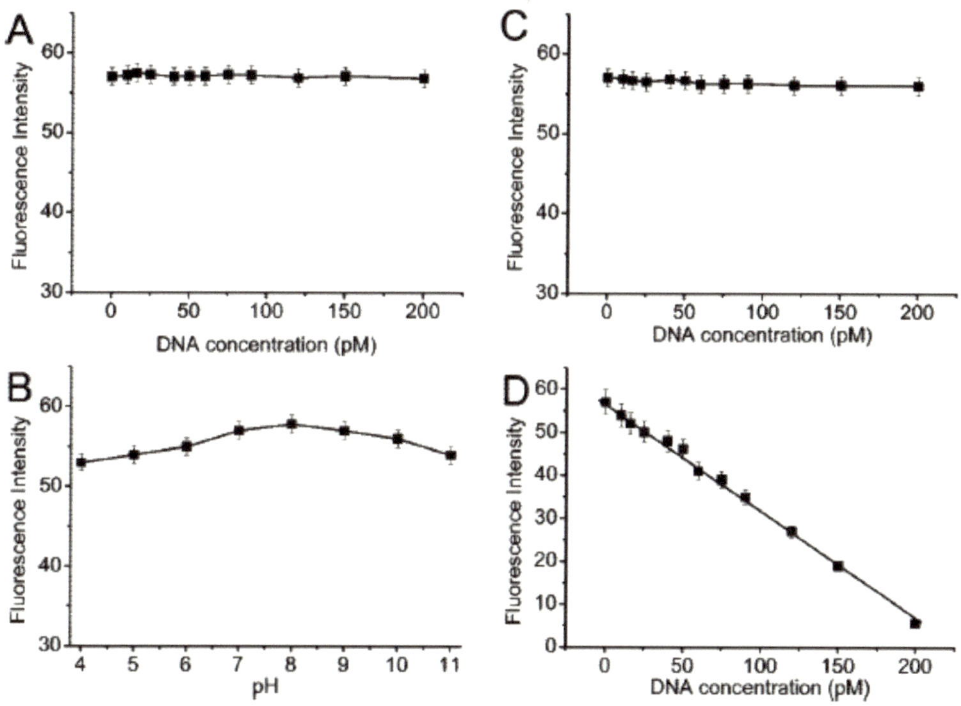

Fig. 18. Plot of supernatant fluorescence intensity versus DNA concentration: (A) effect of cDNA target concentration on fluorescence intensity of QD-DNA probe; (B) fluorescence of QD-DNA probe at different pH values from 4.0 to 11.0; (C) noncDNA target with concentration from 0 to 200 pM incubated with CNT-DNA and QD-DNA probes (no fluorescence changes in supernatant); and (D) marked fluorescence decrease found for CNT-DNA/QD-DNA mixture solution after the incubation with cDNA target. (With permission from American Chemical Society)

stable in the presence of oligonucleotides at different pH values. From the plot of fluorescence intensity versus DNA concentration (Fig. 18D), their increasing concentrations caused a linear reduction in the fluorescence intensity of QDs at the target 1 concentration range of 0~200 pM. Furthermore, noncDNA target was used as a control, and after addition of noncDNA target, little fluorescence change can be observed (Fig. 18C), indicating no CNT-QD assembly can be formed in the presence of noncDNA . In this experiment, the available DNA detection range is from 0 to 200 pM, and according to the measurement results on the samples of gradually diluted DNA targets, the available good repeatability limit of detection is as low as ~0.2 pM (Fig. 18D).

6.8 Simultaneous three DNA target detection based on CNT-QD assembly system (System 2)

Three QDs with different emission wavelengths at 510, 555, and 600 nm were used as probes to detect three target DNA molecules simultaneously (called QD_{510}, QD_{555}, and QD_{600} probes). Their spectrally resolved fluorescence signal is displayed in Fig. 19A. As can be seen, the fluorescence signal can be split into three bands with emission wavelengths of 510 (QD_{510}), 555 (QD_{555}), and 600 nm (QD_{600}). Plotting the fluorescence intensities for the three wavelengths, that is, $\lambda = 510$, $\lambda = 555$, and $\lambda = 600$ nm, respectively, against the concentration of DNA targets results in the linear decrease of fluorescence intensity given in Fig. 19B-F. In Fig. 19B, we used a noncDNA target as control experiment, no fluorescence changes of the supernatant after incubation of noncDNA can be observed, indicating no CNT-QD hybrid formation in this system. By adding three cDNA targets all together to the CNT-DNA and QD-DNA probe solution (six-probe solution), we found a fluorescence decrease at the three wavelengths (510, 555, and 600 nm; Figure 19C), indicating CNT-QD hybrids formed in all the probes corresponding to three different wavelengths. In Figure 19D, only one cDNA target (sequence shown in Scheme 3, system 2, CNT-QD_{510}) was incubated with the six-probe solution, and we found fluorescence decreased only at 510 nm, and no fluorescence changes were found at 555 and 600 nm, indicating there is only CNT-QD_{510} hybrid formed, but no CNT-QD_{555} and CNT-QD_{600} hybrids. Similar results can be seen from Fig. 19E and F. Only CNT-QD_{555} and CNT-QD_{600} hybrids were formed upon the addition of only corresponding cDNA target, respectively. Briefly, we demonstrated the CNT-QD system can be used as simultaneous multicomponent detection system with detection limits of ~0.2 pM for each DNA target.

From Fig. 19C-F and Fig. 18D, we noticed that fluorescence intensity is 60 at [DNA] = 0, but on the other hand, fluorescence intensity was 0 at [DNA] = 225 pM. Although at the fluorescence intensity= 0, the amount of target DNA is 2 μL x 225 pM = 4.5×10^{-16} mol. In the CNT-DNA probe and QD-DNA solution, [DNA] = 1.0 nM and volume = 50 μL, so the DNA amount in the CNT-DNA probe or QD-DNA probe solution is 50 μL x 0.1 nM = 5×10^{-15} mol. From Figure 18D, if we assumed all QD particles were binding to the CNT surface at the point of fluorescence intensity = 0, we concluded that the QD number should be equal to the DNA amount (i.e., 4.5×10^{-16}), so the number of DNA molecules on each QD particle is: 5×10^{-15} mol/4.5×10^{-16} = 11. That is to say, there are 11 DNA molecules on each QD, and only 1 DNA molecule performed DNA hybridization to produce the CNT-QD hybrid. But we could not calculate the number of DNA probes on each CNT, since the lengths of the CNT were not known.

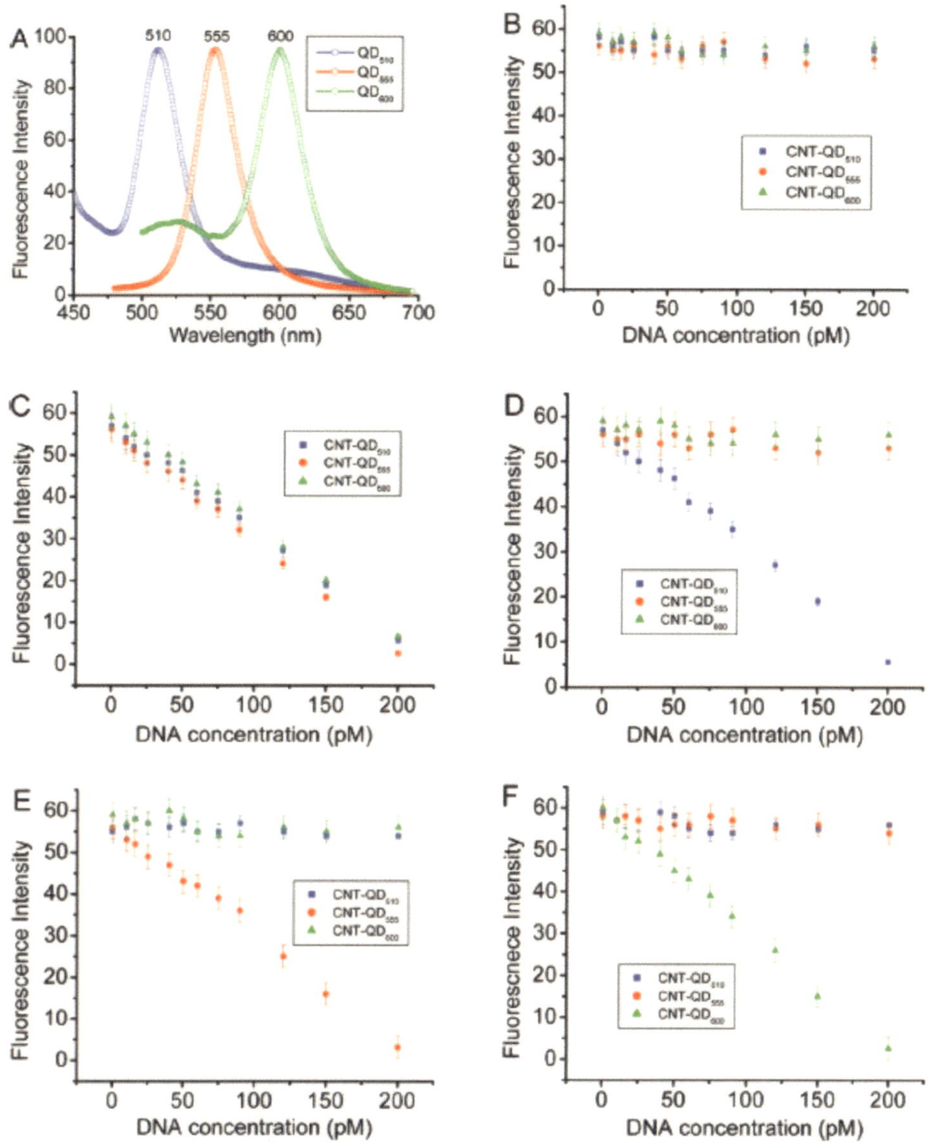

Fig. 19. Sensitivity and linearity analysis of the three-component CNT-QD DNA detection system by plotting fluorescence intensity against target DNA concentration. (A) QD probes with three different fluorescence wavelengths at 510, 555, and 600 nm; (B) plotting of supernatant fluorescence intensity against DNA concentration upon incubation of noncDNA target with CNT-DNA and QD-DNA probes; (C) fluorescence decreased at the three peaks 510, 555, and 600 nm upon the adding three cDNA targets to the six-probe solution; (D) only the 510 nm band decreased by adding the only DNA target corresponding to CNT-QD510 system; and (E, F) fluorescence decreases at 555 and 600 when cDNA corresponding to CNT-QD_{555} and CNT-QD_{600} system, respectively, is added. (With permission from American Chemical Society)

According to Fig. 18D and Fig. 19C-F, the QD fluorescence intensity (*F*) scales with the DNA concentration [DNA] (pM) through the following:

$$F = -26.7\,[\text{DNA}] + 60 \tag{3}$$

6.9 CNT-QD assembly for antigen detection via antigen-antibody (Ag-Ab) immunoreaction (System 3)

The purpose of the CNT-QD assembly system is to ultrasensitively detect target biomolecules, trying to confirm that the CNT-QD ultrasensitive detection system can be used not only for a cDNA target, but also for an antigen-antibody system. The formation of the CNT-QD hybrid via immunoreaction of an antigen with CNT-Ab and QD-Ab probes can be demonstrated from the fluorescence decrease of unbound QD fluorescence, as shown in Fig. 20. Similar to System 1, we used the QD particles with fluorescence emission wavelength at 570 nm. Upon incubating with target BRCAA1 antigen, CNT-QD hybrid was removed by a simple centrifugation step, and supernatant QD fluorescence intensity decreased linearly at an antigen concentration ranging from 0 to 1.0 nM (Fig. 20a). By measuring the gradually diluted samples, we found that we can obtain good repeatability results within the concentration scope of more than ~0.01 nM, which is also located within the scope of linearity between the QDs fluorescence and BRCAA1 concentration. For the control experiment, in the absence of the target antigen, no fluorescence decrease was observed from Fig. 20b.

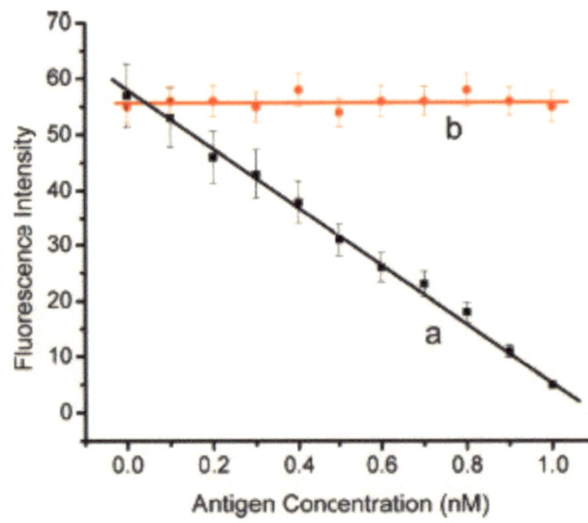

Fig. 20. CNT-QD assembly system for antigen detection. (a) Linear plot of fluorescence intensity in supernatant against antigen concentration and (b) control experiment were conducted in the absence target antigen. (With permission from American Chemical Society)

From Fig. 20, we noticed that the fluorescence intensity is 0 at the point of the target [Ag] = 1.1 nM; that is, the amount of antigen is 1.1 nM x 10 μL = 1.1×10^{-14} mol. In CNT-Ab and QD-Ab probes, the antibody amount is 1.0 nM x 50 μL = 5×10^{-14} mol. The antibody number on each QD particle is (5×10^{-14} mol) / (1.1×10^{-14} mol) = 4.5, so we concluded that 4.5 antibody molecules are on each QD particle, and only 1 antibody on QD bound to the antigen to form the CNT-QD hybrid.

According to Fig. 20, the QD fluorescence intensity (F) scales with the antigen concentration [Ag] (nM) through the following:

$$F = -51.8\,[Ag] + 57 \quad (4)$$

As a comparison, the ELISA detection method for BRCAA1 antigen was performed according to the testing procedure for antigen-antibody interaction. The novel CNT-QD hybrids method reported herein was compared with ELISA (Cui et al., 2004a; Medintz et al., 2005; Pan et al., 2006a). BRCAA1 antigen samples were especially selected to be used for the immunoassay using the ELISA methods, and the result was ~0.5 nM. In the case of protein detection, BRCAA1 antigen with a concentration < 0.5 nM cannot be detected by using the ELISA methods. In the CNT-QD method, BRCAA1 antigen can be detected with good repeatability at ~0.1 nM below the detection limit. The results suggest the excellent CNT-QD method has a higher sensitivity than ELISA. The protein detection limit is ~0.1 ppm by using a Dot-Blot fluorescent staining method (Yamada et al., 2004). In the case of oligonucleotide detection, the detection limit of the colorimetric polynucleotide detection method (Elghanian et al., 1997) is the same as that of our method. Bio-Barcodes assays have been studied and work comparably well over the 20~700 nM target concentration range (Nam et al., 2002; Nam et al., 2003). Therefore, quantification and detection of DNA/protein can be performed with higher accuracy and sensitivity.

Regarding the potential mechanism, the CNT probe and QD probe can form the nanocomposites under the existence of the complementary target oligonucleotides; the distance between CNT and QD highly depends on the length of the oligonucleotides, as observed in experiments; CNTs can quench the fluorescence signal of QDs; and the dynamic quenching and PL resonance energy transfer between CNTs and QDs should be responsible for the phenomena. When the CNT-QD method is used to detect the DNA or antigen molecules, the ratio of the CNT and QD probes is very important. According to our experience, 1:1 is suitable for almost all detected samples. When the sample concentration is lower than the concentration of the CNT and QD probes, part CNT probe and QD probe will be redundant because the distance between uncomplementary CNT probe and QD probe is far more than 10 nm. Therefore, the quenching degree of the QD probe caused by the CNT probe is much less, and the quenching degree can be detected as a control group. By measuring these gradually diluted samples with known concentration, the dose-effects standard curve can be set up before the samples with unknown concentration are detected. Therefore, the redundant CNT probes do not affect the final result of the detected samples.

Regarding the specificity and efficiency of the CNT-QD method, as is known, oligonucleotide hybridization has been broadly used for genetic diagnosis, chip detection, and so on. Its specificity and hybridization efficiency have been confirmed. Therefore, in this work, we did not focus on investigating the specificity and efficiency of the CNT-QD method. Our further work will evaluate its specificity and dynamic efficiency based on one-, two-, and three-base-pair mismatch probes and various temperature conditions as well as large quantities of background genomic and oligonucleotide existence as nonselective matrix effects.

Regarding the stability of the CNT and QD probes, asODN modified CNTs can enhance markedly the water solubility and dispersability of CNTs as reported (Zheng et al., 2003). We have also observed that the CNT probe is very stable at room temperature in PBS buffer for several months. The asODN-modified CdTe QDs are also water-soluble and very stable. Their PL intensities do not change at room temperature and in a dark environment for almost 2 years, and are almost not affected by different pH values, as reported (Pan et al.,

2008). Conversely, the PL intensity of unmodified QDs can be affected seriously by different pH values.

7. Conclusions and future prospects

In this section, we firstly designed a novel facile strategy to assemble CNTs and QDs in common aqueous solution based on a simple electrostatic interaction between amine-terminated CNTs and carboxyl-capped CdSe QDs. Individual QD was separately coated on the CNT surface to form CNT-QD nanocomplexes; as a result, PL of QDs can be quenched by CNT. Increasing of the CNT concentration led to the decrease of the PL intensity of QDs. The Stern-Volmer equation and doublelogarithmic equation were successfully used to calculate the quenching constant (K_{SV} and K_b) of CNT/QD systems, indicating that the interparticle space is the main parameter that determines the PL quenching constant, which can be explained by dynamic quenching and energy-transfer quenching mechanisms. Furthermore, quenching constants (K_{SV} and K_b) show little changes by varying pH values from 4.0 to 10.4, indicating a high quenching stability in different pH environments. Our results demonstrated that the dependent emissive properties of QDs coupled with CNTs opens a straightforward methodology for investigating the interaction between fluorescent molecules and other nanomaterials and further application in optical chemical sensors.

Based on the above novel facile strategy to assemble CNTs and QDs, we developed a novel and efficient DNA/protein detection method for biomolecules (such as oligonucleotide and antigen), that is based on DNA hybridization and antibody-antigen immunoreaction by using asODN-labeled CNTs and QDs as the molecular probes. The CNTs and QDs probes functionalized with alkyloligonucleotide can be assembled into a nanohybrid structure upon the addition of a target oligonucleotide. This strategy based on oligonucleotide hybridization assembly can be used to prepare multicomponent assembly materials comprising differently shaped oligonucleotide-functionalized nanomaterials. We have also developed a promising nanoscale CNTs probe and QDs probe for the direct, rapid, inexpensive, and sensitive detection and quantification of DNA/protein. Our probe combines the DNA hybridization and antigen-antibody interaction and is versatile and capable of simultaneous processing of multiple samples. The established method has great potential in applications such as ultrasensitive pathogen DNA or antigen or antibody detection, molecular imaging, and photoelectrical biosensors.

Based on the effects of MWNTs on the PL properties of CdSe QDs showed that CNTs could suppress the PL of QDs through both dynamic and energy transfer quenching mechanisms, a novel ultrasensitive DNA or antigen detection strategy by the CNT-QD assembly was designed for the direct, rapid, inexpensive, and sensitive detection and quantification of DNA/protein. Due to the DNA hybridization and antigen-antibody interaction and is versatile, simultaneous processing of multiple samples and higher assay throughput can be achieved. Nanotechnology provides a great opportunity to analytical chemists to develop better sensing strategies, but also relies on modern analytical techniques to pave its way to practical applications.

To sum up, nanomaterials are opening new horizons in the development of biosensor devices for simultaneous detection and measurement of specific multi-DNAs and antigens. Those biosensor devices could be useful for diagnosing and monitoring infectious diseases, monitoring the pharmokinetics of drugs, detecting cancer and disease biomarkers, analyzing breath, urine and blood for drugs of abuse, detecting biological and chemical warfare agents, and monitoring pathogens in food, among other conceivable applications. The unique and attractive properties of nanomaterials can markedly improve the sensitivity, selectivity, specificity and rapidity of biomolecules detection, offer the promising capability of detecting or manipulating atoms and molecules, and have great potential in the development of the miniaturizability or portability analytical system.

8. Acknowledgements

This work is supported by the National Key Basic Research Program (973 Project) (2010CB933901 and 2010CB93302), National 863 Hi-tech Project (2007AA022004), Important National Science & Technology Specific Projects (2009ZX10004-311), National Natural Scientific Fund No.20803040), Special project for nano-technology from Shanghai (No.1052nm04100), New Century Excellent Talent of Ministry of Education of China (NCET-08-0350), Shanghai Science and Technology Fund (10XD1406100) and Shanghai Jiao Tong University Innovation Fund for Postgraduates.

9. References

Abu-Salah, K.M., Alrokyan, S.A., Khan, M.N., Ansari, A.A., 2010. Nanomaterials as analytical tools for genosensors. Sensors 10, 963-993.

Alivisatos, A.P., 1996. Semiconductor clusters, nanocrystals, and quantum dots. Science 271, 933.

Alivisatos, P., 2003. The use of nanocrystals in biological detection. Nature Biotechnology 22, 47-52.

Ao, L., Gao, F., Pan, B., He, R., Cui, D., 2006. Fluoroimmunoassay for antigen based on fluorescence quenching signal of gold nanoparticles. Analytical chemistry 78, 1104-1106.

Asefa, T., Duncan, C.T., Sharma, K.K., 2009. Recent advances in nanostructured chemosensors and biosensors. Analyst 134, 1980-1990.

Cataldo, F., 2008. Medicinal chemistry and pharmacological potential of fullerenes and carbon nanotubes. Springer Verlag.

Chan, W.C.W., Maxwell, D.J., Gao, X., Bailey, R.E., Han, M., Nie, S., 2002. Luminescent quantum dots for multiplexed biological detection and imaging. Current Opinion in Biotechnology 13, 40-46.

Chan, W.C.W., Nie, S., 1998. Quantum dot bioconjugates for ultrasensitive nonisotopic detection. Science 281, 2016.

Chen, R.J., Bangsaruntip, S., Drouvalakis, K.A., Wong Shi Kam, N., Shim, M., Li, Y., Kim, W., Utz, P.J., Dai, H., 2003. Noncovalent functionalization of carbon nanotubes for highly specific electronic biosensors. Proceedings of the National Academy of Sciences of the United States of America 100, 4984.

Cheung, C.L., Wong, S.S., Joselevich, E., Woolley, A.T., Lieber, C., 1998. Covalently functionalized nanotubes as nanometer-sized probes in chemistry and biology. Nature 394, 52-55.

Clapp, A.R., Medintz, I.L., Mauro, J.M., Fisher, B.R., Bawendi, M.G., Mattoussi, H., 2004. Fluorescence resonance energy transfer between quantum dot donors and dye-labeled protein acceptors. Journal of the American Chemical Society 126, 301-310.

Cooper, M.A., 2002. Optical biosensors in drug discovery. Nature Reviews Drug Discovery 1, 515-528.

Cui, D., 2007. Advances and prospects on biomolecules functionalized carbon nanotubes. Journal of Nanoscience and Nanotechnology, 7 4, 1298-1314.

Cui, D., Han, Y., Li, Z., Song, H., Wang, K., He, R., Liu, B., Liu, H., Bao, C., Huang, P., 2009. Fluorescent magnetic nanoprobes for in vivo targeted imaging and hyperthermia therapy of prostate cancer. Nano Biomed Eng 1, 61-74.

Cui, D., Jin, G., Gao, T., Sun, T., Tian, F., Estrada, G.G., Gao, H., Sarai, A., 2004a. Characterization of BRCAA1 and its novel antigen epitope identification. Cancer Epidemiology Biomarkers & Prevention 13, 1136.

Cui, D., Pan, B., Zhang, H., Gao, F., Wu, R., Wang, J., He, R., Asahi, T., 2008. Self-assembly of quantum dots and carbon nanotubes for ultrasensitive DNA and antigen detection. Analytical chemistry 80, 7996-8001.

Cui, D., Tian, F., Kong, Y., Titushikin, I., Gao, H., 2004b. Effects of single-walled carbon nanotubes on the polymerase chain reaction. Nanotechnology 15, 154.

Cui, D., Zhang, H., Sheng, J., Wang, Z., Toru, A., He, R., Gao, F., Cho, H.S.C.S., Huth, C., Hu, H., 2010. Effects of CdSe/ZnS Quantum Dots covered Multi-walled Carbon Nanotubes on Murine Embryonic Stem Cells. Nano Biomed Eng 2, 246-256.

Dai, H., 2002. Carbon nanotubes: synthesis, integration, and properties. Accounts of chemical research 35, 1035-1044.

Drbohlavova, J., Adam, V., Kizek, R., Hubalek, J., 2009. Quantum dots¡ªcharacterization, preparation and usage in biological systems. International Journal of Molecular Sciences 10, 656.

Edgar, R., McKinstry, M., Hwang, J., Oppenheim, A.B., Fekete, R.A., Giulian, G., Merril, C., Nagashima, K., Adhya, S., 2006. High-sensitivity bacterial detection using biotin-tagged phage and quantum-dot nanocomplexes. Proceedings of the National Academy of Sciences of the United States of America 103, 4841.

Elghanian, R., Storhoff, J.J., Mucic, R.C., Letsinger, R.L., Mirkin, C.A., 1997. Selective colorimetric detection of polynucleotides based on the distance-dependent optical properties of gold nanoparticles. Science 277, 1078.

Fan, L.J., Jones Jr, W.E., 2006. Studies of Photoinduced Electron Transfer and Energy Migration in a Conjugated Polymer System for Fluorescence Turn-On¡± Chemosensor Applications. The Journal of Physical Chemistry B 110, 7777-7782.

Gao, X., Yang, L., Petros, J.A., Marshall, F.F., Simons, J.W., Nie, S., 2005. In vivo molecular and cellular imaging with quantum dots. Current Opinion in Biotechnology 16, 63-72.

Genin, E., Carion, O., Mahler, B., Dubertret, B., Arhel, N., Charneau, P., Doris, E., Mioskowski, C., 2008. CrAsH- Quantum Dot Nanohybrids for Smart Targeting of Proteins. Journal of the American Chemical Society 130, 8596-8597.

Grzelczak, M., Correa-Duarte, M.A., Salgueirino-Maceira, V., Giersig, M., Diaz, R., Liz-Marzan, L.M., 2006. Photoluminescence Quenching Control in Quantum Dot-Carbon Nanotube Composite Colloids Using a Silica-Shell Spacer. Advanced Materials 18, 415-420.

Guldi, D.M., Rahman, G.M.A., Sgobba, V., Kotov, N.A., Bonifazi, D., Prato, M., 2006. CNT-CdTe versatile donor-acceptor nanohybrids. Journal of the American Chemical Society 128, 2315-2323.

Guldi, D.M., Zilbermann, I., Anderson, G., Kotov, N.A., Tagmatarchis, N., Prato, M., 2005. Nanosized inorganic/organic composites for solar energy conversion. Journal of Materials Chemistry 15, 114-118.

Guo, Y., Shi, D., Cho, H., Dong, Z., Kulkarni, A., Pauletti, G.M., Wang, W., Lian, J., Liu, W., Ren, L., 2008. In vivo Imaging and Drug Storage by Quantum©\Dot©\Conjugated Carbon Nanotubes. Advanced Functional Materials 18, 2489-2497.

Han, M., Gao, X., Su, J.Z., Nie, S., 2001. Quantum-dot-tagged microbeads for multiplexed optical coding of biomolecules. Nature Biotechnology 19, 631-635.

Hatchett, D.W., Josowicz, M., 2008. Composites of intrinsically conducting polymers as sensing nanomaterials. Chemical reviews 108, 746-769.

Hu, L., Zhao, Y.L., Ryu, K., Zhou, C., Stoddart, J.F., Gruner, G., 2008. Light-Induced Charge Transfer in Pyrene/CdSe-SWNT Hybrids. Advanced Materials 20, 939-946.

Huang, P., Wang, K., Pandoli, O., Zhang, X., Gao, F., Shao, J., You, X., He, R., Song, H., Cui, D., 2010. Fluoroimmunoassay for antigen based on fluorescence quenching between quantum dots and gold nanoparticles (Proceedings Paper).

Hwang, S.H., Moorefield, C.N., Wang, P., Jeong, K.U., Cheng, S.Z.D., Kotta, K.K., Newkome, G.R., 2006. Dendron-tethered and templated CdS quantum dots on single-walled carbon nanotubes. Journal of the American Chemical Society 128, 7505-7509.

Iijima, S., 1991. Helical microtubules of graphitic carbon. Nature 354, 56-58.

Jianrong, C., Yuqing, M., Nongyue, H., Xiaohua, W., Sijiao, L., 2004. Nanotechnology and biosensors. Biotechnology advances 22, 505-518.

Kamat, P.V., 2007. Meeting the clean energy demand: Nanostructure architectures for solar energy conversion. The Journal of Physical Chemistry C 111, 2834-2860.

Katz, E., Willner, I., 2004. Biomolecule-Functionalized Carbon Nanotubes: Applications in Nanobioelectronics. ChemPhysChem 5, 1084-1104.

Krishna, M.V.R., Friesner, R., 1991. Quantum confinement effects in semiconductor clusters. The Journal of Chemical Physics 95, 8309.

Kuhn, H., 1994. Reflections on biosystems motivating supramolecular engineering. Biosensors and Bioelectronics 9, 707-717.

Kumar, C.S.S.R., 2007. Nanomaterials for biosensors. Vch Verlagsgesellschaft Mbh.

Li, Q., Sun, B., Kinloch, I.A., Zhi, D., Sirringhaus, H., Windle, A.H., 2006a. Enhanced self-assembly of pyridine-capped CdSe nanocrystals on individual single-walled carbon nanotubes. Chemistry of materials 18, 164-168.

Li, W., Gao, C., Qian, H., Ren, J., Yan, D., 2006b. Multiamino-functionalized carbon nanotubes and their applications in loading quantum dots and magnetic nanoparticles. J. Mater. Chem. 16, 1852-1859.

Li, Z., Huang, P., He, R., Lin, J., Yang, S., Zhang, X., Ren, Q., Cui, D., 2010a. Aptamer-conjugated dendrimer-modified quantum dots for cancer cell targeting and imaging. Materials Letters 64, 375-378.

Li, Z., Huang, P., Lin, J., He, R., Liu, B., Zhang, X., Yang, S., Xi, P., Ren, Q., 2010b. Arginine-Glycine-Aspartic Acid-Conjugated Dendrimer-Modified Quantum Dots for Targeting and Imaging Melanoma. Journal of Nanoscience and Nanotechnology 10, 4859-4867.

Lin, Y., Taylor, S., Li, H., Fernando, K.A.S., Qu, L., Wang, W., Gu, L., Zhou, B., Sun, Y.P., 2004. Advances toward bioapplications of carbon nanotubes. J. Mater. Chem. 14, 527-541.

Liu, C., 2009. Research and development of nanopharmaceuticals in China. Nano Biomedicine and Engineering 1, 1-12.

Liu, Z., Tabakman, S., Welsher, K., Dai, H., 2009. Carbon nanotubes in biology and medicine: in vitro and in vivo detection, imaging and drug delivery. Nano research 2, 85-120.

Maurel, V., Laferri¨re, M., Billone, P., Godin, R., Scaiano, J., 2006. Free radical sensor based on CdSe quantum dots with added 4-amino-2, 2, 6, 6-tetramethylpiperidine oxide functionality. The Journal of Physical Chemistry B 110, 16353-16358.

Medintz, I.L., Clapp, A.R., Mattoussi, H., Goldman, E.R., Fisher, B., Mauro, J.M., 2003. Self-assembled nanoscale biosensors based on quantum dot FRET donors. Nature materials 2, 630-638.

Medintz, I.L., Uyeda, H.T., Goldman, E.R., Mattoussi, H., 2005. Quantum dot bioconjugates for imaging, labelling and sensing. Nature materials 4, 435-446.

Nam, J.M., Park, S.J., Mirkin, C.A., 2002. Bio-barcodes based on oligonucleotide-modified nanoparticles. Journal of the American Chemical Society 124, 3820-3821.

Nam, J.M., Thaxton, C.S., Mirkin, C.A., 2003. Nanoparticle-based bio-bar codes for the ultrasensitive detection of proteins. Science 301, 1884.

Niemeyer, C.M., 2001. Nanoparticles, proteins, and nucleic acids: biotechnology meets materials science. Angew. Chem. Int. Ed 40, 4128-4158.

Nirmal, M., Brus, L., 1999. Luminescence photophysics in semiconductor nanocrystals. Accounts of chemical research 32, 407-414.

Oh, E., Hong, M.Y., Lee, D., Nam, S.H., Yoon, H.C., Kim, H.S., 2005. Inhibition assay of biomolecules based on fluorescence resonance energy transfer (FRET) between quantum dots and gold nanoparticles. Journal of the American Chemical Society 127, 3270-3271.

Ornberg, R.L., Harper, T.F., Liu, H., 2005. Western blot analysis with quantum dot fluorescence technology: a sensitive and quantitative method for multiplexed proteomics. Nature Methods 2, 79-81.

Pan, B., Ao, L., Gao, F., Tian, H., He, R., Cui, D., 2005. End-to-end self-assembly and colorimetric characterization of gold nanorods and nanospheres via oligonucleotide hybridization. Nanotechnology 16, 1776.

Pan, B., Cui, D., Gao, F., He, R., 2006a. Growth of multi-amine terminated poly (amidoamine) dendrimers on the surface of carbon nanotubes. Nanotechnology 17, 2483.

Pan, B., Cui, D., He, R., Gao, F., Zhang, Y., 2006b. Covalent attachment of quantum dot on carbon nanotubes. Chemical physics letters 417, 419-424.

Pan, B., Cui, D., Ozkan, C.S., Ozkan, M., Xu, P., Huang, T., Liu, F., Chen, H., Li, Q., He, R., 2008. Effects of carbon nanotubes on photoluminescence properties of quantum dots. The Journal of Physical Chemistry C 112, 939-944.

Pan, B., Cui, D., Xu, P., Ozkan, C., Feng, G., Ozkan, M., Huang, T., Chu, B., Li, Q., He, R., 2009. Synthesis and characterization of polyamidoamine dendrimer-coated multi-walled carbon nanotubes and their application in gene delivery systems. Nanotechnology 20, 125101.

Pan, B., Gao, F., He, R., Cui, D., Zhang, Y., 2006c. Study on interaction between poly (amidoamine) dendrimer and CdSe nanocrystal in chloroform. Journal of colloid and interface science 297, 151-156.

Pandey, P., Datta, M., Malhotra, B., 2008. Prospects of nanomaterials in biosensors. Analytical Letters 41, 159-209.

Patolsky, F., Gill, R., Weizmann, Y., Mokari, T., Banin, U., Willner, I., 2003. Lighting-up the dynamics of telomerization and DNA replication by CdSe-ZnS quantum dots. Journal of the American Chemical Society 125, 13918-13919.

Peng, X., Manna, L., Yang, W., Wickham, J., Scher, E., Kadavanich, A., Alivisatos, A., 2000. Shape control of CdSe nanocrystals. Nature 404, 59-61.

Pradhan, N., Goorskey, D., Thessing, J., Peng, X., 2005. An alternative of CdSe nanocrystal emitters: pure and tunable impurity emissions in ZnSe nanocrystals. Journal of the American Chemical Society 127, 17586-17587.

Robel, I., Bunker, B.A., Kamat, P.V., 2005. Single-Walled Carbon Nanotube-CdS Nanocomposites as Light-Harvesting Assemblies: Photoinduced Charge-Transfer Interactions. Advanced Materials 17, 2458-2463.

Sheehan, P.E., Whitman, L.J., 2005. Detection limits for nanoscale biosensors. Nano letters 5, 803-807.

Sheeney-Haj-Ichia, L., Basnar, B., Willner, I., 2005. Efficient generation of photocurrents by using CdS/carbon nanotube assemblies on electrodes. Angewandte Chemie International Edition 44, 78-83.

Shi, D., Lian, J., Wang, W., Liu, G., He, P., Dong, Z., Wang, L.M., Ewing, R.C., 2006. Luminescent carbon nanotubes by surface functionalization. Advanced Materials 18, 189-193.

Si, H.Y., Liu, C.H., Xu, H., Wang, T.M., Zhang, H.L., 2009. Shell-Controlled Photoluminescence in CdSe/CNT Nanohybrids. Nanoscale research letters 4, 1146-1152.

Sittampalam, G.S., Kahl, S.D., Janzen, W.P., 1997. High-throughput screening: advances in assay technologies. Current Opinion in Chemical Biology 1, 384-391.

Slowing, I.I., Trewyn, B.G., Giri, S., Lin, V.S.Y., 2007. Mesoporous silica nanoparticles for drug delivery and biosensing applications. Advanced Functional Materials 17, 1225-1236.

So, H.M., Won, K., Kim, Y.H., Kim, B.K., Ryu, B.H., Na, P.S., Kim, H., Lee, J.O., 2005. Single-walled carbon nanotube biosensors using aptamers as molecular recognition elements. Journal of the American Chemical Society 127, 11906-11907.

Spichiger-Keller, U.E., 1998. Chemical Sensors and Biosensors for Medical and BiologicalApplications. Data Processing 1, 1.

Strehlitz, B., Nikolaus, N., Stoltenburg, R., 2008. Protein detection with aptamer biosensors. Sensors 8, 4296-4307.

Turner, A.P.F., 2000. Biosensors--sense and sensitivity. Science 290, 1315.

Turner, A.P.F., Karube, I., Wilson, G.S., 1987. Biosensors: fundamentals and applications. Oxford University Press, USA.

Valcarcel, M., Cardenas, S., Simonet, B., 2007. Role of carbon nanotubes in analytical science. Analytical chemistry 79, 4788-4797.

Valcarcel, M., Simonet, B., Cardenas, S., 2008. Analytical nanoscience and nanotechnology today and tomorrow. Analytical and Bioanalytical Chemistry 391, 1881-1887.

Valcarcel, M., Simonet, B.M., Cardenas, S., Suarez, B., 2005. Present and future applications of carbon nanotubes to analytical science. Analytical and Bioanalytical Chemistry 382, 1783-1790.

Vamvakaki, V., Chaniotakis, N.A., 2007. Carbon nanostructures as transducers in biosensors. Sensors and Actuators B: Chemical 126, 193-197.

Wang, X., Guo, X., 2009. Ultrasensitive Pb2+ detection based on fluorescence resonance energy transfer (FRET) between quantum dots and gold nanoparticles. Analyst 134, 1348-1354.

Xing, Y., Chaudry, Q., Shen, C., Kong, K.Y., Zhau, H.E., Chung, L.W., Petros, J.A., O'Regan, R.M., Yezhelyev, M.V., Simons, J.W., 2007. Bioconjugated quantum dots for multiplexed and quantitative immunohistochemistry. Nature Protocols 2, 1152-1165.

Yamada, N., Ozawa, S., Kageyama, N., Miyano, H., 2004. Detection and quantification of protein residues in food grade amino acids and nucleic acids using a Dot-Blot fluorescent staining method. Journal of agricultural and food chemistry 52, 5329-5333.

Yang, H., Guo, Q., He, R., Li, D., Zhang, X., Bao, C., Hu, H., Cui, D., 2009. A quick and parallel analytical method based on quantum dots labeling for ToRCH-related antibodies. Nanoscale research letters 4, 1469-1474.

Yang, H., Li, D., He, R., Guo, Q., Wang, K., Zhang, X., Huang, P., Cui, D., 2010a. A Novel Quantum Dots-Based Point of Care Test for Syphilis. Nanoscale research letters 5, 875-881.

Yang, R., Tang, Z., Yan, J., Kang, H., Kim, Y., Zhu, Z., Tan, W., 2008. Noncovalent assembly of carbon nanotubes and single-stranded DNA: an effective sensing platform for probing biomolecular interactions. Analytical chemistry 80, 7408-7413.

Yang, W., Ratinac, K.R., Ringer, S.P., Thordarson, P., Gooding, J.J., Braet, F., 2010b. Carbon Nanomaterials in Biosensors: Should You Use Nanotubes or Graphene? Angewandte Chemie International Edition 49, 2114-2138.

Zhang, F., Ali, Z., Amin, F., Riedinger, A., Parak, W.J., 2010. In vitro and intracellular sensing by using the photoluminescence of quantum dots. Analytical and Bioanalytical Chemistry 397, 935-942.

Zhang, X., Guo, Q., Cui, D., 2009. Recent advances in nanotechnology applied to biosensors. Sensors 9, 1033-1053.

Zheng, M., Jagota, A., Semke, E.D., Diner, B.A., McLean, R.S., Lustig, S.R., Richardson, R.E., Tassi, N.G., 2003. DNA-assisted dispersion and separation of carbon nanotubes. Nature materials 2, 338-342.

Zhong, W., 2009. Nanomaterials in fluorescence-based biosensing. Analytical and Bioanalytical Chemistry 394, 47-59.

11

Electrochemical Biosensing with Carbon Nanotubes

Francesco Lamberti[1], Monica Giomo[1] and Nicola Elvassore[2]
[1]*Department of Chemical Engineering, University of Padova*
[2]*Department of Chemical Engineering, University of Padova*
VIMM - Venetian Institute of Molecular Medicine, Padova
Italy

1. Introduction

The nanobioelectrochemistry is a new interdisciplinary field which aims to combine the purposes of bionanotechnology with electrochemistry methodology. It focuses on the study of the electron transfer (ET) kinetics that occur at biointerfaces during redox reactions Chen et al. (2007). The ET results in a electron current that can be easily quantified allowing accurate and high sensitive measurements. These properties are extremely relevant on biological field in which the lacking of quantitative measurement is often a bottle neck in developing new processes.

The majority of biosensors is based on one or more bioactive molecules used in conjunction with an electrode. A redox reaction could be detected electrochemically by three different measurements: i) direct redox of a molecule involved in biological environments; ii) redox of a small mediator species that shuttles between the bioactive molecule and the electrode; iii) direct ET between the biomolecule redox site and the electrode (Fig. 1). Bioactive molecules are referred as such enzymes that require cofactors (as FAD or NADH) for catalytic activity. These bioactive molecules ensure high specificity due to structure recognition between enzymatic protein and substrate and high sensitivity due to high redox catalytic efficiency of cofactors. This latter mechanism of bioelectrochemical sensor is less common because it requires an intimate coupling between electrodes and biomolecules preserving their biological activity. On the other hand, direct electron transfer mechanism has intrinsic advantages with respect to the other two mechanisms because the electrochemical signal can be quantitatively related to a biological phenomenon without signal dissipation generated by the additional mediator. It is well known that the mediator can either react on the electrode or diffuse away in the bulk solution leading to a general sensitivity lowering. In this context, achieving direct ET could not be straightfarward and for this reason, modification of biomolecules or electrode surfaces through the use of novel nanostructured materials as mediator and the engineering of biointerfaces has been reported Hartmann (2005); Hernandez-Santos et al. (2002); Kohli et al. (2004); Wang (2005). The integration at nanoscale length is of paramount importance for reducing the probability of mediator charge dissipation at the interface towards the bulk. The new interface realized comprehending the system nanostructure/biomolecule can be defined as the nanostructured biointerface.

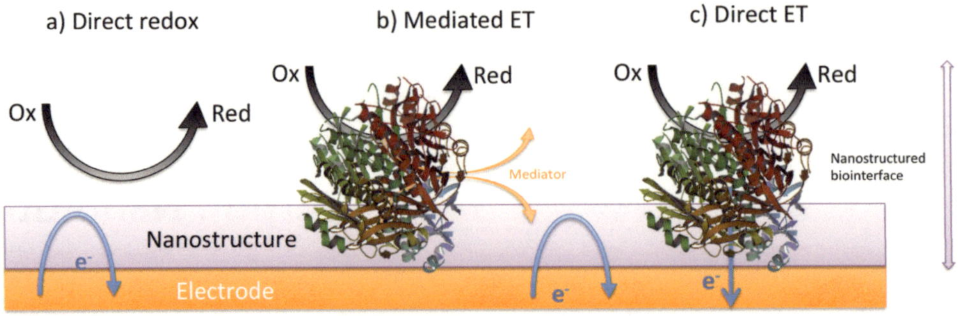

Fig. 1. Scheme of electron transfer (ET) processes on nanostructured electrode surfaces. The system biomolecule + nanostructure forms the nanostructure biointerface. a) Direct redox reaction involving bioactive molecules present in solution; b) mediated ET between electrode surface and nanostructured biointerface; c) direct ET between electrode surface and nanostructured biointerface.

Among nanomaterials, carbon nanotubes could be a perfect solution to overcome the efficiency limitation described above and for this reason are widely used for fabricating the functional biointerfaces enhancing the sensors response. CNTs are well-ordered, nanomaterials with a high aspect ratio; typical lengths are from several hundred nanometers for single-walled carbon nanotubes (SWCNTs) and several to hundreds of nanometers for coaxial multi-walled carbon nanotubes (MWCNTs) Dresselhaus et al. (1996); Smart et al. (2006). Their use is justified by recent studies that demonstrated that CNTs can enhance the electrochemical reactivity of CNT electrochemical systems Musameh et al. (2002); Zhao et al. (2002) and the ET rates of biomolecules Gooding et al. (2003); Yu et al. (2003), accumulate commonly used biomolecules Wang, Kawde & Musameh (2003), and alleviate surface fouling effects for molecules absorption in presence of complex media Musameh et al. (2002). To take advantage of these remarkable properties, CNTs need to be chemically functionalised following oxidation protocols in order to obtain ordered nanostructure interface. Vertical alignment of oxidized nanotubes on electrodes shown to be one of the most exciting and promising strategy of modification of electrodes with CNTs.

Among various attractive characteristics, it was mainly the electric properties of carbon nanotubes that stimulated large scale industrial production of CNT-based materials. However, the electronic response of individual nanotubes is reported to be sensitive to various parameters, such as the synthesis method, defects, chirality, diameter and degree of crystallinity Dresselhaus et al. (2005). It is known that solids with high aspect ratios can produce three-dimensional networks when incorporated into polymer materials. When added as well-dispersed fillers, they provide a conductive path through the composite. Therefore, carbon nanotubes were shown to increase both thermal and electric conductivities of polymers at low percolation thresholds (up to a few weight percents).

On the other hand, carbon nanotubes can be linked to metal or semiconductive surfaces in order to enhance sensitivity, specificity and usage as sensors as we will report further.

Carbon nanotubes may interact with biological environments such as proteins, DNA or neurochemicals. Electronic properties strongly affect the efficiency of interaction of nanotubes with biosystems because of nanotubes defects, metal impurities from fabrication and percentage of doping in bulk materials may alter metabolites detection while biosensing.

Recent studies demonstrated that CNTs enhance electrochemical reactivity Musameh et al. (2002); Zhao et al. (2002) but a chemical functionalization is needed in order to take advantage of these fundamental properties. As will be discussed further, covalent and non covalent functionalization methods are reported to modify interactions between electrodes and CNTs. Finally, it is note worthy to consider the induced field effect in developing carbon nanotubes based biosensors: when polarizing carbon nanotubes modified electrodes in saline media, current flows very close to the electrodic surface thus perturbing linked biosystem Alivisatos (2003). This would lead to modification of theoretical electronic nanotubes properties for application in which a single nanotube or a small quantity of CNTs is needed (Field-Effect transistors Dastagir et al. (2007); Martinez et al. (2009), nanoprobes Burns & Youcef-Toumi (2007); Wong et al. (1998), patch clamp Mazzatenta et al. (2007)).

The aim of this chapter is to review the most relevant contributions in the development of electrochemical biosensors based on carbon nanotubes (CNTs) particularly focusing on the modification of properties and possible applications arising from when a vertically alignment of CNTs is chosen. Therefore, after a brief introduction on the origin of the peculiar electronic properties of nanotubes, an in-depth study on preparation, characterization and biosensing application on vertically aligned SWCNTs modified electrodes is shown. Finally a section focused on future perspectives is provided in which we will analyze the possibility to modify existent materials with CNTs forward a bulk modification strategy. Also a section on the possibility to integrate the CNTs electrochemical devices in microfluidic platforms is presented. This latter technology allows to diminish the average dimensions of the substrates reducing cost and time of analysis and to enhance the selectivity while performing experimental investigations in high-throughput fashion.

2. Electrochemical biosensing with carbon nanotubes

The nanostructure of the nanobiointerface has several fundamental requirements: i) the thickness has to be comparable with respect to average biomolecule systems; ii) it could posses an intrinsic high conductivity to diminish any added resistance; iii) it could be chemically functionalised to assemble the nanostructured with the bioactive molecule realizing the nanobiointerface. Carbon nanotubes (CNTs) respond to all of these required features because of their tunable dimensions, their good electric properties and their easy chemistry.

Recently carbon nanotubes (CNTs) have also been incorporated into electrochemical sensors Britto et al. (1999); Campbell et al. (1999); Che et al. (1998); Luo et al. (2001); Wang et al. (2001). CNTs offer unique advantages including enhanced electronic properties, a large edge plane / basal plane ratio and a rapid electrode kinetics. In general, CNT-based sensors have higher sensitivities, lower limits of detection and faster electron transfer kinetics then traditional carbon electrodes. Many variables need to be tested and then optimized in order to create a CNT-based sensor. The performance can depend on the synthesis method of the nanotube, CNT surface modification, the method of electrode attachment and the addition of electron mediators.

Electrochemistry implies the transfer of charge from one electrode to another one. This means that at least two electrodes constitute an electrochemical cell to form a closed electrical circuit. However, a general aspect of electrochemical sensors is that the charge transport within the transducer part of the whole circuit is always electronic. By the way, the charge transport in the sample can be electron-based, ionic, or mixed. Due to the curvature of carbon graphene sheet in nanotubes, the electron clouds change from a uniform distribution around the C-C backbone in graphite to an asymmetric distribution inside and outside the cylindrical sheet of the nanotube. When the electron clouds are distorted, a rich π-electron conjugation forms outside the tube, therefore making the CNT electrochemically active Meyyappan (2005). Electron donating and withdrawing molecules such as NO_2, NH_3, and O_2 will either transfer electrons to or withdraw electrons from single-walled carbon nanotubes (SWCNTs). Thereby giving SWCNTs more charge carriers or holes, which increase or decrease the SWCNT conductance Meyyappan (2005).

Recent studies demonstrated that CNTs can enhance the electrochemical reactivity of important biomolecules Andreescu et al. (2008); Erokhrin et al. (2008); Musameh et al. (2002); Zhao et al. (2002), and can promote the electron-transfer reactions of proteins (including those where the redox center is embedded deep within the glycoprotein shell) Gooding et al. (2003); Yu et al. (2003). In addition to enhanced electrochemical reactivity, CNT-modified electrodes have been shown to be useful to accumulate important biomolecules (e.g., nucleic acids) Wang, Kawde & Musameh (2003) and to alleviate surface fouling effects (such as those involved in the NADH oxidation process) Musameh et al. (2002). The remarkable sensitivity of CNT conductivity to the surface adsorbates permits the use of CNTs as highly sensitive nanoscale sensors. These properties make CNTs extremely attractive for a wide range of electrochemical biosensors ranging from amperometric enzyme electrodes to DNA hybridization biosensors. To take advantages of the remarkable properties of these unique nanomaterials in such electrochemical sensing applications, the CNTs need to be properly functionalized and immobilized. There are different ways for confining CNT onto electrochemical transducers. Most commonly, this is accomplished using CNT coated electrodes Liu et al. (2008); Luong et al. (2004); Vairavapandian et al. (2008); Wang, Kawde & Musameh (2003); Wang, Musameh & Lin (2003) or using CNT / binder composite electrodes Rubianes & Rivas (2003); Sljukic et al. (2006); Wang & Musameh (2003). The CNTs driven electrocatalytic effects and the increasing use of modified CNTs for electroanalytical applications have been recently reviewed Vairavapandian et al. (2008).

Among the traditionally used electrode materials such as graphite, gold or mercury, CNTs showed better behavior than the others which also have good conducting ability and high chemical stability. CNT-based electrochemical transducers offer substantial improvements in the performance of amperometric enzyme electrodes, immunosensors and nucleic-acid sensing devices. The greatly enhanced electrochemical reactivity of hydrogen peroxide and NADH near the proximity or on the CNT-modified electrodes makes these nanomaterials extremely attractive for numerous oxidase- and dehydrogenase-based amperometric biosensors. For example, vertically aligned CNTs structures can act as molecular wires to allow efficient electron transfer between the underlying electrode and the redox centers of enzymes. The CNT transducer can greatly influence for enhancing the response of the biocatalytic reaction product and provide amplification platforms carrying multiple enzyme tags: it is shown that the vertical orientation is required for obtaining high ET results in such experiments.

For this reason the next section will provide a schematic view for the realization of vertically aligned SWCNTs modified electrodes particularly focusing on critical features and open issues.

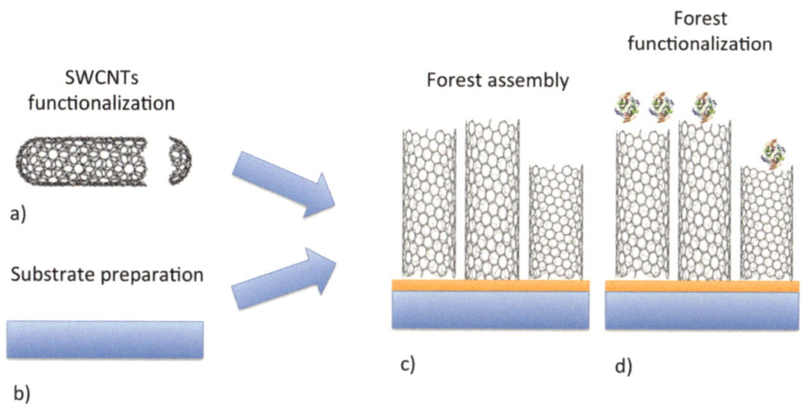

Fig. 2. Steps involving the fabrication of SWCNTs forests based sensors via chemical assembly. a) Substrate preparation; b) SWCNTs functionalization; c) Forest assembling; d) Forest functionalization.

3. Fabrication of vertically aligned SWCNTs modified electrodes

Among carbon nanotubes modified electrodes, surface modification strategy is one of the most used because of the high versatility in chemical modification of substrates and further functionalization, facile impact, low cost and low wastes. In particular, the most promising approach is to develop self-assembled monolayer of CNTs perpendicular oriented to the surface of the electrode in order to realize a forest of carbon nanotubes Diao & Liu (2010). There are a lot of works in which it is demonstrated that the vertical alignment is a good choice for assembling because it can enhance the performance of many nanotube-based devices such as emitters in panel displays Bonard et al. (1998); De Heer et al. (1995), nanoprobes as tips protrusion for optimum high resolution images collection Wong et al. (1998) and in the electrochemical biosensing field because of the good conductivity of the nanotubes, their small diameter and high aspect ratio Chou et al. (2005); Diao & Liu (2005); Diao et al. (2002); Gooding et al. (2003); Yu et al. (2006). It has also been found that a vertical orientation enhance the electron transfer reaction rates at the electrodes with respect to a random dispersion of nanotubes on the surface Chou et al. (2005). Finally forests can be simply functionalised with enzymes and specific redox reaction involving biomolecules can be achieved Liu et al. (2005); Patolsky et al. (2004).

In this chapter we will focus only on chemical self-assembly technique for forests production because, with respect to other conventional approaches (such as CVD, arc-discharge and laser ablation) it shows highly flexibility in topographical control of nanotubes vertical assembly in term of CNTs superficial density and micro and nanopatterns. In addition, conventional approaches may produce forests with endless nanotubes, randomly curled and highly tangled also requiring expensive experimental setups.

The general scheme for obtaining a biosensor based on forests of SWCNTs (the most used in this field of research) via chemical assembly, is presented in Fig. 2. Many steps are involved in the fabrication of a forest of SWCNTs:

1. Substrate preparation
2. SWCNTs functionalization

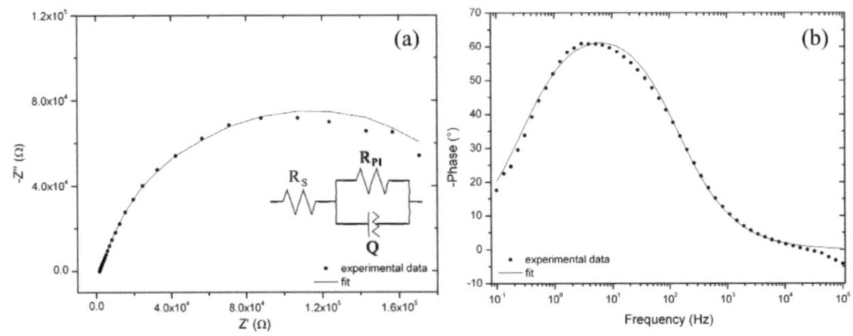

Fig. 3. Typical EI spectra for a gold electrode after electrochemical pretreatment protocol. In panel (a) it is shown the Nyqvist plot representing the semicircle typical of a clean polarized electrode in solution whereas the inset represents the model used for fitting experimental values (R_S is the uncompensated resistance solution, R_{P1} and Q the associated polarization resistance and real capacitance respectively of the gold electrode); panel (b) shows the Bode phase plot: a resonance peak for the circuit at ca. 10 Hz is found.

3. Forest assembling
4. Forest functionalization

3.1 Substrate preparation

Different substrates are chosen for preparing SWCNTs forests: different metals (gold De Heer et al. (1995); Lamberti et al. (2010); Nan et al. (2002); Patolsky et al. (2004); Sheeney-Haj-Ichia et al. (2005) and silver Wu et al. (2001)) or other materials (silicon Yu et al. (2007; 2006), glass Chattopadhyay et al. (2001); Jung et al. (2005), Nafion film Wei et al. (2006)). Substrates cleanness is of paramount importance for achieving an optimum assembling of nanotubes: organic molecules or oxides adsorbed on the surface would affect the efficiency of nanotube coupling. Normally metal substrates (gold in particular) undergo a rigid cleanness protocol that foresees first a mechanical polishing with alumina or similar, then a chemical treatment in strong mixtures like piranha solution and/or an electrochemical treatment step Carvalhal et al. (2005). It is note worthy that it is the electrochemical cleanness the most important step in the protocol: for example we have found that electrochemical impedance spectra (EIS) of clean gold surfaces after electrochemical step reveal that polycristalline gold surfaces are the cleanest and smoothest with a lesser amount of oxides (Fig. 3). Also it is known that impurities affect polycristallinity, adhesion of metals on substrates and reproducibility: for example, physical enhanced chemical vapor deposition (PECVD) deposition allows to obtain clean films because depositions are performed in UHV Lamberti et al. (2011).

3.2 SWCNTs functionalization

Carbon nanotubes prior to being covalently linked to the surface, need to be purified and functionalised because metal nanoparticles and carbonaceous impurities byproducts are still present from production step. Also it is well known that sp_2 carbon atoms of sidewall of nanotubes are more stable than the sp_2 ending atoms: for this reason, it was demonstrated by Liu et al. Liu et al. (1998) that any chemical attack of the tubes would start from the ends of the tubes and proceeds shortening the nanotubes from the defects produced shortening the tubes.

There are many purification strategies in literature such as mixed acids bath Chou et al. (2005); Diao et al. (2002), piranha solution Liu et al. (1998); Ziegler et al. (2005), ozonation Rauwald et al. (2009), phosphomolybdic acid Warakulwit et al. (2008), persulfate solution Liu et al. (2007), pyrolysis Gu et al. (2002) and electron beam irradiation Rauwald et al. (2009). Carbon nanotubes are not only cut and purified when treated in these oxidant conditions but carboxylic functions would create at damage sites: in particular it has been shown by our group that a distribution of oxygenated species would be present on the surface of nanotubes, from carboxylic functions to aldehydes Lamberti et al. (2010). We also showed that nanotubes can be shortened in a controlled way by monitoring the temperature of oxidation bath instead of the time as people normally do: in such circumstances we can monitor not only the length of the nanotubes (obtained by AFM with a standard deviation of some tens of nm maximum) but also the distribution of oxygenated species that appear at the sidewalls by XPS and Raman measurements (Fig. 4). Carboxylated nanotubes may be easily functionalized thanks to chemical versatility of -COOH group. As we will discuss after, acid groups are of paramount importance for coupling nanotubes to the electrode surfaces and to link proteins and other biomolecules in fabricating biosensors.

As final conclusion we can assume that the oxidative cutting by prolonged sonication in strong oxidation has several advantages such as ease of operation, simple equipment and no special requirements. Moreover, this oxidative method allows to obtain short nanotubes in fast times.

3.3 Forest assembling

Surface condensation method (Fig. 5 a) and b)) is the most used fabrication technique for assembling SWCNTs forests in literature because the coupling efficiency is resulted the best Diao & Liu (2010). It develops in forming firstly the self-assembled monolayer (SAM) of small bridge molecules on the substrate (alkanethiols or mercaptoalcohols) and then a reaction involving carbodiimide (CDI) reactions took place at the acid functions. SWCNTs are dissolved in dimethylformamide (DMF) or dimethyl sulfoxide (DMSO) and reaction take place at 60 °C. Usually three different CDIs are used depending on the type of solvent used for dissolving nanotubes: DCC (N,N′-Dicyclohexylcarbodiimide), EDAC (N-(3-Dimethylaminopropyl)-N′-ethylcarbodiimide) and DIPC (N,N′-Diisopropylcarbodiimide) (Fig. 5 d)). As we can see in Fig. 5 c), CDI reaction provides to the formation of an amidic/estheric bond between the C atom of the nanotube and the N (or O) atom of the SAM on the electrode. The active intermedium reagent is the O-acyl isourea that is transformed into amide when a nucleophilic substitution happens in presence of an amine or an alcohol. That is the problem because the O-acyl isourea can undergo to a intramolecular rearrangement: this byproduct, N-acyl urea, is totally stable in most common solvents (typical 5%-10% wt is dissolved). For this reason people try to minimize this unwanted reaction by stabilizing O-acyl isourea adding to reaction solution some pyridine.

Also, the urea molecules deriving from CDIs are not always soluble in common solvents: the corresponding urea for DCC, Dyciclohexylurea (DCU) is totally unsoluble in most of organic solvents and filtration is necessary to remove it; the corresponding urea for EDAC, the EDAU, is soluble in water and for this reason it is used for peptide synthesis, whereas DIPU, the corresponding urea for DIPC, is quite soluble in organic solvents and the remains are normally removed by rinsing with solvent.

It is note worthy that, for this reason, forests fabricated with DCC reagent (almost all) can

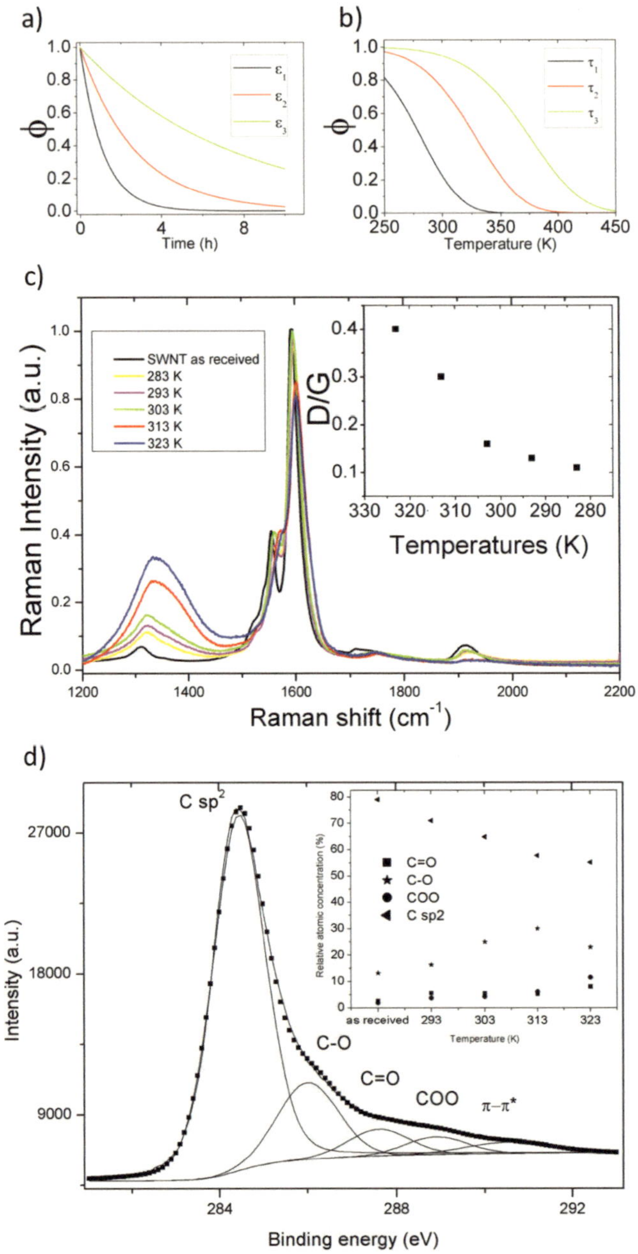

Fig. 4. a) Time-dependence of the fraction of oxidized carbon (ϕ); b) temperature-dependence of ϕ; c) Raman spectra (exciting line 633 nm) of SWCNTs oxidized at five different temperatures. Inset reports the ratio of the D/G peak intensities as a function of temperature and the dotted line indicates the D/G ratio for the pristine nanotubes; d) X-ray photoemission spectrum of carboxylated carbon nanotubes. The spectrum reveals that there are many partially oxygenated species in addition to the -COO species. The inset graph shows the temperature dependence of the C sp^2 component and the corresponding enhancing of the oxygenated species.

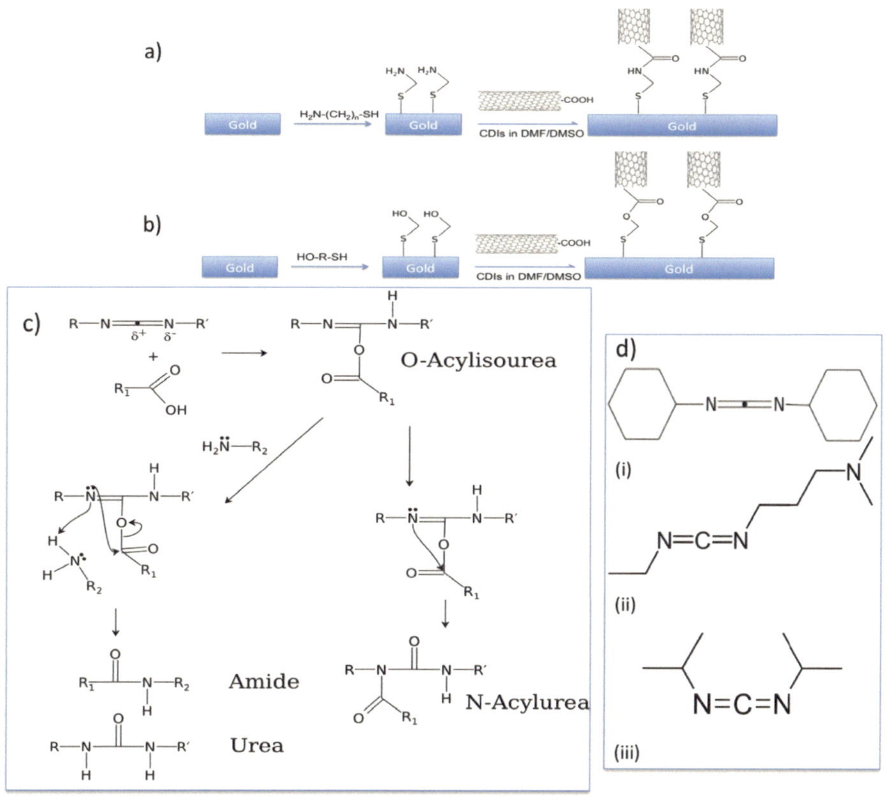

Fig. 5. a) Surface condensation method described within the text using alkanethiols or b) mercaptoalcohols. c) Chemical scheme of reactions occuring when forming amidic/estheric group in fabricating SWCNTs forests: N-Acylisourea is the unwanted by-product of the reaction that is almost unsoluble in all common solvents; d) structural formula of the existing cardodiimmides: (i) DCC (N,N'-Dicyclohexylcarbodiimide), (ii) EDAC (N-(3-Dimethylaminopropyl)-N'-ethylcarbodiimide) and (iii) DIPC (N,N'-Diisopropylcarbodiimide).

not be clean: so in our work Lamberti et al. (2010), we used DIPC instead of common DCC in order to avoid unwanted reagents on samples. In Fig. 6, AFM 3D images of SWCNTs forests at different height are presented. Vertical aligned SWCNTs have been prepared through amide or esther formation using DCC or EDAC on gold substrates Chou et al. (2005); Diao & Liu (2005); Diao et al. (2002); Gooding et al. (2007; 2003); Huang et al. (2006); Nan et al. (2002); Nkosi & Ozoemena (2008); Ozoemena et al. (2007); Patolsky et al. (2004), silicon Huang et al. (2007) and glass substrates Bonard et al. (1998); Jung et al. (2007; 2005). For substrates different from gold the only modification in the technique is the modification of the substrate: typically silicon surface are treated with alkylaminotrimethoxysilane Huang et al. (2007), a compound that contains an amino group. Before amination, substrate need to be modified with hydroxyl groups in order to realize a organosilane SAM on the surface Ulman (1996).

Other methodologies were used for fabricating SWCNTs forests. First of all, the "Au-S bonding" production scheme that implies that the thiols were previously covalently attached to the acidic ending groups of the nanotube and afterwards CNTs were put in contact

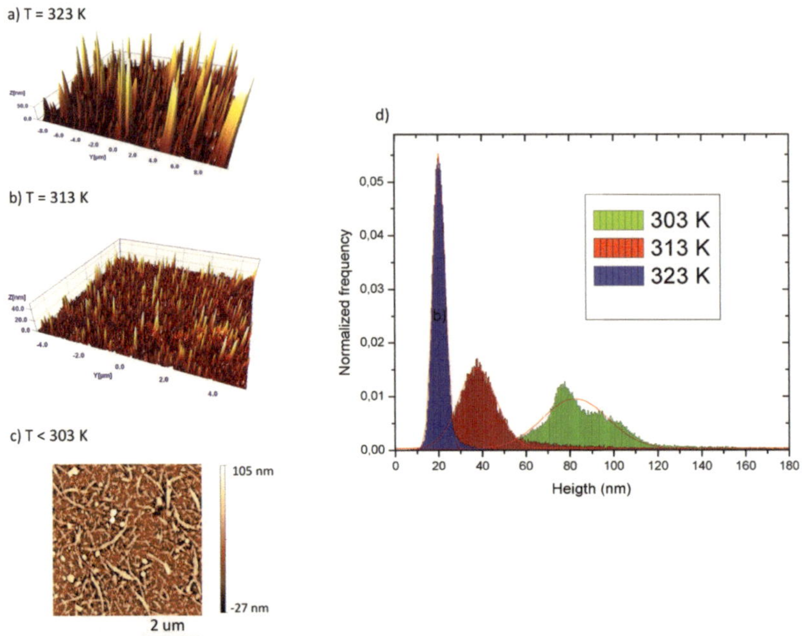

Fig. 6. Atomic force microscopy 3D image of SWNTs forests built at different temperature conditions: a) at 313; b) at 323; c) the AFM 2D image of randomly dispersed carboxylated SWNTs treated at 283 K oxidation process; d) normalized frequency of height of SWNTs measured with AFM for three different temperatures. At 283 and 293 K there is no evidence of SWNT forests.

with gold substrates allowing the formation of Au-S bond Liu et al. (2000). Van der Waals interactions between sidewalls of nanotubes are recognized as the main forces that can prevent the nanotubes to horizontal deposition. The coupling efficiency was relatively low and in order to improve the surface coverage of nanotubes people switch to surface condensation strategy and the electrostatic interaction strategy.

This latter takes advantage of the electrostatic forces between carboxylated SWCNTs and surface. This is possible because carboxylated nanotubes after oxidation are negatively charged ($-COO^-$) allowing electrostatic attraction between them and positively charged surfaces. Papadimitrakopoulos et al. Wei et al. (2008; 2006); Yu et al. (2005; 2006) develop several works in which using a metal-assisted self-assembly technique modifying electrodes with Fe^{3+}. This strategy was used to fabricate carbon nanotubes forests on different substrates such as silicon Chattopadhyay et al. (2001), Nafion-modified silicon Wei et al. (2006), glass Chattopadhyay et al. (2001), gold Wei et al. (2007) and graphite Yu et al. (2003).

4. Forest characterization

Once forests have been fabricated on different substrates, there is the problem to characterize these nanometric structures. Here we summarize the most relevant contribution to the characterization techniques used in literature: Atomic Force Microscopy (AFM) imaging, Raman spectroscopy and X-ray Photoelectron Spectroscopy (XPS). Also Fourier transform infrared spectroscopy (FTIR) Diao et al. (2002) and Quartz Crystal Microbalance (QCM)

Chattopadhyay et al. (2001) are used but they are not widely used as the above methods. Table 1 shows the principal characterization techniques used in literature and the results that can be obtained.

Technique	Result
AFM	Morphology and surface coverage
Raman	Composition and SWCNTs orientation
XPS	Composition and SWCNTs degree of difectuality
FTIR	Information of surface functionalities
QCM	Information on mass change during the assembling

Table 1. List of the main techniques used for characterizing SWCNTs forests. Effective results of each technique is also presented.

4.1 AFM imaging

AFM is considered one of the most wanted technique for the characterization of forests because of AFM provides direct imaging of the nanostructures fabricated on electrodes. AFM imaging analysis can provide the topography of the forest, understanding if nanotubes are deposited in bundles or individually by the deconvolution of the AFM tip by simple geometry considerations Diao et al. (2002); Gooding et al. (2003); Liu et al. (2000); Yu et al. (2003). AFM data also provide informations about the surface coverage, the surface distribution and height of the forests evaluating the coupling efficiency Chattopadhyay et al. (2001); Diao & Liu (2005); Diao et al. (2002); Gooding et al. (2003); Lamberti et al. (2010; 2011); Liu et al. (2005; 2000); Patolsky et al. (2004); Yu et al. (2007). Despite the fact AFM can reach a subnanometric resolution, roughness of the substrates, not ideally surfaces and thermal noise would limit AFM to identify forests whom average height is not higher than a few nanometers.

4.2 Raman spectroscopy

Carbon nanotubes are Raman active Ajayan (1999): SWCNTs are identified by radial breathing modes at Raman shift ca. 200 cm^{-1}. Also Raman spectroscopy is very sensible: only one nanotube, in principle, can be detectable. For this reason, Raman scattering can be used for monitoring or confirming the presence of nanotubes on samples but it does not give any information about the surface coverage or surface distribution of nanotubes. Nevertheless, Papadimitrakopoulos et al. Chattopadhyay et al. (2001) shown that using polarized Raman spectroscopy it would be possible to obtain informations about the orientation of the nanotubes on samples: the intensity was the highest when the polarization of the incident laser is perpendicular to the substrate and the lowest when parallel giving evidence to the vertical alignment of the deposited nanotubes.

Raman scattering also can give informations about quantity of defects as we have previously reported Lamberti et al. (2010): D-band (Raman shift 1330 cm^{-1}) intensity enhances its value increasing oxidation temperature of nanotubes i.e. the number of defects enhances when nanotubes are treated in higher oxidative temperature conditions.

4.3 XPS

X-ray Photoelectron Spectroscopy is mostly used for studying surfaces. It provides informations about the chemical bonds involving the atoms that are present on the surface of a material. For this reason, carbon atom would have a different XPS peak if linked to an N

atom or an O atom. Also, different hybridization of the same atom give different XP spectra. So, by coupling this technique to Raman spectroscopy we can obtain different informations about oxygenated species at the sidewalls when nanotubes are in an oxidized form. Lamberti et al. Lamberti et al. (2010) shown that increasing nanotube oxidation temperature the relative percentage of C sp^2 component diminishes and also, -COO contribution enhances whereas a distribution of oxygenated species is always present (phenols, pyrans, ethers, anhydrides).
XPS can also provide informations about the quantity of bonds in comparison to C-C sp^2. Therefore, it can also quantitatively describe the defect density in CNTs but if Raman measurements are available, informations about defects density in the CNTs surface can be better detected because of Raman spectroscopy is recognized as one the most suitable technique for characterizing carbon based materials. As a final conclusion, Raman and XP spectroscopies provide informations about the quantity and quality of defects in oxidized nanotubes respectively.

5. Electrochemical characterization

The knowledge of electrochemical properties of carbon nanotubes modified electrode is of paramount importance for obtaining informations about the applicability of such electrodes in sensing, biosensing, nanoelectronic devices, field emitters and nanoprobes. The goal is to study the charge transfer (i.e. the electrons' flow) between the electrode and redox species and the electrochemical response of vertical alignment in function of its features such as height and surface density. Redox species can be free-moving in solution or covalently bonded to the SWCNTs.

5.1 Redox species dissolved in solution

Before discussing the works reported in literature for this kind of study, it is note worthy to understand what steps are included in the electron path from the redox species to the underlying collecting electrode. We can assume three different steps as defined in Fig. 7: (i) the electron transfer between redox center and forest, (ii) the electron flow across the nanotube, (iii) the electron jump between nanotube and electrode. In this context, Diao et al. Diao & Liu (2005) proposed a charge transfer model based on tunneling process.
It is noted that the heterogeneous electron transfer (HET) at the open ends of SWCNTs should be remarkably more rapid than that at the sidewalls Cui, Lee, Raphael, Wiler, Hetke, Anderson & Martin (2001); Li et al. (2002). It was shown that the bridge molecule SAM blocks the electron transfer between the redox probe in solution and the underlying electrode. Also it has been demonstrated that the vertical alignment could promote the ET though there is an insulating monolayer in between them Diao et al. (2002) as described in the typical behavior in Fig. 8 (panel 1.A). This phenomenon is also confirmed in EI spectra: in Fig. 8 (panel 1.B) it is shown a Nyqvist plot representing the impedance behavior of electrodes modified with an insulating molecule and a SWCNTs forest. The insulation increases the associated polarization resistance of the circuit whereas the forest assembling decreases it.
From this preliminary consideration scientists try to study the effect in heterogeneous electron transfer (HET) kinetics in samples in which the surface coverage and height vary. Gooding et al. Chou et al. (2005); Liu et al. (2005) investigated the ET kinetics of $Fe(CN)_6^{3-/4-}$ at SWCNTs forests: he found that ET occurs much easier in vertically aligned carbon nanotubes with respect to randomly dispersed SWCNTs. Also, he correlates XPS results in determining

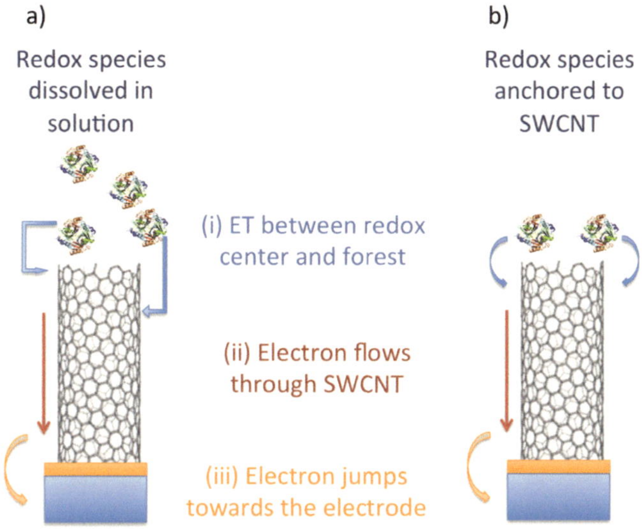

Fig. 7. Schematic representation of ET phenomena occurring in two different situations involving SWCNTs forests: with the redox center free-diffusing in solution and directly linked to the open ends of the nanotube.

that the oxygenated species at the open ends are responsible for high ET. Gooding has also shown Gooding et al. (2003) that aligned nanotubes acts as molecular wires and added no significant electrical resistance to the electron transfer process. Willner's group Patolsky et al. (2004); Sheeney-Haj-Ichia et al. (2005) assumed that defect on the sidewalls may introduce local damage in p-conjugation lowering the ET rate but because of the short height of the forests, the electrical resistance is very short and for this reason negligible. Lamberti et al. in a very recent work Lamberti et al. (2011) gave a good contribution in this sense: our work allows to show that SWCNTs forests fabricated with nanotubes of different heights have significantly different ET kinetic, i.e. shorter is the forest, higher is the ET and of course, smaller is the resistance. Fitting electrochemical impedance spectra with Randles modified cell model, let us to know ET dynamics of system. This result is motivated because forests produced with our temperature-controlled method, allow to obtain nanotubes with a narrower height distribution and more sensible data can be collected.

The electron transfer dynamics between SWCNTs and substrate is the most important step. Diao and colleges Diao & Liu (2005) shown that the adsorbed SWCNTs act as many "electron relay stations" that mediate electrons between the metal electrode and redox centers. Also it is noted that the linking bonds at the ends of the tubes ensure an high electron transfer between nanotubes and surfaces Chidsey et al. (1990); Cui, Primak, Zarate, Tomfohr, Sankey, Moore, Moore, Gust, Harris & Lindsay (2001); Cui et al. (2002); Finklea & Hanshew (1992); Koehne et al. (2004) but Gooding et al. Chou et al. (2009) reported that the electron-transfer rate at the nanotube-modified electrodes decayed exponentially with distance when the chain of the molecule bridge is increased. In this context our recent study on the combined effect on height and surface density of nanotubes give an overall point of view of HET dynamics in SWCNT forests Lamberti et al. (2011). This work starts from the result of Willner's group Patolsky et al. (2004) that found that mixing cysteamine (CYS) layer with 2-mercaptoethanol (ME) molecule to modify gold substrate, would enhance following nanotubes coupling efficiency.

It is shown that esther reaction in these experimental conditions is slower than the amide formation. Taking advantage of this result, we tried to control the surface density by choosing different ratios concentration of CYS/ME determining that a slow amount of ME is necessary to obtain highest values of ET kinetics. AFM images shown that samples with a low relative concentration of CYS would present SWCNTs forests with a broader height distribution: angle contact measurements reveal that nanotubes coupling efficiency is strongly linked to interface processes. The final aim of this study was to actually conclude the study on HET dynamics in SWCNTs forests, determining the optimum conditions for fabricating forests as used in particular for biosensing.

5.2 Redox species anchored to forests

Carbon nanotubes can be used for directly reaching redox centre in species in which ET is very low: in such circumstances, high-sensitivity electrochemical and bioelectrochemical sensing can be performed.
Redox enzymes Gooding et al. (2003); Liu et al. (2005); Patolsky et al. (2004), electroactive complexes Nkosi & Ozoemena (2008); Ozoemena et al. (2007); Yu et al. (2008) and ferrocene Flavel et al. (2009; 2008); Gooding et al. (2007); Yu et al. (2007) were rightly attached to vertical aligned SWCNTs. Ferrocene, a molecular redox probe usually used for ET studies because of its ideal Nerstian behavior, was attached by Shapter and coworkers Yu et al. (2007) an they found that the presence of nanotubes in samples modified with ferrocene with or without the forest, improve the ET. This result is very important because it suggests that vertical alignment of carbon nanotubes can transport electrons. Moreover, other works show that orientation of nanotubes affect ET properties: for randomly dispersed carbon nanotubes the ET is low with respect to vertical alignment due to difficulty of hopping for electrons from one nanotube to an other Gooding et al. (2007).
The possibility of anchoring enzymes or biomolecules would open the way to making biosensing with carbon nanotubes forests: Gooding works and Willner's are the milestones in this sense Gooding et al. (2003); Patolsky et al. (2004). Gooding et al. Gooding et al. (2003) were the first who tried to attach an enzyme to the open ends of a nanotube: they studied the ET kinetics in vertical aligned SWCNTs on which were immobilized microperoxidase (MP-11) and even if ET efficiency is dependent on the spacer thickness, this is negligible when this is in the submicrometric range as in the case of carbon nanotube spacers. Nevertheless, Willner and colleges Patolsky et al. (2004) anchored a reconstructed Glucose Oxidase (apo-GOx) to the opened ends of nanotubes and they found that nanotubes with an average shorter length would increase ET in oxidizing glucose dissolved in solution (Fig. 8). This is in contrast to the previous thesis of Gooding work in which he found that no contributes are to be recognized to length of nanotubes and further work has to be done for solving the dispute.
Willner's work also shown the possibility of using nanotubes as ET mediators for oxidizing or reducing species in solution, as demonstrated by Ozoemena et al. Ozoemena et al. (2007) who oxidize dopamine taking advantage of linking an iron complex to the open ends.
As final conclusion, we can assume that vertical alignment is the best choice for obtaining high direct ET of species in solution or anchored to the forests with respect to other strategies of nanotubes modification. Also they can act as charge transfer mediators for performing redox reactions of species in solution.

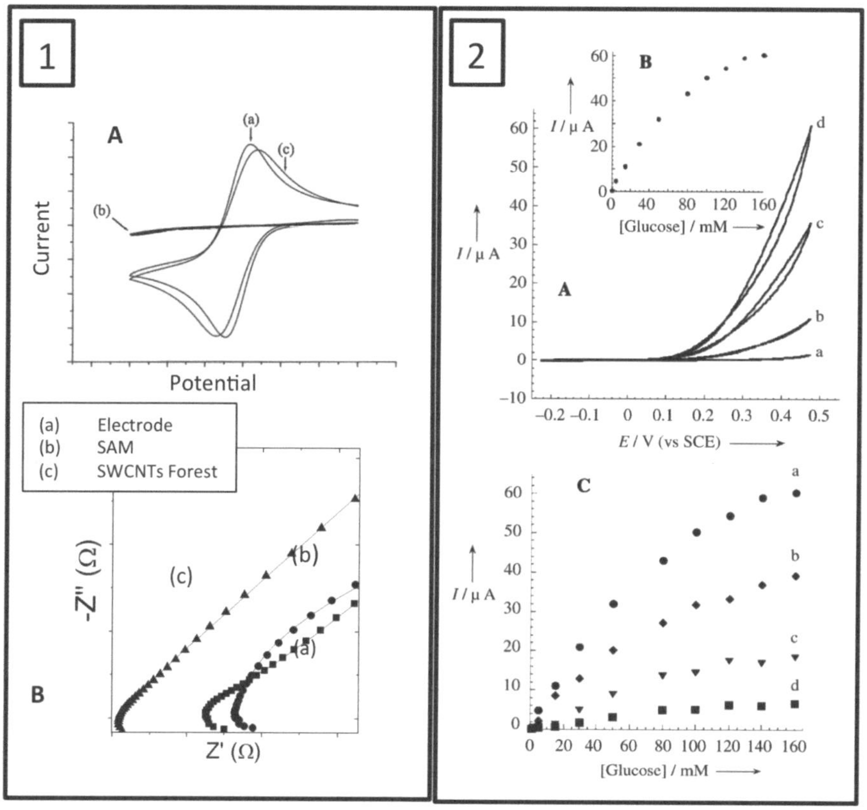

Fig. 8. Schematic view of the electrochemical measurements performed with SWCNTs forests in presence of a redox species. A. Redox species dissolved in solution: typical i) CV and ii) EIS showing the oxidation and the reduction of a redox probe on (a) an electrode, (b) a modified electrode with SAM and (c) a SWCNTs forest modified electrode. B. Redox species anchored to SWCNTs. A) Cyclic voltammograms corresponding to the electrocatalyzed oxidation of different concentrations of glucose by the GOx reconstituted on the 25 nm long FAD-functionalized CNTs assembly: a) 0 mm glucose. b) 20 mm glucose, c) 60 mm glucose, d) 160 mm glucose. Data recorded in phosphate buffer, 0.1 m, pH 7.4, scan rate 5 mVs^{-1}. B) Calibration curve corresponding to the amperometric responses of the reconstituted GOx/CNTs (25 nm) electrode (at E = 0.45 V) in the presence of different concentrations of glucose. C) Calibration curves corresponding to the amperometric responses (at E = 0.45 V) of reconstituted GOxÐCNTs electrodes in the presence of variable concentrations of glucose and different CNT lengths as electrical connector units: a) about 25 nm SWCNTs. b) about 50 nm SWCNTs. c) about 100 nm SWCNTs. d) about 150 nm SWCNTs. With permission to Patolsky et al. (2004).

6. Conclusions and future perspectives

SWCNTs vertical alignment has been demonstrated as a powerful tool for biosensing taking advantage of peculiar properties of carbon nanotubes, their chemistry and straightforward integration with biological environments. Anyway some fabrication features (need of a bridge molecule and coupling efficiency) limits the widespread application of the bidimensional strategy of materials modification.

In such direction, carbon nanotubes based electrodes should address industrial application by the realization of 3D modified materials that can increase active sites number by a bulk doping of materials or eliminate the strong dependence from commonly used noble metals based electrodes. Small quantities of carbon nanotubes could be necessary to realize conductive materials from insulating samples as it is already shown for available synthetic polymers MacDonald et al. (2005): a low degree of doping is needed in order to maintain unchanged wanted bulk properties of starting material such as biocompatibility or stiffness. In such cases, CNTs are suited for biocompatible doped materials since a low level of doping is needed with respect to commonly used fillers like graphite, metals or conductive polymeric structures particles. Also the ability of nanotubes to align following an applied external field creating an electronic percolation path would probably enhance the conductivity of doped materials and in principle further diminish the concentration of dopants. Once realized, carbon nanotube bulk doped biocompatible based devices could be inserted in human body for the realization of drug delivery systems, in vivo biosensors or tissue replacements.

SWCNTs based sensors are actually developed only to measurements in static liquid environments. SWCNTs based sensors integrated in fluidic systems would potentially allow to perform continuous on-line monitoring of multiple-analyte in order to control bioprocesses: for instance combination of glucose and lactate measurements can be related to oxygen-dependent metabolic activity. By enzyme functionalization of forests these measurements can be performed and integrated nanobiosensors can be realized: temporal sequence of dynamic processes and a high-throughput could be achieved. Integrating an electrochemical detector module into microfluidic platforms is preferable because of its inherent portability, the easy of fabrication of the microelectrodes and the lowest costs if compared with other commercial detection systems.

Electrochemical measurements only detect the electrical properties of analyte species undergoing redox reactions, so they are limited to electroactive species. The specific electrode potential can be employed to filter out compounds other than the analyte being detected. In combination with capillary electrophoresis separation, electrochemical detection often provides very good detection limits in microfluidics. Electrochemical detectors for detecting metabolic activity at the extracellular, single-cell level have recently been reviewed Yotter & Wilson (2004) and integration with carbon nanotubes based electrodes is possible.

7. References

Ajayan, P. (1999). Nanotubes from carbon, *Chemical reviews* 99(7): 1787–1800.

Alivisatos, P. (2003). The use of nanocrystals in biological detection, *Nature Biotechnology* 22(1): 47–52.

Andreescu, S., Njagi, J. & Ispas, C. (2008). *The New Frontiers of Organic and Composite Nanotechnology*, ELSEVIER.

Bonard, J., Salvetat, J., Stockli, T., de Heer, W., Forró, L. & Châtelain, A. (1998). Field emission from single-wall carbon nanotube films, *Applied physics letters* 73(7): 918–920.

Britto, P., Santhanam, K., Rubio, A., Alonso, J. & Ajayan, P. (1999). Improved charge transfer at carbon nanotube electrodes, *Advanced Materials* 11(2): 154–157.

Burns, D. & Youcef-Toumi, K. (2007). Shortening carbon nanotube-tipped afm probes, *International Journal of Nanomanufacturing* 1(6): 799–809.

Campbell, J., Sun, L. & Crooks, R. (1999). Electrochemistry using single carbon nanotubes, *Journal of the American Chemical Society* 121(15): 3779–3780.

Carvalhal, R., Sanches Freire, R. & Kubota, L. (2005). Polycrystalline gold electrodes: A comparative study of pretreatment procedures used for cleaning and thiol self-assembly monolayer formation, *Electroanalysis* 17(14): 1251–1259.

Chattopadhyay, D., Galeska, I. & Papadimitrakopoulos, F. (2001). Metal-assisted organization of shortened carbon nanotubes in monolayer and multilayer forest assemblies, *J. Am. Chem. Soc* 123(38): 9451–9452.

Che, G., Lakshmi, B., Fisher, E. & Martin, C. (1998). Carbon nanotubule membranes for electrochemical energy storage and production, *Nature* 393(6683): 346–349.

Chen, D., Wang, G. & Li, J. (2007). Interfacial bioelectrochemistry: Fabrication, properties and applications of functional nanostructured biointerfaces, *The Journal of Physical Chemistry C* 111(6): 2351–2367.
URL: *http://pubs.acs.org/doi/abs/10.1021/jp065099w*

Chidsey, C., Bertozzi, C., Putvinski, T. & Mujsce, A. (1990). Coadsorption of ferrocene-terminated and unsubstituted alkanethiols on gold: electroactive self-assembled monolayers, *Journal of the American Chemical Society* 112(11): 4301–4306.

Chou, A., Bocking, T., Singh, N. K. & Gooding, J. J. (2005). Demonstration of the importance of oxygenated species at the ends of carbon nanotubes for their favourable electrochemical properties, *Chemical Communications* 7: 842.
URL: *http://xlink.rsc.org/?DOI=b415051a*

Chou, A., Eggers, P., Paddon-Row, M. & Gooding, J. (2009). Self-assembled carbon nanotube electrode arrays: effect of length of the linker between nanotubes and electrode, *The Journal of Physical Chemistry C* 113(8): 3203–3211.

Cui, X., Lee, V. A., Raphael, Y., Wiler, J. A., Hetke, J. F., Anderson, D. J. & Martin, D. C. (2001). Surface modification of neural recording electrodes with conducting polymer/biomolecule blends, *J Biomed Mater Res* 56(2): 261–72.

Cui, X., Primak, A., Zarate, X., Tomfohr, J., Sankey, O., Moore, A., Moore, T., Gust, D., Harris, G. & Lindsay, S. (2001). Reproducible measurement of single-molecule conductivity, *Science* 294(5542): 571.

Cui, X., Primak, A., Zarate, X., Tomfohr, J., Sankey, O., Moore, A., Moore, T., Gust, D., Nagahara, L. & Lindsay, S. (2002). Changes in the electronic properties of a molecule when it is wired into a circuit, *The Journal of Physical Chemistry B* 106(34): 8609–8614.

Dastagir, T., Forzani, E., Zhang, R., Amlani, I., Nagahara, L., Tsui, R. & Tao, N. (2007). Electrical detection of hepatitis c virus rna on single wall carbon nanotube-field effect transistors, *Analyst* 132(8): 738–740.

De Heer, W., Chatelain, A. & Ugarte, D. (1995). A carbon nanotube field-emission electron source, *Science* 270(5239): 1179.

Diao, P. & Liu, Z. (2005). Electrochemistry at chemically assembled single-wall carbon nanotube arrays, *J. Phys. Chem. B* 109(44): 20906–20913.

Diao, P. & Liu, Z. (2010). Vertically aligned Single-Walled carbon nanotubes by chemical assembly - methodology, properties, and applications, *Advanced Materials* 22(13): 1430–1449.
URL: *http://doi.wiley.com/10.1002/adma.200903592*

Diao, P., Liu, Z., Wu, B., Nan, X., Zhang, J. & Wei, Z. (2002). Chemically assembled single-wall carbon nanotubes and their electrochemistry, *ChemPhysChem* 3(10): 898–991.

Dresselhaus, M., Dresselhaus, G. & Eklund, P. (1996). *Science of fullerenes and carbon nanotubes*, Academic Press, New York.

Dresselhaus, M., Dresselhaus, G., Saito, R. & Jorio, A. (2005). Raman spectroscopy of carbon nanotubes, *Physics Reports* 409(2): 47–99.

Erokhrin, V., Kumar Ram, M. & Yavuz, O. (2008). *The New Frontiers of Organic and Composite Nanotechnology*, ELSEVIER.

Finklea, H. & Hanshew, D. (1992). Electron-transfer kinetics in organized thiol monolayers with attached pentaammine (pyridine) ruthenium redox centers, *Journal of the American Chemical Society* 114(9): 3173–3181.

Flavel, B., Yu, J., Ellis, A. & Shapter, J. (2009). Electroless plated gold as a support for carbon nanotube electrodes, *Electrochimica Acta* 54(11): 3191–3198.

Flavel, B., Yu, J., Shapter, J. & Quinton, J. (2008). Electrochemical characterisation of patterned carbon nanotube electrodes on silane modified silicon, *Electrochimica Acta* 53(18): 5653–5659.

Gooding, J., Chou, A., Liu, J., Losic, D., Shapter, J. & Hibbert, D. (2007). The effects of the lengths and orientations of single-walled carbon nanotubes on the electrochemistry of nanotube-modified electrodes, *Electrochemistry Communications* 9(7): 1677–1683.

Gooding, J., Wibowo, R., Liu, J., Yang, W., Losic, D., Orbons, S., Mearns, F., Shapter, J. & Hibbert, D. (2003). Protein electrochemistry using aligned carbon nanotube arrays, *Journal of the American Chemical Society* 125(30): 9006–9007.

Gu, Z., Peng, H., Hauge, R., Smalley, R. & Margrave, J. (2002). Cutting single-wall carbon nanotubes through fluorination, *Nano letters* 2(9): 1009–1013.

Hartmann, M. (2005). Ordered mesoporous materials for bioadsorption and biocatalysis, *Chem. Mater* 17(18): 4577–4593.

Hernandez-Santos, D., Gonzalez-Garcia, M. & Garcia, A. (2002). Metal-nanoparticles based electronalysis, *Electroanalysis* 14(18): 1225–1235.

Huang, X., Im, H., Yarimaga, O., Kim, J., Jang, D., Lee, D., Kim, H. & Choi, Y. (2006). Electrochemical behavior of needle-like and forest-like single-walled carbon nanotube electrodes, *Journal of Electroanalytical Chemistry* 594(1): 27–34.

Huang, X., Ryu, S., Im, H. & Choi, Y. (2007). Wet chemical needlelike assemblies of single-walled carbon nanotubes on a silicon surface, *Langmuir* 23(3): 991–994.

Jung, M., Choi, T., Joo, W., Kim, J., Han, I. & Kim, J. (2007). Transparent conductive thin films based on chemically assembled single-walled carbon nanotubes, *Synthetic Metals* 157(22-23): 997–1003.

Jung, M., Jung, S., Jung, D., Ko, Y., Jin, Y., Kim, J. & Jung, H. (2005). Patterning of single-wall carbon nanotubes via a combined technique (chemical anchoring and photolithography) on patterned substrates, *The Journal of Physical Chemistry B* 109(21): 10584–10589.

Koehne, J., Li, J., Cassell, A., Chen, H., Ye, Q., Ng, H., Han, J. & Meyyappan, M. (2004). The fabrication and electrochemical characterization of carbon nanotube nanoelectrode arrays, *Journal of Materials Chemistry* 14(4): 676–684.

Kohli, P., Wirtz, M. & Martin, C. (2004). Nanotube membrane based biosensors, *Electroanalysis* 16(1-2): 9–18.

Lamberti, F., Agnoli, S., Meneghetti, M. & Elvassore, N. (2010). Nanotubes Oxidation Temperature Controls the Height of Single-Walled Carbon Nanotube Forests on Gold Micropatterned Thin Layers, *Langmuir* 26(13): 11344–11348.

Lamberti, F., Giomo, M. & Elvassore, N. (2011). Heterogeneous electron transfer dynamics for swcnts forests on patterned gold layers with different height and density, *Submitted to ACS Nano* .

Li, J., Cassell, A., Delzeit, L., Han, J. & Meyyappan, M. (2002). Novel three-dimensional electrodes: electrochemical properties of carbon nanotube ensembles, *The Journal of Physical Chemistry B* 106(36): 9299–9305.

Liu, J., Chou, A., Rahmat, W., Paddon-Row, M. N. & Gooding, J. J. (2005). Achieving direct electrical connection to glucose oxidase using aligned swcnt arrays, *Electroanal.* 17(1): 38–45.

Liu, J., Rinzler, A. G., Dai, H., Hafner, J. H., Bradley, R. K., Boul, P. J., Lu, A., Iverson, T., Shelimov, K., Huffman, C. B., Rodriguez-Macias, F., Shon, Y.-S., Lee, T. R., Colbert, D. T. & Smalley, R. E. (1998). Fullerene pipes, *Science* 280(5367): 1253–1256.

Liu, L., Zhang, F., Xi, F. & Lin, X. (2008). Highly sensitive biosensor based on bionanomultilayer with water-soluble multiwall carbon nanotubes for determination of phenolics, *Biosensors and Bioelectronics* 24(2): 306–312.

Liu, Y., Gao, L., Sun, J., Zheng, S., Jiang, L., Wang, Y., Kajiura, H., Li, Y. & Noda, K. (2007). A multi-step strategy for cutting and purification of single-walled carbon nanotubes, *Carbon* 45(10): 1972–1978.

Liu, Z., Shen, Z., Zhu, T., Hou, S., Ying, L., Shi, Z. & Gu, Z. (2000). Organizing single-walled carbon nanotubes on gold using a wet chemical self-assembling technique, *Langmuir* 16(8): 3569–3573.

Luo, H., Shi, Z., Li, N., Gu, Z. & Zhuang, Q. (2001). Investigation of the electrochemical and electrocatalytic behavior of single-wall carbon nanotube film on a glassy carbon electrode, *Analytical chemistry* 73(5): 915–920.

Luong, J., Hrapovic, S., Wang, D., Bensebaa, F. & Simard, B. (2004). Solubilization of multiwall carbon nanotubes by 3-aminopropyltriethoxysilane towards the fabrication of electrochemical biosensors with promoted electron transfer, *Electroanalysis* 16(1-2): 132–139.

MacDonald, R. A., Laurenzi, B. F., Viswanathan, G., Ajayan, P. M. & Stegemann, J. P. (2005). Collagen-carbon nanotube composite materials as scaffolds in tissue engineering, *Journal of Biomedical Materials Research Part A* 74A(3): 489–496. URL: *http://doi.wiley.com/10.1002/jbm.a.30386*

Martinez, M. T., Tseng, Y. C., Ormategui, N., loinaz, I., Eritjia, R. & Bokor, J. (2009). abel-free dna biosensors based on functionalized carbon nanotube field effect transistors, *Nano Letters* 9: 530–536.

Mazzatenta, A., Giugliano, M., Campidelli, S., Gambazzi, L., Businaro, L., Markram, H., Prato, M. & Ballerini, L. (2007). Interfacing neurons with carbon nanotubes: electrical signal transfer and synaptic stimulation in cultured brain circuits, *The Journal of neuroscience* 27(26): 6931.

Meyyappan, M. (2005). *Carbon Nanotubes: Science and Application*, CRC, Usa.

Musameh, M., Wang, J., Merkoci, A. & Lin, Y. (2002). Low-potential stable NADH detection at carbon-nanotube-modified glassy carbon electrodes, *Electrochemistry Communications* 4(10): 743–746.

Nan, X., Gu, Z. & Liu, Z. (2002). Immobilizing shortened single-walled carbon nanotubes (swnts) on gold using a surface condensation method, *Journal of colloid and interface science* 245(2): 311–318.

Nkosi, D. & Ozoemena, K. (2008). Self-assembled nano-arrays of single-walled carbon nanotube-octa (hydroxyethylthio) phthalocyaninatoiron (ii) on gold surfaces: Impacts of swcnt and solution ph on electron transfer kinetics, *Electrochimica Acta* 53(6): 2782–2793.

Ozoemena, K., Nyokong, T., Nkosi, D., Chambrier, I. & Cook, M. (2007). Insights into the surface and redox properties of single-walled carbon nanotube–cobalt (ii) tetra-aminophthalocyanine self-assembled on gold electrode, *Electrochimica acta* 52(12): 4132–4143.

Patolsky, F., Weizmann, Y. & Willner, I. (2004). Long-range electrical contacting of redox enzymes by swcnt connectors, *Angewandte Chemie* 116(16): 2165–2169.

Rauwald, U., Shaver, J., Klosterman, D., Chen, Z., Silvera-Batista, C., Schmidt, H., Hauge, R., Smalley, R. & Ziegler, K. (2009). Electron-induced cutting of single-walled carbon nanotubes, *Carbon* 47(1): 178–185.

Rubianes, M. & Rivas, G. (2003). Carbon nanotubes paste electrode, *Electrochemistry Communications* 5(8): 689–694.

Sheeney-Haj-Ichia, L., Basnar, B. & Willner, I. (2005). Efficient generation of photocurrents by using cds/carbon nanotube assemblies on electrodes, *Angewandte Chemie International Edition* 44(1): 78–83.

Sljukic, B., Banks, C., Salter, C., Crossley, A. & Compton, R. (2006). Electrochemically polymerised composites of multi-walled carbon nanotubes and poly(vinylferrocene) and their use as modified electrodes: Application to glucose sensing, *Analyst* 131(5): 670–677.

Smart, S., Cassady, A., Lu, G. & Martin, D. (2006). The biocompatibility of carbon nanotubes, *Carbon* 44(6): 1034–1047.

Ulman, A. (1996). Formation and structure of self-assembled monolayers, *Chemical reviews* 96(4): 1533–1554.

Vairavapandian, D., Vichchulada, P. & Lay, M. D. (2008). Preparation and modification of carbon nanotubes: Review of recent advances and applications in catalysis and sensing, *Analytica chimica acta* 626(2): 119–129.

Wang, J. (2005). Carbon-Nanotube based electrochemical biosensors: A review, *Electroanalysis* 17(1): 7–14.
URL: *http://doi.wiley.com/10.1002/elan.200403113*

Wang, J., Kawde, A. & Musameh, M. (2003). Carbon-nanotube-modified glassy carbon electrodes for amplified label-free electrochemical detection of DNA hybridization, *Analyst* 128(7): 912–916.

Wang, J., Li, M., Shi, Z., Li, N. & Gu, Z. (2001). Electrocatalytic oxidation of 3,4-dihydroxyphenylacetic acid at a glassy carbon electrode modified with single-wall carbon nanotubes, *Electrochimica Acta* 47(4): 651–657.

Wang, J. & Musameh, M. (2003). Carbon nanotube/teflon composite electrochemical sensors and biosensors, *Analytical chemistry* 75(9): 2075–2079.

Wang, J., Musameh, M. & Lin, Y. (2003). Solubilization of carbon nanotubes by Nafion toward the preparation of amperometric biosensors, *Journal of the American Chemical Society* 125(9): 2408–2409.

Warakulwit, C., Majimel, J., Delville, M.-H., Garrigue, P., Limtrakul, J. & Kuhn, A. (2008). Controlled purification, solubilisation and cutting of carbon nanotubes using phosphomolybdic acid, *Journal of Material Chemistry* 18(34): 4056–4061.

Wei, H., Kim, S., Kim, S., Huey, B., Papadimitrakopoulos, F. & Marcus, H. (2007). Patterned forest-assembly of single-wall carbon nanotubes on gold using a non-thiol functionalization technique, *Journal of Materials Chemistry* 17(43): 4577–4585.

Wei, H., Kim, S., Kim, S., Huey, B., Papadimitrakopoulos, F. & Marcus, H. (2008). Site-specific forest-assembly of single-wall carbon nanotubes on electron-beam patterned siox/si substrates, *Materials Science and Engineering: C* 28(8): 1366–1371.

Wei, H., Kim, S., Marcus, H. & Papadimitrakopoulos, F. (2006). Preferential forest assembly of single-wall carbon nanotubes on low-energy electron-beam patterned nafion films, *Chemistry of materials* 18(5): 1100–1106.

Wong, S., Woolley, A., Odom, T., Huang, J., Kim, P., Vezenov, D. & Lieber, C. (1998). Single-walled carbon nanotube probes for high-resolution nanostructure imaging, *Applied physics letters* 73: 3465.

Wu, B., Zhang, J., Wei, Z., Cai, S. & Liu, Z. (2001). Chemical alignment of oxidatively shortened single-walled carbon nanotubes on silver surface, *J. Phys. Chem. B* 105(22): 5075–5078.

Yotter, R. & Wilson, D. (2004). Sensor technologies for monitoring metabolic activity in single cells - Part II: Nonoptical methods and applications, *IEEE Sensors Journal* 4(4): 412–429.

Yu, J., Mathew, S., Flavel, B., Johnston, M. & Shapter, J. (2008). Ruthenium porphyrin functionalized single-walled carbon nanotube arrays a step toward light harvesting antenna and multibit information storage, *Journal of the American Chemical Society* 130(27): 8788–8796.

Yu, J., Shapter, J., Johnston, M., Quinton, J. & Gooding, J. (2007). Electron-transfer characteristics of ferrocene attached to single-walled carbon nanotubes (swcnt) arrays directly anchored to silicon (1 0 0), *Electrochimica Acta* 52(21): 6206–6211.

Yu, X., Chattopadhyay, D., Galeska, I., Papadimitrakopoulos, F. & Rusling, J. (2003). Peroxidase activity of enzymes bound to the ends of single-wall carbon nanotube forest electrodes, *Electrochemistry Communications* 5(5): 408–411.

Yu, X., Kim, S., Papadimitrakopoulos, F. & Rusling, J. (2005). Protein immunosensor using single-wall carbon nanotube forests with electrochemical detection of enzyme labels, *Molecular BioSystems* 1(1): 70–78.

Yu, X., Munge, B., Patel, V., Jensen, G., Bhirde, A., Gong, J. D., Kim, S. N., Gillespie, J., Gutkind, J. S., Papadimitrakopoulos, F. & Rusling, J. F. (2006). Carbon nanotube amplification strategies for highly sensitive immunodetection of cancer biomarkers, *Journal of the American Chemical Society* 128(34): 11199–11205.

Zhao, Q., Gan, Z. & Zhuang, Q. (2002). Electrochemical sensors based on carbon nanotubes, *Electroanalysis* 14(23): 1609–1613.

Ziegler, K. J., Gu, Z., Peng, H., Flor, E. L., Hauge, R. H. & Smalley, R. E. (2005). Controlled oxidative cutting of single-walled carbon nanotubes, *Journal of American Chemical Society* 127: 1541–1547.

12

Carbon Nanotubes as Suitable Electrochemical Platforms for Metalloprotein Sensors and Genosensors

M. Pacios[1], I. Martín-Fernández[2], R. Villa[2], P. Godignon[2], M. Del Valle[1], J. Bartrolí[1] and M.J. Esplandiu[3]
[1]*Grup de Sensors i Biosensors, Departament de Química, Facultat de Ciències, Edifici C-Nord, Universitat Autònoma de Barcelona, Barcelona,*
[2]*Centro Nacional de Microelectrónica (CSIC), Campus UAB, Barcelona,*
[3]*Centro de Investigación en Nanociencia y Nanotecnología, CIN2 (CSIC-ICN), Campus UAB, Barcelona,*
Spain

1. Introduction

In the last decade we have been witnesses of the immense scientific production concerning cabon nanotubes (CNTs). Apart from their exceptional electrical and mechanical properties, CNTs provide interesting electrochemical properties which arise from the combination of their good electrical conductivity, their nanometer size, high aspect ratio, structure, electrochemical stability and surface chemistry (McCreery, 2008; Esplandiu, 2009). A vast number of papers have claimed electrocatalytical properties on CNTs which were correlated with enhanced electron transfer rates, increased signal currents and decreased overpotentials, (Wang, 2005; Gooding, 2005; Banks, 2006). However the fundamentals of such special electroactivity of CNTs are still in debate. In some cases, it has been observed an enhanced electrochemical response of CNT to particular redox systems which was explained by the presence of some metal impurities from the catalyst involved in the CNT growth (Jones, 2007). On the other side, it is well known that edge defects and oxygen functionalities at the surface play a major role in the electrochemical behavior of graphitic materials (McCreery, 2008). Under this context, there has been a large amount of work that has explained the special electrocatalytical properties of the CNTs linked to that of Highly Oriented Pyrolytic Graphite (HOPG). These studies are based on the fact that CNT properties are not so different from HOPG and exhibit similar electrochemical anisotropy. Open ends of the CNT or locations on the tube axis where graphitic sheets terminate (edge defects) exhibit higher electron kinetics (similar to edge-HOPG) whereas CNT walls have lower electron transfer kinetics (similar to basal HOPG) (Moore, 2004; Banks, 2005). However such issue should not lead us to completely disregard the intrinsic electrochemical activity of the CNT walls (Dumitrescu, 2009; Salinas-Torres, 2011). Summed up to all these aspects, CNTs exhibit high specific surface area not only for the electron transfer process itself but also for covalent or non-

covalent (bio)molecule fucntionalization (Katz, 2004; Wang, 2005; Tasis, 2006). Moreover, CNTs represent a versatile material which can be tailored in different geometrical electrode arrangements for exploiting their more electroactive sites (Esplandiu, 2009). All these ingredients make CNT systems ideal candidates for amplified signal transduction in the (bio)electrochemical sensing field.
Accordingly, an overwhelming number of electrochemical biosensing systems on CNT electrodes have been studied such as enzymatic/redox protein biosensors, genosensors or immunosensors which have boosted the potentialities of CNTs as suitable platforms for transducing biorecognition events into useful electroanalytical signals.

1.1 Enzymatic/protein CNT sensors

Electrochemical enzymatic/protein sensors are based on the immobilization of redox proteins on an electrode. In most of the cases the electrode itself can replace physiological partners or natural electron acceptors, providing the driving force to energize the reaction and a sensor to measure the response. For instance, in the case of glucose oxidase (GOx), which catalyses the reaction glucose + $O_2 \rightarrow$ glucolactone +H_2O_2, with oxygen acting as its natural electron acceptor to regenerate the enzyme, the electrode can take the role of its natural partner oxygen (Patolksy, 2004; Cai, 2004). Consequently the main goals when preparing redox enzymatic sensors are to preserve the biomolecule activity when immobilized on the electrode and also to guarantee a good electrochemical communication between the redox active center and the electrode to facilitate the enzyme turnover (Guiseppi-Elie, 2002; Wang, 2005; Gooding, 2005; Kim, 2007; Katz, 2004). However, such goals are rather difficult to achieve since proteins, when adsorbed on electrode surfaces, undergo denaturation with the consequent loss of their electrochemical activity. Moreover, some enzymes contain the redox active centre buried deep inside the protein and it is necessary to utilize mediators to accomplish an electrochemical response (Cai, 2004; Willner, 2004). Therefore, the nature and properties of the surface of the electrode are crucial in dictating how the electron exchange takes place. In this sense, not only the electrochemical reactivity of the surface is important, but also the disposition of the protein and its redox active center on the surface. This in turn strongly depends on the electrode charge density, its hidrophilicity/hidrophobicity or the strength of the interaction.
All the aspects mentioned above have to be considered and optimized in order to get a detectable, reliable and quantifiable electroanalytical response of the enzymatic process.
Under this context, CNTs have been highly appreciated as suitable platforms for anchoring redox enzymes/protein without the loss of the bioactivity and also for allowing a more direct electron transfer with the protein electroactive center and thus avoiding the use of redox mediators (Willner, 2004; Cai, 2004; Esplandiu, 2009). For instance, GOx represents a very interesting model system which is known to have the redox active center (flavin adenine dinucleotide, FAD) deep inside the peptide environment (Willner, 2005; Liu, 2005). Many nanotube modified electrodes have shown to exhibit direct electron transfer to glucose oxidase and at the same time preserving its bioactivity. The direct electron transfer was attributed to the nanotubes being able to penetrate the protein and get close to the FAD center due to their nanometric size (Guiseppi-Elie, 2002).
Additionally CNTs can play an amplification role for the recognition and transduction events due to their high specific area that allows the analyte to be accumulated (Wang, 2005). Moreover, CNTs have been also shown to promote a dramatic decrease in the

overpotential of some important species (H_2O_2, NADH) involved in electrocatalytic enzymatic reactions such as oxidases or dehydrogenases (Wang, 2003; Wang 2005, Rubianes, 2006). For instance, the bioactivity of a CNT electrode modified with GOx can be monitored by the reduction or oxidation of H_2O_2. The detection of H_2O_2 has been carried out at such low overpotentials that interferences with other analytes can be minimized (Wang, 2003; Rubianes, 2006).

Direct electron transfer has also been observed with many metalloproteins such as cytochrome C, horseradish peroxidase, microperoxidase (MP-11), myoglobin, hemoglobin, catalase, azurin, etc, immobilized on different CNT modified electrodes (Gooding, 2003; Yu, 2003; Zhao, 2006; Zhao, 2006a; Zhang, 2007; Li, 2006; Zhang, 2004; Esplandiu, 2009a; Pacios, 2009. Palangsuntikul, 2010; Kumar, 2011). Myoglobin (Mb), a small water soluble protein involved in the binding and storage of oxygen in cells, has been the subject of numerous studies in redox protein sensors. This protein has a hemin core capable of redox activity and exhibits a catalytical response in presence of some analytes such as oxygen, H_2O_2 or NO (Zhang, 2004; Li, 2006; Zhao, 2006; Pacios, 2009; Esplandiu, 2009a). It is generally difficult for Mb to transfer electrons to a conventional electrode, but CNT electrodes have once again demonstrated their performance as suitable electrode material for such protein and have promoted applications as potential (bio)sensors of oxygen or hydrogen peroxide or as reduction catalyst in batteries and fuel cell systems. Taking advantage of the CNT benefits as electrode substrates for (bio)sensing, some groups have investigated the physisorbed Mb response on randomly dispersed carbon nanotube modified substrates (Zhang, 2004; Li, 2006; Zhao, 2006). In all cases, direct electron exchange between Mb and CNT was observed in a quasi-reversible heme Fe(III)/Fe(II) voltammetry response. However, it has been scarcely exploited the use of forest carbon nanotubes, which are expected to be more suitable platforms for the protein anchorage and for the electron transfer kinetics (due to the high density of fast electron exchange CNT edges facing the redox biomolecules) (Yu, 2003; Esplandiu, 2009).

1.2 Genosensors

A genosensor consists in a substrate modified with specific oligonucleotides (probe DNA) that can detect complementary DNA sequences (target DNA) through hybridization. The determination of nucleic acid sequences from humans, animals, bacteria and viruses constitute the starting point to solve different problems such as food and water contamination caused by microorganisms, detection of generic disorders, tisssue matching, forensic applications, etc. A growing number of research groups in the field of DNA sensors have opted to use electrical and electrochemical detection strategies in an effort to avoid some of the problems associated with the fluorescence protocols (Drummond, 2003, Bonanni, 2009; Bonanni, 2010; Bonanni, 2010a). The electrochemical biorecognition represents a practical approach in which the DNA probe is immobilized on an electrical active surface and the hybridization process is measured as a change in electrical parameters (e.g. current, potential, conductance, impedance, capacitance). Again, the use of CNTs in electrochemical DNA biosensors is quite promising for the reasons given before: CNTs constitute novel platforms for DNA immobilization with important electrochemical transducing properties and high specific surface area which can increase not only the attached DNA amount but also concentrate a great number of enzymes or electroactive nanoparticles to amplify DNA hybridization (Wang, 2003; Wang, 2004; Wang, 2004a; Wang,

2005; He, 2006; Kim, 2007,). For instance it has been observed in some studies an enhanced guanine signal which was attributed to a CNT-induced interfacial accumulation of the analyte (due to its large surface area/volume ratio). Other groups have demonstrated the advantages of CNT edges for increasing the electron transfer rates of guanine and adenine bases by measuring adsorbed DNA on bamboo like multiwalled carbon nanotubes (these ones having higher density of edge planes as compared to hollow type multiwalled CNTs) (Heng, 2005).

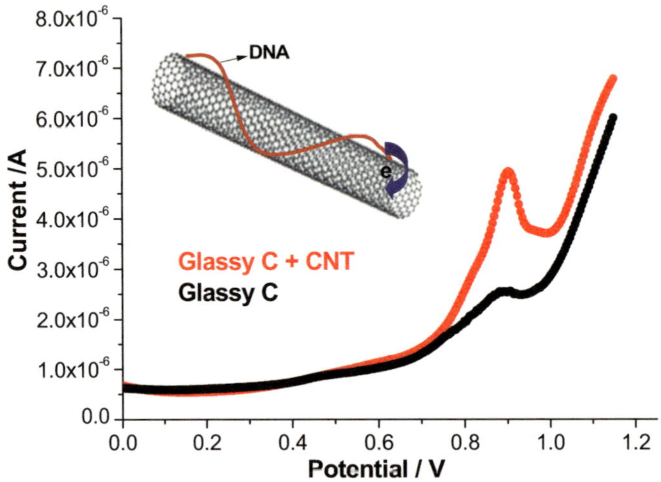

Fig. 1. Differential Pulse Voltammetry (DPV) of a glassy carbon/CNT electrode with physisorbed Polyguanine (Poly-G) (in red) and of a glassy carbon electrode with physisorbed Poly-G (in black). The peak corresponds to the direct oxidation of guanine. Note the enhanced signal of the Poly-G oxidation when it is adsorbed on the CNT support. That is due to the high surface area of CNTs which allow more analyte loading on their surface.

There are many strategies to detect hybridization events; they can be grouped mainly in those based on labeling target DNA or those using a label free approach (Drummond, 2003; Bonanni, 2010a). In the latter case, the more simple strategy is based on the direct oxidation of the DNA bases physisorbed on the CNT material. Although label-free DNA detection protocols are quite desirable since they greatly simplify DNA hybridization assays, such detection schemes can suffer from some drawbacks. On one side, the physisorption of DNA is a multisite attachment which reduces the flexibility of the biomolecule on the electrode support. In the case of hydrophobic electrodes such as CNTs, DNA physisorbs with the bases (apolar region of the molecule) in close contact with the hydrophobic walls whereas the phosphate backbone faces the solution (Carot, 2010). Such orientation of the adsorbed DNA and the reduced flexibility of the adsorbed DNA can decrease the efficiency of the hybridization event. Albeit the direct oxidation does not need the additional use of redox labellings, the process itself also presents the difficulties of the high oxidation potentials at which significant background currents are present (Wang, 2003; Wang, 2004). Among the DNA bases, guanine is the one which exhibits the lower electrochemical overpotential for oxidation, though still remains relative high (close to 1 V). CNTs can help to increase the

CNT signal response from the background noise due to surface area effects (more loading of the probe and target species as observed in Figure 1)(Wang, 2003; Wang, 2004; He, 2006). Despite the potential capacity of CNTs to overcome some of these issues, the strategy is not so practical since DNA oxidation is an irreversible process that impedes the consequent use of the electrode for additional studies (Drummond, 2003) and is limited by the content of DNA bases with low oxidation overpotentials such as guanine. In many cases it is also necessary to manipulate the DNA probe in order to avoid interferences of the guanine oxidation from the probe when monitoring the DNA target (He, 2006). That is why the guanine in single-stranded DNA probe is often replaced by inosine. With that strategy one can reduce the background signal interference derived from single strand DNA probe and consequently increase the hybridization signal derived from double-stranded DNA helix (He, 2006).

There are also many studies that rely on the DNA base oxidation by using flexible probe DNA analyte covalently anchored by a single point linkage (Drummond, 2003; Koehne, 2004). That approach can be achieved by using carbodiimide chemistry which allows establishing an amide bond between -COOH functional groups of the CNT and amine terminated oligonucleotide probes. Such strategy is supposed to increase the hybridization efficiency but requires base oxidation mediators such as ruthenium complexes that mediate the redox process between the guanine base and the electrode. The need of a mediator is due to the fact that in a single point DNA attachment, the guanine bases are not in close contact with the CNT surface like in the multisite physiadsorbed oligonucleotides and consequently the direct base oxidation is not achievable. Again this strategy suffers from the irreversibility of the redox process (Figure 2).

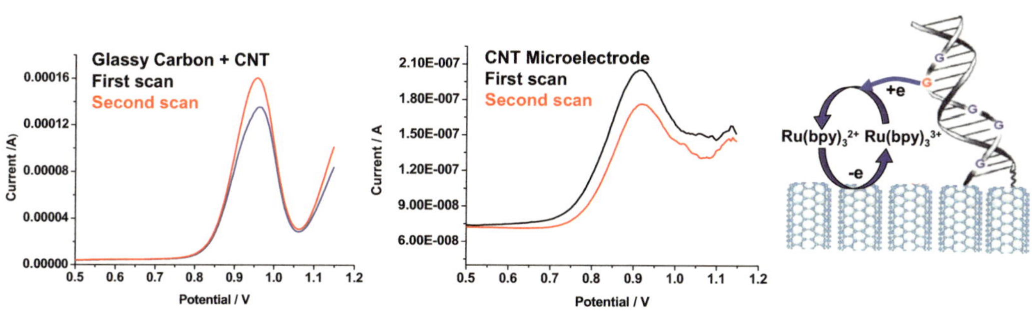

Fig. 2. DPV profiles of covalently attached Poly-G on CNT/glassy carbon electrodes and on CNT microelectrodes. In this case a redox indicator must be used ($(Ru(byp)_3{}^{2+/3+})$) which oxidizes guanine and then exchanges the electron with the electrode. The first scan of the DPV profiles exhibits an enhanced oxidation signal due to the effect of the guanine oxidation but due to the redox irreversibility of guanine there is no contribution of the DNA base in the second scan and consequently the oxidation signal is lower than in the first scan.

Regarding the labelled DNA detection schemes, one common strategy includes the chemical labeling of target DNA sequences with redox active molecules, enzymes, nanoparticles, etc (Fan, 2003; Pividori, 2003; He, 2004a; He, 2006; Bonanni, 2007; Bonanni, 2010). In the case of enzymatic labels bound to the DNA target, the enzyme triggers the catalysis of a redox active reaction under hybridization. For instance by using a horseradish peroxidase-labeled

DNA target, H_2O_2 peroxide is produced which can be detected amperometrically, so the redox reaction of H_2O_2 is an indication of the hybridization process. A variation in such approaches involves a three-component "sandwich" assay, in which the redox label is attached to a synthetic sequence specifically designed to bind an overhang portion of the probe and target DNA. All these approaches need an extra chemical labeling step either in the target DNA or in the synthetic oligonucleotide which makes the process more expensive and effortful (Drummond, 2004).

One alternative way that avoids chemical labeling steps is to use redox active reporter molecules that intrinsically associate with the double helix in a non-covalent manner. These reporter molecules can interact with the DNA either by electrostatic interactions or by intercalative means (in such cases DNA has been considered by some authors as a charge mediator) (Drummond, 2004, Gorodetsky, 2008). Some examples of the former molecules are the highy positively charged cobalt and ruthenium complexes (eg. $Co(phen)_3^{3+}$, $Ru(NH_3)_6^{3+}$). These positively charged complexes can strongly interact with the negatively phosphate backbone. As the hydridization process takes place, proportionally more positively charged complexes bind yielding to a higher signal. The other kind of reporter molecules is redox probes that can be specifically intercalated in double strand DNA. An example of these sorts of molecules is the methlylene blue. These molecules have been also used as reporters on perturbations in base stacking.

The approaches with non-covalent and reversible redox reporters are highly sensitive, simple and do not damage the sample which gives the advantage of further use of the sample. Moreover, the redox molecule reporter strategy together with the functionalization and signal amplification capabilities of the CNTs, the powerful electrochemical detection techniques and the possibilities of sensor miniaturization can provide important stimulus to the field of genosensing and diagnosis.

Under the context of the above introduction, in this chapter we would like to highlight some examples of how carbon nanotubes exhibit important capabilities as biosensor transducers. The first example will be related to the role of CNTs in enzymatic sensors, by modifying the CNT platforms with myoglobin (Mb). Although myoglobin is not an enzyme, it can behave as such in presence of certain analytes such as H_2O_2, O_2 or NO. Myoglobin also has the active redox center embedded deep in the protein network which makes the direct electron transfer in conventional electrodes difficult. We will demonstrate the suitability of CNT platforms for loading myoglobin without losing its activity and also the potentialities of CNT electrodes for enhancing the direct electrochemistry with the redox active center of the protein. Special emphasis will be put on vertically aligned CNT electrodes. We will demonstrate how all these features can exert a big influence on the sensor capabilites of myoglobin when exposed to specific analytes (O_2 or H_2O_2) which encourage the use of such CNT modified systems in clinical diagnosis or in environmental assays.

The second example will be related to the use of CNTs as platforms for genosensors. In this section we will mainly discuss the use of microelectrode arrays made of vertically aligned CNTs as substrates for chemically anchoring of DNA. We will use the strategy of the highly positively charged redox reporters as indicators of the hybridization process which will be monitored by cyclic voltammetry and by other simple and highly sensitive techniques such as chronocoulometry and impedance spectroscopy. We will show that miniaturization of CNT electrodes results crucial when using impedance measurements for detecting DNA biorecognition with positively charge redox reporters.

2. Experimental

2.1 Myoglobin modified CNT electrodes

2.1.1 Working electrodes

Different carbon nanotube electrodes were used as working electrodes: vertically aligned CNTs and CNT epoxy composites. The behahvior of the CNT electrodes were compared in different aspects with other electrodes such as graphite epoxy composites, Au coated silicon electrodes and Highly Oriented Pyrolytic Graphite (HOPG, SPI-3 grade, SPI supplies, PA, USA).

The vertically aligned multi-walled carbon nanotubes were grown on silica substrates and were provided by MER Corporation (Tucson, AZ, USA). The thickness of the resulting nanotube film was around 30-40 μm.

Graphite-expoxy composites were made from graphite powder of 10-20 μm particle size. Carbon nanotube composites were fabricated using multi-walled carbon nanotubes (MWNTs) grown by CVD with lengths between 5-15 μm, diameter of 30 nm and purity of 96-98% (SES Research, Houston, TX, USA). In both cases, Epotek H77 resin and hardener (both from Epoxy Technology, Billerica, MA, USA) were used for the composite fabrication. Composite electrodes were prepared, following the standard protocol in our laboratories (Pumera, 2006; Pacios, 2008). The composite was prepared by first mixing manually the epoxy resin and the hardener in a 20:3 ratio (w:w), respectively. Then the carbon source was added to the epoxy mixture in a 20% in weight proportion. The carbon paste was well mixed during one hour and put in the cavity of the plastic body (around 3.5 mm in depth). Finally the material was cured at 100 °C for 10 hours. Before each use, the surface electrode was polished with emery paper of different grain grades and then with 0.3 μm alumina paper (polishing strips 301044-001, Thermoelectron).

2.1.2 Electrode modification

Horse heart myoglobin (Sigma, MA, USA) was dissolved in 0.1 M phosphate buffer solution (PBS, 137 mM NaCl, 2.7 mM KCl, 10 mM Na_2HPO_4, 2 mM KH_2PO_4, pH= 7.4).

For the covalent immobilization of the proteins on carbon electrodes, *N*-(-3-dimethylaminopropyl)-*N'*-ethylcarbodiimide hydrochloride (EDAC, Sigma, MA, USA) was used.

All electrodes were first oxidized by electrochemical pretreatments in order to increase the oxygen moieties for covalent attachment by carbodiimide chemistry. In the case of the vertical aligned CNT, also a heat treatment (at 470°C) in air was performed. Then, the carbon electrodes were immersed in myoglobin solutions in PBS (3mg/ml) in presence of EDAC (3 mg/ml) during 12 hours. Finally, they were thoroughly washed in PBS for one hour and dried with nitrogen.

2.1.3 Electrochemical behavior and electrocatalytical effects

The electrocatalytic activity of myoglobin in presence of H_2O_2 was studied by using hydrogen peroxide solutions at different concentrations. Such solutions were dearated bubling N_2 for at least 15 minutes.

2.1.4 Instrumentation

All measurements were performed with the normal three-electrode configuration. A Pt wire was used as counter electrode and an Ag/AgCl electrode was acting as reference (in close proximity to the working electrode to minimize the ohmic drop). For the electrochemical characterization we have used Cyclic voltammetry from an electrochemical workstation (IME6, Zahner, Germany).

2.2 DNA modified CNT electrodes

2.2.1 Working electrodes

The microelectrode array electrode with forest CNTs was fabricated as follows (Martin-Fernández, 2010). First, a 150 nm thick Pt layer was patterned by conventional photolithography on silicon wafers that were previously thermally oxidized. A 30 nm thick titanium layer was deposited by sputtering between SiO_2 and Pt to improve the adherance to the substrate. Then a 15 nm thick SiO_2 layer was deposited by PECVD and after that a 4 nm Pt layer was sputtered on top of the SiO_2 which would act as the catalyst layer. The SiO_2 layer between the two Pt ones is to avoid the diffusion of Pt forming the electrode (thick Pt layer) and the catalyst layer. After that, CNT growth was performed in a JetStar 100ST Rapid Thermal CVD from Jipelec in a two step process, the catalyst activation step (800 ^{o}C, 500 sccm H_2 for 5 min) and the CNT growth step (800°C, 1000 sccm CH_4 for 5 min). The final fabrication step consited in introducing the devices in a HF based solution to etch the 15 nm SiO_2 layer on the contact pads after the wafers had been diced. The final array of microelectrodes consisted in four electrodes of 300 μm x 300 μm size as can be depicted in Fig. 3.

Fig. 3. CNT microelectrode with the zoom of two microelectrode patches of 300 μm size each of them. The CNTs were grown by chemical vapor deposition in an upright configuration.

For comparison we also used macroscopic electrodes made of vertically aligned multi-walled carbon nanotubes grown on silica substrates (MER Corporation, Tucson, AZ, USA) and CNT dispersed on glassy carbon electrodes.

2.2.2 Electrochemical pre-treatment of the working electrodes

Electrochemical pretreatment was performed for two reasons, one for cleaning the CNTs by oxidizing the amorphous carbon and the other for generating carboxylic moieties for further covalent functionalization. Three electrochemical treatment procedures were performed with different electrolytes (in 0.1 M HNO_3 , 10 sec, 1 V; in 0.1 M KCl 60 sec, 1.75 V and in 1 M NaOH, 60 sec, 1 V). Treatments in NaOH solutions provided the best electrochemical response based on cylic voltammetry measurements (CV) in presence of a redox benchmark $Fe(CN)_6^{3-/4-}$. Specifically, for such treatment, we observed higher current signal and a decrease in the peak potential difference as indication of the enhancement of the electrochemical performance (increase of the electrode reversibility).

2.2.3 The immobilization of oligonucleotide on electrode

The immobilization of the DNA was performed by following the carbodiimide chemistry. The -COOH terminal groups of the CNTs were activated by immersing in 0.05 M EDAC/0.03M Sulfo-NHS in PBS solution for 15 minutes. These compounds transform the carboxyl group in an amine-reactive ester. After rinsing in PBS, the modified electrode was incubated overnight in 1 μM PolyC-NH_2 oligonucleotide (20 bases). Then the electrodes were immersed for 15 minutes in 0.05% SDS + 0.04 M Hydroxylamine solution to remove non-specific adsorption and to deactivate free -COOH groups.

2.2.4 Hybridization with DNA target

The electrode was immersed for other 15 minutes in 0.01 M PEG in PBS to avoid non-specific adsorption in the consequent hybridization process. After that, the electrodes were incubated for 30 minutes in different concentrations of the complementary strand (from 10^{-10}M to 10^{-6}M in PBS) of PolyG (20 bases). Negative controls were also made with no-complementary oligonucleotide (PolyT).

2.2.5 Instrumentation

All measurements were performed with same three-electrode configuration mentioned above. For the electrochemical characterization we have used different electrochemical techniques: cyclic voltametry, differential pulse voltammetry, chronocoulometry and impedance spectroscopy.

Cyclic voltammetry and Impedance Spectroscopy were obtained with an electrochemical workstation (IME6, Zahner, Germany). Chronocoulometry were performed in Autolab electrochemical workstation.

The cyclic voltammetry and impedance spectroscopy were performed in presence of the redox reporter $Ru(NH_3)_6^{3+}$ at a concentration of 1.8 mM in TRIS buffer. The chronocoulometry was performed either in 10 mM TRIS + 10 mM NaCl or in the same electrolyte with 50μM of $Ru(NH_3)_6^{3+}$.

3. Results

3.3 Myoglobin modified CNT electrodes

3.3.1 Direct electrochemistry of the metalloproteins

As mentioned in the introduction, one of the main issues in electrochemical enzymatic/protein sensors is to achieve the capability of direct electron transfer between the electroactive biomolecule center and the electrode and to keep at the same time the biomolecule bioactivity. This issue is very important to guarantee efficient applications. In this section we will demonstrate how CNT electrodes facilitate the direct electron transfer with myoglobin by using cyclic voltammetry in deaerated solutions of PBS. Cyclic voltammetry is a useful technique to get fast information on the thermodynamic of processes and on the redox kinetics of heterogeneous electron transfer reactions. It allows a rapid location of redox potentials of the electroactive species and convenient evaluation of the effect of the environment upon the redox process. Fig. 4 shows the cyclic voltammograms for the case of myoglobin covalently immobilized on different carbon substrates: vertically aligned CNTs, CNT epoxy composites and graphite composites. These electrodes exhibit an anodic and cathodic peak corresponding to the direct electron transfer between the hemin core of the myoglobin (Fe(III)/Fe(II)) and the carbon based electrodes

(Esplandiu, 2009a). Such electrochemical response is reversible and can be evidenced by the small peak potencial diferences (ΔEp) between the anodic and cathodic wave. The smaller the peak separation, the more facile is the electron transfer process.

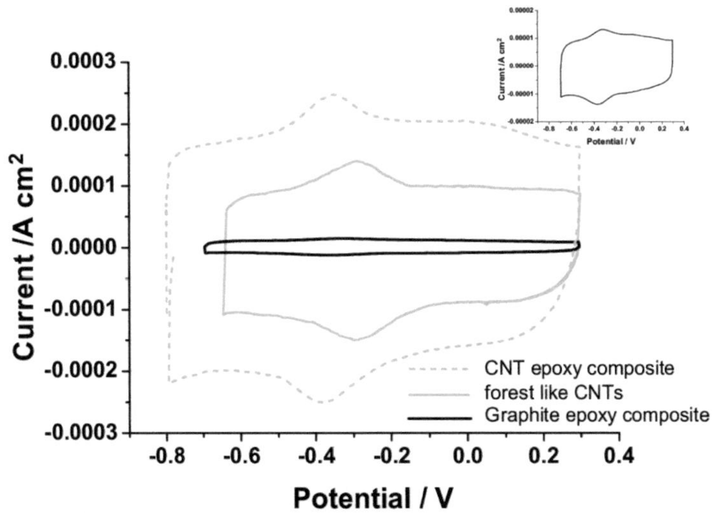

Fig. 4. Cyclic voltammograms for a forest like CNT electrode, CNT and graphite epoxy composite modified with Mb in deareated PBS solutions at a scan rate of 0.1 V/s. A zoom of the graphite composite response is also depicted. Reproduced with permission from (Esplandiu, 2009a). Copyright, 2009, IOP Publishing.

We have used the values of ΔEp at 0.1 V/s in order to extract the electron transfer rate constants using the Laviron method (Laviron, 1979) (Table 1). From such values one can see that the electrochemical response of Myoglobin on CNT electrodes is improved compared to the case of graphite composites. We have found that forest-like CNTs exhibit the smallest peak separation and highest electron transfer rate constant followed by the CNT and graphite epoxy composites (Esplandiu, 2009a).

Carbon electrode	**ΔEp / mV**	**k s^{-1}**	**Γ*r (mol cm^{-2})**
Graphite epoxy composite	43±3	3.1±0.3	6.8 10^{-11}
Carbon nanotube epoxy composite	21±4	7±1	8.0 10^{-10}
Forest like CNTs	8±3	15±4	3.0 10^{-10}

Table 1. Electrochemical parameters of the carbon based electrodes. (Adapted from (Esplandiu, 2009a)

As mentioned in the introduction, edge defects promote high electron transfer rates at the carbon electrode interface and also become the active sites for the covalent attachment. Probably the higher density of edges of the forest CNT electrodes, which are concentrated just at the interface with the electrolyte, could favor a better protein wiring at the CNT substrate and promote the electron exchange. In the case of the CNT composites, the relatively lower electron transfer rate can be rooted in the more heterogeneous complexity of the porous CNT composite. Protein can be anchored on CNT composites through exposed edges on the CNT tips, on defects at the uncoated walls or on defects at insulated polymer coated CNTs. All these characteristics could explain the lower transfer rate as compared with the forest like systems. Although the CNT epoxy composites have lower electron transfer kinetics, they exhibit a higher current signal, which can be due to a higher material porosity providing more surface area for protein immobilization. An estimate of the protein surface coverage (Γ) for every electrode can be made from the peak current (I_p) as a function of the scan rate by using the relation corresponding to a reversible adsorbate electroactive layer, $i_p = 9.39\ 10^5\ n^2\ \nu A\Gamma$. In this relation, n represents the number of exchanged electrons, ν is the scan rate, and A is the electroactive surface area which depends on the roughness of the surface. The true surface area and coverage are very hard to evaluate independently, therefore we have reported the protein coverage per unit of geometrical area, i.e $\Gamma.A/A_{geom}= \Gamma.r$. The values obtained are listed in Table 1, and give an indication of the amount of protein deposited in each electrode. As anticipated, the higher Mb surface coverage on the CNT composite as compared to the other carbon electrodes is probably due to its larger porosity and roughness degree, which facilitates the immobilization of higher amounts of protein and amplifies the signal response (Esplandiu, 2009a).

3.3.2 Electroacatalytic activity of metalloproteins

So far we have proved that direct electron communication is possible between the active redox center of the immobilized protein and the carbon electrode. We have concluded that carbon nanotube materials, either as CNT forest or as CNT composites, exhibit improved electrochemical performance compared to the graphite composites. In the case of the forest CNT, increased electron transfer kinetics is achieved whether in the case of the CNT composites, their high degree of porosity facilitates the immobilization of more amount of protein and therefore the electrochemical signal is amplified.
In this section, we will demonstrate that the protein immobilization process on CNT also preserves the bioactivity of the protein to typical analytes (O_2 and H_2O_2).

3.3.2.1 Electroacatalytic activity in presence of oxygen

The results reported above have shown the reversible behavior of the redox active core of myoglobin in anaerobic conditions. However when the metalloproteins are exposed to oxygen, a strong catalytic effect appears. Figure 5 comparatively shows the remarkable changes in the electrochemical response of myoglobin in absence or in presence of oxygen. In aerobic conditions the electrocatalytic effect is reflected in an increase of the Fe(III) reduction peak accompanied by the disappearance of Fe(II) oxidation peak (Zhang, 2004; Zhao, 2006; Zhao, 2006 a; Zhang, 2007; Li, 2006).
Such catalytic effect can be explained by the following path. As the potential is cathodically scanned, the direct electrochemical reduction of Protein-Fe(III) to Protein-Fe(II) takes place at the carbon electrode. Fe(II) has a very high affinity for oxygen and the latter process is immediately followed by a fast reaction of Protein-Fe(II) with oxygen. The product of Protein-Fe(II)-O_2 could then undergo again electrochemical reduction at the potential of Protein-Fe(III) reduction, producing H_2O_2 and Protein-Fe(II). Protein-Fe(II) again reacts fast with O_2 to continue the cycle. This scheme of reactions is consistent with the large cathodic wave (due to the catalytic reduction of Fe(II)-O_2) and the disappearance of the Fe(II) anodic wave (due to the fast reaction of Fe(II) with oxygen) (Nakajima, 1987; Onuoha, 1997; Zhang, 2002).

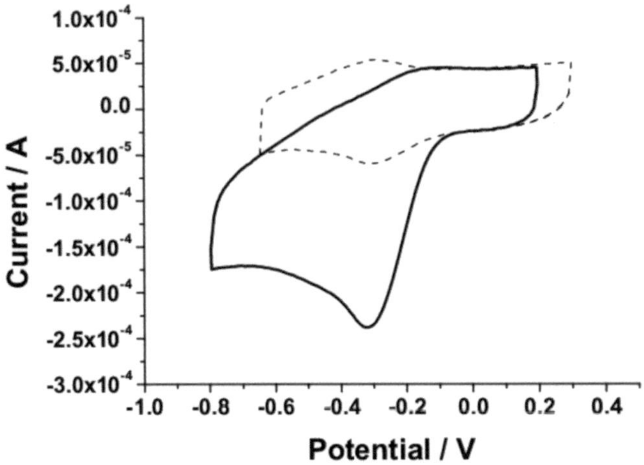

Fig. 5. Cyclic voltammograms for a forest like CNT electrode modified with Myoglobin in deareated PBS solutions (dashed curve) and air saturated PBS solutions (solid curve). Scan rate: 0.1 V/s. Reproduced with permission from (Esplandiu, 2009a). Copyright, 2009, IOP Publishing.

Electrochemical response on different substrates

In this section we will compare the electrocatalytical behavior of Mb in presence of oxygen when immobilized on different electrodes. Firstly, Fig. 6 shows the behavior of the electrochemical response of the different Mb-modified carbon electrodes in an aerated solution. Fig. 6 also shows, as a reference, the voltammograms of protein-free carbon electrodes, where there was no signature of a peak, at least in the scanned potential window.

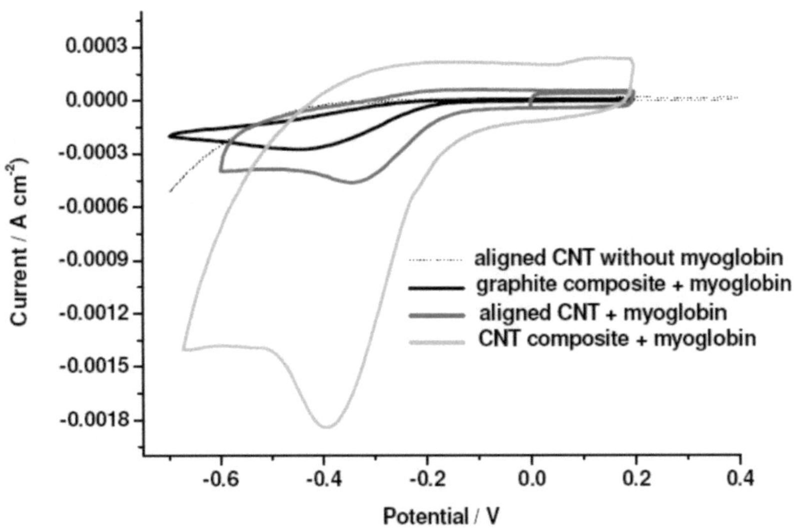

Fig. 6. I/V profiles for the different carbon electrodes in aerated PBS solutions at a scan rate of 0.1V/s. Reproduced with permission from (Esplandiu, 2009a). Copyright, 2009, IOP Publishing.

It can be observed that in terms of current density, the CNT composite exhibits the higher values, followed by the forest like CNTs, and the graphite composite. The reason for that has to do with the protein coverage on the different carbon based systems, as discussed previously. The sensitive response of Mb to oxygen also demonstrates that the immobilization procedure on such carbon systems does not perturb the bioactivity of the protein.

We have also followed the catalytic behavior of Mb in presence of oxygen but when immobilized in other electrodes such gold and HOPG (Pacios, 2009). Figure 7 shows the Mb/electrode response normalized to the same geometric area. Three voltammogram scans are depicted for each electrode material; they were taken with three minutes of difference among them. The poor voltammetric response achieved in the case of HOPG and Au electrodes is remarkable. The oxygen catalytic response progressively disappeared in different i/V runs in the case of Au and HOPG which can reflect either difficulties in anchoring firmly the protein on such electrode surfaces or a progressive denaturation of the protein with the consequent loss of its bioactivity. However, a more stable and reproducible response was kept at the forest CNT substrates. In the results presented in Fig. 7 a, b and c, Mb has been physisorbed on the forest CNT, HOPG and Au electrodes. At this point it is worth mentioning that physisorbed Mb on CNT forest behaves similarly as the covalently attached Mb. The main difference between the system with a higher degree of covalently attached protein and the one with only physisorbed biomolecules is reflected in the long term electrode lifetime. We verified that covalently attached Mb/CNT systems could provide reproducible response up to a month with only a reduction of the electroactivity of about 15%, whereas Mb physorbed on CNT electrodes exhibited stable electroactivity response only for a maximum of one week (Pacios, 2009).

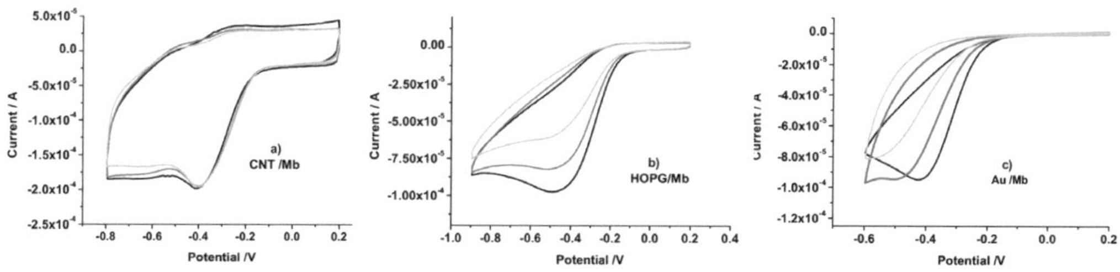

Fig. 7. Oxygen reduction catalytical signal for different electrode surfaces in PBS solutions at 0.1 V/s. Three cyclic repetitive runs are shown for each electrode taken with a difference of three minutes among them (black line: 1st cycle; grey: 2nd cycle; light grey: 3rd cycle). Reproduced with permission from (Pacios, 2009) only plots b and c. Copyright, 2009, American Scientific Publishing.

3.3.2.2 Electrocatalysis of hydrogen peroxide

In this section, we will present the catalytic activity of myoglobin in presence of hydrogen peroxide when carbon-based electrodes are used. By adding aliquots of H_2O_2 to the PBS solution, a considerable increase of the reduction peak was observed at about the same potential at which the electrocatalytic oxygen reduction was taken place. The disappearance of the anodic peak was also detected. All these features can be observed in Fig. 8, in which the behavior of the vertically aligned CNT electrode is taken as a representative example. A reference test using the carbon electrodes in absence of protein but in presence of different concentrations of H_2O_2 showed no obvious cathodic peak response, at least in the potential range under study. The catalytic reduction of H_2O_2 exhibits a behavior similar to the one found with oxygen, which suggests that the reaction mechanisms could be similar. Several plausible explanations have been provided in the literature (Nakajima, 1987; Onuoha, 1997; Zhang, 2002), under the assumption that H_2O_2 can act as an oxidant or reductant agent. It has been suggested that H_2O_2 reacts with MbFe(III) to give an oxyferryl radical (•MbFe(IV)=O) which can react again with H_2O_2 to produce oxygen and Mb(III). When MbFe(III) is reduced at the electrode, it can react fast with oxygen and follow the same path described in the previous section for the oxygen.

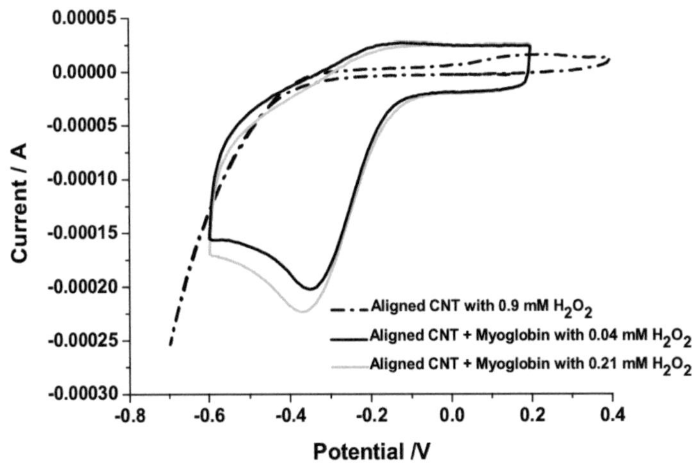

Fig. 8. I/V profiles for the vertically aligned CNT electrode modified with Mb in PBS solutions under different concentrations of H_2O_2 at a scan rate of 0.1V/s. The electrochemical behaviour of CNT in absence of Mb was also included. Adapted from (Esplandiu, 2009a).

In order to have a closer inspection of the electroanalytical parameters, calibration curves were performed on the three carbon electrodes by analizing the current density changes at the catalytic peak as a function of the H_2O_2 concentration (Esplandiu, 2009a). From such calibration curves one would expect a zone of linear response to the analyte that makes possible to extract the sensitivity of the electrochemical response, the limit of detection (LOD) and the range of linear behavior. Table 2 summarizes the electroanalytical characteristics of the myyoglobin graphite composites, CNT composites and CNT forest electrodes.

Electrode	LOD	Sensitivity ($Acm^{-2}M^{-1}$)	Linear range
Graphite Composite	$9. 10^{-5}$ M	0.10	Small
CNT composite	$6. 10^{-5}$ M	0.84	wide
Forest CNT	$5. 10^{-8}$ M	0.47	wide

Table 2. Electroanalytical parameters for the detection of hydrogen peroxide on the different carbon based electrodes.

In the case of the graphite epoxy composites, the limit of detection (LOD) for hydrogen peroxide was of 9.10^{-5} M, the sensitivity, extracted from the curve slope, was of 0.14 A cm^{-2}/M and the linear behaviour extended for a reduced range of concentrations (Fig. 9 a). Above a certain concentration (between 6.10^{-4} and 1.10^{-3} M), the electrode reached saturation and upon further addition of H_2O_2 the current density slightly decreases probably due to a partial protein inactivation in the presence of a higher concentration of peroxide. In the case of the carbon nanotube composite (Fig. 9 b), it was obtained a good calibration curve with a LOD around 6.10^{-5} M, wide range of linear behavior and remarkable high sensitivity of about 0.84 A cm^{-2}/M. Vertically aligned CNT electrode exhibited very low detection limits (LOD around 5.10^{-8} M), a sensitivity of 0.47 A cm^{-2}/M and a very important wide linear concentration range (Fig. 9 c). Among the different analyzed electrode systems, the high signal/noise ratio of the vertically aligned CNT electrode at such low H_2O_2 concentration is remarkable, which could be a consequence of a facilitated electron transfer process of the protein in such environment. In order to have a better quantification of the sensitivity of the different electrodes to the analyte, a normalization of the curves with respect to their electrochemically active protein coverage was made. At the same electroactive protein coverage, the vertically-aligned CNT electrode is almost two times more sensitive than the case of the CNT composite. Thus the higher apparent sensitivity of the CNT composite electrode can be only ascribed to the larger protein coverage due to the higher degree of porosity of this electrode.

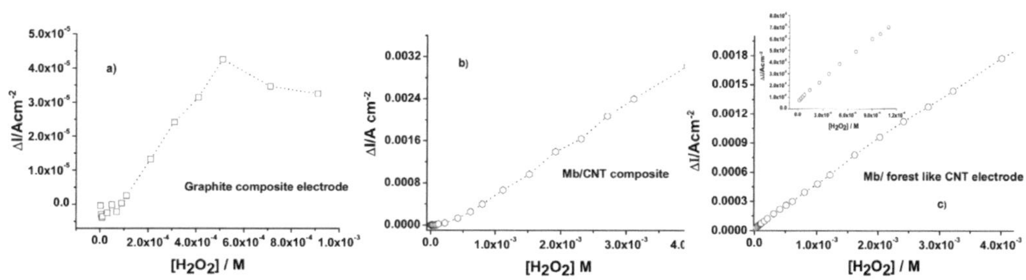

Fig. 9. Calibration curves of hydrogen peroxide on the different Mb/carbon based electrodes. In the forest like system an inset of the curve at the lower concentrations has also been included. Reproduced with permission from (Esplandiu, 2009a). Copyright, 2009, IOP Publishing

These results highlight the good electrochemical performance of electrodes made by a vertical disposition of CNT. The reasons underlying this enhanced performance could be attributed to the fact that this particular configuration results in a more suitable environment for the covalent anchoring of the protein, providing a favourable orientation of the proteins on the tubular structure while keeping a robust protein immobilization. At the same time, the vertical arrangement of CNT facilitates a high surface density of CNT edges, which have been shown to exhibit an improved and fast electron transfer response. All these features can result in a faster electrode response and consequently in a better electron coupling with the protein system.

3.4 DNA modified CNT electrode: Detection of DNA hybridization by using reversible redox indicators

As stated in the introduction, the use of redox reporter molecules for following up the hybridization process is a simple and straightforward method that avoids sample damage or the chemical modification of the DNA. We have used electrostatic groove redox binders based on highly positively charged ruthenium complexes which are expected to accumulate at the electrode interface with the hybridization process. The work has been developed by using mainly forest CNT microelectrode arrays, although the electrochemical response of CNT dispersed on glassy carbon electrodes to the DNA recognition was also followed. PolyC-NH_2, used as probe, was covalently attached to the CNTs and then PolyG was used as target in presence of $Ru(NH_3)_6^{2+/3+}$. We monitored the $Ru(NH_3)_6^{2+/3+}$ electrochemical signal before and after the hybridization process by cyclic voltammetry (CV), chronocoulometry and impedance measurements. We also performed calibration curves by adding increasing concentrations of target PolyG on the probe modified CNT electrode. It will be demonstrated that the combination of CNT microelectrode arrays, the chosen redox indicators and the impedance technique turns out to be a simple and practical methodology for a fast approach to the genosensing process.

3.4.1 Cyclic voltammetry

Cyclic voltammetry is a fast approach to detect the hybridization scheme. Although it does not exhibit the high sensitivity of the other two techniques that will be shown later on, it is very suitable for providing a general trend in the hybridization process. We have found that as the concentration of the target DNA is increased, an augment of the peak current and a decrease in the peak potential difference are observed. Such features indicate an electrostatic surface accumulation of the cationic redox marker due to the increase of the negatively charged oligonucleotide phosphates with the increasing hybridization. Figure 10 a shows this effect for

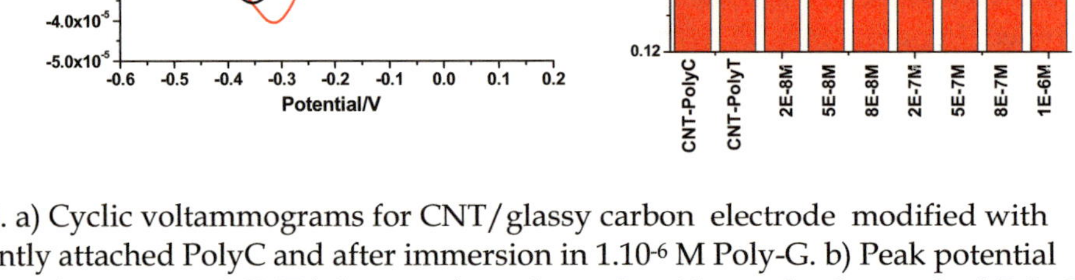

Fig. 10. a) Cyclic voltammograms for CNT/glassy carbon electrode modified with covalently attached PolyC and after immersion in 1.10^{-6} M Poly-G. b) Peak potential difference histogram of CNT/glassy carbon electrode with covalently attached PolyC in presence of non-complementary 1.10^{-6} M polyT and in presence of increasing of complementary PolyG targets. In all the cases the electrodes were blocked with PEG for avoiding non-specific adsorption. The electrolyte was 1.8 mM $Ru(NH_3)_6^{3+}$ in Tris and was deoxygenated with N_2 bubbling for 5 minutes prior to measurements.

a CNT/glassy carbon electrode modified with probe DNA and Fig. 10 b shows the peak potential difference for the the same electrode in presence of increasing aliquots of target oligonuleotides. Such accumulation of the cationic redox marker at the electrode interface makes electron transfer kinetics of the redox probe more favorable. That is sensitively reflected in a peak potential difference decrease. Negative controls in the hybridization process were also obtained by targeting the probe modified electrode with a non-complementary strand such as PolyT.

It is worthy to mention that these features were also found in CNT microelectrode arrays functionalized with PolyC, immersed in increasing concentrations of PolyG strands. These results demonstrate that both CNT platforms are able to anchor the DNA probe and follow the hybridization process. However, this technique can not discriminate between the redox response of the ruthenium species diffusing from the solution and those ones electrostatically adsorb on the phosphate backbone when both of them are present simultaneously. For discrimination of both kinds of species, chronocolumetric measurements can be performed.

3.4.2 Chronocoulometric measurements

Chronocoulometry measurements can help to estimate the accumulated cationic redox marker at the oligonucleotide CNT modified surface as indicative of the hybridization process (Steel, 1998).

In this method a pulse of potential is applied and the response in current is monitorized as a function of the time. From the integration of the current profiles vs time, one can obtain the charge (Q). Chronocoulometry is useful for measuring electrode surface areas, diffusion coefficientes, adsorption of electroactive species and the mechanisms and rate constants for chemical reactions coupled to electron transfer reactions.

The total charge (Q_{total}) measured in response to the potential step comes from three sources (eq. 1) :

1. Charge of the double layer (Q_{dl}) which dominates at shorter times
2. Charge of the adsorbed species (Q_{ads})
3. Charge of the species diffusing at the electrode (Q_{diff}) which dominates at longer times.

$$Q_{TOT} = Q_{dl} + Q_{ads} + Q_{diff} = Q_{dl} + nFA\Gamma_o + \frac{2nFAD_o^{1/2}C_o}{\pi^{1/2}}t^{1/2} \qquad \text{(eq.1)}$$

F is the Faraday constant (96500 C), A is the electrode area, n the number of electrons and C_o is the concentration of the redox species at the bulk electrolyte, D is the diffusion coeficient of the species and Γ_o is the amount of adsorbed species at the interface of the electrode.

In order to estimate the charge corresponding to the electroactive species adsorbed on the surface, one has first to substract the double layer charge (Steel, 1998). Accordingly, we have performed chronocoulometric measurements with the background electrolyte (in absence of the electroactive species that undergo adsorption on the electrode). By extrapolation of a least squares fit from the linear part of the curve charge vs $t^{1/2}$, one can obtain the double layer charge (Q_{dl}) from the the intercept at time zero. Then the chronocoulometric measurements are performed in presence of the redox markers and again the interception at zero time is estimated which corresponds to the charge of the double layer + the charge of the adsorbed amount. Therefore, from both measurements one can get the adsorbed amount by Q_{dl} substraction (Fig. 11). As compared to CV, the chronocoulometry has the advantage that the double layer charge and the charge due to the redox reaction of species adsorbed on the electrode surface can be differentiated from the charge due to the redox reaction of the molecules that diffuse to the electrode. That prevents the confussion over which concentration is being measured, thus symplifying the analysis.

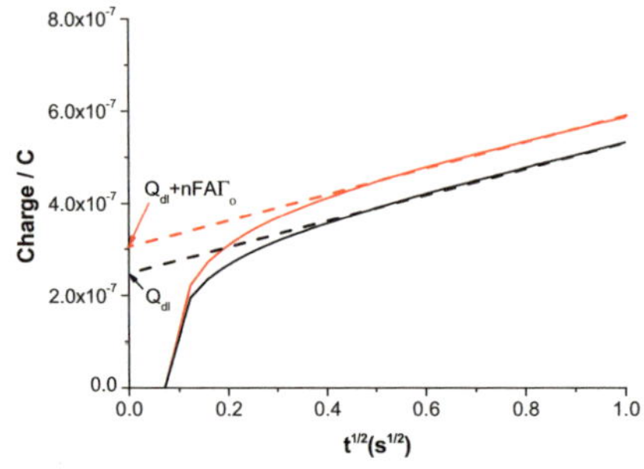

Fig. 11. Illustration of the procedure in order to get the amount of electroactive species adsorbed on an CNT/PolyC electrode. First the chronocoulometric measurements were performed in the background electrolyte, buffer solution of 10 mM TRIS + 10 mM NaCl (black curve) in order to extract the Q_{dl}. Then the solution was changed to 50μM of $Ru(NH_3)_6^{3+}$ in TRIS+NaCl in order to determine the amount of the adsorbed redox species (red curve). Potential pulse from 0.1 V to -0.4V and duration time of 0.5 s.

We have performed such procedure to calculate the charge corresponding to the amount of electroactive species adsorbed on the electrode (Q_{ads}= nFAΓ_0). The values of Q_{ads} have been taken as indicators of the hybridization process (Steel, 1998). In the low ionic strength buffer solution we are using, the trivalent ruthenium species preferentially exchanges with the native monovalent DNA counterion and associates with the phosphate groups in a ratio 1:3. Figure 12 shows the Q_{ads} as a function of increasing aliquots of target DNA for the CNT microelectrode. It can be observed that the probe modified electrode is very sensitive to concentrations below 2.5 10^{-7} M of the target and then a saturation is reached. The limit of detection in such CNT microelectrode system was found to be around 1 nM. Controls with non-complementary strand showed no increase in the charge associated with the electrode surface.

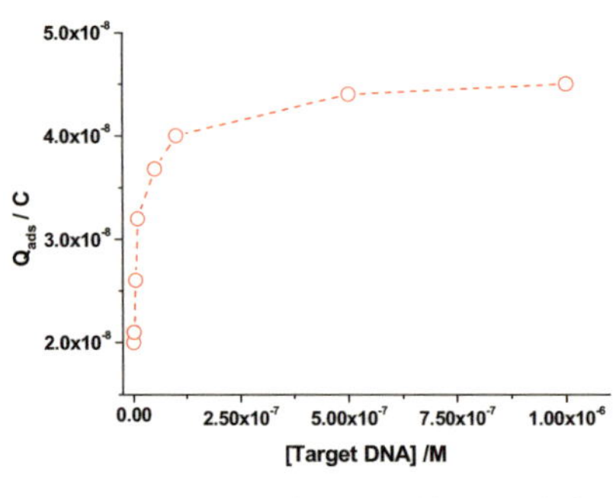

Fig. 12. Q_{ads} on a CNT microelectrode array chemically modified with PolyC and as a function of increasing concetrations of target PolyG.

The use of this technical tool in combination with CNT platforms has been demonstrated to be a quite sensitive strategy to monitor the hybridization process. One can also go a step forward and calculate Γ_o from Q_{ads} to get more details in the hybridization process (Steel, 1998). Although we will not cover such calculation in this chapter, it is worthy to mention that from Γ_o it is possible to get the density of the immobilized DNA on the surface of the electrodes and by calculating the amount of DNA before and after the hybridization, the number of probe and target DNA strands attached to the electrode surface and the hybridization efficiency can be calculated (Steel, 1998).

3.4.3 Electrochemical Impedance Spectroscopy

Electrochemical Impedance Spectroscopy (EIS) is a technique which is used to characterize electrode processes and complex interfaces. This method is based on the application of an AC potential ($E(t) = E_o \cos(\omega + t)$) of small amplitude (typically 10 mV). As a consecuence an AC current is obtained, $I(t) = I_0 \cos(\omega t - \varphi)$. From the relation of both signals the impedance (Z) is obtained ($Z = E(t)/I(t)$). The measurements are carried out at different AC frequencies and thus the name of impedance spectroscopy. Typical frequency range is 0.5 Hz to 100 kHz. Impedance methods allows characterizing the double layer interface at the electrodes and the physicochemical processes of widely differing time constants, sampling electron transfer at high frequency and mass transfer at low frequency. Impedance results are commonly fitted to equivalent circuits of resistors and capacitors, such as the Randles circuit shown in Fig. 13, which is often used to interpret simple electrochemical systems. There are many ways to plot impedance data. Since the function impedance is an AC signal, one can express it in terms of complex number ($Z = Z_{real} + j\, Z_{imag}$) and plot the Z_{real} vs Z_{imag} such kind of representation is called Nyquist plot shown in Fig. 13, which provides visual insight into the system dynamics at the electrochemical interface. Normally such plot exhibits a kind of semicircle profile plus a linear region. The points at which the impedance data cut the real impedance data represent resistance values. In the typical plot of the figure R_{ct} is the charge-transfer resistance, which is inversely proportional to the rate of electron transfer and consequently provides us information about the easiness for electron transfer at the electrode interface (a rough estimation of R_{ct} is related with the diameter of the semicircle); C_d is the double-layer capacitance and can be obtained from the maximum value of impedance data at the semicircle; R_s is the electrolyte resistance and can be extracted from impedance data at the higher frequencies; Z_w is the Warburg impedance and is identified with the linear portion of the impedance spectra that appears at the lower frequencies. The Warburg impedance arises from mass-transfer limitations and can be used to measure effective diffusion coefficients.

Another way to represent the impedance results is by using the Bode Plot where the modulus of the impedance ($\log |Z|$) and the phase angle (φ) between the AC potential and the AC current as a function of the frequency ($\log \omega$) are plotted (Fig. 13). In this plot, the impedance data which are frequency independent represent the behavior of the resistive processes (phase angles close to 0) whereas the ones that are dependent on the frequency are more related to capacitive or diffusive processes (phase angles between -90° or -45°) . Thus the impedance spectra can give us a broad overview of the different processes taking place at the electrochemical interface (capacitive, resistive, diffusion effects) and which one is dominating more at a specific range of frequencies.

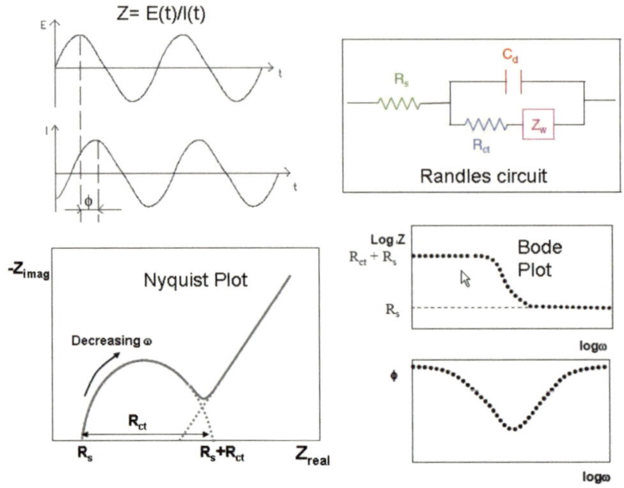

Fig. 13. Impedance function and typical equivalent circuit at the electrochemical interface (the Randles circuit). Different representations of the impedance data (Nyquist and Bode plots)

Impedance spectroscopy is a very versatile tool widely used in different fields (corrosion, semiconductor electrodes, polymers and coatings, batteries, fuel cells, electrode kinetics and mechanisms, biomedical and biological systems, solid-state systems, etc.). Since EIS turns out to be very sensitive to probe the interfacial properties it has become an attractive tool to monitor biorecognition processes at the electrode surface such as in immunosensing or in genosensing fields. This technique also offers the added value of monitoring processes in label free conditions. From the considerable amount of papers dealing with EIS in the electrochemical biosensing area, part of them deals with non-faradaic measurements based on the monitoring of the capacitance values and the other part relies on the used of redox indicators to monitor the changes in the charge transfer resistance.

We have also profited from this technique and performed impedance measurements to follow the hybridization process. Before going in the details of such approach, we would like to mention an important aspect of this study which has to do with the scaling down of the electrode size and its effect on the impedance spectra. Since we are performing impedance spectra in presence of a redox active reporter ($Ru(NH_3)_6^{2+/3+}$, the more approppiate impedance parameter to probe the interfacial changes is the charge transfer resistance (R_{ct}). In contrast to other widely used redox probes such as $Fe(CN)_6^{3-/4-}$, the ruthenium electroactive species is well known to be a more reversible redox pair and also it does not physiadsorb on carbon electrodes, which makes the use of this redox probe quite desirable for many applications. However and due to its high electron kinetics, the R_{ct} analysis of interfacial processes on metallic macroelectrodes becomes impossible. In such cases the impedance spectra is fully dominated by the Warburg impedance, that is, the electrochemical process is diffusion controlled. However, the diffusion control of the electrochemical process can start to decrease when the size of the electrode is reduced. Figure 14 shows this transition from diffusion to electron kinetics control due to a decrease in the electrode area and therefore an increment of the system impedance. The Nyquist plot changes from a straight line (Warburg impedance control by using macroscopic forest CNT electrode) to a semicircle (charge kinetics control by using CNT microelectrode array). Similar changes have been observed in carbon nanofiber electrode arrays (Siddiqui, 2010). Such change in the control process at the electrochemical interface has a real advantage in our case. The impedance detection of the hybridization process based on the charge transfer resistance of the ruthenium reporters is now possible if CNT microelectrodes are used. Under these conditions one can take the R_{ct} parameter as indicative of the hybridization process.

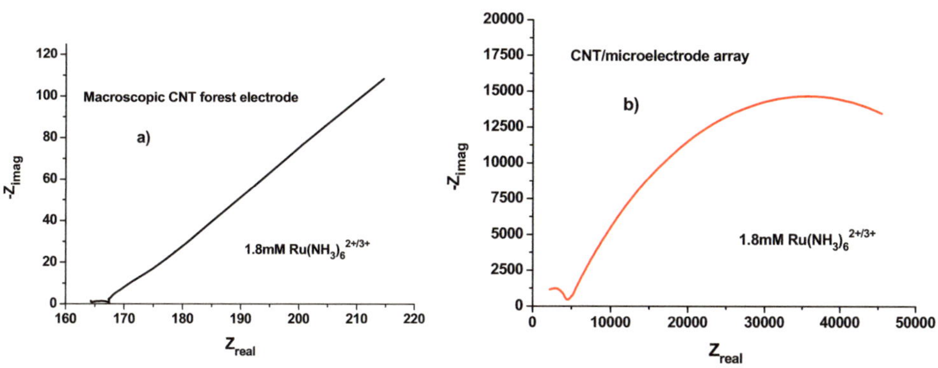

Fig. 14. Impedance spectra of a macroscopic CNT forest electrode (a) and a CNT microelectrode array in presence of 1.8 mM $Ru(NH_3)_6^{2+/3+}$ (b). It can be observed the transition of the impedance spectra from a diffusion-controlled system to a more kinetic-controlled one with the scaling down of the electrode size.

We have performed the monitoring of the hybridization process on the CNT microelectrode array. The parameter R_{ct} was obtained by fitting the impedance spectra to the equivalent circuit depicted in Fig. 13 but without the Warbug parameter and with the double layer capacitance being reresented by a constant phase element (CPE), a parameter that takes into consideration the inhomogeneities of the electrode. The impedance of a CPE is $Z_{CPE} = C^{-1} (j\omega)^{-1}$, where C is the capacitance and n is a fitting parameter that can vary from 0 to 1. When n=1, the constant phase element approaches the value of a pure capacitance. In our case, n was ranging between 0.78-0.73.
The quantitative values of R_{ct} have been plotted in Fig. 15 as a function of increasing target concentrations. As mentioned before, this parameter provides us information about the easiness for electron transfer at the electrode interface, and the trend of this parameter is similar to the findings of the cyclic voltammetry. R_{ct} decreases as the cationic redox marker is accumulated on the interface due to the surface increase of negatively charged phosphates. The electrostatic attraction of the redox marker facilitates the electron exchange as the hybridization process is taken place. The change in resistance is again very sensitive at concentrations below 2.5 10^{-7} M. Above such concentration the charge transfer resistance reaches a plateau. Limits of detection were calculated to be around 1. 10^{-10} M demonstrating the high sensitivity of this technique.
The results of this section allow us to stress the power of the impedance technique in monitoring biorecognition events with a simple hydridization detection scheme. This approach could not have been achieved without the versatility of the CNTs to be disposed in a microelectrode array configuration.

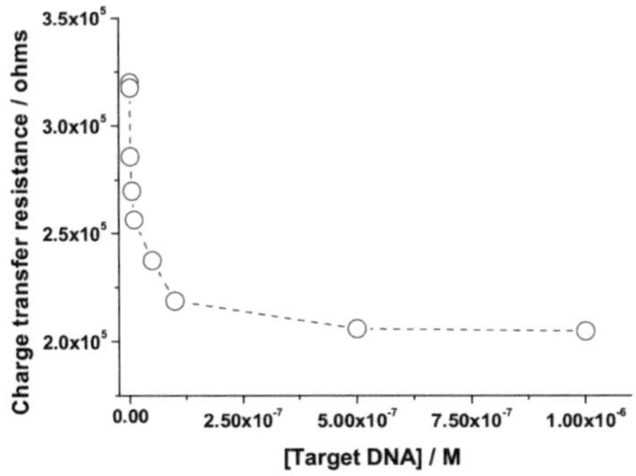

Fig. 15. Impedance spectra of the probe modified CNT microelectrode performed with 1.8 mM $Ru(NH_3)_6^{3+}$ in PBS with increasing concetrations of target DNA. The electrolyte was deoxygenated with N_2 bubble for 5 minutes prior to measurements. The impedance was performed with PBS or 1.8 mM $Ru(NH_3)_6^{3+/2+}$ in TRIS at -0.26 V potential, 10 mV amplitude, from 300 mHz to 100 kHz lower and upper limit.

4. Conclusion and future directions

In this chapter we have tried to highlight the electroanalytical CNT properties for biosensing applications. The remarkable electrochemical properties of CNT arise as a combination of different aspects given mainly by their surface structure, nanometric size and aspect ratio, their high specific area, their electrochemical anisotropy and their versatility to be disposed in different electrode configurations.

We have illustrated two examples that demonstrate the suitability of CNT as electrochemical transducers in enzymatic and DNA sensors. In the first example, we have proven that carbon nanotubes-based materials exhibit an enhanced electrochemical response towards Myoglobin as compared to other electrode substrates. The CNT systems represent an adequate platform for the protein wiring which allows on one side the direct electrochemistry between the hemin active redox center of the protein and on the other side the preservation of the protein bioactivity to different substrates (oxygen, hydrogen peroxide, etc.). We have tested two different CNT electrodes: forest CNT and epoxy CNT composite electrodes. In the former CNT electrode, the enhanced electrochemical signal could arise from a more facilitated electron transfer. This seems to be a consequence of the exposition of a higher density of the more electroactive sites (edge sites), just at the electrode/solution interface. That could facilitate the wiring and electron exchange with Mb.

In the case of the CNT composites, the electrochemical improvement comes from the more porous character of this material and consequently higher surface area for the immobilization of more amount of protein which amplifies the signal response. All these aspects exert influence in the final electroanalytical characteristics such as the remarkable low detection limits in the response of the protein to its specific substrates (case of the Mb/forest CNT electrodes), the high sensitivity due to the large protein loading (case of the CNT composite electrodes) and the wide analyte concentration range. Indeed all these features encourage the use of Mb/CNT electrodes as efficient sensors for oxygen, hydrogen peroxide or nitric oxide with could bring impact in clinical and environmental applications.

The second example reported here is about the use of electrochemical CNT transducers in genosensors. In our studies, we have mainly used a microelectrode array made of vertically aligned CNTs as electrode platform. That electrode system results in a robust platform for single point linkage of probe DNA and for following up the hybridization process with a simple and reversible detection scheme based on electrostatic groove redox binders which are positively charged. Different electrochemical techniques were used such as Cyclic Voltammetry, Chronocoulometry and Impedance Spectroscopy. Chronocoulometry allowed us to discriminate the accumulated redox species at the interface from those diffusing in the electrolyte. The monitoring of the accumulated species provided us a sensitive way to follow the DNA hybridization. On the other side, the combination of the CNT microelectrode array, together with the impedance measurements turned out to be a powerful approach to follow the hybridization process with such highly positively charged redox indicators. It is worthy to mention that without such CNT microelectrode and the scaling down size effects of the electrode on the impedance measurements it should have not been possible to characterize the biorecongition process with that simple but sensitive approach. Again all these results hold promise for the development of simple, and practical genosensors for many applications.

The two examples of biosensing with CNT platforms reported in this chapter encourage the exploitation of novel and miniaturized CNT electrode configurations that can be tunable according to different electrochemical applications. Such applications do not remain restricted to the electrochemical biosensing field but can cover other exciting areas such as the development of electrochemical actuators or electrochemical energy-harvesting devices.

Acknowledgements. This work was financially supported by the Ministry of Education and Science of Spain (Project NAN2004-093006-C05-03 and CTQ-2006-15681-C02-01). M. Pacios acknowledges a FPI predoctoral grant.

5. References

Banks, C.E.; Compton, R.G. (2006) New electrodes for old: from carbon nanotubes to edge plane pyrolytic graphite. *The Analyst*, 131, (1), pp. 15-21.

Bonanni, A.; Del Valle, M. (2010) Use of nanomaterials for impedimetric DNA sensors: a review. *Analytica Chimica Acta*, 678, pp. 7-17.

Bonanni, A.; Esplandiu, M.J.; Del Valle, M. (2007) Signal amplification for impedimetric genosensing using gold-streptavidin nanoparticles, *Electrochimica Acta*, 53 , (11), pp. 4022-4029.

Bonanni, A; Esplandiu, M.J., Del Valle, M. (2009) Impedimetric genosensors employing COOH-modified carbon nanotube screen-printed electrodes, *Biosensors and Bioelectronics*, 4 (9), pp. 2885-2891.

Bonanni, A; Esplandiu, M.J.; Del Valle, M. (2010) Impedimetric genosensing of DNA polymorphism correlated to cystic fibrosis: a comparison among different protocols and electrode surfaces. *Biosensors and Bioelectronics*, 26 (4) pp. 1245-1251.

Cai, C; Chen, J. (2004) Direct electron transfer of glucose oxidase promoted by carbon nanotubes. *Analytical Biochemistry*, 332, pp. 75-83.

Carot, M.L.; Torresi, R.M.; García, C.D.; Esplandiu, M.J.; Giacomelli, C.E. (2010) Electrostatic and hydrophobic interactions involved in CNT biofunctionalization with short ss-DNA. *Journal of Physical Chemistry C*, 114 (10) pp. 4459-4465.

Drummond, T.G.; Hill, M.G.; Barton, J.K. (2003) Electrochemical DNA sensors, *Nature Biotechnology*, 21 (10) pp. 1192-1199.

Dumitrescu, I; Unwin, P.R.; Macpherson, J.V. (2009) Electrochemistry at Carbon Nanotubes: Perspective and Issues. *Chemical Communications*, (45), pp. 6886-6901.

Esplandiu, M. J.; Pacios, M. Cyganek; L. Bartrolí; J. del Valle, M. (2009a) Enhancing the electrochemical response of myoglobin with carbon nanotube electrodes. *Nanotechnology*, 20, pp. 355502.

Esplandiu, M.J. (2009). Electrochemistry on Carbon Nanotube modified surfaces, in: *Chemically Modified Electrodes*, R.C. Alkire, D.M. Kolb, J. Lipkowski, P.N. Ross (Eds.), 57-209, Wiley-VCH Verlag, ISBN 978-3-527-31420-1, Weinheim, Germany.

Fan, C.H.; Plaxco, K.W., Heeger, A.J. (2003) Electrochemical interrogation of conformational changes as a reagentless method for the sequence-specific detection of DNA, *Proceedings of the National Academy of sciences of the United States of America*, 100 (16) pp. 9124-9137.

Gooding, J.J. (2005). Nanostructuring Electrodes with Carbon Nanotubes: A Review on Electrochemistry and Applications for Sensing. *Electrochimica Acta*, 50, (15), pp.3049-3060.

Gooding, J.J.; Wibowo, R.; Liu, J.; Yang, W.; Losic, D; Orbons, S.; Mearns, F.J.; Shapter, J.G.; Hibbert, D.B. (2003) Protein electrochemistry using aligned carbon nanotube arrays. *Journal of the American Chemical Society*, 125, pp. 9006-9007.

Gorodetsky, A.A.; Buzzeo, M.C.; Barton, J.K. (2008) DNA-mediated electrochemistry, *Bioconjugate Chemistry*, 19 (2), pp. 2285-2296.

Guiseppi-Elie, A; Lei, C.; Baughman, R.H. (2002) Direct electron transfer of glucose oxidase on carbon nanotubes. *Nanotechnology*, 13 (5) pp. 559-564.

He, P.; Xu, Y.; Fang, Y. (2006) Applications of carbon nanotubes in electrochemical DNA biosensors, *Microchimica Acta*, 152, pp. 175-186.

He, P.G.; Dai, L.M. (2004a) Aligned carbon nanotube-DNA electrochemical Sensors, *Chemical Communications*, 3, pp. 348-349.

Heng, L.Y.; Chou, A.; Yu, J.; Chen, Y.; Gooding, J.L. (2005) Demonstration of the advantages of using bamboo-like nanotubes for electrochemical biosensor applications compared with single walled carbon nanotubes, *Electrochemistry Communications*, 7 (12), pp. 1457-1462.

Jones, C.P.; Jurkschat, K.; Crossley, A.; Compton, R.G.; Riehl, B.L.; Banks, C.E. (2007) Use of high-purity metal-catalyst-free multiwalled carbon nanotubes to avoid potential experimental misinterpretations, *Langmuir*, 23 (18), pp. 9501-9504.

Katz, E.; Willner, I. (2004). Biomolecule-functionalized carbon nanotubes: applications in nanobioelectronics. *ChemPhysChem*, 5, (8), pp.1084-1104.

Kim, S.N.; Rusling, J.F.; Papadimitrakopoulos. (2007) Carbon Nanotubes for electronic and electrochemical detection of biomolecules. *Advanced Materials*, 19, (20), pp. 3214-3228.

Koehne, J.; Li, J.; Cassell, A.M.; Chen, H.; Ye, Q.; Ng, H.T.; Han, J.; Meyyappan, M. (2004). The fabrication and electrochemical characterization of carbon nanotube nanoelectrode arrays. *Journal of Materials Chemistry*, 14, pp. 676-684.

Kumar, S.A.; Wang, S.F.; Chang, Y.T.; Lu, H.C.; Yeh, C.T. (2011) Electrochemical properties of myoglobin deposited on multi-walled carbon nanotubes/ciprofloxacin film. *Colloids and surfaces B-Intefaces*, 82 (2) pp. 526-531.

Laviron, E. (1979) General expression of the linear potential sweep voltammogram in the case of diffusionless electrochemical systems. *Journal of Electroanalytical Chemistry*, 101, pp. 19-28.

Li, Y.; Lin, X.; Jiang, C. (2006) Fabrication of a Nanobiocomposite Film Containing Heme Proteins and Carbon Nanotubes on a Choline Modified Glassy Carbon Electrode: Direct Electrochemistry and Electrochemical Catalysis. *Electroanalysis*, 18, pp. 2085-2091.

Liu, J.; Chou, A.; Rahmat, W.; Paddon-Row, M.N.; Gooding, J.J. (2005). Achieving direct electrical connection to glucose oxidase using single walled carbon nanotube arrays. *Electroanalysis*, 17, (1), pp. 38-45.

Martín-Fernández, I. (2010) Thesis: RTCVD synthesis of carbon nanotubes and their wafer scale integration into FET and sensor processes, Barcelona, Spain.

McCreery, R.L. (2008). Advanced Carbon electrode materials for molecular Electrochemistry, *Chemical Reviews*, 108 (7) pp. 2646-2687.

Moore, R.R.; Banks, C.E; Compton, R.G. (2004) Basal plane pyrolytic graphite modified electrodes: comparison of carbon nanotubes and graphite powder as electrocatalysts. *Analytical chemistry*, 76, (10), pp. 2677-2682.

Nakajima, R. Yamazaki, I. (1987) The mechanism of oxyperoxidase formation from ferryl peroxidase and hydrogen peroxide. *Journal of Biological Chemistry*, 262, pp. 2576.

Onuoha, A.C. Zu, X. Rusling, J. (1997) Electrochemical Generation and Reactions of Ferrylmyoglobins in Water and Microemulsions. *Journal of the American Chemical Society*, 119, pp. 3979.

Pacios, M. ; del Valle, M. ; Bartrolí, J. ; Esplandiu, M.J. (2008). Electrochemical behavior of rigid carbon nanotube composite electrodes. *Journal of Electroanalytical Chemistry*, 619, pp. 117-124.

Pacios, M.; Del Valle, M.; Bartrolí, J.; Esplandiu, M. J. (2009) Electrocatalyzed O_2 Response of Myoglobin immobilized on Multi-Walled Carbon Nanotube Forest Electrodes. *Journal of Nanoscience and Nanotechnology*, 9, pp. 1-7.

Palangsuntikul R.; Somasundrum, M.; Surareungchai, W. (2010) Kinetic and analytical comparison of horseradish peroxidase on bare and redox modified single-walled carbon nanotubes, *Electrochimica Acta*, 56 (1) pp. 470-475.

Patolsky, F; Weizmann, Y.; Willner, I. (2004) Long-range electrical contacting of redox enzymes by SWCNT connectors. *Angewandte Chemie Int. Ed.*, 43, pp. 2113-2117.

Pividori, M.I.; Merkoci, A.; Alegret, S. (2003) Graphite-epoxy composites as a new transducing material for electrochemical genosensing, *Biosensors and Bioelectronics*, 19 (5), pp. 473-484.

Pumera, M.; Merkoci, A.; Alegret, S. (2006) Carbon nanotube-epoxy composites for electrochemical sensing. *Sensors and Actuators B-Chemical*, 113, (2), pp. 617-622.

Rubianes, M.D.; Rivas, G. (2004) Enzymatic biosensors based on carbon nanotube paste electrodes. *Electroanalysis*, 17 (1), pp. 73-78.

Salinas-Torres, D.; Huerta, F.; Montilla, F.; Morallón, E. (2011) Study of electroactive and electrocatalytic electrodes surfaces of single walled carbon nanotube-modified. *Electrochimica Acta*, 56, pp. 2464-2470.

Siddiqui, S; Arumugam, P.U.; Chen, H.; Li, j.; Meyyappan, M. (2010) Characterization of carbon nanofiber electrode arrays using Electrochemical Impedance Spectroscopy: effect of scaling down the electrode size, *ACS Nano*, 4, (2), pp. 955-961.

Steel, A.B.; Herne, T.M.; Tarlov, M.J. (1998) Electrochemical quantitation of DNA immobilized on gold, *Analytical Chemistry*, 70, (22), pp. 4670-4677.

Tasis, D.; Tagmatarchs, N.; Bianco, A.; Prato, M. (2006) Chemistry of Carbon Nanotubes. *Chemical Reviews*, 106, (3), pp. 1105-1136.

Wang, J. (2005) Carbon-nanotube based electrochemical biosensors: a review. *Electroanalysis*, 17, (1), pp. 7-14.

Wang, J.; Kawde, A.; Musameh, M. (2003) Carbon nanotube modified glassy carbon electrodes for amplified label-free electrochemical detection of DNA hybridization, *The Analyst*, 128, pp. 912-916.

Wang, J.; Li, M.; Shi, Z.; Li, N.; Gu, Z. (2004) Electrochemistry of DNA at single-wall carbon nanotubes, *Electroanalysis*, 16 (1-2), pp. 140-144.

Wang, J.; Liu, G.D.; Jan, M.R. (2004a) Ultrasensitive electrical biosensing of proteins and DNA: carbon nanotube derived amplification of the recognition and transduction events, *Journal of the American Chemical Society*, 126 (10), pp. 2010-3011.

Wang, J.; Musameh, M.; Lin, Y. (2003) Solubilization of carbon nanotubes by nafion: towards the preparation of amperometric biosensors. *Journal of the American Chemical Society*, 125, pp. 2408-2409.

Yu, X.; Chattopadhyay, D.; Galeska, I.; Papadimitrakopoulos, F. ; Rusling, J.F. (2003) Peroxidase activity of enzymes bound to the ends of single-wall carbon nanotube forest electrodes. *Electrochemistry Communications*, 5, (5), pp. 408-411.

Zhang, H. Hu, N. (2007) Assembly of myoglobin layer-by-layer films with poly(propyleneimine) dendrimer-stabilized gold nanoparticles and its application in electrochemical biosensing. *Biosensensors and Bioelectronics* 23, pp. 393-399.

Zhang, L.; Zhao, G.; Wei, X.; Yang, Z. (2004) Electroreduction of Oxygen by Myoglobin on Multi-walled Carbon Nanotube-Modified Glassy Carbon Electrode. *Chem. Lett.* 33, pp. 86-87.

Zhang, Z. Chouchane, S. Magliozzo, R. Rusling, S. (2002) Direct voltammetry and enzyme catalysis with M. Tuberculosis catalase-peroxidase, peroxidases and catalase in lipid films. *Analytical Chemistry*, 74, pp. 163.

Zhao, L.; Liu, H.; Hu, N. (2006) Electroactive films of heme protein-coated multiwalled carbon nanotubes. *J. Colloid Interface Sci.* 296, pp. 204-11.

Zhao, L.; Liu, H.; Hu, N. (2006a) Assembly of layer-by-layer films of heme proteins and single-walled carbon nanotubes: electrochemistry and electrocatalysis. *Anal. Bioanal. Chem.* 384, pp. 414-422.

13

Carbon Nanotube-Mediated Labelling Platforms for Stem Cells

H. Gul-Uludag[1,4], W. Lu[2], P. Xu[1], J. Xing[3,6] and J. Chen[1,4,5]
[1]Department of Biomedical Engineering, University of Alberta, Edmonton, Alberta,
[2]IntelligentNano Inc., Edmonton, Alberta,
[3]Department of Laboratory Medicine and Pathology, University of Alberta, Edmonton, Alberta,
[4]Department of Electrical and Computer Engineering, University of Alberta, Edmonton, Alberta,
[5]National Institute of Nanotechnology,
[6]Cross-cancer Institute, Edmonton, Alberta,
Canada

1. Introduction

The discovery of carbon nanotubes (CNTs) promised to revolutionize biomedical research by offering unique performance attributes inherent in their unique mechanical, electrical, optical and magnetic properties [1-6]. CNTs are highly anisotropic cylindrical nanostructures with lengths ranging from several hundred nanometres to several micrometres, and diameters of 0.4–100 nm. CNTs can be classified as single-walled carbon nanotubes (SWNTs) or multi-walled carbon nanotubes (MWNTs), depending on the number of graphene layers which compose its structure.

CNTs are typically prepared using one of three methods: laser ablation [7], arc-discharge [8], and chemical vapor deposition (CVD) [9]. CVD, the most popular pathway of production, involves reacting a metal catalyst (such as nickel, cobalt with a hydrocarbon feedstock at high temperatures (>800°C) to produce CNT. Both SWNTs and MWNTs, are commercially available in varying degrees of purity. Pristine CNT (non-purified and/or non-functionalised) are completely insoluble in all solvents -- a property which has generated some health concerns in terms of toxicity [10]. Therefore the functionalization of CNT is a key step towards novel biomedical applications. Advanced methods to chemical modification and functionalization have enabled solvation and dispersal of CNTs in water [11-13]. Properly functionalized CNT have a high propensity to cross cell membranes. Various functionalized SWNTs have been shown to transport various bio-molecules across into living cells without cytotoxicity [14, 15]. Nanospearing or nanopenetration molecular delivery, which relies on the penetration of magnetic carbon nanotubes (mCNTs) into cells by magnetic field exposure, was also recently suggested by Cai et al [3]. In this technique, a rotating magnetic field first guided CNTs to spear the cells. In a second step, a static field pulled CNT into the cells. The researchers have achieved unprecedented high molecular delivery efficiency into difficult-to-transfect cells such as primary mammalian neurons and

B lymphocytes with high viability after transfection. In addition to providing a successful molecular delivery platform, CNTs have been also used for successful cellular labelling/tracking in recent years. CNTs coated with peptides have been reported to be successfully delivered and imaged in living human cervical cancer HeLa cells [16]. More interestingly, CNT fluorescence was detected and imaged from living Drosophila larva by using near-infrared (NIR) with no adverse effects on the viability and growth of the larva [17]. These recent findings prompted our group to investigate CNT-mediated labelling platforms for stem cell tracking.

One of potential applications of nanotechnologies in stem cell research is the noninvasive tracking of stem cells and progenitor cells transplanted *in vivo*. Hematopoietic stem/progenitor cells (HSPC) are the most common type of stem cells used for cellular therapies for many decades [18]. Transplantation of HSPC obtained from peripheral blood (PB), bone marrow (BM) or from umbilical cord blood (CB) has been routinely used to rescue bone marrow function following high-dose chemotherapy for non-malignant and malignant hematologic disorders such as leukemias [19, 20]. Labelling HSPC and the consecutive non-invasive tracking *in vivo* following transplantation have become important in understanding HSPC homing/engraftment into BM and the therapeutic efficacy for future stem-cell-based therapies. The fate of transplanted stem cells or progenitor cells is generally assessed by labelling them *in vitro* with a fluorescent dye, thymidine analog, transfected gene such as LacZ, or green fluorescent protein (GFP). However, the methodology for real-time tracking of HSPCs *in vivo* is still lacking, which seriously restricts the progress of research. This chapter describes the evaluation of the uptake efficiency of fluorescein-isothiocyanate-labelled functionalized mCNT (FITC-mCNT) into HSPC and their effect on the HSPC properties.

2. Materials and methods

2.1 Source and isolation of human $CD34^+$ HSPC

Peripheral blood leukapheresis product (LP) was obtained with the patients' informed consent (in accordance with the institutional guidelines approved by the Human Research Ethic Board of the University of Alberta) before cryopreservation. CB was collected immediately after delivery in a sterilized tube containing heparin (1000 IU/ml), and with the informed consent of the mother (in accordance with the institutional guidelines approved by the Health Research Ethics Board of the University of Alberta). Light density cells from CB and LP were obtained by Percoll density gradient centrifugation and enriched for $CD34^+$ cells by immunoaffinity selection with MACS paramagnetic beads (Miltenyi Biotec, Auburn, CA, USA), according to the manufacturer's instructions [21]. The purity of isolated CB and LP $CD34^+$ cells were >95% and >91%, respectively, as determined by fluorescence-activated cell sorter (FACS) analysis.

2.2 Synthesis of FITC-mCNT

Single-walled mCNT containing Ni and Y at the tip, with an average diameter of 1.2–1.5 nm and a length of 2–5 μm, were obtained from Sigma-Aldrich (Ontario, Canada). These specific mCNT were chosen for our experiment because they have been used in magnetic-field-driven biomolecule delivery with very high efficiency [3]. Compared to the reported methods, our synthesis of FITC-mCNT is much simpler and yet returns higher chemical yields [11]. As shown in figure 1, the mCNTs were oxidized to form carboxylic acid groups on the surface [3, 11, 22-24], which were reacted with thionyl chloride and then 2_-(ethylenedioxy)bis(ethylamine) to produce amine terminated nanotubes. The amine was then reacted with FITC to provide FITC-labelled highly water-soluble FITC-mCNT. Infrared (IR), X-ray photoelectron spectroscopy (XPS) and UV–vis spectroscopy were used to validate the chemical reactions of the intermediate and the final FITC-mCNT products.

1. Oxidation of the carbon nanotubes: The first oxidation step was carried out as described previously [3, 11, 22-24]. Briefly, the purchased carbon nanotubes (200 mg) were refluxed with 0.5 M HNO_3 (100 ml) for 48 h to introduce carboxylic groups. After refluxing, the solution was diluted with deionized water, filtered over a 0.2 µm polycarbonate filter (Millipore) and washed several times with deionized water. The sample was collected and dried overnight in a vacuum oven at 80 °C to obtain mCNT 2 (170 mg) (figure 1).

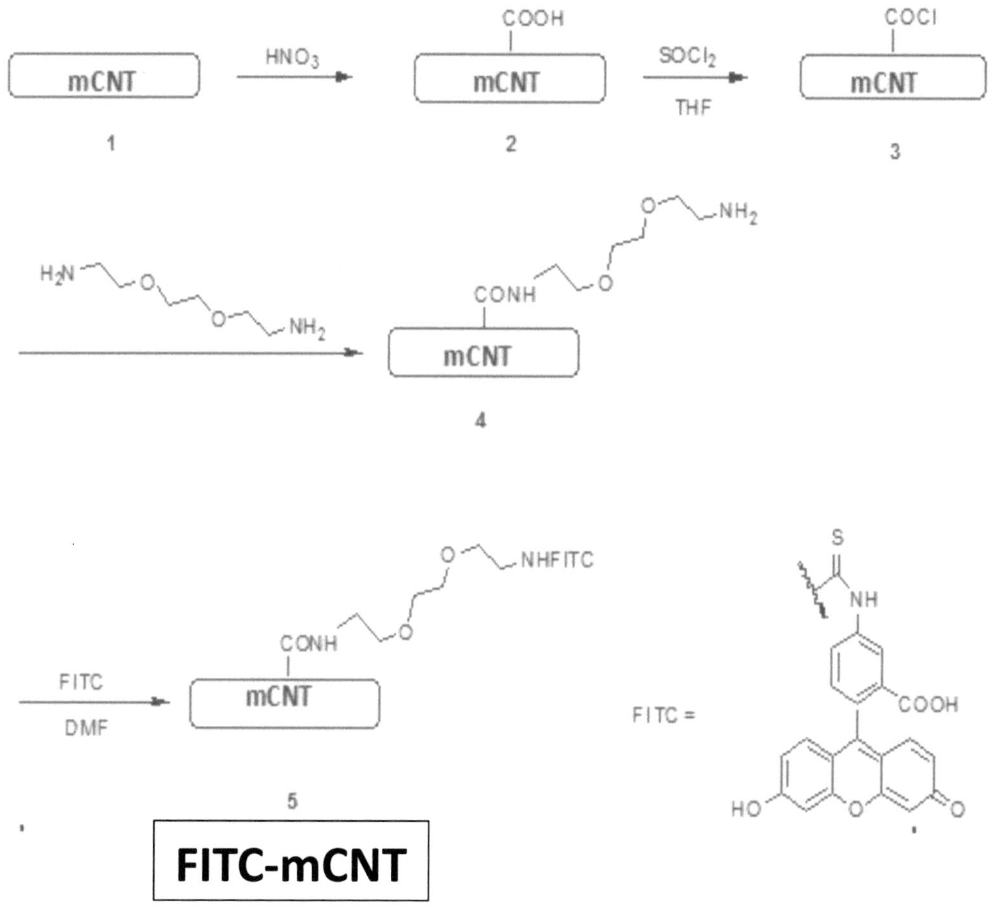

Fig. 1. Schematic illustration of FITC-mCNT synthesis. (1) The mCNTs were oxidized to form carboxylic acid groups on the surface. (2) These nanotubes were reacted with thionyl chloride, (3) and then 2_-(ethylenedioxy)bis(ethylamine) to produce amine-terminated nanotubes. (4) The amine was then reacted with FITCs to form highly water-soluble FITC-mCNTs. (5) Freshly synthesized fluorescent and magnetic FITC-mCNTs were then used in magnetic-field-driven HSPC uptake experiments.

2. Reaction with thionyl chloride to give SWNT-COCl: A suspension of mCNT 2 (100 mg) in 20 ml of $SOCl_2$, together with five drops of dimethylformamide (DMF), was stirred at 70 ◦C for 24 h. The mixture was cooled and centrifuged at 2000 rpm for 30 min. The excess $SOCl_2$ was decanted and the resulting black solid was washed with anhydrous THF (3 × 20 ml) and dried overnight in a vacuum oven at 80°C to give mCNT 3 (78 mg) (figure 1).
3. Coupling with 2-(2-(2-aminoethoxy)ethoxy)ethan amine: The mixture of mCNT 3 (50 mg) and anhydrous 2-(2-(2-aminoethoxy)ethoxy)ethanamine (120 ml) was heated at 100 ◦C for 100 h. During this time, the liquid phase became dark. After cooling, the mixture was poured into methanol (100 ml) and centrifuged to give a black solid, which was washed several times with methanol. The resulting solid was dried overnight in a vacuum oven at 80 ◦C to give mCNT 4 (42 mg) (figure 1).
4. Labeling with FITC: A suspension of the mCNT 4 (25 mg) and FITC (5 mg) in anhydrous DMF (10 ml) was stirred in the dark for 5 h. Then the reaction mixture was poured into anhydrous ethyl ether (40 ml) and centrifuged to give a black solid, which was washed with methanol until TLC (10% MeOH in dichloromethane) showed no free FITC left. The product was dried overnight in a vacuum oven at 80 °C to give mCNT 5 (23 mg) (figure 1).

2.3 Magnetic-field driven HSPC uptake of FITC-mCNT

$CD34^+$ cells with a density of 2.5 to $3*10^5$ cells/plate were placed in poly-L-lysine (10 μg/ml)-coated p35 mm culture dishes and incubated for 45 min at 37 °C, 5% CO2. The magnetic-field-driven delivery method was applied as previously described [3]. Culture dishes were vertically placed into a beaker containing 50 ml serum-free IMDM media with different concentrations of FITC-mCNT or mCNT (10, 20 and 40 μg/ml) and then the beaker was placed on a magnetic stirrer set up at 1200 rpm for 10 min. Culture dishes were then transferred to an Nd–Fe–B permanent magnet for another 10 min. The uptake experiment was terminated by washing the cells with PBS and the cells were incubated for 1, 3, 6, 24 and 48 h at 37◦C, 5% CO2 for uptake efficiency of FITC-mCNT.

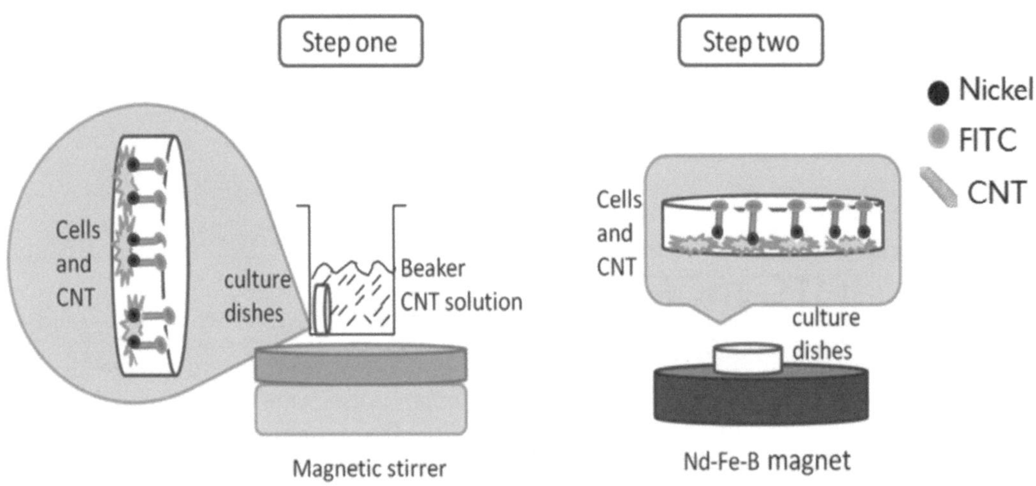

Fig. 2. Schema of magnetic-filed driven delivery method.

2.4 FACS analysis

The cells exposed to FITC-mCNT at different time points and concentrations were collected, extensively washed and then fixed in 1% paraformaldehyde prior to FACS analysis (FACscan, Becton-Dickinson, San Jose, CA, USA) to determine uptake efficiencies.

2.5 Confocal microscopy

$CD34^+$ cells were seeded at a density of $1*10^5$ cells/cm^2 on cover slips previously coated with poly-L-lysine (10 μg/ml) for 45 min at 37 $^\circ$C, 5% CO_2. The cells exposed to 40 μg/ml FITC-mCNT and mCNT (the control) with the nanotube spearing method. Uptake was terminated by washing the cells twice with ice-cold PBS. After 1, 3 and 6 h incubation, the cells were fixed in 3.7% paraformaldehyde, stored overnight at 4 °C and examined under an inverted confocal laser scanning microscope (Carl Zeiss LSM510, Toronto, Canada) equipped with imaging software (LSM 5 Image Browser, Carl Zeiss).

2.6 Cell viability

Cell viability was measured by the trypan blue exclusion assay. $CD34^+$ cells were collected and pelleted by centrifugation at 700 *g* for 5 min. The cells were re-suspended in 25 µl phosphate buffered saline (PBS), mixed with 5 µl of 0.4% trypan blue solution and counted using a hemocytometer.

2.7 Colony-forming unit (CFU) assay

After exposure to the FITC-mCNT solution with a concentration of 40 μg ml−1 for 1, 3 and 6 h, $CD34^+$ cells were plated in triplicate in standard semisolid methylcellulose haematopoietic progenitor culture media (human MethoCult GF; Stem Cell Technologies, Vancouver, BC, Canada) at concentrations of $1*10^3$ /ml These culture plates were incubated at 37 $^\circ$C in 5% CO_2. Colonies were identified and enumerated 14 days later.

3. Results and discussion

In the present study, we examined the magnetic-field-driven uptake of FITC-mCNT into HSPC. FITC-mCNT with both fluorescent and magnetic properties were synthesized and freshly prepared for our uptake experiments as summarized in figure 1. In order to test our fluorescent and magnetic FITC-mCNTs for labelling HSPC, $CD34^+$ cells obtained from LP were exposed to solutions of different concentrations of FITC-mCNT or mCNT alone (10, 20 and 40 μg/ml) in the presence of a magnetic field. The uptake efficiencies of these cells were determined using FACS 1, 3 and 6 h after the uptake. In LP CD34+ cells, the FITC-mCNT uptake started (45% FITC uptake) 1 h after exposure even at the lowest FITC-mCNT concentration (10 μg/ml and reached its maximum efficiency (83% FITC uptake) after 6 h (figure 3(A)).

FITC reached 83%, 90% and 100% in LP CD34+ cells at 6 h after uptake of FITC-mCNT with 10, 20 and 40 μg/ml concentrations, respectively. Although LP CD34+ cells were ~100% FITC-positive at 3 and 6 h after uptake of FITC-mCNT (40 μg/ml), the mean fluorescence (MF) was the highest at 6 h after FITCmCNT uptake (figure 3(B)), which indicates more uptake of FITC-mCNT into $CD34^+$ cells as time increases. Therefore, we further investigated the internalization of the highest concentration of FITC-mCNT (40 μg/ml) into these cells at 24 and 48 h after uptake. LP $CD34^+$ cells were still about 98% FITC-positive 24 and 48 h after FITC-mCNT uptake, though their MF significantly dropped compared to that at 6 h (figure 4), most likely due to FITC degradation over time.

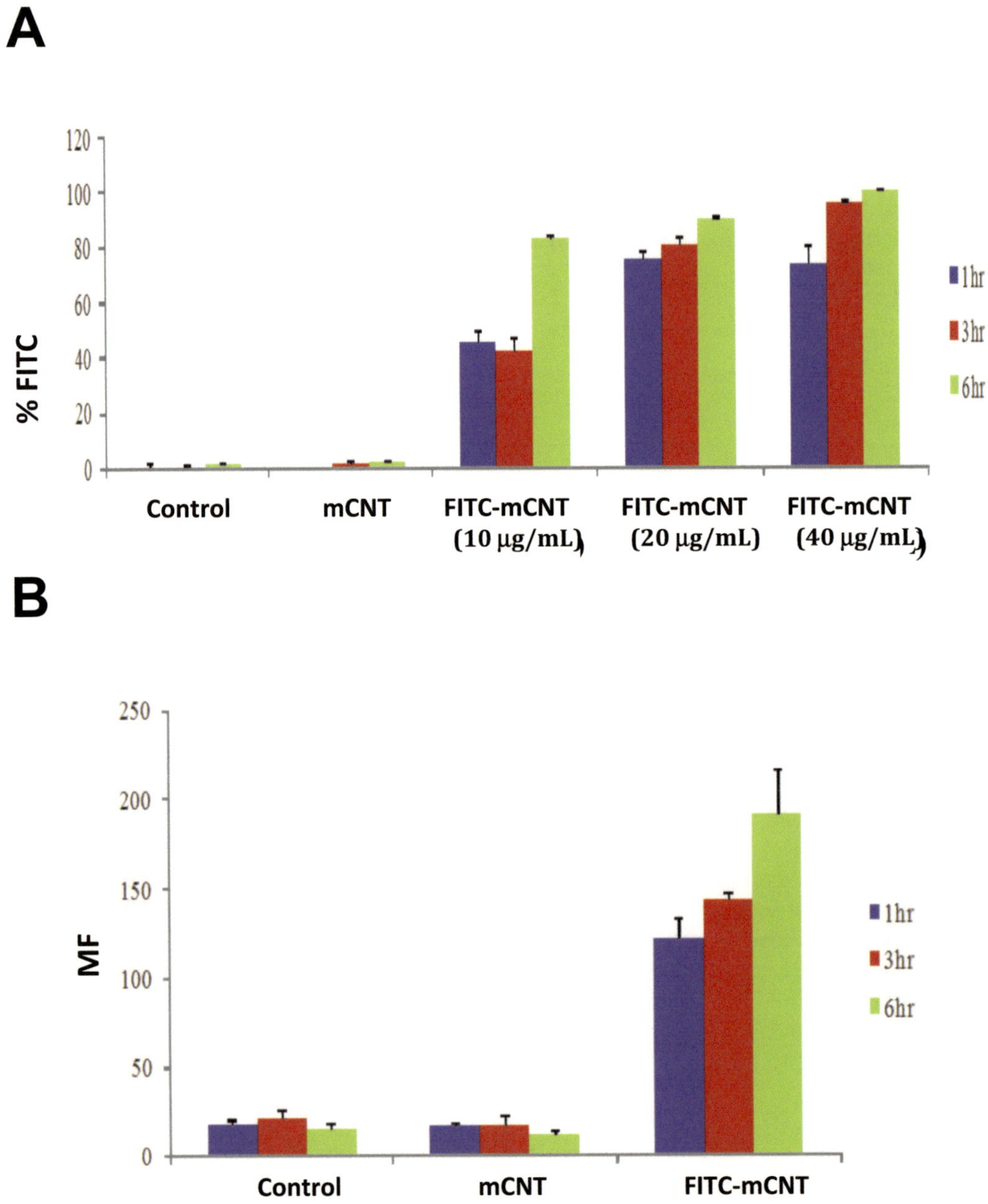

Fig. 3. Efficient uptake of FITC-mCNTs in LP HSPC as shown by FACS analysis. (A) % FITC of LP CD34+ cells exposed to increasing concentrations of FITC-mCNTs or mCNTs (10, 20 and 40 μg/ml) at 1, 3 and 6 h after exposure. (B) Mean fluorescence of LP CD34+ cells exposed to FITC-mCNT (40 μg/ml) at 1, 3 and 6 h after delivery.

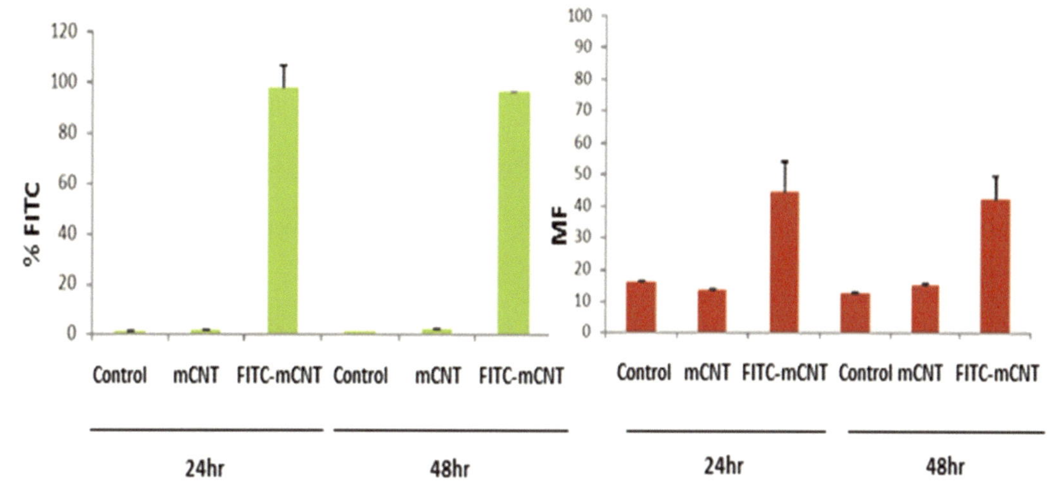

Fig. 4. % FITC and MF of LP CD34+ exposed to FITC-mCNT (40 μg/ml) 24 and 48 h after delivery.

To confirm that even less mature HSPC can efficiently uptake FITC-mCNT, we next studied the internalization of FITC-mCNT (40 μg/mlinto CB-derived HSPC containing a larger immature stem cell fraction ($CD34^+$, CD38−phenotype) than LP HSPC. The efficient uptake started after about 1 h in CB $CD34^+$ cells as demonstrated by ~40% FITC uptake rate shown in figure 5(A), which was lower compared to the result in LP $CD34^+$ cells. FITC-mCNT uptake reached 90% about 6 h after its exposure to the cells, similar to the result in LP $CD34^+$ cells, based on the FACS analysis (figure 5(A)). Confocal analysis also confirmed the efficient internalization of FITC-mCNT into CB CD34+ cells (figure 5(B)), similar to confocal data obtained in THP-1 and MCF-7 cells [25]. FITC-mCNT uptake was saturated at 6 h after its exposure to CB $CD34^+$ cells similar to LP $CD34^+$ cells (figure 5(B)). However, we did not observe a fluorescence signal by confocal analysis 1 h after FITC-mCNT exposure to cells, most probably due to a weak fluorescence signal (data not shown). The HSPC that were not exposed to FITC-mCNT (or the control) did not show any evidence of green fluorescence by confocal analysis, which is consistent with the FACS data in figure 5(B). These data suggest that FITC-mCNT are efficiently internalized by HSPC in a time and concentration-dependent manner. In addition, this efficient mCNT delivery method does not depend on a targeted cell type because this method employs a physical magnetic force for delivery, unlike a biological mechanism like viral. Compared with other nanomaterial uptake using native or modified supermagnetic iron oxide (SPIO) [26, 27], FITC-mCNT labelling or uptake is more efficient for stem cell labelling.

Our rapid FITC-mCNT labelling of HSPC might offer a solution for the difficulty of tracking the movement of transplanted HSPC in patients.

To investigate the cytotoxicity of mCNT, we studied the HSPC viability after FITC-mCNT uptake by trypan blue exclusion assay. With the exposure of FITC-mCNT to various concentrations (10, 20 and 40 μg/ml, the viability of LP CB $CD34^+$ cells was not compromised compared to the control even 6 h after delivery when the uptake rate reached its peak in these cells (figure 6(A)). As expected, the viability of CB $CD34^+$ cells was also not affected by the uptake of FITC-mCNT (figure 6(B)).

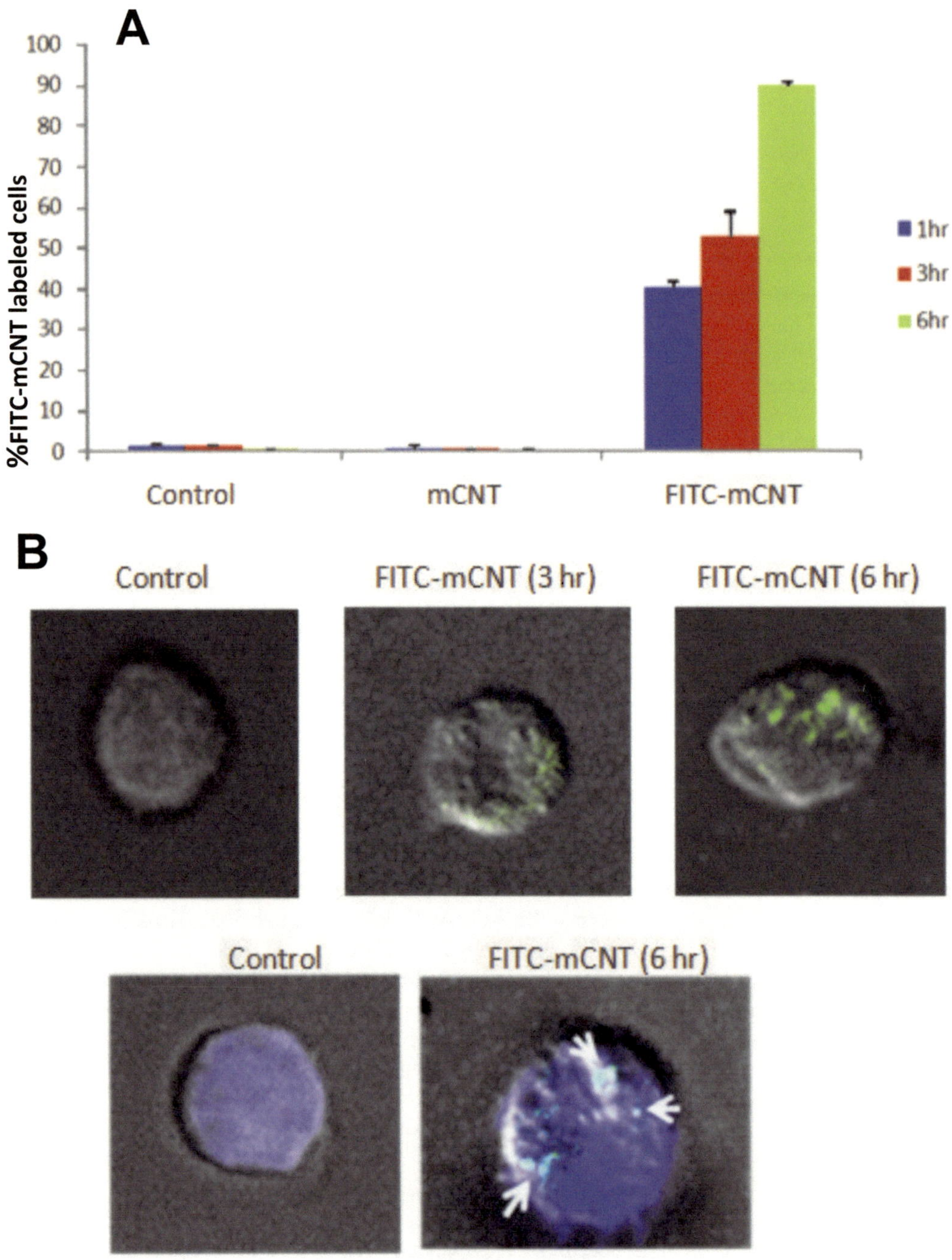

Fig. 5. FITC-mCNTs are also taken up effectively into CB HSPC. (A) % FITC of CB CD34+ cells labelled with FITC-mCNT (40 μg/ml). (B) Confocal microscopy of FITC-mCNT-labelled CB CD34+ cells at 3 and 6 h after delivery. Blue (DAPI): nucleus.

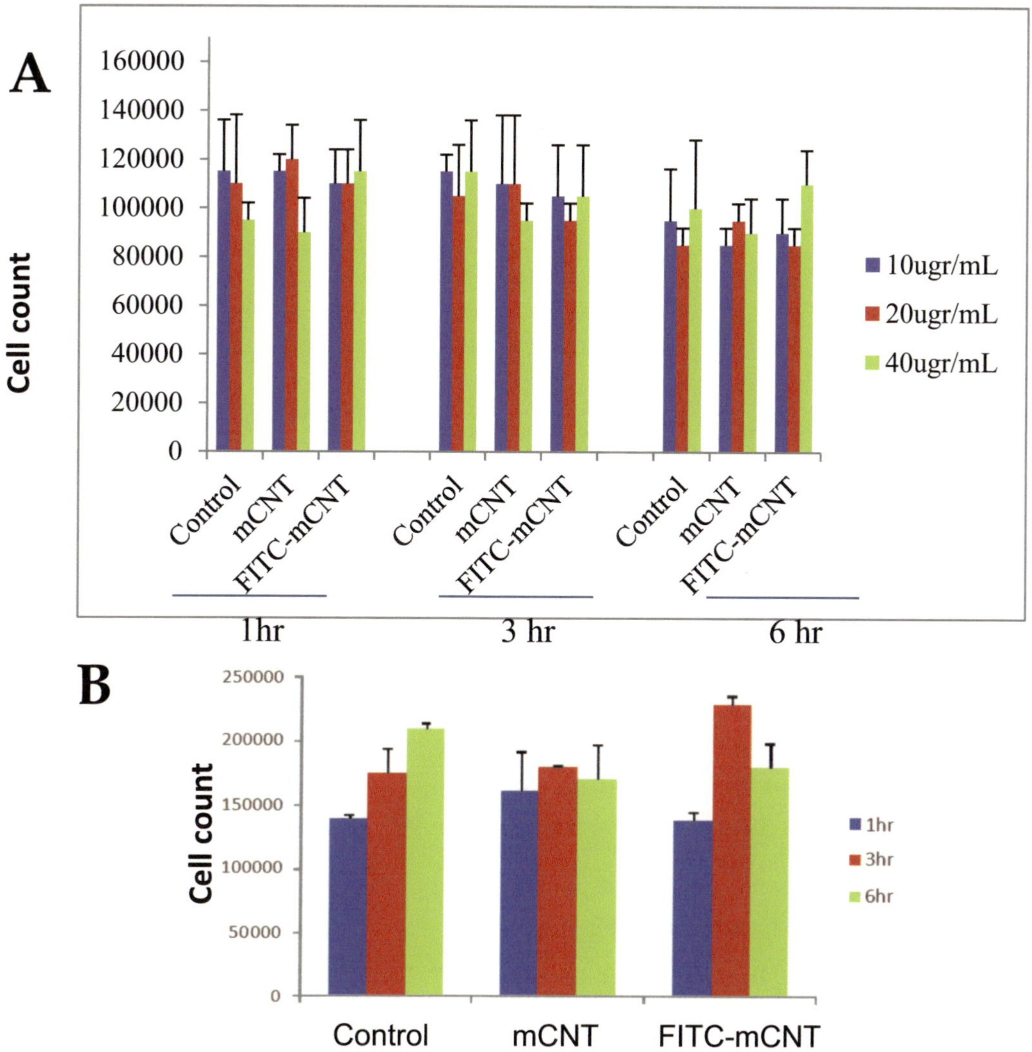

Fig. 6. FITC-mCNTs show no adverse effect on cell viability of HSPC cell count of FITC-mCNT-labelled (A) LP CD34+ cells and (B) CD34+ cells

It has been reported that CNT has no adverse effect on macrophages. It was also not cytotoxic and has no significant effect on adipogenic, osteogenic or chondrogenic differentiation of hMSC [28]. To investigate the long-term cytotoxicity effect of our magnetic-field-driven FITC-mCNT uptake and their impact on the differentiation of HSPC, we performed a colony unit assay (CFU) on FITC-mCNT-labelled HSPC 1, 3 and 6 h after uptake. After 14 days, colonies were identified and enumerated. No evidence was observed of cytotoxicity nor was the differentiation affected in FITC-mCNT labelled HSPC because there was no difference in overall colony number or type (CFU-GM: colony-forming unit of granulocyte/macrophage; BFU-E: burst forming unit of erythrocyte; CFU-GEMM: colony-forming unit of granulocyte macrophage-erythroid megakaryocyte) between the FITC-mCNT labelled and the control HSPC (figure 7). These observations suggest that FITC-mCNT internalization is not only efficient and safe, but it also does not alter the HSPC's properties, similar to the effect of COOH-functionalized single-walled CNT in human mesenchymal stem cells [28].

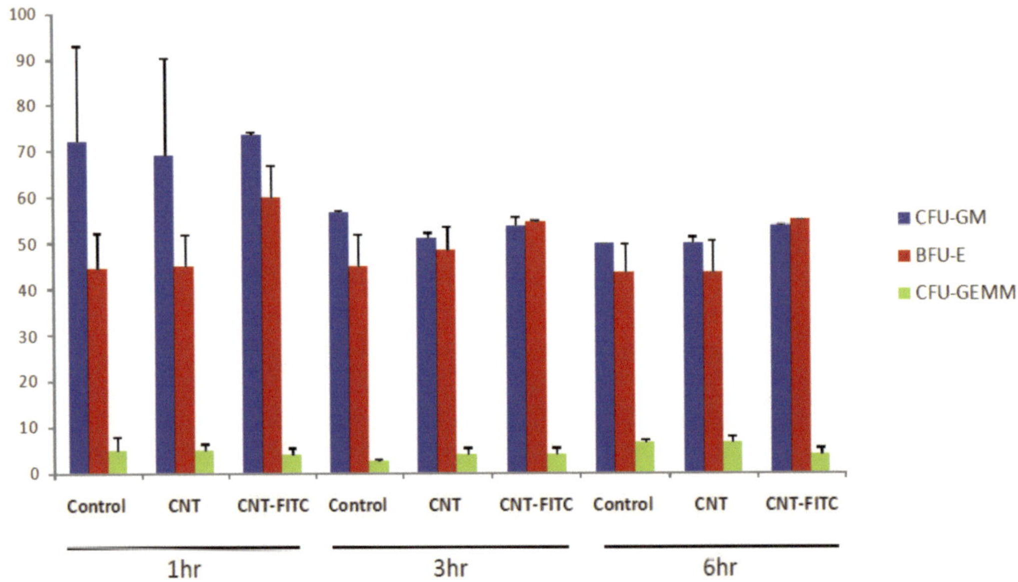

Fig. 7. Internalization of FITC-mCNTs did not affect the differentiation of HSPC. Colony-formation unit of FITC-mCNT-labelled HSPCs at 1, 3 and 6 h after the internalization by clonogenic assay.

3. Conclusion and future outlook

In the current study, a highly efficient and safe mCNT-mediated labelling method for stem cells has been presented. For the efficacy of future stem cell-based therapies, it is crucial to image stem cells and their final location *in vivo*. Detection by magnetic resonance imaging (MRI) may serve as a noninvasive *in vivo* method to study the fate of transplanted stem cells labelled with magnetic nanoparticles. However, the detection of mCNT by standard MRI would be difficult due to its low resolution. Thus, mCNT modified with a contrast agent such as gadolinium chelates could be used for *in vivo* detection and tracking of stem cells using MRI. A promising new approach to non-invasive imaging by exploiting the near-infrared (NIR) optical properties of mCNT might also create significant advances for *in vivo* monitoring of transplanted stem cells. In addition to their use for stem cell labelling and tracking, mCNT could also serve as a perfect platform to deliver genes into hard-to-transfect stem cells. Many strategies for enhancing stem cell transplantation, proliferation and differentiation require the delivery of biomolecules into the cytosolic or nuclear compartments. However, transfection efficiency in stem cells is very low -- a consequence of the low division frequency of the target cells and the quiescent nature of most primitive stem cells. Though viral vectors are capable of efficiently transporting recombinant DNAs into a cell, the undesirable consequences of a viral integration process (i.e. haphazard activation or silencing of host genes), as well as the immunogenicity of viral particles, have recently raised safety concerns about viral vectors [29]. mCNTs that cross the cell membrane through mechanisms other than endocytosis might be an interesting approach to deliver biomolecules within stem cells. Therefore, systemic studies must assess the toxicological impacts of nanotubes on a biological system and evaluate the effects of CNT on the function of stem cells in vivo.

4. Acknowledgments

The authors would like to acknowledge funding supports from the Canadian Institutes of Health Research (CIHR) and the Industrial Research Assistance Program (IRAP) program of the National Research Council (NRC), Canada. The authors would also like to acknowledge the valuable discussions with Dr Eric Swanson, the Director of the National Research Council (IRAP program), Edmonton, Canada.

5. References

[1] Klumpp C, Kostarelos K, Prato M and Bianco A 2006 *Biochim. Biophys. Acta-Biomembr.* 1758 404–12
[2] Pantarotto D, Singh R, McCarthy D, Erhardt M, Briand J P, Prato M, Kostarelos K and Bianco A 2004 *Angew. Chem. Int. Edn* 43 5242–6
[3] Cai D, Mataraza J M, Qin Z H, Huang Z, Huang J, Chiles T C, Carnahan D, Kempa K and Ren Z 2005 *Nat. Methods* 2 449–54
[4] Bekyarova E, Ni Y, Malarkey E B, Montana V, McWilliams J L, Haddon R C and Parpura V 2005 *J. Biomed. Nanotechnol.* 1 3–17
[5] Davis J J, Coles R J and Hill H A O 1997 *J. Electroanal. Chem.* 440 279–82
[6] Besteman K, Lee J O, Wiertz F G M, Heering H A and Dekker C 2003 *Nano Lett.* 3 727–30
[7] Thess A, Lee. R, Nikolaev. P, Diah. H, Petit. P,Robert.J, Xu.C,Fischer.J.E and Samalley.R.E 1996 Science 273 483–7
[8] Bethune.D,Kiang.C,Beyers.R 1993 *Nature* 363 605-607
[9] Cassell AM, Raymakers JA, Kong J, Dai HJ 1999 *J Phys Chem B* 103 6484–92
[10] Colvin VL 2003 *Nat Biotechnol* 21 1166-1170
[11] Bianco A, Kostarelos K, Partidos CD, Prato M 2005 *Chem Commun* 5 571-577
[12] Kostarelos K, Lacerda L, Partidos CD, Prato M, Bianco A 2005 *J Drug Deliv SciTechnol* 15 41-47
[13] Bianco A 2004 *Expt Opin Drug Deliv* 1 57-65
[14] Kam N W S and Dai H J 2005 *J. Am. Chem. Soc.* 127 6021–6
[15] Kam N W S, Liu Z A and Dai H J 2006 *Angew Chem Int. Edn* 45 577–81
[16] Chin S F, Baughman R H and Pantano P 2007 *Exp. Biol. Med.* 232 1236–44
[17] Leeuw T K, Reith R M, Simonette R A, Harden M E, Cherukuri P T, Tsyboulski D A, Beckingham K M and Weissman R B 2007 *Nano Lett.* 7 2650–4
[18] Copelan E A 2006 *New Engl. J. Med.* 354 1813–26
[19] Lu L, Shen R N and Broxmeyer H E 1996 *Crit. Rev. Oncol.Hematol.* 22 61–78
[20] Kolb H J, Simoes B and Schmid C 2004 *Curr. Opin. Oncol.* 16 167–73

[21] Gul H, Marquez-Curtis L A, Jahroudi N, Lo J, Turner A R and Janowska-Wieczorek A 2009 *Stem Cells Dev.* 18 831–8

[22] Hirsch A 2002 *Angew.Chem. Int. Edn* 41 1853–9

[23] Baker S E, Cai W, Lasseter T L, Weidkamp K P and Hamers R J 2002 *Nano Lett.* 2 1413–17

[24] Balasubramanian K and Burghard M 2005 *Small* 1 180–92

[25] Lu W, Gul H, Xu P, Ang W T, Xing J, Zhang J and Chen J 2009 *Proc. IEEE/NIH Life Science Systems and Application Workshop* pp 173–5

[26] Bulte J W and Kraitchman D L 2004 *NMR Biomed.* 17 484–99

[27] Bulte J W *et al* 2001 *Nat. Biotechnol.* 19 1141–7

[28] Mooney E, Dockery P, Greiser U, Murphy M and Barron V 2008 *Nano Lett.* 8 2137–43

[29] Uchida N, Sutton R E, Friera A M, He D, Reitsma M J, ChangW C, Veres G, Scollay R andWeissman I L 1998 *Proc. Natl Acad. Sci. USA* 95 11939–44

14

MWCNT Used in Orthopaedic Bone Cements

Nicholas Dunne and Ross W. Ormsby
School of Mechanical & Aerospace Engineering,
Queen's University of Belfast, Ashby Building, Belfast,
UK

1. Introduction

This chapter discusses the use of carbon nanotube (CNT) based nanocomposites for biomedical applications, particularly in the area of orthopaedic bone cement used in joint replacement surgery.

The chapter initially introduces total joint replacements and poly methyl methacrylate (PMMA) bone cement. The associated issues and drawbacks with the use of these PMMA bone cements in terms of mechanical and thermal properties are then discussed in detail. The application of various MWCNT types (in terms of chemical functionality) at various weight loadings in augmenting some of the issues described is then presented. The next section of this chapter discusses the biological response to the various nanocomposite bone cements with MWCNT. The chapter concludes by discussing issues of CNT interaction with the body, and outlines the current trends in tagging and tracking the movement of MWCNT.

2. The hip joint

The hip joint (Figure 1) is a synovial ball and socket joint allowing for rotation about three perpendicular axes. It is constructed of the femoral head and the acetabulum of the pelvic bone. The femoral head and acetabulum are covered by cartilage. In a healthy hip joint the cartilage acts as a protective cushion to allow smooth movement of the joint, thus reducing friction and to some extent absorb shock. The presence of the synovial membrane secretes synovial fluid into the joint in order to nourish and lubricate the articulating cartilage (Martini and Bartholomew, 2000). The hip joint is responsible for the transfer of weight from the leg to the body, and as such, can be under substantial mechanical stresses.

2.1 Potential problems with the hip joint

Problems with the hip joint can arise due to cartilage damage within the joint caused by disease, trauma, or congenital conditions. This can lead to the surrounding tissues becoming inflamed, causing considerable pain.

Arthritis (joint inflammation) is the main cause of hip joint degradation (Havelin *et al.* 2003; Malchau *et al.* 2002). There are more than one hundred rheumatic diseases that can cause chronic pain, stiffness, and swelling in the synovial joints. The Arthritis Research

Campaign (ARC 2002) reported that in the UK, 206 million working days were lost due to arthritis and joint related conditions. The National Institute of Arthritis and Musculoskeletal and Skin Diseases (NIAMS 2004) stated that two of the most common forms of arthritis are osteoarthritis and rheumatoid arthritis. Primary osteoarthritis is a result of the gradual eroding of the cartilage layer (Figure 2a). It most commonly affects those over the age of 60 (ARC 2002) and remains the most common cause for primary joint surgery (94% of patients in 2005 (NJR 2006)). Congenital conditions such as a deformed joint or defective cartilage can result in osteoarthritis; however obesity, joint fracture, ligament tears, or other injury can damage cartilage, resulting in secondary osteoarthritis. It is noteworthy that while increased occurrences of osteoarthritis are indicative of an aging population, obesity is currently a major risk factor of osteoarthritis (ARC 2002). Overall, it is clear that osteoarthritis is the most common indication for joint replacement irrespective of age (Furnes *et al.* 2005; Karrholm *et al.* 2008; NJR 2006).

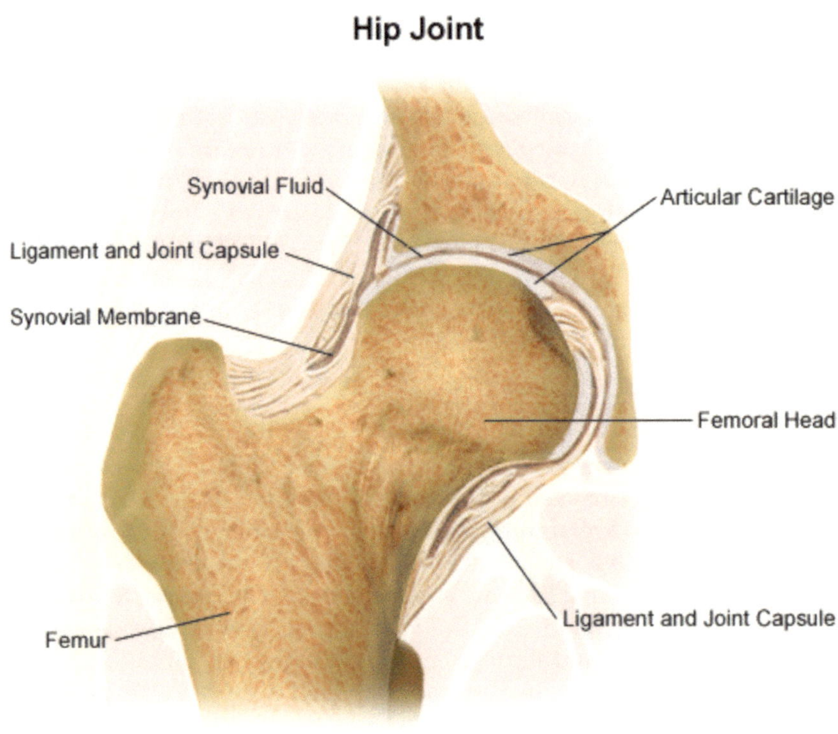

Fig. 1. Anatomy of a healthy hip (Martini and Bartholomew, 2000).

Rheumatoid arthritis is a chronic inflammatory disease of the joints whereby the synovium within the joint becomes inflamed. This inflammatory process damages the surrounding bone and cartilage (Figure 2b). Rheumatoid arthritis most commonly occurs during middle age of adulthood; however the disease can affect children and young adults as well. Rheumatoid arthritis usually affects joints symmetrically and most frequently attacks the hands, wrists, elbows, shoulders, knees and elbows.

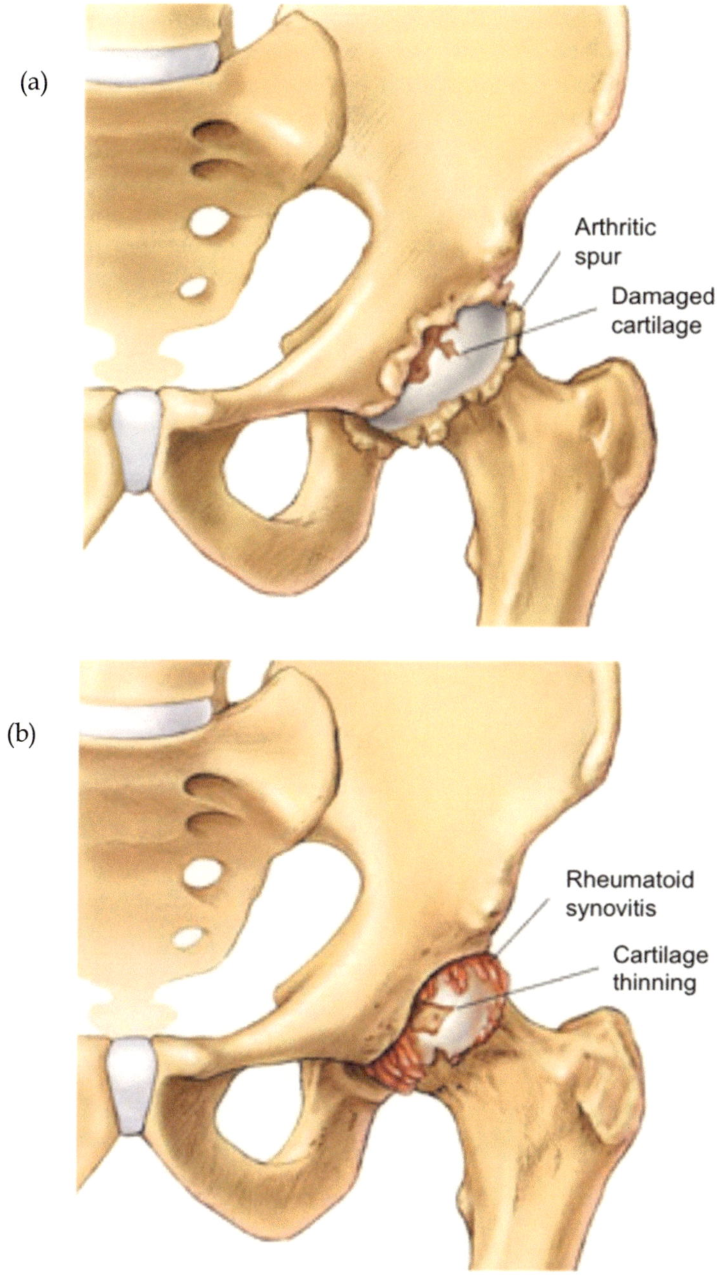

Fig. 2. Illustration of (a) Osteoarthritis and (b) Rheumatoid arthritis of the hip (Zimmer Inc., 2010)

2.2 Primary joint replacement

It is estimated that more than 29% of the population in the UK, are affected by arthritis and joint pain (ARC 2002). If partial damage of the joint has occurred, it may be possible to repair or replace just the damaged areas; if the entire joint is damaged, however, a total joint replacement (TJR) may be necessary to relieve pain and to maintain function of the joint (Prendergast 2001). When replacing a total joint, the diseased or damaged parts are removed and artificial parts, i.e. prostheses or implants, are fitted. Due to the associated risks of surgery, in addition to high financial cost (in 1999-2000, hip and knee replacements alone cost the UK's health and social services £405 million (ARC 2002)), TJR is considered the last resort after failure of non-surgical treatment (Felson *et al.* 2000). TJR may be performed on a variety of joints, including hip, knee, ankle, shoulder, elbow, fingers and wrist. However, hip replacements are by far the most common, as reported, for example, in Norway between 1987 and 2004 (Furnes *et al.* 2005). Figure 3 shows an example of total hip replacement (THR) components.

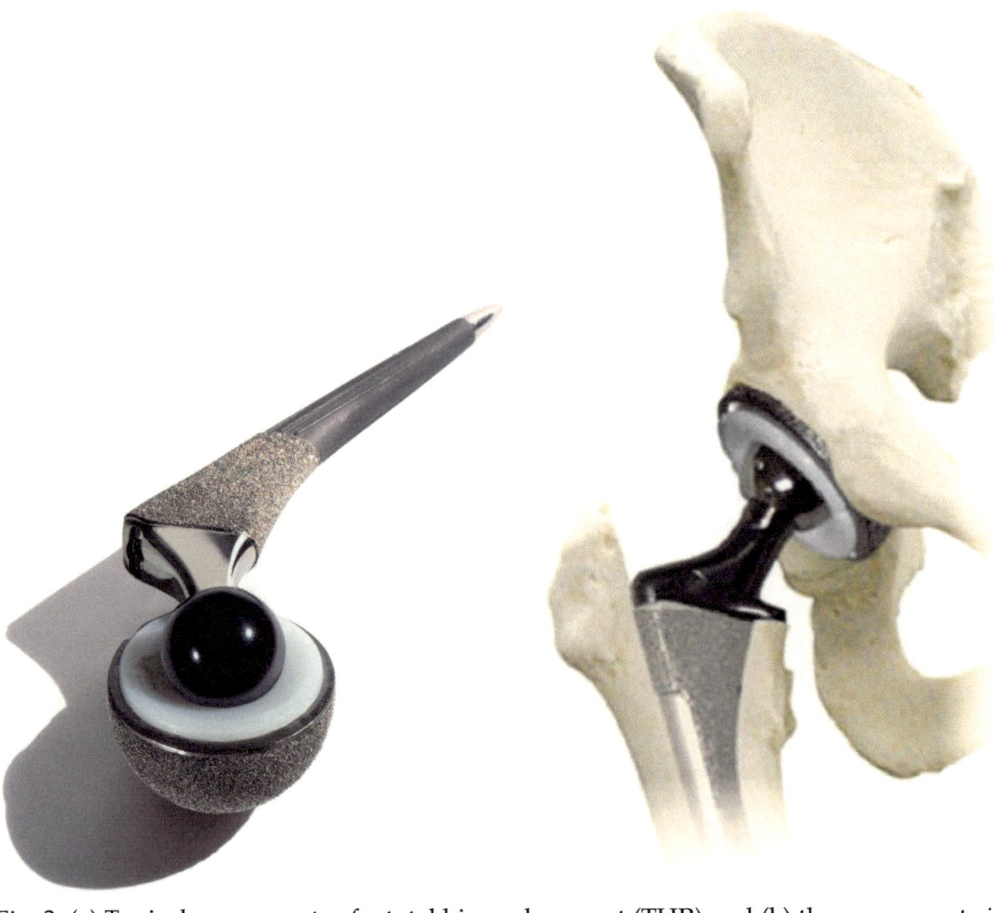

Fig. 3. (a) Typical components of a total hip replacement (THR) and (b) the components *in vivo* (Smith and Nephew Inc., 2008)

During TJR, the most commonly used method of implant fixation is with a load transferring grout-like material, typically an acrylic based bone cement. The major advantage of these cemented joint replacements is the reduced operation recovery time: once polymerised the cement is capable of bearing load and offers immediate stability (figure 4).

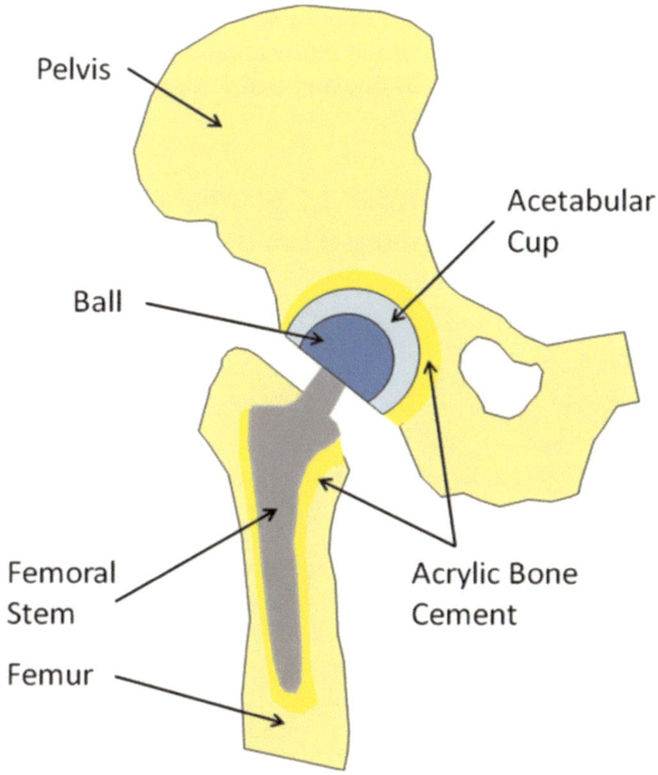

Fig. 4. Schematic diagram of a cement TJR.

However, if the cement mantle becomes loose, the surrounding bone may resorb and ultimate failure of the implant may occur. Uncemented implants were introduced to overcome these shortcomings, for example, cement wear particles, in addition to residual monomer and the highly exothermic polymerisation causing cellular necrosis to the surrounding bone. Uncemented implants typically use a roughened porous surface to promote bone growth around the prosthesis (Hungerford and Jones 1988). However, the bone cavity produced during the operation needs to be precise to ensure the implant is initially held in place through an interlocking mechanical fixation between the implant and the bone. It is also essential that the surrounding bone is healthy to enable this technique to be successful. In addition, the recovery time is long as the bone is required to regenerate. A combination of cemented and uncemented implants is also employed and often termed a 'hybrid'. More recently, resurfacing arthroplasty has been introduced, where less of the bone is removed compared with conventional TJR. Resurfacing procedures not only require the removal of less bone, but cause fewer complications during revision surgeries because the femoral canal is retained intact (Amstutz *et al.* 1998). On average, the number of primary arthroplasties in developed nations is increasing each year (Furnes *et al.* 2005; Karrholm *et al.* 2008; NJR 2006). Figure 5 demonstrates the proportion of cemented, uncemented and hybrid replacements. It is obvious from this graph just how much more popular cemented procedures are. It should be noted that the total number of TJRs performed in England and Wales is significantly greater than Sweden and Norway. In 2004, for example, 48,987 THRs were recorded in England and Wales, compared to just 13,366 in Sweden and 7,061 in Norway. It should also be noted that the population of England and Wales (approximately 55m) is significantly greater than Sweden and Norway (approximately 13m). This equates to approximately 1 in every 1100 people in England and Wales, and 1 in every 650 people in Sweden and Norway.

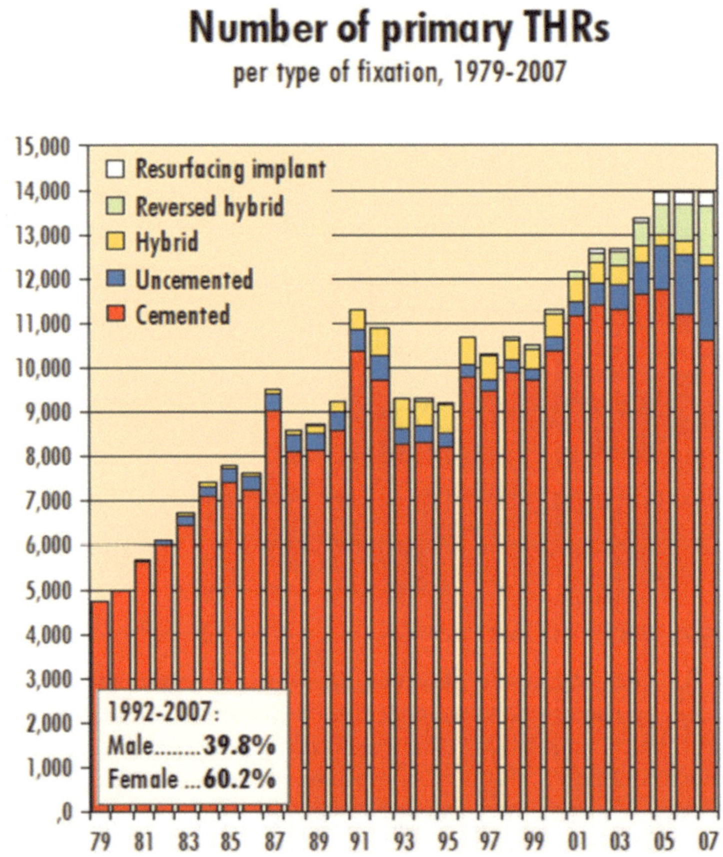

Fig. 5. Number and fixation type of primary THRs preformed in Sweden from 1979 to 2007. (Karrholm *et al.* 2008)

Fully cemented TJR remain the most frequently used implant fixation procedure with 51% of primary THR in England and Wales being cemented in 2005, compared with 54% in 2004. In contrast, there was a slight increase in the application of bone cement for other fixation surgical procedures (NJR 2006). Alternative studies have shown that use of primary cemented TJRs over the last 10 years has remained consistent, whilst the application of cementless implants has almost doubled over the same period (Karrholm *et al.* 2008). This may be partly explained by the increase in the number of TJRs required for a younger age group (<60 yrs and <55 yrs for Sweden and England/Wales respectively). This cohort received more surgical procedures involving uncemented and hybrid implants (Karrholm *et al.* 2008; NJR 2006). It is important to note that whilst there is a slight decrease in the use of cemented implants in THR procedures; bone cement is still required for the majority of implant procedures. In 2005, cement was used for the fixation of 73% of femoral stems and 53% of acetabular cups, in England and Wales (NJR 2006).

2.3 Acrylic bone cement

Poly (methylmethacrylate) (PMMA) has been used in orthopaedics since the early 1960s (Charnley 1960). It was first introduced by Sir John Charnley and Dr Dennis Smith. Also known as acrylic bone cement, it acts as a grouting agent for the fixation of artificial joints as well as the treatment of spinal compression fractures (vertebroplasty). In TJR, bone cement fills the space between prosthesis and bone and acts as an elastic buffer, therefore transferring mechanical load on the implant to the bone. This function of distributing stresses is critical for implant longevity. If the external stresses exceed the ability of the cement to transfer the load, a fracture results (Kuehn *et al.*, 2005).

Acrylic bone cement is a two phase system, consisting of a polymer powder and monomer liquid. The powder phase primarily consists of spherical PMMA beads (82–89 wt. %), in addition to an inorganic radiopacifying agent, usually barium sulphate or zirconium dioxide (10 - 15 wt. %). The powder component also contains benzoyl peroxide (BPO; 0.5–2.6 wt. %), which catalyses polymerisation. The liquid phase is largely MMA monomer (98 wt. %), with 2 wt. % N, N-Dimethyl-*p*-toluidene (DmpT) which accelerates the polymerisation. From a chemical point of view, MMA is an ester of methacrylic acid with a polymerisable double bond. When the liquid and powder phases are mixed, the initiator (BPO) reacts with the accelerator (DmpT) to form free radicals in what is known as the 'initiation reaction'. These free radicals initiate polymerisation of MMA into PMMA by adding to the polymerisable double-bond of the monomer molecule. Temperatures during this reaction can reach up to 110°C. During polymerisation, the bone cement is worked into a 'dough' phase that can be moulded or injected. In a relatively short amount of time (10 - 15 minutes) the bone cement hardens to ca. 90% of its final mechanical properties (Kuehn *et al.*, 2005). Although current revision rates of cemented TJR are low, improved mechanical and thermal properties are required to further reduce subsequent surgeries of cemented arthroplasties, and increase the longevity of the implant. With 88.7% of current cemented implants expected to last at least 14 years (Karrholm *et al.* 2008), this would mean that more physically active patients would have to undergo a number of revision surgeries in their lifetime. Furthermore, younger patient populations are more likely to impose heavier, more complex loadings on the implant, as they would wish to continue pursuing an active lifestyle.

3. Composition and polymerisation reaction

3.1 Composition

Acrylic bone cement, as mentioned is primarily composed of poly methylmethacrylate (PMMA). Most commercial acrylic bone cements comprise of a two part self-curing acrylic polymer, usually formulated as a 2:1 powder to liquid ratio. These components are mixed immediately prior to implantation during surgery and delivered directly to the implant site. The compositions of the main commercial bone cements are summarised in Table 1.0, showing variations in chemical composition. Other cements may also contain antibiotics (e.g. gentamicin sulphate (Lewis 2003; Hendriks *et al.* 2004)) in order to improve the body's response to the implant, reducing risk of subsequent infection and implant rejection.

Constituent	CMW-1	CMW-3	Palacos R	Simplex P	Zimmer LVC
POWDER COMPONENTS					
Benzoyl peroxide (BPO)	2.60	2.20	0.5-1.6	1.19	0.75
Barium sulphate ($BaSO_4$)	9.10	10.00	-	10.00	10.00
Zirconium dioxide (ZrO_2)	-	-	14.85	-	-
Chlorophyll	-	-	200 ppm	-	-
PMMA	88.30	87.80	-	16.55	89.25
PMMA-Methacrylic acid (P(MMA/MA))	-	-	83.55-84.65	-	-
PMMA-*styrene* copolymers P(MMA/ST)	-	-	-	82.26	-
LIQUID COMPONENTS					
NN Dimethyl P Toluidine (DmpT)	0.40	0.99	2.13	2.48	2.75
Hydroquinone	15-20 ppm	15-20 ppm	64 ppm	75 ppm	75 ppm
Mehtylmethacrylate (MMA)	98.66	98.07	97.87	97.51	97.25
Ethanol	0.92	0.92	-	-	-
Ascorbic Acid	0.02	0.02	-	-	-
Chlorophyll	-	-	267 ppm	-	-
Gentamicin sulphate	-	-		-	-

Table 1. Compositions of six commercial formulations of bone cement (Lewis 1997). The compositions are given in percent (w/w) except where stated otherwise.

3.2 Polymerisation reaction

PMMA is an amorphous polymer, which is plasticised on the addition of the monomer methyl methacrylate (MMA). When bone cement is mixed two processes occur, firstly the monomer is absorbed by the PMMA beads and secondly, a free radical polymerisation reaction occurs (Kuehn *et al.* 2005). This reaction is shown schematically in Figure 6 below.

During this reaction the DmpT causes the BPO to decompose leaving a benzoyl radical, and a benzoyl anion (Figure 6a). These benzoyl radicals then initiate the polymerisation of the MMA by combining and forming an active centre (Figure 6b). These active centers then combine with multiple molecules to form a polymer chain (Figure 6c). This reaction forms a viscous fluid allowing the polymerising cement to be moulded as required, i.e. this is the stage when the surgeon would inject the bone cement into the prepared bone canal prior to implanting the stem. As the monomer begins to polymerise, the cement hardens around the stem, holding it in place. This reaction is highly exothermic, an example of a temperature plot of bone cement during polymerisation is shown in Figure 7. The heat energy produced during polymerisation is 57 kJ per mole MMA, resulting in temperatures, which can exceed 100°C. These elevated temperatures can cause cellular bone necrosis which can ultimately contribute to aseptic loosening (Dunne and Orr 2002; Stanczyk and van Rietbergen 2004; Kuehn *et al.* 2005). It should be noted though that the polymerisation temperatures experienced *in vivo* have been much lower (between 40-47 °C) at the bone interface (Toksvig-Larsen *et al.*, 1991). This is due to the reduced thicknesses of bone cement mantle, the presence of blood circulation, and the dissipation of heat through the implant and

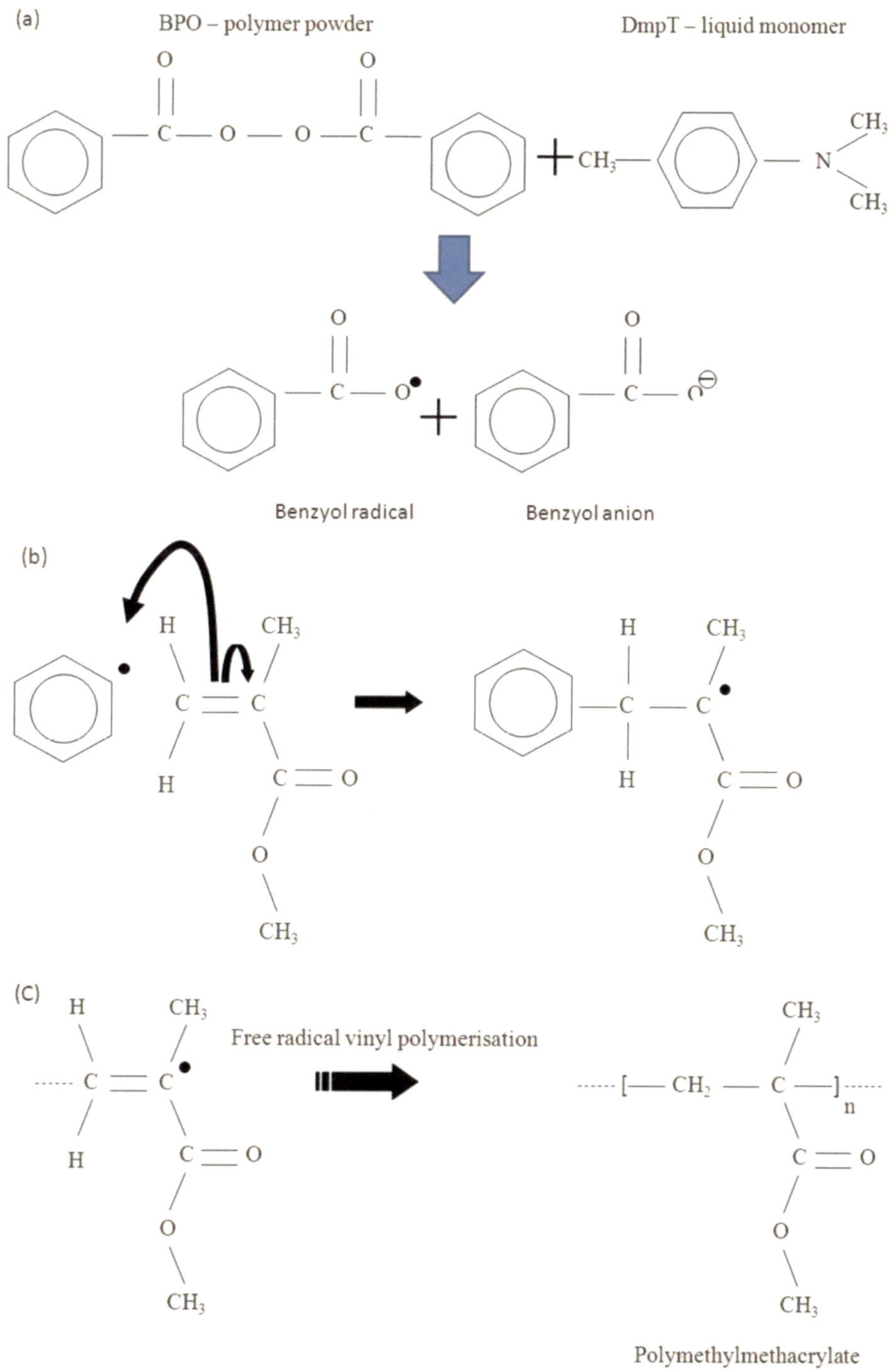

Fig. 6. (a) Schematic diagram showing the decomposition of BPO leaving a benzoyl radical, and a benzoyl anion; (b) How these benzoyl radicals initiate polymerisation of MMA; (c) formation of a polymer chain.

surrounding tissue (Kuehn *et al.* 2005). It has been shown that volumetric shrinkage can occur due to thermal contraction on cooling and the changing density as polymerisation progresses (Gilbert *et al.* 2000; Kuehn *et al.* 2005). Gilbert *et al.* (2000) reported that volumetric shrinkage as a result of density variation, due to the exothermic polymerisation was between 5.1 % and 6.5 % depending on mixing method employed and type of cement. Both shrinkage mechanisms have been identified as factors which influence the levels of residual stresses within the cement (Gilbert *et al.* 2000; Orr *et al.* 2003).

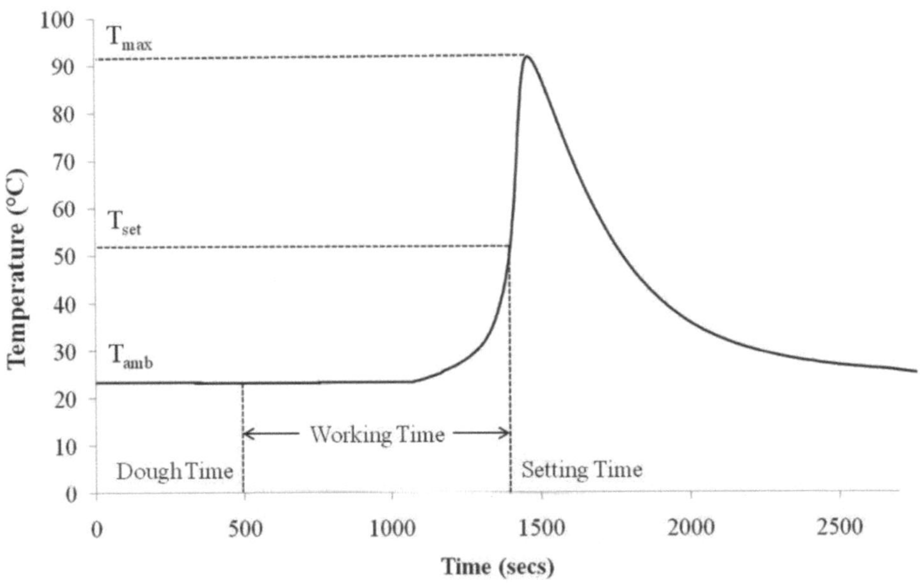

Fig. 7. A typical curing curve for acrylic bone cement where T_{max} is the maximum temperature reached, T_{set} is the setting temperature and T_{amb} is the ambient temperature.

As illustrated in Figure 7, the time that has elapsed after initial mixing when the cement takes a homogeneous dough-like state is known as the 'dough time'. This point may be identified with temperature or, average molecular weight of the polymer. However, as specified in the British Standard BS 7253 (ISO 5833:2002), it is the point at which the cement will no longer stick to powderless surgical gloves (typically 2-3 minutes after initial mixing). The time from the end of dough time until the cement can no longer be manipulated, is defined as the working time. During an operation this is the time during which the surgeon must insert the stem and adjust its position. Finally, the setting time is the time from the onset of mixing until the surface temperature reaches one half of the maximum temperature, as described in ISO 5833:2002.

4. Current issues with acrylic bone cement

4.1 Mechanical properties

The main role of bone cement is to transfer load between bone and the metallic prosthesis. Several studies have shown that the composition of acrylic bone cement significantly influences the mechanical properties of the cement (including Harper and Bonfield 2000; Lewis 2000). It is during the polymerisation process that numerous cement properties, for example viscosity, setting time, maximum cure temperature, and volumetric shrinkage etc, can be determined. These material characteristics may influence a cemented TJR performance. It has also been shown that the variability in the mechanical static and dynamic properties of commercial bone cements is significant, with greater relative differences reported in fatigue properties (Lewis 1997; Harper and Bonfield 2000). Harper and Bonfield (2000) found that there was some correlation between the static and fatigue strengths, however the ranking of the different cements tested did not match exactly. Mechanical properties are known to be affected by: cement composition, size and morphology of the PMMA beads, molecular weight, cement mixing technique, and the powder-liquid ratio (Harper and Bonfield 2000; Lewis 1997). The variation in tensile strength, for example, is reported to vary between 24–49 MPa for five different commercial bone cement formulations, depending on the mixing technique, specimen age and test conditions (Lewis 1997).

4.2 Thermal properties

As mentioned previously, *in vivo* temperatures during the exothermic polymerisation of bone cement can cause thermal necrosis (tissue death) of the bone cells and impaired local blood circulation, which can lead to early failure through aseptic loosening of the implant (Huang *et al.* 2005). It has been reported that for epithelial cellular death to occur, an exposure time of 1 s is required for temperatures above 70 °C, 30 seconds for temperatures greater than 55 °C, and approximately five hours for temperatures greater than 45 °C (Starke *et al.* 2001). Collagen protein molecules are denatured at 45 °C, and experience irreversible damage at 60 °C if held at these temperatures for an hour. It has also been reported that thermal necrosis occurs in bone tissue when exposure is greater than 1 minute for temperatures above 50 °C and denaturation of sensory nerves occurs for temperatures above 45 °C if exposure exceeds 30 minutes. The amount of heat generated during polymerisation is dependent on the amount of reacting monomer, however the maximum temperature reached is also dependent upon the rate of heat dissipation. *In vitro* testing completed by Stanczyk and van Rietbergen (2004) suggested that the tips of bone trabeculae protruding into setting cement may experience temperatures in excess of 70 °C. In the 1960s, upon first use of bone cement in TJR, Charnley believed that while temperatures of ~100 °C could be reached during polymerisation, in the presence of a metallic prosthesis, which would act as a heat sink, there was a reduction in the peak temperature experienced *in vivo*. (Charnley 1960). Since then, numerical simulations and *in vitro* studies of thermal necrosis and peak exotherms in TJR, have helped establish two methods which may assist the reduction of thermal necrosis: (a) the use of thin cement mantle layers, and (b) pre-cooling of the bone surface (Chandler *et al.* 2006; Fukushima *et al.* 2002).

An additional potential adverse consequence of using standard acrylic bone cements is the leaching of residual liquid monomer into the surrounding tissue, which may cause inflammation, chemical necrosis and even death. Average levels of residual monomer can be as high as 5%, however local concentrations may be as high as 15 %, increasing the likelihood of chemical necrosis (Stanczyk and van Rietbergen 2004). Vacuum mixing of acrylic bone cement has been associated with reduced levels of residual monomer as mixing bone cement at reduced pressures increases monomer polymerisation (Bettencourt *et al.* 2001).

There is considerable variation in the chemical composition of different brands of cement (Table 2.1). Often this difference involves more than one of the basic constituents, making it difficult to draw any conclusions regarding the effect of composition on mechanical properties of the cement. It is accepted that the intrinsic properties of the monomer units and the high molecular weight dictate their subsequent mechanical properties such as craze strength, creep resistance and fatigue performance (Sauer and Richardson 1980; Hull and Clyne 1996a). Lewis (2003) reviewed the effect of molecular weight on fatigue performance of bone cement, reporting that increasing the molecular weight of either the powder or the fully cured cement improves the fatigue performance of acrylic bone cement, assuming all other parameters remain fixed. Lewis (2003) suggested that this increase in mechanical performance was related to the increase in polymer chain entanglement due to increased molecular weight which in turn, increased the resistance of the bone cement to craze formation and lead to subsequent increased fatigue crack propagation resistance (Sauer and Richardson 1980; Lewis 2000). Deb *et al.* (2003) reported that increasing the quantity of initiator and activator increased the peak temperature reached during polymerisation reaction and, in addition, lowered the setting time. The content of the residual monomer in the cured bone cement specimens was additionally determined, and it was reported that the highest concentrations of initiator and activator provided the lowest content of residual monomer. However, the concentration of these compounds within the cement must be controlled as they have detrimental health implications when released into the patient. It has also been shown that the type of activator used in polymerisation may significantly influence the fatigue life and fracture toughness of bone cement due to changes in the molecular weight of the resulting polymer (Deb *et al.* 2003). Residual MMA can result after incomplete polymerisation, and is known to influence the mechanical properties and fatigue performance of acrylic bone cement by acting as a plasticiser (Vallo *et al.* 1997; Lewis and Janna 2004). Unreacted MMA is also a possible source of toxicity in the surrounding tissue with possible effects such as hypotension, tissue irritation and alveolar lesions. It has been seen that complete polymerisation and therefore minimal residual MMA content, can be ensured by selecting a suitable initiator activator ratio without significantly affecting fracture toughness (Hasenwinkel *et al.* 2002). Alternatively, the presence of residual monomer can reduce the amount of shrinkage of the bone cement, assuming no other sources of shrinkage occur (Gilbert *et al.* 2000).

4.3 Fatigue failure of bone cement - *In vivo* analysis

Within a cemented implant femur, four main areas of weakness have been recognized as potential failure initiation sites, and can be identified as the: (1) cement, (2) bone-cement interface, (3) cement-prosthesis interface and (4) host bone. Jasty *et al.* (1991) used fractographic analysis to examine *ex-vivo* femoral components, reporting evidence of de-bonding at the cement-prosthesis interface in the majority of the TJR investigated. Partial or complete fracture of the cement mantle was frequently coupled with de-bonding at the cement-prosthesis interface. In the early stages of failure, micro-cracking was evident at the cement-bone interface, although these micro-cracks were considered to be non-critical events as there was no evidence to suggest they were associated with complete fracture across the cement mantle. Fatigue damage accumulation is therefore common prior to overall loosening of the implant *in vivo*. Cemented hip replacements typically experience final failure after several fracture sites have developed, although single, longitudinal cement fractures causing loosening, and subsequent failure have also been recorded *in vivo* (Jasty *et al.* 1991; Topoleski *et al.* 1990). As mentioned previously, fractographic analysis performed on *ex-vivo* specimens of failed bone cement has allowed *in vivo* failure mechanisms to be observed and, as a result, several groups have demonstrated that *in vitro* testing can replicate the micro-mechanisms of failure *in vivo* (Topoleski *et al.* 1990/1993; Verdonschot and Huiskes 1997a; Murphy and Prendergast 2002).

Typically, fracture surfaces were identified with a stepped or irregular fatigue region, this region then evolved into a flat, rapid fracture region (Topoleski *et al.* 1990). This stepped or irregular surface can be attributed to the coalescence of micro-cracks that have formed ahead of the crack tip during crack propagation. Initiating micro-cracks (Figure 8) were believed to exist as a result of internal defects such as pores, aggregates of the radiopaque agent, inclusions from the bone at the bone-cement interface, residual stress-induced cracks at the cement-prosthesis interface, and implant design (Jasty *et al.* 1991; Bishop *et al.* 1996; McCormack and Prendergast 1996; Orr *et al.* 2003; Prendergast 2001b; Murphy and Prendergast 2002).

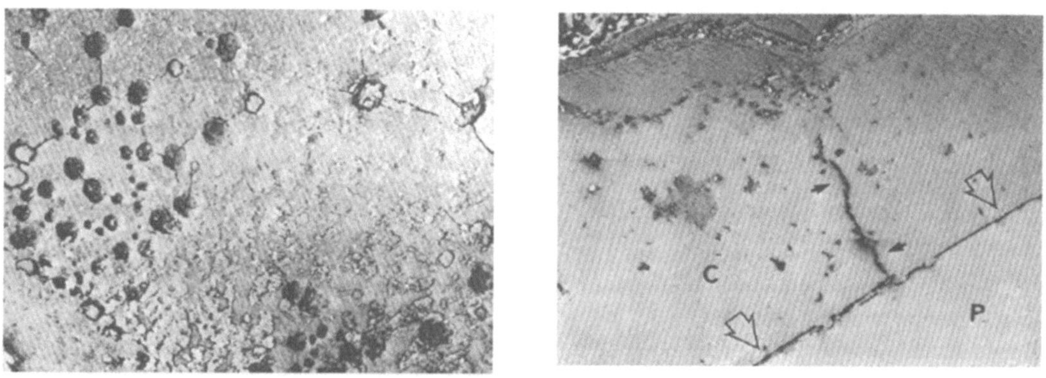

Fig. 8. Scanning electron micrographs showing (a) micro fractures through pores near distal end of prosthesis and (b) an incomplete fracture through the cement mantle originating at the cement-prosthesis interface, Jasty *et al.* (1991).

McCormack *et al.* (1998) used experimental and finite element modeling of the cemented construct to complete statistical analysis of micro-crack accumulation. Representation of micro-crack initiation and propagation was achieved for a longitudinal cross-section of the implanted construct. This allowed for the modeling of the bone-cement and cement-prosthesis interactions. The damage accumulation was found to vary significantly over different regions of the cement mantle. It was reported that more significant cracking occurred at the lateral side of the cement mantle compared with the medial side, however an increased rate of crack formation was seen at the distal end (*cf.* proximal), with more cracks initiating from the bulk of the cement (*cf.* the interfaces). It was also noted that a greater incidence of cracks originating from the bone-cement interface was seen to occur compared to the cement-prosthesis interface. Alternative studies have reported that the location at which a fracture initiates depends on the type of loading applied to the specimen. McCormack and Prendergast (1999) reported a greater occurrence of fatigue cracks initiated from pores within the cement mantle when examining cement under bending loads. Under torsional loads, they observed that cracks initiated most often at the interfaces (McCormack *et al.* 1999). Additionally, as previously mentioned in a study by Jasty *et al.* 1991, *ex-vivo* observations reported evidence of cracks initiating from the cement-prosthesis interface and from voids within the bulk of the cement, suggesting both crack growth scenarios may be important. Prendergast (2001b) confirmed the dependence of crack initiation on loading type, and also suggested that in order to reduce this damage accumulation, the volume of cement stressed to a critical degree must be minimised. There is some disagreement as to the predominant source of fatigue damage accumulation; *in vitro* test specimens demonstrate the benefits of a reduced porosity within the cement (Dunne *et al.* 2003). However, the associated complexity of the *in vivo* cement mantle, in addition to the stress singularities introduced at sharp corners of the implant, and the cement interfaces, may over-shadow the stress amplifying effect of porosity (Janssen *et al.* 2005b). It is however, widely accepted that porosity has a marked influence on damage accumulation, as pores have the potential to act as initiation sites or aid crack coalescence. It should be noted that failure of the cemented construct is not just influenced by damage accumulation and final fracture of the cement mantle. Wear particles associated with the breakdown of cement during de-bonding and fracture, may be transported throughout the implant, leading to an immune-response and the development of osteolysis. Resultant bone degradation will ultimately lead to aseptic loosening of the implant (Anthony *et al.* 1990). A further feature of bone cement is its ability to creep under sustained loading (either static or fatigue), which may contribute to damage development over time. The relationship between creep and damage accumulation is complex. Creep is thought to promote stress relaxation within the cement mantle, reducing the damage accumulation rate (Stolk *et al.* 2004). However, it may also serve to increase levels of implant migration, but the magnitude of this has been shown to be insignificant (Verdonschot and Huiskes 1997b).

4.4 Mechanisms of failure in bone cement – *In vitro* analysis

4.4.1 Fatigue crack initiation

In vitro strain measurements completed by O'Connor *et al.* (1996) have shown the variation in stresses within the cement mantle. The presence of stress raisers within the cement mantle (e.g. porosity) have the potential to sufficiently raise stresses and cause fatigue crack initiation, and subsequent failure. Whilst extensive research has been conducted on the factors that cause crack initiation, a full understanding of the micromechanics involved is yet to be achieved (Lewis 2003). *In vitro* studies by Orr *et al.* (2003) suggested that cracks may be present prior to the loading of the implant, i.e. once the cement has polymerised. Whether cracks are more likely to occur at the bone-cement interface or at the cement-prosthesis interface is a source of discussion. Bishop *et al.* (1996) reported that pores may be more likely to occur at the cement-prosthesis interface due to the presence of a temperature gradient. If a prosthesis conditioned at room temperature is implanted into bone at body temperature, the polymerisation process will begin at the bone-cement interface, hence the cement-prosthesis interface will polymerise later. Porosity at the cement-prosthesis interface due to cement shrinkage will cause reductions in the static and dynamic properties. In contrast, Orr *et al.* (2003) reported that voids and micro-cracks are more likely to occur at the cement-bone interface due to the presence of residual stresses caused by thermal shrinkage around the metallic implant, with a small proportion of cracks initiating from pores within the mantle. McCormack and Prendergast (1999) proposed that initial levels of new crack initiation are higher early on in the loading history, due to stress relief occurring at regions of stress intensity. Furthermore, McCormack and Prendergast (1999) suggested that crack growth rate is the same for all types of micro-cracks, whether "pre-loaded" (i.e. cracks formed as a result of stress relief during cement shrinkage or, from regions of high stress concentration) or 'load-initiated' (i.e. cracks formed due to fatigue loading). As a result it is believed that pre-loaded cracks play a critical role in the aseptic loosening process and thus, the overall failure of cement mantle. Any improvement made to the mixing process (reduction in porosity) and to the level of shrinkage during polymerisation, may then, in theory, impede levels of damage accumulation.

4.4.2 Fatigue crack growth

Fatigue crack growth can propagate in two different phases and are typically observed as a flat, rapid fracture region proceeded by an irregular, or stepped fatigue region (Topoleski *et al.* 1990). The stepped or irregular region, is representative of the early stages of slow crack growth, and may be accounted for in polymers by a process known as "discontinuous crack growth" (DCG). DCG refers to a single burst of fatigue crack advance after several hundred fatigue cycles (Takemori 1984). At high stress intensity factors (ΔK), striated growth usually occurs at the crack tip. Striations refer to the growth bands visible on a fracture surface whose spacing is equivalent to crack growth rate per stress cycle. Striations are orientated perpendicular to the direction of crack growth. Striations are often confused with DCG bands, with the main differences being that the DCG band spacing is significantly greater than the crack length increment per cycle and these bands arise at low ΔK values. Once crack initiation has occurred, it will be in the DCG regime, and the mechanisms by which the crack develops throughout this regime will ultimately determine the fatigue crack growth resistance of the material (Takemori 1984). Fractographic studies have shown that DCG is of relevance for acrylic bone cement with distinct bands being observed in *ex-vivo* samples (Jasty *et al.* 1991; Topoleski *et al.* 1990). DCG is a function of the testing and specimen preparation conditions, environmental effects and compositional changes, all factors that influence the fracture properties of the polymer (Takemori 1984). Changes to the bone cement that influence these factors must be carefully considered with respect to their affect on the overall structural performance of the cement. During the early stages of fatigue crack propagation, DCG band formation is favoured by the development and growth of crazes ahead of the crack tip (Skibo *et al.* 1977). Crazes are identifiable as dense arrays of fibrils inter-dispersed with elongated voids that appear ahead of the crack tip, effectively reducing the density of the polymer in that region. Crazes are generally perpendicular to the applied stress, which result in inelastic deformation by craze widening in the local principal stress direction. In amorphous glassy polymers (e.g. PMMA) brittle fractures occur through crazing and crack propagation (Scheirs 2000b). Crazing has been reported in detail for PMMA by Pulos and Knauss (1998a/b/c). Pulos and Knauss (1998a/b/c) described how damage (identifiable with crazing) ahead of the fatigue crack tip may occur over many cycles, causing a sudden jump in the crack. As mentioned previously, crazing is prominent in polymers ahead of the crack tip at low ΔK values and DCG may occur within this craze zone once the maximum opening of the craze zone reaches a critical value (Figure 9).

Crazing is a common form of polymer deterioration. Crazing is often a precursor to crack growth and failure, however in thermoplastics the presence of crazing can aid fracture toughening. In these cases, the mechanism of crazing enables polymers to absorb energy through the matrix (in-elastic) deformation. This is possible because the energy used to initiate crazing, and crack growth is large and allows the energy to be dissipated over a large area (Luo *et al.* 2004; Topoleski *et al.* 1990). Alternatively, in thermosetting plastics, crazing may lower the strength of the polymer and lead to premature failure (Scheirs 2000b). Crack propagation through a polymer may also be retarded through crack bridging and, to some extent, micro-cracking; this effect may influence both static and cyclic failure. Previous work in the literature suggests that secondary cracks are present in cement failure (a result of tensile stress relief) (Verdonschot and Huiskes 1997b), therefore consideration of these toughening mechanisms may be appropriate. Non-uniform extension of a crack tip can result in un-cracked ligaments. It should be noted that micro-cracks only act as a toughening mechanism when they are constrained; otherwise they are detrimental to the fracture toughness as they propagate and develop into long cracks (Nalla *et al.* 2004).

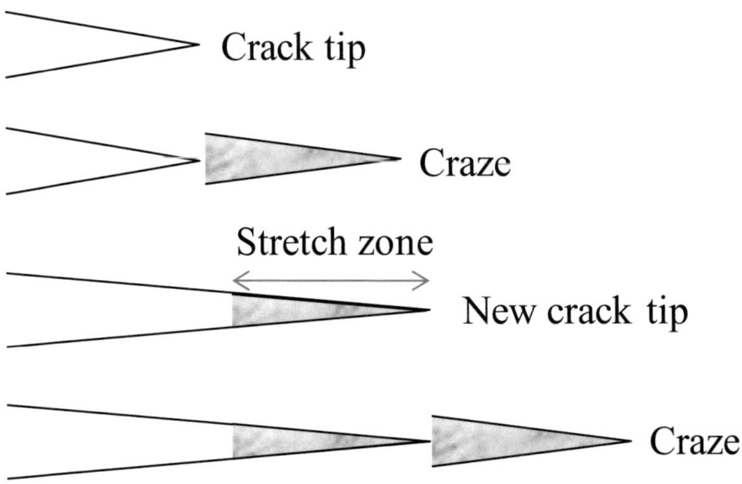

Fig. 9. Illustration of discontinuous band growth (DCG) within the craze zone, ahead of a crack tip.

4.5 Effect of residual stresses and cement shrinkage

It is well accepted that residual stresses are generated within the cement mantle following polymerization and have a direct influence on the stress distribution at the cement-prosthesis interface. Knowledge and understanding of these processes may allow a more accurate prediction of load transfer and in-service conditions in the cement mantle (Nuno and Amabili 2002; Nuno and Avanzolini 2002; Orr et al. 2003; Roques *et al.* 2004). Stresses due to shrinkage, exist in cement surrounding the stem in both the longitudinal and the hoop direction (Roques *et al.* 2004). These stresses will be at their greatest immediately post-operatively, with a reduction occurring with time as stress relaxation, and creep occur (Verdonschot and Huiskes 1997a; Nuno and Amabili 2002; Nuno and Avanzolini 2002; Stolk *et al.* 2004). This may create more favourable stress distributions at the interfaces between the cement-bone, and cement-prosthesis, as finite element modelling of bonded and un-bonded stems predicts that an increase in compressive stresses at these regions may occur (Verdonschot and Huiskes 1997a). The presence of pores at either interface has been attributed to volume shrinkage of the cement during polymerization. Some polymerization shrinkage will occur while the cement is still viscous and hence can be accommodated by flow (Orr *et al.* 2003). However, as polymerization progresses, this flow may not accommodate shrinkage and cracks can initiate in high stress areas as a mechanism of stress relief. While residual stresses do exist in fully polymerized bone cement, the additional presence of porosity, high stress concentrations or excessive heat generated during polymerization may still be required for large cracks to initiate. As residual stresses alone may not be enough to generate cracks (Lennon and Prendergast 2002). The ultimate tensile strength of various bone cements range between 24 and 49 MPa (Lewis 1997), whereas residual stresses of between 2.5 MPa (Nuno and Avanzolini 2002) and 12.6 MPa (Orr *et al.* 2003) have been reported. The direction in which the cement will shrink is of great significance. Orr *et al.* (2003) reported this has a direct relation to the levels of micro-cracking that may occur as a result of shrinkage stress, although the use of acoustic emission has provided evidence for the shrinkage of the cement onto the femoral implants (Roques *et al.* 2004). Furthermore, shrinkage is known to be affected by the volume fraction of monomer content (Gilbert *et al.* 2000).

4.6 The role of porosity

Lewis (1997) identified four main reasons why porosity occurs in bone cements;

i. The entrapment of air between the polymer powder and monomer liquid as the powder is wetted by the monomer upon mixing,
ii. Evaporation of the liquid monomer during polymerisation,
iii. Entrapment of air during mixing,
iv. Entrapment of air upon transfer of the dough into the cement gun (depending on mixing method).

Materials engineering principles and the relevant literature explain that the presence of pores within the cement mantle act as stress raisers, which may then act as crack initiation sites under applied loads. Many researchers have demonstrated that reduced porosity allows for improved compressive, flexural and fatigue properties of acrylic bone cement (Lewis 1997; Murphy and Prendergast 2000; Dunne and Orr 2002; Dunne *et al.* 2003), therefore the level of porosity (both macro- and micro-pores) should be minimised. Pores of diameter ≥1 mm are deemed macro-pores, and are generally introduced during the mixing process when air is trapped within the cement mixture. These macro-pores are often cited as being the cause of low fatigue life for test specimens as crack initiation is often associated with a single pore. Micro-pores have diameters ≤1 mm, and may be established due to the evaporation of the liquid monomer during the polymerisation process and/or entrapment of air during mixing (Dunne *et al.* 2003). It is often observed that multiple smaller pores (≤1 mm) in close proximity are more detrimental than one larger pore (≥2 mm); this is often a result of the type of mixing method employed (Murphy and Prendergast 2000). A multiple pore arrangement is typically observed for hand mixed specimens (Figure 10) where the combined interaction of the pores produces a stress concentration large enough to cause fatigue crack initiation. In contrast, vacuum mixing (Figure 11) usually generates a smaller distribution of pores. There is evidence to suggest that hand mixed cement reduces the level of shrinkage that cement experiences due to the high level of porosity introduced during mixing, (Dunne *et al.* 2003), as only the cement shrinks during polymerisation and not the voids (Kuehn *et al.* 2005). However, porosity may be beneficial to reduce residual stresses prior to loading; this benefit could be outweighed by the adverse effects observed for fatigue crack initiation and propagation. When the propagating crack tip reaches a pore, failure is considered to occur instantly across the void area, effectively causing the crack propagation rate to increase. Conversely though, there are studies that suggest that pores act as crack "blunters" thereby increasing the fatigue life of the bone cement (Topoleski *et al.* 1993), although crack acceleration into the void must also occur due to the local stress concentration at such a defect.

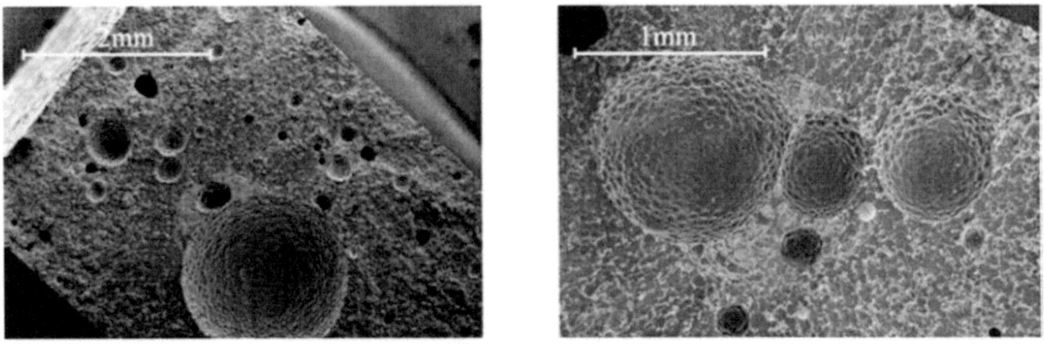

Fig. 10. Scanning electron micrograph of a hand-mixed fracture surface for commercial bone cement showing a large pore with a large number of small pores (Murphy and Prendergast 2000).

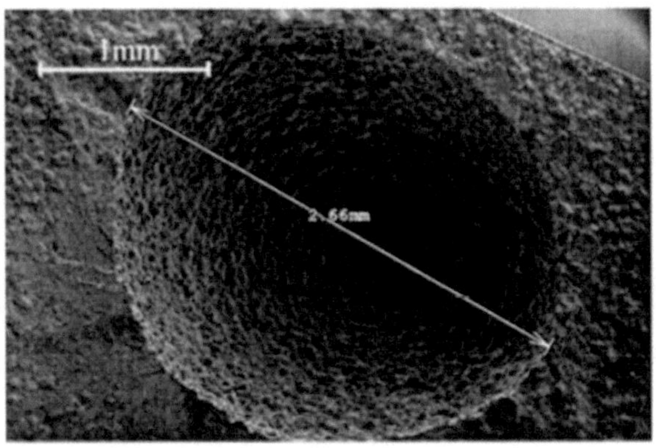

Fig. 11. Scanning electron micrograph of a vacuum mixed fracture surface of commercial bone cement, with a larger pore size compared to hand mixed cement (Murphy and Prendergast 2000).

4.7 Presence of radiopaque agent

Radiopaque agents are included in bone cement formulations (approximately 10-15 % wt.) to allow the cement to be distinguishable from the surrounding body tissues on radiographs. Barium sulphate ($BaSO_4$), Zirconium dioxide (ZrO_2) and iodine-containing copolymers are a few of the possible radiopaque agents used, however $BaSO_4$ is most commonly used. It has been reported that $BaSO_4$ particles do not influence the polymerisation reaction or handling properties of bone cement (Pascual *et al.* 1996). Additional studies reviewed by Lewis (2003) shown that radiopacifiers may have a positive effect on the fatigue life of acrylic bone cement, although this depended on the particle size and morphology, with the inclusion of "nanoparticles" of $BaSO_4$ (~100 nm in diameter) leading to significant increases in the fatigue life (Ginebra *et al.* 2002). This improvement in fatigue life was attributed to crack tip blunting, possibly due to the increased number of $BaSO_4$ particles encountered by the crack tip. This is in agreement with Vallo *et al.* (1997) who proposed that it is the interactions between the crack front and secondary phase particles that account for an increase in toughness; such a mechanism would involve 'crack pinning' and, in effect, an increase in crack length. Conversely, detrimental effects on bending strength, bending modulus and impact strength have been reported after increasing the loading of radiopaque agents (Liu *et al.* 2001). These reductions have been linked to limited bonding between the $BaSO_4$ particles and the host polymeric matrix (Molino and Topoleski, 1996). Furthermore, it has been observed that large agglomerations of $BaSO_4$ can act as fatigue crack initiation sites with the potential of causing overall failure (Kurtz *et al.* 2005).

4.8 Micromechanical analysis of fatigue failure

When examining the fatigue life of PMMA bone cement, Topoleski *et al.* (1993) suggested that micro-crack propagation occurred primarily through the inter-bead matrix, in addition to micro-crack formation ahead of the crack tip, as is modeled in Figure 12. Topoleski *et al.* (1993) also stated that the PMMA beads themselves may experience cleavage or crazing during rapid fracture. Other works by Murphy and Prendergast (2002) suggested that micro-cracks propagate primarily through the inter-bead matrix, but indicated that crack arrest could occur within pre-polymerised beads.

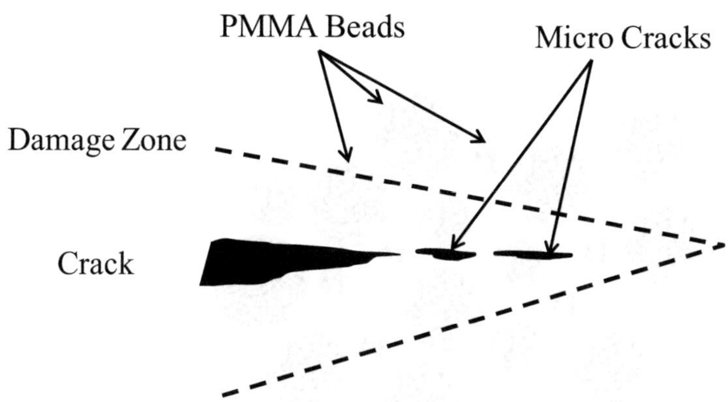

Fig. 12. Schematic of the proposed model of fatigue crack propagation and damage formation of Topoleski *et al.* (1993).

In relation to the porosity that remains after mixing, it is well established that the lower the porosity, the better the static and dynamic properties of cement (Dunne and Orr 2002). Murphy and Prendergast (2000/2002) suggested that pore initiated fractures may be linked to mechanical stress concentration, caused by adjacent PMMA beads at the pore surface, as seen in Figure 13. Topoleski *et al.* (1993) also suggested that pores situated within the fatigue crack damage zone act as micro-crack nucleation sites (see Figure 14), effectively increasing the area of the fatigue damage zone ahead of the crack tip. Conversely, the presence of porosity promotes levels of micro-crack initiation and could be considered to increase rates of crack propagation. Hence porosity can be seen as being both destructive and constructive. Finite element (FE) analysis showed that pores contributed to both fatigue crack acceleration and deceleration, depending on the location of the pores within the stress field, irrespective of size (Janssen *et al.* 2005a). Crack retardation was prominent when pores existed near to the propagating crack, but not close enough to initiate the crack deviating from its original path. The presence of the pore in this scenario reduced the stress in the cement by causing the formation of secondary cracking initiating at the pore.

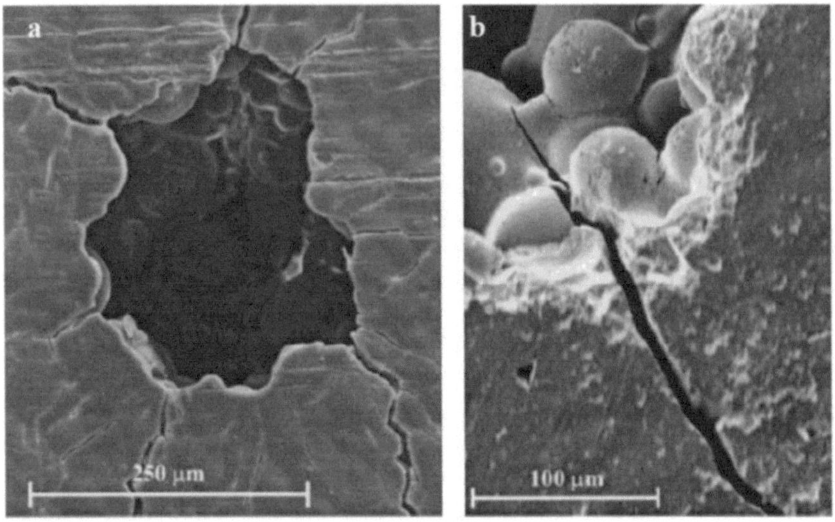

Fig. 13. SEM micrograph of (a) microcracks propagating from a pore and (b) a crack initiation site at a stress concentration between PMMA beads (Murphy and Prendergast 2002).

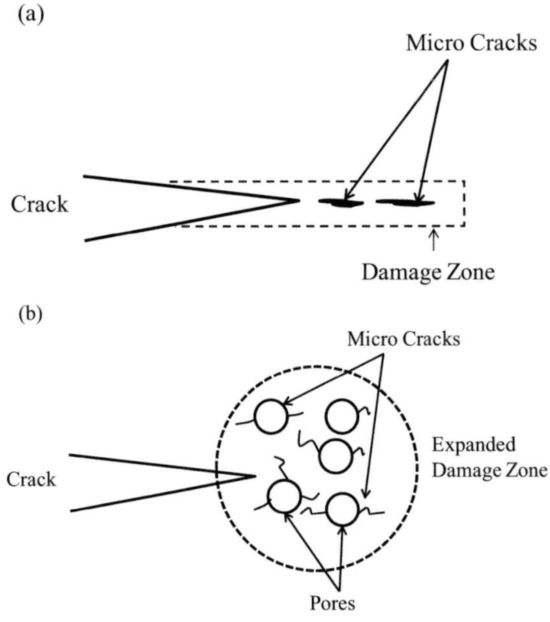

Fig. 14. Expansion of the fatigue crack due to porosity at the crack tip, as suggested by Topoleski *et al* (1993). (a) Damage zone remains linear when no pores are at the crack tip. (b) Expansion of damage zone as a result of micro-crack nucleation (i.e. pores).

In general, it is understood that crack propagation characteristics of bone cement at a microstructural level are heavily reliant on whether the failure regime is a fatigue crack or a fast 'impact' fracture (Prendergast 2001b). Fatigue crack propagation is directly influenced by bone cement microstructure, porosity, residual stresses and agglomerations of radiopaque agent. An impact or fast failure however, does not have a dependence on these same parameters. For any polymer, the success in achieving improved mechanical performance remains in the materials' potential to inhibit or delay crack initiation and improve its resistance to crack propagation.

5. Developments in acrylic bone cement

5.1 Mechanical properties

The intrinsic mechanical properties of acrylic bone cement (such as strength, fracture toughness and fatigue crack propagation resistance) in addition to the presence of extrinsic factors such as porosity, agglomerates of radiopaque agents and other such stress concentrations may limit its long-term survival (Lewis 2003). Within the current literature there have been many attempts to improve the fatigue performance of Acrylic bone cement. Most studies have tried to control the extrinsic factors, in particular porosity (Norman *et al.* 1995; Murphy and Prendergast 2000; Lewis 2003) by means of vacuum mixing or centrifugation. However, this does not address the underlying intrinsic factors which can be broadly categorised into two areas: (a) mechanical studies, focusing on improving mechanical performance, and (b) biological studies where the focus may be on the effect of bioactive inclusions or the addition of antibiotics.

5.2 Mechanical performance

A significant portion of the literature is directed towards discussing the potential to increase mechanical performance of acrylic bone cement *via* reinforcement with fibres or secondary phase particles: for example, carbon (Pilliar *et al.* 1976; Robinson *et al.* 1981; Pal and Saha 1982; Wright and Robinson 1982; Saha and Pal 1986), polyethylene (Yang *et al.* 1997; Narva *et al.* 2005), titanium (Topoleski *et al.* 1998; Kotha *et al.* 2006), hydroxyapatite (HA) (Serbetci *et al.* 2004), glass beads (Shinzato *et al.* 2000), glass flake (Franklin *et al.* 2005), glass fibres (Narva *et al.* 2005), and steel fibers (Kotha *et al.* 2004). Mechanical properties that have been reported to improve include: compressive, tensile, and bending strength, elastic modulus, fracture toughness and fatigue resistance, when compared to cement without reinforcement. In addition to the mechanical improvements provided by these fillers, further benefits have been identified with respect to the peak temperature reached during polymerisation. As mentioned previously, high temperatures experienced *in vivo* can cause thermal necrosis of the bone cells surrounding the cement mantle, in addition to the coagulation of blood, which can potentially lead to aseptic loosening of the implant, and ultimately implant failure (Lewis 1997). Reduced *in situ* polymerisation temperatures have been observed for, but not limited to, steel, carbon fibres (CF) and multiwalled carbon nanotube (MWCNT) reinforced bone cement (Pilliar *et al.* 1976; Saha and Pal 1986; Kotha *et al.* 2004, Marrs *et al.*, 2006). A number of researchers have investigated adding CF as a reinforcing agent using clinically applicable cement mixing techniques for both *in vitro* testing (Robinson *et al.* 1981; Pal and Saha 1982; Wright and Robinson 1982; Saha and Pal 1986;) and for *in vivo* applications (Pilliar *et al.* 1976). Pilliar *et al.* (1976) reported that the inclusion of randomly oriented CF (0.6 cm) improved fatigue performance, tensile strength, Young's modulus and impact resistance (i.e. indicative of toughness), compared to cement without reinforcement. The thermal properties were also observed for the two cement types; the dough time and the setting time were unaffected by the addition of CF, whilst the maximum curing temperature was lowered for the cement with added CF (53 °C compared to 57 °C). Interestingly these research groups found that the addition of CF increased the viscosity of the cement above the required level by ASTM standards (ASTM F451–99a), meaning that use of these cements in a clinical setting would not be ideal. Fractographic analysis identified poor distribution of CF, and evidence of poor CF-PMMA bonding, although fibre pullout was noted. This CF-reinforced cement was used *in vivo* with no detrimental mechanical or biological response observed after 18 months. During the 1980s, problems associated with the high starting viscosity of the cement, and subsequent reduced levels of intrusion, were investigated *in vitro* following the development of low viscosity cement. Robinson *et al.* (1981) confirmed that CF-reinforcement of a commercial cement increased fracture toughness of both regular and low viscosity cements. However, the low viscosity cements (both reinforced and conventional) displayed a reduction in fracture toughness when compared to the equivalent regular viscosity cement. For the reinforced cement, Wright and Robinson (1982) reported a decreased crack growth rate versus the unreinforced cement). Saha and Pal (1986) investigated the effect of mechanically, or hand-mixed CF-reinforced bone cements, and found that mechanical mixing provided superior performance. They attributed this to the improved dispersion of CFs throughout the cement matrix.

Investigations concerning the addition of titanium (Topoleski *et al.* 1998; Kotha *et al.* 2006) or CF (Pilliar *et al.* 1976; Saha and Pal 1986) to bone cement suggested that commercially viable mixing methods are indeed possible. Topoleski *et al.* (1998) used SEM analysis to confirm that before cement failure, a good bond between the fibers and the host matrix existed, although subsequent damage led to evidence of fiber de-bonding, plastic deformation and ductile rupture of the fibers. Topoleski *et al* (1998) also reported that the presence of the fibers prevented crack propagation (Topoleski *et al.* 1998). Additionally, fiber based additives have been shown to dissipate energy associated with static crack propagation, resulting in improved fracture toughness of acrylic bone cement, through crack diversion and crack tip blunting (Gilbert *et al.* 1995). Orientation and dispersion of the fibers, in addition to good interfacial bonding, were all identified to have a positive effect on improving mechanical properties due to reinforcement (Gilbert *et al.* 1995; Yang *et al.* 1997).

5.3 Biological performance

Bone cement is a biologically inert component of the implant construct. Conventional acrylic bone cement does not normally promote bone ingrowth. Several studies however have attempted to improve the biological performance of bone cement. These have included the incorporation of bioactive agents, such as HA based powders, glass ceramic particles or glass beads (Lee *et al.* 1997; Mousa *et al.* 2000; Shinzato *et al.* 2000). Each of these additives has been reported to enhance the biocompatibility of bone cement, thus reducing the formation of fibrous tissue at the bone-bone cement interface. Mousa *et al.* (2000) used apatite-wollastonite glass ceramic (AW-GC) particles to reduce the amount of monomer required for polymerisation, which lead to a reduction in the peak exotherm, and thermally induced bone necrosis as well as decreasing the levels of cement shrinkage. Similar results have been reported for cements containing glass beads, which have also been shown to improve bioactivity (i.e. osteoconductivity) compared with HA powder (Shinzato *et al.*

2000). Additionally, it has been reported that many of these bioactive cement composites exhibited no detrimental influence on mechanical performance, and in some cases improvements were observed (Mousa *et al.* 2000; Shinzato *et al.* 2000).

Concerns regarding biological performance include the use of antibiotics, which are integrated to reduce risk of infection and associated revision (Kuehn *et al.* 2005); many antibiotic-loaded cements are currently commercially available and, for those containing gentamicin sulphate, are believed to not cause any adverse affect on the fatigue performance (Baleani *et al.* 2003).

6. Polymer matrix composites

The mechanical success of any polymer composite is governed by the successful transfer of load between the matrix and the reinforcement. This transfer of load is dependent upon the volume fraction, dispersion, orientation of the reinforcing phase, the host matrix-reinforcement interface and the individual mechanical properties of the phases that are present (Gilbert *et al.* 1995; Hull and Clyne 1996b; Yang *et al.* 1997). Within fibre-reinforced composites four main microstructural regions exist: (1) the matrix, (2) the fibre, (3) the interface, and, in some composite systems, (4) the interphase. An interphase may be present if a mechanical or chemical interaction takes place between the polymer matrix and the reinforcing phase (examples includes adsorption of the polymer onto the surface of the reinforcing agent in particulate-reinforced polymers, inter-diffusion of the components during blending and chemical reactions at the polymer/fibre interface) (Pukanszky 2005).

Fibres aligned in the direction of applied load are particularly effective at reinforcing composites. Corresponding mechanical properties which effect failure performance may be identified, with a complex interaction between individual phase properties, the interface strength between the host polymer and the reinforcing fibres and the composite microstructure. Polymers are the most common form of composite matrix and are often reinforced with a low fraction of fillers such as glass, CF or Aramid. This results in composites of high specific strength and modulus as the low levels of additives allows a more homogeneous dispersion (Callister 2000a). Of these three reinforcements, CF composites often exhibit the best resistance to fatigue failure due to superior mechanical properties as well as the higher thermal conductivity of carbon fibres which assists in the dissipation of heat during cyclic loading (Scheirs 2000a). In compression, the mechanical performance of fibre-reinforced composites is dependent on the interaction between the host polymer matrix and the fibre. For optimum reinforcement, the matrix would provide lateral stabilisation to the fibre preventing subsequent buckling. Alternatively, the tensile behaviour is governed by the tensile strength of the fibre additive (Hull and Clyne 1996a).

Fatigue failure in polymer composites is commonly characterised by a gradual reduction in stiffness (Scheirs 2000c). Without reinforcement, fatigue failure typically occurs perpendicular to the applied load; in contrast, the presence of fibres generally results in a diffuse damage zone due to the combination of a number of sub-critical failure modes and crack shielding mechanisms. In general, crack propagation through fibre-reinforced polymers may be considered as a multi-faceted interaction between the polymer matrix, the fibre reinforcement and the associated interface/interphase regions. A combination of mechanisms may occur and subsequently, fibre inclusions may impede crack growth by three main mechanisms (Mandell *et al.* 1980; Sauer and Richardson 1980):

i. Debonding of interface/interphase between fibre and matrix - as a crack approaches, failure of the interface occur serving to blunt the crack tip and reduce crack propagation.
ii. Crack bridging - transferring load across a given matrix crack, reducing the crack.
iii. Fibre pullout, subsequent to crack bridging, may also absorb energy due to matrix deformation and/or interface friction.

7. Carbon nanotubes

It was in 1980 that Sumio Iijima first recorded an 'onion-shaped particle' in the order of 0.8 - 1 nm in diameter. It was not until five years later that Iijima realised that this 'onion-like structure' was the fullerene C_{60}, which was believed to be discovered by Kroto, Heath, O'Brien, Curl and Smalley in 1985, (Figure 15).

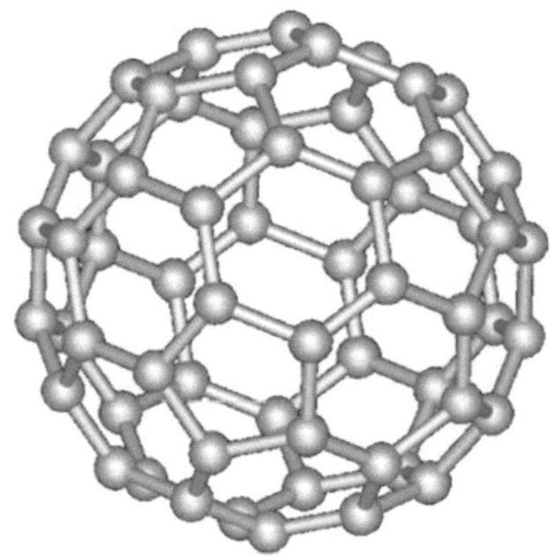

Fig. 15. Carbon C_{60} molecule (Iijima, 1991).

Although Curl *et al.* (2001) later confirmed that it was Osawa who first documented the concept in 1970. It was in 1991, whilst working as an electron microscopist that Iijima's study of soot deposited on the cathode during the arc-evaporation synthesis of fullerenes led to the sighting of a needle-shaped material. What was originally described as *"microtubules of graphitic carbon"* is now commonly known as carbon nanotube (CNT). Whilst being considered an accidental discovery, Iijima believes it was the *"power of serendipity"*. Initially, Iijima produced individual tubes of graphitic carbon with diameters of 4-30 nm and a length of up to 1 µm using arc-discharge evaporation methods (Iijima 1991). At present, CNT can be synthesised *via* electric arc discharge (Iijima 1991; Shi *et al.* 2000), laser ablation (Zhang *et al.* 1998; Zhang and Iijima 1998) or, more commonly chemical vapour deposition (Sinnott *et al.* 1999; Andrews *et al.* 2002). Extensive research has been conducted on these processing methods by Andrews *et al.* (2002), and Thostenson *et al.* (2001).
CNT can occur as either single-walled nanotube (SWCNT) or multiwalled nanotube (MWCNT) structures. SWCNT consist of a single graphene sheet rolled up as a seam-free tube. They can be thought of as a linearly extended fullerene (Ajayan 1997; Iijima 1991/2002; Baughman *et al.* 2002). SWCNT usually exist as agglomerations due to the van der Waals forces between each tube, with diameters on average between 0.7-2 nm, whilst their lengths are often 5-30 µm (Ajayan 1997; Colbert 2003). MWCNT can be described as an array of

SWCNT that are concentrically arranged inside one another with an internal diameter as small as 2.2 nm. The distance between the individual SWCNT that constitute MWCNT (or the graphite inter-layer separation) is typically 0.34 nm (Ajayan 1997; Iijima 1991/2002) (Figure 16). Iijima (1991) used electron diffraction to establish that the crystal axis of the graphene tubes consisted of carbon-atom hexagons arranged in a helical manner about the tube axis (Figure 17). The ends of CNT are closed off by the presence of pentagonal carbon rings near the tip regions, whilst deformations and imperfections of the cylinder occur as pentagons or heptagons within the main structure of the tube (Ajayan 1997).

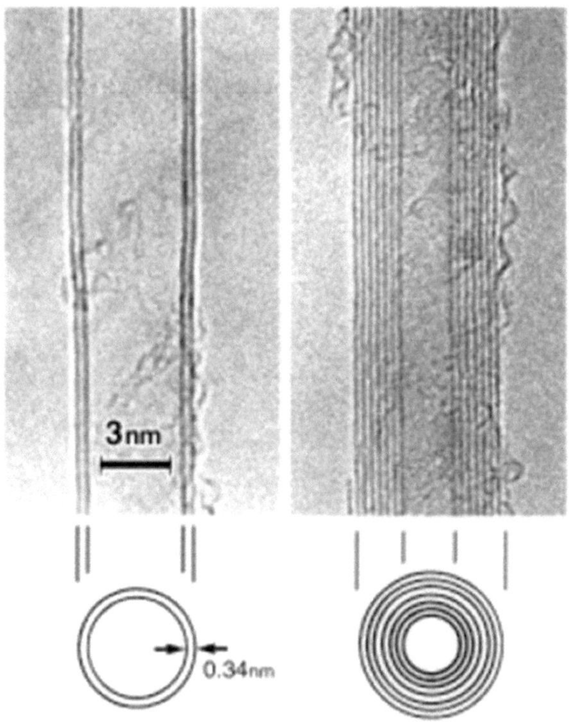

Fig. 16. Transmission electron micrograph (TEM) of MWCNT (Iijima 1991).

SWCNT, are known to be stiff and exceptionally strong (high Young's modulus and high tensile strength). Furthermore, SWCNT can stretch beyond 20% of their original length and bend over double without kinking (Baughman *et al.* 2002; Colbert 2003). However, the mechanical properties of individual CNT, whether SWCNT or MWCNT are the subject of much research with a significant variation in recorded properties existing. The use of CNT in various matrices can greatly enhance mechanical properties. Wong *et al.* (1997) provided an insight into the potential uses of CNT. Atomic force microscopy (AFM) was employed to determine the elasticity, strength and toughness of individual silicon carbide nanorods (SiC NRs) and MWCNT that were attached to molybdenum disulphide surfaces. The average bending strength of the MWCNT was 14.2 ± 8.0 GPa; with the maximum bending strength being substantially smaller than that of the SiC NRs at 53.4 GPa. In contrast, whilst both nanostructures exhibited high values for the Young's modulus (highlighting their suitability as reinforcing agents in ceramic, metal and polymer matrix composites) the Young's modulus for the MWCNT was almost double that of the SiC NRs (1.28 ± 0.59 TPa and ~600 GPa, respectively). The Young's modulus value for the in-plane modulus of highly orientated pyrolytic graphite was recorded at 1.06 TPa (Blakslee *et al.* 1970) and is believed to be the largest known for a bulk material. Wong *et al.* (1997) concluded that while the stiffer MWCNT had a lower ultimate strength, the elastic buckling displayed by the MWCNT (i.e. the energy storing capabilities of CNT before failure) showed them to be the "tougher" nanostructure. More recently, Demczyk *et al.* (2002) investigated the direct failure of individual MWCNT under tension using TEM. While the mode of failure, either ductile or brittle, could not be determined, results confirmed a tendency to fail *via* a mode now known as 'telescopic failure', with initial failure observed in the outermost walls followed by a 'sword in sheath effect' of the inner cylinders. TEM observations also confirmed the elastic capabilities of MWCNT during deformation in bending, even after being highly distorted.

Fig. 17. Helical arrangement of carbon atom hexagons that make up a graphene sheet in a MWCNT (Iijima 1991).

It is clear that CNT offer significant potential to improve the properties for many existing materials; the challenge remains for the superior properties exhibited by CNT individually to be successfully applied and optimised in practical applications such as nanocomposites. Not only do the properties of the CNT themselves vary due to impurities during processing (Baughman *et al.* 2002), but on addition of CNT to a matrix, the problem becomes multi-faceted: CNT dimensions, dispersion, alignment, concentration, CNT-matrix interface/adhesion and choice of matrix are some of the many issues that govern the final properties of the composite material. At present, production of high purity SWCNT still remains costly and of a low yield: purification reduces yield further, adds to cost and damages the structure (Andrews *et al.* 2002). Low manufacturing costs and high yields of aligned MWCNT are now possible however, making them the preferred choice in bulk composite material development.

8. Properties of CNT-reinforced polymers

Since the discovery of CNT in 1991, research incorporating them into various matrix materials has been plentiful. This increase in research can be attributed to the fact that CNT have extremely high aspect ratios (typically >150:1), modulus and low density. The addition of CNT to polymer composite materials has enhanced mechanical (Wong *et al.* 1997; Andrews *et al.* 2002), electrical (Baughman *et al.* 2002; Colbert 2003) and thermal properties (Kim *et al.* 2001; Baughman *et al.* 2002; Colbert 2003), in matrices including polycarbonate (Ding *et al.* 2003; Eitan *et al.* 2006), polystyrene (Andrews *et al.* 2002; Thostenson and Chou 2003; Park *et al.* 2005) ultrahigh molecular weight polyethylene (Ruan *et al.* 2003) and PMMA (Hwang *et al.* 2004; Marrs *et al.* 2005).
Extensive literature has explored the effects of CNT on mechanical properties of various polymer matrices. It has been shown that the addition of CNT can increase the toughness of polymer matrices due to crack bridging, changes in morphology of the matrix and the additional energy required for de-bonding and nanotube pullout (Dalton *et al.* 2003; Ruan *et al.* 2003; Andrews and Weisenberger 2004). As with other polymer composites, increases in modulus may be identified with stress transfer from the matrix to the CNT. It should be noted though that the presence of agglomerations of CNT can have significant adverse influence on the mechanical properties of CNT-polymers, acting as stress concentrations and fracture initiation points within the composite microstructure. This effect was recorded by Marrs *et al.* (2005), who incorporated MWCNT at various levels of loading in to PMMA bone cement. They characterised the fatigue, quasi-static tensile and bend properties for these MWCNT-PMMA nanocomposites. They found that the optimal performance was for the addition of 2 wt% MWCNT. They also report that loadings above this resulted in reduced mechanical properties, although results were still superior when compared to pure PMMA (Marrs *et al.* 2005). Cadek *et al.* (2004) reported an increase in Young's modulus by a factor of two after the addition of 0.6 vol% CNT to poly (vinyl alcohol), an effect which they attributed to nanotube diameter and resultant surface area. Varying levels of reinforcement as a result of a change in nanotube diameter suggested that MWCNT of smaller diameter provided optimal reinforcement due to increased surface area. Whilst the enhancing effects of CNT with respect to tensile strength and modulus have been recorded (Ruan *et al.* 2003), it has been established that theoretical predictive models such as Rule of Mixtures approach or the Halpin-Tsai model predict superior reinforcement capabilities than experimental data provides. Part of this discrepancy may be due to factors such as poor interfacial bonding, inhomogeneous dispersion, and CNT quality (Andrews *et al.* 2002; Fisher *et al.* 2002; Andrews and Weisenberger 2004). Poor adhesion/bonding at the interface between the host polymer and the CNT may have a detrimental effect on the mechanical properties of CNT-composites (Andrews *et al.* 2002). Nanotube pullout experiments, using atomic force microscopy, may be used to determine the force required to separate a single CNT from a polymer matrix (Barber *et al.* 2003; Baroud *et al.* 2004). It is widely accepted that a strong interfacial adhesion between the reinforcing nanotube and the polymer matrix leads to the effective transfer of load (Cooper *et al.* 2002; Barber *et al.* 2003; Goh *et al.* 2003, Marrs *et al.*,

2007). Barber *et al.* (2003) measured the average interfacial stress required to remove a single MWCNT from a polyethylene-butene matrix as 47 MPa. Comparing this value to the 10 MPa measured for poorly bonded interfaces in other fibre-reinforced polymers, Barber *et al.* (2003) suggested that this enhancement was due to the presence of covalent bonding at the interface between CNT and the host polymer, potentially due to a chemical interaction between the polyethylene-butene matrix and defects on the surface of the nanotube. Moreover, it is believed that the mechanical properties of the polymer immediately surrounding the nanotube may be enhanced when compared to the bulk of the polymer. Nano pullout test results by Barber *et al.* (2003) showed no evidence of the polymer yielding, even at pullout stresses that were ten times higher than average tensile strength of the polymer matrix. This may be explained using differential scanning calorimetry (DSC) measurements to determine the polymer crystallinity. Cadek *et al.* (2004) compared polymer crystallinity for PVA after the addition of 0.6 vol. % CNT and measured a linear increase in crystallinity with increasing volume fraction of CNT, suggesting a crystalline polymer coating is formed at the nanotube surface. Observations of the fracture surface using SEM confirmed that polymer wetting of the nanotube surface was achieved (Figure 18); diameters of the nanotubes in the composite were larger than the as-received CNT (Ding *et al.* 2003).

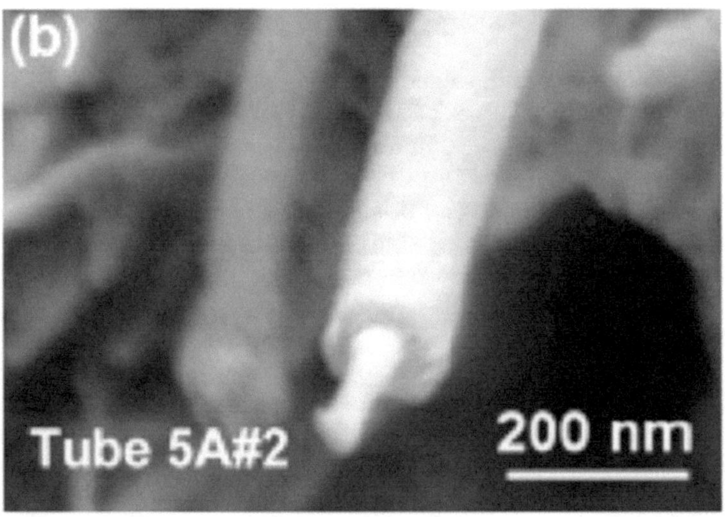

Fig. 18. High-resolution SEM image highlighting MWCNT coated with a polymer sheath protruding from a MWCNT-polycarbonate fracture surface, as observed by Ding *et al.* (2003).

The interface between the host matrix and reinforcement phase, in addition to the interphase region may play a pivotal role in optimising the mechanical performance of a polymer composite (Gilbert *et al.* 1995; Hull and Clyne 1996b; Yang *et al.* 1997; Eitan *et al.* 2006). When investigating the effect of MWCNT on the crystallinity of PVA and PVA nanocomposites Ryan *et al.* (2006) used dynamic mechanical analysis (DMA) to confirm significant improvements in the Young's modulus. Ryan *et al.*(2006) also confirm the findings reported by Marrs *et al.* (2005), reporting that limits in the amount of CNT added to the polymer exist for achieving optimal mechanical performance. Ryan *et al.* (2006) explained that improvements in mechanical properties are seen for lower fractions of CNT with detrimental effects introduced at higher levels of loadings due to the higher incidence of CNT agglomerations. Of further interest is the degree of crystallinity present at the CNT-polymer interface: reported for CNT-reinforced PVA, the large increases in Young's modulus were attributed to the variations in crystallinity. A consequence of the ability of CNT to act as nucleation sites for crystals in both the solution and melt/solid-state phase (Coleman *et al.* 2004; Ryan *et al.* 2006). It is the presence of a well-bonded crystalline interface between the polymer matrix and the nanotubes that may account for the improved mechanical properties due to the increased levels of stress transfer (Cadek *et al.* 2004; Coleman *et al.* 2004). This would allow for failure/crack deflection to occur at the matrix-crystalline interface rather than at the nanotube interface. It has been highlighted that the presence of CNT agglomerations, particularly seen with the use of SWCNT, limits the ability of PVA to act as crystal nucleation sites and, as a result inferior mechanical performance are not seen. Similar findings have been reported for other semi-crystalline polymers (Hull and Clyne 1996a) such as UHMWPE (Ruan *et al.* 2003), polypropylene (Leelapornpisit *et al.* 2005; Seo *et al.* 2005) and polyamide (Chao *et al.* 2006). These reports highlighted the need for careful selection of the processing method that allows for optimal levels of crystallisation at the nanotube-matrix interface (Coleman *et al.* 2004). To date, limited work has been published regarding the crystallisation of PMMA at the nanotube-matrix interface: Coleman *et al.* (1998) proposed that PMMA would be unable to bond to CNT due to the spatial arrangement of the polymer. Alternative mechanisms have been sought to improve the interface and interphase properties of CNT- amorphous polymers. The most commonly employed approach is the covalent attachment of chemically functional groups to the CNT at the defect sites. This is completed in order to achieve similar polymer sheathing effects around the nanotube. In chlorinated polypropylene, Coleman *et al.* (2004) reported that the thickness of the polymer sheath surrounding the nanotubes depends on the volume occupied by the functional groups. Upon failure, fracture was seen to occur away from the CNT-polymer matrix interface. Immobilisation of the polymer chains in the region surrounding the nanotubes was proposed as an additional reinforcement mechanism associated with MWCNT-reinforcement of amorphous polycarbonate. Functionalisation of the MWCNT surface led to an increase in thickness of this interphase region, subsequently improving load transfer capabilities between the matrix and the nanotubes (Eitan *et al.* 2006). This suggestion was highlighted by Jia *et al.* (1999), they reported that PMMA can bond with CNT, although this was dependent on the processing methods utilised. Using *in situ* polymerisation, initiated using the free radical initiator, Azobisisobutyronitrile (AIBN), to form the polymer, additions of MWCNT (both unfunctionalised and carboxyl functionalised) resulted in nanocomposites of improved mechanical properties (in particular, tensile strength, toughness and hardness), with the carboxyl functionalised nanotubes out-performing their unfunctionalised counterparts. High interfacial strengths were associated with bonding between the open π-bonds of the CNT (believed to be initiated by the AIBN) and the open bonds in the PMMA possibly creating a C-C bond between the PMMA and the CNT. Jia *et al.* (1999) also reported that higher loadings of MWCNT led to a more brittle polymer with reduced toughness and tensile strength. Velasco-Santos *et al.* (2003) demonstrated the use of an amorphous polymer matrix is potentially advantageous over semi-crystalline polymers like PVA; clear reasoning behind this is not given although the presence and nature of the interphase region may be an explanation.

9. Mechanisms of failure of CNT-polymers

Andrews *et al.* (2002) reported that CNT within polymer matrices under tensile stress may align themselves parallel to the direction of the applied load, enabling crack bridging behind a crack tip. Andrews *et al.* (2002) explained that this phenomenon reduced the stress concentration in the region surrounding the crack tip, and ultimately reduced crack propagation. Key reinforcement mechanisms that have been identified in CNT-polymer systems are nanotube pullout, bending of nanotubes (often due to surface defects such as iron oxide catalyst inclusions from the CNT production process) and telescopic fracture of nanotubes (Andrews *et al.* 2002; Cooper *et al.* 2002; Demczyk *et al.* 2002; Ding *et al.* 2003; Hwang *et al.* 2004). Such phenomena are schematically illustrated in Figure 19.

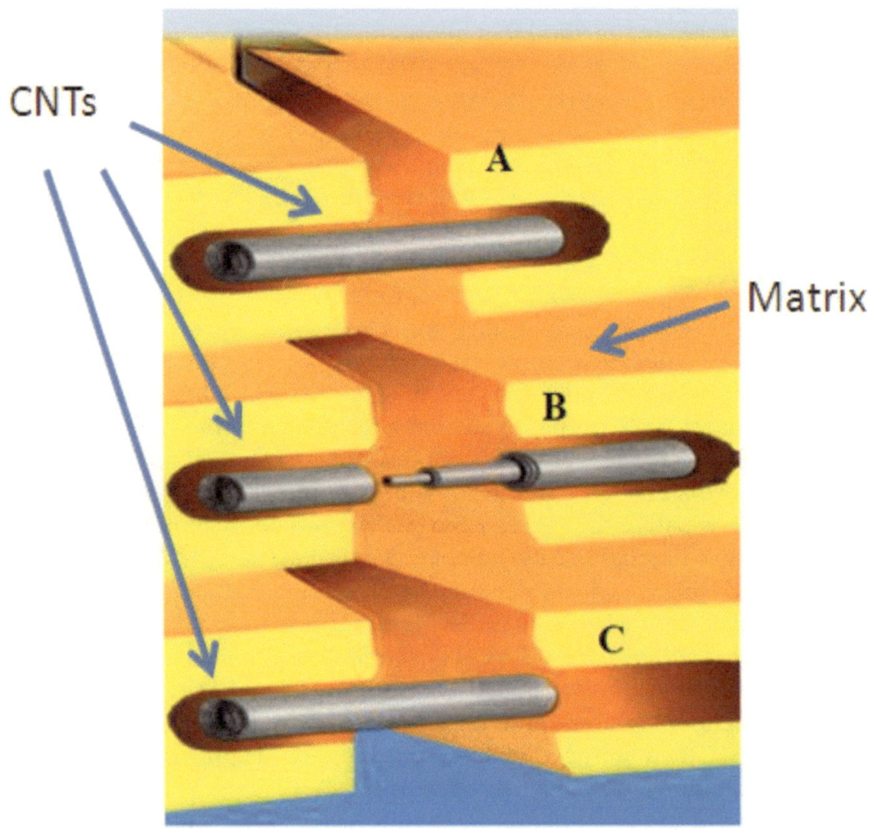

Fig. 19. Schematic illustration of (A) a CNT bridging a crack, (B) telescopic failure of a CNT and (C) CNT fibre pullout experienced in CNT-reinforced polymers (Sinnott and Andrews 2001).

Crack bridging of the matrix may arise in CNT-containing polymers. Hwang *et al.* (2004) directly observed nanotube pullout. (Figure 20b) and failure of the graphene layers resulting in telescopic failure (Figure 20c).

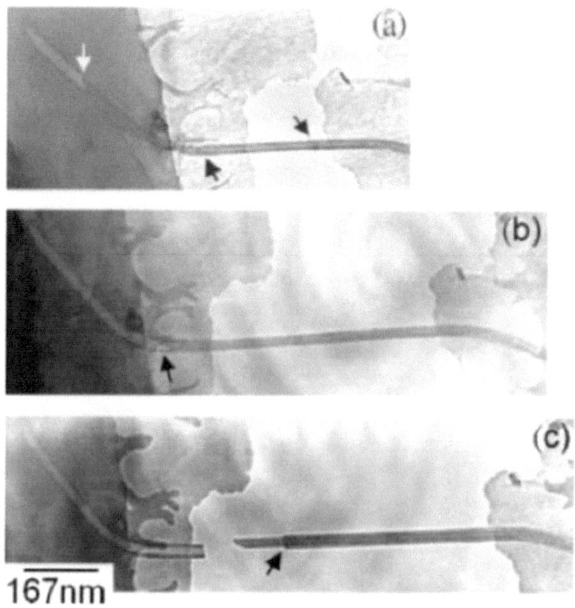

Fig. 20. TEM images of MWCNT-containing PMMA showing (a) breaking of graphene layers, (b) MWCNT pullout and (c) final telescopic failure of the MWCNT (Hwang *et al.* 2004).

10. Biocompatibility of CNT

The Royal Society and the Royal Academy of Engineering, UK, published a report (2004) discussing the associated ethical, health and safety, and social implications of nanotechnology (The Royal Society and the Royal Academy of Engineering 2004). With an increased interest in the use of nanotechnology, the Government later published its own report (*'Characterising the potential risks posed by engineered nanoparticles'*, November 2005) and follow-up studies (*'First quarterly update on the Voluntary Reporting Scheme for engineered nano-scale materials'*, December 2006). The full report addressed many issues concerning the potential use of nanotechnology and CNT. Concerns have been raised that the properties that promote the use of nanoparticles in certain applications may also have health implications, such as their high aspect ratios, surface reactivity and their ability to cross cell membranes (The Royal Society and the Royal Academy of Engineering 2004; Kagan *et al.*, 2005; Fadeel *et al.*, 2007). The report highlighted that the main risks associated with CNT stem from their high surface to volume ratio to which a target organ may be exposed, in addition to the chemical reactivity of the surface, the physical dimensions of the nanoparticles and their solubility. Speculation surrounding the use of CNT has equated their effect on health to that of asbestos (due to their similar size and shape). CNT are therefore suspected as being potentially carcinogenic, and additionally, may cause inflammation or functional changes to proteins due to their large surface area. However, it has been argued that no new risks to health have been introduced as a result of the increasing use of nanoparticles as part of composite materials, and that most concerns derive from the possibility of detached, or 'free' nanoparticles and nanotubes from the matrix (The Royal Society and the Royal Academy of Engineering 2004). It is believed that, if airborne, the likelihood of CNT existing as individual fibres is improbable as electrostatic forces cause the CNT to agglomerate which reduces their ability to be inhaled into the deeper areas of the lungs. However, when investigating of the inhalation of stable non-purified SWCNT aerosols in mice, Shvedova *et al.* (2008) reported that the chain of pathological events was realised through an early inflammatory response and oxidative stress culminated in the development of multifocal granulomatous pneumonia and interstitial fibrosis (Figure 21).

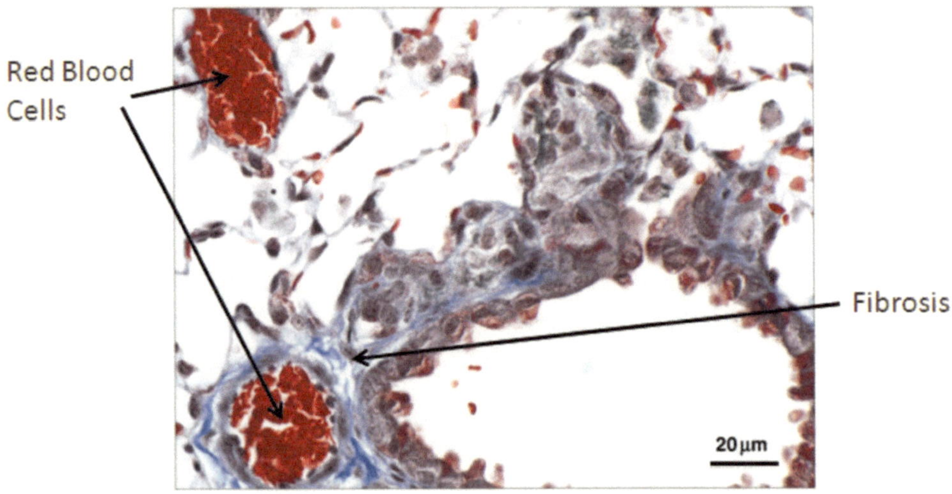

Fig. 21. Representative image of lung section from the SWCNT inhalation study depicting granuloma formation on day 28 post treatment. Fibrosis is indicated by blue staining in this Masson's Trichrome stained section of the lung (Shvedova *et al.* 2008).

Smart *et al.* (2006) reviewed the often conflicting findings pertaining to the cytotoxicity and biocompatibility of CNT. They concluded that, as-received (i.e. untreated, or unfunctionalised) CNT exhibited some degree of toxicity (observed both *in vitro* and *in vivo*) with detrimental effects associated with the presence of transition metal ions, used as catalysts in the CNT production. Smart *et al.* (2006) also reported that CNT that have been chemically functionalised have yet to demonstrate toxicity effects. It is highlighted that the tendency for CNT to aggregate may impact the reported results, although quantification of this fact has yet to be investigated. With research into the use of CNT, and nanotechnology ever increasing, the uncertainty regarding toxicity has been brought to public attention. As a result, it has been recommended that further research is necessary regarding the biological impacts of nanoparticles and nanotubes, including their exposure pathways within the body, and that methodologies for *in situ* monitoring should also be developed (The Royal Society and the Royal Academy of Engineering 2004). Recent studies have addressed the issue of CNT uptake by different cell types. While the results seem to be controversial, it is apparent that the presence or absence of specialised signals determined the recognition and subsequent interactions of CNT with cells. Overall, pristine CNT carrying no recognisable signals were poorly taken-up whereas CNT modified chemically (e.g. oxidatively modified and functionalised) or by adsorbed macromolecules (e.g. proteins, lipids) were more readily recognised and engulfed by cells (Shvedova *et al.*, 2010). Several *in vitro* studies support the concept that pristine CNT are not readily taken up by lung cells. Davoren *et al.* (2007) reported no measurable uptake of CNT in A549 cells (a human alveolar type II cancer cell line). Likewise, Herzog *et al.* (2007) reported no uptake of CNT in either A549 cells or BEAS-2B cells (a human bronchial epithelial cell line). Lastly, no evidence of uptake of CNT was reported after electron microscopic evaluation of exposed RAW 264.7 cells (mouse peritoneal macrophage cell line) (Shvedova *et al.*, 2005). In contrast, functionalistion of SWCNT with a phospholipid signal, phosphatidylserine, made CNT recognisable *in vitro* by different phagocytic cells, including murine RAW264.7 macrophages, primary monocyte-derived human macrophages and dendritic cells, and microglia from rat brain (Figure 22) (Shvedova *et al.*, 2009).

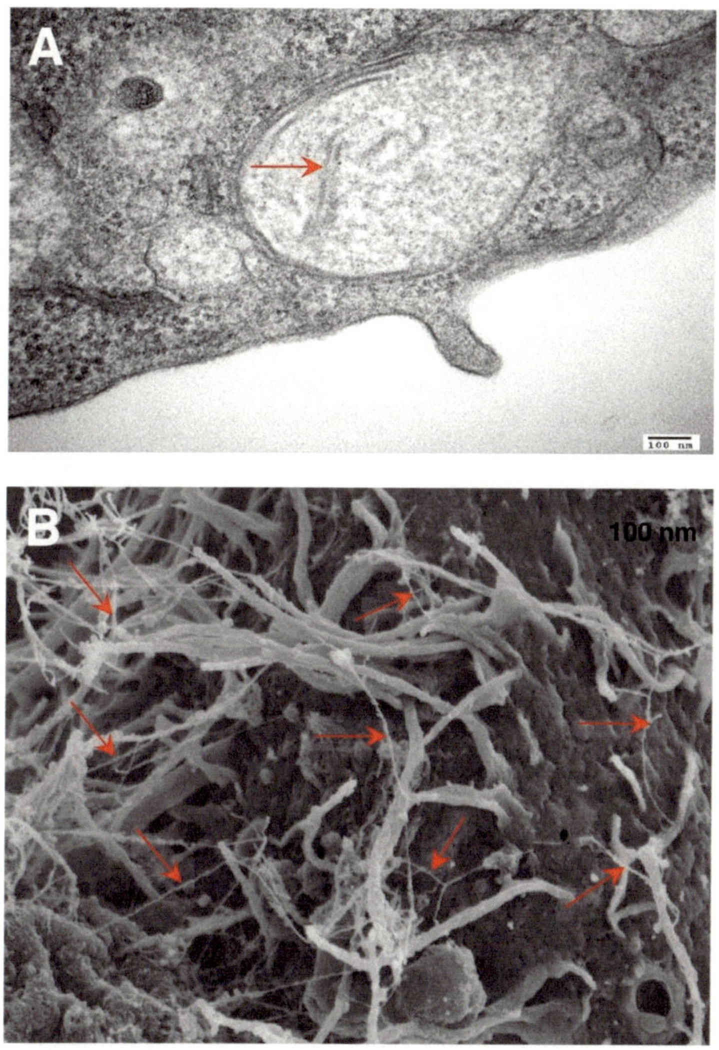

Fig. 22. Representative transmission electron micrograph (A) and scanning electron micrograph (B) of RAW264.7 macrophages with engulfed PS-coated SWCNT. Arrows indicate SWCNT. (Shvedova *et al.*, 2009).

11. CNT-reinforced biomaterials

Nanotechnology in biomaterials is not a new idea (Hrkach *et al.*, 1997). Nanomaterials have been used as implant coatings, bulk materials, drug delivery, actuators, diagnostic tools and devices (Sinha and Yeow, 2005). When biomaterials incorporating nanomaterials are studied, much of the emphasis is on the interaction between the biological tissue and the biomaterial at a molecular level. Using the interface between bone, and a metallic implant as an example, a positive biological interaction is essential if a good fixation is to be obtained. Chun *et al.* (2004) examined this interaction by coating titanium (Ti) substrates with helical rosette self-assembled organic nanotubes (HRN). HRN display chemical and structural similarities to various constituents of bone (Figure 23). Chun *et al.* (2004) found that the HRN-coated titanium displayed enhanced interaction with the naturally-occurring nanostructures' constituents such as collagen fibres and HA; this was measured as a function of the cell adhesion of human fetal osteoblas (hFoBs) cells (Figure 24).

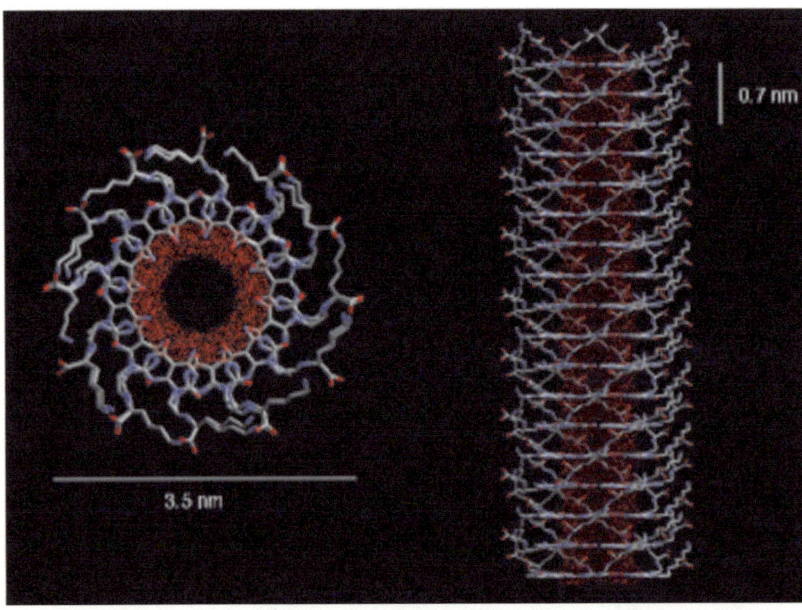

Fig. 23. Diagram of helical rosette nanotube (HRN) (Chun *et al.*, 2004).

Webster *et al.* (2004) incorporated carbon nanofibres (CNF) into polycarbonate-urethane (PCU) for neural or orthopaedic prosthetic devices. They reported that this material had the potential to increase neural and osteoblast functions, as cell attachment increased with CNF loading. Additionally they stated that the functions of cells that contributed to glial scar-tissue formation for neural prostheses (astrocytes) and fibrous-tissue encapsulation for bone implants (fibroblasts) decreased on the PCU composites containing increasing amounts of CNFs. In this manner, this study provided the first evidence that CNF formulations may interact with neural and bone cells, which is important for the design of successful neural probes and orthopaedic implants. Furthermore, Webster *et al.* (2004) summarised that using nanotechnology in biological systems may be potentially feasible as biological systems are governed by molecular behaviour at the nanoscale, and therefore the properties of which are accustomed to high levels of interaction at this nanoscale. This study by Webster *et al.*

highlighted the potential for CNT to be used in PMMA bone cement to encourage cell growth at the bone-cement interface with the aim of reducing aseptic loosening by enhancing the mechanical interlock in the cancellous bone.

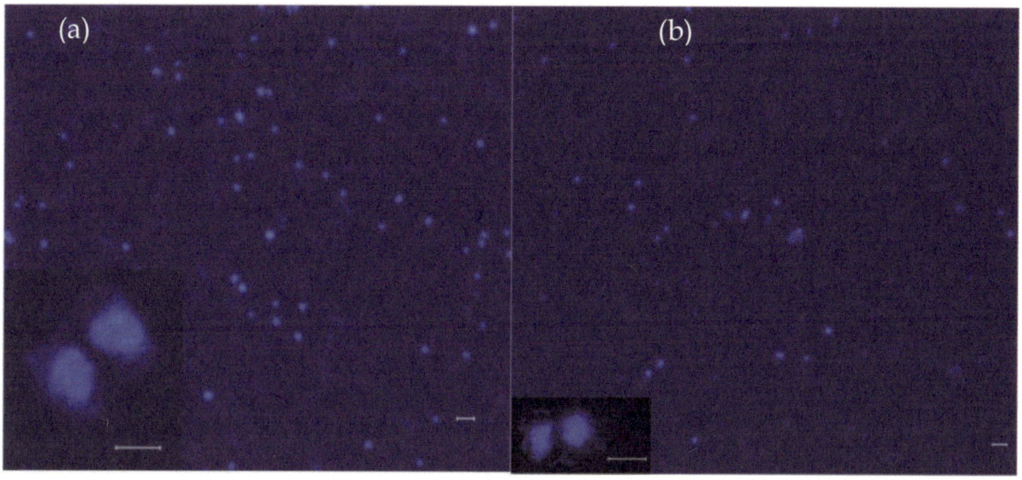

Fig. 24. Fluorescently stained cells on Ti substrates. (a) HRN coated Ti. (b) Uncoated Ti. (Magnification: 20×; inset magnification 200×). Scale bars = 60 μm, inset bars = 50 and 100 μm for (a) and (b), respectively (Chun *et al.*, 2004).

CNT exhibit many unique mechanical, thermal, and electrical properties. However, their potential use for bioengineering applications and medical materials is almost wholly dependent on their biocompatibility. Cui *et al.* (2005) investigated the effect of SWCNT on human HEK293 cells (human embryo kidney cells). Results showed that SWCNT can inhibit HEK293 cell proliferation, inducing cell apoptosis (programmed cell death as controlled by the nuclei in normally functioning cells) and decreasing cellular adhesive ability in a dose and time-dependent manner. Their results also showed that HEK293 cells initiated active responses such as secretion of small 'isolation' proteins to isolate the cells attached to the SWCNT from the rest of the cell mass; a response that offers potential for medical chemistry and disease therapy.

Synthetic bone scaffolds is an area where the biocompatibility of materials used, such as polymers or peptide fibers, is still an issue where possible rejection by the body is feasible. CNT offer mechanical advantages over the polymers or peptide fibers currently used in bone scaffolds. Zhao *et al.* (2005), investigated the use of chemically functionalised SWCNT as a scaffold material for the growth of artificial bone, they identified the potential for the self-assembly of HA on the surface of SWCNT. They suggested that this was possibly due to the presence of negatively charged functional groups on the SWCNT that attract the calcium cations present in HA (Zhao *et al.* 2005). The group also proposed that it is the high tensile strength, high degree of flexibility, and low density of CNT that make these materials ideal for the production of bone. The diameters of SWCNT used in the study by Zhao *et al.* (2005) are of similar order and magnitude to the triple helix collagen fibres within bone, and as such can act as scaffolds for the nucleation and growth of HA.

The potential for CNT to be used within bioengineering applications is by no means endless, however while many more applications could be discussed, the following papers offer further information on the use of nanotechnology for biomedical applications: Sinha and

Yeow (2005), Webster *et al.* (2004) and Bellare *et al.* (2002). Investigations concerning the cytotoxic response of CNT-containing materials have reported encouraging results confirming their potential use in orthopaedic applications (Smart *et al.* 2006); however, many questions remain unanswered and as yet, the understanding of the toxicity and biocompatibility of CNT-reinforced materials is not fully established.

12. CNT-reinforced bone cement

12.1 Mechanical properties

MWCNT offer the potential to augment mechanical properties of PMMA bone cement due to their strength and aspect ratio. The addition of MWCNT to PMMA bone cement has been shown to significantly improve the static mechanical properties (Marrs *et al.*, 2006 and Marrs, 2007; Ormsby *et al.*, 2010a; Ormsby *et al.*, 2010b), and the fatigue performance of MMA-co-Sty copolymer based bone cement (Marrs *et al.*, 2006). Marrs *et al.* (2006) investigated the influence of unfunctionalised MWCNT in PMMA based bone cements. They reported moderate improvements (13-24 %) in the static properties when 2 wt. % MWCNT was incorporated into PMMA bone cement. Marrs (2007) reported significant improvements (>300 %) in the dynamic properties of methyl methacrylate-styrene copolymer (MMA-co-Sty), a chief component of commercial bone cement when unfunctionalised MWCNT (2 wt. %) were added. However, both studies (Marrs *et al.*, 2006 and Marrs, 2007) used non-clinically relevant methods to ensure optimal dispersion of the MWCNT. The MWCNT were dispersed through a molten matrix of pre-polymerised commercial bone cement powder using stainless steel counter rotating rotors in a mixing chamber at 220 °C. The two materials were heated and subjected to high-shear mixing. Once the molten composite had cooled and hardened, it was crushed into pellets and hot pressed under vacuum to form films. These films were subsequently machined into testing specimen. Each specimen was then annealed at 125°C for a minimum of 15 h to alleviate any residual stresses that formed during machining.

The uniform distribution of CNT within the polymer matrix is critical for maximising the interfacial bond between the CNT and polymer matrix and therefore achieving optimal improvements in mechanical properties (Andrews *et al.*, 2002; Marrs, 2007). It has also been reported that alignment and optimum dispersion of the CNT is important in the context of improving the thermal properties of a nanocomposite (Xie *et al.*, 2005). The CNT must create a well dispersed, overlapping network facilitating the transport of electrons, phonons, and heat energy.

Many processing techniques have been employed to uniformly disperse CNT within polymer matrices (Xie *et al.*, 2005; Andrews *et al.*, 2002). The two most commonly used techniques involve (i) *in situ* dispersion (sonication of the CNT in solution) and (ii) high temperature shear mixing. These techniques are primarily used to separate the entanglements and agglomerations of the as-produced CNT, and secondly to disperse the individual CNT throughout the matrix. Andrews *et al.* (2004) stated that these techniques produce more favourable results when small concentrations of CNT are used, however, mixing higher concentrations of CNT (>5 wt. %) increases the viscosity of the mixture irrespective of the state of the polymer. Andrews *et al.* (2002) postulated that an elevated viscosity hinders effective dispersion of the CNT into the polymer matrix, therefore, the energy induced into the mixing process must be increased, but, at the risk of shortening the CNT or irreversibly damaging the matrix material. Moreover, it has been reported that the efficacy of MWCNT reinforcement is largely dependent on the level of loading of MWCNT, the dispersion of these MWCNT and the peak stress of dynamic loading cycle (Marrs *et al.*, 2006; Marrs, 2007).

Ormsby *et al.* (2010a) addressed the limitations of the studies by Marrs *et al.*, (2006; Marrs, 2007) by incorporating unfunctionalised (MWCNT-UNF) and carboxyl (MWCNT-COOH) functionalised MWCNT (0.1 wt. %) into PMMA bone cement using three different preparation techniques. CNT were either added to the liquid MMA component of the cement via magnetic stirring or ultrasonic disintegration, or dry blended with the polymer powder component. A contemporary vacuum mixing system was subsequently used to mix the bone cement following the normal protocol for a joint replacement surgical procedure. Improvements in static mechanical properties and thermal properties of the MWCNT-PMMA nanocomposite cement were observed (Ormsby *et al.*, 2010a). Ormsby *et al.*, (2010a) demonstrated that adding MWCNT (0.1 wt. %) to the polymer powder or liquid monomer components prior to cement mixing with a proprietary mixing system, improved the mechanical properties of the resultant cement, provided the appropriate method for incorporating the MWCNT was used (≈21 %). This was a significant finding because mechanical failure of the bone cement mantle remains a major problem in joint replacement surgery (Topoleski *et al.*, 1990). Like typical fibre-reinforced composites, mechanical failure of PMMA bone cement is believed to take place in three phases, (1) crack initiation due to an initial imperfection in material stability, (2) slow crack growth, and (3) rapid propagation to fracture (Figure 25a) (Topoleski *et al.*, 1995). Although mixing the cement under the application of a vacuum and injecting the cement into the surgical site using a closed delivery system have improved the mechanical performance of the cement, residual material voids and poor surgical technique can contribute to weak or thin regions within the cement mantle causing these regions to be more susceptible to mechanical failure (Marrs *et al.*, 2006).

Ormsby *et al.*, (2010a) reported that adding MWCNT to the liquid monomer by magnetic stirring had an overall negative effect on the mechanical performance of the bone cement. This was largely attributed to the poor dispersion of MWCNT in the liquid monomer and resulting in MWCNT agglomerations within the cement matrix (Figure 25b).

These agglomerations acted as stress concentrations within the cement, providing a mechanism for premature failure of the cement when subjected to load. In contrast, dry blending MWCNT in the polymer powder or disintegrating the MWCNT in the liquid monomer using ultrasonic agitation suitably disentangled the nanotubes and homogenously dispersed the MWCNT in the resulting nanocomposite (Ormsby *et al.*, 2010a). Andrews and Weisenberger (2004) also reported that ultrasonic disintegration was an effective method for dispersion of MWCNT at low levels (<5 wt. %) of concentration (Andrews and Weisenberger, 2004). Marrs *et al.* (2006) stated that care is needed when dispersing MWCNT within a polymer matrix, and reported the adverse effects of sporadic, inadequately dispersed, clumps of MWCNT, particularly at levels of loading greater than 5 wt.%.

The presence of well-dispersed MWCNT in PMMA cement with their anticipated strong nanotube-matrix bonding and high tensile properties, suggests that a percentage of the MWCNT would be orientated with their longitudinal axis perpendicular to the crack wave. Such MWCNT were effective in bridging the initial crack and preventing crack propagation, further enhancing the longevity of the cement mantle (Figures 25c and 25d), by improvement in mechanical properties. These improvements are clinically beneficial for the use of reinforced PMMA bone cement in TJR due to a reduction in the rate of crack propagation. This effect may be most important for improperly placed femoral implants with thinner cement mantle layers, which continues to be cited as a factor that may reduce implant longevity (Morscher and Wirz, 2002). Additionally PMMA dental prostheses (dentures) are also known to fail prematurely through thin connectors due to impact and fatigue loading (Ormsby *et al.*, 2010a). There could be an application for MWCNT inclusion in PMMA dental prostheses, enhancing the functionality of denture-based acrylic materials when subjected to fatigue loading (Polyzois *et al.*, 1996).

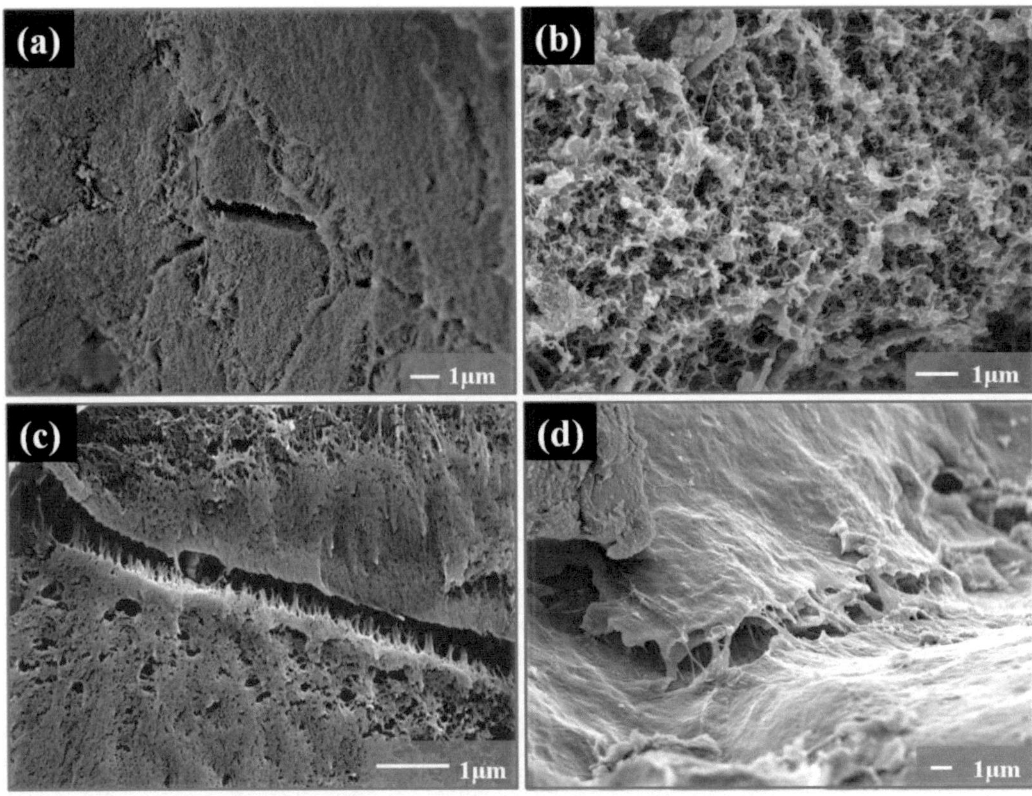

Fig. 25. SEM images showing (a) A large pore on the short rod chevron notched fracture surface of the control cement (X 300). (b) Unfunctionalised MWCNT dry blended in the PMMA polymer powder cement showing an agglomeration of barium sulphate, which was the fracture initiation point for this specimen (X 150). (c) Functionalised MWCNT disintegrated in the MMA liquid monomer by ultrasonication, MWCNT can be seen to bridge a micro-crack across the cement surface, X 5,000, (d) Functionalised MWCNT ultrasonically disintegrated within the MMA liquid monomer, MWCNT can be seen to bridge a micro-crack on the cement surface, X15,000 (Ormsby *et al.*, 2010a).

The filler/matrix interface in fibre-reinforced polymer composites is critical in controlling load transfer from the matrix to the fibre, failure mechanisms, and degradation (Ormsby *et al.*, 2010a). Gojny *et al.* (2003) reported that functionalisation of MWCNT led to reduced agglomeration and improved interaction between the nanotubes and the polymer resin. Ormsby *et al.*, (2010a) used MWCNT that were surface modified with a carboxyl grouping, as it has been reported that the static mechanical properties of PMMA polymer resin can be significantly improved with this arrangement (Pande *et al.*, 2008). Ormsby *et al.* (2010a) observed that surface modification of the MWCNT with carboxyl groups did not result in significant improvements in the compressive or bend properties of the PMMA cement on a consistent basis. However, the fracture toughness of the PMMA cement was significantly enhanced (p-values<0.001) when the MWCNT were surface modified with carboxyl groups. It is unclear currently as to whether the improvements in performance of the MWCNT-PMMA cements are a direct consequence of good MWCNT dispersion within the PMMA matrix, providing mechanical reinforcement, or is due to a chemical interaction between the MWCNT and PMMA matrix (Ormsby *et al.*, 2010a). Eitan *et al.*, (2006) used strain dependent Raman spectroscopy to show that there is load transfer from the matrix to the nanotubes, and that the efficiency of the load transfer is improved by surface modification of the MWCNT.

It is also interesting to observe that significant improvements in fracture toughness did not correlate to improvements of the same magnitude for strength and modulus of the different cement combinations tested. Ormsby *et al.* (2010a) postulated that the methods adopted for specimen preparation, specimen configuration and the different modes of loading employed during testing could account for this. It has been reported that different loading regimes evaluate differing reinforcement mechanisms within the specimen microstructure, therefore dissimilar responses are expected (Lewis and Mladsi, 2000; Wagner and Chu, 2006). Wagner and Chu (2006) also found distinctions in mechanical properties when testing three dental core ceramic based materials. They found significant differences in the biaxial flexural strength, but reported no significant difference for the indentation fracture toughness for the materials tested (Wagner and Chu, 2006).

Subsequent to this investigation, Ormsby *et al.* (2010b) also published a study investigating the efficacy of adding different concentrations of MWCNT to PMMA bone cement of varying functionality as a means of improving MWCNT dispersion and thus augmenting the mechanical properties of the PMMA bone cement further. The bone cement was prepared using the optimal method for MWCNT incorporation, as determined in their previous study (Ormsby *et al.*, 2010a). Ormsby *et al.* (2010b) reported that adding MWCNT at low loadings (≤0.25 wt. %) to MMA monomer, prior to cement mixing with a proprietary mixing system, improved the mechanical properties of the resultant nanocomposite cement. Adding carboxyl and amine functional groups enhanced the dispersion of the MWCNT within the cement matrix and potentially increased the interaction between the carbon nanotubes and the cement, thereby improving the mechanical integrity of the resultant nanocomposite cement. These improvements in mechanical strength are potentially significant as mechanical failure of the bone cement mantle remains a prevalent issue in total joint replacement surgery often leading to revision surgical procedures. Adding MWCNT at higher loadings (≥0.5 wt. %) provided a negative effect on the mechanical performance of the nanocomposite cement. This was attributed to poor dispersion of MWCNT resulting in agglomerations forming within the cement matrix. In contrast, low loadings (≤0.25 wt. %) of MWCNT were more readily disentangled by the application of ultrasonic energy and homogenously dispersed in the resulting nanocomposite. The presence of well-dispersed MWCNT in PMMA cement with their anticipated strong nanotube-matrix bonding and high tensile properties, suggests that a percentage of the MWCNT would be orientated with their longitudinal axis perpendicular to the crack wave. Such MWCNT were effective in bridging the initial crack and preventing crack propagation, further enhancing the mechanical integrity of the cement mantle. These improvements could have clinical benefits for the application of MWCNT-PMMA nanocomposite cement in TJR surgery, due to a reduction in the rate of crack propagation through the reinforced nanocomposite cement mantle. This effect may have greatest significance for misaligned femoral implants resulting in areas of thinner cement mantle thickness, which continues to be cited as a main factor of cement mantle failure (Ormsby *et al.*, 2010b).

Gojny *et al.*, (2003) also reported that the addition of chemical functional groups to the MWCNT can provide a negative charge to the MWCNT and thus reduced agglomeration and improve interaction between the nanotubes and the host polymer. The results of this study by Ormsby *et al.*, (2010b) concurred with the findings of Gojny *et al.* (2003). The PMMA bone cement with MWCNT-UNF exhibited least significant improvements (p-value<0.1) for all mechanical properties measured. This reduced improvement in mechanical properties was attributed to poor dispersion of MWCNT within the cement matrix, resulting in the occurrence of MWCNT agglomerations. The MWCNT-UNF provided a degree of mechanical reinforcement at lower loading (≤0.25 wt. %), largely due to the reduced tendency for MWCNT agglomerations. MWCNT-COOH provided the most significant (p-value<0.001) improvements in all mechanical properties of the PMMA cement. It is proposed these significant improvements are a result of a homogenous dispersion of the MWCNT within the PMMA matrix aided by the negatively charged carboxyl groups. This homogeneous dispersion in tandem with interfacial interactions between the functionalised MWCNT and PMMA matrix could provide improved mechanical properties of the resultant nanocomposite. The bone cements incorporating amine functionalised MWCNT (MWCNT-NH_2) also improved mechanical properties. These improvements were less significant p-value<0.01 when compared with the addition of MWCNT-COOH. It is postulated that this is due to the lower level of functional groups present on the MWCNT-NH_2 when compared with the MWCNT-COOH (that is 0.5% vs. 4.0%, functional groups, respectively). This lower concentration of MWCNT-NH_2 functional groups may result in a more heterogeneous dispersion of the MWCNT within the cement matrix, therefore resulting in a less successful transfer of stress through the cement mantle.

12.2 Thermal properties

PMMA bone cement is produced by a free radical reaction on mixing the polymer powder and liquid monomer constituents. The polymerisation reaction is a highly exothermic chemical reaction and as a consequence the peak temperatures typically reach 80-100 °C. It has been reported that polymerizing bone cement has the potential to cause thermal necrosis of the surrounding bone cells, which is one of the mitigating factors for aseptic loosening of an implant fixed with PMMA bone cement (Dunne and Orr, 2002).

Reducing the polymerisation reaction of PMMA bone cement, therefore lowering the extent of thermal necrosis has been investigated by many research groups. Meyer *et al.* (1973) reported reducing the temperature (22 °C) prior to bone cement mixing had a significant influence on the polymerisation reaction of the PMMA cement. They concluded that mixing PMMA cement at a temperature of 4°C resulted in a peak temperature (T_{max}) of 53 °C, while mixing the same cement at 37 °C increased the peak temperature to 125 °C. Meyer *et al.* (1973) also investigated the effects of pre-chilling the femoral prosthesis prior to implantation into the bone cavity; they found adopting this approach did not influence the peak temperature. Larsen *et al.* (1991) also investigated the effects of pre-chilling the femoral prosthesis, however, they reported a 5 °C reduction in the peak temperature at the bone–

cement interface. Additionally, Lidgren *et al.* (1987) found using chilled water to pulse-lavage the bone cavity prior to cement delivery had a significant effect on the extent of the polymerisation reaction, the peak temperature was subsequently reduced from 59 °C to 45 °C. The mixing method used to prepare the PMMA bone cement prior to delivery into the bone cavity also has a role in its polymerisation reaction. Dunne and Orr (2002) reported the level of heat generated for bone cement prepared under the application of a vacuum was significantly reduced when compared to the same cement prepared under atmospheric conditions using a bowl and spatula mixing arrangement. Other methods can be used to reduce the degree of polymerisation reaction of PMMA bone cement, such as altering the compositions or constituents of the cement. However, this can have a significant influence on the mechanical and handling performance of the bone cement (Lewis *et al.*, 2007).

CNT incorporation has previously been reported to improve the thermal properties of a range of polymers, including polyethylene (McClory et al., 2010), polyurethane (Marrs *et al.*, 2006), polystyrene (Andrews and Weisenberger, 2004), polyvinyl alcohol and methyl methacrylate-styrene copolymer (Xie *et al.*, 2005).

Andrews and Weisenberger (2004) proposed that the thermal property improvements for CNT-polymer composites are a function of CNT type, degree of dispersion, CNT loading, CNT alignment and polymer matrix. Xie *et al.* (2005) reported a significant improvement (≈125%) in the thermal conductivity of an epoxy when 1.0 wt. % SWCNT powder was added. Choi *et al.* (2003) observed an increase (≈300%) in the thermal conductivity of an epoxy for a SWCNT loading of 3.0 wt. %. The thermal properties of PMMA bone cement have been modified with MWCNT by Ormsby *et al.*, (2010a). Incorporating either unfunctionalised or carboxyl functionalised MWCNT into the PMMA powder or liquid monomer prior to mixing both components together had a significant effect on the exothermic polymerisation reaction. It was observed that maximum temperature and the setting properties exhibited during polymerisation were significantly reduced by the inclusion of 0.1wt. % (unfunctionalised or carboxyl functionalised) MWCNT into the PMMA cement, irrespective of the method of introduction. Other studies have also reported reductions in the thermal properties of PMMA cement on addition of 5-15 wt. % steel fibres (Kotha *et al.*, 2002). Dunne and Orr (2002) reported that reduction of the polymerisation exotherm will decrease the likelihood of residual stresses developing within the cement mantle, which can cause premature failure of the cement when subjected to mechanical loading.

The importance of minimising the bone cement exothermic reaction has been stressed, as it may result in a permanent cessation of blood flow and bone tissue necrosis, which shows no sign of repair after 100 days (Moritz and Henriques, 1947; Feith, 1975; Eriksson and Alberksson, 1983; Mjoberg *et al.*, 1984). The cumulative TNI (Thermal Necrosis Index) has been used previously to assess the level of irreparable damage bone cement caused by heat generation (Moritz and Henriques, 1947; Dunne and Orr, 2002). If TNI exceeds one there is the possibility of thermal damage to the living tissue cells. The thermal necrosis index is typically calculated at two temperatures; >44 °C and >55 °C, chosen because the temperature threshold for impaired bone regeneration has been reported to be in the range of 44-47 °C (Moritz and Henriques, 1947; Eriksson and Alberksson, 1983). The incorporation of MWCNT to PMMA based bone cement may reduce the incidence of polymerisation induced hot spots and thermal necrosis of the surrounding tissue adjacent to the cement mantle, which is believed to be observed radiographically (Linder, 1977). Reducing the occurrence of such tissue damage may improve the mechanical integrity of the cement-bone interface, thereby promoting implant longevity.

Ormsby *et al.*, (2010a) report that the incorporation of unfunctionalised or carboxyl functionalised MWCNT assisted in the dissipation of the heat produced during the exothermic polymerisation reaction of PMMA bone cement, irrespective of the method of introduction. With unfunctionalised MWCNT, this reduction was not below the levels necessary to prevent thermal tissue damage as the TNI was greater than one. In contrast, surface modification of the MWCNT with carboxyl groups and subsequent addition to the liquid monomer using magnetic stirring did reduce the TNI values at >44 °C and >55 °C to levels below one.

In a subsequent study by Ormsby *et al.*, (2011) the incorporation of unfunctionalised, amine, and carboxyl functionalised MWCNT at increasing wt. % assisted in the dissipation of the heat produced during the polymerisation of PMMA bone cement. It was observed that any effect on the reaction exotherm was dependant on MWCNT loading. The greater reductions in exotherm were reported for the highest level of MWCNT loading (1.0 wt. %). Saha and Pal (1986) reported a similar finding when examining carbon fibre reinforced bone cement. The greatest reductions in peak exothermic temperature were associated with the highest levels of carbon fibre. It is important to note that the types of MWCNT used within the study by Ormsby *et al.*, (2011) had thermal conductivity values of >3000Wm^{-1} k^{-1}. It was therefore proposed that the MWCNT act as a heat sink within the PMMA bone cement and therefore assist in the dissipation of the heat generated during the polymerisation reaction (Ormsby *et al.*, 2011). This behaviour is also a function of the extent of MWCNT dispersion and distribution throughout the PMMA bone cement matrix, such that uniform dispersion of MWCNT within the cement will dissipate the thermal energy throughout the cement matrix. This is further aided by the interconnectivity of MWCNT entanglements and the very large surface area of MWCNT (600-1000 m^2/g) (Peigney *et al.*, 2009). Bonnet *et al.* (2007) found a similar effect on the addition of 7 vol. % of SWCNT to PMMA reporting a 55 % increase in the thermal conductivity. It is therefore hypothesised that the thermal conductivity of the PMMA bone cement described here will have also increased due to MWCNT addition.

It has been stated that for composites incorporating CNT to be thermally conductive, they must form a percolated network of overlapping or touching CNT for the transport of heat energy (Marrs, 2007). Therefore bone cements with relatively poor levels of MWCNT dispersion (≥0.5 wt. %) within the PMMA matrix, due to agglomerations (Figure 26), demonstrated the greatest reduction in thermal properties.

It is possible to use this theory to explain why the MWCNT of different chemical functionality provided differing thermal properties. The addition of MWCNT-UNF and MWCNT-NH_2 provided more significant reductions in the polymerisation reaction when compared to the MWCNT-COOH. It is suggested by Ormsby *et al.*, (2011) that this may be due to a less homogeneous dispersion of the MWCNT-UNF and MWCNT-NH_2 within the cement in comparison to the improved dispersion of the MWCNT-COOH.

Ormsby *et al.* (2010b) added MWCNT of various chemical functionality at increasing loadings to PMMA cement and assessed the mechanical properties of the resultant composites. They reported significant improvements in mechanical properties at low levels of MWCNT loading (≤0.5 wt. %). Ormsby *et al.* (2010b) showed that MWCNT-COOH provided the greatest improvement in mechanical properties, due to the improved MWCNT dispersion associated with improved interfacial interactions between these MWCNT and PMMA through enhanced van der Waals attraction and hydrogen bonding.

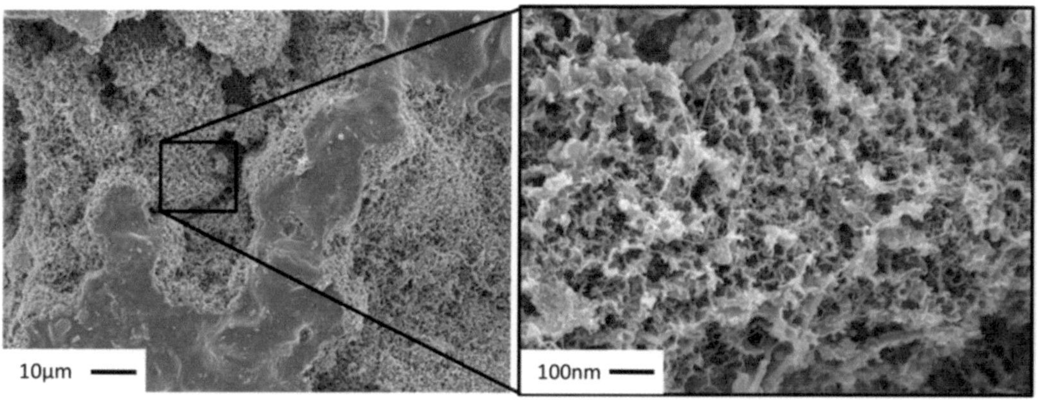

Fig. 26. SEM image of 1.0 wt% MWCNT filled PMMA bone cement showing MWCNT-UNF.

It is noteworthy that the MWCNT inclusion altered the rate of PMMA polymerisation. A slower rate of polymerisation extended the time taken for the bone cement to fully polymerise, which in turn reduced the T_{max} and TNI values. It is postulated that the presence of MWCNT in the cement not only altered the kinetics of the polymerisation reaction, but additionally played a role dissipating heat energy. Incorporation of carboxyl and amine functionalised MWCNT had a greater influence on the polymerisation reaction of the bone cement used in this study, compared to the unfunctionalised (Ormsby *et al.*, 2011).

12.3 Rheology properties

The efficacy of PMMA bone cement in anchoring a TJR is affected by many fundamental characteristics. Among these are the rheological, polymerisation, and handling properties, whose significance is two-fold (Ormsby *et al.*, 2011). Firstly, the ease with which the cement flows into the intramedullary bone canal facilitates the controlled positioning of the prosthesis. This is critical as it has been reported that initial prosthesis position is a contributory factor to the longevity of the cemented implant (Jones *et al.*, 1992; Lewis and Carroll, 2001). Secondly, the rheological properties of the cement may play an important role in the development of pores in the cement during mixing and delivery. Such pores may act as sites for the initiation of cracks, which, in turn, can cause or contribute to aseptic loosening of the prosthesis (Jones *et al.*, 1992).

To date, there have been limited studies examining the viscoelastic properties of PMMA bone cement, with oscillatory shear rheometry (OSR) being the most common method employed. Harper *et al.* (2000) observed that the complex viscosity (η^*) of VersaBond ™ and Palacos® R cements increased from 1000 Pa.s at 2.5 min to 5000 Pa.s at 6 min. They defined this sharp increase in η^* as the onset of cure. Spiegelberg and McKinley (1998) determined the critical gel time of Simplex P™ cement as 9.7 min. Farrar and Rose (2001) investigated the initial polymerisation reaction of several commercial bone cements. They examined η^* over a range of temperatures and concluded the polymerisation of bone cement is strongly dependent on temperature. Ormsby *et al.*, (2011) have assessed the influence of differing MWCNT (unfunctionalised, carboxyl functionalized or amine functionalised) on the rheological properties and cure kinetics of the polymerising PMMA bone cement. They investigated how the differing MWCNT systems influenced the time at which the onset of polymerisation occurred, as well as the time at which polymer gelation occurred. Ormsby *et al.*, (2011) found that MWCNT addition significantly influenced the rheological behaviour of the polymerising cement. For each cement investigated, η^* increased with time. Ormsby *et al.*, (2011) explained this trend applying the Krieger–Dougherty equation (Equation 1) (Krieger and Dougherty, 1959), which describes the viscosity of a concentrated suspension (η).

$$\eta^* = \eta_s \left(1 - \frac{\varphi}{\varphi_m}\right)^{-[\eta]\,\varphi_m} \tag{1}$$

where, η_s is the viscosity of the suspending medium, φ is the phase volume of the particles in the suspension, φ_m is the maximum packing fraction of those particles, and η is the intrinsic viscosity. This equation may be used to comment on the variation of the polymerizing bone cement's η^* as a function of time, although its application is limited as it primarily applies to Newtonian suspensions. During the initial stages of mixing the powder and liquid components, the high initial viscosity is attributed primarily to the swelling of the polymer beads within the cement powder (Lewis and Carroll, 2001). As elapsed time from start of mixing increases and swelling causes both φ_m and η to decrease. Thus, η^* increases with t, a trend observed in the present results of the studies of Lewis and Carroll, (2001), and Ormsby *et al.*, (2011).

Ormsby *et al.*, (2011) found that the incorporation of chemically functionalised MWCNT (MWCNT-COOH and MWCNT-NH_2) into PMMA bone cement significantly extended the onset of cure. This effect was more pronounced as MWCNT loading was increased. Indicating the time delay before the onset of cure for these composite cements is in part due to the role the functional groups play in altering the polymerisation reaction, in addition to physically preventing cross-linking of the polymer chain. The onset of cure of the PMMA cements with MWCNT-UNF addition was also delayed, but to a lesser extent. In all cases, MWCNT addition to PMMA bone cement prevented macro-gelation from occurring.

It was also significant to observe that on addition of MWCNT-COOH, gel-times increased up to a loading of ≤0.5 wt. %, but at 1.0 wt. % the gel-time decreased, compared to the control sample. This finding is commonly reported for heavily-filled polymers, as the cement may exhibit solid-like properties from the onset of mixing. Therefore, initially the filled bone cement will have a higher viscosity than the control, but the actual onset of polymerisation may not occur until much later in the reaction. Lalko *et al.* (2009) reported a similar behavior after incorporating increasing fractions of functionalised CNT into polycarbonate.

Lower loadings of MWCNT-COOH (≤0.5 wt. %) did extend gel-times, when compared to the control (MWCNT free bone cement), again supporting the hypothesis that the reduced rate of polymerisation is due to chemical interactions between functional groups present on the surface of the MWCNT and the polymer matrix, as the time before gelation occurs increased with level of loading (and thus concentration of functional groups). Ormsby *et al.*, (2011) have suggested that the physical presence may indeed affect the rate of polymerisation, but the functional chemical groups may be the predominant influence.

This hypothesis is supported by the theory that if the functional groups on the MWCNT were indeed interrupting the polymerisation reaction by terminating polymer chains via formation of covalent bonds, then the onset of cure (sudden increase in complex viscosity) would never occur and the cement would never reach a hardened state. It is noteworthy though, for this to occur, the MWCNT loading would need to be significantly higher than 1.0 wt. % (Ormsby *et al.*, 2011). Interestingly, the gel-times remain relatively unchanged for the bone cement with MWCNT-UNF, with no clear pattern evident irrespective of MWCNT-UNF loading. These results would indicate that the prolonged time before the onset of cure experienced in the bone cements with MWCNT-COOH is dependent on the carboxyl functional group.

13. Summary and conclusions

As the number of primary TJR continues to increase each year and, even with the reported decrease in the proportion of cemented TJR performed, PMMA bone cement is still required for the majority of TJR procedures. At present with longer life expectancy and younger patient populations requiring TJR, an increase in cemented revisions seems inevitable. Aseptic loosening is continually cited as being the most common indication for revision. It is well established that for cemented implants a number of factors contribute to aseptic loosening, of which, fatigue damage of the cement mantle has been observed *in vivo*. Therefore, a crucial requirement exists for the development of new technologies and biomaterials for the treatment of traumatic injuries and chronic diseases, which allow less tissue damage and more tissue regeneration and are conducive to rapid patient recovery. Particularly for biomaterials and devices designed to replace a degenerated or diseased joint, bone structure, many questions need to be answered. Such devices and implants would benefit significantly from availability of a material that is multi-functional and can meet the biomechanical and biological requirements.

The conventional biomaterials available today are reaching their maximum capabilities, notwithstanding their successful application in treating and preventing different medical conditions. There is a need for the development of new biomaterials which must satisfy several requirements ranging from physical, mechanical, biological, toxicological and other characteristics, depending on the final clinical application.

Carbon is chemically inert and CNT not only demonstrate superior mechanical, chemical and electrical properties, but also have the potential to be biocompatible particularly when appropriately functionalised. Also, encapsulation of other materials within CNT could potentially create applications for therapeutic use in medicine. Incorporation of MWCNT into PMMA based orthopaedic bone is a case in point, whereby a high degree of MWCNT-polymer matrix interaction has been shown to increase the fracture resistance during mechanical loading. Furthermore, it has been reported that MWCNT-PMMA bone cement leads to increased viscosity and reduced polymerisation temperatures. Reducing the temperature generated during polymerisation could reduce the thermal cellular necrosis experienced *in vivo*, reducing the probability of aseptic loosening. Furthermore, a reduction in the exotherm of bone cement will reduce residual stresses within the cement mantle as a consequence of excessive shrinkage.

To fully exploit the use of MWCNT in PMMA bone cement further development and research is required. In particular a detailed investigation of the biocompatibility of the MWCNT composite cements is required. This would require exposing human osteoblast cells to the composite MWCNT-PMMA bone cements, ultimately leading to *in vivo* cell work. This would provide a clearer indication of the MWCNT composite cements potential integration into the body.

Regardless of this interest, there are many issues and limitations to be considered. The field of nanomaterials for biomedical and bioengineering applications is still very much in its infancy and many difficult questions remain unanswered, including manufacturing, safety and regulatory issues. Preliminary investigations substantiate the enormous potential of MWCNT systems for biomedical and bioengineering applications either as a structure, coating, scaffold or composite; although most of these are only at laboratory-scale and *in vitro* testing. There is a major requirement for interdisciplinary collaboration and exchange of knowledge at many levels to effectively address the current issues, before being able to fully understand and explore the true potential of CNT for biomedical and bioengineering applications.

14. References

American Society for Testing and Materials (ASTM), Specification F451-95. Standard specification for acrylic bone cement, 1996 Annual Book of ASTM Standards, Vol 13.01. Philadelphia, American Society for Testing and Materials (1996) pp. 49-45.

Ajayan, P. M. (1997). "Carbon nanotubes: novel architecture in nanometer space." Progressin Crystal Growth and Characterization of Materials 34(1-4): 37-51.

Apple, D. (2003). "A pioneer in the quest to eradicate world blindness." Bulletin of the World Health Organization 81(10): 756-757.

Andrews, R., Jacques, D., Qian, D. L. and Rantell, T. (2002). "Multiwall carbon nanotubes: synthesis and application." Accounts of Chemical Research 35(12): 1008-1017.

Andrews, R. and Weisenberger, M. C. (2004). "Carbon nanotube polymer composites." Current Opinion in Solid State and Materials Science 8(1): 31-37.

Anthony, P. P., Gie, G. A., Howie, C. R. and Ling, R. S. (1990). "Localised endosteal bone lysis in relation to the femoral components of cemented total hip arthroplasties." Journal of Bone and Joint Surgery - British Volume 72-B(6): 971-979.

Baleani, M., Cristofolini, L., Minari, C. and Toni, A. (2003). "Fatigue strength of PMMA bone cement mixed with gentamicin and barium sulphate vs. pure PMMA." Proceedings of the Institution of Mechanical Engineers, Part H (Journal of Engineering in Medicine) 217: 9-12.

Barber, A. H., Cohen, S. R. and Wagner, H. D. (2003). "Measurement of carbon nanotube polymer interfacial strength." Applied Physics Letters 82(23): 4140-4142.

Baroud, G., Samara, M. and Steffen, T. (2004). "Influence of mixing method on the cement temperature-mixing time history and doughing time of three acrylic cements for vertebroplasty." Journal of Biomedical Materials Research Part B-Applied Biomaterials 68B(1): 112-116.

Baughman, R. H., Zakhidov, A. A. and De Heer, W. A. (2002). "Carbon nanotubes – the route toward applications." Science 297(5582): 787-792.

Bellare, A., Fitz, W., Gomoll, A., Turell, M. B., Scott, R. and Thornhill, T. (2002). "Using nanotechnology to improve the performance of acrylic bone cements." The Orthopaedic Journal at Harvard Medical School Volume 4.

Bishop, N. E., Ferguson, S. and Tepic, S. (1996). "Porosity reduction in bone cement at the cement-stem interface." Journal of Bone and Joint Surgery-British Volume 78B(3): 349-356.

Blakslee, O. L., Proctor, D. G., Seldin, E. J., Spence, G. B. and Weng, T. (1970). "Elastic constants of compression-annealed pyrolytic graphite." Journal of Applied Physics 41(8): 3373-82.

Bonnet, P., Sireude, D., Garnier, B., Chauvet, O. (2007) Thermal properties and percolation in carbon nanotube-polymer composites. Journal of Applied Physics Letters 91: 201910-1–201910-3.

BSI, British Standards Institution (1993). Non-metallic materials for surgical implants. Part 1. Specification for acrylic resin cement. BS 7253. ISO 5833.

Cadek, M., Coleman, J. N., Ryan, K. P., Nicolosi, V., Bister, G., Fonseca, A., Nagy, J. B., Szostak, K., Beguin, F. and Blau, W. J. (2004). "Reinforcement of polymers with carbon nanotubes: The role of nanotube surface area." Nano Letters 4(2): 353-6.

Callister, W. D. (2000a). Chapter 17: Composites. Materials Science and Engineering: An Introduction, John Wiley & Sons, Inc.: 520-561.

Chao, G., Hailin, Z., Yanping, W., Watts, P. C. P., Hao, K., Xiaowen, C. and Yan, D. (2006). "In situ polymerization approach to multiwalled carbon nanotubes-reinforced nylon 1010 composites: Mechanical properties and crystallization behavior." Polymer 47(1): 113-22.

Chandler, M., Kowalski, R. S. Z., Watkins, N. D., Briscoe, A. and New, A. M. R. (2006). "Cementing techniques in hip resurfacing." Proceedings of the Institution of Mechanical Engineers Part H-Journal of Engineering in Medicine 220(H2): 321-331.

Chun, A. L., Moralez, J. G., Fenniri, H. and Webster, T. J. (2004). "Helical rosette nanotubes: A more effective orthopaedic implant material." Nanotechnology 15(4): 234-9.

Colbert, D. T. (2003). "Single-wall nanotubes: A new option for conductive plastics and engineering polymers." Plastics, Additives and Compounding 5(1): 18-25.

Coleman, J. N., Cadek, M., Blake, R., Nicolosi, V., Ryan, K. P., Belton, C., Fonseca, A., Nagy, J. B., Gun'ko, Y. K. and Blau, W. J. (2004). "High-performance nanotubereinforced plastics: Understanding the mechanism of strength increase." Advanced Functional Materials 14(8): 791-798.

Coleman, J. N., Curran, S., Dalton, A. B., Davey, A. P., McCarthy, B., Blau, W. and Barklie, R. C. (1998). "Percolation-dominated conductivity in a conjugated polymercarbon-nanotube composite." Physical Review B (Condensed Matter) 58(12): 7492-5.

Cooper, C. A., Cohen, S. R., Barber, A. H. and Wagner, H. D. (2002). "Detachment of nanotubes from a polymer matrix." Applied Physics Letters 81(20): 3873-5.

Cui, D., Tian, F., Ozkan, C. S., Wang, M. and Gao, H. (2005). "Effect of single wall carbon nanotubes on human HEK293 cells." Toxicology Letters 155(1): 73-85.

Curl, R. F., Smalley, R. E., Kroto, H. W., O'Brien, S. and Heath, J. R. (2001). "How the news that we were not the first to conceive of soccer ball C60 got to us." Journal of Molecular Graphics and Modelling 19(2): 185-186.

Demian, H. W. and McDermott, K. (1998). "Regulatory perspective on characterization and testing of orthopedic bone cements." Biomaterials 19(17): 1607-1618.

Dalton, A. B., Collins, S., Munoz, E., Razal, J. M., Ebron, V. H., Ferraris, J. P., Coleman, J. N., Kim, B. G. and Baughman, R. H. (2003). "Super-tough carbonnanotube fibres." Nature 423(6941): 703.

Davoren, M., Herzog, E., Casey, A., Cottinerau, B., Chambers G., and Byrne H.J., 2007. In vitro toxicity evaluation of single walled carbon nanotubes on human A549 lung cells, Toxicol in Vitro 21, 438–448.

Deb, S., Lewis, G., Janna, S. W., Vazquez, B. and San Roman, J. (2003). "Fatigue and fracture toughness of acrylic bone cements modified with long-chain amine activators." Journal of Biomedical Materials Research Part A 67A(2): 571-577.

Demczyk, B. G., Wang, Y. M., Cumings, J., Hetman, M., Han, W., Zettl, A. and Ritchie, R. O. (2002). "Direct mechanical measurement of the tensile strength and elastic modulus of multiwalled carbon nanotubes." Materials Science and Engineering A 334(1-2): 173-178.

Ding, W., Eitan, A., Fisher, F. T., Chen, X., Dikin, D. A., Andrews, R., Brinson, L. C., Schadler, L. S. and Ruoff, R. S. (2003). "Direct observation of polymer sheathing in carbon nanotube-polycarbonate composites." Nano Letters 3(11): 1593-7.

Dunne, N.J., Orr, J.F., (2002). Curing characteristics of acrylic bone cement. Journal of Material Science 13(1), 17-22.

Dunne, N. J., Orr, J. F., Mushipe, M. T. and Eveleigh, R. J. (2003). "The relationship between porosity and fatigue characteristics of bone cements." Biomaterials 24(2): 239-245.

Eitan, A., Fisher, F. T., Andrews, R., Brinson, L. C. and Schadler, L. S. (2006). "Reinforcement mechanisms in MWCNT-filled polycarbonate." Composites Science and Technology 66(9): 1159-1170.

Eriksson, A.R., Alberksson, T. (1983) Temperature threshold levels for heat-induced bone tissue injury-A vital microscopy in the rabbit. Journal of Prosthetic Dentistry 50(1): 101-107.

B. Fadeel, V. Kagan, H. Krug, A. Shvedova, M. Svartengren and L. Tran. (2007) There's plenty of room at the forum: potential risks and safety assessment of engineered nanomaterials, Nanotoxicology 1 (2), 73–84.

Farrar, D.F., Rose, J. (2001) Rheological properties of PMMA bone cement during curing. Biomaterials 22: 3005–3013.

Feith, R. (1975) Side effect of acrylic cement implanted into bone. Acta Orthopaedica Scandinavica 214;161.

Fisher, F. T., Bradshaw, R. D. and Brinson, L. C. (2002). "Effects of nanotube waviness on the modulus of nanotube-reinforced polymers." Applied Physics Letters 80(24): 4647.

Franklin, P., Wood, D. J. and Bubb, N. L. (2005). "Reinforcement of poly(methyl methacrylate) denture base with glass flake." Dental Materials 21(4): 365-370.

Fukushima, H., Hashimoto, Y., Yoshiya, S., Kurosaka, M., Matsuda, M., Kawamura, S. and Iwatsubo, T. (2002). "Conduction analysis of cement interface temperature in total knee arthroplasty." Kobe Journal of Medical Science 48: 63-72.

Gilbert, J. L., Hasenwinkel, J. M., Wixson, R. L. and Lautenschlager, E. P. (2000). "A theoretical and experimental analysis of polymerization shrinkage of bone cement: A potential major source of porosity." Journal of Biomedical Materials Research 52(1): 210-218.

Gilbert, J. L., Ney, D. S. and Lautenschlager, E. P. (1995). "Self-reinforced composite poly(methyl methacrylate): static and fatigue properties." Biomaterials 16(14): 1043-1055.

Gilbert, J. L., Hasenwinkel, J.M., Wixson, R.L., and Lautenschlager, E.P.A. (2002) Theoretical and experimental analysis of polymerisation shrinkage of bone cement: A potential major source of porosity. Journal of Biomedical Material Research 52(1): 210-218.

Ginebra, M. P., Albuixech, L., Fernandez-Barragan, E., Aparicio, C., Gil, F. J., San Roman, J., Vazquez, B. and Planell, J. A. (2002). "Mechanical performance of acrylic bone cements containing different radiopacifying agents." Biomaterials 23(8): 1873-1882.

Granchi, D., Cenni, E., Savarino, L., Ciapetti, G., Forbicini, G., Vancini, M., Maini, C., Baldini, N. and Giunti, A. (2002). "Bone cement extracts modulate the osteoprotegerin/osteoprotegerin-ligand expression in Mg63 osteoblast-like cells." Biomaterials 23(11): 2359-2365.

Goh, H. W., Goh, S. H., Xu, G. Q., Pramoda, K. P. and Zhang, W. D. (2003). "Crystallization and dynamic mechanical behavior of double-C60-end-capped poly(ethylene oxide)/multiwalled carbon nanotube composites." Chemical Physics Letters 379(3-4): 236-41.

Harper, E.J., German, M.J., Bonfield, W., Braden, M., Dingeldein, E., Wahlig, H. (2000) A comparison of VersaBondTM, a modified acrylic bone cement with improved handling properties, to Palacos R, Simplex P. Trans Sixth World Biomater Cong, Kamuela (Big Island). 540.

Hendriks, J. G. E., Van Horn, J. R., Van Der Mei, H. C. and Busscher, H. J. (2004). "Backgrounds of antibiotic-loaded bone cement and prosthesis-related infection." Biomaterials 25(3): 545-556.

Herzog, E., Casey, A., Lyng, F.M., Chambers, G., Byrne, H.J., and Davoren, M. (2007). A new approach to the toxicity testing of carbon-based nanomaterials. The clonogenic assay, Toxicol Lett 174, 49–60.

Huang, K. Y., Yan, J. J. and Lin, R. M. (2005). "Histopathologic findings of retrieved specimens of vertebroplasty with polymethylmethacrylate cement - case control study." Spine 30(19): E585-E588.

Hull, D. and Clyne, T. W. (1996a). Fibres and matrices. An Introduction to Composite Materials. Clarke, D. R., Suresh, S. and Ward, I. M., Cambridge University Press: 9-38.

Hull, D. and Clyne, T. W. (1996b). General introduction. An Introduction to Composite Materials. . Clarke, D. R., Suresh, S. and Ward, I. M., Cambridge University Press: 1-8.

Hwang, G. L., Shieh, Y.-T. and Hwang, K. C. (2004). "Efficient load transfer to polymergrafted multiwalled carbon nanotubes in polymer composites." Advanced Functional Materials 14(5): 487-491.

Iijima, S. (1991). "Helical microtubules of graphitic carbon." Nature 354: 56-58.

Iijima, S. (2002). "Carbon nanotubes: past, present, and future." Physica B: Condensed Matter 323(1-4): 1-5.

Janssen, D., Aquarius, R., Stolk, J. and Verdonschot, N. (2005a). "The contradictory effects of pores on fatigue cracking of bone cement." Journal of Biomedical Materials Research - Part B Applied Biomaterials 74(2): 747-753.

Janssen, D., Stolk, J. and Verdonschot, N. (2005b). "Why would cement porosity reduction be clinically irrelevant, while experimental data show the contrary." Journal of Orthopaedic Research 23(4): 691-697.

Jasty, M., Maloney, W. J., Bragdon, C. R., Oconnor, D. O., Haire, T. and Harris, W. H. (1991). "The initiation of failure in cemented femoral components of hip arthroplasties." Journal of Bone and Joint Surgery-British Volume 73(4): 551-558.

Jia, Z., Wang, Z., Xu, C., Liang, J., Wei, B., Wu, D. and Zhu, S. (1999). "Study on poly (methyl methacrylate)/carbon nanotube composites." Materials Science & Engineering A (Structural Materials: Properties, Microstructure and Processing) A271(1-2): 395-400.

Jones, P.R., Hukins, D.W.L., Porter, M.L., Davies, K.E., Hardinge, K., Taylor, C.J. (1992) Bending and fracture of the femoral component in cemented total hip replacement. Journal of Biomechanical Engineering 14 (5): 444.

Kagan, V.E., Tyurina, Y.Y., Tyurin, V.A., Konduru, N.V., Potapovich A.I., and Osipov A.N. (2006). Direct and indirect effects of single walled carbon nanotubes on RAW 264.7 macrophages: role of iron, Toxicol Lett 165 (1), 88–100.

Keener, J. D., Callaghan, J. J., Goetz, D. D., Pederson, D. R., Sullivan, P. M. and Johnston, R. C. (2003). "Twenty-five-year results after charnley total hip arthroplasty in patients less than fifty years old - a concise follow-up of a previous report." Journal of Bone and Joint Surgery-American Volume 85A(6): 1066-1072.

Khor, K. H., Buffière, J. Y., Ludwig, W., Toda, H., Ubhi, H. S., Gregson, P. J. And Kim, P., Shi, L., Majumdar, A. and McEuen, P. L. (2001). "Thermal transport measurements of individual multiwalled nanotubes." Physical Review Letters 87(21): 215502-1.

Krieger, I.M., Dougherty, T.J. (1959) A mechanism for non-Newtonian flow in suspensions of rigid spheres. Transaction of the Society of Rheology 3: 137–152.

Konnecke, K. and Rehage, G. (1981). "Crystallization and stero association of steroregular PMMA." Colloid and Polymer Science 259: 1062-1069.

Kotha, S. P., Li, C., McGinn, P., Schmid, S. R. and Mason, J. J. (2006). "Improved mechanical properties of acrylic bone cement with short titanium fiber reinforcement." Journal of Materials Science: Materials in Medicine 17(8): 743-8.

Kotha, S. P., Li, C., Schmid, S. R. and Mason, J. J. (2004). "Fracture toughness of steelfiber-reinforced bone cement." Journal of Biomedical Materials Research - Part A 70(3): 514-521.

Kuehn, K. D., Ege, W. and Gopp, U. (2005a). "Acrylic bone cements: compositions and properties." Orthopedic Clinics of North America 36: 17-28.

Kurtz, S. M., Villarraga, M. L., Zhao, K. and Edidin, A. A. (2005). "Static and fatigue mechanical behavior of bone cement with elevated barium sulfate content for treatment of vertebral compression fractures." Biomaterials 26(17): 3699-3712.

Larsen, S.T., Franzen, S., Ryd, L. (1991) Cement interface temperature in hip arthroplasty. Acta Orthopaedica Scandinavica 62(2): 102-105.

Lalko, M.P., Rakesh, L., Hirschi, S. (2009) Rheology of polycarbonate reinforced with functionalized and unfunctionalized single-walled carbon nanotubes. Journal of Thermal Analysis and Calorimetry 95(1): 203-206.

Lee, R. R., Ogiso, M., Watanabe, A. and Ishihara, K. (1997). "Examination of hydroxyapatite filled 4-META/MAA-TBB adhesive one cement in in vitro and in vivo environment." Journal of Biomedical Materials Research 38(1): 11-16.

Leelapornpisit, W., Ton-That, M.-T., Perrin-Sarazin, F., Cole, K. C., Denault, J. and Simard, B. (2005). "Effect of carbon nanotubes on the crystallization and properties of polypropylene." Journal of Polymer Science, Part B: Polymer Physics 43(18): 2445-2453.

Lennon, A. B. and Prendergast, P. J. (2002). "Residual stress due to curing can initiate damage in porous bone cement: experimental and theoretical evidence." Journal of Biomechanics 35(3): 311-321.

Lewis, G. (1997). "Properties of acrylic bone cement: state of the art review." Journal of Biomedical Materials Research 38(2): 155-182.

Lewis, G. (2000). "Relative roles of cement molecular weight and mixing method on the fatigue performance of acrylic bone cement: simplex-p versus osteopal." Journal of Biomedical Materials Research 53(1): 119-130.

Lewis, G. (2003). "Fatigue testing and performance of acrylic bone-cement materials: stateof-the-art review." Journal of Biomedical Materials Research Part B-Applied Biomaterials 66B(1): 457-486.

Lewis, G., Carroll, M., 2001. Rheological properties of acrylic bone cement during curing and the role of the size of the powder particles., Journal of Biomedical Material Research (App Biomaterials) 63, pp. 191 - 199.

Lewis, G., Janna, S. and Carroll, M. (2003). "Effect of test frequency on the in vitro fatigue life of acrylic bone cement." Biomaterials 24(6): 1111-17.

Lewis, G. and Janna, S. I. (2004). "Effect of fabrication pressure on the fatigue performance of cemex acrylic bone cement." Biomaterials 25(7-8): 1415-1420.

Lewis, G., Van Hooy-Corstjens, C. S. J., Bhattaram, A. and Koole, L. H. (2005). "Influence of the radiopacifier in an acrylic bone cement on its mechanical, thermal, and physical properties: Barium sulfate-containing cement versus iodine-containing cement." Journal of Biomedical Materials Research - Part B Applied Biomaterials 73(1): 77-87.

Lidgren, L., Bodelind, B., Möller, J. (1987) Bone cement improved by vacuum mixing and chilling. Acta Orthopaedica 58(1): 27-32.

Linder, L. (1977) Reaction of bone to the acute chemical trauma of bone cement. Journal of Bone Joint Surgery 59(1): 82-87.

Liu, C., Green, S. M., Watkins, N. D., Gregg, P. J. and McCaskie, A. W. (2001). "Some failure modes of four clinical bone cements." Proceedings of the Institution of Mechanical Engineers, Part H (Journal of Engineering in Medicine) 215(H4): 359- 66.

Luo, W.-B., Yang, T.-Q. and Wang, X.-Y. (2004). "Time-dependent craze zone growth at a crack tip in polymer solids." Polymer 45(10): 3519-3525.

Mandell, J. F., Huang, D. D. and McGarry, F. J. (1980). Crack propagation modes in injection molded fibre reinforced thermoplastics. Cambridge, Massachusetts, School of Engineering, Massachusetts Institute of technology: 1-34.

Marrs, B., Andrews, R., Rantell, T. and Pienkowski, D. A. (2006). "Augmentation of acrylic bone cement with multiwall carbon nanotubes." Journal of Biomedical Materials Research Part A 77(A): 269-276.

Marrs, B., Andrews, R., Rantell, T., Pienkowski, D. (2006) Augmentation of acrylic bone cement with multiwall carbon nanotubes. Journal of Biomedical Material Research 77(2): 269-276.

McCormack, B. A. O. and Prendergast, P. J. (1996). "Interface failure in implants cemented with different bone-cements - a fracture mechanics approach". 2nd International Symposium on Computer Methods in Biomechanics and Biomedical Engineering, Swansea, Gordon and Breach.

McCormack, B. A. O. and Prendergast, P. J. (1999). "Microdamage accumulation in the cement layer of hip replacements under flexural loading." Journal of Biomechanics 32(5): 467-75.

McCormack, B. A. O., Walsh, C. D., Wilson, S. P. and Prendergast, P. J. (1998). "A statistical analysis of microcrack accumulation in PMMA under fatigue loading: applications to orthopaedic implant fixation." International Journal of Fatigue 20(8): 581-593.

McClory, C., McNally, T., Brennan, G., Erskine, J. (2007) Thermosetting polyurethane multiwall carbon nanotube composites. Journal of Applied Polymer Science 105: 1003-1011.

McClory, C., McNally, T., Baxendale, M., Pötschke, P., Blau, W., Ruether, M. (2010) Electrical and rheological percolation of PMMA/MWCNT composites as a function of CNT geometry and functionality. European Polymer Journal 46 (5): 854-868.

McNally, T., Potschke, P., Halley, P., Murphy, M., Martin, D., Bell, S.E.J. (2005) Polyethylene multiwall carbon nanotube composites by melt blending. Polymer. 46 8222-8232.

Meyer, P.R., Lautenschlager, E.P., Moore, B.K. (1973) On the setting properties of acrylic bone cement. Journal of Bone Joint Surgery 55(A): 149-156.

Mjoberg, B., Petterson, H., Roseqvist, R., Rydholm, A. (1984) Bone cement, thermal injury and radiolucent zone. Acta Orthopaedica Scandinavica 55: 597-600.

Molino, L. N. and Topoleski, L. D. T. (1996). "Effect of BaSO4 on the fatigue crack propagation rate of PMMA bone cement." Journal of Biomedical Materials Research 31(1): 131-137.

Moritz, A.R., Henriques, F.C. (1947). Studies of thermal injury-The relative importance of time and surface temperature in the causation of cutaneous burns. American Journal of Pathology 23: 695-720.

Mousa, W. F., Kobayashi, M., Shinzato, S., Kamimura, M., Neo, M., Yoshihara, S. and Nakamura, T. (2000). "Biological and mechanical properties of PMMA-based bioactive bone cements." Biomaterials 21(21): 2137-2146.

Murphy, B. P. and Prendergast, P. J. (2000). "On the magnitude and variability of the fatigue strength of acrylic bone cement." International Journal of Fatigue 22(10): 855-864.

Murphy, B. P. and Prendergast, P. J. (2002). "The relationship between stress, porosity, and nonlinear damage accumulation in acrylic bone cement." Journal of Biomedical Materials Research 59(4): 646-654.

Nalla, R. K., Kruzic, J. J. and Ritchie, R. O. (2004). "On the origin of the toughness of mineralized tissue: microcracking or crack bridging?" Bone 34(5): 790-798.

Narva, K. K., Lassila, L. V. J. and Vallittu, P. K. (2005). "Flexural fatigue of denture base polymer with fiber-reinforced composite reinforcement." Composites Part A: Applied Science and Manufacturing 36(9): 1275-1281.

Norman, T. L., Kish, V., Blaha, J. D., Gruen, T. A. and Hustosky, K. (1995). "Creep characteristics of hand-mixed and vacuum-mixed acrylic bone-cement at elevated stress levels." Journal of Biomedical Materials Research 29(4): 495-501.

Nuno, N. and Amabili, M. (2002). "Modelling debonded stem-cement interface for hip implants: Effects of residual stresses." Clinical Biomechanics 17(1): 41-48.

Nuno, N. and Avanzolini, G. (2002). "Residual stresses at the stem-cement interface of an idealized cemented hip stem." Journal of Biomechanics 35(6): 849-852.

O'Connor, D. O., Burke, D. W., Jasty, M., Sedlacek, R. C. and Harris, W. H. (1996). "In vitro measurement of strain in the bone cement surrounding the femoral component of total hip replacements during simulated gait and stair-climbing." Journal of Orthopaedic Research 14(5): 769-777.

Ormsby, R., McNally, T., Mitchell, C.A., Dunne, N. (2010a) Incorporation of Multiwall Carbon Nanotubes to Acrylic Based Bone Cements: Effects on Mechanical and Thermal Properties. Journal of Mechanical Behaviour of Biomedical Materials 3(2): 136-145.

Ormsby, R., McNally, T., Mitchell, C.A., Dunne, N. (2010b) Influence of multiwall carbon nanotube functionality and loading on mechanical properties of PMMA/MWCNT bone cements. Journal of Material Science - Materials in Medicine 21(8): 2287-2292.

Ormsby, R., McNally, T., Mitchell, C.A., Halley, P., Martin, D., Dunne, N. (2011). Thermal and Rheological properties of PMMA/MWCNT bone cements. Carbon, DOI:10.1016/j.carbon.2011.02.063.

Orr, J. F., Dunne, N. J. and Quinn, J. C. (2003). "Shrinkage stresses in bone cement." Biomaterials 24(17): 2933-2940.

Pal, S. and Saha, S. (1982). "Stress relaxation and creep behaviour of normal and carbon fibre reinforced acrylic bone cement." Biomaterials 3: 93-96.

Pascual, B., Vazquez, B., Gurruchaga, M., Goni, I., Ginebra, M. P., Gil, F. J., Planell, J. A., Levenfeld, B. and Roman, J. S. (1996). "New aspects of the effect of size and size distribution on the setting parameters and mechanical properties of acrylic bone cements." Biomaterials 17(5): 509-516.

Park, J. U., Cho, S., Cho, K. S., Ahn, K. F., Lee, S. J. and Lee, S. J. (2005). "Effective insitu-preparation and characteristics of polystyrene-grafted carbon nanotube composites." Korea-Australia Rheology Journal 17(2): 41-45.

Peigney, A., Laurent, C., Flahaut, E., Bacsa, R.R., Rousset, A. (2009) Specific surface area of carbon nanotubes and bundles of carbon nanotubes. Carbon 39(4): 507-514.

Pilliar, R. M., Blackwell, R., Macnab, I. and Cameron, H. U. (1976). "Carbon fiber reinforced bone cement in orthopedic surgery." Journal of Biomedical Materials Research 10(6): 893-906.

Prendergast, P. J., and Murphy, B., (2000). "On the magnitude and variability of the fatigue strength of acrylic bone cement." International Journal of Fatigue 22(10): 855-864.

Prendergast, P. J. (2001a). Chapter 35: Bone prostheses and implants. Bone Mechanics Handbook. Cowin, S. C., CRC Press LLC.

Prendergast, P. J. (2001b). "The functional performance of orthopaedic bone cement." Key Engineering Materials 198-199: 291-300.

Pukanszky, B. (2005). "Interfaces and interphases in multicomponent materials: past, present, future." European Polymer Journal 41(4): 645-662.

Pulos, G. C. and Knauss, W. G. (1998a). "Nonsteady crack and craze behavior in PMMA under cyclical loading. I. Experimental preliminaries." International Journal of Fracture 93(1-4): 145-59.

Pulos, G. C. and Knauss, W. G. (1998b). "Nonsteady crack and craze behavior in PMMA under cyclical loading. II. Effect of load history on growth rate and fracture morphology." International Journal of Fracture 93(1-4): 161-85.

Pulos, G. C. and Knauss, W. G. (1998c). "Nonsteady crack and craze behavior in PMMA under cyclical loading. III. Effect of load history on cohesive force distribution on the craze." International Journal of Fracture 93(1-4): 187-207.

Robinson, R. P., Wright, T. M. and Burstein, A. H. (1981). "Mechanical properties of poly(methyl methacrylate) bone cements." Journal of Biomedical Materials Research 15(2): 203-208.

Roques, A., Browne, M., Taylor, A., New, A. and Baker, D. (2004). "Quantitative measurement of the stresses induced during polymerisation of bone cement." Biomaterials 25(18): 4415-4424.

Ruan, S. L., Gao, P., Yang, X. G. and Yu, T. X. (2003). "Toughening high performance ultrahigh molecular weight polyethylene using multiwalled carbon nanotubes." Polymer 44(19): 5643-54.

Ryan, K. P., Cadek, M., Nicolosi, V., Walker, S., Ruether, M., Fonseca, A., Nagy, J. B., Blau, W. J. and Coleman, J. N. (2006). "Multiwalled carbon nanotube nucleated crystallization and reinforcement in poly (vinyl alcohol) composites." Synthetic Metals 156(2-4): 332-335.

Saha, S. and Pal, S. (1986). "Mechanical characterization of commercially made carbonfibre-reinforced polymethylmethacrylate." Journal of Biomedical Materials Research 20(6): 817-826.

Sauer, J. A. and Richardson, G. C. (1980). "Fatigue of polymers." International Journal of Fracture 16(6): 499-532.

Scheirs, J. (2000a). Chapter 15: Failure of fibre-reinforced composites. Compositional and Failure Analysis of Polymers. A Practical Approach, John Wiley & Sons, Ltd.: 449-481.

Scheirs, J. (2000b). Chapter 12: Mechanical failure mechanisms of polymers. Compositional and Failure Analysis of Polymers. A Practical Approach, John Wiley & Sons, Ltd.: 304-362.

Seo, M.K., Lee, J.-R. and Park, S.-J. (2005). "Crystallization kinetics and interfacial behaviors of polypropylene composites reinforced with multi-walled carbon nanotubes." Materials Science and Engineering A 404(1-2): 79-84.

Serbetci, K., Korkusuz, F. and Hasirci, N. (2004). "Thermal and mechanical properties of hydroxyapatite impregnated acrylic bone cements." Polymer Testing 23(2): 145-55.

Shannon, B. D., Klassen, J. F., Rand, J. A., Berry, D. J. and Trousdale, R. T. (2003). "Revision total knee arthroplasty with cemented components and uncemented intramedullary stems." Journal of Arthroplasty 18(7): 27-32.

Shi, Z., Lian, Y., Liao, F. H., Zhou, X., Gu, Z., Zhang, Y., Iijima, S., Li, H., Yue, K. T. and Zhang, S.-L. (2000). "Large scale synthesis of single-wall carbon nanotubes by arc-discharge method." Journal of Physics and Chemistry of Solids 61(7): 1031-1036.

Shinzato, S., Kobayashi, M., Mousa, W. F., Kamimura, M., Neo, M., Kitamura, Y., Kokubo, T. and Nakamura, T. (2000). "Bioactive polymethyl methacrylate-based bone cement: Comparison of glass beads, apatite- and wollastonite-containing glass ceramic, and hydroxyapatite fillers on mechanical and biological properties." Journal of Biomedical Materials Research 51(2): 258-272.

Shvedova, A.A., Kisin, E.R., Mercer, R. Murray, Johnson V.J., and Potapovich A.I., 2005. Unusual inflammatory and fibrogenic pulmonary responses to single-walled carbon nanotubes in mice, Am. J. Physiol 289 L698–L708

Shvedova, A.A., Fabisiak, J.P., Kisin, E.R., Murray, A.R., Roberts J.R., and Tyurina Y.Y. 2008. Sequential exposure to carbon nanotubes and bacteria enhances pulmonary inflammation and infectivity, Am J Respir Cell Mol Biol 38 (5), 579–590.A.A.

Shvedova, A.A., Kisin, E.R., Porter, D., Schulte, P., Kagan, V.E., Fadeel, B., Castranova, V. 2009. Mechanisms of pulmonary toxicity and medical applications of carbon nanotubes: Two faces of Janus? Pharmacology and Therapeutics 121 (2), 192-204.

Sinha, N. and Yeow, J. T. W. (2005). "Carbon nanotubes for biomedical applications." IEEE Transactions on Nanobioscience 4(2): 180-195.

Sinnott, S. B. and Andrews, R. (2001). "Carbon nanotubes: synthesis, properties, and applications." Critical Reviews in Solid State and Materials Sciences 26(3): 145-249.

Sinnott, S. B., Andrews, R., Qian, D., Rao, A. M., Mao, Z., Dickey, E. C. and Derbyshire, F. (1999). "Model of carbon nanotube growth through chemical vapor deposition." Chemical Physics Letters 315(1-2): 25-30.

Skibo, M. D., Hertzberg, R. W., Manson, J. A. and Kim, S. L. (1977). "On the generality of discontinuous fatigue crack growth in glassy polymers." Journal of Materials Science 12(3): 531-42.

Smart, S. K., Cassady, A. I., Lu, G. Q. and Martin, D. J. (2006). "The biocompatibility of carbon nanotubes." Carbon 44(6): 1034-1047.

Spiegelberg, S.H., McKinley, G.H. (1998) Characterization of the curing process of bone cement with multi-harmonic shear rheometry. In: Trans 24th Annual Meeting of the Society of Biomaterials, San Diego, CA. 283.

Starke, G. R., Birnie, C. and van den Blink, P. A. (2001). Numerical modelling of cement polymerisation and thermal bone necrosis. Computer methods in biomechanics and biomedical engineering, Gordon & Breach, London.

Stanczyk, M. and van Rietbergen, B. (2004). "Thermal analysis of bone cement polymerisation at the cement-bone interface." Journal of Biomechanics 37(12): 1803-1810.

Stolk, J., Verdonschot, N., Murphy, B. P., Prendergast, P. J. and Huiskes, R. (2004). "Finite element simulation of anisotropic damage accumulation and creep in acrylic bone cement." Engineering Fracture Mechanics 71(4-6): 513-528.

Suresh, S. (1998). Chapter 12 Fatigue crack growth in noncrystalline solids. Fatigue of Materials, Cambridge University Press: 408-431.

Takemori, M. T. (1984). Polymer fatigue. Annual Review of Materials Science. Vol.14, Annual Reviews: 171-204.

The Royal Society and the Royal Academy of Engineering (2004). Nanoscience and nanotechnologies: opportunities and uncertainties. London, The Royal Society and the Royal Academy of Engineering.

Thostenson, E. T. and Chou, T.-W. (2003). "On the elastic properties of carbon nanotubebased composites: modelling and characterization." Journal of Physics D: Applied Physics 36(5): 573-582.

Thostenson, E. T., Zhifeng, R. and Tsu-Wei, C. (2001). "Advances in the science and technology of carbon nanotubes and their composites: A review." Composites Science and Technology 61(13): 1899-912.

Topoleski, L. D. T., Ducheyne, P. and Cuckler, J. M. (1990). "A fractographic analysis of in vivo poly(methyl methacrylate) bone-cement failure mechanisms." Journal of Biomedical Materials Research 24(2): 135-154.

Topoleski, L. D. T., Ducheyne, P. and Cuckler, J. M. (1993). "Microstructural pathway of fracture in poly(methyl methacrylate) bone-cement." Biomaterials 14(15): 1165-1172.

Topoleski, L. D. T., Ducheyne, P. and Cuckler, J. M. (1998). "Flow intrusion characteristics and fracture properties of titanium-fibre-reinforced bone cement." Biomaterials 19(17): 1569-1577.

Vallo, C. I., Cuadrado, T. R. and Frontini, P. M. (1997). "Mechanical and fracture behaviour evaluation of commercial acrylic bone cements." Polymer International 43(3): 260-268.

Velasco-Santos, C., Martinez-Hernandez, A. L., Fisher, F., Ruoff, R. and Castano, V. M. (2003). "Dynamical-mechanical and thermal analysis of carbon nanotube-methylethyl methacrylate nanocomposites." Journal of Physics D (Applied Physics) 36(12): 1423-8.

Verdonschot, N. and Huiskes, R. (1997a). "Acrylic cement creeps but does not allow much subsidence of femoral stems." Journal of Bone and Joint Surgery-British Volume 79B(4): 665-669.

Verdonschot, N. and Huiskes, R. (1997b). "The effects of cement-stem debonding in THA on the long-term failure probability of cement." Journal of Biomechanics 30(8): 795-802.

Webster, T. J., Waid, M. C., McKenzie, J. L., Price, R. L. and Ejiofor, J. U. (2004). "Nano-biotechnology: carbon nanofibres as improved neural and orthopaedic implants." Nanotechnology 15(1): 48-54.

Wong, E.W., Sheehan, P.E. and Lieber, C.M. (1997). "Nanobeam mechanics: elasticity, strength and toughness of nanorods and nanotubes." Science 277(5334): 1971-5.

Wright, T.M. and Robinson, R.P. (1982). "Fatigue crack propagation in poly methylmethacrylate bone cements." Journal of Materials Science 17(9): 2463-8.

Xie, X.L., Mai, Y.W., Zhou, X.P. (2005) Dispersion and alignment of carbon nanotubes in polymer matrix: A review. Material Science Engineering 49(4): 89–112.

Yang, J.M., Huang, P.Y., Yang, M.C. and Lo, S.K. (1997). "Effect of MMA-g- UHMWPE grafted fiber on mechanical properties of acrylic bone cement." Journal of Biomedical Materials Research 38(4): 361-369.

Zhang, Y., Gu, H. and Iijima, S. (1998). "Single-wall carbon nanotubes synthesized by laser ablation in a nitrogen atmosphere." Applied Physics Letters 73(26): 3827-9.

Zhang, Y. and Iijima, S. (1998). "Microscopic structure of as-grown single-wall carbon nanotubes by laser ablation." Philosophical Magazine Letters 78(2): 139-44.

Zhao, B., Hu, H., Mandal, S. K. and Haddon, R. C. (2005). "A bone mimic based on the self-assembly of hydroxyapatite on chemically functionalized single-walled carbonnanotubes." Chem. Mater. 17(12): 3235-3241.

Part 3

CNTs as Chemical Sensors

15

Carbon Nanotubes in Electrochemical Sensors

M. Mazloum-Ardakani and M.A. Sheikh-Mohseni
Department of Chemistry, Faculty of Science, Yazd University
I.R. Iran

1. Introduction

Nanoscale materials, defined as having at least one dimension less than 100 nm, have received steadily growing interest due to their unique properties and application potential. Typical examples include zero-dimensional nanoparticles, one-dimensional nanowires, and two-dimensional graphenes. Because of the quantum confinement of electrons in one or more dimensions, novel electrical, optical, and magnetic properties can be achieved in nanostructures. Nowadays carbon nanotubes (CNTs) are among the most promising nanomaterials. Depending on the chirality along the graphene sheet, either semiconducting or metallic electronic states are created.

There are numerous studies on carbon nanotubes applications. For example, CNTs has been widely used in structure of different sensors. Various advantages of CNTs as sensor materials have been shown for analysis of diversified chemicals of food quality, clinical and environmental interest. High thermal conductivity, remarkable mechanical properties, chemically stability and high surface to volume ratio of CNTs is very appealing for sensing applications.

Electrochemical sensors (ECS) have been proven as an inexpensive and simple analytical method with remarkable detection sensitivity, reproducibility, and ease of miniaturization rather than other instrumental analysis methods. Since CNTs possess interesting electrochemical properties, contributed by the activity of edge-plane-like graphite sites at the CNT ends, it can be used for construction of electrochemical sensors (CNT-ECS). CNT-ECS exhibit low detection limit, high sensitivity and fast response due to the signal enhancement provided by high surface area, low overvoltage, and rapid electrode kinetics.

Because of importance of electrochemical sensors and CNTs, this chapter is focused on advantages and applications of CNTs in electrochemical sensors. At first, CNTs are introduced and different electrochemical sensors are defined based on the type of output signal.

Afterwards, different methods are investigated and classified for preparation of electrochemical sensors based on CNTs. These strategies have been applied for the immobilization of CNTs on electrochemical transducers. The one common way is coating the electrode surface with a CNT suspension and in another approach the CNT is mixed with mineral oil to fabricate paste electrode. Some other methods, such as self- assembly and polymerization, are used less frequently.

Finally, the advantages and applications of CNTs are investigated in different ECSs (potentiometry and voltammetry). In potentiometric sensors, CNTs similar to other nanostructure materials have been proposed as alternative solid-state transducers that can improve the detection limit and selectivity coefficient of the sensor. On the other hand, there are many electroanalytical methods using voltammetric and amperometric techniques with CNT-modified electrodes in the recent literature. In the case of electrocatalysis sensors the electrocatalytic activity of the electrode was improved by using CNTs.

2. Electrochemical sensors

A chemical sensor is a small device that can be used for direct measurement of the analyte in the sample matrix. Ideally, such a device is capable of responding continuously and reversibly and does not perturb the sample. By combining the sample handling and measurement steps, sensors eliminate the need for sample collection and preparation. Chemical sensors consist of a transduction element covered by a chemical or biological recognition layer. This layer interacts with the target analyte, and the chemical changes resulting from this interaction are translated by the transduction element into electrical signals. Electrochemical sensors represent an important subclass of chemical sensors in which an electrode is used as the transduction element. Such devices hold a leading position among sensors presently available, have reached the commercial stage, and have found a vast range of important applications in the fields of clinical, industrial, environmental, and agricultural analyses.
There are different methods for classification of electrochemical sensors. One of them is based on output signal from electrochemical sensor. In this manner electrochemical sensors classified to two major classes: potentiometric and voltammetric sensors.

2.1 Potentiometric sensors

Potentiometric sensors or so-called ion-selective electrodes (ISEs) have been the subject of continuous research efforts. This group of chemical sensors is characterized as simple in preparation, robust in operation and moderately selective in analytical performance. Some kinds of ISEs become routine devices in analytical laboratories (Mazloum-Ardakani et al., 2006).
Ideally, potentiometric sensors detect the ion activity in the sample. This makes it a unique class of chemical sensors that may be very useful in bioavailability or speciation studies. The Nernst Equation is normally used to describe the ideal response of such a cell:

$$EMF = K + (RT/ z_I F) \ln a_I \quad (1)$$

where EMF is the electromotive force (the observed potential at zero current), K is a constant potential contribution that often includes the liquid-junction potential at the reference electrode, a_I is the sample activity for the ion I with charge z, and R, T, and F are the gas constant, absolute temperature, and Faraday constant, respectively (Wang, 2006). Note that the ion activity effectively describes the so-called free, or uncomplexed, concentration of the analyte, which is often the relevant driving force in chemical or biochemical reactions. This quantity may be orders of magnitude smaller than the total analyte concentration if a complexing agent is present in the sample.

2.1.1 Lower detection limit

In general, the lower detection limit (LOD) is defined as the concentration of the analyte at which the signal is increased relative to the background level by three times the standard deviation of the noise (Currie, 1995). According to IUPAC recommendations, the definition of the LOD in potentiometry is unique (Guilbault et al., 1976). The potentiometric response or the EMF is a linear function of the logarithm of the activity of the free (uncomplexed) ions a_I in solution. Its slope is described by the Nernst Equation (1) as $59.2/z_I$ mV decade^{-1} at 25°C (Fig. 1). Below the LOD, it has a constant value, which is ideally defined by the response of the sensor to another (interfering) ion J. The potentiometric lower LOD is defined as the cross section of the two linear parts of the response function (Fig. 1). However, typical noise during potentiometric measurements may be as low as 0.06–0.08 mV. The LOD according to the general definition used in analytical chemistry (three times the standard deviation of this noise) is therefore orders of magnitude lower than that given by the definition for potentiometry (Midgley, 1979).

2.1.2 Improvement of the lower detection limit

In most cases described so far, the lower detection limit of solvent polymeric membrane-based ion-selective electrodes (ISEs) lies in the micromolar range (Sokalski et al., 1999). In their usual design, the electric contact between the sensing membrane and the inner reference electrode is accomplished with an electrolyte solution of the ion to be measured (the primary ion). It has been established that this setup may cause severe biases because the primary ion activities near the membrane can be bigger by orders of magnitude than in the bulk of the sample (Bakker & Pretsch, 2002). This artifact, caused by zero-current ion fluxes induces less than optimal lower detection limits of the ISEs. The reason for the enhanced primary ion activity in the stagnant layer near the membrane induces by concentration gradients in the membrane. Such gradients in the membrane arise either because of an increase of the concentration of the primary ions on the inner membrane side or because of a decrease on the outer membrane side. The first mechanism is due to a coextraction of the primary ion together with its counter ion from the inner solution into the membrane, while the second one is a consequence of the partial replacement of primary ions by interfering ions from the sample on the outer membrane side (Sokalski et al., 1999).

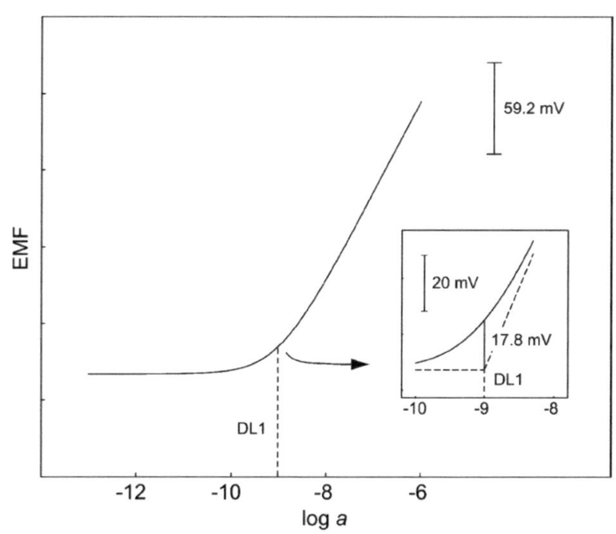

Fig. 1. The traditional definition of the lower LOD of potentiometric sensors is defined as the cross-section of both extrapolated linear portions of the calibration curve (DL).

There are various ways for improvement of LOD for potentiometric sensors (Mazloum-Ardakani et al., 2008). The basic of them is elimination of ion flux from inner solution. Some of these remedies have been brought below.
It was shown that ISEs with low detection limits can be fabricated for alkali metals by adding an ion-exchange resin to the inner solution of the membrane (Qin et al., 2000). In another way low detection limit ISEs can be obtained by applying a small current across the ion-selective membrane. Such constant-current (galvanostatic) systems were shown to yield detection limits down to the picomolar range (Pergel et al., 2001). In other work, it was recognized that rotating electrodes can be very useful in potentiometry if concentration polarizations near the membrane are relevant. The measuring range and detection limit of potentiometric sensors was shifted to lower concentrations with increasing rotational speeds (Ye & Meyerhoff, 2001).

2.1.3 Solid contact ion selective electrodes

In another important way the using of solid-contact (coated wire) electrodes prevent the ion release from the inner side of the membrane, compared to the traditional liquid-contact electrodes. Therefore the LOD and linear response range improved in this case. However, the original coated wire electrodes suffer from potential instabilities which have been mainly attributed to the lack of a well-defined redox couple and also the formation of a thin aqueous layer at the membrane–metal interface. Numerous works has been done to improve the stability of solid state electrode, which include the modification of electrode and the utilization of ion-to-electron transducers. In the last years different research teams have focused their investigations on different materials that are able to convert effectively the ionic signal through the ion-selective membrane into an electronic signal (ion to electron transducer) (Bobacka et al., 2008).

2.2 Carbon nanotubes in potentiometric sensors

Electroactive conducting polymers have been the most commonly used as ion to electron transducer in solid contact potentiometric sensors due to their ability to generate a high redox capacitance that confers a high stability to the recorded signal (Michalska, 2006). However, carbon materials including three-dimensionally ordered macroporous carbon, SWCNT and MWCNT have been used to facilitate the ion to electron transfer in solid-state ISE (Lai et al., 2007; Parra et al., 2009). Carbon-based materials are outstanding transducers, insensitive to oxygen and light, and very hydrophobic, which make them more advantageous compared to the conducting polymers in some applications. Furthermore, carbon nanotubes did not show any evidence of the formation of a water layer between the ion selective membrane and the transducer that see in the case of conducting polymer (Lai et al., 2007). The transduction mechanism of carbon nanotubes in solid-contact ions elective electrode has been studied by Crespo et al. (2009), and they proposed an asymmetric capacitor model to illustrate why carbon nanotubes can act as ion to electron transducer in solid-state ISEs.
The nanostructured characteristics of CNTs that are associated to their high surface to volume ratio, together with the inert character of the carbon structures, seem to be the intrinsic reasons for their outstanding transducing properties. Specifically, SWCNT display some distinctive characteristics. All the atoms in SWCNT are located at the surface; therefore, the current flows at the surface and small variations of the local chemical environment can be detected. For this reason, SWCNT are also a suitable material for ultrasmall sensors.

From a practical point of view, ensemble of carbon nanotubes are easy to handle and can easily be deposited on different electrode surfaces. There are different strategies for utilization of CNTs in construction of solid contact ion selective electrodes. We classified these methods in two categories. These classes are carbon paste-CNT (CPE-CNT) and other solid electrode (except carbon paste)/CNT/polymeric membrane (SE/CNT/PM). The other solid electrodes may be glassy carbon and gold electrodes.

2.2.1 Carbon paste-CNT ion selective electrodes (CPE-CNT-ISE)

Due to advantages of carbon paste electrodes (CPEs) such as renewability, stable response, low ohmic resistance, and no need for internal solution, they have recently attracted attention to use as electrochemical sensors (Wang, 2006). To date, most CPE-based potentiometric sensors reported are based on incorporation of a selective agent into the carbon paste. The base of carbon pastes is usually a mixture of powdered graphite and mineral oil binder. Nowadays, carbon nanotubes have been used in carbon paste electrodes for preparation of different electrochemical sensors because of their special properties.

For construction of CPE-CNT-ISE four or five components is needed: powdered graphite, paraffin, CNT, ionophore and a suitable salt as additive (if necessary). General procedure for preparation of CPE/CNT-ISE is as flowing: the components with appropriate weight mixed well usually with a mortar and pestle until a uniformly wet paste was obtained. Often different amount of components was used and several carbon paste electrodes prepared because the weight of components must be optimized. Sometimes the ionophore may be dissolved in a suitable solvent and then mixed with other component of carbon paste. This is doing for better homogenizing of components of the paste.

After construction of carbon paste, the paste was then packed into the end of an appropriate tube (usually: 3-5 mm internal diameter and 3-10cm long). This tube may be glassy or plastic. A copper wire was inserted into the opposite end to establish electrical contact. When necessary, a new surface was obtained by pushing an excess of paste out of the tube, which was then polished with paper. The electrode was finally conditioned by soaking in a solution of primary ion. The time of conditioning and the concentration of primary ion solution must be optimized.

In some of cases the room-temperature ionic liquids was used instead of paraffin for binder agent in carbon paste or it was used as fifth or sixth components of carbon paste. Incorporation of mineral oil gives CPEs some disadvantages. Mineral oil is not component-fixed since it is derived from refining of petroleum and processing of crude oil. As a result, contaminants or matrix components may unpredictably influence detection and analysis. In addition, the mechanical stability of CPEs rests somewhere between that of membrane electrodes and solid electrodes. Therefore ionic liquids have been used in carbon paste electrodes. These sensors are, in turn, called carbon ionic liquid electrodes (CILEs) (Safavi et al., 2007). Ionic liquids are a good choice as binders in carbon paste electrodes due to their chemical stability, low vapor pressure, low toxicity, low melting temperature, high ionic conductivity and good electrochemical and thermal stability (Maleki et al., 2006).

Using CNT in the composition of the carbon paste (for preparation of potentiometric sensors) not only improves the conductivity of the sensor, but also increases the transduction of the chemical signal to electrical signal. By increasing the conductivity, the dynamic working range and response time of the sensor improve. If the transduction property of the sensor increases, the potential response of the sensor improves to Nernstian values. Also, a wider linear response range and higher sensitivity was reached when the

CNTs were present in the composition of the electrode. However, with increasing the amount of the CNTs higher than a certain value in the composition of the carbon paste long response time and lower sensitivity for the potentiometric sensor obtained. It is probably due to the high surface area formed on the electrode surface that may offer special opportunities for the capturing ions on the electrode surface (Faridbod et al., 2010).
Table 1 shows different parameters of one carbon paste-CNT potentiometric sensor in comparison of some other potentiometric sensor without CNT. As can be seen detection limit, linear range and slope is improved in the case of carbon paste-CNT rather than other electrodes.

Slope (mVdecade^{-1})	Detection limit (mol L^{-1})	Linear range (mol L^{-1})	Reference
29.3	2.5×10^{-9}	5.0×10^{-9} to 1.0×10^{-4}	CPE-CNT Khani et al., 2010
32.6	8.9×10^{-7}	1.0×10^{-6} to 1.0×10^{-1}	Mahajan et al., 2009
34	1.0×10^{-8}	1.0×10^{-8} to 1.0×10^{-3}	Yu et al., 2007
58.8	1.5×10^{-7}	3.2×10^{-7} to 3.2×10^{-4}	Abu-Shawish, 2009
29.6	6.5×10^{-7}	1.0×10^{-6} to 1.0×10^{-1}	Ion et al., 2009
28.7	4.5×10^{-7}	5×10^{-6} to 1×10^{-2}	Lu et al., 2003
30.2	5.0×10^{-8}	1.0×10^{-7} to 1.0×10^{-2}	Rofouei et al., 2009
30.0	4.4×10^{-8}	7.0×10^{-8} to 1.0×10^{-1}	Gupta et al., 2007
30.0	5.0×10^{-8}	1.0×10^{-7} to 1.0×10^{-2}	Bakhtiarzadeh & Ghani, 2008

Table 1. Different parameters of some Hg (II)-potentiometric sensors.

2.2.2 Solid electrode (except carbon paste)/CNT/polymeric membrane ion selective electrodes

Solid electrodes with extended anodic potential windows have attracted considerable analytical interest. Of the many different solid materials that can be used as working electrodes, the most often used are carbon, platinum and gold. Silver, nickel, and copper can also be used for specific applications. In this book the carbon paste electrode not considered in solid electrode and discussed in previous section.
Among solid electrodes the glassy carbon electrode (GCE) has been very popular because of its excellent mechanical and electrical properties, wide potential window, chemical inertness (solvent resistance), and relatively reproducible performance. The material is prepared by means of a careful controlled heating program of a premodeled polymeric (phenolformaldehyde) resin body in an inert atmosphere (Adams, 1969). The carbonization process proceeds very slowly over the 300–1200 °C to ensure the elimination of oxygen, nitrogen, and hydrogen. The structure of glassy carbon involves thin, tangled ribbons of cross-linked graphite-like sheets. Because of its high density and small pore size, no impregnating procedure is required. However, surface pretreatment is usually employed to create active and reproducible glassy carbon electrodes and to enhance their analytical performance. Such pretreatment is usually achieved by polishing (to a shiny "mirror-like" appearance) with successively smaller alumina particles (down to 0.05μm). The electrode should then be rinsed with deionized water before use. Additional activation steps, such as electrochemical, chemical, heat, or laser treatments, have also been used to enhance the

performance. The improved electron transfer capability has been attributed to the removal of surface contaminants, exposure of fresh carbon edges, and an increase in the density of surface oxygen groups (that act as interfacial surface mediators) (Wang, 2006).

Because of special properties of GCE it has been used very as transducer for construction of solid contact ion selective electrodes. There are different ways for utilization and deposition of CNT on solid electrodes. In one method, a solution of CNT was prepared by introducing appropriate amount of CNT into DMF (usually in concentration of 1 mg/ml) and sonication of this solution. Then suitable amount of DMF-CNT solution (about 5 µl) was placed directly onto the GCE surface and dried at room temperature to form a CNT film at the GCE surface. This method can be carried out in some more details as flowing: CNTs were deposited by spraying an aqueous dispersion containing 10^{-2} wt.% of the CNTs and 1% wt. of sodium dodecyl sulphate (SDS), thereby coating the GC with a layer of CNTs. Before deposit, the dispersion was homogenized using a sonicator. The CNTs were deposited in successive steps. The dispersion of CNTs was sprayed; the layer of CNTs was then dried, thoroughly washed with water, and dried again. The process was repeated several times.

After deposition of CNTs on GCE surface (GCE/CNT), suitable amount (10-100 µl) of polymer membrane solution deposit onto the electrode by drop casting. The electrode maintain under dry ambient conditions for appropriate time for example for one day (GCE/CNT/polymeric membrane). The membrane solution has been prepared by four components: a suitable polymer (e.g. poly vinyl chloride (PVC)), plasticizer (e.g. di butyl phthalate (DBP)), ionophore and additive (if it is necessary). The type and the amount of these components must be optimizing to potentiometric sensor show best response. The conditioning of the prepared sensor is necessary before use (Mazloum-Ardakani et al., 2005).

The layer of CNTs on the solid electrode (GC or gold electrode) functions as a transducer providing a high degree of stability in the potentiometric signal, converting the ionic current in the polymeric membrane to the electronic current that flows through the solid electrode conducting rod.

This fabrication of solid-contact ISEs based on CNT requires the additional step to produce the intermediate layer, which makes this type of electrode more complicated to prepare than coated wire electrode. Recently, a facile method has been reported to fabricate single piece solid-contact electrode by adding CNTs into the polymeric membrane solution directly and sonication of it (Zhu et al., 2009). Then this solution subsequently, is drop cast on the solid electrode disk and the solvent is evaporated overnight at room temperature. With the aid of polymeric dispersants CNTs facilitating the ion-to-electron transfer can be well-dispersed in different polymer matrices, which lead to a general approach to prepare all solid-state ion-selective sensors by simple one-step drop casting method.

For combination of properties of conducting polymers and carbon nanotubes a composite of them can be used as ion-to-electron transducer in solid contact ion selective electrodes. In one of the reported methods, negatively charged carbon nanotubes were used as dopants in the electrochemical synthesis of the one conducting polymer. This conducting polymer doped with CNT (CP/CNT) electropolymerized on a solid electrode (CP/CNT/GCE) as transducer and the sensing polymeric membrane applied at CP/CNT/GCE (Mousavi et al., 2009).

2.3 Voltammetric sensors

In voltammetry a time-dependent potential is applied to an electrochemical cell, and the current flowing through the cell is measured as a function of that potential. A plot of current as a function of applied potential is called a voltammogram, providing quantitative and qualitative information about the species involved in the oxidation or reduction reaction. Direct voltammetry of more substance at bare electrodes takes place in high over potential and at nearly potential close to other (Mazloum-Ardakani et al., 2010a). In addition, the direct oxidation results electrode surface contamination (fouling) due to the adsorption of oxidized products. These cases result poor sensitivity, poor selectivity and unstable analytical signals. To avoid the above obstacles in electroanalytical methods, modified electrodes have been developed (Kalimuthu & John, 2009). Chemically modified electrodes represent a modern approach to electrode systems. These electrodes rely on the placement of a reagent onto the surface, to impart the behavior of that reagent to the modified surface. The modification of electrode surfaces can solve many electroanalytical problems, and may form the basis for new analytical applications and different sensing devices. There are four principle enhancement techniques for voltammetric sensors and modified electrodes, namely selective preconcentration, permselectivity, selective recognition and electrocatalysis. Among different voltammetric sensors, in this section we discussed only the voltammetric sensors based on electrocatalysis.

2.3.1 Electrocatalysis voltammetric sensors

Often the desired redox reaction at the bare electrode involves slow electron transfer kinetics and therefore occurs at an appreciable rate only at potentials substantially higher than its thermodynamic redox potential. Such reactions can be catalyzed by attaching to the surface a suitable electron transfer mediator. Electrocatalytic reactions play a central role in electrochemistry and a vital role in sensing and energy-related applications. The function of the mediator is to facilitate the charge transfer between the analyte and the electrode. In most cases the mediated reaction sequence (e.g., for an oxidation process) can be described as below:

$$M_{red} \rightarrow M_{ox} + ne$$

$$M_{ox} + A_{red} \rightarrow M_{red} + A_{ox}$$

where M represents the mediator (or modifier) on the electrode surface and A is analyte (Bard & Faulkner, 2001). Hence, the electron transfer takes place between the electrode and mediator and not directly between the electrode and the analyte. The active form of the catalyst is electrochemically regenerated. The net results of this electron shuttling are a lowering of the overvoltage of analyte oxidation to the formal potential of the mediator and an increase in current density.

Different derivatives of hydroquinone (Mazloum-Ardakani et al., 2010b) and transition metal complexes (Amini et al., 2001; Janda et al., 1996; Shahrokhian et al., 2004) are well known as electron mediators or modifiers in the electrocatalytic determination of some important compounds by voltammetric sensors. These modifiers have a high electron transfer rate constant. Usually, modifier in the absence of analyte exhibits a well-behaved redox reaction in cyclic voltammetry technique. But in the presence of analyte the anodic peak current in cyclic voltammogram increased drastically and the cathodic peak eliminated at suitable potential scan rate (low scan rates). This mechanism is called electrocatalysis or EC′ (Bard & Faulkner, 2001). Different electrochemical techniques such as cyclic voltammetry, chronoamperometry and hydrodynamic voltammetry can be used for investigation of electrocatalysis. These techniques will provide the means to determine different kinetic parameters such as rate of electron transfer between electrode and modifier,

rate of chemical reaction between modifier and analyte and transfer coefficient (α) for oxidation of modifier and analyte (Mazloum-Ardakani et al., 2009).
Since differential pulse voltammetry has high current sensitivity and good resolution, it has been used for quantitative analysis by electrocatalysis voltammetric sensors. For obtaining the calibration curve differential pulse voltammetry apply for different concentrations of analyte. Then the peak current of differential pulse voltammograms plot versus concentrations of analyte. The most important merit of figures for one voltammetric sensor are linear range of calibration curve and detection limit. These can affect by different parameters. For example, even the factors that increase the current sensitivity can improve the detection limit of the sensor (Beitollahi et al., 2009).
Utilization of even substrate with modifier that improves the electrocatalysis of analyte will increase the sensitivity and thus the detection limit and other factors of the sensor will be better. Nanomaterials can be used in structure of the voltammetric sensors for this purpose. There are different methods for modification of the electrode and utilization of nanomaterials (e.g. carbon nanotubes) in voltammetric sensors. In the next section we classified these methods and explain them in details.

2.4 Carbon nanotubes in electrocatalysis voltammetric sensors

Carbon nanotubes represent an increasingly important group of nanomaterials with unique geometric, mechanical, electronic, and chemical properties (Baughman, 2002). The unique properties of carbon nanotubes make them also extremely attractive for the tasks of surface modification in voltammetric sensors. More recent studies have demonstrated that carbon nanotubes-modified electrodes can promote the electrochemical reactivity of important analytes and impart resistance against surface fouling in electrocatalysis voltammetric sensors (Wang & Musameh, 2003). The electrocatalytic activity of carbon nanotubes has been attributed to the presence of edge plane defects at their end caps (Banks et al., 2004).
Modification of the electrode and utilization of carbon nanotubes for preparation of electrocatalysis voltammetric sensors can be divided into two major groups: modification of the bulk (such as carbon paste and carbon ceramic electrodes) and modification of the surface of the electrode (such as adsorption and preparation of polymer film). The most usual methods for using of carbon nanotubes in electrocatalysis voltammetric sensors have been discussed as flowing: modification of the bulk of carbon paste and carbon ceramic by both modifier and carbon nanotubes; modification of the surface of other solid state electrodes (except carbon paste and carbon ceramic e.g. glassy carbon) by drop casting of a solution of carbon nanotube at electrode surface and then modification with modifier; modification of the surface of glassy carbon electrode by a polymer and carbon nanotubes simultaneously.

2.4.1 Carbon nanotubes in bulk modification

There are tow methods for bulk modification of electrodes in electrocatalysis voltammetric sensors. In one of them carbon paste has been used as matrix for incorporating carbon nanotubes and modifier. General procedure for preparation of carbon paste electrode was described in section 2.2.1. In electrocatalysis voltammetric sensors three component has been used: carbon powder, paraffin and modifier. The amount of modifier is 0.5-10% w/w of total weight commonly.
The subtle electronic properties of carbon nanotubes suggest that they have the ability to promote electron transfer when used as the electrode material in electrochemical reactions.

These properties provide a new manner of electrode surface modification for designing new voltammetric sensors and novel electrocatalytic materials (Beitollahi et al., 2008). Therefore carbon nanotubes incorporate as a forth component in carbon paste to improve the electrocatalytic properties of the electrocatalysis voltammetric sensors. The effect of carbon nanotubes on the response of a one electrocatalysis voltammetric sensor has been described below as an example.

We made one electrochemical sensor based on carbon paste electrode modified with 3,4-dihydroxybenzaldehyde-2,4-dinitrophenylhydrazone incorporating carbon nanotubes, which makes the modified electrode highly sensitive for simultaneous electrochemical detection of norepinephrine, acetaminophen and tryptophan (Mazloum-Ardakani et al., 2011). Fig. 2 depicts the cyclic voltammetry responses for the electrochemical oxidation of 0.3 mM norepinephrine at unmodified carbon paste (curve b), modified carbon paste with only modifier (curve e), modified carbon paste with only carbon nanotubes (curve d) and modified carbon paste with both modifier and carbon nanotube (curve f). As it is seen, the peak potential of the norepinephrine at carbon paste modified with carbon nanotube is lower than unmodified carbon paste which indicates the catalysis effect of carbon nanotube. However, modified carbon paste with modifier and carbon nanotube (curve f) shows much higher anodic peak current for the oxidation of norepinephrine compared to modified

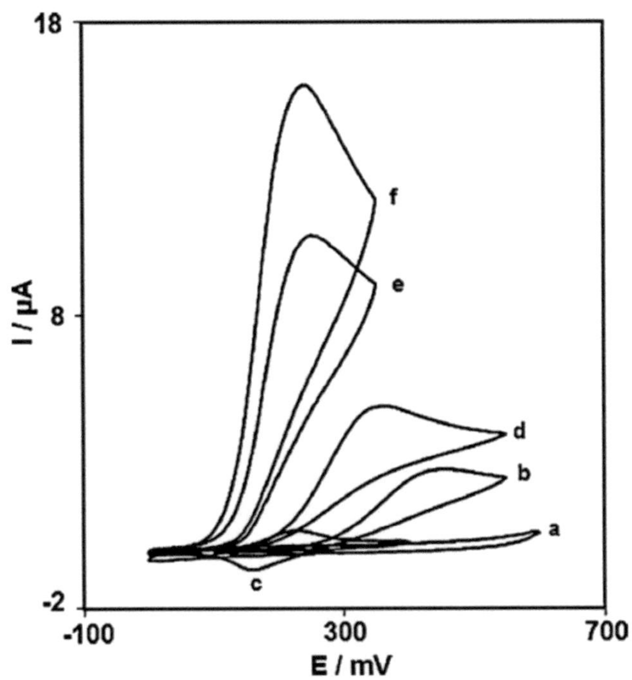

Fig. 2. Cyclic voltammograms of (a) unmodified carbon paste and (c) modified carbon paste with modifier in 0.1 M phosphate buffer solution (pH 7.0); (b) as (a) +0.3 mM norepinephrine; (d) as (b) at the surface of modified carbon paste with only carbon nanotubes; (e) as (b) at the surface of modified carbon paste with modifier; (f) as (b) at the surface of modified carbon paste with modifier and carbon nanotube. All scan rates are 10 mVs^{-1}

carbon paste with only modifier (curve e), indicating that the combination of carbon nanotube and the modifier has significantly improved the performance of the electrode toward norepinephrine oxidation. Also the linear range and detection limit of carbon paste based electrocatalysis voltammetric sensor will be improved by using of carbon nanotubes. For example the detection limit and linear range of a carbon paste sensor without carbon nanotubes for determination of dopamine reported as 1.0×10^{-6} M and 2.0×10^{-6}–1.5×10^{-3} M, respectively (Safavi et al., 2006). However these are 8.7×10^{-8} and 1.0×10^{-7}–9.0×10^{-4} at our reported carbon paste sensor based on carbon nanotubes (Mazloum-Ardakani et al., 2009).

Another way for bulk modification of electrode is using of carbon ceramic as matrix which can hold modifier and carbon nanotubes simultaneously. The carbon ceramic electrodes prepared by sol–gel techniques are interesting materials in the fields of electrochemical sensors and biosensors owing to their renewability, polishibility, chemical stability, physical rigidity, porosity, and permeability in electroanalysis (Lin & Brown, 1997).

The advantages and utilization of carbon nanotubes in carbon ceramic based sensors is similar to carbon paste sensors. The general procedure for construction of modified carbon ceramic sensors with carbon nanotubes is as flowing (Tsionsky et al., 1994).

The mixture solution of methyltrimethoxysilane, methanol and hydrochloric acid (each of them with appropriate concentration) has been stirred to ensure uniform mixing, after which suitable amount of graphite powder (e.g. 1.0 g in 1 mL of solution) and carbon nanotubes (e.g. 0.01 g in 1 mL of solution) has been added and the mixture shaken for about 5 min. The mixture has been added to Teflon or glassy tube with about 3 mm inner diameter. Then it is dried for about 48 h at room temperature.

2.4.2 Carbon nanotubes in surface modification

One common way for modification of the surface of electrode with carbon nanotube was described in section 2.2.2. Suitable amount of a solution of carbon nanotube in a solvent (e.g. 5 µl of carbon nanotube in DMF with concentration of 1 mg/ml) was placed directly onto the solid electrode (e.g. glassy carbon) surface and dried at room temperature to form a carbon nanotube film at the electrode surface. After that, usually the electrochemical deposition has been applied for immobilization of modifier at the glassy carbon electrode. This procedure is as flowing.

Electrochemical activation of the glassy carbon electrode has been performed by continuous potential cycling from about -1.1 to 1.6 V at sweep rate of 100 mVs^{-1} in sodium bicarbonate solution (e.g. 0.1 M) until a stable voltammogram obtain. After rinsing with doubly distilled water, the activated electrode was modified subsequently as follows. It was placed in a buffer solution containing modifier (in concentration of 0.1 to 1 mM) and was modified by cycling the potential around the redox peak of modifier for several cycles or several minutes (Mazloum-Ardakani et al., 2007). Finally, the electrode was rinsed thoroughly with water and the GCE/CNT/mod is prepared.

In another way for modification of the surface of the electrode the carbon nanotube dispersed in conducing polymer film at the surface. Recently, conducting polymer/CNTs composites have received significant interest because the incorporation of CNTs into conducting polymers can lead to new composite materials possessing the properties of each component with a synergistic effect that would be useful in particular application (Chen et al., 2000). Polypyrrole and polyaniline can be used for fabrication of CNT/PPy and CNT/PANI nanocomposite electrodes due to the ease in the preparation through copolymerization by a chemical or electrochemical approach and the resulting nanocomposites exhibits high conductivity and stability (Guo et al., 2005).

The SEM image of three different composite of CNT/conducting polymer has been shown in fig. 3. This figure indicates the difference between conducting polymer film and composite of CNT/polymer. The SEM images of the PPY-CNT, PANI-CNT and PEDOT-CNT composites show commonly a three-dimensional (3D) network composed of interconnected fibrils with similar diameters. The diameter of the as-received CNTs was in the range of 10–30 nm, while the fibrils in the composites were of 30–60 nm in width. This difference indicates that the conducting polymer formed a uniform coating on the surface of individual CNTs. It can then be concluded that the 3D network was formed with the CNTs serving as the backbone, thus greatly enhancing the mechanical properties of the composite film (Peng et al., 2007).

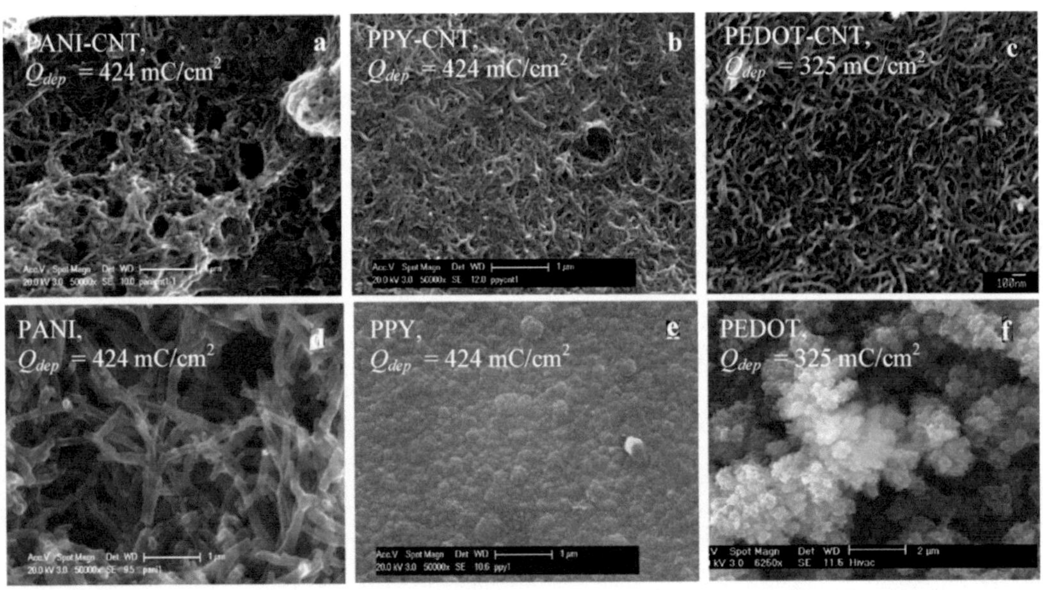

Fig. 3. SEM images of the surfaces of CNT composite films with polyaniline (PANI), polypyrrole (PPY) and poly[3,4-ethylenedioxythiophene] (PEDOT) (a–c) and of their pure polymer (d–f); Q_{dep} is deposition charges.

As an example for using of composite of CNT and conducting polymer in voltammetric sensors, we can state a reported sensor based on PANI/CNTs composite. This PANI/CNTs composite modified electrode fabricated by galvanostatic electro polymerization of aniline on MWCNTs-modified gold electrode, exhibits enhanced electrolytic behavior to the reduction of nitrite and facilitates the detection of nitrite at an applied potential of 0.0 V. A linear range from 5.0×10^{-6} to 1.5×10^{-2} M for the detection of sodium nitrite has been observed at the PANI/MWCNTs-modified electrode with a detection limit of 1.0 μM (Guo et al., 2005).

In another example of surface modification by conducting polymer/CNT composite the films of overoxidized polypyrrole directed single-walled carbon nanotubes (SWNTs) have been electrochemically coated onto glassy carbon electrode (Li et al., 2007). For preparation of this composite, electroactive monomer pyrrole was added into the solution containing sodium dodecyl sulfate and SWNTs. Then, electropolymerization was proceeded at the surface of glassy carbon electrode and a novel kind of conducting polymer/SWNTs composite film with the orientation of SWNTs were obtained correspondingly. Finally, this obtained polypyrrole (PPy)/ SWNTs film modified electrode was oxidized at a potential of +1.8 V. This oxidized composite film modified glassy electrode exhibited excellent electrocatalytic properties for some species such as nitrite, ascorbic acid, dopamine and uric acid; and was used as a new sensor for practical applications. Compared with previous CNTs modified electrodes, in this reported work SWNTs were oriented towards the outside of modified layer, which made the film easily conductive. Moreover, this proposed film modified electrode was more stable, selective and applicable.

3. Conclusion and future directions

There are numerous studies on carbon nanotubes applications because of interesting and unique properties of them. One important application of CNTs is using of them in structure of different sensors. Since CNTs possess interesting electrochemical properties, it can be used for construction of electrochemical sensors specially potentiometric and voltammetric sensors.

In potentiometric sensors, CNTs have been proposed as alternative solid-state transducers that can improve the detection limit and selectivity coefficient of the sensor by facilitating the ion to electron transfer and elimination of the ion flux from the internal solution. Carbon paste and solid electrode/CNT/polymeric membrane are tow way for utilization of CNTs in potentiometric sensors that discussed in detail.

The unique properties of carbon nanotubes make them extremely attractive for the tasks of surface modification in voltammetric sensors. CNT-modified electrodes can improve the electrocatalytic activity of the modifier for electroanalysis of important analytes and impart resistance against surface fouling in electrocatalysis voltammetric sensors. Modification of the electrode with a modifier and utilization of carbon nanotubes for preparation of electrocatalysis voltammetric sensors was divided and discussed into tow major groups: modification of the bulk and modification of the surface of the electrode.

In the future, synthesis of novel carbon nanotubes for example by functionalization of them for improving of their properties, and using of them in construction of different electrochemical sensors will open a new window to electrochemical sensors field. Also, application of CNTs in microelectrodes and nanosensors can be interesting.

So far we have discussed the one sensor and one analyte approach, however, arrays of independent electrodes can offer much more analytical information and thus hold a great potential for many practical applications. The use of multielectrode or multisensor arrays takes advantage of the partial selectivity of an individual electrode, by combining several electrodes and examining the relative responses of all the sensors together.

4. Acknowledgment

The authors would like to thank Yazd University Research Council.

5. References

Abu-Shawish, H.M. (2009). A mercury (II) selective sensor based on N,N_-bis(salicylaldehyde) phenylenediamine as neutral carrier for potentiometric analysis in water samples. *J. Hazard. Mater.* Vol.167, pp.602–608, ISSN 0304-3894

Adams, R. N. (1969). *Electrochemistry at Solid Electrodes*, CRC press, ISBN 0824710053, New York

Amini, M.K.; Shahrokhian, S.; Tangestaninejad, S. & Mirkhani V. (2001). Iron (II) Phthalocyanine-Modified Carbon-Paste Electrode for Potentiometric Detection of Ascorbic Acid. *Anal. Biochem.*, Vol.290, No.2, pp.277-282, ISSN 0003-2697

Bakhtiarzadeh, F. & Ghani, S.A. (2008). An ion selective electrode for mercury (II) based on mercury (II) complex of poly(4-vinyl pyridine). *J. Electroanal. Chem.* Vol.624, pp.139–143, ISSN 1572-6657

Bakker, E.; Pretsch, E. (2002). Peer Reviewed: The new wave of ion-selective electrodes. *Anal. Chem.*, Vol.74, No.15, pp.420A-426A, ISSN 0003-2700

Banks, C.E.; Moore, R.R. & Compton, R.G. (2004). Investigation of modified basal plane pyrolytic graphite electrodes: definitive evidence for the electrocatalytic properties of the ends of carbon nanotubes. *Chem. Commun.*, Vol.16, No. 16, pp.1804-1805, ISSN 1359-7345

Bard, A.J. & Faulkner, L.R. (2001). Electrochemical Methods Fundamentals and Applications (Third Edition), ISBN 0-471-04372-9, John Wiley & Sons, New York

Baughman, R. H.; Zakhidov, A. & De Heer, W.A. (2002). Carbon nanotubes--the route toward applications. *Science*, Vol.297, No.5582, pp.787-792, ISSN 0036-8075

Beitollahi, H.; Mazloum-Ardakani, M.; Naeimi, H. & Ganjipour, B. (2009). Electrochemical characterization of 2, 2′-[1, 2-ethanediylbis (nitriloethylidyne)]-bis-hydroquinone-carbon nanotube paste electrode and its application to simultaneous voltammetric determination of ascorbic acid and uric acid. *J. Solid State Electrochem.*, Vol.13, pp.353–363, ISSN 1432-8488

Beitollahi, H.; Mazloum-Ardakani, M.; Ganjipour, B. & Naeimi, H. (2008). Novel 2,2′-[1,2-ethanediylbis(nitriloethylidyne)]-bis-hydroquinone double-wall carbon nanotube paste electrode for simultaneous determination of epinephrine, uric acid and folic acid. *Biosens. Bioelectron.*, Vol.24, pp.362–368, ISSN 0956-5663

Bobacka, J.; Ivaska, A. & Lewenstam, A. (2008). Potentiometric Ion Sensors. *Chem. Rev.*, Vol.108, No.2, pp.329–351, ISSN 0009-2665

Chen, G.Z.; Shaffer, M.S.P.; Coleby, D.; Dixon, G.; Zhou, W.; Fray, D.J. & Windle, A.H. (2000). Carbon nanotubes and polypyrrole composite: coating and doping. *Adv. Mater.* Vol.12, pp.522–526, ISSN 0935-9648

Crespo, G.A.; Macho, S.; Bobacka, J. & Rius, F.X. (2009). Transduction mechanism of carbon nanotubes in solid-contact ion-selective electrodes. *Anal. Chem.*, Vol.81, No.2, pp.676–681, ISSN 0003-2700

Currie, L.A. (1995). Nomenclature in evaluation of analytical methods including detection and quantification capabilities (IUPAC Recommendations 1995). *Pure Appl. Chem.*, Vol.67, No. 10, pp. 1699-1723, ISSN 0033-4545

Faridbod, F.; Ganjali, M.R.; Larijani, B.; Hosseini, M. & Norouzi, P. (2010). Ho^{3+} carbon paste sensor based on multi-walled carbon nanotubes: Applied for determination of holmium content in biological and environmental samples. *Mater. Sci. Eng. C*, Vol.30, No.4, pp.555–560, ISSN 0928-4931

Guilbault, G.G.; Durst, R.A.; Frant, M.S.; Freiser, H.; Hansen, E.H.; Light, T.S.; Pungor, E.; Rechnitz, G.; Rice, N.M.; Rohm, T.J.; Simon, W. & Thomas, J.D.R. (1976). Recommendations for nomenclature of ion-selective electrodes. *Pure Appl. Chem.*, Vol.48, No.1, pp.127-132, ISSN 0033-4545

Guo, M.; Chen, J.; Li, J.; Tao, B. & Yao, S. (2005). Fabrication of polyaniline/carbon nanotubes composite modified electrode and its electrocatalytic property to the reduction of nitrite. *Anal. Chim. Acta.* Vol.532, pp.71–77, ISSN 0003-2670

Gupta, V.K.; Singh, A.K.; Khayat, M.A. & Gupta, B. (2007). Neutral carriers based polymeric membrane electrodes for selective determination of mercury (II). *Anal. Chim. Acta* Vol.590, pp.81–90, ISSN 0003-2670

Ion, A.C.; Ion, I.; Stefan, D.N. & Barbu, L. (2009). Possible mercury speciation in urine samples using potentiometric methods. *Mater. Sci. Eng. C*, Vol.29, pp.1–4, ISSN 0928-4931

Janda, P.; Weber, J.; Dunsch, L. & Lever, A.B.P. (1996). Detection of Ascorbic Acid Using a Carbon Fiber Microelectrode Coated with Cobalt Tetramethylpyridoporphyrazine. *Anal. Chem.*, Vol.68, No.6, pp.960-965, ISSN 0003-2700

Kalimuthu, P. & John, S.A. (2009). Modification of electrodes with nanostructured functionalized thiadiazole polymer film and its application to the determination of ascorbic acid. *Electrochim. Acta*, Vol.55, No.1, pp.183-189, ISSN 0013-4686

Khani, H.; Rofoueia, M.K.; Arabb. P.; Gupta, V.K. & Vafaei, Z. (2010). Multi-walled carbon nanotubes-ionic liquid-carbon paste electrode as a super selectivity sensor: Application to potentiometric monitoring of mercury ion (II). *J. Hazard. Mater.* Vol.183, pp.402–409, ISSN 0304-3894

Lai, C.; Fierke, M.A.; Stein, A. & Buhlmann, P. (2007). Ion-selective electrodes with three dimensionally ordered macroporous carbon as the solid contact. *Anal. Chem.*, Vol.79, No. 12, pp.4621–4626, ISSN 0003-2700

Li, Y.; Wanga, P., Wang, L. & Lin, X. (2007). Overoxidized polypyrrole film directed single-walled carbon nanotubes immobilization on glassy carbon electrode and its sensing applications. *Biosens. Bioelectron.*, Vol.22, pp.3120–3125, ISSN 0956-5663

Lin, J. & Brown, C.W. (1997). Sol-gel glass as a matrix for chemical and biochemical sensing. *Trends Anal. Chem.*, Vol.16, No.4, pp.200-211, ISSN 0165-9936

Lu, J.; Tong, X. & He, X. (2003). A mercury ion-selective electrode based on a calixarene derivative containing the thiazole azo group. *J. Electroanal. Chem.*, Vol.540, pp.111–117, ISSN 1572-6657

Mahajan, R.K.; Puri, R.K.; Marwaha, A.; Kaur, I. & Mahajan, M.P. (2009). Highly selective potentiometric determination of mercury (II) ions using 1-furan-2-yl-4-(4-nitrophenyl)-2-phenyl-5H-imidazole-3-oxide based membrane electrodes. *J. Hazard. Mater.* Vol.167, pp.237–243, ISSN 0304-3894

Maleki, N.; Safavi, A. & Tajabadi, F. (2006). High-performance carbon composite electrode based on an ionic liquid as a binder. *Anal. Chem.*, Vol.78, No.11, pp.3820-3826, ISSN 0003-2700

Mazloum-Ardakani, M.; Beitollahi, H.; Amini, M.K.; Mirkhalaf, F. & Mirjalili, B.B.F. (2011). A highly sensitive nanostructure-based electrochemical sensor for electrocatalytic determination of norepinephrine in the presence of acetaminophen and tryptophan. *Biosens. Bioelectron.*, Vol.26, pp.2102–2106, ISSN 0956-5663

Mazloum-Ardakani, M.; Beitollahi, H.; Sheikh-Mohseni, M.A.; Naeimi, H. & Taghavinia, N. (2010a). Novel nanostructure electrochemical sensor for electrocatalytic determination of norepinephrine in the presence of high concentrations of acetaminophene and folic acid. *Appl. Catal. A: Gen.*, Vol.378, pp.195-201, ISSN 0926-860X

Mazloum-Ardakani, M.; Beitollahi, H.; Sheikh Mohseni, M.A.; Benvidi, A.; Naeimi, H., Nejati-Barzoki, M. & Taghavinia, N. (2010b). Simultaneous determination of epinephrine and acetaminophen concentrations using a novel carbon paste electrode prepared with 2,2′-[1,2 butanediylbis(nitriloethylidyne)]-bis-hydroquinone and TiO_2 nanoparticles. *Colloids Surf. B*, Vol.76, pp.82-87, ISSN 0927-7765

Mazloum-Ardakani, M.; Beitollahi, H.; Ganjipour, B.; Naeimi H. & Nejati M. (2009). Electrochemical and catalytic investigations of dopamine and uric acid by modified carbon nanotube paste electrode. *Bioelectrochemistry*, Vol.75, pp.1–8, ISSN 1567-5394

Mazloum-Ardakani, M.; Sheikh Mohseni, M.A. & Salavati-Niasari, M. (2008). Novel thiocyanate-selective electrode based on binuclear molybdenum complex of bis-N,O-bidentate Schiff base. *Can. J. Anal. Sci. Spect.*, Vol.53, No.4, pp.179-188, ISSN 1205-6685

Mazloum-Ardakani, M.; Rahimi, P.; Ebrahimi Karami, P.; Zare, H.R. & Naeimi, H. (2007). Electrocatalytic oxidation of cysteine by quinizarine at glassy carbon electrode. *Sens. Actuators B*, Vol.123, pp.763–768, ISSN 0925-4005

Mazloum-Ardakani, M.; Jalayer, M.; Naeimi, H.; Heidarnezhad, A. & Zare, H.R. (2006). Highly selective oxalate-membrane electrode based on 2,2_-[1,4-butandiyle bis(nitrilo propylidine)]bis-1-naphtholato copper(II). *Biosenso. Bioelectron.*, Vol.21, pp.1156–1162, ISSN 0956-5663

Mazloum-Ardakani, M.; Sadeghi, A. & Salavati-Niasari, M. (2005). Highly selective thiocyanate membrane electrode based on butane-2,3-dione bis(salicylhydrazonato)zinc(II) complex. *Talanta*, Vol.66, pp.837–843, ISSN 0039-9140

Michalska, A. (2006). Optimizing the analytical performance and construction of ion-selective electrodes with conducting polymer-based ion-to-electron transducers. *Anal. Bioanal. Chem.*, Vol.384, No.2, pp.391–406, ISSN 1618-2642

Midgley, D. (1979). Limit of detection in analysis with ion-selective electrodes. *Analyst*, Vol.104, pp.248-257, ISSN 0003-2654

Mousavi, Z.; Bobacka, J.; Lewenstam, A. & Ivaska, A. (2009). Poly (3,4-ethylenedioxythiophene) (PEDOT) doped with carbon nanotubes as ion-to-electron transducer in polymer membrane-based potassium ion-selective electrodes. *J. Electroanal. Chem.*, Vol.633, No.1, pp.246–252, ISSN 1572-6657

Parra, E.J.; Crespo, G.A.; Riu, J.; Ruiz, A. & Rius, F.X. (2009). Ion-selective electrodes using multi-walled carbon nanotubes as ion-to-electron transducers for the detection of perchlorate. *Analyst*, Vol.134, pp.1905–1910, ISSN 0003-2654

Peng, C.; Jin, J. & Chen, G.Z. (2007). A comparative study on electrochemical co-deposition and capacitance of composite films of conducting polymers and carbon nanotubes. *Electrochim. Acta* Vol.53, pp.525–537, ISSN 0013-4686

Pergel, E.; Gyurcsanyi, R. E.; Toth, K. & Lindner, E. (2001). Picomolar detection limits with current-polarized Pb^{2+} ion-selective membranes. *Anal. Chem.*, Vol.73, pp.4249-4253, No.17, ISSN 0003-2700

Qin, W.; Zwickl, T. & Pretsch, E. (2000). Improved Detection Limits and Unbiased Selectivity Coefficients Obtained by Using Ion-Exchange Resins in the Inner Reference Solution of Ion-Selective Polymeric Membrane Electrodes. *Anal. Chem.*, Vol.72, No.14, pp.3236-3240, ISSN 0003-2700

Rofouei, M.K.; Mohammadi, M. & Gholivand, M.B. (2009). Mercury (II) selective membrane electrode based on 1,3-bis(2-methoxybenzene)triazene. *Mater. Sci. Eng. C*, Vol.29, pp.2154–2159, ISSN 0928-4931

Safavi, A.; Maleki, N.; Honarasa, F.; Tajabadi, F. & Sedaghatpour, F. (2007). Ionic liquids modify the performance of carbon based potentiometric sensors. *Electroanalysis*, Vol.19, pp.582-586, ISSN 1040-0397

Safavi, A.; Maleki, N.; Moradlou, O. & Tajabadi, F. (2006). Simultaneous determination of dopamine, ascorbic acid, and uric acid using carbon ionic liquid electrode. *Anal. Biochem.*, Vol.359, pp.224–229, ISSN 0003-2697

Shahrokhian, S.; Souri, A. & Khajehsharifi, H. (2004). Electrocatalytic oxidation of penicillamine at a carbon paste electrode modified with cobalt salophen. *J. Electroanal. Chem.*, Vol.565, No.1, pp.95-101, ISSN 1572-6657

Sokalski, T.; Zwickl, T.; Bakker, E. & Pretsch, E. (1999). Lowering the detection limit of solvent polymeric ion-selective electrodes. 1. Modeling the influence of steady-state ion fluxes. *Anal. Chem.*, Vol.71, No.6, pp.1204-1209, ISSN 0003-2700

Tsionsky, M.; Gun, G.; Glezer, V. & Lev, O. (1994). Sol-gel-derived ceramic-carbon composite electrodes: introduction and scope of applications. *Anal. Chem.*, Vol.66, No.10, pp.1747-1753, ISSN 0003-2700

Wang, J. (2006). *Analytical Electrochemistry* (Third Edition), ISBN-13 978-0-471-67879-3, John Wiley & Sons, Hoboken, New Jersey

Wang, J. & Musameh, M. (2003). Carbon nanotube/teflon composite electrochemical sensors and biosensors. *Anal. Chem.*, Vol.75, No.9, pp.2075-2079, ISSN 0003-2700

Ye, Q. & Meyerhoff, M.E. (2001). Rotating electrode potentiometry: lowering the detection limits of nonequilibrium polyion-sensitive membrane electrodes. *Anal. Chem.*, Vol.73, No.2, pp.332-336, ISSN 0003-2700

Yu, X.; Zhoua, Z.; Wang, Y.; Liu, Y.; Xie, Q. & Xiao, D. (2007). Mercury (II)-selective polymeric membrane electrode based on the 3-[4-(dimethylamino)phenyl]-5-mercapto-1,5-diphenylpentanone. *Sens. Actuators B*, Vol.123, pp.352–358, ISSN 0925-4005

Zhu, J.; Qin, Y. & Zhang, Y. (2009). Preparation of all solid-state potentiometric ion sensors with polymer–CNT composites. *Electrochem. Commun.*, Vol.11, pp.1684–1687, ISSN 1388-2481

16

Application of Carbon Nanotubes Modified Electrode in Pharmaceutical Analysis

Lingbo Qu[1,2] and Suling Yang[1]
[1]*Department of Chemistry, Zhengzhou University, Zhengzhou 450001,*
[2]*Chemistry and Chemical Engineering School, Henan University of Technology, Zhengzhou 450001, PR China*

1. Introduction

The development of electrochemical sensors has attracted considerable attention as a low-cost method to the sensitive detection of a variety of pharmaceutical analytes. Since the discovery of carbon nanotubes (CNTs) in 1991 [1], research on CNTs has grown rapidly. In recent years, CNTs have also been used as electrode modified materials because CNTs offer unique advantages including enhanced electronic properties, a large edge plane/basal plane ratio, and electron transfer reactions [2]. Thus, CNTs-based sensors generally have higher sensitivities in a low concentration or in the complex matrix, lower limits of detection, and faster electron transfer kinetics than traditional carbon electrodes. Many factors need to be investigated in order to create an optimal CNTs-based sensor. Electrode performance can be influenced by the pretreatment of the nanotube, CNTs surface modification, the method of electrode attachment, and the addition of electron mediators. With the further development of CNTs and nanotechnology, studies on preparation, properties and application of CNTs-based modified electrodes have still been a hot topic attracting lots of researchers in the world. This article is presented on the application of CNTs modified electrode in different pharmaceutical analytes, which mainly includes the electrochemical studies on weak basic pharmaceuticals, weak acidic pharmaceuticals and other related small biological molecules. The physical and catalytic properties make CNTs ideal for use in sensors. Most notably, CNTs display high electrical conductivity, chemical stability, and mechanical strength.

2. CNTs modified electrode used in analysis of weak basic pharmaceuticals

Caffeine (3, 7-dihydro-1, 3, 7-trimethyl-1H-purine-2, 6-dione) and theophylline (3, 7-dihydro-1, 3-dimethyl-1H-purine-2, 6-dione), are two important active alkaloids that are widely distributed in beverages and plant products mainly including tea, coffee bean, cocoa and cola nuts. They are known to have many pharmacological effects, such as gastric acid secretion, diuretic, cardiac stimulant, and stimulant of central nervous system [3]. However, appropriate dosing is crucial because of the serious side adverse reactions in the presence of high concentrations of these compounds as the risk factors for asthma, kidney malfunction and cardiovascular diseases [4].

Nafion, a perfluorinated sulphonated cation exchanger with properties of excellent antifouling capacity, chemical inertness and high permeability to cations, has been extensively employed as an electrode modifier. CNTs can be homogeneously dispersed in Nafion solution because of the hydrophobic side chains and polar head groups of Nafion. Nafion/CNTs composite thin film-modified electrodes have their attractive effects in electroanalytical applications. Recently, we [5, 6] reported that Nafion/CNTs -modified electrode was made, and used as a sensor in the electrochemical determination of caffeine and theophylline. This sensor can ameliorate the problems of high overpotential, slow electrode reaction, and low sensitivity, which occurs at conventional electrodes.
Figure 1 displays the characterization of the Nafion/CNTs composite film on the glassy carbon electrode (GCE) by using scanning electron microscopy (SEM). It is obvious that the Nafion/CNTs composite film was uniformly coated on the electrode surface and formed a spaghetti-like porous reticular formation. The special surface morphology offered a much larger real surface area than the apparent geometric area.

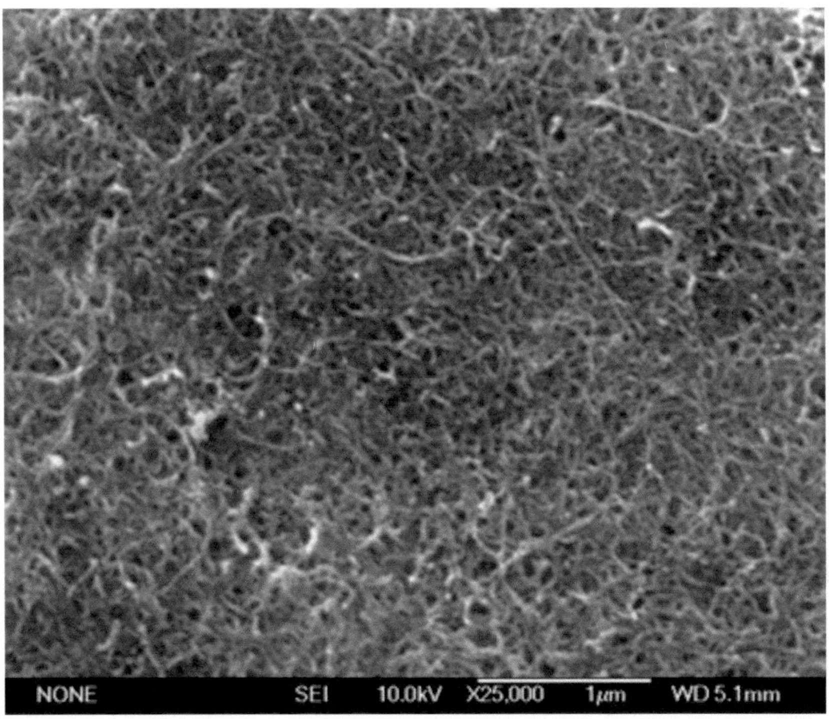

Fig. 1. SEM image of Nafion/CNTs composite film on glassy carbon electrode.

The experimental results demonstrated that caffeine and theophylline can be effectively accumulated at Nafion/CNTs composite film-modified electrode and produce a sensitive anodic peak in a 0.01 mol/L H_2SO_4 medium, respectively (Figure 2). Under the same conditions, no anodic peak of caffeine and theophyline was observed at the bare GCE. Especially at the Nafion/CNTs nanocomposite -modified electrode, the peak current was significantly higher than those at the CNTs/GCE or the Nafion/GCE. The oxidation process of caffeine and theophyline at Nafion/CNTs/GCE or Nafion/GCE is irreversible. Compared with the Nafion/GCE, the oxidation potential at the Nafion/CNTs/GCE was negatively shifted. This phenomenon may be an evidence of catalytic effect of CNTs toward caffeine and theophyline oxidation. The reasons for the notable sensitivity of the determination at the Nafion/CNTs/GCE may be summarized as follows:(1) the Nafion/CNTs/GCE contains the cation exchanger of Nafion which has selective cation exchange enriched ability due to the electrostatic interaction. (2) CNTs display attractive characteristics, such as much larger specific surface area, excellent adsorptive ability and catalytic ability. Without a doubt, the synergetic functions of Nafion and CNTs make contributions to the higher current response [7-12].

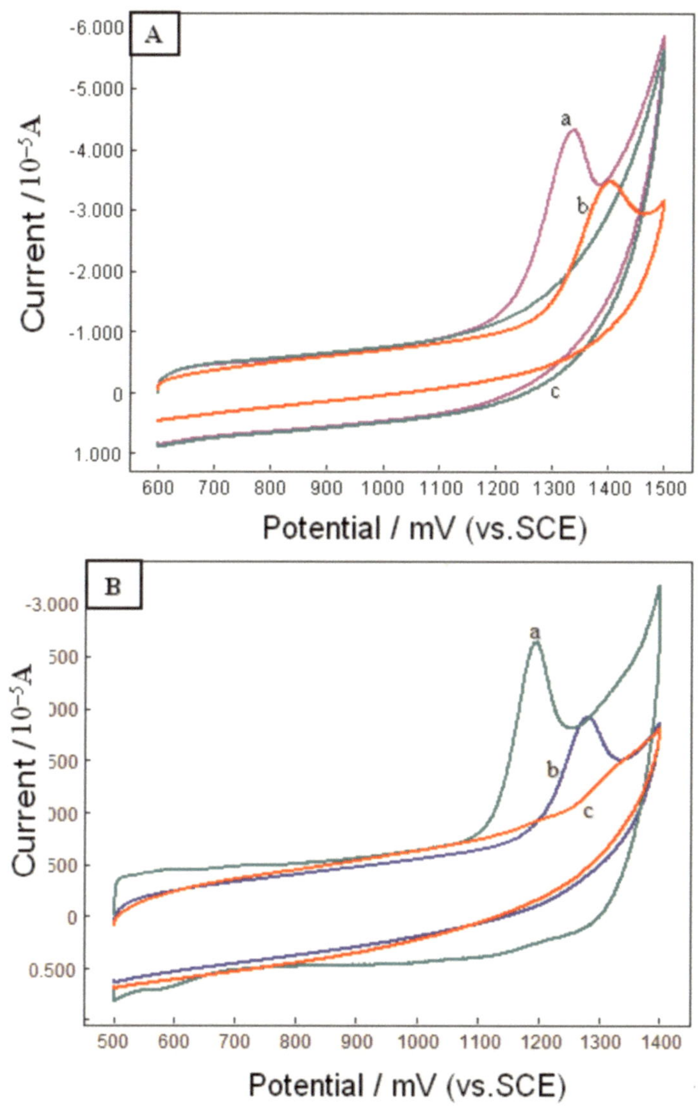

Fig. 2. Cyclic voltammograms of the Nafion/CNTs-modified GCE (a), Nafion-modified GCE (b), the bare GCE (c) in 0.01 mol/L H_2SO_4 medium containing: A. 3.0 ×10^{-5} mol/L caffeine; scan rate 80 mV/s; B. 2.0 × 10^{-5} mol/L theophyline; scan rate 100 mV/s.

Under the suitable conditions, the anodic peak current was linear to caffeine concentration in the range of $6.0 \times 10^{-7} - 4.0 \times 10^{-4}$ mol/L with a limit of detection of 2.3×10^{-7} mol/L; theophylline in the range of $8.0 \times 10^{-8} - 6.0 \times 10^{-5}$ mol/L with a detection limit of 2.0×10^{-8} mol/L.

Regeneration and reproducibility are the two vital characteristics for the modified electrode, which should be investigated for analytical determination. The same Nafion/CNTs/GCE was used for five times successive measurement of 2.0×10^{-5} mol/L TP. After each measurement, the surface of the Nafion/CNTs/GCE was regenerated by successively scan cycle between 500 and 1400 mV in 0.01 mol/L H_2SO_4 medium solutions for three cycles. The relative standard deviation (RSD) of the peak current was 2.4% (n = 5), which revealed the good regeneration and reproducibility.

This newly exploited method was successfully used to determine caffeine in beverage samples and theophylline in drug samples. The detected results of caffeine in beverage samples are shown in Table 1. The results are in agreement with the value obtained employing UV-vis spectroscopic method [13] as well as the regulation of the American Beverage Association [14] in the range of 4.3×10^{-4} - 8.7×10^{-4} mol/L. Caffeine concentration of energy drinking water is in agreement with the declared content (i.e., 200 mg/L).

Sample	DPV value	UV-vis spectroscopic value
Cola beverage1	$6.2 \pm 0.53 \times 10^{-4}$ mol/L	$6.1 \pm 0.29 \times 10^{-4}$ mol/L
Cola beverage 2	$5.3 \pm 0.16 \times 10^{-4}$ mol/L	$5.2 \pm 0.53 \times 10^{-4}$ mol/L
Energy drinking water	200 ± 10 mg/L	198 ± 14 mg/L
Green tea	28.9 ± 0.42 mg/g	29.6 ± 0.25 mg/g

Table 1. Results obtained in determination of caffeine in beverage samples and tea using the DPV (proposed) and UV-vis spectroscopic methods (n = 5) .

The content of TP in theophylline sustained-release tablet was calculated to be 0.0921g per tablet (the declared content was 0.1 g per tablet); the content of TP in aminophylline injection was calculated to be 0.1976 g per ampoule (the declared content was 0.2 g per ampoule). The determined contents of TP were in agreement with the declared contents of TP in real samples. The results demonstrated that the proposed methods could be efficiently used for the determination of caffeine and TP.

3. CNTs modified electrode used in analysis of weak acidic pharmaceuticals

Ascorbic acid (AA) is widely present in many biological liquids, medicines, fruits and beverages. It is one of the most important soluble vitamins and plays a significant role in the biological functioning, such as the supplement of inadequate dietary intake , wound healing [15, 16], prevention as well as treatment of common cold, mental illness and infertility [17]. Uric acid (UA) is one of the principal end products of purine metabolism in human body. Abnormal levels of UA are symptoms of several diseases, like gout, hyperuricaemia and Lesch-Nyhan syndrome [18]. Hence, monitoring the concentration of UA in biological fluids has their clinical significance. Generally, AA and UA always coexist in biological fluids such as blood and urine, and AA has a close oxidation peak potential of UA, which results in poor selectivity determination of AA or UA in real samples on conventional electrodes. Therefore, it is essential to exploit more sensitive, selective and simple methods for the simultaneous determination of AA and UA. Recently, our group has realized the simultaneous determination of AA and UA using Nafion/CNTs composite film-modified electrode [19].

Figure 3A demonstrated the cyclic voltammetry (CV) curves of a mixture of UA and AA (containing 8.0×10^{-5} mol/L UA and 1.0×10^{-3} mol/L AA) (a), 1.0×10^{-3} mol/L AA (b) and without AA and UA (c) in 0.1 mol/L NaCl (pH 6.5) at Nafion/CNTs/GCE, respectively.

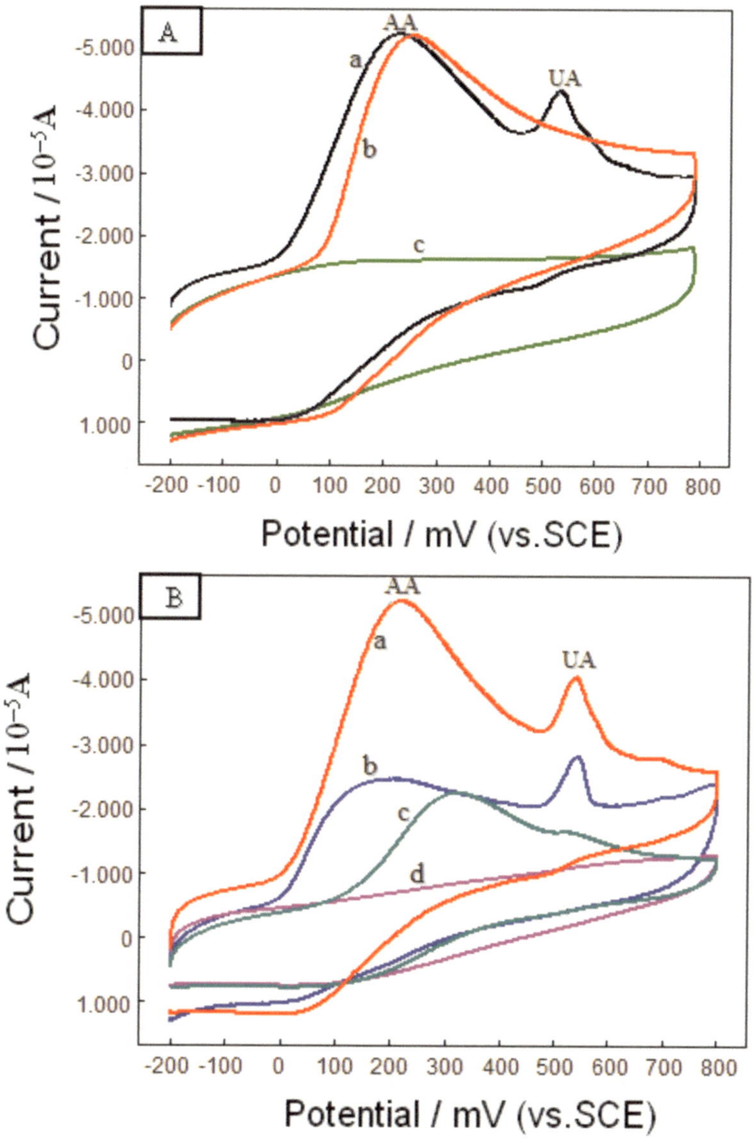

Fig. 3. (A) CV curves of the mixture containing 1.0×10^{-3} mol/L AA and 8.0×10^{-5} mol/L UA (a), 1.0×10^{-3} mol/L AA (b) and without AA and UA (c) in 0.1 mol/L NaCl solution (pH 6.5) at the Nafion/CNTs/GCE; (B) CV curves of the mixture containing 1.0×10^{-3} mol/L AA and 8.0×10^{-5} mol/L UA in 0.1 mol/L NaCl solution (pH 6.5) at the Nafion/CNTs/GCE (a), CNTs/GCE (b), bare GCE (c) and Nafion/GCE (d); scan rate: 100 mV/s, rest time: 3 s

Figure 3A-a showed two anodic peaks at around the potential of 534 and 214 mV, which attributed to the oxidation of UA and AA with a 320 mV separation of both peaks. Figure 3B revealed CV response of a mixture of UA and AA (containing 8.0×10^{-5} mol/L UA and 1.0×10^{-3} mol/L AA) in 0.1 mol/L NaCl (pH 6.5) at Nafion/CNTs/GCE (a), CNTs/GCE (b), bare GCE (c) and Nafion/GCE (d). Under the same conditions, no anodic peak of AA or UA was observed at the Nafion/GCE. At the bare GCE, UA and AA exhibited an overlapped and broad anodic peak extended over a potential region of 98–580 mV with the mixed potential at 310 mV. However, CV for the Nafion/CNTs/GCE and CNTs/GCE showed two anodic peaks with a separation of about 320 mV towards UA and AA, which was broad enough for their simultaneous electrochemical determination. Nevertheless，at the Nafion/CNTs nanocomposite-modified electrode, the peak currents were significantly higher than those at the CNTs/GCE or the bare GCE.

Under the optimized conditions, the peak currents of AA and UA were proportional to their concentration at the ranges of 8.0×10^{-5} to 6.0×10^{-3} mol/L and 6.0×10^{-7} to 8.0×10^{-5} mol/L, respectively. The proposed method was used for the detection of AA and UA in real samples, such as Vitamin C Injection and urine sample.

The total value of AA in Vitamin C Injection was 242 g/L. The total value of AA was in agreement with the declared content (i.e., 250 g/L). The total UA concentrations detected in urine sample was 4.37×10^{-3} mol/L, which was consistent with the containing level of a healthy human. The results demonstrated that the proposed methods could be efficiently used for the determination of AA and UA.

Paeonol (2-hydroxyl-4-methoxyacetophone, PN) is a major phenolic component of Cortex Moutan. It is known to have anti-aggregatory, anti-oxidant and anti-inflammatory activities [20]. Paeonol has been used in the treatment of arthritis and suppress ADP- or collagen-induced human blood platelet aggregation in a dose-dependent manner due to its analgesic, anti-pyretic, and anti-bacterial properties [21]. Many researchers have given increasing attention to the pharmacokinetic activities of paeonol. Therefore, the technique of quantitative determination of paeonol is crucial to the evaluation and popularization of traditional Chinese medicines as well as drug products containing paeonol [22, 23]. We have carried out the determination of paeonol in pharmaceutical and biological samples using Nafion/CNTs-modified electrode as a sensitive voltammetric sensor [24].

As shown in figure 4, paeonol exhibited an anodic peak on the Nafion/CNTs/GCE (Figure 4-a), Nafion/GCE (Figure 4c) and a bare GCE (Figure 4-b) in 0.1 mol/L phosphate buffer solution (pH 7.0). According to Wang's report [25], paeonol was easy to yield compact insulated polymer arylether on the bare electrode surface, which could prevent the progression of the electrode reaction on the electrode surface. Nafion, a perfluorinated sulfonate polymer is a good cation-exchanger provided with adsorption ability and high surface area. It maybe effectively prevents the macro-polymer arylether adsorbing on the modified electrode surface and avoid anode fouling by forming a dense film with a low permeability. Thus, when Nafion was present, the anodic peak current (i_{pa}) was effectively separated from the background current (curve c). When Nafion and CNTs were combined together, the peak potential moved in the negative direction slightly at about 836 mV (vs. SCE) with the obvious enhancement of the peak current (curve a). This phenomenon maybe related to the influence of CNTs on the electron transfer rate, the effective electrode area, as well as the accumulation amount of paeonol. The resulting composite materials coated electrode exhibited more sensitive response to paeonol, meaning that composite materials could cooperate with each other to enhance the voltammetric response of paeonol. In addition, it was observed that paeonol also produced a slight cathodic peak at about 440 mV (vs. SCE) on the Nafion/CNTs/GCE, which indicated that Nafion/CNTs/GCE was much more sensitive than Nafion/GCE in improving paeonol response under the same conditions.

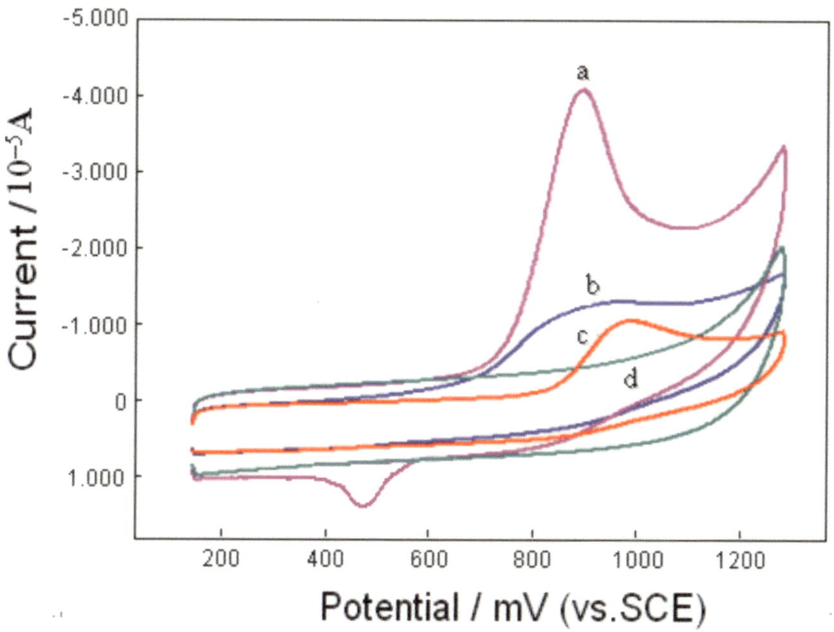

Fig. 4. Cyclic voltammograms on the Nafion/CNTs-modified GCE (a), the bare GCE (b), Nafion-modified GCE (c) containing 8.0×10^{-5} mol/L paeonol and Nafion/CNTs-modified GCE (d) without paeonol in 0.1 mol/L NaH_2PO_4- Na_2HPO_4 (pH 7.0) medium; scan rate 100 mV/s.

The relationship between the anodic peak current and the concentration of paeonol was studied using differential pulse anodic stripping voltammetry. Under the optimum instrumental conditions, the anodic peak current was proportional to paeonol concentration in the range of 6.0×10^{-7}- 6.0×10^{-5} mol/L with a detection limit of 4.0×10^{-7} mol/L. The Nafion/CNTs/GCE was applied to the determination of paeonol in Liuweidihuang Wan and Cortex Moutan and the results were listed in Table 2. Total value of paeonol in Liuweidihuang was 3.20 ± 0.75 mg/g, which was close to the revealed value of 3.22 mg/g [26]. Total value of paeonol in cortex moutan was 20.05 ± 0.50 mg/g, which was consistent with the reported value of 20.61 ± 0.986 mg/g [27]. The accuracy of the method was evaluated by its recovery during spiked experiments. Confirming those quantitative and reproducible results of this method, the direct determination of paeonol in spiked Liuweidihuang Wan, Cortex Moutan, urine and plasma samples were carried out, and the results were also displayed in Table 2. The recoveries of paeonol from the drug matrices and biological samples, such as urine and plasma demonstrated that this proposed method could be applied to the detection of paeonol in pharmaceutical and biological samples with excellent sensitivity and selectivity.

Sample	Original (μmol/L)	Spike (μmol/L)	Found (μmol/L)	Average recovery (%)	RSD of recovery (%)
LiuweidihuangWan[a]	7.71 ± 0.19	40	47.52 ± 0.52	98.7	2.3
		8	15.83 ± 0.24	101.5	2.0
		2	9.63 ± 0.17	96.0	1.5
Cortex Moutan[b]	48.26 ± 0.24	40	88.10 ± 0.21	99.6	2.0
		8	57.92 ± 0.23	108.2	3.2
		2	50.30 ± 0.14	102.0	2.3
Urine[c]	—	80	79.80 ± 0.31	99.8	2.6
		20	20.25 ± 0.10	101.3	1.7
		2	1.99 ± 0.22	99.5	1.9
Plasma[c]	—	80	81.20 ± 0.42	101.5	3.0
		20	21.00 ± 0.19	105.0	3.7
		2	1.96 ± 0.20	98.0	2.2

Table 2. Determination of the content of paeonol in pharmaceutical and biological samples ∗ on the Nafion/CNTs/GCE.
a Dilution factor: 1/100.
b Dilution factor: 1/10.
* Number of samples assayed: 5.

4. CNTs with methylene blue composite film-modified electrode for the simultaneous voltammetric detection of dopamine and uric acid in the presence of high concentration of ascorbic acid

Dopamine (DA) as one of the most important catecholamines, is an significant neurotransmitter and plays a vital role in the central nervous, renal, hormonal and cardiovascular systems [28, 29]. Inadequate DA-containing neurons may cause neurological disorders such as schizophrenia and Parkinson's disease [30]. Therefore, determining the concentration of this neurochemical is of great clinical importance. Uric acid (UA) is one of the principal end products of purine metabolism in the human body. Hence, monitoring the concentration of UA in biological fluids has their clinical significance. Electrochemical methods can be used to determine DA and UA due to their electrochemical active. But the significant problem encountered with the detection of DA or UA using electrochemical methods arising from the primary interference of ascorbic acid (AA). Generally, AA, DA and UA always coexist in biological fluids, and AA has an overlapping oxidation peak of DA and UA at bare electrodes, which results in poor selectivity determination of DA or UA in real samples. Therefore, it is essential to eliminate the interference of AA by using suitable film-modified electrode for the selective determination of DA and UA.

Through the adsorption of methylene blue (MB) onto the CNTs, our group has successfully prepared a sensitive voltammetric sensor CNTs/MB for the selective determination of DA and UA [31]. In order to ascertain the formation of MB-CNTs nanostructure, UV-vis spectral measurements, atomic force microscope (AFM), scanning electron microscopy (SEM) and electrochemical impedance spectra (EIS) were carried out.

As shown in figure 5, the absorption spectra of water-dispersed solutions of free MB (curve a), CNTs (curve b) and the MB-CNTs adduct (curve c) were displayed. The UV-vis spectrum of the CNTs dispersed in aqueous solution exhibited a strong absorbance at 263 nm. The spectrum of free MB in aqueous solution displayed two strong absorption peaks at 294 and 665 nm (curve a), characteristic of the MB monomer in solution. The chemisorption of MB onto the CNTs was evident from the spectrum of the MB-CNTs composite (curve c), which was similar to that of CNTs. However, a close inspection of the spectrum of CNTs and the MB-CNTs adsorptive nanostructure revealed that there was a change in the spectrum of CNTs after its adsorption of MB. A new peak appeared at 690 nm was observed, which was due to the absorbance of the aggregation of MB molecules onto the CNTs [32].

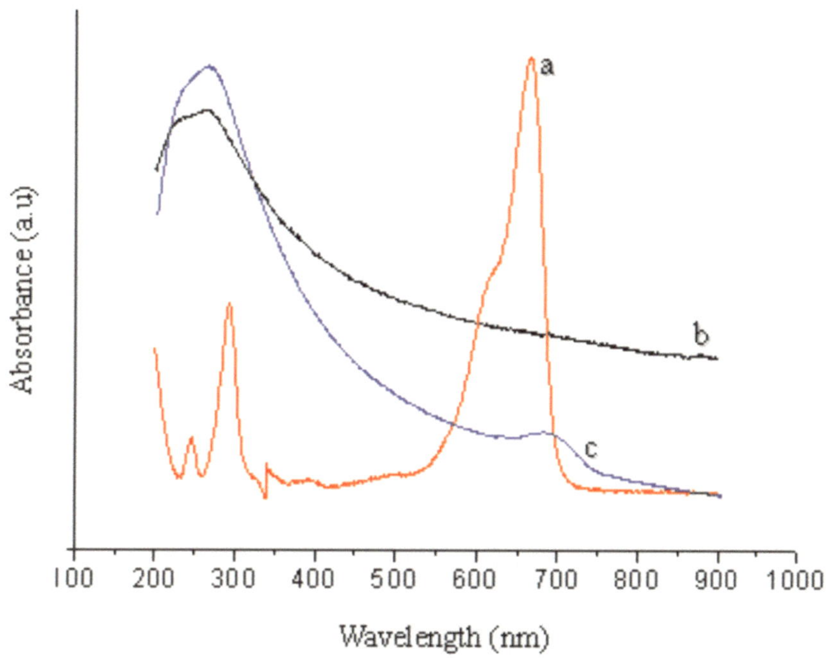

Fig. 5. Absorption spectrum of aqueous dispersion of MB (a), aqueous dispersion of CNTs (b), and aqueous dispersion of MB-CNTs (c).

Figure 6 presented the morphology of CNTs (Figure 6 A, B) and MB-CNTs (Figure 6 C, D) powder samples on a mica sheet, obtained by an atomic force microscope (AFM). The powder samples showed the existence of CNTs in a network-like structure with an unorderly arrangement. The adsorption of MB on CNTs seemed to break the intertwined network of CNTs into individual carbon nanotubes. In addition, some breaking and shortening of CNTs into smaller ones were obvious. The average diameter of individual strand of CNTs was about 38.05 nm. The average diameter of individual strand of MB-CNTs was about 62.03 nm, which was nearly twice that of CNTs indicating clearly that the individual strand of CNTs was entirely encapsulated or wrapped by the MB to form a tubular composite. Moreover, the MB-CNTs sample could be readily dispersed in water to give a black-colored solution, but the CNTs couldn't be easily dissolved or dispersed in aqueous solution. This observation again indicated an adsorption of MB on CNTs.

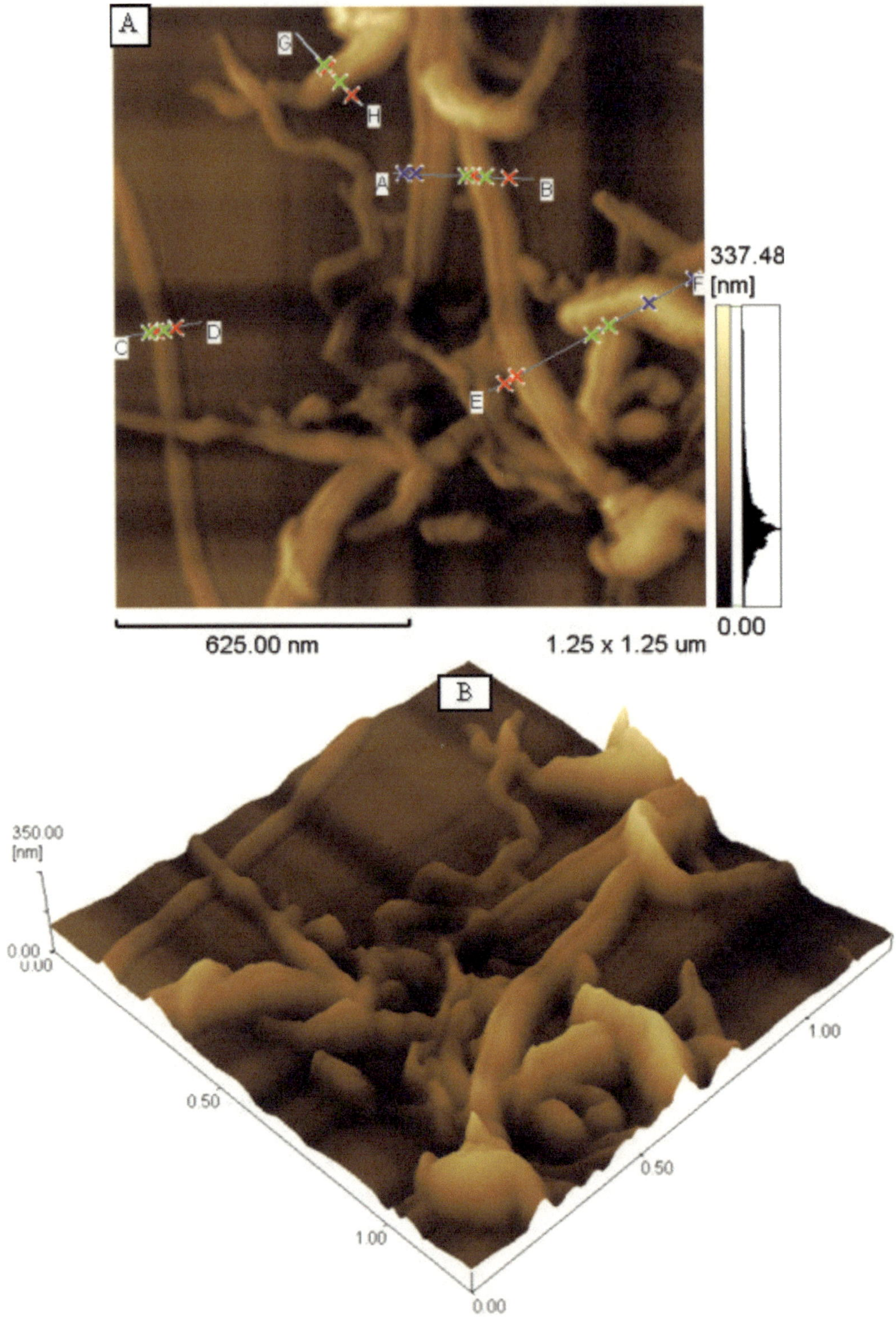
A
G
H
A
B
C
D
E
F
337.48
[nm]
0.00
625.00 nm
1.25 x 1.25 um
B
350.00
[nm]
0.00
0.00
0.50
1.00
0.00
0.50
1.00

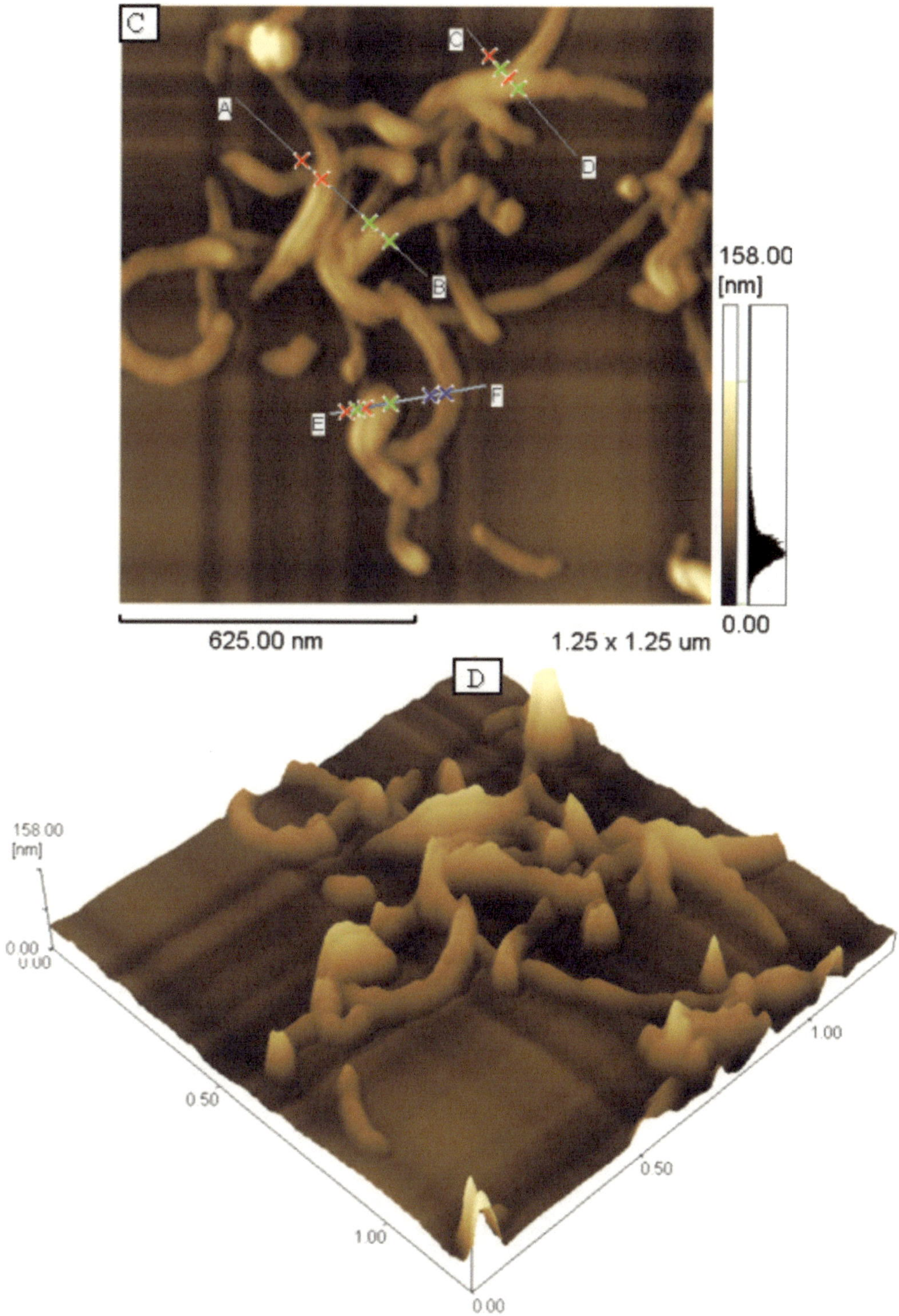

Fig. 6. AFM image of CNT (A, B) and MB-CNTs (C, D) power samples on a mica sheet.

Figure 7 demonstrated cyclic voltammograms of the CNTs/GCE (curve a) and MB-CNTs/GCE (curve b) immersed into 0.1 mol/L pure phosphate buffer (pH 3.0) containing no MB. There was no any redox wave of CNTs/GCE in the potential range of -400–600 mV. In contrast to CNTs/GCE, a pair of redox waves of the MB-CNTs/GCE was observed at 95 and 27 mV, suggesting that electroactive MB can adsorb onto the CNTs and form an electrochemically functional nanostructure.

Figure 8 showed the results of the electrochemical impedance spectra (EIS) of a bare GCE, CNTs/GCE and MB-CNTs/GC electrode in a solution containing 0.1 mol/L KCl and 1.0 mmol/L $Fe(CN)_6{}^{3-/4-}$. The Nyquist plot exhibited a poor small semicircle portion at higher frequencies, and a line at lower frequencies for the bare GCE (Figure 8A), which was similar to the characteristic of a diffusion limiting step of the electrochemical process. However, at CNTs/GCE, the Nyquist plots displayed a clear bigger semicircular feature in addition to the linear feature (Figure 8B-a). The semicircle portion, as observed at higher frequencies, was associated with a process that was limited by electron transfer. The linear features observed at lower frequencies were attributed to diffusion-limited electron transfer. Adsorption of MB on the CNTs to fabricate MB-CNTs/GCE caused a remarkably decrease of the semicircle portion at higher frequencies to form a nearly-straight line (Figure 8B-b). This may be demonstrated that MB-CNTs film introduced an advantage to the interfacial electron transfer. Because of the positive charge of MB in the film, the $[Fe(CN)_6]^{4-/3-}$ probe could arrive to the surface of the electrode promptly. The impedance changes of the different modified electrodes revealed that MB had adsorbed on the CNTs.

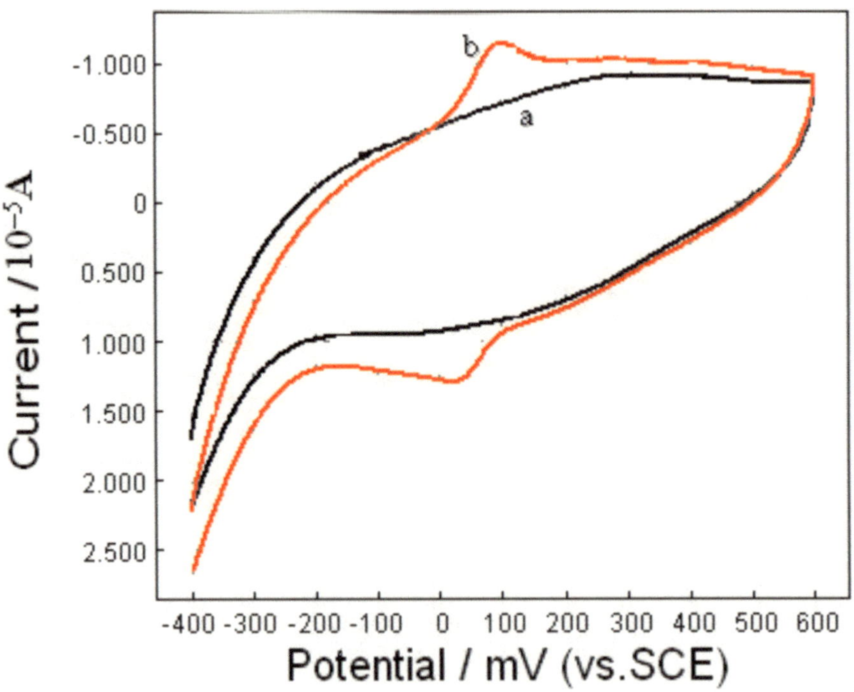

Fig. 7. Cyclic voltammograms of the CNTs/GCE (a) and MB-CNTs/GCE (b) in 0.1 mol/L phosphate buffer (pH 3.0); scan rate 80 mV/s

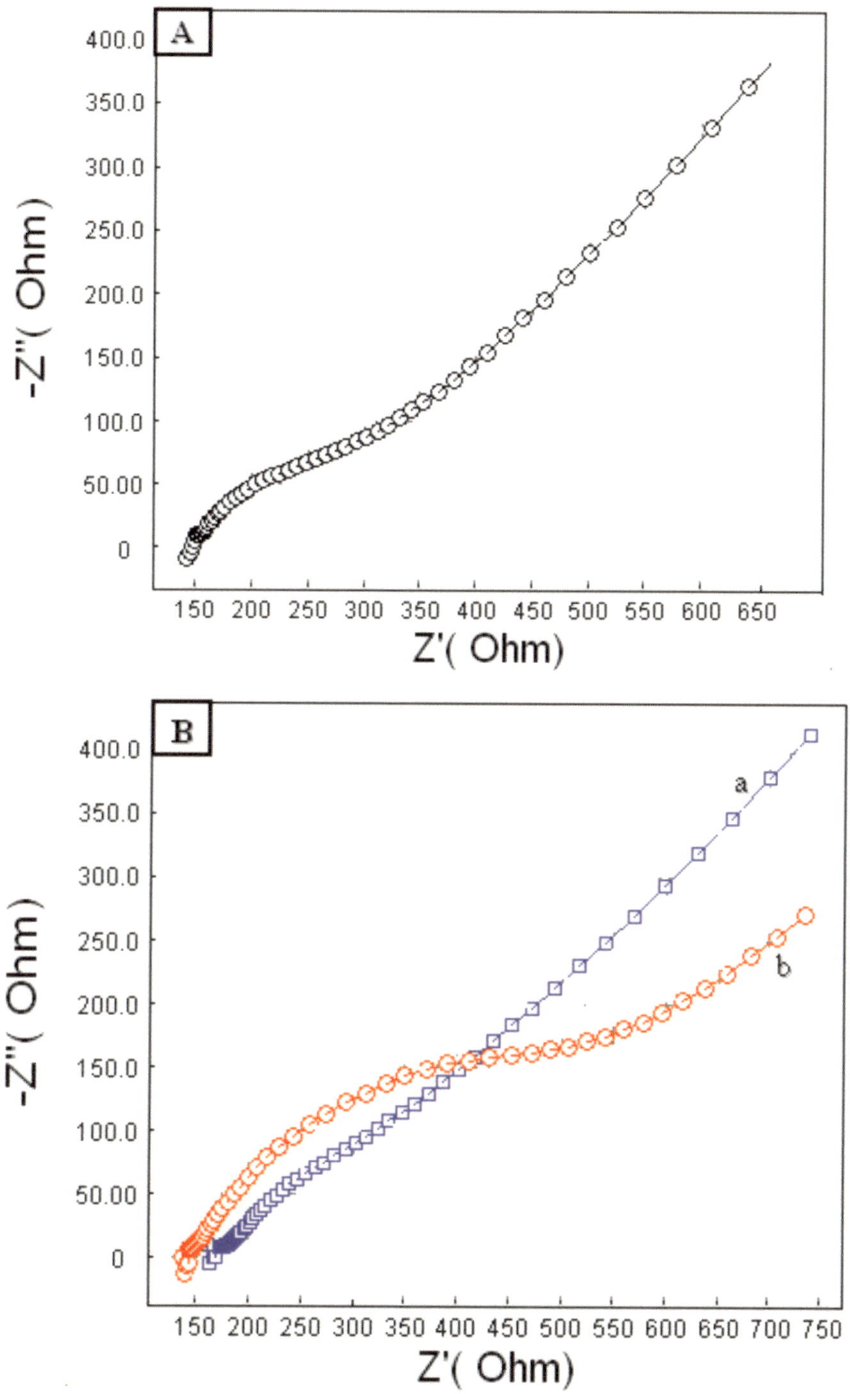

Fig. 8. (A) Nyquist diagram of EIS at bare GCE; (B) Nyquist diagrams of EIS at MB-CNTs/GCE (a) and CNTs/GCE (b). EIS condition: frequency range: 100 kHz–1 Hz; signal amplitude: 5 mV; solution: 1.0 mmol/L $Fe(CN)_6{}^{3-/4-}$ in 0.1 mol/L KCl.

The modified electrode showed excellent electrocatalytic activity toward dopamine (DA) and uric acid (UA) in 0.1 mol/L phosphate solution medium (pH 3.0). Figure. 9A demonstrated the cyclic voltammetry (CV) curves of a mixture of 3.0×10^{-4} mol/L DA and 6.0×10^{-4} mol/L UA (a), 3.0×10^{-4} mol/L DA (b), 6.0×10^{-4} mol/L UA (c), and buffer solution without DA and UA (d) in 0.1 mol/L phosphate solution (pH 3.0) at MB-CNTs/GCE, respectively. Figure. 9A-a showed two anodic peaks at around the potential of 429 and 603 mV, which attributed to the oxidation of DA and UA with a 174 mV separation of both peaks, which was broad enough for their simultaneous electrochemical determination of DA and UA. Figure 9B revealed CV responses of a mixture of 3.0×10^{-4} mol/L DA and 6.0×10^{-4} mol/L UA in 0.1 mol/L phosphate solution (pH 3.0) at MB-CNTs/GCE (a), CNTs/GCE (b), bare GCE (c) and MB/GCE (d). Under the same conditions, poor anodic peaks of DA and UA were observed at the bare GCE. At the MB/GCE, DA and UA exhibited an overlapped and broad anodic peak extended over a potential region of 450–630 mV with the mixed potential at 530 mV. CV for CNTs/GCE showed two anodic peaks with a separation of about 144 mV towards DA and UA. Comparing MB-CNTs/GCE with CNTs/GCE, a remarkable increase in redox peak currents of DA was observed with the anodic potential shift negatively (more than 30 mV), which revealed that MB in the matrix of CNTs/GCE can act as an efficient electron mediator for the electrocatalytic oxidation of DA，besides CNTs can enhance the electron-transfer rate and make more DA participate in the electrochemical reaction due to their accumulation and catalytic ability. The anodic (E_{pa}) and cathodic peak potentials (E_{pc}) of DA were at about 429 mV and 336 mV (vs. SCE), respectively, and the peak currents ratio of i_{pa}/i_{pc} was about 1.0, which showed that the electrode reaction of DA was almost reversible.

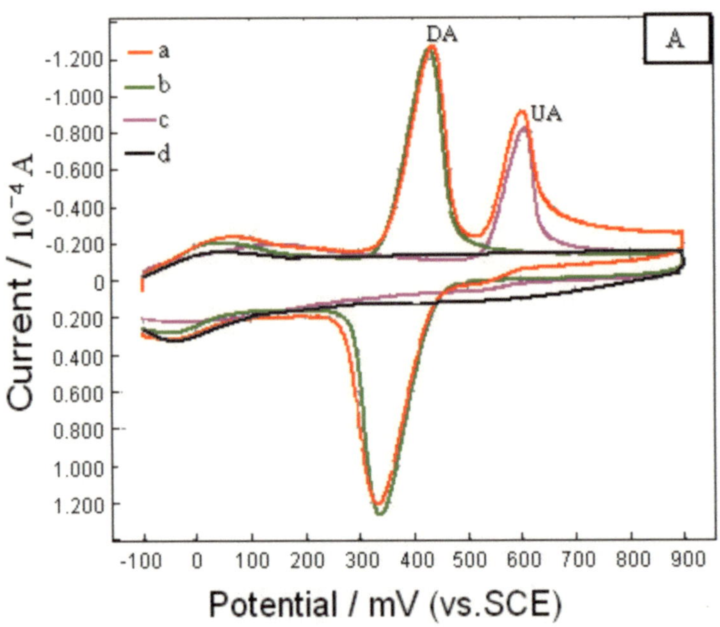

Fig. 9. (A) CV curves of the mixture containing 3.0×10^{-4} mol/L DA and 6.0×10^{-4} mol/L UA (a), 3×10^{-4} mol/L DA (b), 6×10^{-4} mol/L UA (c), and buffer solution without DA and UA (d) in 0.1 mol/L phosphate solution (pH 3.0) at the MB-CNTs/GCE.

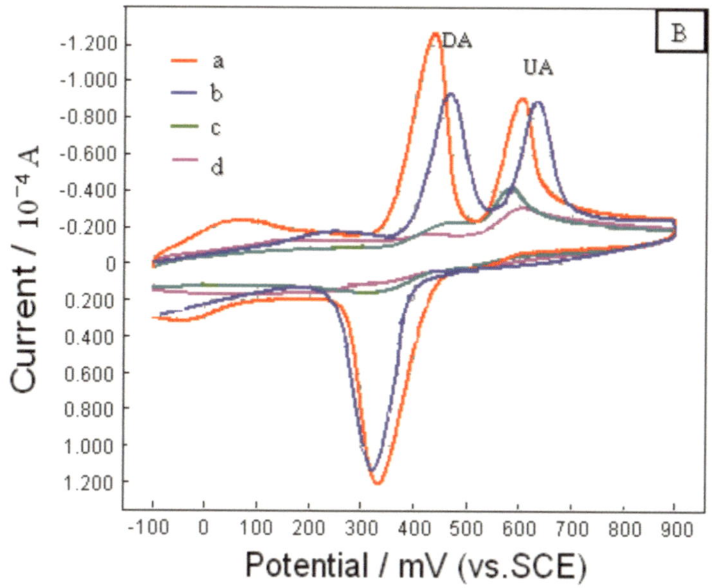

Fig. 9. (B) CV curves of the mixture containing 3.0×10^{-4} mol/L DA and 6.0×10^{-4} mol/L UA in 0.1 mol/L phosphate solution (pH 3.0) at different modified electrodes: MB-CNTs/GCE (a), CNTs/GCE (b), bare GCE (c), and MB/GCE (d); scan rate 80 mV/s.

Generally, as the electroactive substance, ascorbic acid (AA) always coexists with DA and UA. The oxidation peak potential of AA is very close to that of DA and UA, which results in poor selectivity determination of DA or UA in real samples on conventional electrodes. Therefore, it is essential to exploit more sensitive, selective and simple methods for the segregative determination of AA, DA and UA. Figure 10 demonstrated the differential pulse anodic stripping voltammetry (DPASV) curves of the mixture of 3.0×10^{-4} mol/L DA and 6.0×10^{-4} mol/L UA (a), 3.0×10^{-3} mol/L AA, 3.0×10^{-4} mol/L DA and 6.0×10^{-4} mol/L UA (b), 1.0×10^{-3} mol/L AA, 3.0×10^{-4} mol/L DA and 6.0×10^{-4} mol/L UA (c), 1.0×10^{-3} mol/L AA (d) and 0.1 mol/L phosphate buffer solution (pH 3.0) (e) in 0.1 mol/L phosphate buffer solution (pH 3.0) on MB-CNTs/GCE. Figure 10-d showed an anodic peak at around 200 mV, which was attributed to the oxidation of AA, the separation potential of AA and DA was about 219 mV, indicating broad enough separation to eliminate the interference of AA and realize the simultaneous electrochemical determinations of DA and UA in the mixed solution.

Under the optimized conditions, In the presence of 1.0 mmol/L AA and 10.0 μmol/L UA, the anodic peak current was linear to the concentration of DA in the range of 0.4−10.0 μmol/L with a detection limit of 0.2 μmol/L DA. The anodic peak current of UA was linear to the concentration in the range of 2.0−20.0 μmol/L and 20.0−200.0 μmol/L with a lowest detection limit of 1.0 μmol/L in the presence of 1.0 mmol/L AA and 1.0 μmol/L DA.

The repeatability and stability are the vital characteristics for the modified electrode, which should be investigated for analytical determination. The same MB-CNTs/GCE was used for five times successive measurement, and the relative standard deviation (RSD) of the peak current was 3.5 % for 3.0×10^{-4} mol/L DA and 4.0 % for 6.0×10^{-4} mol/L UA. In addition, five freshly prepared MB-CNTs/GCE were used to measure 3.0×10^{-4} mol/L DA in the same condition. All five electrodes exhibited similar current responses and a relative standard deviation of 5.6% was obtained. These findings revealed that the electrochemical behavior of the MB-CNTs/GCE was highly repeatable. The stability of the MB-CNTs/GCE was studied by determining the steady-state response current of 3.0×10^{-4} mol/L DA every day after preparation, the RSD of steady-state response current was 6.5%. When not in use, the sensor was stored in 0.1 mol/L PBS buffer (pH 7.0) at 4 °C. The results showed that the steady-state response current only decreased by 12% after 7 days, indicating that the MB-CNTs/GCE electrode was considerably stable.

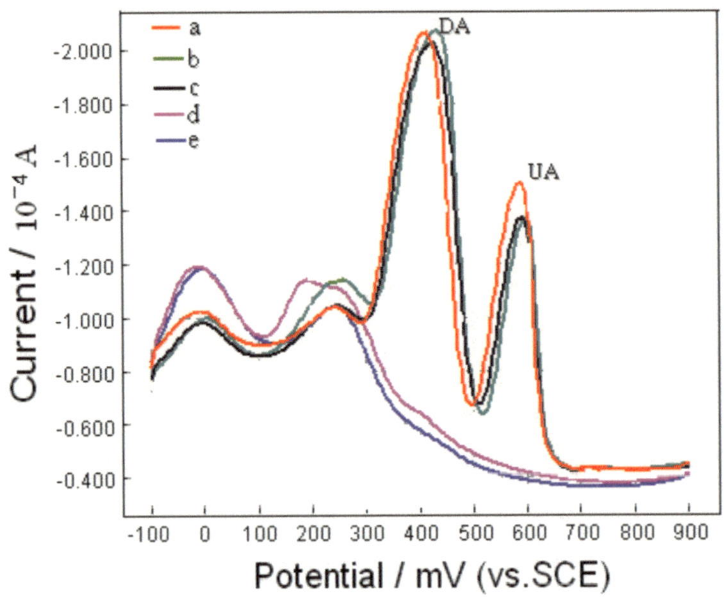

Fig. 10. Differential pulse anodic stripping voltammograms of the mixed solution of 3.0×10^{-4} mol/L DA and 6.0×10^{-4} mol/L UA (a), 3.0×10^{-3} mol/L AA, 3.0×10^{-4} mol/L DA and 6.0×10^{-4} mol/L UA (b), 1.0×10^{-3} mol/L AA, 3.0×10^{-4} mol/L DA and 6.0×10^{-4} mol/L UA (c), 1.0×10^{-3} mol/L AA (d) and 0.1 mol/L phosphate buffer solution (pH 3.0) (e) on the MB-CNTs/GCE; instrument parameters: accumulation time: 40 s, accumulation potential: 100 mV, pulse amplitude: 50 mV, pulse increment: 4 mV, pulse width: 40 ms and pulse period: 120 ms.

The MB-CNTs/GCE electrode was used in real samples analysis, such as Dopamine Hydrochloride Injection and urine. The detected results were shown that the total value of DA in Dopamine Hydrochloride Injection was 10.48 mg/mL, which was in agreement with the declared content (i.e., 10.0 mg/mL). The total UA concentrations detected in urine sample was 4.54×10^{-3} mol/L, consistent with the containing level of a healthy human.

5. Gold nanoparticles/ethylenediamine/carbon nanotubes-modified glassy carbon electrode as the voltammetric sensor for selective determination of rutin in the presence of ascorbic acid

Rutin (3′, 4′, 5, 7-tetrahydroxyflarone-3β-d-rutinoside), as a kind of the most abundant bioactive flavonoid called as vitamin p, is widely present in multivitamin preparations and more than 70 herbal remedies. As a natural flavone derivative, rutin has a wide range of

physiological activities including anti-inflammatory, hemostat, antibacterial anti-tumor and anti-oxidant [33–35]. It is always used clinically as the therapeutical medicine [36, 37]. For example, It can be applied to the treatment of diseases, such as capillary bleeding by diluting the blood, reducing capillary permeability and lower blood pressure [38]. Hence, it is necessary to develop simple, sensitive, economical and efficient techniques for the determination of rutin. Recently, we reported a preparation and application of AuNPs/en/CNTs composite film-modified glassy carbon electrode (AuNPs/en/CNTs/GCE) [39].
SEM can effectively prove the surface morphologies of the modified electrode. The morphology of the CNTs film (Figure 11-a) showed a network-like structure. Compared with the CNTs film, the SEM image of AuNPs/en/CNTs films (Figure 11-b) displayed many observable light dots which were due to the assembly of AuNPs. SEM image (Figure 11-b) confirmed that gold nanoparticles were typically bound on CNTs with fairly even distribution, although a few aggregates were observed.

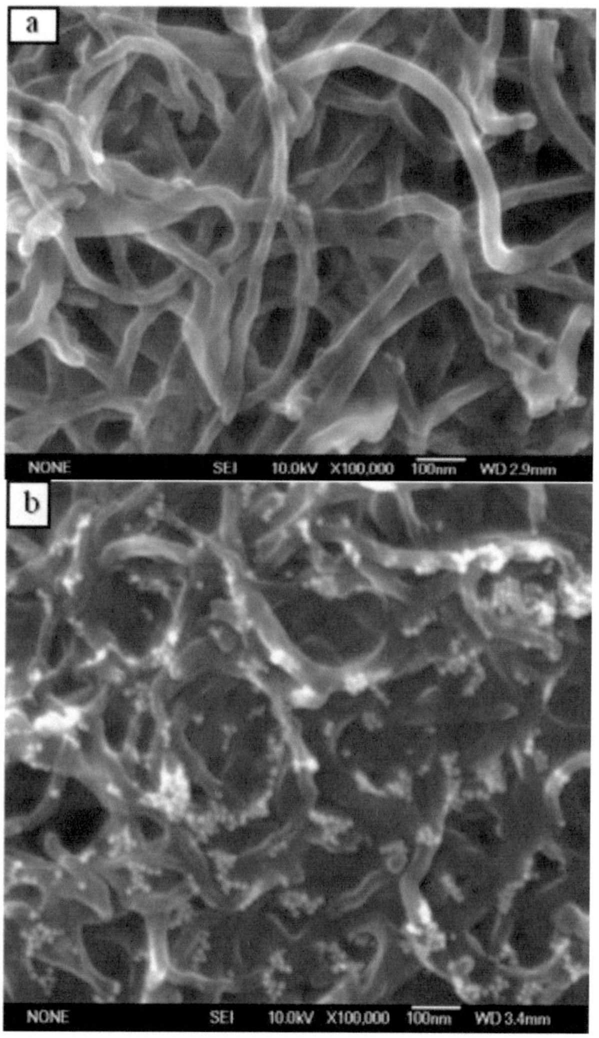

Fig. 11. SEM images of CNTs film (a), and AuNPs/en/CNTs films (b)

Experiments revealed that the redox peak currents of rutin could be remarkably enhanced on AuNPs/en/CNTs/GCE, meaning good electrocatalytic activity for the oxidation of rutin. Figure 12 showed the cyclic voltammograms of rutin on different modified electrodes. Rutin didn't display any redox peaks at the bare GCE (a), which demonstrated the weaker adsorption and slower electrochemical reaction of rutin on the GCE surface. However, there were well-defined redox peaks on the en/CNTs/GCE (b), CNTs/GCE (c) and AuNPs/en/CNTs/GCE (d) in 0.1 mol/L phosphate buffer solution (pH 3.5). But the heights of the redox peaks were clearly higher in the case of AuNPs/en/CNTs/GCE than that of the redox peaks on the en/CNTs/GC or CNTs/GCE. The anodic (E_{pa}) and cathodic peak potentials (E_{pc}) were at about 487 mV and 432 mV (vs. SCE), respectively, and the ratio of i_{pa}/i_{pc} was about 1.2, which showed that the electrode reaction was almost reversible. Nano-gold and CNTs can enhance the electron-transfer rate and make more rutin participate in the electrochemical reaction due to their accumulation and catalytic ability.

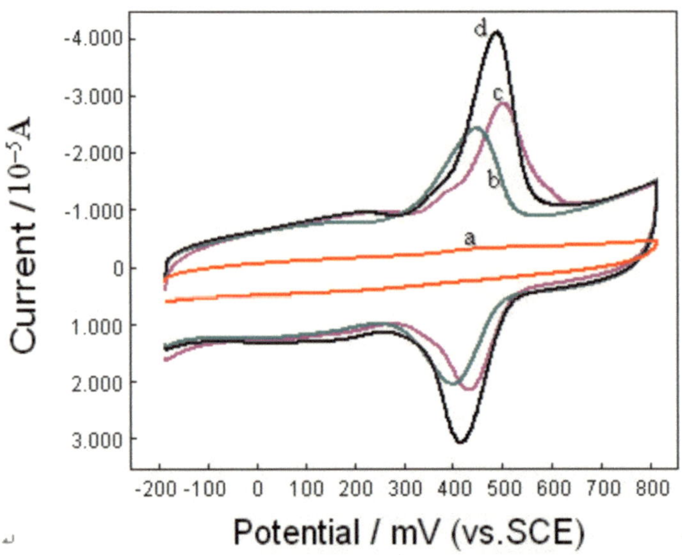

Fig. 12. Cyclic voltammograms of 8.0×10^{-5} mol/L rutin in 0.1 mol/L phosphate buffer (pH 3.5) on the different electrodes: the bare GCE (a), en/CNTs /GCE (b), CNTs/GCE (c) and AuNPs/en/CNTs/GCE (d); scan rate 100 mV/s.

Generally, as the electroactive substance, ascorbic acid (AA) always coexists in the Compound Rutin Tablets. The oxidation peak potential of AA is very close to that of rutin, which results in poor selectivity determination of AA or rutin in real samples on conventional electrodes. Therefore, it is essential to exploit more sensitive, selective and simple methods for the segregative determination of AA and rutin. Figure 13 demonstrated the cyclic voltammetry (CV) curves of 1.0×10^{-4} mol/L AA (a), the mixture of 2.0×10^{-5} mol/L rutin and 1.0×10^{-4} mol/L AA (b), 2.0×10^{-5} mol/L rutin (c) and without AA and rutin (d) in 0.1 mol/L phosphate buffer solution (pH 3.5) on AuNPs/en/CNTs/GCE. Figure 13 b showed two anodic peaks at around 186 mV and 487 mV, which were attributed to the oxidation of AA and rutin with a 301 mV separation of both peaks, indicating broad enough separation for the simultaneous electrochemical determinations of rutin and AA in the mixed solution.

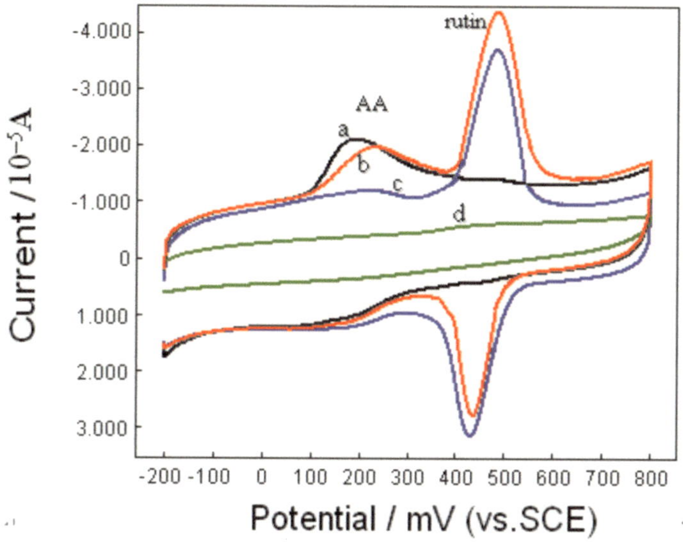

Fig. 13. Cyclic voltammograms of 1.0×10^{-4} mol/L AA (a), the mixed solution of 1.0×10^{-4} mol/L AA and 2.0×10^{-5} mol/L rutin (b), 2.0×10^{-5} mol/L rutin (c) and 0.1 mol/L phosphate buffer solution (pH 3.5) (d) on the AuNPs/en/CNTs/GCE; scan rate 100 mV/s.

Under optimized conditions, the anodic peak current was linear to the rutin concentration in the range of 4.8×10^{-8} mol/L − 9.6×10^{-7} mol/L. The regression equation was: $i_{pa} = 2.3728\ C_{rutin} - 0.1782$ (i_{pa}: 10^{-5}A, C_{rutin}: μmol/L, $r = 0.9973$). The detection limit of 3.2×10^{-8} mol/L was obtained. Regeneration and reproducibility are two important characteristics for the modified electrode, which should be investigated. The same modified GCE was used for six times successive measurements of 2.0×10^{-5} mol/L rutin. After each measurement, the surface of the AuNPs/en/CNTs/GCE was regenerated by successively cycling between -200 mV and 800 mV (vs. SCE) in 0.1 mol/L phosphate buffer solution (pH3.5) for six cycles. The relative standard deviation (RSD) of the anodic peak current was 4.3%, which suggested good regeneration and reproducibility of the modified electrode.

This method was used for the determination of rutin in the Compound Rutin Tablets and Rutin Tablets. The contents of rutin in the Compound Rutin Tablets and Rutin Tablets were calculated to be 20.0 ± 0.87 mg and 19.4 ± 0.54 mg per tablet, respectively (the declared content of rutin was 20 mg per tablet). In order to test the accuracy of the proposed method, the conventional method of HPLC was employed to determine the contents of rutin in the Compound Rutin Tablets and Rutin Tablets (the contents of rutin in the Compound Rutin Tablets and Rutin Tablets were 18.9 ± 0.47 mg and 18.1 ± 0.57 mg per tablet, respectively). The quantitative results obtained by HPLC were in agreement with the data determined by the proposed electrochemical method, indicating that the method was selective and suitable for rutin determination in real samples.

6. Preparation of yttrium hexacyanoferrate/carbon nanotube/Nafion nanocomposite film-modified electrode: Application to the electrocatalytic oxidation of L-cysteine

L-cysteine (L-CySH), a sulfur-containing molecule, is one of the most important amino acids. It is widely present in many medicines, food and biological tissues, such as cysteine protease, vasopressin and anti-diuretic hormone [40]. It plays a significant role in biological systems, playing a role in folding and defolding mechanisms [41]. An inadequate dietary intake of L-CySH may cause number of clinical problems, for instance, liver damage, skin lesions, and slowed growth [42]. Therefore, the sensitive detection of L-CySH is clinically significant. Nevertheless, some traditional methods, such as the chromatographic methods, are time-consuming, expensive, and require complicated preconcentration, multisolvent extraction and trained technicians. In contrast, electrochemical methods are characterized by their simplicity, high sensitivity, good stability, low-cost instrumentation, small scale and on-site monitoring [43]. However, although L-CySH is an electroactive compound, there are some drawbacks in electroanalysis, for example, large overpotential, low sensitivity and sluggish electron-transfer kinetics at conventional electrodes. These obstacles can result in oxide formation and fouling of the electrode surface [44]. To overcome these obstacles, we have successfully prepared YHCFNP, and the electrocatalytic oxidation and amperometric determination of L-CySH were accomplished at a modified electrode compose of a mixture of YHCFNP and CNTs [45].

As can be seen in figure 14, when YHCFNP and carbon nanotube were mixed together, they were uniformly dispersed. In this case, CNTs can be used as the carrier and entanglement (diameter: 10 – 20 nm, length: 1–2 μm) to fix YHCFNP on the electrode surface.

Fig. 14. Typical SEM images of the mixed YHCFNP and CNTs

Figure 15 shows the cyclic voltammograms of different electrodes with and without 0.5 mmol/L L-CySH in 0.1 mol/L PBS (pH 6.82). As figure 15A reveals, at the YHCFNP/CNTs/Nafion-modified GCE, there is a pair of well-defined redox peaks at 152 mV and 250 mV at a scan rate of 20 mV/s without L-cysteine. Under the same conditions, there was no redox peak at the CNTs/Nafion-modified GCE and bare GCE.

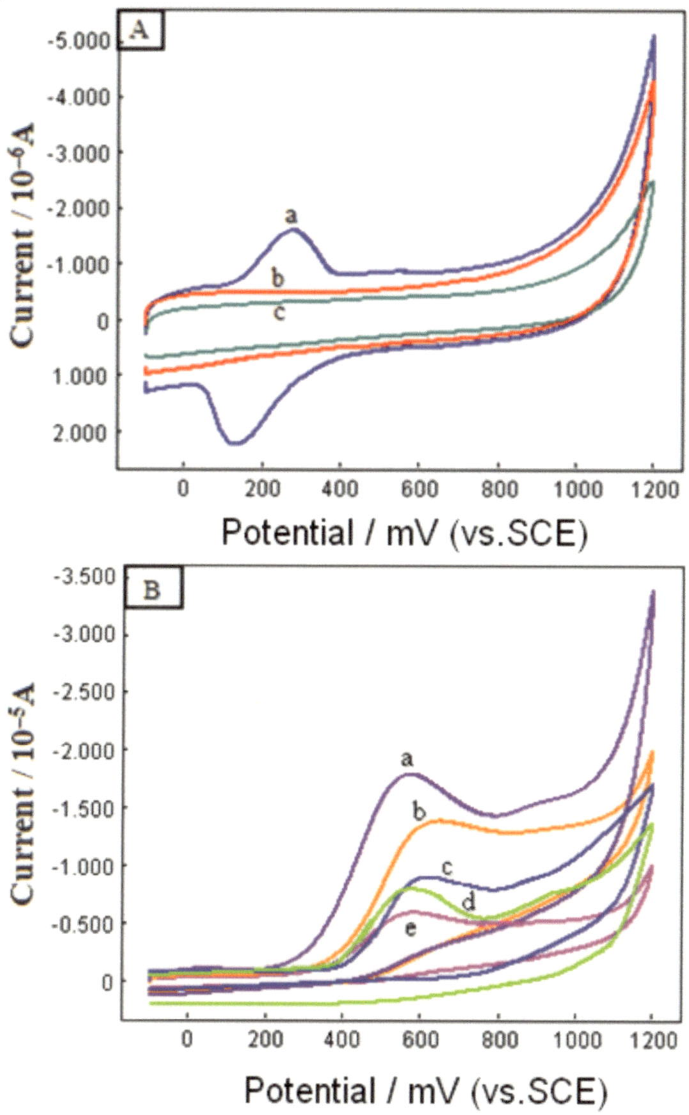

Fig. 15. A. Cyclic voltammograms of different electrodes in 0.1 mol/L PBS (pH 6.82) without L-cysteine: YHCFNP/CNTs/Nafion-modified GCE (a), CNTs/Nafion-modified GCE (b), GCE (c) ; B. Cyclic voltammograms of different electrodes in mol/L PBS (pH 6.82) with 0.5 m mol/L L-cysteine: YHCFNP/CNTs/Nafion-modified GCE (a), CNTs/Nafion-modified GCE (b), CNTs-modified GCE (c), YHCFNP-modified GCE (d), GCE (e). Scan rate 20 mV/s.

Figure 15 B depicts the cyclic voltammograms of different electrodes with the addition of 0.5 m mol/L L-CySH in 0.1 mol/L PBS (pH 6.82). L-cysteine produced a weak anodic peak at the bare GCE (c), which demonstrates the slower electrochemical reaction of L-cysteine at the GCE surface. This phenomenon may be related to electrode fouling caused by the deposition of this compound and its oxidation products on the electrode surface. At the CNTs-modified GCE (c) , there was a well-defined anodic peak , and the peak current was clearly higher than that at the bare GCE. Indeed, in a number of cases, the L-cysteine molecules under study can interact with carbon nanotube in a way that a well-polished "traditional" carbon electrode cannot. This phenomenon is evidence of the catalytic effect of CNTs toward L-cysteine oxidation. In most cases, the electrocatalytic activity of carbon nanotube is attributed to edge plane like-sites/defects, which may occur at the ends, and along the tube axis [46]. Recently it has been revealed that additionally, metallic impurities remaining from the fabrication processes can be the origin for certain analytes [47–49]. Because of the excellent evenly dispersing capacity of Nafion for CNTs, a stronger anodic peak of L-CySH was shown at the CNTs/Nafion-modified GCE (b). At the YHCFNP-modified GCE (d), a well-defined anodic peak appeared. The peak current was obviously higher than that at the bare GCE, which demonstrated that the YHCFNP-modified GCE can improve the electrochemical reaction of L-cysteine on the electrode surface. In addition, a well-defined anodic peak appeared at the YHCFNP/CNTs/Nafion-modified GCE (a), and the height of the anodic peak was clearly higher than that at the CNTs/Nafion-modified GCE. The anodic potential (E_{pa}) was about 570 mV. In contrast, the anodic potential at the CNTs/Nafion-modified GCE was 642 mV, indicating a negative shift of about 72 mV. The experimental results indicate that the electrooxidation of L-CySH is remarkably improved by the YHCFNP/CNTs/Nafion-modified GCE, which may result from the high dispersion of YHCFNP/CNTs nanocomposite with high surface area and good electronic properties.

Under the optimum experimental conditions, the electrochemical response to L-cysteine at the YHCFNP/CNTs/Nafion-modified GCE was fast (within 4 s). Linear calibration plots were obtained over the range of 0.20–11.4 μmol/L with a low detection limit of 0.16 μmol/L. The YHCFNP/CNTs/Nafion-modified GCE exhibited several advantages, such as high stability and good resistance against interference by ascorbic acid and other oxidizable amino acids.

7. Concluding remarks

Nowadays, the study of CNTs modified electrode is the forefront subject that offers enormous possibilities in pharmaceutical analysis. The facility of the preparation and functionalization of CNTs makes them very useful for the development of modified electrodes with specific detection of medicine molecules, as well as for the fabrication of third generation sensors (no mediator is needed). As far as the modified electrode is concerned, the CNTs coated and polymer embedded electrodes are more widely used than the CNTs paste electrode and CNTs intercalated electrode in the determination of pharmaceutical analysis owing to the different dispersants. Thus, to explore and use the materials as dispersants with good and green friendly properties is an important field. Moreover, CNTs intermingled with other nonmaterials (such as nano Au particles, nano YHCF, *etc*) were modified on conventional electrodes. These nanocomposites provide a synergic effect which results in the improvement in the response property of modified electrodes. So, CNTs hybrid with other nanomaterials used as modified material was also an important part. In addition, further research on the mechanism of the electrochemical reaction between medicine and modified electrode is a very important aspect in relation to CNTs modified electrode. Although CNTs modified electrode used in pharmaceutical analysis is receiving increasing interest for sensor construction in recent years, the study should involve the combination of nanomaterials, analysis and life sciences.

8. References

[1] S. Iljima, Nature. 354 (1991) 56.
[2] Y.R. Wang, P. Hu, Q.L. Liang, G.A. Luo, Y.M. Wang, Chin. J. Anal. Chem. 36 (2008) 1011.
[3] D. M. Graham, Nutr. Rev. 1978, 36, 97.
[4] A. Turnball, J. Br. Nutr. Found. Nutr. Bull. 1981, 6, 153.
[5] S.L. Yang, R. Yang, G. Li, L.B. Qu, J.J. Li, L.L Yu, J. Electroanal. Chem. 639 (2010) 77.
[6] S.L. Yang, R. Yang, G. Li, J.J. Li, L.B. Qu, J. Chem. Sc. 122 (2010) 919.
[7] Y.C. Tsai, J.M. Chen , S.C. Li , F. Marken, Electrochem. Commun. 6 (2004) 917.
[8] D. Sun, Z.M. Sun, J. Appl. Electrochem. 38 (2008)1223.
[9] Y.C. Tsai, J.M. Chen, F. Marken, Microchim. Acta. 150 (2005) 269.
[10] K.B. Wu, S.S. Hu, Microchim. Acta. 144 (2004) 131.
[11] W.S. Huang, C.H. Yang, S.H. Zhang, Anal. Bioanal. Chem. 375 (2003) 703.
[12] R.X. Guo, Q.Xu, D.Y. Wang, X.Y. Hu, Microchim. Acta. 161 (2008) 265.
[13] A. Belay, K. Ture, M. Redi-Abshiro, A. Asfaw, Food. Chem. 108 (2008) 310.
[14] American Beverage Association, Nutrition&Health, Ingredients, Caffeine, http://minisites.ameribev.org/products/pdf/caffeine-levels.pdf, last accessed October, 2009.
[15] B.A. Fox, A.G. Cameron, Food Science, Nutrition and Health, 15th Ed. Edward Arnold, London. 1989.
[16] Combs GF, The Vitamins: Fundamental Aspects in Nutrition and Health, 2nd ed., Academic Press, San Diego. 1992.
[17] O.Arrigori, C.D. Tullio, Biochim. Biophys. Acta. 1569 (2002) 1
[18] J.M. Zen, J.J. Jou, G. Ilangovan, Analyst. 123 (1998) 1345
[19]G. Li, S.L. Yang, L.B. Qu, R. Yang, J.J. Li, J. Solid. State. Electrochem. 15 (2011) 161.
[20] S.H. Kim, S.A. Kim, M.K. Park, Int. Immunopharmacol. 4 (2004) 279.
[21] X.A. Wu, H.L. Chen, X.G. Chen, Biomed. Chromatogr 17 (2003) 504.
[22] P. Drasar, J. Moravcova, J. Chromatogr. A. 812 (2004) 3
[23] T.Y.K. Chan, Drug Saf. 17 (1997) 209.
[24] S.L. Yang, L.B. Qu, R. Yang, J.J. Li, L.L. Yu, J. Appl. Electrochem. 40 (2010) 1371.
[25] Y. Wang, J. Wu, D. Li, Chin. J. Anal. Chem. 34 (2006) 1331.
[26] Y.L. Chang, B. Liu, Zhongguo Zhong Yao Za Zhi. 31 (2006) 653.
[27] K. Yu, Y.W. Wang, Y.Y. Cheng, J. Pharm. Biomed. Anal. 40 (2006) 1257.
[28] R.M. Wightman, L.J. May, A.C. Michael, Anal. Chem. 60 (1988) 769A.
[29] A. Liu, I. Honma, H.S. Zhou, Biosens. Bioelectron. 21 (2005) 809.
[30] J.W. Mo, B. Ogorevc, Anal. Chem. 73 (2001) 1196.
[31] S.L. Yang, G. Li, R. Yang, M.M. Xia, L.B. Qu, J. Solid. State. Electrochem. DOI 10.1007/s10008-010-1210-x.
[32] T. Sagara, K. Niki, Langmuir. 9 (1993) 831.
[33] R.M. Gene, C. Cartana, T. Adzet, E. Marin, T. Panella, S. Canigueral, Planta. Med. 62 (1996) 232.
[34] A. Hasan, I. Ahmad, Fetoterapia. 67 (1996) 182.
[35] R. Ramanathan, W.P. Das, C.H. Tan, Int.. J. Oncol. 3 (1993) 115.
[36] W.Q. Sun, J.F. Sheng, Handbook of Natural Active Constituents, Chinese Medicinal Science and Technology Press, Beijing, 1998, 2240

[37] J.E.F. Reynolds, Martindale. The Extra Pharmacopoeia, 31 ed., The Royal Pharmaceutical Society, Council of the Royal Pharmaceutical Society of Great Britain, London, 1996, 1679.
[38] J. van der Geer, J.A.J. Hanraads, R.A. Lupton, J. Sci. Commun. 163 (2000) 51.
[39] S.L. Yang, L.B. Qu, G. Li, R. Yang, C.C. Liu, J. Electroanal. Chem. 645 (2010) 115.
[40] Dorland's Illustrated Medical Dictionary, 27th. d., E.J. Taylor, d.;W.B. Saunders: Philadelpia, 1988.
[41] D.Voet, J. G. Voet, Biochemistry, 2nd ed.; John Wiley & Sons: New York, 1995, 1263.
[42] W. Dro¨ge, H. P. Eck, S. Mihm, Immunol. Today. 13 (1992) 211.
[43] O.A. Sadik, W.H. Land, J. Wang, Electroanalysis. 15 (2003) 1149.
[44] Z.F. Chen, H.Z. Zheng, C. Lu, Y.B. Zu, Langmuir. 23 (2007) 10816.
[45] L.B. Qu, S.L. Yang, G. Li, R. Yang, J.J Li, L.L Yu, Electrochim. Acta. 56 (2011) 2934
[46] C.E. Banks, T.J. Davies, G.G. Wildgoose, R.G. Compton, Chem. Commun. 7 (2005) 829.
[47] J. Kruusma, N. Mould, K. Jurkschat, A. Crossley, C.E. Banks, Electrochem. Commun. 9 (2007) 2330.
[48] M. Merisalu, J. Kruusma, C.E. Banks, Electrochem. Commun. 12 (2010) 144.
[49] J. Kruusma, V. Sammelselg, C.E. Banks, Electrochem. Commun. 10 (2008) 1872.

17

Single-Walled Carbon Nanotube Network Gas Sensor

Sunglyul Maeng
Department of Electrical & Electronic Engineering
Woosuk University
Republic of Korea

1. Introduction

Gas sensors are generally defined as devices to detect or measure concentration of (bio-) chemicals in gaseous state. The demands of highly sensitive gas sensors for industry, environmental monitoring, safety, biomedicine, and pharmaceutics have provoked the intensive research interests in micro-electromechanical systems (MEMS) and nanotechnology. The nanotechnology in conjunction with MEMS has created huge potential to build highly sensitive, low cost, compact gas sensors with low power consumption. (Graf et al.,2006; Panchanpakesan et al.,2006; Udrea et al.,2007; Maeng et al., 2008-b)

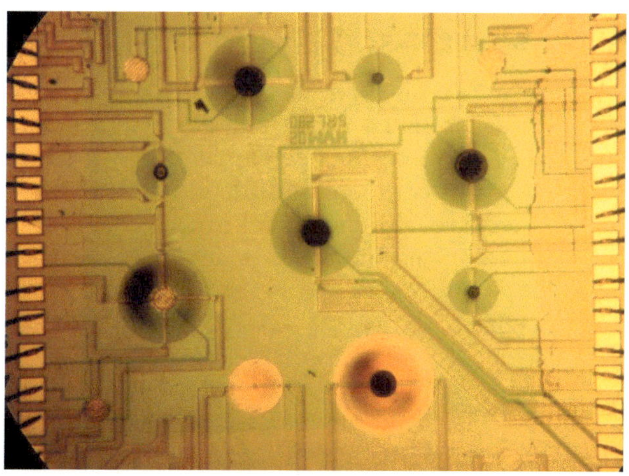

Fig. 1. An example of MEMS sensor system integrated with nanostructured sensing materials. The darkish areas are covered with multi-walled CNTs.

There exist various types of MEMS gas sensors: two terminal resistance type, field-effect transistor (FET) type, capacitance type, and surface acoustic wave (SAW) type, etc. Among them, the two terminal resistance type gas sensors are most widely used and investigated as the manufacturing process is very cost effective. In the manufacturing of two terminal resistance type gas sensors, semiconducting materials such as SnO_2, ZnO and WO_3 are used as sensing elements. Among them, nanocryatalline SnO_2 has been

focused in gas sensor applications as this material was first commercialized for industrial resistance type gas sensors. (Göpel & Schierbaum,1995; Cheng et al.,2004) As the demand for sensors both highly sensitive and very compact increases, however, the problems of this material have been revealed. The conventional semiconducting materials need high temperature operation for enhancement of the sensitivity. If the grain size of the sensing elements becomes very small, the coalescence of the grains is inevitable and this leads to the change of the base-line resistance of the sensors. (Shukla et al.,2003) Another problem is the thickness and uniformity control of the semiconducting sensing materials. The conventional sensors adopt thick layer of sensing materials and the control of the thickness uniformity is not difficult. In the MEMS gas sensors, ultrathin layer of nanostructured sensing materials is needed as this facilitates both the high sensitive detection of gases and the low power operation of the sensors. (Udrea et al., 2009) The thickness uniformity control of extremely thin films is not as easy as the thick film cases. Recent development of metal oxide semiconducting nanowires, nanobelts, and nanosheets overcomes the problem of coalescence of materials by thermal heating and accelerates the progress of MEMS gas sensors. (Park et al., 2010; Moon et al., 2010; Maeng et al., 2011) However, the thickness uniformity control problem still remains.
The use of semiconducting CNTs as gas sensing elements instead of metal oxide nano-materals proposed couple of decades before and have been studied intensively due to their inherent properties such as high strength, high electrical and thermal conductivity and high surface-to-volume ratio. In this chapter, the recent development of two terminal resistance type single-walled carbon nanotube (SWCNT) network gas sensors will be reviewed.

2. Individual SWCNT gas sensor

The early work by Kong et al. showed that the electrical resistance of individual semiconducting SWCNTs dramatically changes when exposed to gaseous molecules such as nitrogen dioxide, ammonia, and oxygen under applied gate voltages. These FET type SWCNT gas sensors exhibited very good sensitivity and fast response time at room temperature. The electrical properties of individual metallic SWCNTs in various chemical environments were also investigated under gate modulation. The resulting electrical conductance changes turned out to be negligible compared to the semiconducting counterparts. (Kong, J. Et al.(2000))
It is difficult to pick up an individual semiconducting SWCNT from as-grown samples, which are mixtures of both metallic and semiconducting SWCNTs. Furthermore, the alignment of an individual semiconducting SWCNT between pre-patterned electrodes is very complicated process. Therefore, the individual semiconducting FET type SWCNT gas sensor is impractical. More practical way of making semiconducting SWCNT sensors is to use mat or network of SWCNTs, which consist of both metallic and semiconducting SWCNTs, as sensing elements.

3. SWCNT network gas sensor fabricated by direct CVD

Direct growth of SWCNTs network from selectively arranged catalyst by chemical vapor deposition (CVD) has been atempted by various authors.(Qi et al., 2003; Stardermann, et al.,2004; Wongwiriyapan et al., 2006)

Qi et al. fabricated SWCNT FET sensors by using direct CVD growth technique at 900ºC and reported that all the devices showed highly sensitive chemical gatings, which indicates that the charge transfer is predominantly through semiconducting SWCNTs. This cannot be easily understood. The nanotubes formed at the devices are mixture of both semiconducting and metallic SWCNTs as the chirality control of SWCNTs is impossible for present day CVD technology. Since the average conductance of a metallic SWCNT is about two orders of magnitude higher than that of semiconducting counterpart(Liang et al.,2007), the main charge transfer would be through metallic SWCNTs for many of the devices.

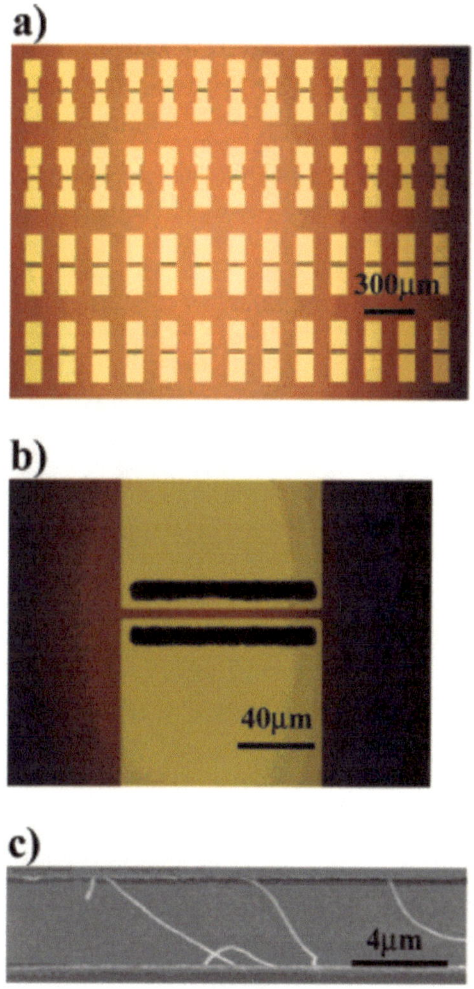

Fig. 2. (a) Optical image of an array of SWCNT network devices. (b) Optical image of one device. The black regions contain catalyst patterned on top of opposing Mo source and drain electrodes. (c) Scanning electron microscopy (SEM) image of a SWCNT network bridging two opposing Mo electrodes in a device. Excepted from Qi et al.,2003. © American Chemical Society

Stadermann et al. grew SWCNT networks directly onto silicon dioxide substrates using CVD. Gold was deposited onto the surface and patterned to form the electrodes. The devices were tested as FETs by applying sufficient back gate voltage (10V) to show conductance changes between 1% and 50%. These data demonstrate that, in a network of SWCNT between two electrodes, a large part of the current is carried by a small number of highly conductive metallic connections, and the connections involving semiconducting tubes, even though much more numerous, carry a comparable or smaller portion of the current.

Wongwiriyapan et al. synthesized SWCNT film directly on a substrate by thermal CVD. Vertically aligned SWCNTs with a high density were grown at 750 °C, while horizontally lying SWCNT networks with a low density were grown in the temperature range 800–950 °C. Figure 3 shows the varieties of SWCNT networks depending on the growth temperatures. Figure 5. in conjunction with Figure 4. shows the relationship between the surface coverage of SWCNTs and the normalized sensor response: the SWCNT networks with the lowest density exhibited the highest normalized sensor response. Here the normalized sensor response is defined as $\Delta G/G_0$, where G_0 and G represent the conductance values of the junction *before* and *after* the exposure to target gas respectively and $\Delta G \equiv G - G_0$. The *sensitivty* of a sensor at certain concentration values of target gas is defined as the maximum achievable normalized sensor response. The poor sensitivities for high density SWCNT network sensor can be attribute to the conduction of carriers through predorminantly metallic pathways in the network.

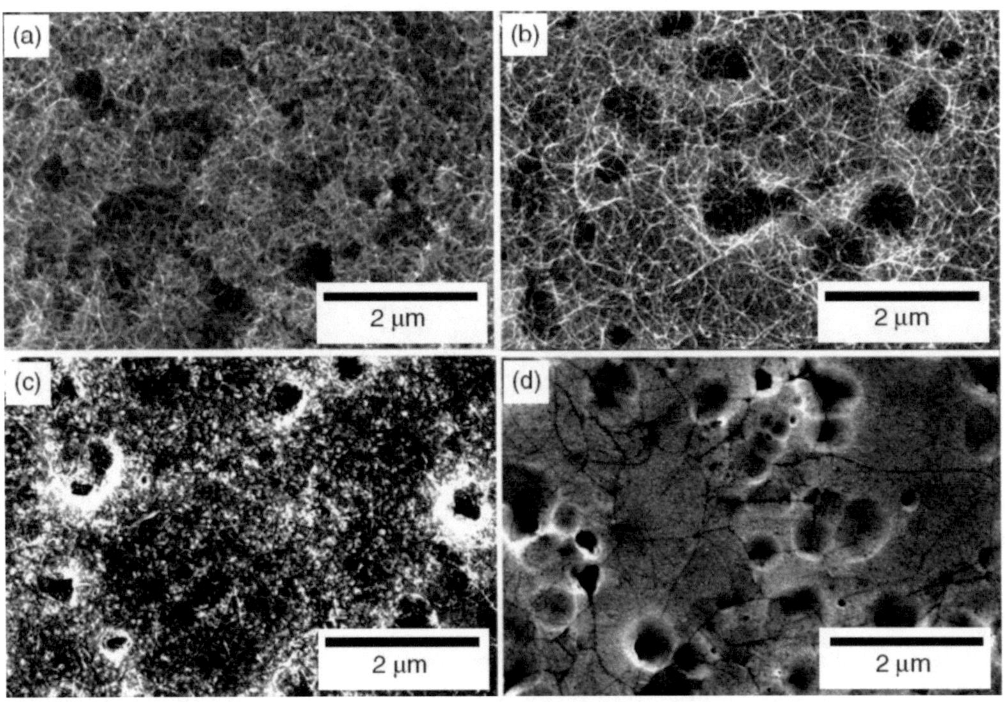

Fig. 3. SEM image of SWCNT network synthesized at different teperatures: (a)750°C, (b)850°C, (c)900°C and (d)950°C. Excepted from Wongwiriyapan et al., 2006. © IOP Publishing Ltd

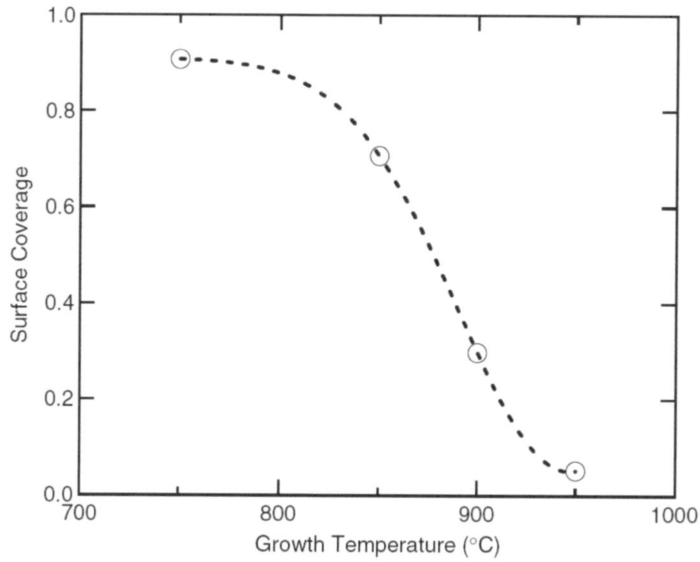

Fig. 4. Surface coverage of SWCNTs synthesized at different growth temperatures. Excepted from Wongwiriyapan et al., 2006. © IOP Publishing Ltd

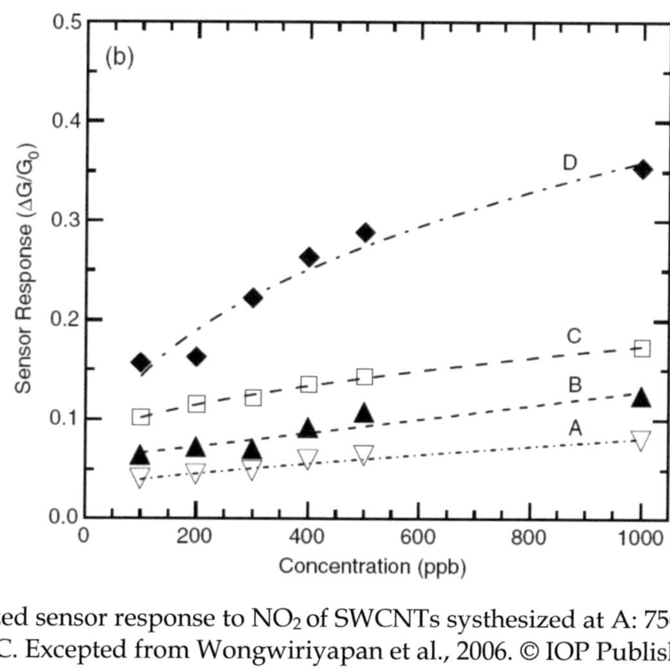

Fig. 5. Normalized sensor response to NO_2 of SWCNTs systhesized at A: 750°C, B:850°C, C:900°C, D:950°C. Excepted from Wongwiriyapan et al., 2006. © IOP Publishing Ltd

This direct growth of SWCNT networks by CVD exhibits complexity with low yield in manufacturing point of view. Furthermore, the SWCNT network sensors can be fabricated on conventional bulky sensor platforms only. The maximum temperature achieved by MEMS gas sensor platform is limited below 700°C at present. (Haque et al., 2008-a, 2008-b)

4. Drop-deposited SWCNT network gas sensor

The most widely reported SWCNT gas sensor is drop-deposited SWCNT network sensor. (Li, et al.,2003; Lu et al.,2006) In this case, the purified SWCNTs are generally dispersed in solutions (surfactants) and then drop-deposited onto interdigitated area of electrodes pre-patterned on the insulating substrate. Figure 6 shows the SEM image of a SWCNT network device fabricated by the drop-deposition method. Figure 7 shows the normalized sensing response of a SWCNT network sensor for various concentrations of NO_2 gas at room temperature. Here 10 minute UV light illumination was introduced to accelerate NO_2 desorption.

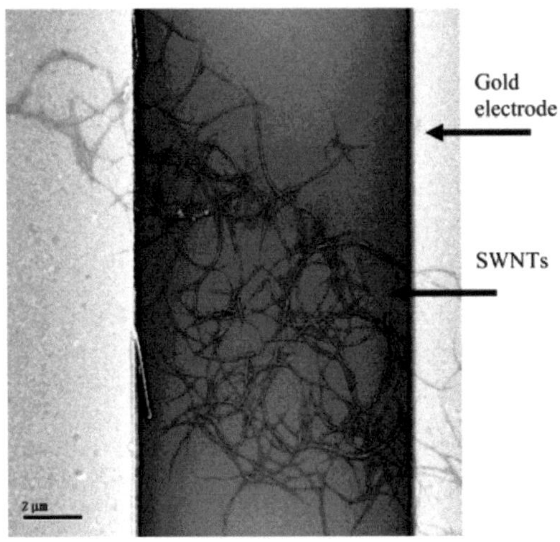

Fig. 6. SEM image of SWCNT network bridging two gold electrodes. Excepted from Li et al., 2003. © American Chemical Society

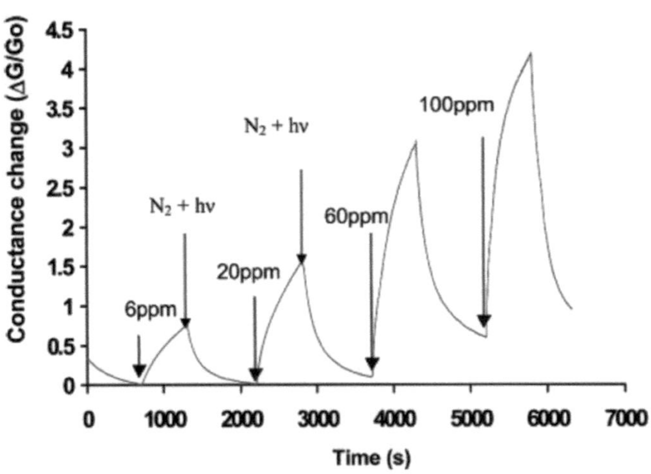

Fig. 7. Normalized sensor response of a drop-deposited SWCNT network sensor upon exposure to various concentration of NO_2 gas. Excepted from Li et al., 2003. © American Chemical Society

This type of fabrication of network sensor is preferred at the research level as it is very simple. But, as shown in Figure 6 the deposited SWCNT network exhibits non-uniformity, which leads to poor reproducibility of the sensor. Thus, this method is not adequate for mass-production.

5. Vacuum filtered SWCNT network gas sensor

In order to enhance non-local uniformity of the SWCNT network sensors, Cho et al. used vacuum filtration method. In this process, 30 nm average pore size AAO filter membrane was installed inside the vacuum filteration system. Then, vacuum filtration of SWCNT solution containing 1% surfactant (SDS) was achieved as shown in Figure 8(a). The SWCNT network deposited on the AAO, then were transferred onto sensor platform as shown in Figure 8(b). By changing the weight of SWCNTs added to the solution, the thickness of the SWCNT network was controlled. As shown in Figure 9, it was revealed that the lower the SWCNT density, the higher the sensitivity of the SWCNT network sensor. The study by Cho et al. gives important information: The SWCNT network sensors must be very thin in order to get high sensitivity. (Cho et al., 2006) Hu et al. reported ultrathin uniform SWCNT network deposition by vacuum filtration method. However, no gas sensors have ever reported by them. (Hu et al., 2004)

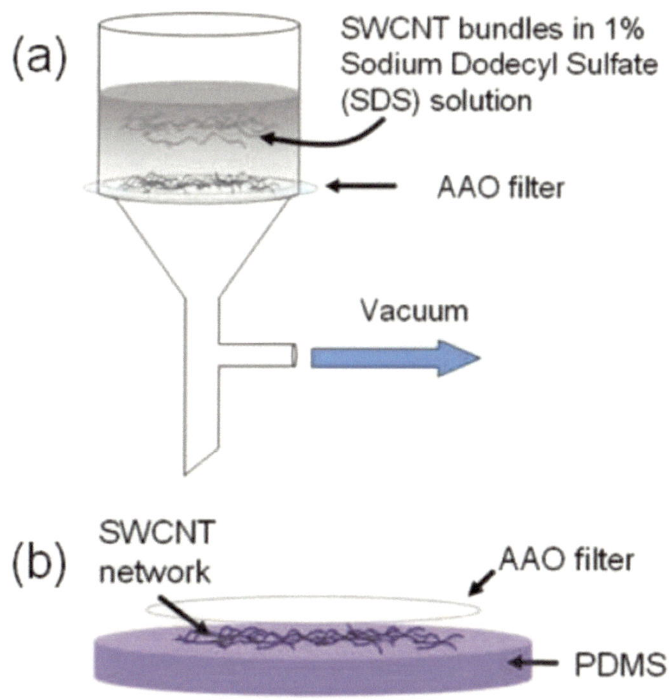

Fig. 8. Schematic illustration of (a) the vacuum filtration method using AAO filter membrane and (b) PDMS mold transfer of SWCNT network. Excepted from Cho et al.,2006. © Materials Research Society

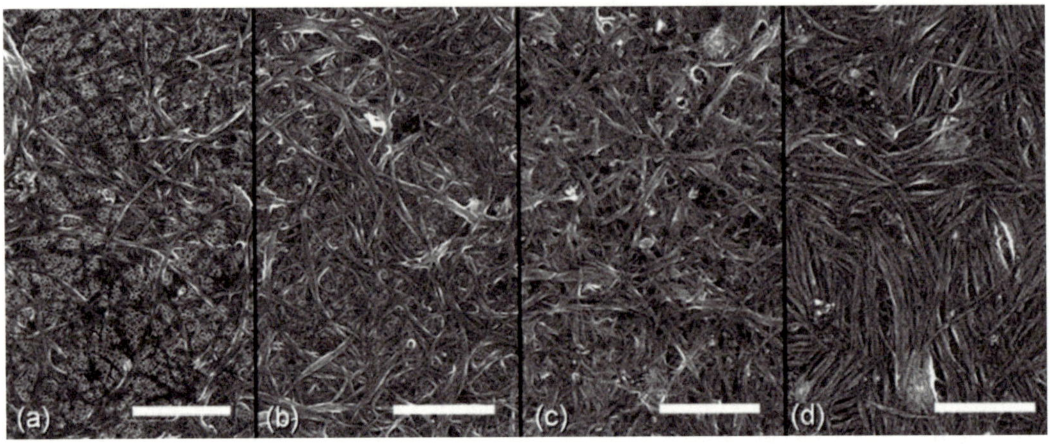

Fig. 9. SEM images of the SWCNT network formed on the filter membrane. The initial SWCNT densities used to form each of the thin film were (a) 0.04mg/ml, (b)0.08mg/ml, (c)0.12mg/ml, and (d)0.16mg/ml. The tool bar all indicate 2 μ m. Excepted from Cho et al.,2006. © Materials Research Society

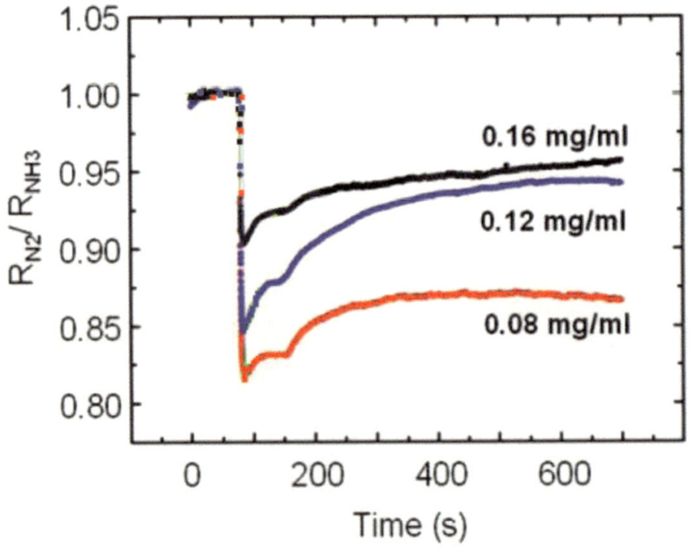

Fig. 10. Normalized sensor responses to the 50% NH_3 gas diluted in N_2 for various sensors with different nanotube densities. Excepted from Cho et al.,2006. © Materials Research Society

6. AC dielectrophoretically assembled SWCNT network gas sensor

To form very uniform and thin SWCNT network, AC dielectrophoresis technique was introduced.(Suehiro et al., 2005; Lee et al., 2006) By using the technique,FET and resistance type sensors were fabricated and tested for various gases.

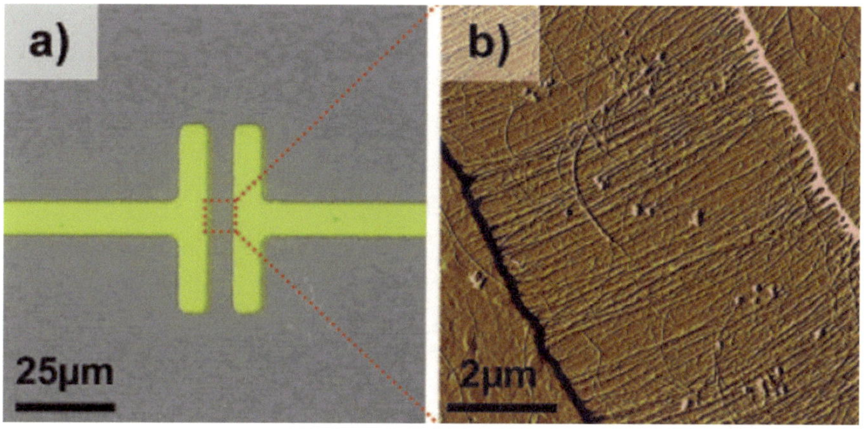

Fig. 11. (a) Optical image of a SWCNT network sensor. (b) AFM image of SWCNT network deposited between the Au gap by ac dielectrophoresis. Excepted from Lee et al.,2006. © American Chemical Society

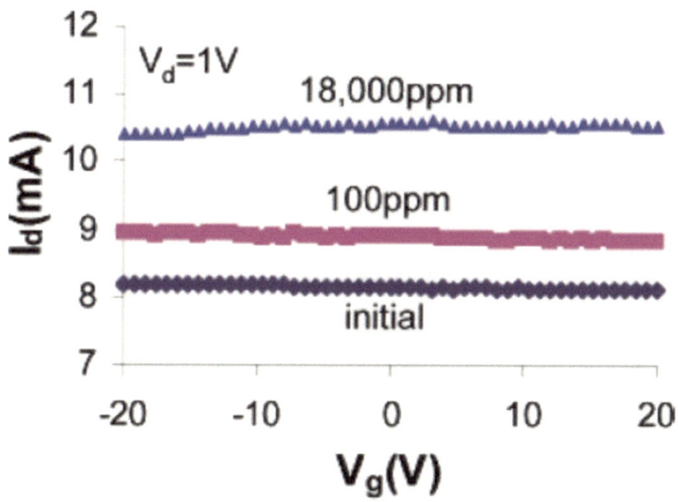

Fig. 12. Dependence of channel current upon gate bias of a SWCNT network FET fabricated by ac dielectrophoresis. The FET characteristics were measured at different $SOCl_2$ concentrations. Small on/of ratio (~1) and vertical shift confirms that the charge transfer between SWCNT network and $SOCl_2$ occurs mainly through metallic pathways. Excepted from Lee et al.,2006. © American Chemical Society

Figure 11. exhibits a SWCNT network FET device fabricated by Lee et al. and Figure 12. shows its perforamance under both gate modulation and gas adsorption. From the observation that gate voltage does not modulate the drain current Lee et al. reached a conclusion that the semiconducting nanotubes do not mainly contribute to the signal. They attributed this to the high density of the nanotube network. If the nanotube density is too high, the signal transduction occurs through metallic pathways, which leads to poor sensitivity.

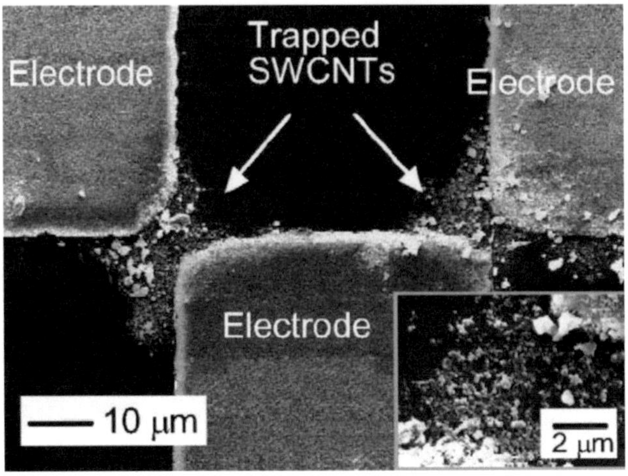

Fig. 13. SEM image of SWCNT network device fabricated by ac dielectrophresis for 3 hours. The inset is magnified image near an electrode corner. Excepted from Suehiro et al.,2005. © Elsevier B.V.

Figure 13. exhibits a SWCNT network device fabricated by Suehiro et al.. By monitoring impedence of the devices during the ac dielectrophresis Suehiro et al. controlled the assembly of SWCNT and reached a conclusion that the semiconducting SWCNT *do* contribute to the sensing. Figure 14 exhibits normalized sensor response of a SWCNT network sensor fabricated by Suehiro et al..

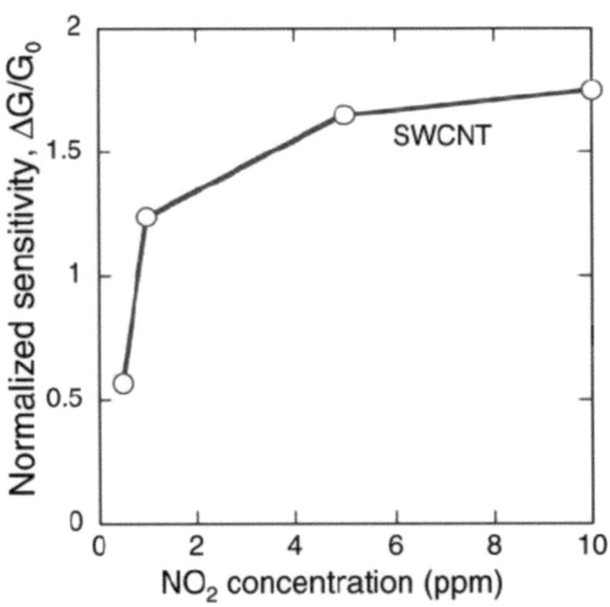

Fig. 14. The dependence of NO_2 gas concentration upon the normalized sensor response of a SWCNT network gas sensor fabricated by ac dielectrophresis. Excepted from Suehiro et al.,2005. © Elsevier B.V.

7. SWCNT network gas sensor fabricated by surface-programmed assembly technique

Recently, ultra uniform and thin SWCNT network sensors fabricated by 'surface-programmed assembly' technique were reported by several authors.(Wang et al.,2009; Tran et al.,2008; Maeng et al.,2008-a) As this technique enables very precise control of nanotube density, it is expected that semiconducting nanotubes contribute to the sensing mechanism as is predicted by percolation theory which leads to high sensitivity. In addition, this technique is expected to facilitate wafer scale production of SWCNT sensor.

Wang et al. reported SWCNT sensors fabricated by using self-assembled monolayer (SAM) of hydrophilic 3- aminoprophiltrimethysilane(APS) as shown in Figure 15. In this process very thin SWCNT network patterns were selectively formed on the functionalized region. From the DMMP detection experiments, they revealed that the sensitivity reaches a maximum when networks of overlapped nanotubes disappear, i.e. the nanotubes are assembled to form a monolayer network. This conclusion supports the observations by Cho et al. that the thinner the SWCNT network the higher the sensitivity of the sensor.(Cho et al., 2006) It is further revealed that SWCNT monolayer with maximum connectivity is ideal for making highest sensitivity gas sensors.

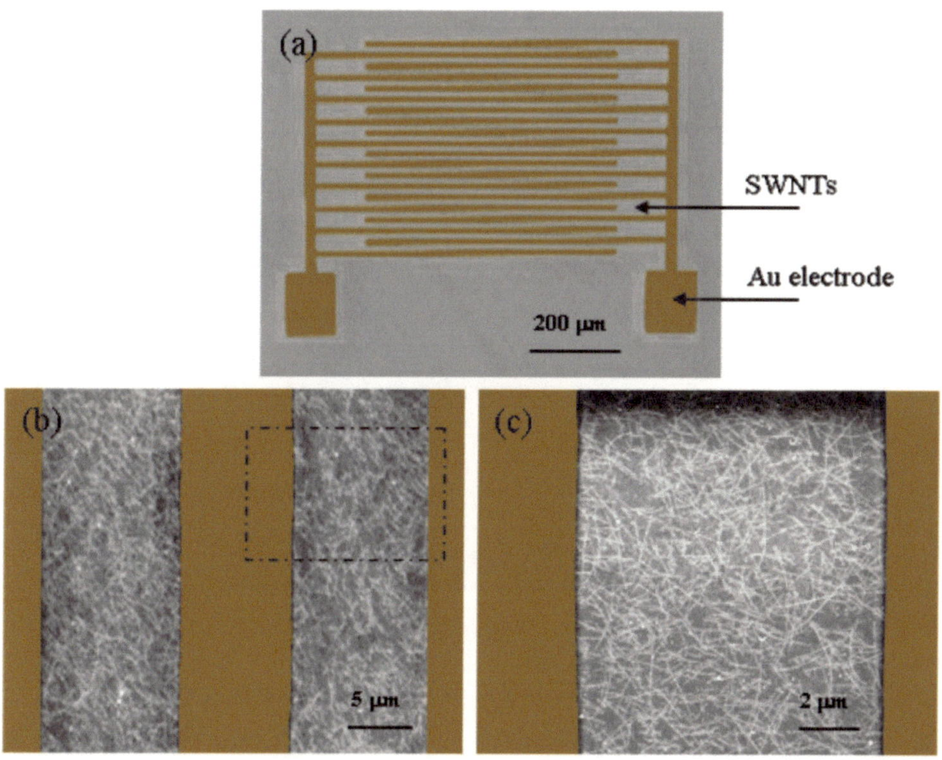

Fig. 15. (a) Optical image of the electrode arrays, where deposited SWCNT networks are separated by Au electrodes. (b) SEM image of SWCNT networks bridging Au electrodes deposited on Si/SiO2 substrate. (c) An enlarged SEM image of (b). Excepted from Wang et al.,2006. © IOP Publishing Ltd

Tran et al. fabricated SWCNT sensors on both functionalized substrate with hydrophilic monolayer of 3-aminoprophilethoxysilane(APTES) and bare substrate as shown in Figure 16.

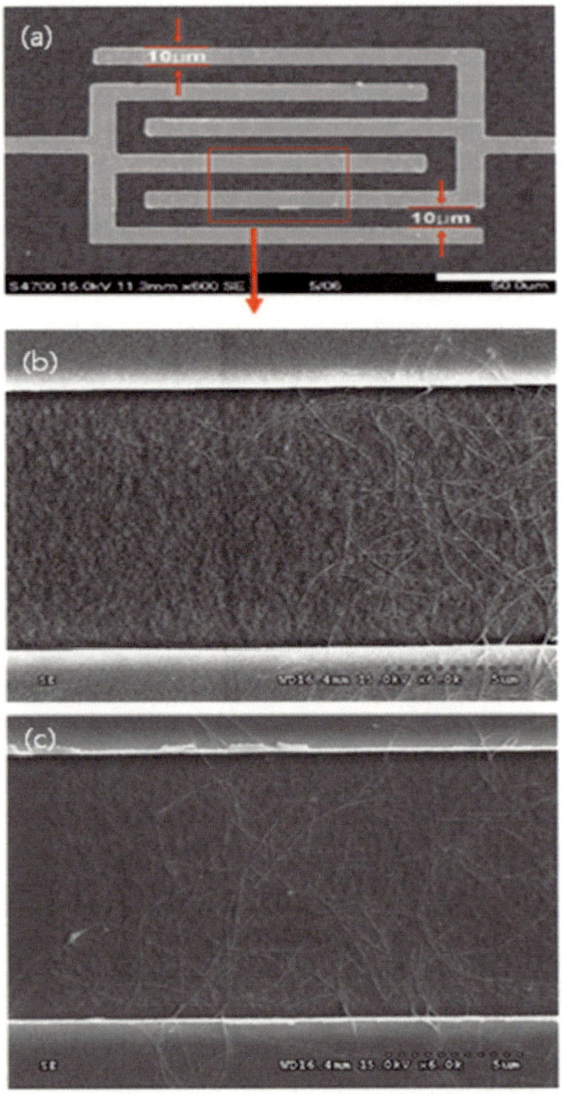

Fig. 16. SEM image of (a) the sensor device which consist of interdigitate electrodes and SWCNTs (b) the SWCNT networks formed on bare surface (c) the SWCNt networks formed on the APTES-treated surface. Excepted from Tran et al.,2008. © Elsevier B.V.

Then, they tried to compare NO_2 detection response of the two sensors as shown in Figure 17. The response time of the functionalized substrate-based gas sensor is very fast (a few seconds), while that of the bare substrate-based one is very slow(hundreds of seconds). It is conjectured that the interaction between APTES and sidewall of SWCNTs gives easier accessibility to the gas molecules. The sensitivity of the sensor on APTES- treated substrate is also shown to be higher than that on bare substrate.

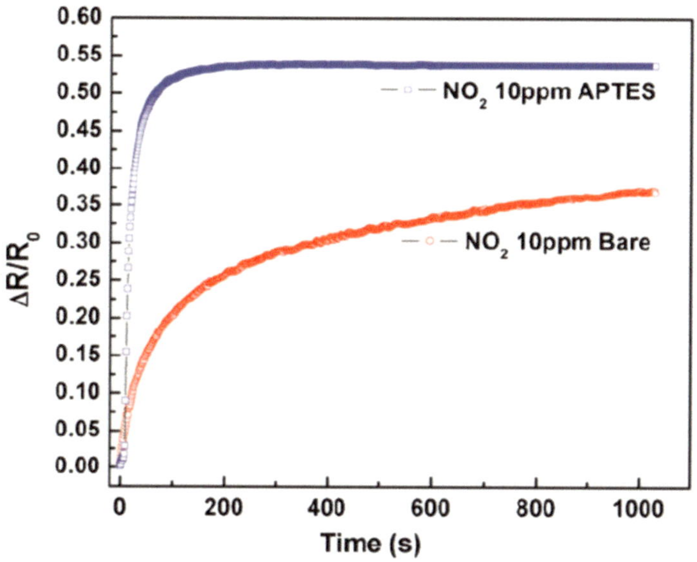

Fig. 17. Comparision of the normalized sensor reponses to 10 ppm NO_2 gas between the SWCNT sensor on a bare and APTES-treated surface. Excepted from Tran et al.,2008. © Elsevier B.V.

Maeng et al. reported highly uniform SWCNT monolayer sensors fabricated by using hydrophobic octadecyltrichlorosiline (OST) surface functionalization. This time, SWCNTs are selectively assembled on non-functionalized or bare surface regions of the substrates. The atomic force microscopy (AFM) image of a SWCNT junction shows a low-density monolayer of SWCNT network between the Au/Ti electrodes. (Figure 18)

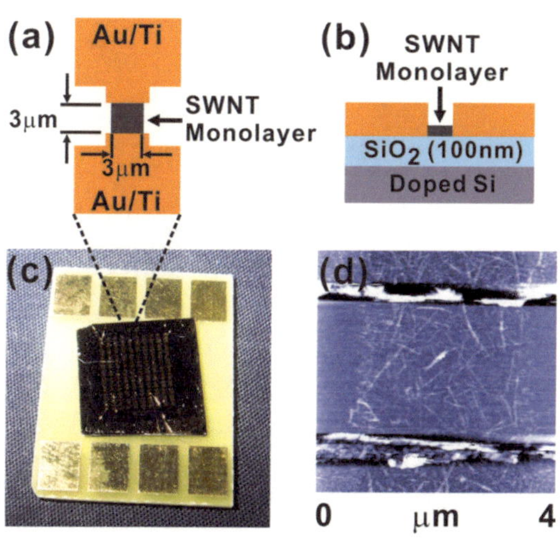

Fig. 18. Schematic illustrations of (a) top view and (b) side-view depicting the structure of the SWCNT network sensor. (c) Optical image of the 10 x 10 array of SWCNT network sensors. (d) AFM topography image of a SWCNT sensor. Excepted from Maeng et al.,2008-a. © American Institute of Physics

Figure 19 shows the normalized responses of typical SWCNT monolayer network sensor fabricated on bare surface to various NO_2 gas concentrations at room temperature. If we compare Figure 19 with Figure 17, it seems that SWCNT monolayer sensor formed on bare substrate exhibits higher sensitivity than that formed on APTES-functionalized substrate for the same concentration of NO_2. This is contradictory to the observation by Tran et al.. It is likely that Tran et al. could not optimize the sensor fabrication conditions and further researches are necessary to determine whether or not SWCNT monolayer sensor formed on bare substrate assumes higher sensitivity than that formed on APTES-functionalized substrate. Considerably high sensitivity of the sensor indicates that semiconducting SWCNTs are involved in the sensing as percolation theory describes.

Figure 20(a) shows the responses of various SWCNT monolayer network sensors to 500 ppb NO_2 gas at room temperature. Figure 20(b) shows the dependence of sensitivity upon base-line conductance. Unexpectedly, the sensitivity of SWCNT network sensors linearly proportional to the inverse of based-line conductance as shown in Figure 20(b).

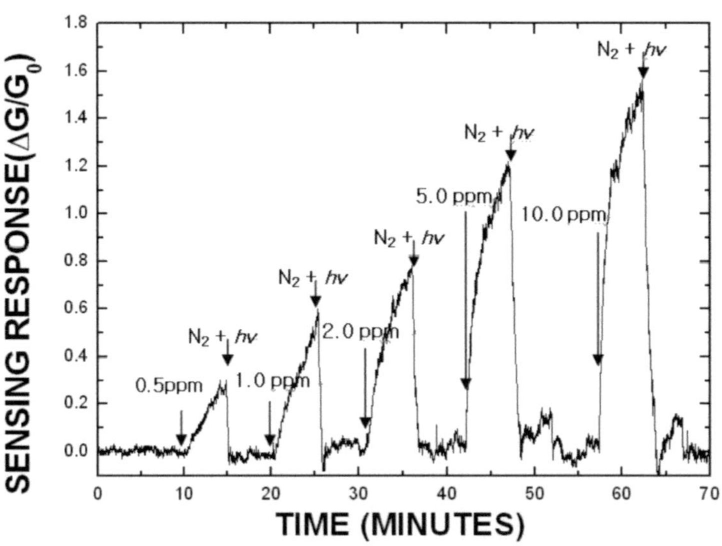

Fig. 19. The dependence of normalized sensing response of a SWCNT network sensor to the concentration of exposed NO_2 gas at room temperature. Excepted from Maeng et al.,2008-a. © American Institute of Physics

In order to construct a conduction model of the SWCNT network sensor, Suehiro et al. assumed that the base-line conductance $G_0 = N_t\ g_0$, where N_t is the total number of SWCNTs and g_0 is the average conductance of one SWCNT. It was implicitly assumed that all SWCNT had the same electrical property. After NO_2 adsorption, the conductance of individual SWCNT is further assumed to be increases by Δg on average, which leads to the total increase of sensor conductance $\Delta G = N_t\ \Delta g$. From this relations, they further derived the relation that normalized sensor response $\Delta G/G_0 = \Delta g/\ g_0$. However, they gave no thought to the fact that the random network both follows percolation theory and consists of very different types of SWCNTs. According to the percolation theory, $G_0 \propto (N_t - N_c)^v$ when all the nanotubes have the similar electrical conductance. Here

$N_c = 1/\pi\,(4.236/L_{tube})^2$ is the critical number corresponding to percolation threshold, and $v \sim 1.94$ is the critical exponent. (Stauffer, 1985) If some of the constituents have totally different electrical conductances as was the case of SWCNT network, the calculation becomes a bit complicated.

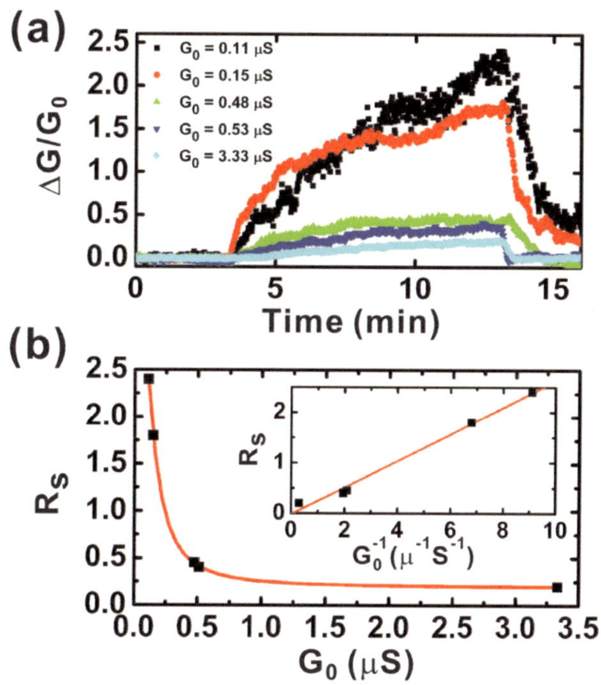

Fig. 20. (a) Normalized sensor response $\Delta G/G_0$ of SWCNT network sensor to 500 ppb NO_2 gas. (b) Sensitivity R_s vs base-line conductance G_0 graph. (Inset) Sensitivity R_s lineraly proportional to the inverse of the base-line conductance. Excepted from Maeng et al.,2008-a. © American Institute of Physics

8. Theoretical explanation of the inverse proportionality of sensitivity to the base-line conductance of SWCNT network sensor based on the percolation theory

The sensing characteristics shown in Figure 20 can be attributed to both the mixing nature of metallic and semiconducting SWCNTs and the randomly assembled network. Even though the SWCNT monolayer network is so uniformly deposited that each devices have equal number of SWCNTs and fixed ratio of metallic SWCNT/semiconducting SWCNT, the fluctuations of the base-line conductance can occur depending on the numbers of metallic SWCNTs which contact with electrodes. If the width of the electrode is comparable to the size of nanotubes as was in the case of the work by Maeng et al. the numbers may differ from device to device. Now, it should be explained how the difference of base-line conductance of devices influence the sensitivity of the devices. $G(c)$, the sheet conductance of a device at a gas concentration c, can be described as:

$$G(c) = g_s(c)P_s + g_m(c)P_m \tag{1}$$

where $g_m(c)$ and $g_s(c)$ are average conductance per a metallic-only pathway and that including semiconducting nanotubes at a gas concentration c, respectively. While the SWCNTs treated by ac dielectrophoresis are aligned with preferential directions, those by surface programmed assembly method are randomly oriented. This enables the application of the percolation theory in estimating possible conduction pathways in a SWCNT monolayer network.

By using the percolation theory, $P_t = P_0(N_t - N_c)^\nu$, $P_m = P_0(N_m - N_c)^\nu$, the relation $\Delta g_m / \Delta g_s \ll 1$, and the relation $g_s(0) / g_m(0) \ll 1$, where P_t is the total conduction paths, P_m is the number of metallic-only paths, P_o is a constant, N_m is the number of metallic SWCNTs in the network, $\Delta g_m = g_m(c) - g_m(0)$, and $\Delta g_s = g_s(c) - g_s(0)$, it is obtained

$$\frac{\Delta G}{G_o} \approx \frac{\Delta g_s P_t}{G_o} - \frac{\Delta g_s}{g_m(0)} \times [1 - \frac{g_s(0)}{g_m(0)} \{(\frac{\varepsilon - \delta}{1 - \delta})^\nu - 1\}]$$

$$\approx \frac{\Delta g_s P_t}{G_o} - \frac{\Delta g_s}{g_m(0)} \tag{2}$$

where $\varepsilon \equiv N_t / N_m = 1.5 \sim 3$ and $\delta \equiv N_c / N_m$.

For a specific gas concentration, the achievable maximum value of $\Delta g_s \equiv \Delta g_{s,max}$ is a constant. Furthermore, if it is assumed that the SWCNTs are dispersed so uniformly that the total number P_t of all conduction paths is a constant, the sensitivity for the specific gas concentration can have the form

$$\text{Sensitivity } R_s \approx a/G_0 - \beta \tag{3}$$

where $a \equiv \Delta g_{s,max} P_t$ = constant and $\beta \equiv \Delta g_{s,max} / g_m(0)$ = constant. (Maeng et al., 2008-a) This result explains the linear proportionality of sensitivity to the inverse of the initial conductance G_0.

9. Future prospects of SWCNT network gas sensor

To secure the reproducibility of the SWCNT monolayer network sensor, the size of the electrodes must be large enough so that the numbers of metallic SWCNTs contacting with electrodes are averaged out. The SWCNT monolayer sensor is very promising not only in the perspective of sensitivity but also in the perspective of sensor operation. Even though the response time of semiconducting SWCNT sensor is about one order of magnitude faster than conventional metal oxide sensors, the recovery time of the sensor is about one order of magnitude slower than the metal oxide counterparts. This slow recovery time of the SWCNT sensors are due to irreversible binding of gas molecules on the SWCNT surface and is a primary hurdle for most SWCNT sensors that have appeared in the literature to date. MEMS-based metal oxide sensors usually adopt microheaters to enhance gas adsorption rate at moderate temperature range (200 ~ 400ºC) and the desorption rate at higher temperature range. The integration of microheater with SWCNT network sensor will be also very useful for acceleration of desorption rate of adsorbed species even though some authors suggest to use UV source for the same purpose. (Li et al., 2003; Karthigeyan, et al., 2008; Maeng, et al., 2008-a) The use of microheaters is highly recommendable as the characteristics of SWCNT devices are affected by humidity. (Sung et al, 2006) Carbon nanotubes are also known to exhibit higher sensitivity at moderately elevated temperature range.(Valentini, et al.,2003) It is, thus, desirable for SWCNTs network to become monolayer so that the temperature of microheater and the nanotubes coincides during the operation. In this regards, the attempt to fabricate mass-production scale SWCNT sensors by depositing SWCNT monolayer onto mocroheaters of the MEMS sensor platforms is much needed and to be explored in the future.

10. Conclusion

In this chapter, an attempt has been made to provide overview of resistance and FET type SWCNT network gas sensors. In the thick SWCNT network sensor, the conduction of charge carriers occurs primarily through metallic-only pathways and the sensor sensitivity is poor. As the thickness of the sensor is reduced, the sensitivity tends to be enhanced. There exist a critical point where main conduction begin to occur through semiconducting-metallic mixed pathways. The critical point is determined by the density of SWCNT network. When the network becomes non-overlapping monolayer, the semiconducting nanotubes contribute to the electrical conduction so significantly that sensitivity reaches a maximum value. The behavior of the sensor sensitivity can be explained satisfactorily by the percolation theory at this condition.

11. Acknowledgment

This work was supported by the RIC program of MKE in Woosuk university.

12. References

Cheng, B.; Russell, J.M.; Shi, W.; Zhang, L. & Samulski, E.T. (2004). Large-scale, solution-phase growth of single-crystalline SnO_2 nanorods. *Journal of the American Chemical Society*, Vol.126, No.19, 5972–5973Göpel, W. & Schierbaum, K.D. (1995). SnO_2 sensors: current status and future prospects. *Sensors and Actuators B*, Vol.26-27, 1-12

Cho, C.-W.; Lim, C.-H.; Woo, C.-S.; Jeon, H.-S.; Park, B.; Ju, H.; Lee, C.-J.; Maeng, S.; Kim, K.-C.; Kim, S.H. & Lee, S.-B.(2006). Highly flexible and transparent single walled carbon nanotube network gas sensors fabricated on PDMS substrates. *Proceedings of 2006 MRS Spring Meeting*, San Francisco, Ca., USA, April 17-21, 2006, # 0922-U11-10

Graf, M.; Frey, U.; Taschini, S.& Hierlemann. A. (2006). Micro hotplate-based sensor array system for the detection of environmentally relevant gases. *Analytical Chemistry*, Vol.78, No.19, 6801-6808

Haque, M.S.; Teo, K.B.K.; Rupensinghe, N.L.; Ali, S.Z.; Haneef, I.; Maeng, S.; Park, J.; Udrea, F. & Milne, W.I. (2008-a). On-chip deposition of carbon nanotubes using CMOS microplates. *Nanotechnology*, Vol. 19, 025607

Haque, M.S.; Ali, S.Z.; Guha, P.K.;Oei, S.P.; Park, J.; Maeng, S.; Teo, K.B.K.; Udrea, F. & Milne, W.I. (2008-b). Growth of carbon nanotubes on fully processed silicon-on-insulator CMOS substrates. *Journal of Nanoscience and Nanotechnology*, Vol.8, No.11,5667-5672

Hu, L.; Hecht, D.S. & Grüner, G.(2004). Percolation in transparent and conducting carbon nanotube networks. *Nano Letters*, Vol.4, No.12, 2513-2517

Karthigeyan, A.; Minami, N. & Iakoubovskii, K.(2008). Highly sensitive, room-temperature gas sensors prepared from celluose derivative assisted dispersions of single-wall carbon nanotubes. Japanese Journal of Applied Physics, Vol.47, No.9, 7440-7443

Kong, J.; Franklin, N.R.; Zhou, C.; Chapline, M.G.; Peng, S.; Cho, K. & Dai, H. (2000). Nanotube molecular wires as chemical sensors, *Science*, Vol.687, No.5453, 622-625, 0036-8075

Lee, C. Y.; Baik, S.; Zhang, J.; Masel, R.I.; & Strano, M.S. (2006). Charge transfer from metallic signle-walled carbon nanotube sensor arrays. *Journal of Physical Chemistry B*, Vol.110, 11055-11061

Li, J.; Lu, Y.; Ye, Q.; Cinke, M.; Han, J. & Meyyappan, M. (2003). Carbon nanotube sensors for gas and organic vapor detection. *Nano Letters*, Vol.3, No.7, 929-933

Liang, X.L.; Wang, S.; Duan, X.J.; Zhang, Z.Y.; Chen, Q.; Zhang, J.; Peng, L.-M.(2007). A comparative study on SWCNT and DECNT field-effect transistors. *Journal of Nanoscience and Nanotechnology*, Vol.7,No.4/5, 1568-1572. 1533-4880

Lu, Y.; Partridge, C.; Meyyappan, M. & Li, J.(2006).A carbon nanotube sensor array for sensitive gas discrimination using principal component analysis. *Journal of Electroanalytical Chemistry*, Vol.593, 105-110, 0022-0728

Maeng, S.; Moon, S. E.; Kim, S.H.; Lee, H.-Y.; Park, S.-J.; Kwak, J.-H.; Park, K.-H.; Park, J.; Choi, Y.; Udria, F.; Milne, W.I.; Lee, B.Y.; Lee M. & Hong, S.(2008-a) Highly-sensitive NO_2 sensor arrays based on undecorated single-walled carbon nanotube monolayer junctions. *Applied Physics Letters*. Vol.93, No.11, 113111. 0003-6951

Maeng, S.; Guha, P.; Udrea, F.; Ali, Z.; Santra, S.; Gardner,J.W.; Park, J.; Kim, S.-H.; Moon, S.E.; Park, K.-H.; Kim, J.-D.; Choi, Y.; Milne, W.I.(2008-b). SOI CMOS-based smart gas sensor system for ubiquitous sensor networks. *ETRI Journal* Vol.30, No.4, 516-525

Maeng, S.; Moon, S.-E. & Kim, S.-W.(2011). Synthesis of novel tin oxide nanoplatelets and their NO_2 sensing properties, *Materials and Manufacturing Processes*, in progress

Moon, S.E.; Lee, H.-Y.; Lee, J.-W.; Choi, N.-J.; Park, S.-J.; Kwak, J.-W.; Park, K.-H.; Kim, J.; Cho, G.-H.; Lee T.-H.; Maeng, S.; Udrea, F.; Milne, W.I. (2010). Low power consumption and high sensitivity carbon monoxide gas sensor using indium oxide nanowire, *Journal of Nanoscience and Nanotechnology*, Vol.10, No.5, 3189-3192

Panchanpakesan, B.; Cavicchi, R.; Semancik, S. & DeVoe, D.L. (2006). Sensitivity, selectivity and stability of tin oxide nanostructures on large area arrays of microhotplates. *Nanotechnology*, Vol.17, 415-425

Park, S.J.; Park, J.; Lee, H.Y.; Moon, S.E.; Park, K.-H.; Kim, J.; Maeng, S.; Udrea, F.; Milne, W.I.; Kim, G.T. (2010). High sensitive NO_2 gas sensor with low power consumption using selectively grown ZnO nanorods, *Journal of Nanoscience and Nanotechnology*, Vol.10, No.5, 3385-3388

Qi, P.; Vermesh, O.; Grecu, M. ; Javey, A. ; Wang, Q. ; Dai, H. ;Peng, S. & Cho, K.J.(2003). Toward large array of multiplex functionalized carbon nanotube sensors for highly sensitive and selective molecular detection, *Nano Letters*, Vol.3, No.3, 347-351

Stauffer, G.(1985). *Introduction to percolation theory*. Taylor & Francis Ltd., 0850663156, London

Shukla, S. ; Patil, S. ; Kuiry, S.C. ; Rahman, Z. ; Du, T. ; Ludwig, L. ; Parish, C. & Seal, S.(2003). Synthesis and characterization of Sol-Gel derived nanocrystalline tin oxide thin film as a hydrogen gas sensor. *Sensor and Actuators B*, Vol.76, 343-353

Stadermann, M.; Papadakis, S.j.; Falvo, M.R.; Novak, J.; Snow, E.; Fu, Q.; Liu, J.; Fridman, Y.; Boland, J.J.; Superfine, R. & Washburn, S. (2004). Nanoscale study of conduction through carbon nanotube networks. Physical Review B, vol.69, 201402, 0163-1829

Suehiro, J.; Zhou, G.; Imakiire, H.; Ding, W. & Hara, M.(2005). Controlled fabrication of carbon nanotube NO2 gas sensor using dielectrophoretic impedence measurement. *Sensors and Actuators B*, Vol.108, 398-403. 0925-4005

Sung, D.; Hong, S.; Kim, Y.-H.; Park, N.; Kim, S. H.; Maeng, S. & Kim, K.-C.(2006). *Ab initio* study of the effect of water adsorption on the carbon nanotube field-effect transistor. Applied Physics Letters, Vol.89, 243110. 0003-6951

Tran, T.H.; Lee, J.-W.; Lee, K.; Lee, Y.D. & Ju, B.-K. (2008) The gas sensing properties of single-walled carbon nanotubes deposited on an aminosilane monolayer. *Sensors and Actuators B*, Vol.129, 67-71. 0925-4005

Udrea, F.; Maeng, S.; Gardner, J.W.; Park, J.; Haque,M.S.; Ali, S.Z.; Guha, P.K.; Vieira, S.M.C.; Kim, H.Y.; Lee, S.Y.; Kim,S.H.; Choi, Y.; Kim, K.C.; Moon, S.E.; Milne, W.I. (2007) Three technologies for a smart miniaturized gas sensor : SOI CMOS, micromachining and CNTs-Challenges and performance. *Technical Digest of International Electron Devices Meeting (IEDM)*, 10-12 December 2007, Washington, DC, USA. 831-834

Udrea, F.; Santra, S.; Guha, P.K.; Ali, S.Z.; Covington, J.A.; Milne, W.I.; Gardner, J.W. & Maeng, S. (2009). Nanotubes and nanorods on CMOS substrates for gas sensing. *International Symposium on Olfaction and Electronic Nose* (*ISEON*), Brescia, Italy, 15-17 April, 2009

Valentini, L.; Armentano, I. & Kenny, M.(2003). Sensors for sub-ppm NO_2 gas detection based on carbon nanotube thin films. Applied Physics Letters, Vol.82, No.6, 961-963. 0003-6951

Wang, Y.; Zhou, Z.; Yang, Z.; Chen, X.; Xu, D. & Zhang, Y. (2009). Gas sensors based on deposited single- walled carbon nanotube networks for DMMP detection. *Nanotechnology*, Vol.20, 345502

Wongwiriyapan, W.; Honda, S.; Konishi, H.; Mizuta, T.; Ohmori, T.; Kishimoto, Y.; Ito, T.; Maekawa, T.; Suzuki, K.; Ishikawa, H.; Murakami, T.; Kisoda, K.; Harima, H.; Oura, K. & Katayama.(2006). Influence of the growth morphology of single-walled carbon nanotubes on gas sensing performance. *Nanotechnology*, Vol.17, 4424-4430.

18

Ammonia Sensors Based on Composites of Carbon Nanotubes and Titanium Dioxide

Marciano Sánchez and Marina Rincón
Centro de Investigación en Energía, Universidad Nacional Autónoma de México
México

1. Introduction

Design of composite materials for ammonia (NH_3) sensing is important because of two main reasons: (1) NH_3 is the most common substitute for chlorofluorocarbons (CFCs) in cooling systems, and (2) most sensors show long recovery times at room temperature due to the tendency of ammonia to strongly interact with many substrates. Multiwalled carbon nanotubes (MWCNTs) have been used to sense polar molecules like carbon monoxide (CO), carbon dioxide (CO_2), NH_3, water (H_2O), and ethanol (C_2H_5OH) (Ong et al., 2002; Valentini et al., 2004; Varghese et al., 2001), as well as non polar gases like helium (He) and nitrogen (N_2) (Adu et al., 2001). For carbon nanotubes, studies have shown that O_2 molecules are electron acceptors with substantial adsorption energies and charge transference, while NH_3, N_2, CO_2, methane (CH_4), H_2O, hydrogen (H_2) and argon (Ar) are electron donors (Zhao et al., 2002). Adsorption in CNTs is determined by adsorption energy and availability of sites, with typically four different adsorption sites: external surface, grooves between CNTs on the bundle outside, pores inside CNTs, and interstitial channel between adjacent tubes inside the bundle (Stan & Cole, 1998; Williams & Eklund, 2000). Additionally, theoretical studies carried out by Jhi et al. about the electronic and magnetic properties of oxidized CNTs indicated their high potential as gas sensors. These studies demonstrated that the sensing mechanism of as prepared CNTs is more related to oxygen doping than to intrinsic properties, and depends on the structural defects caused by the synthesis methods (Jhi et al., 2000).

With the aim of increasing the response of CNTs to certain gases, some studies about substitutional functionalization of CNTs have been reported (Peng & Cho, 2003). Although boron (B) and nitrogen (N) doping improve the response to gases like CO and H_2O, B_xC_y nanotubes show stronger chemisorption, in contrast with the substitutional functionalization with nitrogen (Villalpando-Paez et al., 2004). Oxidation of CNTs in acid solutions to graft oxygen functional groups [i.e., carboxyl (COOH), hydroxyl (OH), and carbonyl (CO)] on CNT walls have been widely used to diversify the sensing options. Another strategy has been the fabrication of compound materials based on CNTs and metallic oxides. Espinoza et al. reported the use of metallic oxides-CNT composites based on commercial tin (SnO_2) and tungsten (WO_3) oxide powders, as well as sol gel TiO_2 materials, in the sensing of NO_2 and CO at room temperature and 150°C. Their results indicated a better performance for CNT/SnO_2 and CNT/WO_3 than for CNT/TiO_2, for the titania composite the sensor response was barely sizable (Espinoza et al., 2007).

Throughout this chapter, we will present the synthesis and performance of compound materials based on MWCNT/TiO_2 tested as resistor and capacitor ammonia sensors. The design aims for low cost room temperature sensors with good reversibility and fast recovery times. For sensor fabrication, both materials were prepared separately (*ex-situ*) and deposited in a multilayer configuration. By means of spacial and spectroscopic resolution techniques we were able to correlate sensor fabrication with performance. We will demonstrate that the *ex situ* method combined with substantial chemical oxidation of pretreated CNTs provide the best composite material with synergistic properties for ammonia sensing. Variations of capacitance were as high as 150%, while changes in resistance were one order of magnitude lower.

2. Experimental

2.1 Film elaboration techniques

MWCNT functionalization

MWCNTs were obtained from Nanostructured & Amorphous Materials Inc. (90 wt.%, outer diameter<10 nm, length: 5-15 μm, tangled). To introduce oxigenated functional groups, CNTs were refluxed at 100°C for 6 h in acid solutions of 0.5 M sulphuric acid (H_2SO_4) with variable concentrations of nitric acid (HNO_3). Functionalized CNTs were labelled according to the concentration of HNO_3 (2.5M, 7.5M, 12.5M) as CNT-2.5, CNT-7.5 and CNT-12.5, CNTs without treatment were labeled as CNT-WT, composites were labeled as CMP with the number of the HNO_3 concentration used for chemical treatment. CNT inks were prepared mixing 10 mg of CNTs with 40 μL of triton X-100 in 5 mL of deionized water, the triton/CNT weight ratio was 0.6.

Sensor fabrication (ex-situ method)

To prepare TiO_2 films, 100 mL sol gel baths containing 92 mL of isopropanol, 0.1 mL of hydrochloric acid (HCl), and 8 mL of titanium isopropoxide were kept at room temperature (27°C) for 24-48 h. Films were deposited by dip coating on glass (Corning) and indium tin oxide (ITO) 1x1.5 in substrates, using 8 dipping/withdrawing cycles at 30 mm/min speed, and air annealing for 5 min at 400°C after each immersion, and at 500°C for 1 h after the last cycle. After TiO_2 deposition, 0.5 mL of CNT ink was drop casted on top and was dried at room temperature for 24 h. The two layer film was annealed in air for 30 min at 400°C, finally another TiO_2 film (8 cycles) were deposited to make a sandwich like configuration; the whole system was annealed at 400°C in air for 1 h.

2.2 Characterization

The changes in chemical composition caused by the functionalization of CNTs were followed by thermogravimetric analysis (TGA, TA Instruments Q500). TGA studies were carried out under 5 mL/min oxygen flow and using 5°C/min heating rate in the 25-800°C temperature range. Particle size and zeta potential (ζ) were measured by dynamic light scattering (DLS, Zetasizer Nano, Malvern Instruments) as a function of CNTs functionalization. Film crystallinity was monitored by X-ray diffraction (Rigaku Dmax 2200, CuKα radiation, λ=1.5405 Å), using the Debye-Scherer equation (Cullity, 1978) for crystallite size determination. Atomic force microscopy was used for surface microstructural studies (Nanosurf Easyscan, Nanosurf AG, Switzerland). The experimental setup for sensing trials for both platforms is shown in Figure 1, basically, nitrogen was passed through a bubbler containing ammonia and was injected into a flask where the resistance or capacitance of the sensors changed as they were exposed to ammonia; the baseline was measured in dry air and the sensor in the picture correspond to a resistor sensor.

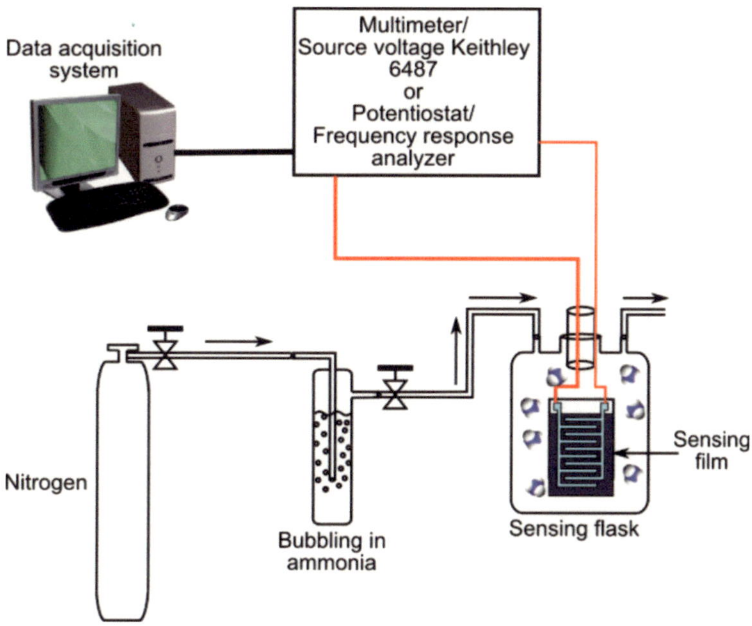

Fig. 1. Experimental setup used for sensing trials

For capacitor sensors, variations of capacitance were followed by electrochemical impedance spectroscopy (EIS) measurements in an Autolab PGSTAT302N potentiostat/galvanostat unit (Eco Chemie), using the configuration depicted in Figure 2. A perturbation of 5-10 mV was applied at open circuit potential in the frequency range from 10^5 to 10^{-3} Hz. Complex non linear least squares (CNLS) fitting of the experimental data was done with the Zsimpwin software (Princeton Applied Research). All the sensing trials were done at room temperature (27°C).

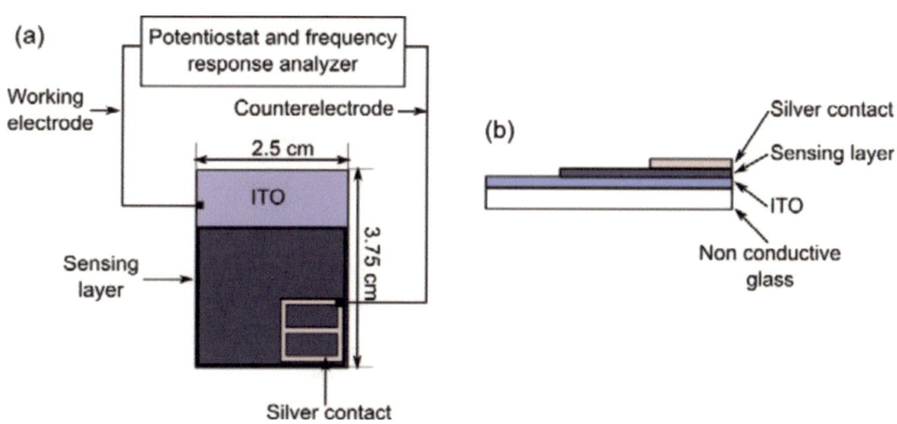

Fig. 2. Capacitor sensor configuration: (a) top view; (b) side view

3. Results and discussion

3.1 CNT characterization

Figure 3 shows the TGA curves of CNTs exposed to various acid treatments. From the variation of mass vs. temperature [Figure 3(a)], or mass/temperature ratio vs. temperature [Figure 3(b)] figures it is easy to appreciate the ~15 wt.% of metal and impurities content of as received CNTs (CNT-WT) as well as its removal by all the acid treatments. The higher oxidation temperature indicates that the treatment used for CNT-2.5 does not damage the nanotube surface, but as the aggressiveness of the functionalization increases, a sizable weakening of the CNT structures is observed, in addition to the grafting of oxygenated functional groups, as evidenced in Fourier transform infrared studies (Sánchez & Rincón, 2009). DLS characterization of CNT aqueous inks prepared with a triton/CNT ratio of 0.6 and sonicated for 30 and 90 min is presented in Figure 4. Ultrasonic treatments of 90 min produced a better dispersion, especially for CNT-WT and CNT-2.5, however, for CNT-7.5 and CNT-12.5 the differences between 30 and 90 min sonication times are smaller [Figure 4(a)]. The absolute value of ζ [Figure 4(b)] shows a continuous increase as the intensity of the chemical treatment does, up to ~18 mV, which is lower than the characteristic values of stable solutions (±30 mV), this increase confirms the presence of functional groups, however in the long term these solutions would show a tendency to precipitate.

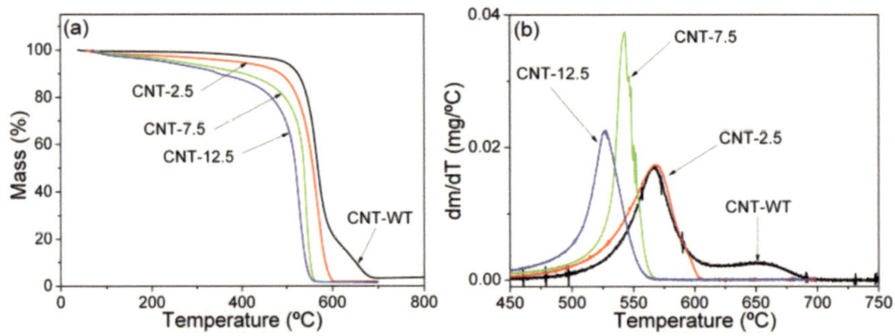

Fig. 3. Thermogravimetric analysis of CNTs in a 5 mL/min oxygen flow and heating ratio of 5°C/min: (a) mass vs. temperature; (b) mass/temperature ratio vs. temperature

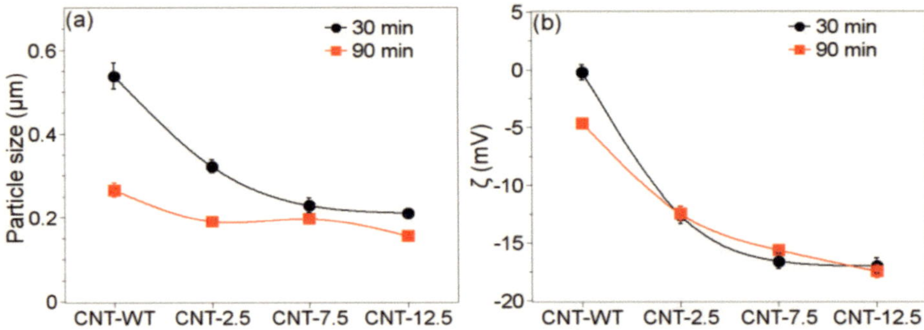

Fig. 4. Dynamic light scattering characterization of CNT inks based on sonication time and functionalization intensity: (a) particle size vs CNT functionalization; (b) ζ potential vs. CNT functionalization

3.2 Structural characterization

The effect of CNT functionalization on the topography of the composite is shown in Figure 5. Figures 5(a-b) show the surface and edge detection of CMP-WT composite. Several longitudinal and tangled formations of CNTs bundled in ropes are evident. In contrast, Figures 5(c-d) show images of CMP-12.5 composite, where CNT/titania grains are slightly smaller, more dispersed, and with shorter ropes. Analysis at a lower scale (not shown) confirmed that a compact and thin layer (~20 nm) of titania was formed covering all the surface including carbon nanotubes, with an increasing thickness as the intensity of chemical treatment does.

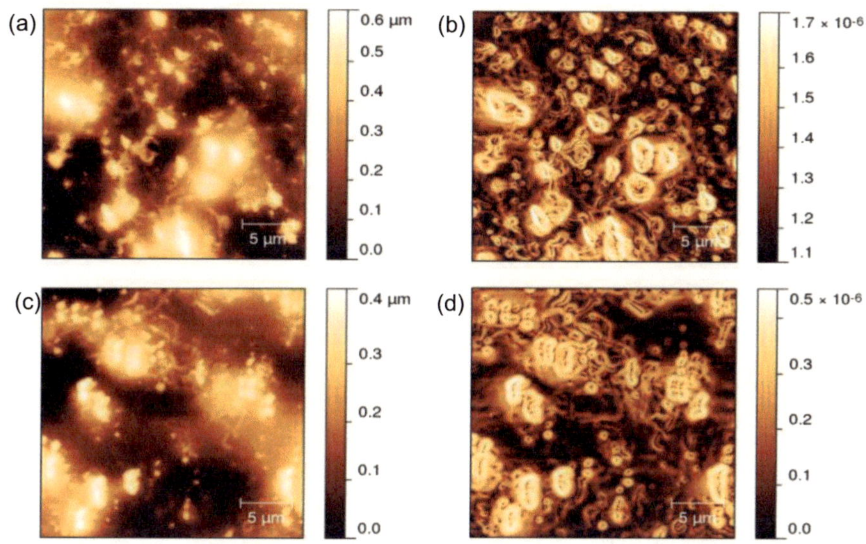

Fig. 5. AFM studies of TiO_2/CNT/TiO_2 composites deposited on glass substrates: (a) topography of CMP-WT; (b) edge detection on Fig. (a) showing CNT ropes; (c) topography of CMP-12.5; (d) edge detection on Fig. (c)

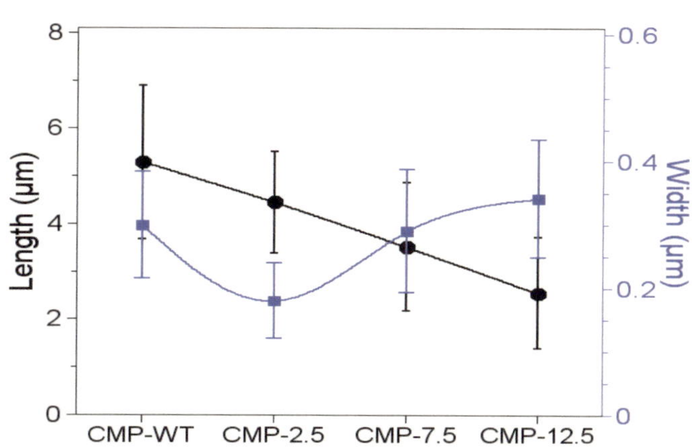

Fig. 6. Length and width of CNT ropes of TiO_2/CNT/TiO_2 composites deposited on glass substrates

Statistical analysis of the length and width of CNT ropes in TiO_2/CNT composites deposited on insulator glass substrates is shown in Figure 6. This graph confirms that the chemical treatment reduces the length of ropes from ~5 µm for CMP-WT to ~2.5 µm for CMP-12.5. A slight increase in the width is observed as the intensity of functionalization increases, which can only be due to a thicker titania layer, given that thinner CNT ropes are obtained at stronger functionalizations.
X-ray diffraction studies (Figure 7) of films prepared in a multilayer configuration (system TiO_2/CNT/TiO_2) shows only the anatase phase and broadening of its peaks. Crystallite size determination using the peak at $2\theta \approx 48°$ (to avoid interference from carbon diffraction at $2\theta = 25°$) shows a decrease from ~25 nm for CMP-WT to ~10 nm for CMP-12.5 (a reduction of ~60%), according to some studies that report the confined growing of titania grains attached to functionalized CNTs which are smaller than the ones growing away from nanotubes (An et al., 2007; Hieu et al., 2008; Song et al., 2007; Yu et al., 2007). This figure also confirms that CNTs with more functional groups are covered with a thicker titania layer.

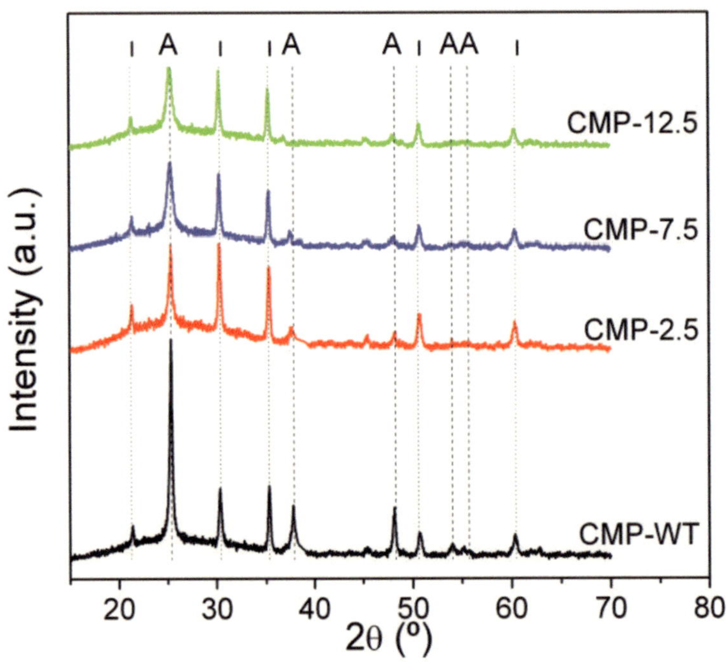

Fig. 7. X-ray diffraction patterns of multilayer systems TiO_2/CNT/TiO_2 deposited on ITO substrates. A: anatase; I: ITO.

3.3 Electrical characterization

3.3.1 Resistor sensors

In this section the performance of CNTs and TiO_2/CNT films as resistor ammonia sensors is presented. Figures 8-10 compare the effect of CNT functionalization on the dynamical response to 1 vol.% ammonia of CNT (Figure 8), CNT/TiO_2 (Figure 9), and TiO_2/CNT/TiO_2 (Fig. 10). Some of these figures show passivation of the most reactive sites during the first pulse, and reversible responses in the subsequent pulses.

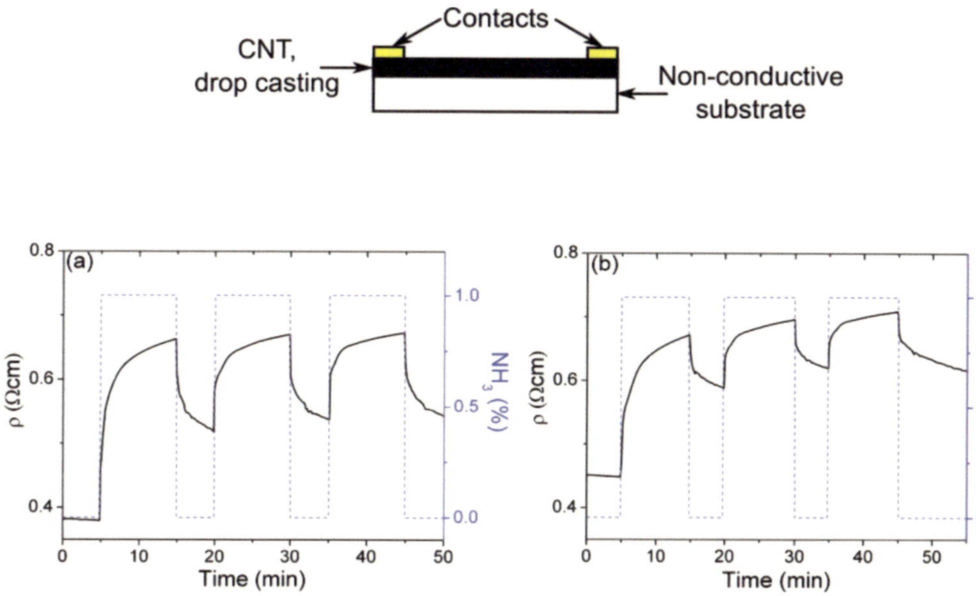

Fig. 8. Dynamical reponse to ammonia of CNT resistor films deposited by drop casting: (a) CNT-WT; (b) CNT-2.5

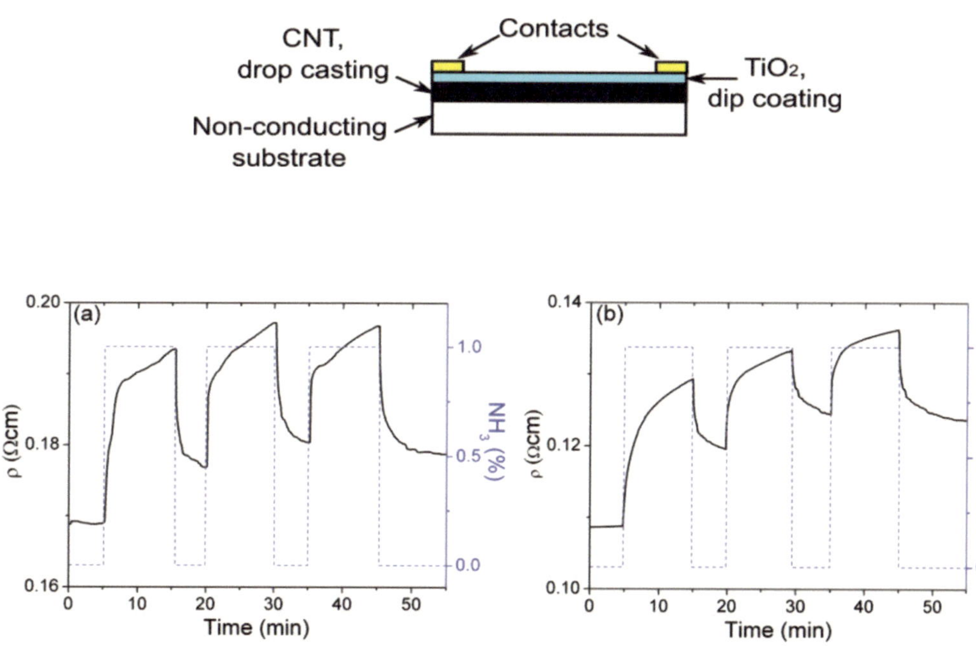

Fig. 9. Dynamical response to ammonia of compound films of CNTs (drop casting) and TiO_2 (dip coating): (a) CNT-WT/TiO_2; (b) CNT-2.5/TiO_2

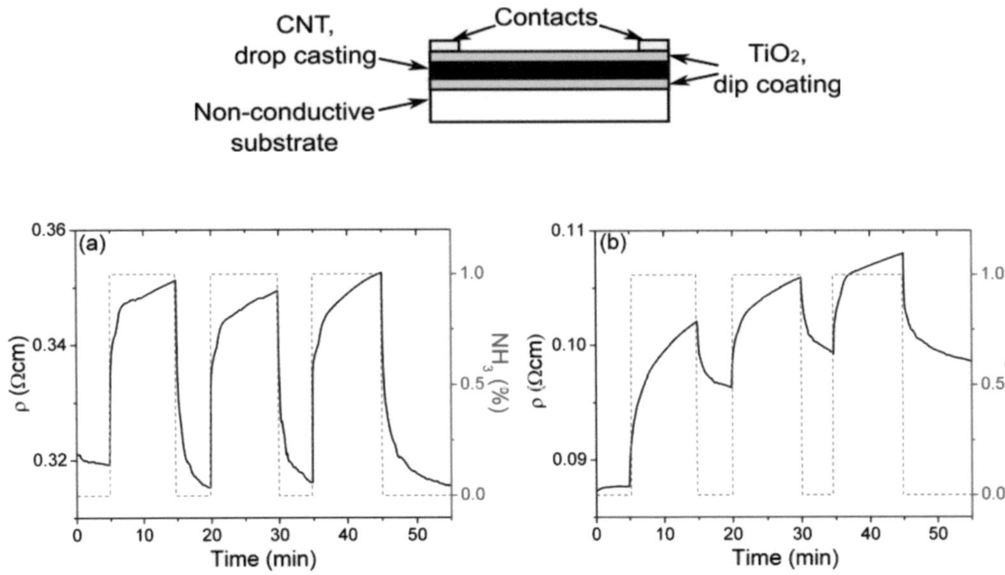

Fig. 10. Dynamical response to ammonia of compound films of CNTs (drop casting) and TiO_2 (dip coating): (a) TiO_2/CNT-WT/TiO_2; (b) TiO_2/CNT-2.5/TiO_2

To quantify the effect of the inclusion of titania layers and the use of functionalized CNTs Figure 11 shows the comparison of the resistivity in air, the sensor response (S_R) (equation 3), the adsorption and desorption times, and the reversibility (equation 4).

$$S_R = R_{ammonia} / R_{air} \tag{3}$$

$$R = \Delta R_{desorption} / \Delta R_{adsorption} \tag{4}$$

The resistivity in air [Figure 11(a)] shows a decrease up to ~80% when CNTs are covered with titanium dioxide, affecting also the value of S. In titania-free materials, the effect of functionalization is to increase the resistivity in air, while in those containing titania the resistivity decreases and this drop is more notorious in sensors with a higher number of titania layers. With respect to the adsorption/desorption times a clear reduction is observed in the time required for adsorption in functionalized materials, whereas the effect of titania layers is difficult to appreciate. For desorption times, the presence of TiO_2 layers causes up to one order decrease (~90%) and it is far more important than CNT functionalization. The faster sensors correspond to a combination of functionalized CNTs and the presence of TiO_2 layers, see systems CNT-2.5/TiO_2 and TiO_2/CNT-2.5/TiO_2.
Some observations from AFM studies can help to explain the electrical behavior of these composites. Dip coating is a technique that applies strong pressures on the surface as films are deposited (Brinker & Hurd, 1994), therefore it results in the deposition of compact and continuous titania films as was observed by AFM and aids to the formation of new paths for charge carriers. These processes are enhanced if functionalized CNTs are used, giving composites with lower resistivity. There are also some studies that report a larger work function for the TiO_2 than for CNT (Ou et al., 2006; Wang et al., 2005), opening the posibillity for electronic transfer from CNT to TiO_2 (Figure 12), increasing the composite conductivity through the increase of titania conductivity and dedoping of CNTs.

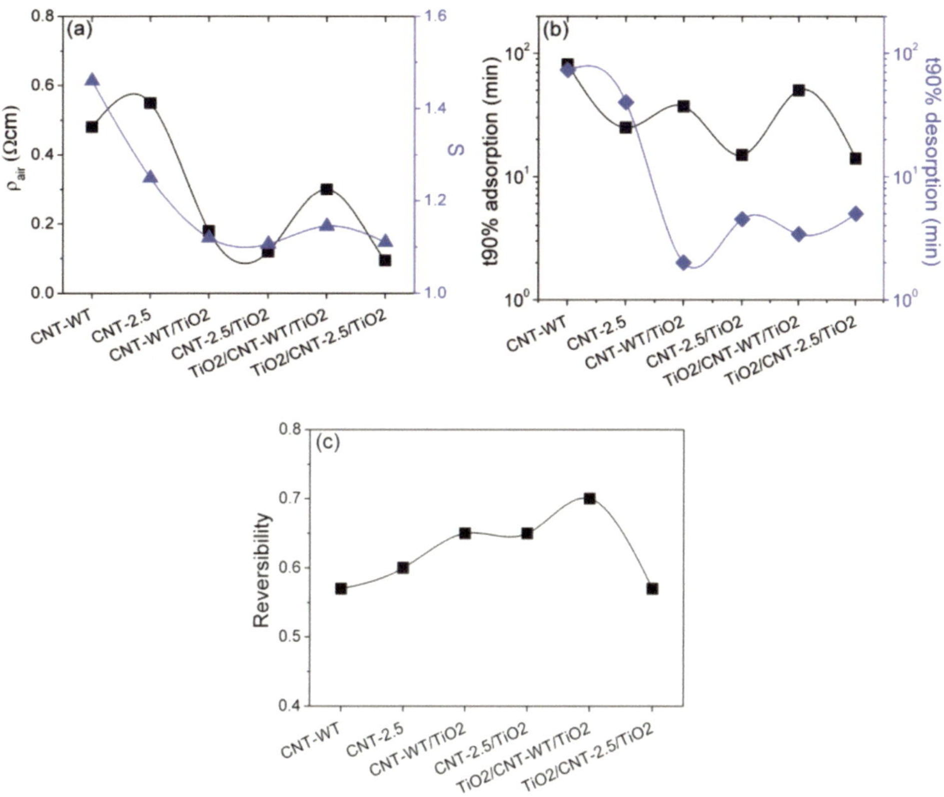

Fig. 11. Characterization of single and multilayer films: (a) resistivity in air and sensor response; (b) adsorption and desorption times; (c) reversibility

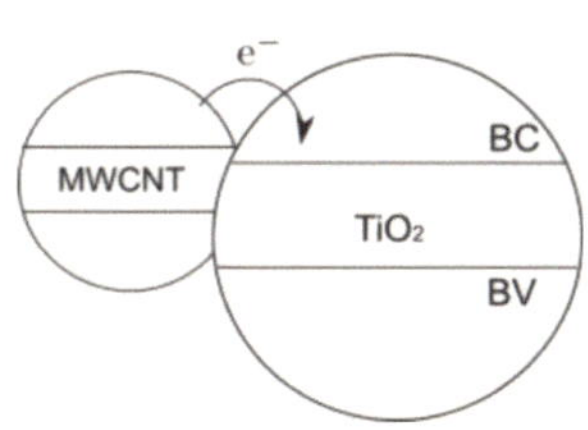

Fig. 12. Electronic transference from MWCNTs to the conduction band of TiO_2

3.3.2 Capacitor sensors

EIS studies of multilayer systems TiO_2/CNT-X/TiO_2 prepared by dip coating and drop casting using CNT-WT and CNT-2.5 are presented in Figures 13 and 14, respectively. Cole-Cole plots are shown in Figures 13(a) and 14(a), the variations of phase angle vs frequency are shown in Figures 13(b) and 14(b), and the results of fitting experimental data with equivalent circuits are presented in Tables 1 and 2. The equivalent circuit used to fit the experimental data of TiO_2/CNT-WT/TiO_2 composite [Figure 13(a)] consists of a resistance R_0 in series with three subcircuits: higher frequencies (R_1C_1), intermediate frequencies (R_2Q_2), and lower frequencies (R_3C_3). The value of R_1 (few ohms) decreases as the material is exposed to ammonia, in contrast to the values of R_2 and of R_3 (several kohms) which show an increase. Capacitance values do not show a unique tendency either, C_1 increases, C_2 it is almost constant, and C_3 decreases with ammonia exposure. Differences among capacitances values are of several orders of magnitude suggesting that adsorption processes are taking place in different sites.

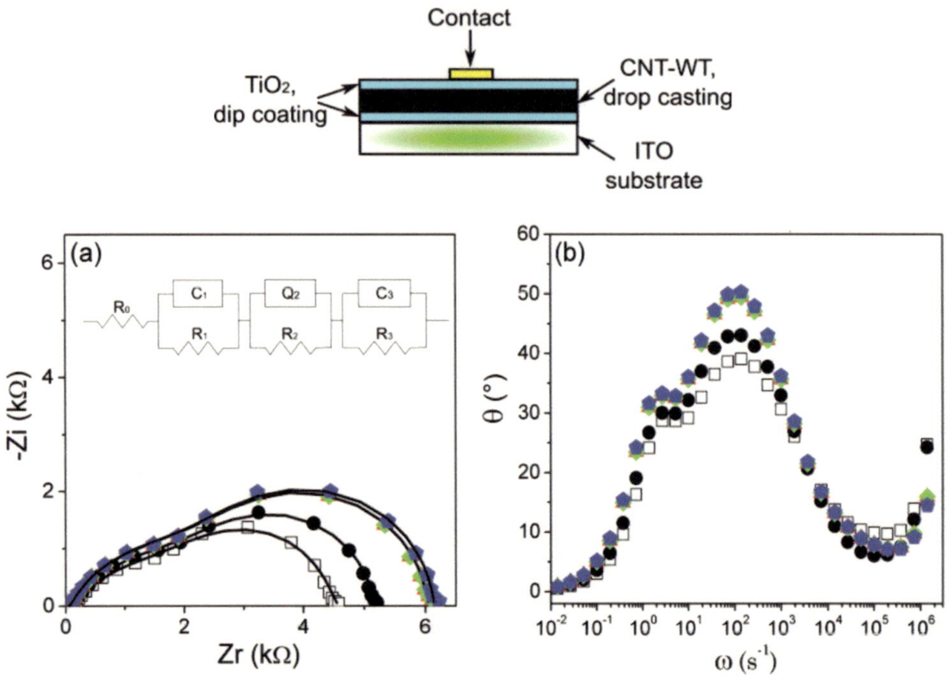

Fig. 13. EIS studies of the composite TiO_2/CNT-WT/TiO_2 with CNTs deposited by drop casting and TiO_2 layers prepared by dip coating: (a) Cole-Cole plot; (b) phase angle vs angular frequency. Colors code: white square: air; black circle: NH_3 1 vol.%; red triangle: NH_3 4 vol.%, 5 min; green diamond: NH_3 4 vol.%, 10 min; blue pentagon: NH_3 4 vol.%, 15 min. Markers are experimental data and lines are the fitting results.

In contrast to the multilayer sensor based on non-functionalized CNTs, EIS results of TiO_2/CNT-2.5/TiO_2 required only two subcircuits to fit the experimental data depicted in Figure 14. The subcircuit at higher (R_1C_1) and lower (R_2Q_2) frequencies are evident in air, but during ammonia exposure only the process at lower frequencies dominates. R_1 values have been related to TiO_2 grain boundaries in TiO_2-CNT composites, although the value of ~10^1 Ω is lower than the typical values (~10^3 Ω) (Sánchez et al., 2009), this difference could be related to processes taking place at the CNT/TiO_2 interface, supporting the improved connectivity of the materials and the direction of charge transfer from CNT=>TiO_2. With respect to the subcircuits at intermediate and lower frequencies in composites based on CNT-WT, changes in capacitances and resistances when the films are exposed to ammonia suggest the presence of two carbon types or two different adsorption sites. In contrast, for composites based on CNT-2.5, amorphous carbon and metal catalysts are removed during functionalization, defining the surface of CNTs as the main adsorption site. Reduction of R_2 in air as well as in ammonia, suggest doping of both TiO_2 and CNTs, because the electronic transference from CNTs to TiO_2 might be facilitated by the formation of covalent bonds during functionalization. Given the tendencies described, we propose that subcircuit 1 is the

Element	Air	1%, 5 min	4%, 5 min	4%, 10 min	4%, 15 min
$R_0(\Omega)$	51	46	46	46	46
$R_1(\Omega)$	53	56	37	26	23
C_1(nF)	3.1	2.1	4.3	8.3	11.0
$\tau_1(\mu s)$	0.16	0.11	0.16	0.21	0.25
$\omega_1(s^{-1})x10^6$	6.08	8.5	6.3	4.6	3.9
$R_2(k\Omega)$	2.78	2.7	2.9	2.8	2.8
$Q_2^0(\mu Ss^{n2})$	35	24	23	22	22
n_2	0.57	0.6	0.7	0.7	0.7
$C_2(\mu F)$	6.0	5.1	5.6	5.9	6.2
τ_2(ms)	17	14	16	17	18
$\omega_2(s^{-1})$	60	72	62	60	57
$R_3(k\Omega)$	1.7	2.4	2.9	3.1	3.3
$C_3(\mu F)$	75	65	64	64	62
τ_3(s)	0.12	0.15	0.19	0.20	0.20
$\omega_3(s^{-1})$	7.8	6.4	5.4	5.0	4.8
S_R		+1.13	+1.30	+1.32	+1.34

Table 1. CNLS fitting results of the system TiO_2/CNT-WT/TiO_2, relaxation times and sensor response. The positive value of S_R is to indicate an increase in the resistance

response of titania grain boundaries, subcircuit 2 is the response of CNTs, and subcircuit 3 is related to the response of impurities in CNT-WT, these correlations are summarized in Table 3. The removal of impurities (i.e. elimination of the third subcircuit) and the decrement in R values reduce the relaxation times up to three orders of magnitude, from seconds for composites with CNT-WT to miliseconds for composites with CNT-2.5. Another benefit of the functionalization is that the capacitance response computed from equation 5 increases up to ~2.5 (i.e., 150%, Table 2) in composites with CNT-2.5, which is about one order higher than the response of resistors [~1.2, i.e., 20%, Figure 11(a), Tables 1 and 2].

$$S_C = C_{ammonia}/C_{air} \quad (5)$$

Element	Air	1%, 5 min	4%, 5 min	4%, 10 min	4%, 15 min
$R_0(\Omega)$	26				
$R_1(\Omega)$	31				
C_1(nF)	44				
$\tau_1(\mu s)$	1.3				
$\omega_1(s^{-1})x10^5$	7.3				
$R_2(\Omega)$	168	219	220	222	222
$Q_2^0(\mu Ss^{n2})$	55	19	15	14	14
n_2	0.6	0.7	0.8	0.8	0.8
$C_2(\mu F)$	1.4	2.5	2.9	3.2	3.4
τ_2(ms)	0.23	0.54	0.64	0.69	0.76
$\omega_2(s^{-1})x10^3$	4.2	1.8	1.5	1.4	1.3
S_R		+1.106	+1.097	+1.104	+1.112

Table 2. CNLS fitting results of the system TiO_2/CNT-2.5/TiO_2, relaxation times and sensor response. The positive value of S_R is to indicate an increase in the resistance

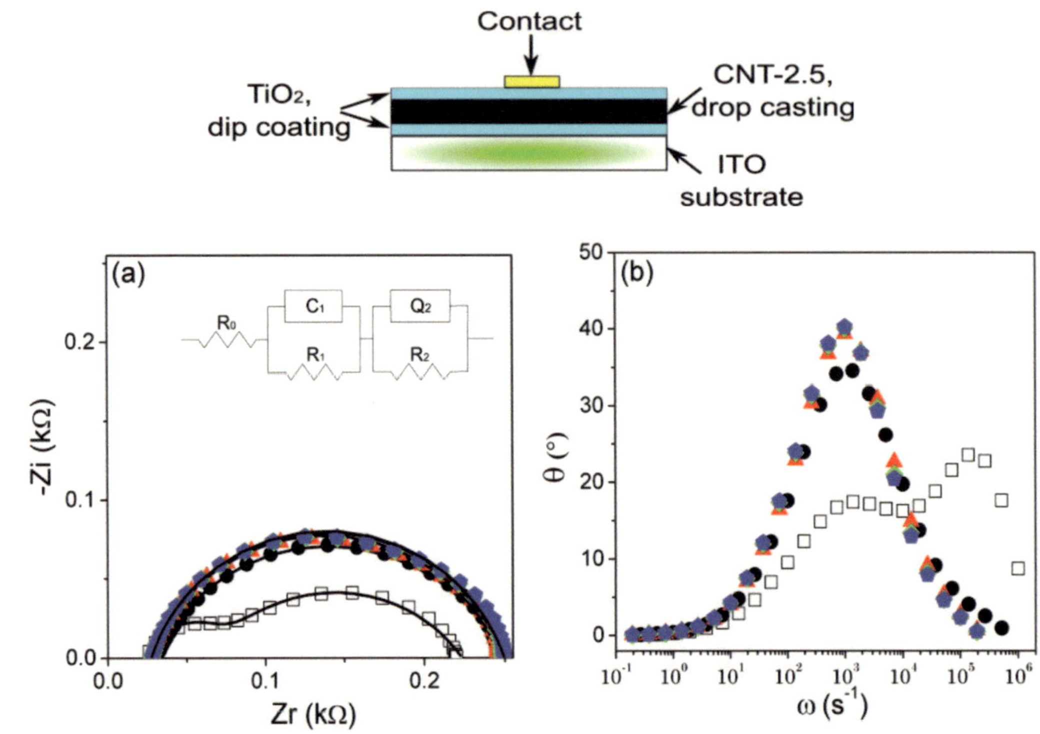

Fig. 14. EIS studies of TiO_2/CNT-2.5/TiO_2 composite with CNTs deposited by drop casting and TiO_2 layers prepared by dip coating: (a) Cole-Cole plot; (b) phase angle vs. angular frequency. Colors code:white square: air; black circle: NH_3 1 vol.%; red triangle: NH_3 4 vol.%, 5 min; green diamond: NH_3 4 vol.%, 10 min; blue pentagon: NH_3 4 vol.%, 15 min. Markers are experimental data and lines are fitting results.

CMP-WT	CMP-2.5	Correlation
R_0	R_0	Resistance due to external connectors
R_1C_1	R_1C_1 (only in air)	TiO_2 grain boundaries
R_2Q_2	R_2Q_2	Carbon nanotubes
R_3C_3		Amorphous carbon and impurities

Table 3. Correlation of the equivalent circuits elements with the adsorption sites of the multilayered composites CMP-WT and CMP-2.5

4. Conclusions and future directions

Design of composite materials based on combinations of nanoparticled titanium dioxide and multiwalled carbon nanotubes, using small amounts of nanotubes for low cost room temperature ammonia sensors have been demonstrated. For sensor fabrication, *ex situ* synthesis of TiO_2 and functionalization of CNTs were carried out separately and deposited in a multilayer configuration. Composites were tested as resistors and capacitors during ammonia sensing based on CNTs functionalization and number of titania layers. Results indicated that the use of titania layers in combination with the substantial chemical oxidation of CNT surface produced a better material with synergistic properties for sensing applications.

5. Acknowledgment

We acknowledge the felloship provided by the Consejo Nacional de Ciencia y Tecnología, CONACYT-México (M. Sánchez), we thank to M.L. Ramón García for the XRD analysis and to P. Altuzar Coello for TGA analysis.

6. References

Adu, C.K.W.; Sumanesekera, G.U.; Pradhan, B.K.; Romero, H.E. & Eklund, P.C. (2001), Carbon nanotubes: a thermoelectric nano-nose, *Chemical Physics Letters*, Vol. 337, pp. 31-35.

An, G.; Ma, W.; Sun, Z.; Liu, Z.; Han, B.; Miao, S.; Miao, Z. & Ding, K. (2007), Preparation of titania/carbon nanotube composites using supercritical ethanol and their photocatalytic activity for phenol degradation under visible light irradiation, *Carbon*, Vol. 45, pp. 1795-1801.

Brinker, C.J. & Hurd, A.J. (1994), Fundamentals of sol-gel dip coating, *Journal de Physique III*, Vol. 4, pp. 1231-1242.

Cullity, B.D. (1978). *Elements of X-Ray Diffraction*, Addison-Wesley, Massachussets, USA.

Espinoza, E.H.; Ionescu, R.; Chambon, B.; Bedis, G.; Sotter, E.; Bittencourt, C.; Felten, A.; Pireaux, J.J.; Correig, X. & Llobet, E. (2007), Hybrid metal oxide and multiwall carbon nanotube films for low temperature gas sensing, *Sensors and Actuators B: Chemical*, Vol. 127, pp. 137-142.

Hieu, N.V.; Duy N.V.; Huy, P.T.; Chien, N.D.; Thamilselvan, M. & Yi, J. (2008), Mixed SnO_2/TiO_2 included with carbon nanotubes for gas-sensing application, *Physica E*, Vol. 41, pp. 258-263.

Jhi, S.H.; Louie, S.G. & Cohen M.L. (2000), Electronic properties of oxidized carbon nanotubes, *Physical Review Letters*, Vol. 85, pp. 1710-1713.

Ong, K.G.; Zeng, K. & Grimes C.A. (2002), A wireless, passive carbon nanotube-based gas sensor, *IEEE Sensors Journal*, Vol. 2, pp. 82-88.

Ou, Y.; Lin, J.; Fang, S. & Liao, D. (2006), MWNT-TiO_2: Ni composite catalyst: A new class of catalyst for photocatalytic H_2 evolution from water under visible light illumination, *Chemical Physics Letters*, Vol. 429, pp. 199-203.

Peng, S. & Cho, K. (2003), Ab initio studies of doped carbon nanotube sensors, *Nano Letters*, Vol. 11, pp. 57-60.

Sánchez, M. & Rincón M.E. (2009), Sensor response of sol-gel multiwalled carbon nanotubes-TiO_2 composites deposited by screen-printing and dip-coating techniques, *Sensors and Actuators B: Chemical*, Vol. 140, pp. 17-23.

Sánchez, M.; Rincón M.E. & Guirado-López R.A. (2009), Anomalous sensor response of TiO_2 films: electrochemical impedance spectroscopy and ab initio studies, *The Journal of Physical Chemistry C*, Vol. 113, pp. 21635-21641.

Song, H.; Qiu, X.; Li, F.; Zhu, W. & Chen, L. (2007), Ethanol electro-oxidation on catalysts with TiO_2 coated carbon nanotubes as support, *Electrochemistry Communications*, Vol. 9, pp. 1416-1421.

Stan, G. & Cole, M.W. (1998), Hydrogen adsorption in nanotubes, *Journal of Low Temperature Physics*, Vol. 110, pp. 539-544.

Valentini, L.; Cantalini, C.; Armentano, I.; Kenny, J.M.; Lozzi, L. & Santucci, S. (2004), Highly sensitive and selective sensors based on nanotubes thin films for molecular detection, *Diamond and Related Materials*, Vol. 13, pp. 1301-1305.

Varghese, O.K.; Kichambre, P.D.; Gong, D., Ong, K.G.; Dickey, E.C. & Grimes, C.A. (2001), Gas sensing characteristics of multi-wall carbon nanotubes, *Sensors and Actuators B: Chemical*, Vol. 81, pp. 32-41.

Villalpando-Páez, F.; Romero, A.H.; Muñoz-Sandoval, E.; Martínez, L.M.; Terrones, H. & Terrones, M. (2004), Fabrication of vapor and gas sensors using films of aligned CN_x nanotubes, *Chemical Physics Letters*, Vol. 386, pp. 137-143.

Wang, W.; Serp, P.; Kalck, P. & Faria, J.L. (2005), Photocatalytic degradation of phenol on MWNT and titania composite catalysts prepared by a modified sol-gel method, *Applied Catalysis B*, Vol. 56, pp. 305-312.

Williams, K.A. & Eklund, P.C. (2000), Monte Carlo simulations of H_2 physisorption in finite-diameter carbon nanotube ropes, *Chemical Physics Letters*, Vol. 320, pp. 352-358.

Yu, H.; Quan X.; Chen, S. & Zhao, H. (2007), TiO_2-multiwalled carbon nanotube heterojunction arrays and their charge separation capability, *The Journal of Physical Chemistry C*, Vol. 111, pp. 12987-12991.

Zhao, J.; Buldum, A.; Han, J. & Lu, J.P. (2002), Gas molecule adsorption in carbon nanotubes and nanotube bundles, *Nanotechnology*, Vol. 13, pp. 195-200.

Part 4

Health Hazard Potentials of CNTs

19

Carbon Nanotubes – Interactions with Biological Systems

Joana Reis[1], Fernando Capela-Silva[2], José Potes[2], Alexandra Fonseca[3], Mónica Oliveira[3], Subramani Kanagaraj[4] and António Torres Marques[5]

[1]Universidade de Évora,
[2]ICAAM, Universidade de Évora,
[3]Universidade de Aveiro,
[4]Indian Institute of Technology Guwahati,
[5]Faculdade de Engenharia da Universidade do Porto,
[1,2,3,5]Portugal
[4]India

1. Introduction

Carbon nanotubes (CNT) are highly versatile materials, with an enormous potential for biomedical applications. Their properties are dependent upon production process and may be modified by subsequent chemical treatment.

Carbon nanotubes can be used to improve polymers' composites mechanical properties. Its tailoring allows for the creation of anisotropic nanocomposites (Kanagaraj et al, 2007; Koerner et al, 2004; Pulskamp et al, 2007; Sen et al, 2004). Due to their semi - conductive behaviour, its usage may provide electrical stimulation (Grunlan et al, 2004; Huang et al, 2003). The use of CNT as translocators in drug-delivery systems or in image diagnosis has also been suggested (Bianco et al, 2005; Cherukuri et al, 2004). High tumour accumulation of single-walled CNT (SWCNT) has been described, anticipating the possibility of further therapeutic uses (Liu et al, 2007). There are several studies on gas, temperature, pressure, glucose, chemical force and resonator mass sensors based on CNT (Barone et al, 2005; Barone et al, 2005b; Collins et al, 2000; Hrapovic et al, 2004; Kong et al, 2000; Lee et al, 2007; Lin et al, 2005; Perez et al, 2005; Wood et al, 1999; Yan et al, 2007; Yang et al, 2006; Yun et al, 2007).

In face of recent studies, special attention has been drawn into promising orthopaedic use of CNT for improving tribological behaviour and material mechanical properties. However, and considering the conductive properties of CNT the range of orthopaedic application may broaden up, since it is known that electrical fields as small as 0,1 mV/cm may enhance osteoblastic proliferation locally (Brighton et al, 1992). CNT based electrodes could be considered for integrating implantable orthopaedic devices. CNT have been reported to have direct and distinct effects on osteoblasts and osteoclasts metabolic functions (Narita et al, 2009; Sirivisoot et al, 2007; Tutak et al, 2009).

CNT have been discovered in 1991 (IIjima, 1991), but seem to have been around for quite a long time, since they were detected in gas combustion streams like the ones in normal households stoves (Murr et al, 2004). The fact that CNT are small enough to be inhaled has raised the question of lung reaction to their presence. The impact on the skin of handlers and the environmental consequences of mass production are also pertinent interrogations, as it is the possibility of secondary organ dissemination.

2. Health hazards

2.1 Respiratory toxicity

Some authors described strong cytotoxic effects on guinea pig alveolar macrophages of SWCNT and, at a smaller extent, of multi-walled carbon nanotube (MWCNT), when compared to fullerenes (C60). The same authors also describe impairment of phagocytic activity (Jia et al, 2005). Cytotoxicity comparable to asbestos-particles induced on murine macrophages has been described by Soto (Soto et al, 2005). Experiments conducted by Magrez on three lung-tumor cell lines suggest CNT led to proliferation inhibition and cell death, although CNT showed less toxicity than carbon black nanoparticles and carbon nanofibers (Magrez et al, 2006). Davoren et al. assessed SWCNT cytotoxicity on a distinct lung-carcinoma cell line (A549) and describe SWCNT concentration - dependent toxicity and the protective effect of serum (Davoren et al, 2007).

Another study, conducted by Sharma, concluded that SWCNT induced oxidative stress in rat lung cells (Sharma et al, 2007). The same oxidative stress related changes are described by Herzog et al. in primary bronchial epithelial cells and A549 cells but the study points out that the length of the response is strongly dependent on the dispersion medium used (Herzog et al, 2009). Pulskamp also describes oxidative stress in two cell lines (rat macrophages NR8383 and human A549) cultured in contact with CNT. However, when comparing purified SWCNT and commercial CNT their findings suggested the biological effects were associated with the metal traces. They also describe puzzling divergent results between MTT (3-(4,5-Dimethylthiazol-2-yl)-2,5-diphenyltetrazolium bromide) and WST (water soluble tetrazolium salt, 2-(4-Iodophenyl)-3-(4-nitrophenyl)-5-(2,4-disulfophenyl)-2H-tetrazolium, monosodium salt) viability assays, both dependent on the activity of mitochondrial dehydrogenases (Pulskamp et al, 2007). These discrepancies can only be explained based on interactions of non-soluble formazan crystals in MTT with CNT, as opposed to WST soluble final formazan salt.

The results of *in vitro* studies that suggest toxicity are partially supported by several *in vivo* studies but, once again, with often divergent conclusions.

Huczko instilled intratracheally a soot of CNT in guinea pigs and later measured tidal volume, breathing frequency, pulmonary resistance and bronchoalveolar fluid cell and protein content. These authors concluded that working with soot containing CNT was probably not a health hazard, but no histopathological study was referred (Huczko et al, 2001). Lam et al. conducted studies in mice and concluded the SWCNT could be toxic if they reached the lungs; Warheit et al. conducted a similar study in rats, describing granuloma formation and considering it as probably resulting from aggregates of CNT (Lam et al, 2004; Warheit et al, 2004).

Muller et al. compared asbestos, carbon black and MWCNT effects when instilled in the trachea of rats, at different doses. These authors described dose-dependent persistent inflammation and granuloma formation, more significant with MWCNT than with carbon black but less extensive than with asbestos (Muller et al, 2005).

Shvedova et al. described unusual acute inflammatory response, early granulomatous reaction and progressive fibrosis in mice exposed to SWCNT, leading to the conclusion of CNT intrinsic toxicity. This study used a technique of pharyngeal aspiration instead of the intratracheal instillation used in the previous studies, and allowed aerosolization of fine SWNCT particles. These particles were associated with fibrogenic response in the absence of persistent local inflammation, suggesting health risks for workers (Shvedova et al, 2005). However, a more recent study describes significant changes in deposition pattern and pulmonary response when SWCNT are more evenly disperse in the suspension prior to pharyngeal aspiration (Mercer et al, 2008).

More recently, inhaled MWCNTs migration to the subpleura and associated increased number of pleural mononuclear cells and subpleural fibrosis was described in mice (Ryman-Rasmussen et al, 2009), further advising caution and appropriate security measures when handling CNT.

Wang et al. presented a study with dispersed SWCNT (DSWCNT) supporting data from previous reports, in the sense that they describe *in vitro* and *in vivo* stimulation of lung fibroblasts proliferation and collagen deposition, and metalloproteinase 9 increased expression, in the absence of inflammation (Wang et al, 2010).

It has also been hypothesized, and demonstrated for other types of nanoparticles, that following inhalation, nanoparticles may reach the central nervous system (CNS)(Elder et al, 2006). Nanoparticles enter the nervous system by transcytosis and are presented to neuron cells (Zensi et al, 2009). Studies showing that inhaled gold nanoparticles accumulate in olfactory bulb of rats and reach the cerebral cortex, as well as the lung and thereof other organs such as esophagus, tongue, kidney, aorta, spleen, septum, heart and blood (Yu et al, 2007).

These observations suggest that if there are high doses of nanoparticles in the air they can enter into the CNS via the olfactory nerve during accidental or prolonged environmental or occupational exposure to humans, and that nanoparticles may exert their effects not only on respiratory tract and neighboring organs but spread to distant organs.

2.2 Epidermal/dermal toxicity

Several studies have also been conducted on epidermal/dermal toxicity of CNT. Functionalized 6-aminohexanoic acid-derivatized SWCNT may cause dose-related rise in inflammatory cytokines (Zhang et al, 2007).

MWCNT induction inflammatory pathways may be similar to those of combustion-derived metals (Ding et al, 2005) and cause decreased cell viability, changes on metabolic, cell signalling, stress and cytoskeletal protein expression (Witzmann & Monteiro-Riviere, 2006). Other authors report presence of chemically unmodified MWCNT in cytoplasmic vacuoles of cultured human keratinocytes and induction of the release of interleukin 8 in a time dependent manner (Monteiro-Riviere et al, 2005) and SWCNT inhibition of HEK9293 cells growth through induction of apoptosis and decreased cell adhesion has also been described (Cui et al, 2005).

Patlolla et al. describe dose and time-dependent cytotoxicity, genotoxicity and induction of apoptosis by purified MWCNT in normal human dermal fibroblasts cells. The MWCNT used in this study had been treated for extraction of metal (Fe) impurities and then, by treatment with sulfuric/nitric acid, functionalized in very high degree. The authors report that 2 to 7% of final weight was due to carboxyl groups (Patlolla et al, 2010).

3. Biological response and mechanisms of toxicity

Whilst assessing *in vitro* cytotoxicity of SWCNT on fibroblasts and trying to bring some light on the issue of how the removal of catalytical metal would influence the toxicity, Tian et al. concluded that the refined SWCNT were more toxic, inducing significant changes on cytoskeleton and cell morphology, probably because of the enhancement of the hydrophobic character by the refinement treatment, the toxicity seemingly directly related to surface area (Tian et al, 2006). Sayes et al. reported decreased SWCNT cytotoxicity in dermal fibroblasts with higher functionalization density (Sayes et al, 2006). However, other authors compared pristine and oxidized MWCNT effects on human T lymphocytes and described increased toxicity of oxidized CNT, with high doses, even if oxidation increased solubility (Bottini et al, 2006).

Koyama et al. reported time-dependent changes in T lymphocytes by measuring CD4 and CD8, associated with local granuloma formation after subcutaneous implantation in mice, although overall toxicological changes were in absolute lower then with asbestos (Koyama et al, 2006). These results might seem somehow in conflict with the findings by Dumortier et al. that concluded that functionalized SWCNT did not affect B and T lymphocytes viability. However, the authors emphasized that absence of functional changes was only observed in the CNT funcionalized via the 1,3-dipolar cycloaddion reaction, in non - oxidized nanotubes (Dumortier et al, 2006). Brown et al. conducted *in vitro* studies that suggested monocytic cells' response is strongly dependent of morphology and state of aggregation of the CNT. Long, straight well-dispersed nanofilaments induced the production of more TNF-α and ROS than highly curved and entangled aggregates; incomplete uptake or frustrated phagocytosis of CNT was also described (Brown et al, 2007). Barillet et al. showed that short (0.1-5 nm) and long (0.1-20 nm) CNT, and the presence of metal residues, induced different cell response and toxicity (Barillet et al, 2010).

The same mechanisms of frustrated phagocythosis, increased production of proinflammatory cytokines and oxidative stress apparently justified the *in vivo* findings described by several authors. They conducted studies with longer implantation times and these effects may eventually lead to carcinogenesis (Fraczek et al, 2008; Poland et al, 2008; Takagi et al, 2008).

Whilst several authors describe low cellular uptake of SWCNT (Barillet et al, 2010; Davoren et al, 2007; Herzog et al, 2007), others describe high cellular uptake (Jia et al, 2005; Pulskamp et al, 2007). As suggested by Shvedova et al. and Mercer et al., the degree of uptake and the cell response may be dependent on the carrier of the CNT (Mercer et al, 2008; Shvedova et al, 2009). Protein adsorption to CNT surface and its subsequent structural change may trigger phagocytosis, and Barillet et al. suggest CNT may induce membrane damage by mechanical action (Barillet et al, 2010).

Bihari et al. reported SWCNT thrombogenic and platelet activation effects in mice and point out this could cause possible systemic problems, along with hindering the use of these materials for drug delivery (Bihari et al, 2010).

These recent studies stress the need of careful re-evaluation and research before enlarging the field of CNT application. However, as Fraczek et al, Lacerda et al, Schipper et al and Sitharaman et al, among others, findings suggest, CNT behaviour *in vivo* depends on their length, functionalization and degree of agglutination (Fraczek et al, 2008; Lacerda et al, 2008; Schipper et al, 2008; Sitharaman et al, 2008). Up to now, several studies report low or undetected liver and systemic toxicity in mice, although CNT presence has been shown in the liver, lung and faeces after intravenous injection (Deng et al, 2007; Fraczek et al, 2008; Schipper et al, 2008).

4. Mechanisms of interaction of CNT

The questions related to possible interactions between CNT and various dye markers, pointing out the difficulties in the interpretation of the obtained results are raised by several authors, pointing out the need for careful interpretation of the results (Casey et al, 2007; Davoren et al, 2007; Pulskamp et al, 2007).

The commonly used MTT assay, used to assess cell viability and proliferation, has been described to falsely lower results due to attachment of insoluble formazan to CNT (Pulskamp et al, 2007).

Guo et al. describe SWCNT dose-dependent adsorption and depletion of over 14 amino acids and vitamins from RPMI cell culture medium. This implies that indirect mechanisms of toxicity may influence the results of *in vitro* studies, since some of these molecules are essential for cell viability and proliferation. SWCNT cause dose-dependent adsorption of culture medium amino acids and vitamins, showing higher affinity for planar aromatic or conjugated structures, and for positively charged solutes (Guo et al, 2008).

Functionalization of SWCNT and MWCNT with terminal or surface specific groups alters solubility and protein adsorption, including of cytokines IL6 and IL8, in a dose-dependent manner (Tian et al, 2006). In the absence of specific chemical affinity between the nanotube surface and the protein, one cause of interference would be the seizing of the molecule inside the nanotube, dependent on molecule size, unless CNT are functionalized with specific groups that promote chemical binding. CNT's active surface issues are equally important, as in a composite CNT surface available for interaction is reduced because nanotubes are embedded in a matrix.

There are several possible mechanisms of interaction. Molecule adsorption is probably strongly dependent on charge and molecule size, and also on the CNT surface available for interaction.

The authors explored protein adsorption to non-functionalized and functionalized multi-walled CNT (MWCNT) and to ultra high molecular weight polyethylene (UHMWPE)/ MWCNT composite and with UHMWPE polymer alone.

Two different proteins were chosen, bovine serum albumin (BSA, Promega) and histone (Histone, calf thymus, Merck). Histones are a group of small proteins, with molecular weights varying from around 21 500 Dalton to 11 200 Dalton; at neutral pH, histones are positively charged (Panyim & Chalkley, 1971). Bovine serum albumin (BSA) has a molecular weight of around 66 700 Dalton and its isoelectric point is 4.7, thus being negatively charged at pH 7, due to the domination of acidic groups over amine groups (Tsargorodskaya et al, 2004).

Solutions of both proteins (concentration 200 μg/mL) were prepared through agitation in PBS (without calcium and magnesium), and the pH adjusted to 7 with HCl 1 N.

MWCNT (range of diameter 60-100 nm, length of the tubes 5–15 μm), non-functionalized and functionalized with carbonyl, carboxyl and hydroxyl groups were added to the solutions (n=6), with a concentration of 100 μg/mL, and mixed by vortexing. Bulk samples of composite (0.2% MWCNT) and polymer were also incubated in the solutions, maintaining the same weight/volume rate.

After 12 hours at room temperature, solutions were filtered using 0.2 μm polyethersulfone low protein binding syringe filters (VWR). Initial albumin and histone solutions were also filtered.

Protein content in the filtrates was assessed, in triplicate, by the bicinchoninic acid assay (BCA Protein Assay, Calbiochem), accordingly to the manufacturer's instructions. PBS was used as blank. The protein content in solutions incubated with the materials is expressed in percentage of histone and BSA filtered solutions, assumed as 100%.

The normal distribution was verified by the Kolmogorov-Smirnov test, homogeneity of variance by the Levene test and means compared ANOVA (Tukey test). The statistical analysis was done using software OriginPro 8 (OriginLab Corporation, USA).

Figure 1 shows the percentages of BSA and histone left in filtrate after incubation with functionalized and non-functionalized CNT, assuming filtered solutions of both proteins as 100%; bars represent mean and error bars represent standard deviation.
When considering BSA and histone adsorption to UHMWPE and MWCNT/UHMWPE composite, no differences were found between both materials and protein solutions.
A statistically significant (at a 0.05 level) higher amount of histone was retained by functionalized CNT, when comparing to non-functionalized CNT. A significantly higher amount of histone was retained in both groups of CNT when comparing to BSA. The difference in protein adsorption between the two groups of CNT incubated in BSA solution was not significant.

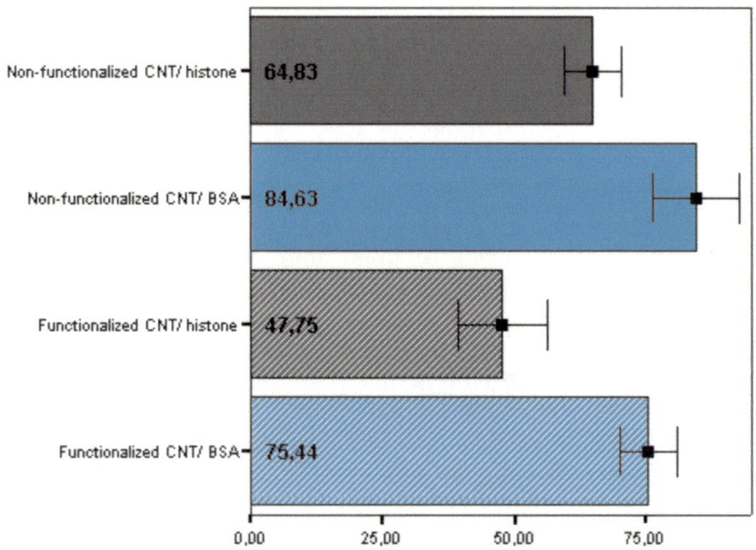

Fig. 1. Percentage of bovine serum albumin (BSA) and histone after incubation with CNT.

The results show that protein charge and size are paramount for the interaction with CNT, likewise CNT surface functional groups and the surface area available for interaction. Smaller, positive charged particles are more likely to bind in significant amounts to both functionalized and non-functionalized CNT, as the present results suggest, and in agreement with the conclusions from previous studies (Guo et al, 2008). Functionalizing CNT may further enhance adsorption.
The need to consider possible interactions between CNT and the substances evaluated and used in assays is, thus, clear. Positive advances are being pursued and achieved in better understanding and predicting the interactions between CNT and other nanoparticles and biological molecules (Xia et al, 2010). As it is also the authors' opinion, and as recent literature emphasizes, there is an urgent need for standardization of assays involving CNT (Ren et al, 2010).

5. Potential medical applications of CNT

CNT may be used in orthopedics as mechanical reinforcement, to tailor surface properties and provide a nanostructured surface that promotes bone cell adhesion and function, or by exploring CNT conductive properties in the view of the development of smart implants.

Few studies are available on the cytocompatibility of composites using CNT as reinforcement. Chlopek et al. reported cytocompatibility of MWCNT similar to that of polysulfone, after culturing osteoblasts and fibroblasts in with polysulfone alone and with polysulfone plus MWCNT. These authors described a slight increase of collagen I production, in the absence of induction of IL6 and free radicals (Chlopek et al, 2006).

George et al. studied the adhesion behaviour of osteoblast-like cells MG63, primary osteoblasts and A549 cells on MWCNT surfaces. They determined that all cell types were capable of attaching and proliferating but not able to penetrate the mesh work. These findings suggested cell spreading and migration were affected (George et al, 2006). However, in a more recent study conducted by Meng et al., fibroblasts had improved growth and collagen synthesis on a nanofibrous scaffold made of composite MWCNT/ polyurethane, when comparing to controls and polyurethane alone (Meng et al, 2008).

Shitaraman et al. described in vivo good tissue response to ultra-short SWCNT, propylene fumarate diacrylate (PPF) composite scaffolds, similar to the response to PPF alone (Shitaraman et al, 2008). Other studies report bone cell growth along CNT and plasma-sprayed carbon nanotube reinforced hydroxyapatite, suggesting CNT conductive properties may be explored (Balani et al, 2007; Zanello et al, 2006).

Sirivisoot's study concluded that osteoblast (bone forming cell) functions (specifically alkaline phosphatase activity and calcium deposition) are significantly greater on MWCNT grown by chemical vapor deposition on anodized porous Ti than on anodized Ti without CNT and currently-used Ti for up to 21 days (Sirivisoot et al, 2007).

Usui et al. implanted highly crystalline MWCNT, with purity around 98%, in subperiosteal and tibial defects in mice. These authors describe MWCNT incorporation into bone marrow tissue and bone matrix. When combined with rhBMP-2, MWCNT-collagen composites accelerated bone formation when compared to collagen and rhBMP-2 alone after implantation in dorsal musculature of mice (Usui et al, 2008).

The results of a study conducted by Tutak et al also describe a rise in alkaline phosphatase activity, collagen I synthesis and total protein content in MC3T3-E1 cells grown on thin film SWCNT substrates, along with a decrease in the number of viable cells, assessed by the MTT assay (Tutak et al, 2010). The authors suggest this may be due to initial toxicity and release of cytoplasmatic growth factors by injured cells.

Osteoblast response is probably dependent on surface energy density and the roughness of the SWCNT surface, both cell adhesion and proliferation being affected. Hydrophilic and medium rough films provoked higher cell adhesion and proliferation (Tutak et al, 2010).

There is, also, growing evidence that MWCNT may inhibit osteoclast differentiation and activity in vivo and *in vitro* and do not affect osteoblasts negatively (Narita et al, 2009).In this study, the authors implanted rhBMP-2/collagen/MWCNT or rhBMP-2/collagen composites in the dorsal musculature of mice. They reported a significantly lower number of osteoclasts in the neo-formed ectopic bone in the rhBMP-2/collagen/80n-MWCNT group than in the rhBMP-2/collagen group.

Yadav et al described the use of MWCNT for reinforcement of hydroxyapatite and gelatin composites with the aim of improving mechanical properties of a potential nanocomposite for bone replacement. The authors describe absence of skin, kidney and liver changes after subcutaneous injection in mice of composites with 0, 1, and 2% non-functionalized MWCNT. However, in the group that received injection of a 4% MWCNT, mild changes were observed, such as hydropic changes and ballooning of hepatocytes and glomerular abelation and swelling of renal tubules. Although a significant increase in the flexural strength of the composite was achieved through the addition of the MWCNT, the authors describe a lowering of the Young modulus, and final elasticity is still far from the one of natural bone (Yadav et al, 2009).

UHMWPE is one of the mostly used materials in biomedical applications such as acetabular cups as bearing surface in total hip arthroplasty. Coupled with a metal or ceramic femoral head, UHMWPE has shown excellent resistance to wear. However, despite the success of the total joint arthroplasty, wear is the major obstacle limiting the long-term performance of the UHMWPE implants. It has been estimated that billions of particles are produced yearly from the surface of a total hip replacement (Kurtz et al, 1999). Although UHMWPE components are in no imminent danger of wearing during patient's lifetime, osteolysis and loosening of the implants are attributed to the debris generated from the articulating surface. Moreover, the mechanical and physical properties of UHMWPE still need to be improved; in particular, enhancing friction and wear resistance can satisfy the material requirement in joint replacement surgeries (Jacobs et al, 2002).

Ruan et al (2003) reported an enhancement of toughness in UHMWPE films with the addition of 1 wt% MWCNT. Their results revealed an increase in strain energy density of ~150 % for the composites as compared with pure UHMWPE. They also reported an increase of ~140 % in ductility and up to 25 % in tensile strength. An analysis by nanoindenter and atomic force microscopy (AFM) of UHMWPE/MWCNT composites has been reported by Wei et al (2006). They have observed a decrease of the friction coefficient with MWCNT content increase.

With such promising results, the authors prepared UHMWPE/CNT composites processed through optimized compression moulding process in order to obtain a suitable reinforced polymeric material for the production of the acetabular cup component used as hip joint implants.

The effective use of carbon nanotubes in composites depends upon its homogeneous dispersion throughout the polymer matrix, without/less destroying their integrity. To prevent the CNT agglomeration and to improve the interfacial bonding between the polymer and the CNT, it was performed a chemical treatment to the MWCNTs following the methodology proposed by Esumi et al (1996). It is believed that the polymer spreads over the MWCNT due to their clean nature and seamless structure of the CNT, leading to an interface coating of the polymer on CNT of about few micron thickness which is due to wettability characteristics of polymer and thus the interfacial bonding between them increases which leads to enhancement of load transfer from polymer to MWCNTs (Ding et al, 2003; Potschke et al, 2002; Kanagaraj et al, 2007). Figure 2 clearly shows the homogeneous distribution of the MWCNT in the polymeric matrix.

The prepared 0.2% MWCNT/UHMWPE composite and pure UHMWPE were subjected to tensile tests for mechanical characterization and some tribological tests, using a ball-on-plate configuration (Kanagaraj et al, 2010). In Figure 3 it can be observed the variation of the wear volume for pure polymer and for composite against the sliding distance

Kanagaraj et al. developed a UHMWPE/MWCNT composite that presents superior wear behaviour (decreased wear volume and wear coefficient) when compared to conventional UHMWPE. Reinforcement of UHMWPE by adding multiwalled carbon nanotubes (MWCNT) allows improvement of mechanical characteristics for biomedical applications.

Fig. 2. SEM pictures of 0.2% MWCNT/UHMWPE composite. The presence of MWCNT is highlighted.

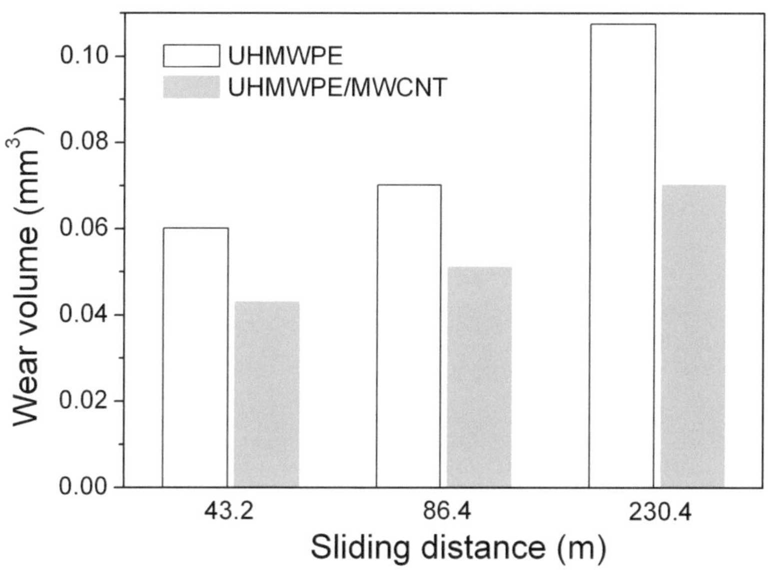

Fig. 3. Wear volume variation for pure polymer and UHMWPE/MWCNT composite against sliding distance.

It is clearly observed that the wear volume of the composite is lower than the one observed for pure polymer, which can be correlated with an increase of toughness of the composite when compared with the pure polymer. The decrease of wear volume was: 28%, 28% and 35 % for sidings' distances of 43.2m, 86.4m and 232.2m, respectively. This decreasing results in a good load transfer effect from the nanotube to the polymer.
The decreasing of wear volume decreasing with incorporation of the MWCNT is consistent with an increasing of toughness for the composite sample as compared with pure polymer. In Table 1 it is shown the mechanical properties for both studied samples (UHMWPE and UHMWPE/MWCNT composite).

	Young's modulus		Tensile strength		Toughness		Breaking elongation	
	GPa	% inc.	MPa	% inc.	J/g	% inc.	%	% inc.
UHMWPE	0.718	0	21.17	0	50.84	0	280.73	0
UHMWPE/CNT	0.754	5.0	25.97	22.7	68.53	34.8	354.7	26.3

Table 1. Mechanical properties of UHMWPE and UHMWPE/MWCNT.

It is clearly observed an increase of the mechanical properties for the composite when compared with pure polymer. The observed increasing in toughness was of about 34.8%.
The developed UHMWPE/MWCNT composite seems to present superior mechanical properties when compared to a conventional UHMWPE, but there is still some controversy when it comes to carbon nanotubes toxicity. To analyze the latter the authors assessed human osteoblast-like MG63 cells viability and proliferation and interleukin-6 (IL-6) production in contact with this composite's particles (Figure 4).

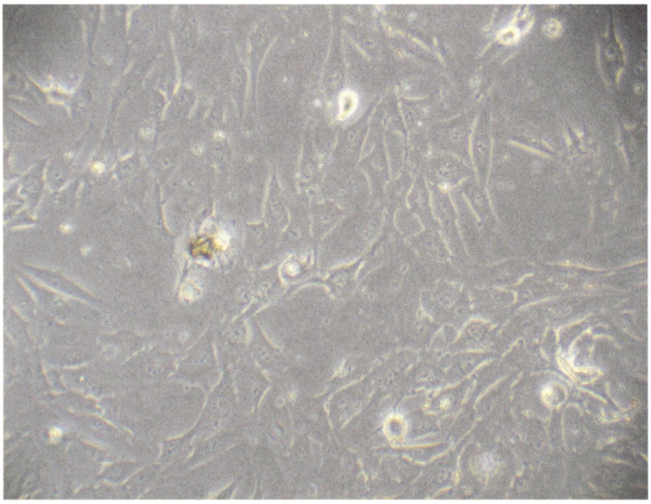

Fig. 4. MG63 after 3 days culture in contact with UHMWPE/MWCNT composite particles.

The results suggest cytocompatibility similar to that of conventional UHMWPE (Table 2). The viability and proliferation were assessed by the WST-1 assay, known to be not influenced by the presence of CNT and the results expressed as percentage of control standard culture plate ± standard deviation. The IL6 production was assessed in culture medium by ELISA assay (Peprotech) and total protein content (Calbiochem) and results expressed as mean± standard deviation.

Assay	UHMWPE/MWCNT	UHMWPE
WST-1	97.92 ± 8.29%	96.19 ± 7.92%
IL6 (pg/mL)	108.99 ± 9.90	92.52 ± 11.02
Total Protein (μg/ mL)	163.29 ± 11.81	137.07 ± 6.17

Table 2. *In vitro* results: MG63 cells after 6 days in contact with UHMWPE/MWCNT composite and UHMWPE

Naresh et al. (2011) has observed hydroxyapatite coating over UHMWPE-CNT composites with time during the simulated body fluid test (SBF), where the test was carried out for a week. The thickness of the hydroxyapatite layer was also increased with time where approximately 55% of weight was increased on 7th day of SBF testing (Figure 5). From the optical images, it is also confirmed (Figure 6).

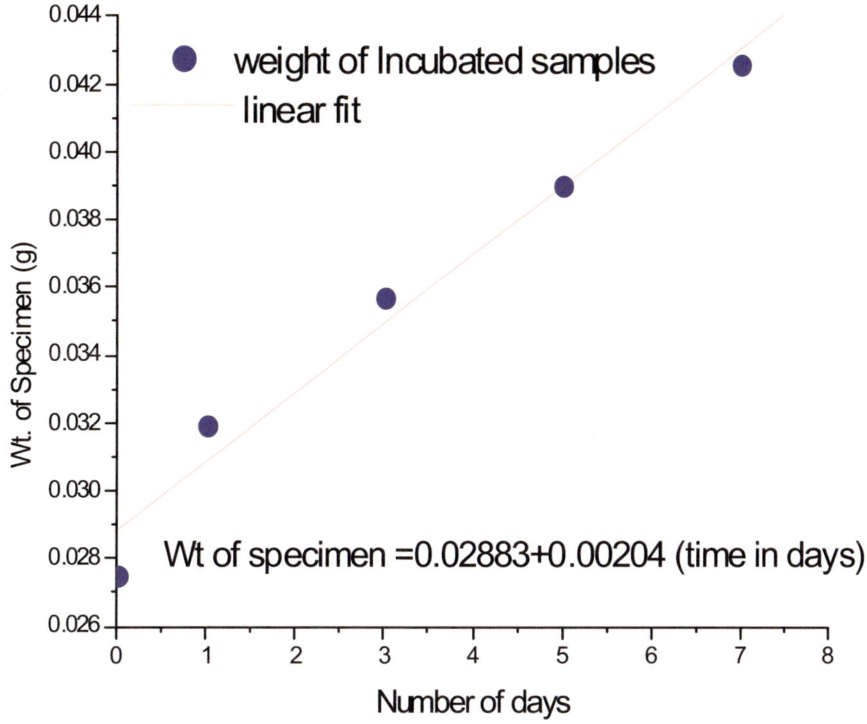

Fig. 5. Hydroxyapatite coating on UHMWPE/CNT composite with time during SBF test.

Fig. 6. Optical microscope images of Hydroxyapatite grown on UHMWPE/CNT 2.0 wt% on day 0, 3 and 5.

As Balani et al. have suggested (2007), the presence of CNT may assist in hydroxyapatite crystals nucleation and growth, promoting higher crystallinity.

Recently, formation of hybrid CNT-protein nanofibers was reported. These nanofibers contained fibronectin that promoted the nucleation of crystalline nuclei and the CNT templates controled the orientation of nuclei and the crystal growth of hydroxyapatite to form flake-like crystals (Wei et al, 2011). Nucleation and crystal growth occur in a biomimetic way.

6. Conclusion and future remarks

The immense potential of CNT for biomedical applications is evident. Either as sensors, drug carriers, imagiology aids, bioelectrodes or reinforcement for composites, these are highly versatile and promising molecules.

The acetabular cup in hip prosthesis and the tribologic surfaces in knee prosthesis is often lined with ultrahigh-molecular-weight polyethylene (UHMWPE). It is accepted that debris particles derived from UHMWPE weight-bearing surfaces are the main cause for debris-associated aseptic loosening. Apart of the sometimes catastrophic consequences for the integrity of the implant, debris particles attract macrophages, causing their activation and the secretion of proinflammatory mediators such as interleukine (IL) 1, IL6, IL17, receptor activator of nuclear factor-κB ligand (RANKL) and proteolytic enzymes.

Better wear behavior and less particle generation would help lowering the numbers of revision arthroplasty procedures due to material failure and aseptic loosening.

Kanagaraj et al. developed a UHMWPE/MWCNT composite that presents superior wear behavior (decreased wear volume and wear coefficient) when compared to conventional UHMWPE. The preliminary *in vitro* studies in human osteoblast-like MG63 show the cell response to particles of this novel composite is similar to that provoked by conventional medical grade UHMWPE. The viability and IL6 production do not show significant differences. More throughout studies, using different cell lines and a deeper look on cell response, can be developed.

However, the research carried out is far from having reached a consensus on CNT toxicity, although further studies are being conducted on the subject. Nevertheless, the main issue is that probably it is being compared what cannot be compared. As neatly indicated by the studies conducted by Herzog et al. (2009) and Barillet et al. (2010), among others, different CNT, produced by different methods, yielding different amounts of metal residues, with varying sizes and surface area, functionalized or not, different degrees of functionalization, dispersed in different mediums, elicit different responses, in different cell lines, assessed by different methodologies! Some of which are known to suffer from interference when CNT are present.

When performing either *in vitro* or *in vivo* studies, a deep knowledge on the type of CNT used and exploring possible interactions with experimental methods is mandatory. A thorough in depth study should be conducted, allowing establishment of experimental methodology guidelines. Studies conducted up to now advice precaution when handling these materials due to possible epidermal and respiratory detrimental effects.

In fact, the same characteristics that make nanomaterials so promising for drug delivery purposes also may be source of concern, if distant spread and accumulation in organs cannot be controlled.

It is clear that CNT surface characteristics are determinant for interaction with other molecules. As inferred from the histone and BSA adsorption assay here presented, adsorption is not just dependent on charge but also on molecule size. The degree in which it occurs is dependent on the surface area available for interaction, and the fact that it can occur in high degree (more than 50% of histone on solution was retained in the functionalized CNT) may have dramatic effects on interaction with cells and cell response, and with assays used to assess this same response. Mechanisms of interaction are slowly being explored and revealed, but tools for predicting adsorption behavior are not yet widely available. Developing these tools and wide spreading their use call upon a interdisciplinary approach, putting together knowledge and technological resources traditionally handled by biologists, chemists, physicists, material science engineers, among others. A non-traditional and unique material demands a non-traditional and custom-made approach, as in the last decades a new era has begun.

The development of knowledge on CNT interactions with biological systems gives hope to fully explore the CNT potential as reinforcement component in composites for orthopedic applications. It makes a better use of their ability to promote biomimetic nucleation and growth of hydroxyapatite crystals and exceptional strength. The growing interest on bone electrophysiology and piezoelectricity, and CNT conductive properties, can anticipate their use in smart implants, able to adapt their performance to the mechanical environment and as constituents of materials that mimic bone natural properties and support osteoblast proliferation and differentiation. CNT applications are almost unlimited, and we can expect to see further research on their application as drug carriers and in imagiology, due to their capacity to cross biological membranes, near-infrared intrinsic fluorescence and biodistribution. The biodistribution and pharmacokinetics may be tuned by controlling the size, the surface chemistry, and the targeting ligand, and CNT can be loaded with a variety of drugs, being a specially promising tool in the fight against cancer.

7. References

Balani, K., Anderson, R., Laha, T., Andara, M., Tercero, J., Crumpler, E. & Agarwal, A., 2007. Plasma-sprayed carbon nanotube reinforced hydroxyapatite coatings and their interaction with human osteoblasts in vitro. *Biomaterials*, Vol. 28, No. 4, (February 2007), pp. 618-624, ISSN: 0142-9612

Barillet, S., Simon-Deckers, A., Herlin-Boime, N., Mayne-L'Hermite, M., Reynaud, C., Cassio, D., Gouget, B. & Carrière, M., 2010. Toxicological consequences of TiO2, SiC nanoparticles and multi-walled carbon nanotubes exposure in several mammalian cell types: an in vitro study. *Journal of Nanoparticle Research*, Vol. 12, No. 1, (January 2010), pp. 61-73, ISSN: 1572-896X

Barone, P.W., Baik, S., Heller, D.A. & Strano, M.S., 2005a. Near-infrared optical sensors based on single-walled carbon nanotubes. *Nature Materials*, Vol. 4, No.1, (January 2005), pp. 86-92, ISSN: 1476-1122

Barone, P.W., Parker, R.S & Strano, M.S., 2005b. In Vivo Fluorescence Detection of Glucose Using a Single-Walled Carbon Nanotube Optical Sensor: Design, Fluorophore Properties, Advantages, and Disadvantages. *Analytical Chemistry*, Vol. 77, No.23, (December 2005), pp. 7556-7562, ISSN: 1520-6882

Bianco, A., Kostarelos, K. & Prato, M., 2005. Applications of carbon nanotubes in drug delivery. *Current Opinion in Chemical Biology*, Vol. 9, No.6, (December 2005), pp. 674-679, ISSN: 1367-5931

Bihari, P., Holzer, M., Praetner, M., Fent, J., Lerchenberger, M., Reichel, C.A., Rehberg, M., Lakatos, S. & Krombach, F., 2010. Single-walled carbon nanotubes activate platelets and accelerate thrombus formation in the microcirculation. *Toxicology*, Vol. 269, No. 2-3, (March 2010), pp. 148-154, ISSN: 0300-483X

Bottini, M., Bruckner, S., Nika, K., Bottini, N., S. Bellucci, Magrini, A., Bergamaschi, A. & Mustelin, T., 2006. Multi-walled carbon nanotubes induce T lymphocyte apoptosis. *Toxicology Letters*, Vol. 160, No. 2, (January 2006), pp. 121-12, ISSN: 0378-4274

Brighton CT, Okereke E, Pollack SR & Clark CC. (1992). In Vitro Bone-Cell Response to a Capacitively Coupled Electrical Field The Role of Field Strength, Pulse Pattern, and Duty Cycle. *Clinical orthopaedics and related research*, Vol. 285, (December 1992), pp. 255-262, ISSN: 1941-7551

Brown, D.M., Kinloch, I.A., Bangert, U., Windle, A.H., Walter, D.M., Walker, G.S., Scotchford, C.A., Donaldson, K. & Stone, V., 2007. An in vitro study of the potential of carbon nanotubes and nanofibres to induce inflammatory mediators and frustrated phagocytosis. *Carbon*, Vol. 45, No. 9, (August 2007), pp. 1743-1756, ISSN: 0008-6223

Casey, A., Herzog, E., Davoren, M., Lyng, F.M., Byrne, H.J. & Chambers, G., 2007. Spectroscopic analysis confirms the interactions between single walled carbon nanotubes and various dyes commonly used to assess cytotoxicity. *Carbon*, Vol. 45, No. 7, (June 2007), pp. 1425-1432, ISSN: 0008-6223

Chang, N., Bellare, A., Choen, R.E., & Spector, M., 2000. Wear behavior of bulk oriented and fiber reinforced UHMWPE. *Wear*, Vol. 241, No 1, (June 2000), pp. 109-117, ISSN: 0043-1648

Cherukuri, P., Bachilo, S.M., Litovsky, S.H. & Weisman, R.B., 2004. Near-Infrared Fluorescence Microscopy of Single-Walled Carbon Nanotubes in Phagocytic Cells. *Journal of the American Chemical Society*, Vol. 126, No.48, (December 2004), pp. 15638-15639, ISSN: 1520-5126

Chlopek, J., Czajkowska, B., Szaraniec, B., Frackowiak, E., Szostak, K. & Béguin, F., 2006. In vitro studies of carbon nanotubes biocompatibility. *Carbon*, Vol. 44, No. 6, (May 2006), pp. 1106-1111, ISSN: 0008-6223

Collins, P.G., Bradley, K., Ishigami, M. & Zettl, A., 2000. Extreme Oxygen Sensitivity of Electronic Properties of Carbon Nanotubes. *Science*, Vol. 287, No. 5459, (March 2000), pp. 1801-1804, ISSN: 1095-9203

Cui, D., Tian, F., Ozkan, C.S., Wang, M. & Gao, H., 2005. Effect of single wall carbon nanotubes on human HEK293 cells. *Toxicology Letters*, Vol. 155, No. 1, (January 2005), pp. 73-85, ISSN: 0378-4274

Davoren, M., Herzog, E., Casey, A., Cottineau, B., Chambers, G., Byrne, H.J. & Lyng, F.M., 2007. In vitro toxicity evaluation of single walled carbon nanotubes on human A549 lung cells. *Toxicology in Vitro*, Vol. 21, No. 3, (April 2007), pp. 438-448, ISSN: 0887-2333

Deng, X., Jia, G., Wang, H., Sun, H., Wang, X., Yang, S., Wang, T. & Liu, Y., 2007. Translocation and fate of multi-walled carbon nanotubes in vivo. *Carbon*, Vol. 45, No. 7, (June 2007), pp. 1419-1424, ISSN: 0008-6223

Ding, L., Stilwell, J., Zhang, T., Elboudwarej, O., Jiang, H., Selegue, J.P., Cooke, P.A., Gray, J.W. & Chen, F.F., 2005. Molecular Characterization of the Cytotoxic Mechanism of Multiwall Carbon Nanotubes and Nano-Onions on Human Skin Fibroblast. *Nano Letters*, Vol. 5, No. 12, (December 2005), pp. 2448-2464, ISSN: 1530-6992

Ding, W., Eitan A., Fisher F.T., Chen, X., Dikin, D.A., Andrews, R., Brinson, L.C., Schadler, L.S. & Ruoff, R.S., 2003. Direct observation of polymer sheathing in carbon nanotube-polycarbonate composites. *Nano Letters*, Vol. 3, No. 11, (November 2003), pp. 1593-2597, ISSN: 1530-6992

Dumortier, H., Lacotte, S., Pastorin, G., Marega, R., Wu, W., Bonifazi, D., Briand, J.-P., Prato, M., Muller, S. & Bianco, A., 2006. Functionalized Carbon Nanotubes Are Non-Cytotoxic and Preserve the Functionality of Primary Immune Cells. *Nano Letters*, Vol. 6, No. 7, (July 2006), pp. 1522-1528, ISSN: 1530-6992

Elder, A., Gelein, R., Silva, V., Feikert, T., Opanashuk, L., Carter, J., Potter, R., Maynard, A., Ito, Y., Finkelstein, J. & Oberdorster, G., 2006. Translocation of inhaled ultrafine manganese oxide particles to the central nervous system. *Environmental Health* Perspectives, Vol. 114, No. 8, (August 2006), pp. 1172-1178, ISSN: 0091-6765

Esumi, K., Ishigami, M., Nakajima, A., Sawada, K., and Honda, H., 1996. Chemical treatment of carbon nanotubes, *Carbon*, Vol. 34, pp. 279-81, ISSN 0008-6223

Fraczek, A., Menaszek, E., Paluszkiewicz, C. & Blazewicz, M., 2008. Comparative in vivo biocompatibility study of single- and multi-wall carbon nanotubes. *Acta Biomaterialia*, Vol. 4, No. 6, (November 2008), pp. 1593-1602, ISSN: 1742-7061

George, J.H., Shaffer, M.S. & Stevens, M.M., 2006. Investigating the cellular response to nanofibrous materials by use of a multi-walled carbon nanotube model. *Journal of Experimental Nanoscience*, Vol. 1, No. 1, (March 2006), pp. 1-12, ISSN: 1745-8099

Grunlan, J.C., Mehrabi, A.R., Bannon, M.V. & Bahr, J.L., 2004. Water-Based Single-Walled-Nanotube-Filled Polymer Composite with an Exceptionally Low Percolation Threshold. *Advanced Materials*, Vol. 16, No. 2, (January 2004), pp. 150-153, ISSN: 1521-4095

Guo, L., Bussche, A.V.D., Buechner, M., Yan, A., Kane, A.G. & Hurt, R.H., 2008. Adsorption of essential micronutrients by carbon nanotubes and the implications for nanotoxicity testing. *Small*,Vol. 4, No. 6, (June 2008), pp. 721-727, ISSN: 1613-6829

Herzog, E., Byrne, H.J., Davoren, M., Casey, A., Duschl, A., Oostingh, G.J., 2009. Dispersion medium modulates oxidative stress response of human lung epithelial cells upon exposure to carbon nanomaterial samples. *Toxicology and Applied Pharmacology*, Vol. 236, No.3, (May 2009), pp. 276–281, ISSN: 0041-008X

Herzog, E., Casey, A., Lyng, F. M., Chambers, G., Byrne, H. J., & Davoren, M. (2007). A new approach to the toxicity testing of carbon-based nanomaterials - The clonogenic assay. *Toxicology Letters*, Vol. 174, No. 1-3, (November 2007), pp. 49–60, ISSN: 0378-4274

Hrapovic, S., Liu, Y., Male, K.B. & Luong, J.H.T., 2004. Electrochemical Biosensing Platforms Using Platinum Nanoparticles and Carbon Nanotubes. *Analytical Chemistry*, Vol. 76, No. 4, (February 2004), pp. 1083-1088, ISSN: 1520-6882

Huang, J.-E., Li, X.-H., Xu, J.-C. & Li, H.-L., 2003. Well-dispersed single-walled carbon nanotube/polyaniline composite films. *Carbon*, Vol. 41, No. 14, (October 2003), pp. 2731-2736, ISSN: 0008-6223

Huczko, A., Lange, H., Calstrokko, E., Grubek-Jaworska, H. & Droszcz, P., 2001. Physiological test of carbon nanotubes: Are they asbestos-like? *Fullerene science and technology*, Vol. 9, No. 2, (April 2001), pp. 251-254, ISSN: 1064-122X

Iijima, S., 1991. Helical microtubules of graphitic carbon. *Nature*, Vol. 354, No. 6348, (November 1991), pp. 56-58, ISSN: 0028-0836

Jacobs, O., Kazanci, M., Cohn, D., & Marom, G., 2002. Creep and wear behaviour of ethylene-butene copolymers reinforced by ultra-high molecular weight polyethylene fibres. *Wear,* Vol. 253, No. 5, (September 2002), pp. 618-625, ISSN: 0043-1648

Jia, G., Wang, H., Yan, Y., Wang, X., Pei, R., Yan, Y., Zhao, B. & Guo, X., 2005. Cytotoxicity of Carbon Nanomaterials: Single-Wall Nanotube, Multi-Wall Nanotube, and Fullerene. *Environmental Science and Technology,* Vol. 39, No. 5, (March 2005), pp.1378-1383, ISSN: 1520-5851

Kanagaraj, S., Mathew, M.T., Fonseca, A., Oliveira, M.S.A., Simões, J.A.O. & Rocha, L.A., 2010. Tribological characterisation of carbon nanotubes/ultrahigh molecular weight polyethylene composites: the effect of sliding distance. *International Journal of Surface Science and Engineering,* Vol. 4, Nos. 4/5/6, pp. 305-321, ISSN: 1749-7868

Kanagaraj, S., Varanda, F.R., Zhil'tsova, T.V., Oliveira, M.S.A. & Simões, J.A.O., 2007. Mechanical properties of high density polyethylene/carbon nanotube composites. *Composites Science and Technology* Vol. 67, No. 15-16, (December 2007), pp. 3071-3077, ISSN: 0266-3538

Koerner, H., Price, G., Pearce, N.A., Alexander, M. & Vaia, R.A., 2004. Remotely actuated polymer nanocomposites - stress-recovery of carbon-nanotube-filled thermoplastic elastomers. *Nature Materials,* Vol. 3, No. 2, (February 2004), pp. 115-120, ISSN: 1476-1122

Kong, J., Franklin, N.R., Zhou, C., Chapline, M.G., Peng, S., Cho, K. & Dai, H., 2000. Nanotube Molecular Wires as Chemical Sensors. *Science,* Vol. 287, No. 5453, (January 2000), pp. 622-625, ISSN: 1095-9203

Koyama, S., Endo, M., Kim, Y.-A., Hayashi, T., Yanagisawa, T., Osaka, K., Koyama, H., Haniu, H. & Kuroiwa, N., 2006. Role of systemic T-cells and histopathological aspects after subcutaneous implantation of various carbon nanotubes in mice. *Carbon,* Vol. 44, No. 6, (May 2006), pp. 1079-1092, ISSN: 0008-6223

Kurtz, S.M., Muratoglu, O.K., Evans, M., and Edidin, A.A., 1999. Advances in the processing, sterilization, and crosslinking of ultra-high molecular weight polyethylene for total joint arthroplasty. *Biomaterials,* Vol. 20, No. 18, (September 1999), pp. 1659-1688, ISSN: 0142-9612

Lacerda, L., Ali-Boucetta, H., Herrero, M.A., Pastorin, G., Bianco, A., Prato, M. & Kostarelos, K., 2008. Tissue histology and physiology following intravenous administration of different types of functionalized multiwalled carbon nanotubes. *Nanomedicine,* Vol. 3, No. 2, (April 2008), pp. 149-161, ISSN: 1549-9634

Lam, C.-W., James, J.T., McCluskey, R. & Hunter, R.L., 2004. Pulmonary Toxicity of Single-Wall Carbon Nanotubes in Mice 7 and 90 Days After Intratracheal Instillation. *Toxicological Sciences* Vol. 77, No.1, (January 2004), pp. 126-134, ISSN: 1096-0929

Lee, H., Yoon, S.W., Kim, E.J., Park, J., 2007. In-Situ Growth of Copper Sulfide Nanocrystals on Multiwalled Carbon Nanotubes and Their Application as Novel Solar Cell and Amperometric Glucose Sensor Materials. *Nano Letters,* Vol. 7, No. 3, (March 2007), pp. 778-784, ISSN: 1530-6992

Lin, Y., Yantasee, W. & Wang, J., 2005. Carbon Nanotubes (CNT) for the Development of Electrochemical Biosensors. *Frontiers in Bioscience,* Vol. 10, No. 1, (January 2005), pp. 492-505, ISSN: 1093-4715

Liu, Z., Cai, W., He, L., Nakayama, N., Chen, K., Sun, X., Chen, X. & Dai, H., 2007. In vivo biodistribution and highly efficient tumour targeting of carbon nanotubes in mice. *Nature Nanotechnology,* Vol. 2, No.1, (January 2007), pp. 47-52, ISSN: 1748-3395

Magrez, A., Kasas, S., Salicio, V., Pasquier, N., Seo, J.W., Celio, M., Catsicas, S., Schwaller, B. & Forro, L., 2006. Cellular Toxicity of Carbon-Based Nanomaterials. *Nano Letters*, Vol. 6, No. 6, (June 2006), pp. 121-1125, ISSN: 1530-6992

Meng, J., Kong, H., Han, Z., Wang, C., Zhu, G., Xie, S. & Xu, H., 2009. Enhancement of nanofibrous scaffold of multiwalled carbon nanotubes/polyurethane composite to the fibroblasts growth and biosynthesis. *Journal of Biomedical Materials Research-Part A* Vol. 88, No. 1, (January 2009), pp. 105-116, ISSN: 1552-4965

Mercer, R.R., Scabilloni, J., Wang, L., Kisin, E., Murray, A.R., Schwegler-Berry, D., Shvedova, A.A. & Castranova, V., 2008. Alteration of deposition pattern and pulmonary response as a result of improved dispersion of aspirated single-walled carbon nanotubes in a mouse model. *American Journal of Physiology - Lung Cellular and Molecular Physiology*, Vol. 294, No. 1, (January 2008), pp. L87-97, ISSN: 1522-1504

Monteiro-Riviere, N.A., Nemanich, R.J., Inman, A.O., Wang, Y.Y. & Riviere, J.E., 2005. Multi-walled carbon nanotube interactions with human epidermal keratinocytes. *Toxicology Letters*, Vol. 155, No. 3, (March 2005), pp. 377-384, ISSN: 0378-4274

Muller, J., Huaux, F., Moreau, N., Misson, P., Heilier, J.-F., Delos, M., Arras, M., Fonseca, A., Nagy, J.B. & Lison, D., 2005. Respiratory toxicity of multi-wall carbon nanotubes. *Toxicology and Applied Pharmacology* Vol. 207, No. 3, (September 2005), pp. 221-231, ISSN: 0041-008X

Murr, L., Soto,K., Esquivel, E., Bang, J., Guerrero, P., Lopez, D. & Ramirez, D., 2004. Carbon nanotubes and other fullerene-related nanocrystals in the environment: A TEM study. *JOM Journal of the Minerals, Metals and Materials Society*, Vol. 56, No.6, (June 2004), pp. 28-31, ISSN: 1543-1851

Naresh K N., Rama S P., & Kanagaraj S., 2011. Biocompatibility and morphology studies on ultrahigh molecular weight polyethylene/multi walled carbon nanotubes nanocomposites. *Proceedings of the 4th Indo-Australian Conference on Biomaterials, Tissue Engineering and Drug Delivery Systems*, Ahmedabad, India, February 10-12

Narita N, Kobayashi Y, Nakamura H, Maeda K, Ishihara A, Mizoguchi T, Usui Y, Aoki K, Simizu M, Kato H, Ozawa H, Udagawa N, Endo M, Takahashi N & Saito N. (2009). Multiwalled Carbon Nanotubes Specifically Inhibit Osteoclast Differentiation and Function. *Nano Letters*, Vol. 9, No. 4, (April 2009), pp. 1406-1413, ISSN: 1530-6992

Panyim, S. & Chalkley, R., 1971. The Molecular Weights of Vertebrate Mistones Exploiting a Modified Sodium Dodecyl Sulfate Electrophoretic Method. *The Journal of Biological Chemistry*, Vol. 246, No. 24, (December 1971), pp. 7557-7560, ISSN: 1083-351X

Patlolla, A., Knighten, B. & Tchounwou, P., 2010. Multi-Walled Carbon Nanotubes Induce Cytotoxicity, Genotoxicity And Apoptosis In Normal Human Dermal Fibroblast Cells. *Ethnicity & Disease*, Vol. 20, No. 1 - Supplement 1, (Spring 2010), pp. S1–65-72, ISSN: 1049-510X

Perez, B., Pumera, M., del Valle, M., Merkoçi, A. & Alegret, S., 2005. Glucose Biosensor Based on Carbon Nanotube Epoxy Composites. *Journal of Nanoscience and Nanotechnology*, Vol. 5, No. 10, (October 2005), pp. 1694-1698, ISSN: 1533-4880

Poland, C.A., Duffin, R., Kinloch, I., Maynard, A., Wallace, W.A.H., Seaton, A., Stone, V., Brown, S., MacNee, W. & Donaldson, K., 2008. Carbon nanotubes introduced into the abdominal cavity of mice show asbestos-like pathogenicity in a pilot study. *Nature Nanotechnology*, Vol. 3, No. 7, (July 2008), pp. 423-428, ISSN: 1748-3395

Potschke, P., Fornes, T.D., & Paul, D.R., 2002. Rheological behaviour of multiewalled carbon nanotube/polycarbonate composites. *Polymer*, Vol. 43, No 11, (May 2002), pp. 3247-3255, ISSN: 0032-3861

Pulskamp, K., Diabaté, S., Krug, H.F., 2007. Carbon nanotubes show no sign of acute toxicity but induce intracellular reactive oxygen species in dependence on contaminants. *Toxicology Letters*, Vol. 168, No. 1, (January 2007), pp. 58-74, ISSN: 0378-4274

Ren, H.-X. Chen, X., Liu, J.-H., Gu, N. & Huang, X.-J., 2010. Toxicity of single-walled carbon nanotube: How we were wrong? *Materials today*, Vol. 13, No. 1-2, (January-February 2010), pp. 6-8, ISSN: 1369-7021

Ruan, S.L., Gao, P., Yang, X.G., and Yu, T.X., 2003. Toughening high performance ultrahigh molecular weight polyethylene using multiwalled carbon nanotubes, *Polymer*, Vol. 44, No. 19 , (September 2003), pp. 5643-5654

Ryman-Rasmussen, J.P., Cesta, M.F., Brody, A.R., Shipley-Phillips, J.K., Everitt, J.I., Tewksbury, E.W., Moss, O.R., Wong, B.A., Dodd, D.E., Andersen, M.E. & Bonner, J.C., 2009. Inhaled carbon nanotubes reach the subpleural tissue in mice. *Nature Nanotechnology*, Vol. 4, No. 11, (November 2009), pp. 695-780, ISSN: 1748-3395

Sayes, C.M., Liang, F., Hudson, J.L., Mendez, J., Guo, W., Beach, J.M., Moore, V.C., Doyle, C.D., West, J.L., Billups, W.E., Ausman, K.D. & Colvin, V.L., 2006. Functionalization density dependence of single-walled carbon nanotubes cytotoxicity in vitro. *Toxicology Letters*, Vol. 161, No.2, (February 2006), pp. 135-142, ISSN: 0378-4274

Schipper, M.L., Nakayama-Ratchford, N., Davis, C.R., Kam, N.W.S., Chu, P., Liu, Z., Sun, X., Dai, H. & Gambhir, S.S., 2008. A pilot toxicology study of single-walled carbon nanotubes in a small sample of mice. *Nature Nanotechnology*, Vol. 3, No. 4, (April 2008), pp. 216-221, ISSN: 1748-3395

Sharma, C.S., Sarkar, S., Periyakaruppan, A., Barr, J., Wise, K., Thomas, R., Wilson, B.L. & Ramesh, G.T., 2007. Single-Walled Carbon Nanotubes Induces Oxidative Stress in Rat Lung Epithelial Cells. *Journal of Nanoscience and Nanotechnology*, Vol. 7, No. 7, (July 2007), pp. 2466-2472, ISSN: 1533-4880

Shvedova, A.A., Kisin, E.R., Mercer, R., Murray, A.R., Johnson, V.J., Potapovich, A.I., Tyurina, Y.Y., Gorelik, O., Arepalli, S., Schwegler-Berry, D., Hubbs, A.F., Antonini, J., Evans, D.E., Ku, B.-K., Ramsey, D., Maynard, A., Kagan, V.E., Castranova, V. & Baron, P., 2005. Unusual inflammatory and fibrogenic pulmonary responses to single-walled carbon nanotubes in mice. *The American Journal of Physiology - Lung Cellular and Molecular Physiology*, Vol. 289, No. 5, (November 2005), pp. L698-708, ISSN: 1522-1504

Shvedova, A.A., Kisin, E.R., Porter, D., Schulte, P., Kagan, V.E., Fadeel, B. & Castranova, V., 2009. Mechanisms of pulmonary toxicity and medical applications of carbon nanotubes: Two faces of Janus? *Pharmacology & Therapeutics*, Vol.121, No. 2, (February 2009), pp. 192–204, ISSN: 0163-7258

Sirivisoot S, Yao C, Xiao X, Sheldon BW & Webster TJ. (2007). Greater osteoblast functions on multiwalled carbon nanotubes grown from anodized nanotubular titanium for orthopedic applications. *Nanotechnology*, Vol. 18, No. 36, (September 2007), pp. 365102 – 6, ISSN: 1361-6528

Sitharaman, B., Shi, X., Walboomers, X.F., Liao, H., Cuijpers, V., Wilson, L.J., Mikos, A.G. & Jansen, J.A., 2008. In vivo biocompatibility of ultra-short single-walled carbon nanotube/biodegradable polymer nanocomposites for bone tissue engineering. *Bone* Vol. 43, No. 2, (August 2008), pp. 362-370, ISSN: 8756-3282

Soto, K.F., Carrasco, A., Powell, T.G., Garza, K.M. & Murr, L.E., 2005. Comparative in vitro cytotoxicity assessment of some manufacturednanoparticulate materials characterized by transmission electron microscopy. *Journal of Nanoparticle Research*, Vol. 7, No. 2-3, (June 2005), pp.145-169, ISSN: 1572-896X

Takagi, A., Hirose, A., Nishimura, T., Fukumori, N., Ogata, A., Ohashi, N., Kitajima, S. & Kanno, J., 2008. Induction of mesothelioma in p53+/- mouse by intraperitoneal application of multi-wall carbon nanotubes. *The Journal of Toxicological Sciences*, Vol. 33, No. 1, (February 2008), pp. 105-116, ISSN:1880-3989

Tian, F., Cui, D., Schwarz, H., Estrada, G.G. & Kobayashi, H., 2006. Cytotoxicity of single-wall carbon nanotubes on human fibroblasts. *Toxicology in Vitro*, Vol. 20, No. 7, (October 2006), pp. 1202-1212, ISSN: 0887-2333

Tsargorodskaya, A., Nabok, A.V. & Ray, A.K., 2004. Ellipsometric study of the adsorption of bovine serum albumin into porous silicon. *Nanotechnology*, Vol. 15, No. 5, (May 2004), pp. 703-709, ISSN 1361-6528

Tutak W, Park KH, Vasilov A, Starovoytov V, Fanchini G, Cai S-Q, Partridge NC, Sesti F & Chhowalla M. (2009). Toxicity induced enhanced extracellular matrix production in osteoblastic cells cultured on single-walled carbon nanotube networks. *Nanotechnology*, Vol. 20, No. 25, (June 2009), pp. 255101 – 255108, ISSN: 1361-6528

Tutak, W., Chhowalla, M. & Sesti, F., 2010. The chemical and physical characteristics of single-walled carbon nanotube film impact on osteoblastic cell response. *Nanotechnology*, Vol. 21, No. 31, (August 2010), pp. 1-6, ISSN 1361-6528

Usui,Y., Aoki, K., Narita, N., Murakami, N., Nakamura, I., Nakamura, K., Ishigaki, N., Yamazaki, H., Horiuchi, H., Kato, H., Taruta, S., Kim, Y.A., Endo, M. & Saito, N., 2008. Carbon Nanotubes with High Bone-Tissue Compatibility and Bone-Formation Acceleration Effects. *Small*, Vol. 4, No. 2, (February 2008), pp. 240 – 246, ISSN: 1613-6829

Wang, A., Lin, R., Polineni, V.K., Essner, A., Stark, C., and Dumbleton, J.H., 1998. Carbon fiber reinforced polyether ether ketone composite as a bearing surface for total hip replacement, *Tribology International*, Vol. 31, No. 11, (November 1998), pp. 661-667, ISSN 0301-679X

Wang, C. and Hwang, L.M., 1996. Transcrystallization of PTFE fiber/PP composites. II. Effect of transcrystallinity on the interfacial strength, *Journal of Polymer Science: Part B: Polymer Physics*, Vol. 34, No 8, (June 1996), pp. 1435-1442, ISSN 1099-0488.

Wang, L., Mercer, R.R., Rojanasakul, Y., Qiu, A., Lu, Y., Scabilloni, J.F., Wu, N., Castranova, V., 2010. Direct Fibrogenic Effects of Dispersed Single-Walled Carbon Nanotubes on Human Lung Fibroblasts. *Journal of Toxicology and Environmental Health, Part A*, Vol. 73, No.5-6, pp. 410–422, ISSN: 1087-2620

Warheit, D. B., Laurence, B.R., Reed, K.L., Roach, D.H., Reynolds, G.A.M. & Webb, T.R., 2004. Comparative Pulmonary Toxicity Assessment of Single-wall Carbon Nanotubes in Rats, *Toxicological Sciences*, Vol. 77, No.1, (January 2004), pp. 117-125, ISSN: 1096-0929

Wei, G., Zhang, Xie, J.L. & Jandt, K. D., 2011. Biomimetic growth of hydroxyapatite on super water-soluble carbon nanotube-protein hybrid nanofibers. *Carbon*, Vol. 49, No. 7, (June 2011), pp. 2216-2226, ISSN: 0008-6223

Wei, Z., Zhao, Y.-P., Ruan, S.L., Gao, P., and Yu, T.X., 2006. A study of the tribological behaviour of carbonnanotube- reinforced ultrahigh molecular weight polyethylene composites. *Surface and Interface Analysis*, Vol. 38, pp. 883-886

Witzmann, F.A. & Monteiro-Riviere, N.A., 2006. Multi-walled carbon nanotube exposure alters protein expression in human keratinocytes. *Nanomedicine*, Vol. 2, No. 3, (September 2006), pp. 158-168, ISSN: 1549-9634

Wood, J.R., Frogley, M.D., Meurs, E.R., Prins, A.D., Peijs, T., Dunstan, D.J. & Wagner, H.D., 1999. Mechanical Response of Carbon Nanotubes under Molecular and Macroscopic Pressures. *The Journal of Physical Chemistry Letters*, Vol. 103, No. 47, (November 1999), pp. 10388-10392, ISSN: 1520-5207

Xia, X.-R., Monteiro-Riviere, N.A. & Riviere, J.E., 2010. An index for characterization of nanomaterials in biological systems. *Nature Nanotechnology*, Vol. 5, No. 9, (September 2010), pp. 671-675, ISSN: 1748-3395

Yadav, S., Bera, T., Saxena, P.S., Maurya, A.K., Garbyal, R.S., Vajtai, R., Ramachandrarao, P. & Srivastava, A., 2010. MWCNTs as reinforcing agent to the Hap–Gel nanocomposite for artificial bone grafting. *Journal of Biomedical Materials Research Part A*, Vol. 93, No. 3, (June 2010), pp. 886-896, ISSN: 1552-4965

Yan, X.B., Chen, X.J., Tay, B.K. & Khor, K.A., 2007. Transparent and flexible glucose biosensor via layer-by-layer assembly of multi-wall carbon nanotubes and glucose oxidase. *Electrochemistry Communications*, Vol. 9, No. 6, (June 2007), pp. 1269-1275, ISSN: 1388-2481

Yang, M., Yang, Y., Liu, Y., Shen, G. & Yu, R., 2006. Platinum nanoparticles-doped sol-gel/carbon nanotubes composite electrochemical sensors and biosensors. *Biosensors and Bioelectronics*, Vol. 21, No.7, (January 2006), pp. 1125-1131, ISSN: 0956-5663

Yu, L. E., Lanry Yung, L.-Y., Ong, C.-N., Tan, Y.-L., Balasubramaniam, K.S., Hartono, D., Shui, G., Wenk, M. R. & Ong, W.-Y., 2007. Translocation and effects of gold nanoparticles after inhalation exposure in rats. *Nanotoxicology*, Vol. 1, No. 3, (January 2007), pp. 235-242, ISSN: 1743-5404

Yun, Y., Bange, A., Shanov, V.N., Heineman, W.R., Halsall, H.B., Dong, Z., Jazieh, A., Tu, Y., Wong, D., Pixley, S., Behbehani, M. & Schulz, M.J., 2007. A Carbon Nanotube Needle Biosensor. *Journal of Nanoscience and Nanotechnology*, Vol. 7, No. 7, (July 2007), pp. 2293-2300, ISSN: 1533-4880

Zanello, L.P., Zhao, B., Hu, H. & Haddon, R.C., 2006. Bone cell proliferation on carbon nanotubes. *Nano Letters*, Vol. 6, No. 3, (March 2006), pp. 562-567, ISSN: 1530-6992

Zensi, A., Begley, D., Pontikis, C., Legros, C., Mihoreanu, L., Wagner, S., Buchel, C., von Briesen, H. & Kreuter, J., 2009. Albumin nanoparticles targeted with Apo E enter the CNS by transcytosis and are delivered to neurons. *Journal of Controlled Release*, Vol. 137, No. 1, (July 2009), pp.78-86, ISSN: 0168-3659.

Zhang, L.W., Zeng, L., Barron, A.R. & Monteiro-Riviere, N.A., 2007. Biological Interactions of Functionalized Single-Wall Carbon Nanotubes in Human Epidermal Keratinocytes. *International Journal of Toxicology*, Vol. 26, No. 2, (March 2007), pp. 103-113, ISSN: 1092-874X

20

Impact of the Carbon Allotropes on Cholesterol Domain: MD Simulation

Zygmunt Gburski, Krzysztof Górny,
Przemysław Raczyński and Aleksander Dawid
Institute of Physics, University of Silesia, Katowice
Poland

1. Introduction

In recent years, immense advancement has been made in the field of nanotechnology. This emerging field will indefinitely become a critical facet of many areas including chemistry, biology, electronics and optics, and will provide unique opportunities for researchers to innovate in unimaginable ways. Nanomaterials, because of their unique mechanical, thermal, optical and electronic properties, have reshaped many segments of modern science and engineering and are increasingly impacting our society, health care, and the environment. Specifically, nanotechnology has great potential in biomedical applications, as mammal/human biology is essentially a very complex system of nano-machines. Nowadays big changes are coming from the marriage of medicine and nanotechnology - the new branch of science called nanomedicine or molecular medicine. A field of utilizing molecular assemblies at the nano-scale of about 100 nm or less for novel and alternative diagnostics and therapeutics, in incredible selectivity and accuracy not achievable through conventional means. One would hope that with the future development of nanomedicine, we will be able to think of today's incurable diseases as curable tomorrow, by looking at a problem at its molecular or even atomic levels and apply medical intervention at the molecular scale.

Presently nanomedicine involves detection of particles (nanobiosensors), drug delivery systems, emulsions, and carriers for delivering vaccines and biomaterials with unusual properties and improved biocompatibility.

Recently, within the realm of nano-scale, carbon nanotubes (CNT) are being tested as medical devices at an increasing rate. Just to mention a few examples: CNTs have been utilized as scaffolds for neuronal and ligamentous tissue growth for regenerative interventions of the central nervous system and orthopaedic sites (Hu et al., 2004), substrates for detecting antibodies associated with human autoimmune diseases with high specificity (Wang et al., 2004), carriers of contrast agent for enhanced magnetic resonance imaging (Sitharaman et al., 2005). When coated with nucleic acids (DNA or RNA), vaccines, and proteins, CNTs have been shown as effective substrates for gene sequencing and as gene and drug delivery vectors to challenge conventional viral and particulate delivery systems (Pantarotto et al., 2004; Kam et al., 2004; Liu et al., 2005; Lu et al., 2004) . Carbon nanotube has been also probed as a vehicle for drug delivery into selected cells (Liu et al., 2008) and as bio-sensor (Wisitsoraat et al., 2010). In this chapter we consider and test the idea that carbon nanotubes might be utilized to remove unwanted molecular aggregates, particularly excess cholesterol, from a living tissues.

Cholesterol is a major sterol component of mammalian cell membranes, it plays important role in maintaining physical and mechanical properties of the membrane. There is a large literature of cholesterol in biomembranes (Róg et al., 2009). Its abundance influences such diverse membrane processes as signal transduction, protein stabilization, protein and lipid sorting, and membrane fusion. Independent of its permanent presence in a cell membrane, cholesterol is transported through the blood as a component of water- soluble carrier aggregates known as lipoproteins. A lipoprotein aggregate is composed of an outer shell of phospholipids, which renders the particle soluble in water; a core of fats called lipid, including cholesterol and a surface apoprotein molecule that allows some tissues to recognize and take up the aggregate core content. Cholesterol can be also found in extracellular medium (lymphatic fluid) of the body. Although cholesterol is essential for the proper functioning of cell membranes, excess cholesterol levels could prove detrimental. Particularly, excess cholesterol may precipitate in forming cholesterol lodgments (domains) in the inner lining of blood vessels. This triggers the subendothelial accumulations of cholesterol-engorged macrophages, called "foam cells", later on leading to the formation of the plaque deposition in atherosclerosis disease (Lusis, 2000). Eventual build up of plaques, cells of inflammation, and blood clotting can block the normal blood flow in the coronary arteries. This is a catastrophic event that stops the flow of nutrients and oxygen to the heart muscle, leading to heart attack (myocardial infarction).
The study of the influence of carbon nanotube on cholesterol is in its infancy. In this report we present our molecular dynamics (MD) investigation of the influence of the carbon allotropes (nanotube, graphene) on the cholesterol molecules: a) embedded in a cell membrane, b) forming a lodgment around selected extracellular domain proteins.

2. Toward the extraction of cholesterol lodgement by carbon allotrope manipulation

The search for new methods for removing of excess cholesterol molecules, precursors of plaque deposition in an early phase of atherosclerosis disease, is a vital subject of molecular medicine. Our recent simulations (Raczyński et al., 2006a, 2007; Gburski et al., 2010) and the material presented here are related to this issue. In this context, we have chosen the carbon allotropes since they are known to be hydrophobic. This is an important feature when it comes to intervention in the biosystem, where water is the inherent component. We made a step towards the investigation of influence of carbon allotropes on cholesterol-contained systems of potential interest in nanomedicine. Particularly, these studies might be related to the quest for future, molecular level treatment of an embryonic phase of artheriosclerosis, celebral hemorrhage, lung diseases, ...*etc.*

2.1 Impact of carbon nanotube on cholesterol in a cell membrane

First we study the influence of carbon nanotube on mammalian cell membrane. The cell membrane separates the intracellular components from the extracellular environment, its architecture is that of phospholipid bilayer. Indispensable component of cell's membrane is cholesterol, one of its primary functions is to guarantee a proper elasticity/stiffness of membrane. Cholesterol molecules are embedded between phospholipids, there is approximately one cholesterol per ten phospholipid molecules in the bilayer (see Fig.1). If we want to use CNT as a medical nanodevice, first we have to know what is the effect of CNT on just mentioned "membrane" cholesterol, *i. e.* those cholesterol molecules which reside within phospholipid bilayer.

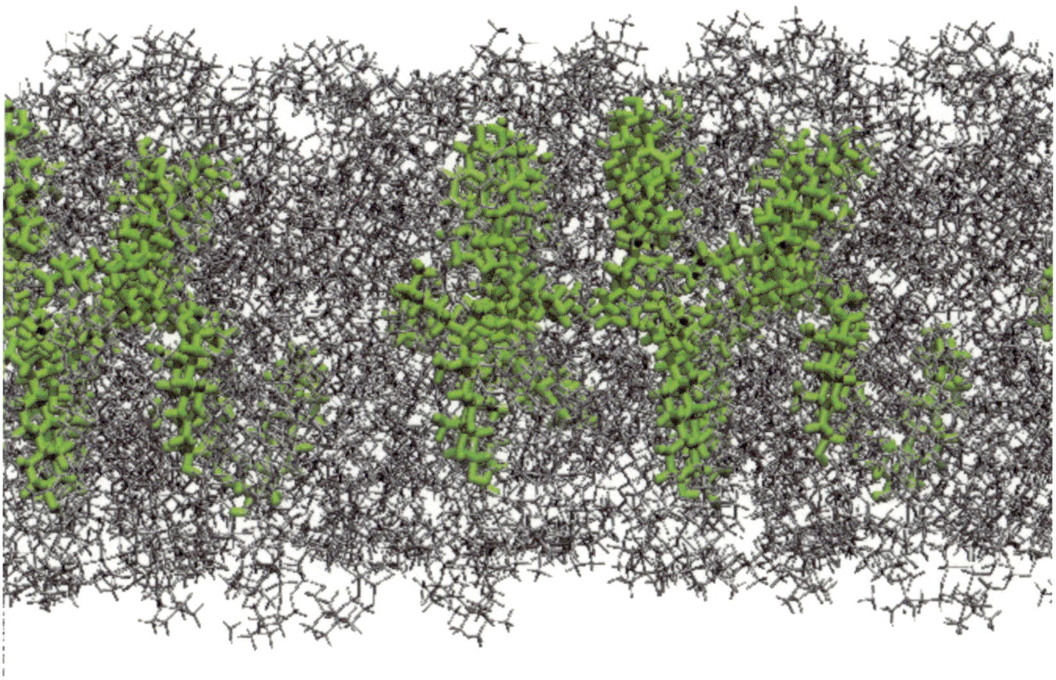

Fig. 1. Fragment of the simulated mammals' cell membrane. The cholesterol molecules (green) are embedded between DMPC phospholipid bilayer (grey)

We consider the cholesterol molecules located between nonpolar hydrophobic fatty acid tails of neighbouring phospholipide molecules which form phospholipide bilayer sheet. The outer side of the bilayer form the polar hydrophilic phosphate heads of phospholipides. We have done MD simulations for this system (without nanotube), for the reference purpose. Next we have placed the carbon nanotube close to the "sea" of phosphate heads and repeated the simulation again. The essential simulation details are given at the end of this section. The dynamical observable of cholesterol molecule has been calculated, namely the mean square displacement $<|\Delta\mathbf{r}(t)|^2>$ of the center of a molecule

$$<|\Delta\mathbf{r}(t)|^2> = <|\mathbf{r}(t)-\mathbf{r}(0)|^2> \quad (1)$$

where $\mathbf{r}$ is the position of the center of mass of a molecule of interest. The comparison of the mean square displacement of the center of mass of cholesterol molecule at T = 309 K with and without the presence of nanotube, is given in Fig. 2. In the absence of nanotube the cholesterol molecule - tighten between phospholipids - can't walk further then 1.1 Å, whereas in the presence of nanotube cholesterol's $<|\Delta\mathbf{r}(t)|^2>$ reaches the saturation at 1.3 Å. We see that the cholesterol molecule somehow "feels" the nanotube, namely it is slightly attracted by nanotube surface because we observe the increase of the maximum distance it can move. However, the plot of $<|\Delta\mathbf{r}(t)|^2>$ function (Fig. 2) tells that cholesterols always remain inside phospholipide bilayer and the nanotube can not pull them out of the cell membrane.

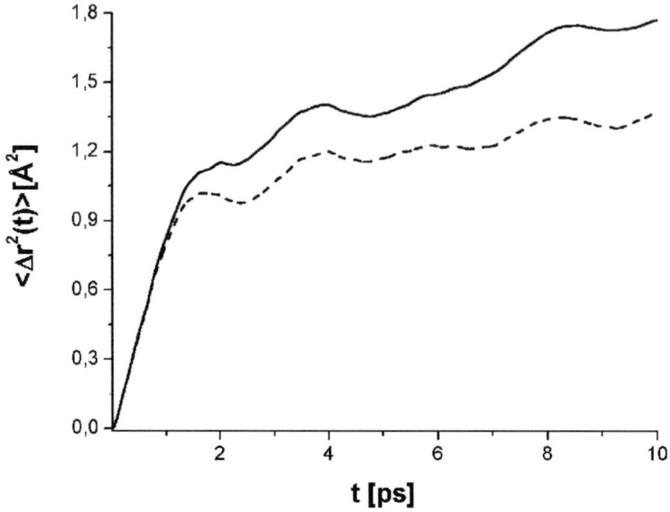

Fig. 2. The mean square displacement of the centre of mass of cholesterol molecule embedded between phospholipids in the presence (solid line) and absence (dashed line) of carbon nanotube.

The slope of the mean square displacement is connected with the translational diffusion coefficient *via* Einstein relation

$$<|\Delta \mathbf{r}(t)|^2> \cong 6Dt \qquad (2)$$

The plot of $<|\Delta \mathbf{r}(t)|^2>$ (see Fig. 2) indicates that the system is in the liquid phase, the slope of the linear part of mean square displacement of cholesterol is $1.4 * 10^{-1}$ Å^2/ps. Hence the value of diffusion coefficient D, calculated from the linear part of $<|\Delta \mathbf{r}(t)|^2>$ using eq. (2) , is $D = 2.4*10^{-6}$ cm^2/s.

Summarizing so far presented results we conclude, that the carbon nanotube only very slightly influences the dynamics of those cholesterols which are located in a cell membrane, between phospholipides. Particularly, what is very important, the carbon nanotube can't pull the cholesterol molecules out of the phospholipide layer.

Essential technical details of MD simulations analysed above.

We have used the standard Lennard-Jones (LJ) interaction potential $V(r_{ij})$ between carbon atoms of the armchair (10,10) nanotube and the atoms (sites) of rigid-body cholesterol $C_{27}H_{45}OH$ and phospholipid. Namely,

$$V(r_{ij}) = 4\varepsilon[(\sigma / r_{ij})^{12} - (\sigma / r_{ij})^{6}], \qquad (3)$$

where r_{ij} is the distance between the atoms i*th* and j*th*, ε is the minimum of potential at the distance $2^{1/6}\sigma$. The cholesterol and phospholipid molecules include lots of atomic sites, but in line with the common procedure for large molecules we treat CH, CH_2 and CH_3 atomic groups as a supersites (pseudoatoms). The L-J parameters for these groups and other atoms involved are taken from (Daura et al., 1988; Kuznetsova & Kvamme, 2002). Moreover, we have included the dipole moments of cholesterol and phospholipid (OH bonds) by putting the charge -0.376 *e* on oxygen and 0.376 *e* on hydrogen atoms of OH bonds (Phelps & Dalby,

1966). The L-J potentials parameters between unlike atoms and pseudoatoms were calculated by the Lorentz-Berthelot rules (Allen & Tildesley, 1989)

$$\sigma_{A-B} = (\sigma_A + \sigma_B)/2, \ \varepsilon_{A-B} = \sqrt{\varepsilon_A \varepsilon_B} \ , \tag{4}$$

where A, B are; C, O, N, S, H, CH, CH_2 and CH_3 atoms or pseudoatoms. The classical equations of motion were integrated by predictor-corrector Adams-Moulton algorithm (Rapaport, 1995). The integration time step was 0.3 fs which ensured total energy conservation within 0.01%. The total simulation time was 1.5 ns.

2.2 Influence of carbon nanotube on cholesterol domain localized on a protein surface

Our next step is to simulate another ensemble, composed of a protein and cholesterols. We have chosen 1KF9 as an example of human extracellular domain protein (Shiffer et al., 2002) since the data required for MD simulation are available for this protein (Protein Data Bank, 2010). This protein was covered by 40 cholesterol molecules. During a preliminary MD simulation we have observed that the cholesterols gather together near protein surface, forming the cholesterol lodgment. Then, we have placed the carbon nanotube near this cholesterol cluster (see Fig. 3).

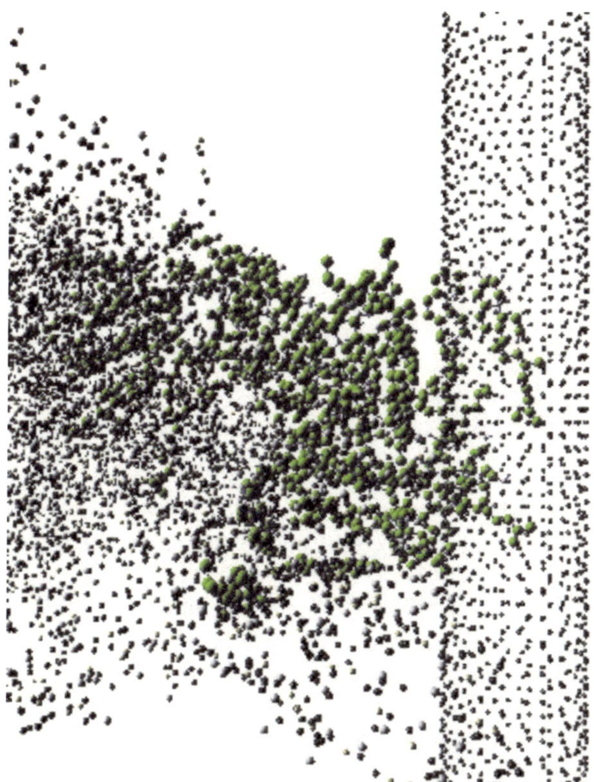

Fig. 3. The snapshot of an equilibrium configuration of the simulated ensemble, showing a fragment of 1KF9 protein and a cholesterol lodgment in the presence of carbon nanotube.

Holding the nanotube immobile in its place we have run simulation again, collecting MD data over 1.5 ns. The radial distribution function g(r) of the centers of mass of cholesterols in the lodgment without nanotube shows two peaks (see Fig. 4). The first one, quite sharp indicates the average near neighbors distance r = 6.1 Å while the second, very broad peak around r ≈ 12.2 Å comes from the farther distance cholesterol neighbors (second shell). There are no other pronounced peaks and for larger intermolecular distance g(r) gradually tends toward zero. The highest peak of g(r) for the lodgment with nanotube (Fig. 4) appears at larger distance r = 7.1 Å comparing to r = 6.1 Å for the lodgment without nanotube. It must be associated with the average centers of mass distance between cholesterols covering nanotube. Note, that in (Raczyński, 2006b) we have calculated g(r) for the single layer of cholesterols surrounding carbon nanotube and the maximum of g(r) was located at r = 7.2 Å. A little peak at r = 5 Å comes from the small number of cholesterols confined between protein and nanotube. The value of g(r) reaches zero around r ≈ 40 Å in both cases, indicating that way the diameter of the cholesterol lodgment.

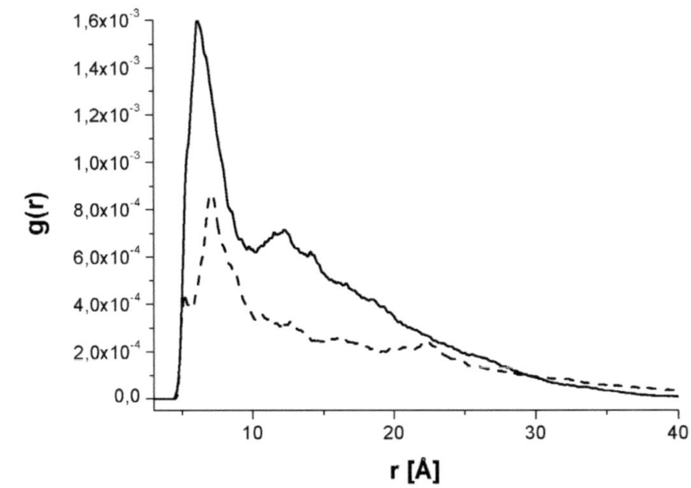

Fig. 4. The radial distribution function g(r) of the centre of mass of cholesterol molecule in the lodgment in the absence (solid line) and presence (dashed line) of the carbon nanotube.

The mean square displacement of the center of mass of cholesterol molecule is shown in Fig. 5. The difference between $<|\Delta \mathbf{r}(t)|^2>$ with and without nanotube is distinct. Without nanotube, the plot of $<|\Delta \mathbf{r}(t)|^2>$ is somehow similar to dense media, the slope of $<|\Delta \mathbf{r}(t)|^2>$ is low hence the translational mobility of cholesterols within the lodgment is very weak. The value of $<|\Delta \mathbf{r}(t)|^2>$ and its slope spectacularly increases when the nanotube is nearby the cholesterol lodgment (cholesterols get mobile). The increase of $<|\Delta \mathbf{r}(t)|^2>$ while nanotube is nearby lodgment reflects a simple fact that the nanotube pulls cholesterols out of the lodgment. The pulled out cholesterols spread all over the nanotube surface forming thin layer covering carbon nanotube. Therefore, removing the nanotube, covered by the intercepted cholesterols, substantially diminishes the number of cholesterol molecules which remain within the lodgment. The process of extraction of the cholesterol lodgment is quite efficient. Our nanotube of 80.5 Å length has pulled out twenty three of the total number of forty cholesterols, reaching 57 % of efficiency (for the snapshot see Fig. 6).

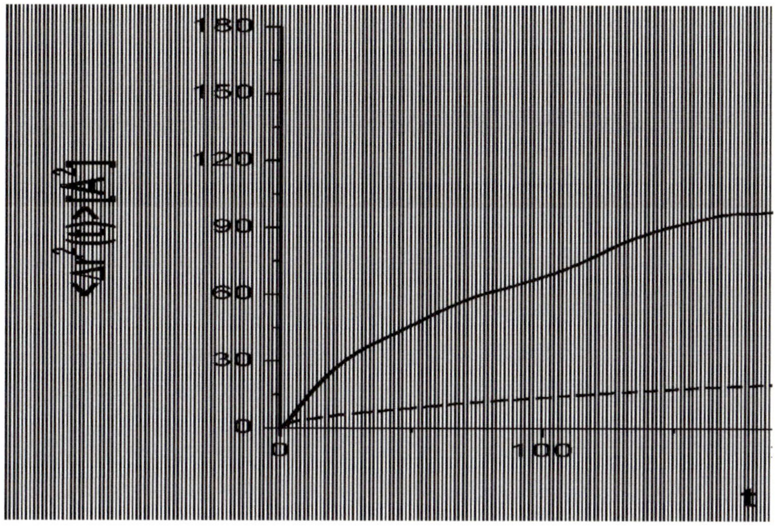

Fig. 5. The mean square displacement of the centre of mass of cholesterol molecule in the lodgment in the absence (dashed line) and presence (solid line) of carbon nanotube.

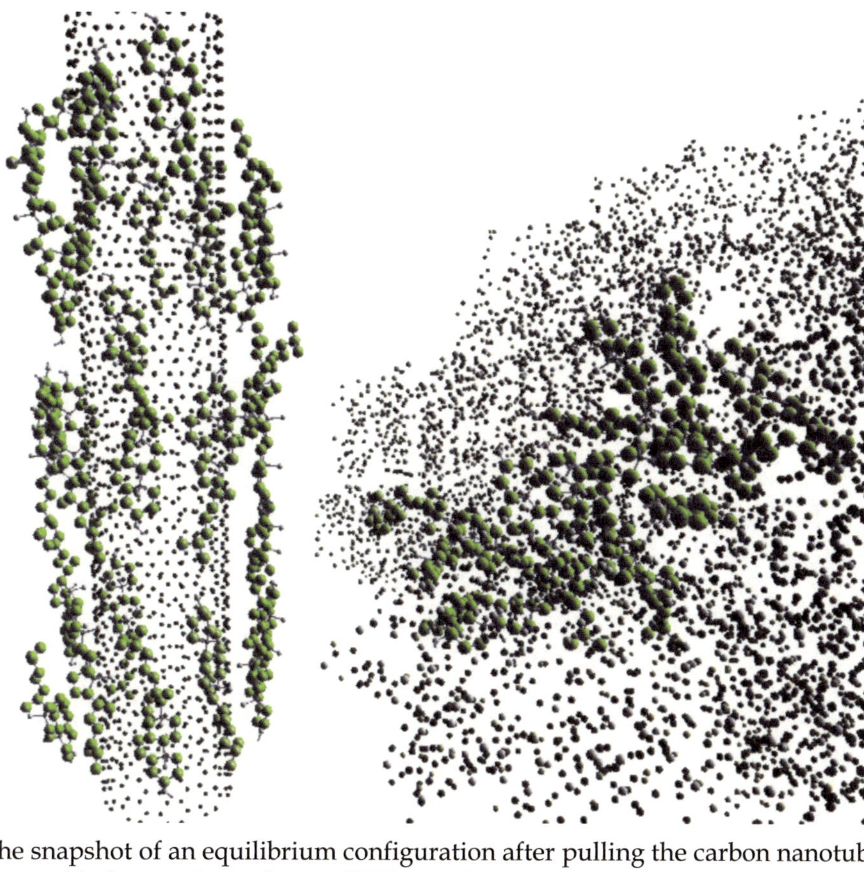

Fig. 6. The snapshot of an equilibrium configuration after pulling the carbon nanotube out of the cholesterol lodgment located over 1KF9 protein

The molecular system discussed above doesn't include water, which is an essential component of living organism. As a next step, we have turned to the study of a more realistic, albeit more complicated fragment of biosystem. Particularly, we made a reconnaissance study, *via* computer simulation, of the influence of the carbon nanotube on the dynamics of cholesterol molecules forming a domain around a selected extracellular protein in a water environment. As an example, 1LQV endothelial extracellular protein was chosen (Protein Data Bank, 2010), see Fig. 7.

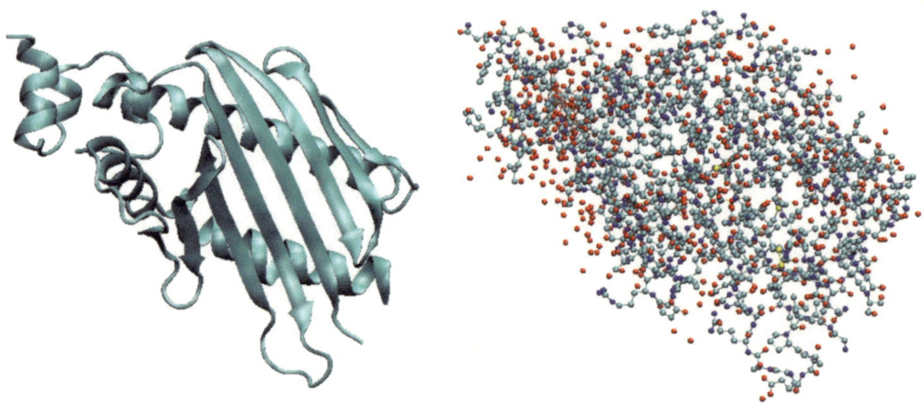

Fig. 7. The endothelial 1LQV protein; ribbon representation (left), atomistic representation (right)

1LQV protein appears in the thin layer of cells named endothelium, this layer forms an interface between circulating blood and the rest of the vessel wall. That is why we selected the human endothelial protein that resides in the innermost layer of a blood artery (see Fig. 8.).

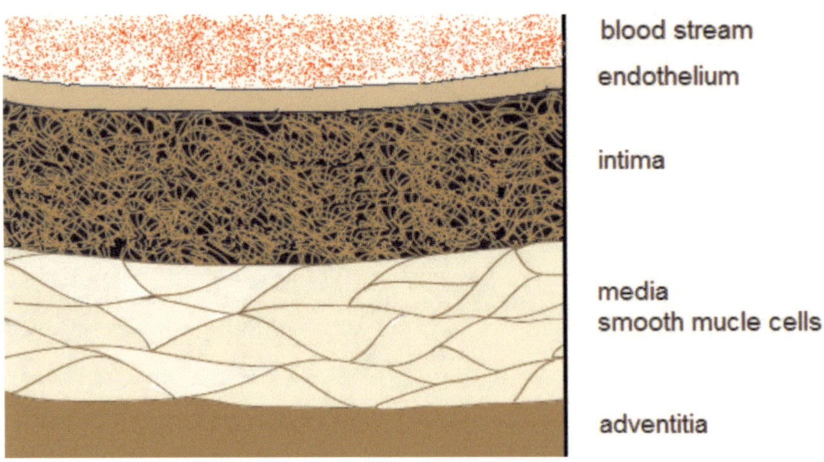

Fig. 8. Structure of a large blood artery

All molecules in the system were modeled on the full atomistic level (simulation details are given at the end of this subsection).
First, the mean square displacement of the center of mass of cholesterol was calculated. From the relation between $<|\Delta\mathbf{r}(t)|^2>$ and the translational diffusion coefficient D (eq. 2), one knows that a nonzero slope of $<|\Delta\mathbf{r}(t)|^2>$ is an indicator of mobility (translational diffusion) of molecules. Fig. 9 shows $<|\Delta\mathbf{r}(t)|^2>$ plots both in the presence and absence of the nanotube. It indicates that in both cases the cholesterol domain is not in the solid state. The diffusion coefficient D of cholesterol, estimated from the linear part of the slope of $<|\Delta\mathbf{r}(t)|^2>$ at the physiological temperature $T \approx 309$ K, is $D = 0.37 * 10^{-6}$ cm^2/s (with nanotube) and is $D = 0.39 * 10^{-6}$ cm^2/s (without nanotube). Note, that these values of D are related to the short time translational dynamics of cholesterol molecule.

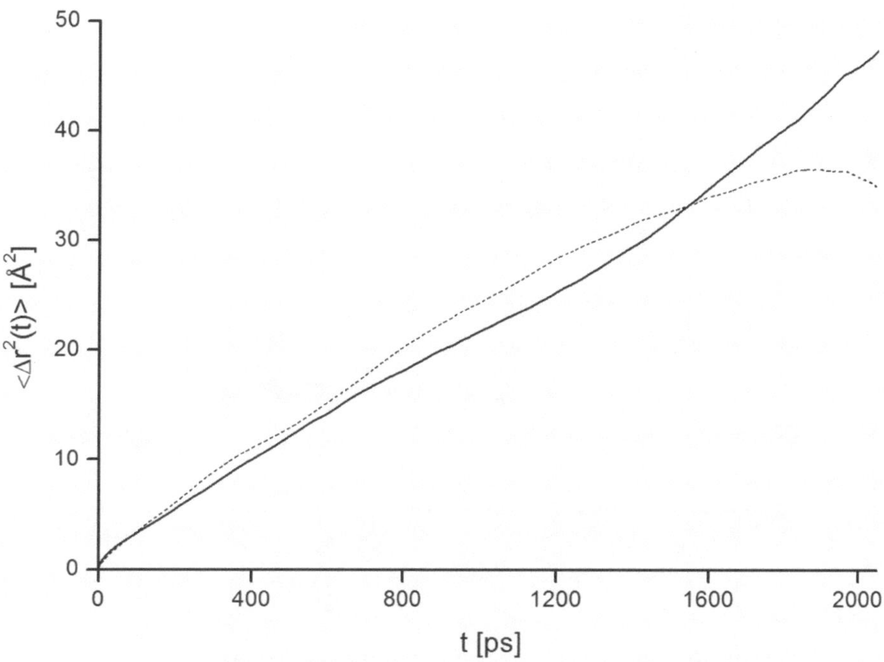

Fig. 9. Mean square displacement of the centre of mass of cholesterol molecule in the domain settled down on the fragment of endothelial 1LQV protein: in the presence (solid line) and absence (dashed line) of carbon nanotube.

If there is no nanotube, the mean square displacement function shows saturation around 35 Å^2. Therefore, although the cholesterols have some mobility, their displacements are restricted by the interaction with protein surface. One can see that saturation of $<|\Delta\mathbf{r}(t)|^2>$ vanishes when the carbon nanotube is placed near the domain. The cholesterol molecules now can migrate farther, the restriction on their translational dynamics, imposed by the protein surface, is overcome by the competitive interaction with the carbon atoms of the nanotube. Some cholesterols move from the domain and build up a thin layer around the carbon nanotube.

To visualize even more the migration of cholesterol molecules from the domain to nanotube we have calculated the radial distribution function of cholesterols with respect to the main axis of the carbon nanotube (denoted $g_{CN\text{-}chol}$ (r) in Fig. 10). Calculations were performed for both initial and final stage of simulation of system with CNT (averaged over 0.2 ns). The radius of nanotube studied is 6.8 Å. Apparently at the final stage some cholesterol molecules approach close to nanotube surface, the first maximum of radial distribution is higher and located nearer the nanotube.

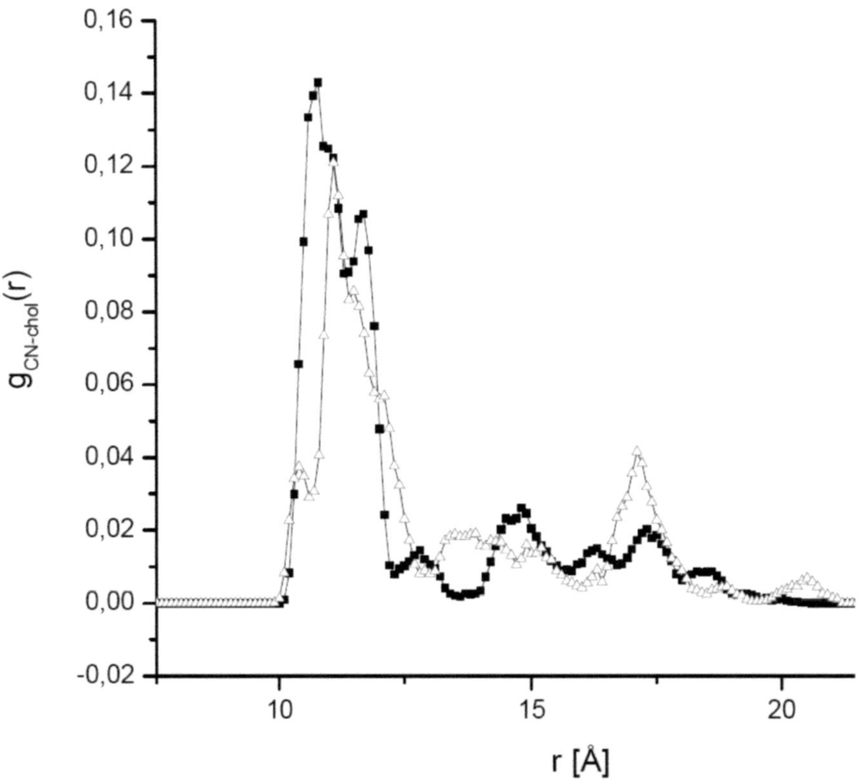

Fig. 10. Radial distribution function of the centers of mass of cholesterols with respect to the main axis of carbon nanotube, calculated at the beginning (open triangles) and the end of simulation run (filled squares).

To show more directly the ability of CNT to remove cholesterol from the domain, steered molecular dynamics simulation (Philips et al., 2005) was carried out. In this simulation external forces were applied to pull out CNT from the domain (see Fig. 11).

The process of removing cholesterols by carbon nanotube is efficient. The nanotube of 60 Å length has pulled out 17 of the total number of 21 cholesterols, reaching 80 % efficiency. The repeating of this action, using clean nanotube, practically removes those remaining cholesterols which have survived the first intervention of the nanotube. The reported ability of the nanotube to extract the cholesterol lodgment at physiological temperature is quite appealing. One would say, this is a kind of nanosurgery (extraction of cholesterol lodgment) made in a computer laboratory.

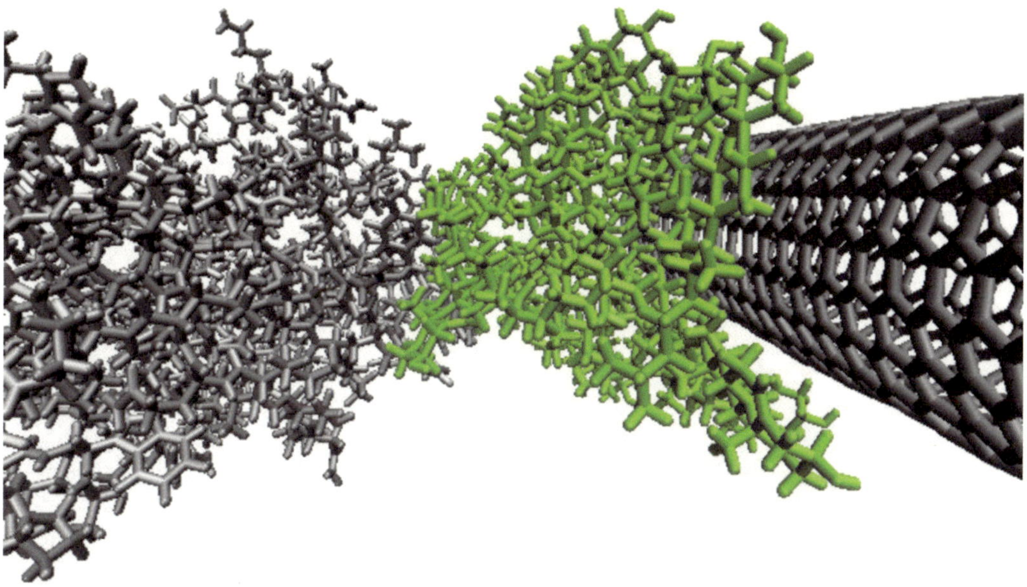

Fig. 11. Final configuration of the simulated ensemble after pulling out the carbon nanotube. Carbon nanotube has removed the majority of cholesterols from the domain originally spread over the surface of 1LQV protein. Water molecules are not shown for the clarity of the picture.

Essential simulation details.
MD simulation of ensemble with 1K9F protein was performed with Lennard-Jones potential between carbon atoms of the armchair (10, 10) nanotube and the atoms (sites) of rigid-body cholesterol $C_{27}H_{45}OH$ and the protein. The equations of motion were integrated by predictor-corrector Adams-Moulton algorithm. The integration time step was 0.3 fs which ensured sufficient total energy conservation. The total simulation time was 1.5 ns.
In case of ensemble with 1LQV protein more sophisticated force field was applied. The molecular dynamics (MD) simulations were performed using the NAMD 2.6 program (Philips et al., 2005) with the all atom CHARMM force field (MacKerell et al., 1998) in NVT (constant number of particles, constant volume and constant temperature) ensemble at the physiological temperature T = 309 K.
The CHARMM force field includes intramolecular harmonic stretching V_{bond}, harmonic bending V_{angle}, torsional $V_{dihedral}$, Van der Waals and Coulombic terms:

$$V_{total} = V_{bond} + V_{angle} + V_{dihedral} + V_{vdW} + V_{Coulomb} \quad (5)$$

The standard NAMD integration algorithm (Brünger-Brooks-Karplus) was used with timestep of 0.5 fs. The ensemble consisting of protein, cholesterol and water molecules was equilibrated for $3*10^6$ time steps with periodic boundary conditions (Rapaport, 1995). After equilibration, the system was simulated for $5*10^6$ time steps (2.5 ns).

2.3 Removing of cholesterol lodgement by graphene sheet

The aim of our next computer experiment was to test whether the graphene, a sheet of carbon just one atom thick, could substantially influence the distribution of cholesterol molecules spread over the protein's surface. For that reason, the graphene sheet (720 carbon atoms) was placed at a distance 2.3 nm from the center of cholesterol lodgment covered 1LQV endothelial protein. At this initial configuration the graphene sheet was separated from the direct interaction with cholesterol molecules by water (see Fig. 12). For the reference purpose we have also done MD simulations for the ensemble without graphene sheet.

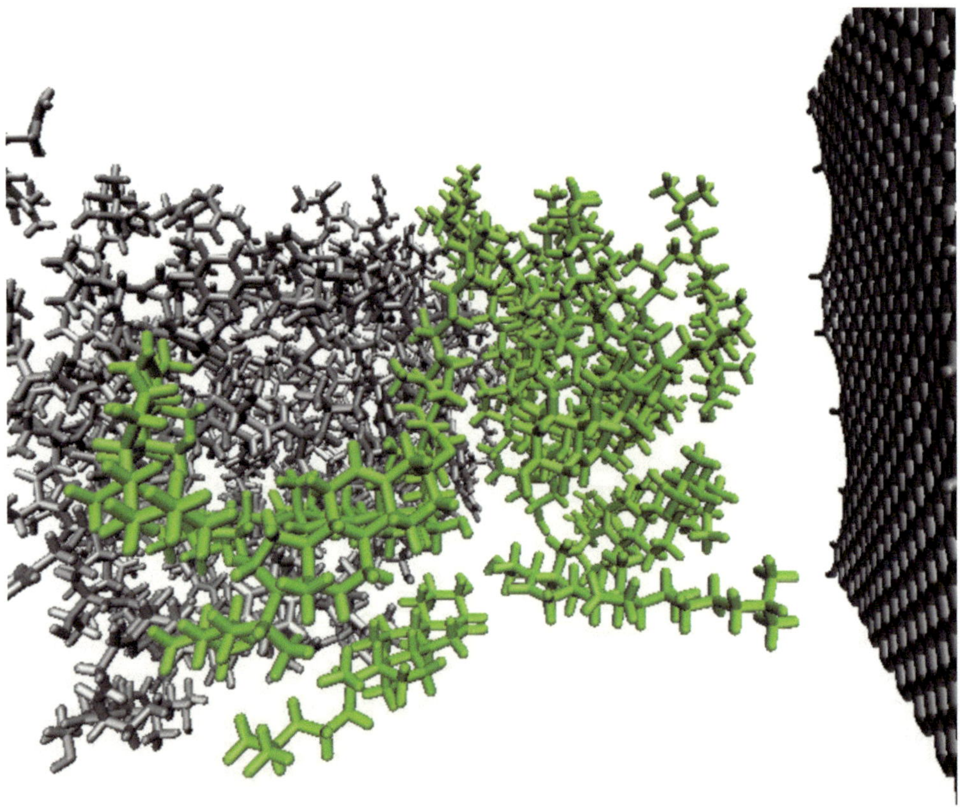

Fig. 12. Snapshot of the initial configuration of the ensemble of 1LQV protein, cholesterol molecules and graphene wall solvated in water, at T = 309 K. Water molecules are not shown to improve the clarity of the picture.

When the graphene is placed nearby the cholesterol domain, the translational mobility of cholesterol molecules significantly increases, reflecting the migration of cholesterols towards the graphene sheet. To visualize the migration of cholesterols, we have also calculated a time evolution of the mean distance between the center of mass of cholesterol lodgment and graphene wall (Fig. 13). It declines with time, clearly showing the process of displacement of a large number of cholesterols from the domain to the graphene sheet.

The snapshot of the instantaneous configuration of the system studied, after completing the migration process, is shown in Fig. 14. Apparently, the large amount of cholesterol molecules was removed from the cluster surrounding the protein.

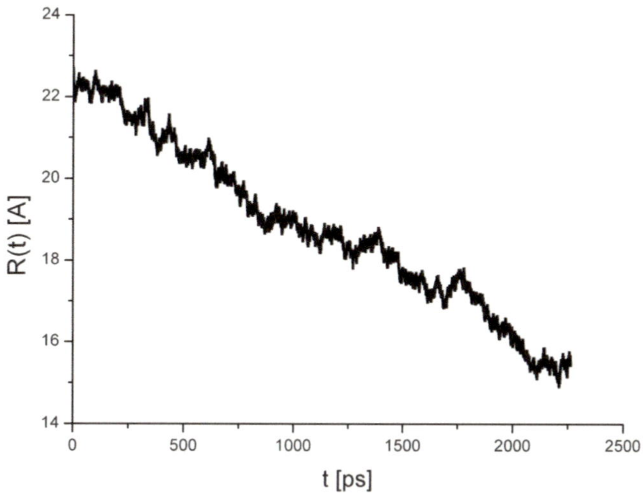

Fig. 13. Mean distance between the center of mass of cholesterols and graphene sheet.

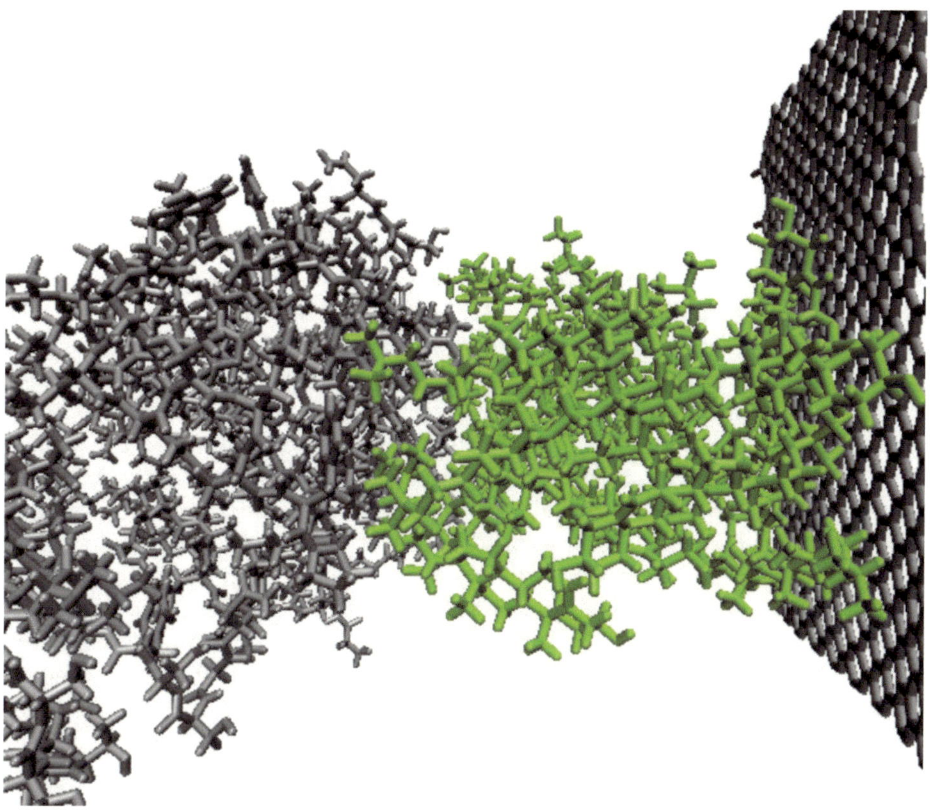

Fig. 14. Snapshot of the final configuration of the ensemble of 1LQV protein, cholesterol molecules and graphene wall solvated in water, at T = 309 K. Water molecules are not shown to improve the clarity of the picture.

At the physiological temperature T= 309 K the graphene sheet substantially influences the dynamics of cholesterol molecules forming a lodgment around a chosen example of endothelial protein (1LQV). In the presence of graphene some cholesterol molecules migrate from the lodgment and settle down on the graphene surface. As a result, the cholesterol lodgment strongly diminishes.
Essential simulation details.
All MD simulations were carried out using NAMD program (Philips et al., 2005) with the all atom CHARMM27 force field (MacKerell et al., 1998; Feller & MacKerell, 2000). The net atomic charges borne by cholesterol were determined in (Hénin & Chipot, 2006). Initially, 21 cholesterol molecules were located near the surface of 1LQV protein, and the system was equilibrated over 10^6 time steps dt (dt=0.5 fs). Next, as usual procedure in case of biological samples, water was added ($9*10^3$ H_2O molecules, TIP3P model (Jorgensen et al., 1983)). All simulations were carried out in the NVT ensemble at the physiological temperature T = 309 K. The results presented here were obtained from a 5 ns production trajectory (10^7 time steps, dt = 0.5 fs) with the periodic boundary conditions.

3. Conclusion

Our MD simulations show that the carbon nanotube can not pull out those cholesterol molecules which are embedded in a cell membrane (phospholipid bilayer). That is exactly what one would appreciate. It means that the presence of nanotube even quite close to the cell membrane doesn't destroy the structure and natural balance of cholesterol - phospholipide system. In other words, the presence of nanotube nearby a cell membrane is neutral for the dynamics of "membrane" cholesterol. Nevertheless, our simulations show that a carbon nanotube can diminish the unwanted cholesterol lodgment assembled on the protein surface. This happens, because the attraction of cholesterols by nanotube prevails cholesterols' tendency to gather together over protein surface (setting up a lodgment). The cholesterol molecules tired out of the lodgment spread all over the nanotube surface, forming thin layer. Quite similar or even stronger removal effect can be achieved by applying graphene sheet instead of carbon nanotube. What we have done could be treated as a kind of nanosurgery made in a computer laboratory. Our simulations suggest that the reported ability of extraction of the cholesterol lodgment by carbon allotropes (nanotube, graphene) might be taken into account while searching for the future medical devices suitable for treatment of the cholesterol lodgment, a precursor of plaque deposition in an early phase of atherosclerosis disease. Note, that both carbon nanotube and graphene are hydrophobic – the very favorable feature in the discussed context. Naturally, more computer simulations and, very importantly, some real life experiments with carbon nanotubes, graphene and cholesterol domains covering selected components of mammals' tissues, are required to farther test the reported observations.

4. Acknowledgment

NAMD and VMD were developed by the Theoretical and Computational Biophysics Group in the Beckman Institute for Advanced Science and Technology at the University of Illinois at Urbana-Champaign.

5. References

Allen M. P. & Tildesley D. J. (1989). *Computer simulation of liquids*, Clarendon Press, ISBN 0198556544, New York, USA

Daura X., Mark A. E., & van Gunsteren W. F. (1998). Parametrization of aliphatic CH_n united atoms of GROMOS96 force field. *Journal of Computational Chemistry* , Vol. 19, No. 5 (April 1998), pp. 535 - 547, ISSN 1089-7690

Feller S. E., & MacKerell Jr A. D. (2000). An Improved Empirical Potential Energy Function for Molecular Simulations of Phospholipids. *J. Phys. Chem. B*, Vol. 104, No. 31 (August 2000), pp. 7510 - 7516, ISSN 1520-6106

Gburski Z., Górny K. & Raczyński P. (2010). The impact of a carbon nanotube on the cholesterol domain localized on a protein surface. *Solid State Communications*, Vol. 150, No. 9-10 (March 2010), pp. 415 - 419, ISSN 0038-1098

Hénin J. & Chipot C. (2006). Hydrogen-bonding patterns of cholesterol in lipid membranes. *Chem. Phys. Letters*, Vol. 425, No. 4 - 6 (July 2006), pp. 329 - 335, ISSN: 0009-2614

Hu H., Ni Y., Montana V., Haddon R. & Parpura V. (2004). Chemically Functionalized Carbon Nanotubes as Substrates for Neuronal Growth. *Nano Letters*, Vol. 4, No. 3 (March 2004), pp. 507 - 511, ISSN 1530-6984

Jorgensen W. L., Chandrasekhar J., Medura J. D., Impey R. W. & Klein M. L. (1983). Comparison of simple potential functions for simulating liquid water. *J. Chem. Phys.*, Vol. 79, No. 2 (July 1983), pp. 926 - 935, ISSN 0021-9606

Kam N. W. S., Jessop T. C., Wender P. A. & Dai H. J. (2004). Nanotube Molecular Transporters: Internalization of Carbon Nanotube-Proein Conjugates into Mammalian Cells. *J. Am. Chem. Soc.* Vol. 126, No. 22 (May 2004), pp. 6850 - 6851, ISSN 0002-7863

Kuznetsova T., & Kvamme B. (2002). Atomistic computer simulations for thermodynamic properties of carbon dioxide at low temperatures. *Energy Conversion and Management*, Vol. 43, No. 18 (December 2002), pp. 2601 - 2623, ISSN 0196-8904

Liu Y.,Wu D. C., Zhang W. D., Jiang X., He C. B., Chung T. S., Goh S. H. & Leong K. W. (2005). Polyethylenimine-Grafted Multiwalled Carbon Nanotubes for Secure Noncovalent Immobilization and Efficient Delivery of DNA. *Angewandte Chemie*. Vol. 44, No. 30 (July 2005), pp. 4782 - 4785, ISSN 1521 - 3773

Liu Z., Chen K. Davis C., Sherlock S., Cao Q., Chen X. & Dai H. (2008). Treatment Drug Delivery with Carbon Nanotubes for *In vivo* Cancer. *Cancer Research*, Vol. 68, No 16 (August 2008), pp. 6652 - 6660, ISSN 0008-5472

Lu Q., Moore J. M., Huang G., Mount A. S., Rao A. M., Larcom L. L. & Ke P. C. (2004). RNA Polymer Translocation with Single-Walled Carbon Nanotubes. *Nano Letters*, Vol. 4, No. 12 (October 2004), pp. 2473 - 2477, ISSN 1530-6984

Lusis, A. J. (2000). Atherosclerosis. *Nature*, Vol. 407, No. 6801 (September 2000), pp. 233 - 241, ISSN 0028-0836

MacKerell Jr A. D., Bashford D., Bellott M., Dunbrack Jr R. L., Evanseck J. D., Field M. J., Fischer S., Gao J., Guo H., Ha S., Joseph- McCarthy D., Kuchnir L., Kuczera K., Lau F. T. K., Mattos C., Michnick S., Ngo T., Nguyen D. T., Prodhom B., Reiher W. E., Roux B., Schlenkrich M., Smith J. C., Stote R., Straub J., Watanabe M., Wiórkiewicz-Kuczera J., Yin D. & Karplus D. (1998). All-Atom Empirical Potential for Molecular Modeling and Dynamics Studies of Proteins. *J. Phys. Chem. B*, Vol. 102, No. 18 (April 1998), pp. 3586 - 3616, ISSN 1520-6106

Pantarotto D., Briand J., Prato M. & Bianco A. (2004). Translocation of bioactive peptides across cell membranes by carbon nanotubes . *Chem. Commun.* No 1 (January 2004), pp. 16 - 17, ISSN 1359-7345

Phelps D. H., & Dalby F. W. (1966). Experimental determination of the electric dipole moment of the ground electronic state of CH. *Phys. Rev. Letters*, Vol. 16, No. 1 (January 1966), pp. 3 - 4, ISSN 0031 - 9007

Phillips J. C., Braun R., Wang W., Gumbart J., Tajkhorshid E., Villa E., Chipot C., Skeel L. & Schulten K. (2005). Scalable molecular dynamics with NAMD. *Journal of Computational Chemistry*, Vol. 26, No. 16 (December 2005), pp. 1781 - 1802, ISSN 1089-7690

Protein Data Bank. (2010). Available from: http://www.rcsb.org/pdb

Raczyński P., Dawid A., Piątek A. & Gburski Z. (2006a). Reorientational dynamics of cholesterol molecules in thin film surrounded carbon nanotube: molecular dynamics simulations. *J. Molecular Structure* Vol. 792 - 793 (July 2006), pp. 216 - 220, ISSN 0022-2860

Raczyński P., Dawid A. & Gburski Z. (2006b). The dynamics of cholesterol in cholesterol-phospholipid assembly localized near nanotube surface: MD study. *J. Molecular Structure*, Vol. 792 - 793 (July 2006), pp. 212 - 215, ISSN 0022-2860

Raczyński P., Dawid A., Sokół M., & Gburski Z. (2007). The influence of the carbon nanotube on the structural and dynamical properties of cholesterol cluster. *Biomolecular Engineering*, Vol. 24, No. 5 (November 2007), pp. 572 - 576, ISSN 1389- 0344

Rapaport D. C. (1996). *The art of molecular dynamics simulation.* (1996). Cambridge University Press New York, ISBN 0521445612, New York, USA

Róg T., Pasenkiewicz-Gierula M., Vattulainen I. & Karttunen M. (2009). Ordering effects of cholesterol and its analogues. *Biochimica et Biophysica Acta (BBA) - Biomembranes*, Vol. 1788, No. 1 (January 2009), pp. 97 - 121, ISSN 0005-2736

Schiffer C. A., Ultsch M., Walsh S Somers, De Vos A. M., & Kossiakoff A. A. (2002). Structure of a Phage Display-derived Variant of Human Growth Hormone Complexed to Two Copies of the Extracellular Domain of its Receptor: Evidence for Strong Structural Coupling between Receptor Binding Sites . *J. Mol. Biol.*, Vol. 316, No 2 (February 2002), pp. 277 - 289, ISSN 00222836

Sitharaman B., Kissell K. R., Hartman K. B., Tran L. A., Baikalov A., Rusakova I., Sun Y., Khant H. A., Ludtke S. J., Chiu W., Laus S., Tóth E., Helm L, Merbach A. E. & Wilson L. J. (2005). Superparamagnetic gadonanotubes are high-performance MRI contrast agents. *Chem. Commun.* No. 31 (July 2005), pp. 3915 - 3917, ISSN 1359-7345

Wang J., Liu G., & Jan M. (2004). Ultrasensitive Electrical Biosensing of Proteins and DNA: Carbon-Nanotube Derived Amplification of the Recognition and Transduction Events. *J. Am. Chem. Soc.*, Vol. 126, No. 10 (March 2004), pp. 3010 - 3011, ISSN 0002-7863

Wisitsoraat A., Sritongkham P., Karuwan C., Phokharatkul D., Maturos T. & Tuantranont A. (2010). Fast cholesterol detection using flow injection microfluidic device with functionalized carbon nanotubes based electrochemical sensor. *Biosensors and Bioelectronics*, Vol. 26, No. 4 (December 2010), pp. 1514 - 1520, ISSN 0956-5663

21

Electric-Field and Friction Effects on Carbon Nanotube-Assisted Water Self-Diffusion Across Lipid Membranes

Niall J. English, José-Antonio Garate and J. M. Don MacElroy
The SEC Strategic Research Cluster and the Centre for Synthesis and Chemical Biology, School of Chemical and Bioprocess Engineering, University College Dublin, Belfield, Dublin 4, Ireland

1. Introduction

The accelerated appearance of applications and theoretical studies of carbon nanotubes (CNTs)[1] in the literature over the last two decades reveals their broad attraction in fields as diverse as nanoelectronics, bionanotechnology and solar energy conversion nanodevices[2-5]. In particular, they provide the basis for simplified models of nanofluidic machines and membrane channels such as aquaporins[6-9] and have been employed in a range of simulation studies at the nanoscale to investigate the transport of water undergoing single-file or highly confined diffusion and permeation in a single CNT or in arrays of CNTs[10-22].

The simulation method of choice in the study of water transport through nanostructures is molecular dynamics (MD) (see, for example, [23-25]) which provides real-time trajectories of all of the atoms within the system under investigation and, through the statistical framework of linear response theory, permits a direct link with macroscopic observables. One of the first studies of water confined within CNTs employing MD, supplemented with an analysis using a continuous-time random walk model (CTRWM), was reported by Berezhkovskii and Hummer[12,13] and demonstrated that water fills a (6,6) CNT spontaneously in a single-file fashion. In later work, MD simulations focused on pressure-driven osmotic water flow in single CNTs and arrays of CNTs[14-16] and in model pores containing charge distributions which are known to be ubiquitous for biological channels[17,18]. Water channels made of double-walled carbon nanotubes embedded in a lipid membrane, where the inner tube is modeled as a (6,6) CNT with a charge distribution emulating the NPA motif of aquaporins[19], have also been investigated, and anomalous observations relating to the dynamics, configurational states and apparent pore-length dependence of the properties of water confined within CNTs have received attention[20-22].

In [26,27], molecular simulations were reported for unidirectional single-file diffusion of water within (5,5) and (6,6) single-walled carbon nanotubes (SWCNT) and normal diffusion in (8,8) and (11,11) SWCNT's across a lipid membrane. These studies employed both equilibrium MD simulation techniques and an adaptation of MD to investigate conditions involving the influence of both static and alternating electric fields. While the fields employed in this work were relatively weak vis-à-vis fields generated by charge distributions, it was observed that the permeation of water for the (5,5) SWCNT embedded in a lipid membrane is enhanced by static electric fields due to a decrease in the fluctuations of the number of water molecules inside the SWCNT. In the larger single-file (6,6) CNT pore, dipole alignment with the imposed external field is more easily achieved due to a greater degree of rotational freedom of the water molecules and this was observed to result in a significant reduction in the water self-diffusion flux[26]. In this work it was also observed that the non-field (6,6) flux results were much higher than those reported by Zhu and Schulten[14], even though the same potential/force field was used in both studies. We suggested that this difference may be due to the array of parallel nanotubes employed by Zhu and Schulten which could cause friction (via long-range electrostatics) between the single files of water. This was confirmed in [27] where it was demonstrated that in these CNT-arrays there is a significant friction effect that retards movement of the single files of water.

In this chapter, we report further studies of non-equilibrium MD simulations of water permeation within (6,6) SWCNT's across a lipid bilayer under the influence of both static and alternating electric fields, taking advantage of GPU-accelerated MD[28] to extend the simulation timescales to longer than was previously possible[26,27], while also investigating single-molecule translational and rotational frequency modes, collective properties and friction in tube-arrays in greater detail, allied with straightforward, mechanistic models to interpret our results.

2. Simulation methodology

All of the GPU-accelerated MD simulations[23-25] were carried out using the NAMDv2.7[28,29] program suite, in conjunction with the CHARMM27 parameter set[30] and the TIP3P model for water[31]. Initially, a periodically replicated single-nanotube system was studied, comprising a (6, 6) armchair SWCNT, with a diameter of 8.20 Å and approximate length of 36.9 Å, embedded in a 1-palmytoil-2-oleoyl-sn-glycero-3-phosphatidylcholine (POPC) membrane patch. Layers of water of around 10Å were added in the +*z* and -*z*-direction to solvate the system. Each atom of the SWCNT was modelled as a neutral sp^2 aromatic carbon with uncapped ends (in a similar way as shown in Fig. 1, although this is for arrays of four CNTs). For all of these systems, the plane of the POPC membrane was placed in the *x-y* plane, and the axis normal to the membrane was along the *z*-direction, with the CNT oriented along this *z*-axis (perpendicular to the plane of the membrane – cf. Fig. 1). The final size of the (three-dimensional-periodic) simulation box was 34.84 x 34 x 66.29 $Å^3$, with total atom numbers of 8,415 (1337 water molecules and 30 phospholipids).

To model the arrays of CNTs, three-dimensional-periodic cells consisting of four (6, 6) SWCNT arrays were constructed with separations of 15, 20 and 25 Å between the geometric central axes of each (see Fig. 1), and inserted along the *z*-axis of a POPC membrane, again solvated with layers of water molecules of around 10 Å in thickness in the +z and -z-directions. In each of these cases, the membrane was also placed in the *x-y* plane, with the principal axis of the SWCNT oriented perpendicular to the membrane. The final sizes for the simulation boxes in the array-systems were 45.75 x 60.32 x 66.68 $Å^3$, 49.36 x 49.21 x 71.27 $Å^3$ and 63.10 x 59.99 x 66.27 $Å^3$ for respective separations of 15, 20 and 25 Å, with total atom numbers of 18,979 (3045 water molecules and 62 phospholipids), 17,901 (3043 water molecules and 54 phospholipids) and 26,309 (4327 water molecules and 88 phospholipids), respectively. All molecular systems were constructed using the VMDv1.86[32] program.

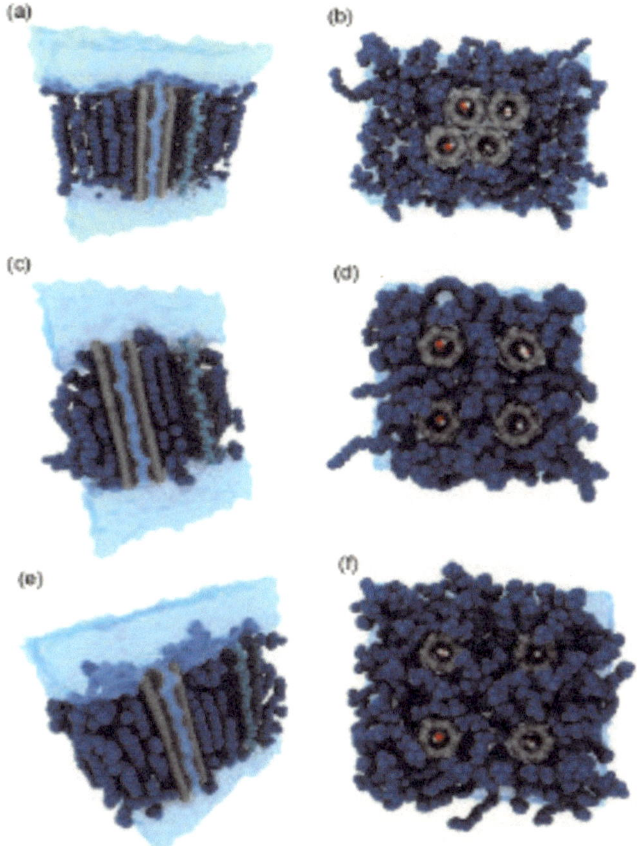

Fig. 1. (a), (c) and (e) Cross-section of the periodic cells simulated for the (6,6) CNT-arrays separated by 15, 20 and 25 Å, respectively. (b), (d) and (f) Plan view (along the z-axis, with the x-y plane in that of the page) of the (6,6) CNT-array systems separated by 15, 20 and 25 Å, respectively.

The Particle Mesh Ewald[33] method was utilised to treat long-range electrostatics to within a relative tolerance of 1×10^{-6} with a cut-off distance of 12 Å applied to real-space Ewald interactions; it was found that a real-space screening coefficient of 0.258 $Å^{-1}$ and (three-dimensional Fourier transform) grid spacings in the 0.9 to 1 Å range in each Cartesian direction led to this desired accuracy level, along with close-to-optimal execution speeds. A cut-off distance of 12 Å was also applied to van der Waals interactions, using a smooth switching function between 10 and 12 Å. Using the r-RESPA method[34], multiple time steps were applied to MD, with 1 fs for bonded interactions, 1 fs for short-range non-bonded interactions (including real-space Ewald interactions) and 2 fs for the full (real- and reciprocal-space) electrostatics evaluation. All production runs were carried out using Langevin dynamics[35] coupled to an NVT reservoir set at 300 K with a damping coefficient of 1 ps^{-1}, representing rather mild coupling which affects diffusion little. The SHAKE[36] algorithm was used to constrain bond lengths to all hydrogen atoms.

Prior to production, an NPT relaxation run was carried out for 1 ns; a Nosé-Hoover approach was applied[37] using anisotropic cell variation, with Langevin dynamics for piston

fluctuation control[38]; here, the set points were 1 atm and 300 K. During this NPT relaxation, the nanotube atoms were fixed, while the lipid bilayer tails and water molecules were allowed to move freely. Following this, the CNT and lipid-head atoms were released slowly, keeping a harmonic constraint thereon with a strength of 1 kcal/mol·Å^2. This strength coefficient was reduced progressively to 1.5, 1.0 and 0.5 kcal/mol·Å^2 every 100 ps for 0.4 ns. After this initial relaxation, during which the pressure and volume had stabilised, MD simulations of each of the (6, 6) CNT-systems (single-tube and array) were carried out for 50 ns in the NVT ensemble (in the absence and presence of external fields – *vide infra*). Three replicates were performed for each system in each case (zero- and in-field).
Use of GPU-accelerated MD implemented in NAMD v2.7[28,29] on an NVIDIA double-precision C2050 platform led to a substantial acceleration (approximately 2.5-fold) vis-à-vis the equivalent number of CPU-cores on quad-core Intel Xeon Harpertown-Stoakley nodes connected via sub-microsecond-latency Infiniband. The acceleration was primarily, though not only, in the evaluation of intermediate-range non-bonded van der Waals and electrostatics interactions, where the computational overhead was reduced approximately six-fold relative to that required with the equivalent number of CPU-cores in a CPU-only approach (on the abovementioned Intel Xeon chips).
Static electric fields (EFs) were applied in the non-equilibrium MD simulations with an intensity of 0.0065 V/Å in the +*z*-direction (*i.e.*, along the CNT-axis). The EF exerts a force over partial atomic charges *ia*:

$$f_{ia} = q_{ia}E \tag{1}$$

Non-equilibrium MD simulations were also carried out in the presence of alternating electric fields[39] along the same *z*-axis (*i.e.*, along the axes of the nanotubes). In this case, the RMS intensity was set to 0.0065 V/Å, the same as that of the peak frequency for the static EF, but the field does change direction every period between the +*z* and –*z* axis, *ipso facto*[40,41]. Three different frequencies were used: 2.45, 50 and 100 GHz. Here, the force exerted over the partial atomic charges is given by:

$$f_{ia} = q_{ia}E_0\cos(\omega t) \tag{2}$$

No constraints were applied to the membrane during MD in either the absence or presence of static and alternating electric fields, while a light 0.5 kcal/mol Å^2 harmonic constraint was applied to the SWCNT atoms.

3. Modelling diffusion and permeation of water in single-file channels

The transport of water through single file channels embedded in lipid bilayers may be analysed from two perspectives (a) self-diffusion of water (e.g. labeled isotopically for direct physical measurement or, for simulation purposes, simply assigned a 'colour') and (b) water permeation under the influence of a pressure gradient (osmotic or otherwise). The two parameters which characterise these transport processes are the diffusive and osmotic permeabilities, p_d and p_f, respectively (see [10] for a review of details). These are defined through the flux expressions:

$$j_d = p_d \Delta c_{tr} \tag{3a}$$

and

$$j_f = \frac{1}{k_B T} p_f \Delta P \tag{3b}$$

where Δc_{tr} is the tracer (or colour) concentration difference across the membrane, P is the trans-membrane liquid pressure difference (due, under normal circumstances, to an osmotic bulk solute concentration difference), T is absolute temperature and k_B is Boltzmann's constant.

In [10] and references cited therein (see also [12-15]) the diffusive permeability has been investigated using a continuous time random walk model (CTRWM) and, in the absence of a chemical potential difference $\Delta\mu$ or an external force, the osmotic permeability was computed by equilibrium MD using a collective diffusion model. In [26, 27], a more general model was developed for the diffusive permeability (which is not restricted in application to single-file pores) and this more general approach is described below. In addition, the collective transport permeability is analysed in this chapter in greater detail using the generalized linear transport theory for slip flow in pores developed in earlier work[42,43].

3.1 The diffusive permeability

To compute the diffusive permeability, p_d, of water in a strongly inhomogeneous system constituting two bulk liquid reservoirs separated by a lipid membrane with embedded carbon nanotubes, the simplest and statistically most accurate approach is to employ a colour diffusion technique. In this approach the flux of 'black' water molecules diffusing from left to right in Fig. 2 is monitored during the course of a long molecular dynamics simulation (a similar analysis applies to 'white' water molecules diffusing in the opposite direction). If the cross-sectional areas in the bulk and in the pore are A and A_p, respectively, and L is the length of the mixing zone between the black reservoir on the left of Fig. 2 and the pore mouth, then at steady state the net flow (particles per unit time) between the black reservoir and the outer perimeter of the CNT is given by

$$j_d = \frac{DA}{L}(c_B(1) - c_B(2)) \tag{4}$$

where $c_B(1)$ is the concentration of black water molecules in the black reservoir and $c_B(2)$ is the concentration of black water molecules just outside the CNT pore mouth.

At the pore mouth a simple reversible rate process takes place for the molecules to either enter the pore or return into the reservoir on the left and the net rate is given by

$$r_{2-2p} = k_f c_B(2) - k_r c_{Bp}(2) \tag{5}$$

where the forward and reverse rate constants k_f and k_r are in units of sec^{-1} and $c_{Bp}(2)$ is the concentration of black water molecules just inside the CNT pore adjacent to the pore mouth. The net rate in Eqn. (5) is a volumetric measure (these rate constants will be determined to a great extent by the hydrogen bond structure and dynamics). Multiplying by a volume V_i (an interfacial volume corresponding to $A_p\delta$ with δ = length scale of the order of the molecular dimensions representing the scale of the concentration inhomogenities) then for steady state conditions we have

$$\begin{aligned} V_i r_{2-2p} &= j_d \\ &= V_i(k_f c_B(2) - k_r c_{Bp}(2)) \\ &= V_i k_f (c_B(2) - \frac{1}{K_p} c_{Bp}(2)) \\ &= \frac{DA}{L}(c_B(1) - c_B(2)) \end{aligned} \tag{6}$$

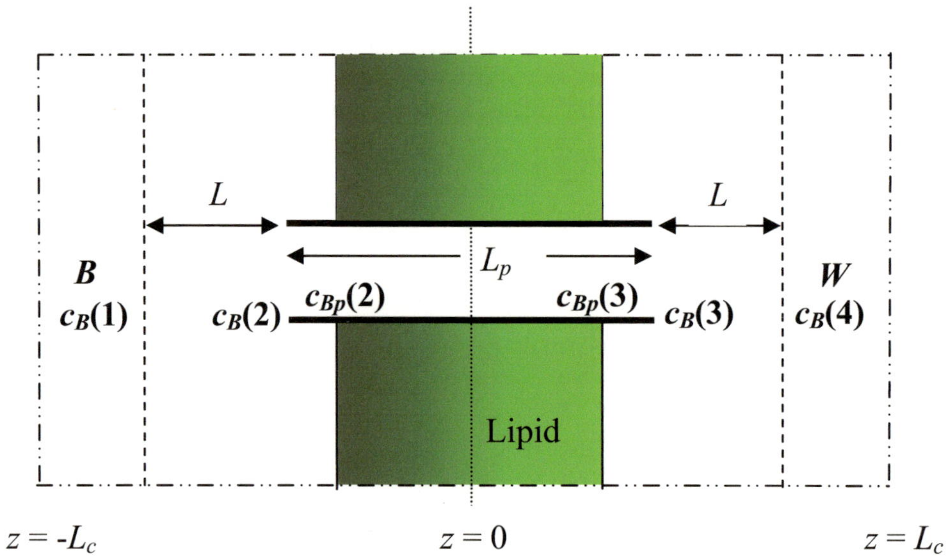

Fig. 2. Schematic of the 'black' and 'white' water reservoirs on either side of the CNT, with the CNT oriented along the z-axis, normal to the membrane patch. The concentration of the black water molecules is specified at the edge of the black reservoir ($c_B(1)$), at the pore mouth ($c_B(2)$), just inside the pore ($c_{Bp}(2)$), with analogous concentrations on the other side of the simulation cell (the positive z-coordinate half), *i.e.*, $c_{Bp}(3)$, $c_B(3)$ and $c_B(4)$. L_c was 3.5 nm in all of the systems simulated.

Eliminating $c_B(2)$ then

$$j_d = \frac{p_1}{K_p}(K_p c_B(1) - c_{Bp}(2)) \tag{7}$$

where K_p is defined as k_f/k_r and p_1 is the partial permeability for diffusive transport from the black reservoir to just inside the CNT:

$$\frac{1}{p_1} = \frac{L}{DA} + \frac{1}{k_f V_i}$$

The steady-state flow along the pore may also be defined using an intra-pore diffusion coefficient, D_p (which is considered to exist in the same sense as proposed in [10, 12-14] but which has a deeper meaning than indicated by the simple CTRWM)

$$j_d = \frac{D_p A_p}{L_p}(c_{Bp}(2) - c_{Bp}(3)) \quad (8)$$

where L_p is the length of the pore and $c_{Bp}(3)$ is the concentration of black water molecules just inside the CNT pore adjacent to the pore mouth near the white particle reservoir.
Also, for the right hand side of the system it follows from Eqn. 7 that

$$j_d = \frac{p_1}{K_p}(c_{Bp}(3) - K_p c_{Bp}(4)) \quad (9)$$

where $c_{Bp}(4)$ is the concentration of black water molecules in the white particle reservoir.
Now combining Eqns. 7, 8 and 9 eliminating the intermediate concentration terms gives

$$j_d = p_d(c_B(1) - c_B(4))$$

where

$$\frac{1}{p_d} = \frac{2L}{DA} + \frac{2}{k_f V_i} + \frac{L_p}{D_p K_p A_p} \quad (10)$$

Eqn. 8 simply states that the permeability is governed by a sum of mass transfer resistance terms due to bulk diffusion, incorporation into the pore mouth and intra-pore diffusion. Normally, one sets $c_B(4) = 0$ (in the simulations the 'black' water molecules entering the 'white' reservoir are simply switched in colour to 'white') and $c_B(1)$ = molecular/molar density of water in the bulk i.e. $c_{H2O} = 33.5\ \text{nm}^{-3}$ or 55.6 mol.litre^{-1}. Therefore

$$j_d = p_d c_{H_2O}$$

The overall permeability can be simply determined from the MD simulations and by selecting different positions for the 'black' and 'white' reservoirs, one can eliminate the bulk diffusion D/L terms.
In the development of the CTRW model reported by Berezhkovskii and Hummer[12], there is the implicit assumption that L is sufficiently small to minimise bulk-phase diffusional resistance while retaining a sub-nanometre range close to the pore mouth ($\delta \sim a$) to ensure that entry/exit kinetics are fully taken into consideration. This condition with the additional assumption that the single file is considered to be densely packed leads to

$$L_p = Na;\ \delta = a$$

where a is a length scale characteristic of the size of the water molecules. Furthermore, the diffusion coefficient within the pore is given by the CTRW model as

$$D_p = \frac{a^2}{2\tau} = \frac{a^2}{2} k_r$$

In our model, we employ the physically reasonable assumption that the intra-pore hopping rate k is equal to k_r, *i.e.*, the reverse rate constant in Eqn. 5. Within the pore, the forward and reverse hopping rate constants are equal. At the pore mouth, however, we distinguish between the forward (pore entry) and the reverse (pore exit) rate constants. For the CNT's under consideration here, one may anticipate $k_r > k_f$ or $K_p < 1$ and if we assume the molecular volume $v_w = aA_p$, then we find

$$\text{Permeation rate} = \frac{k_f a A_p}{2}\frac{1}{N+1}c_{H2O} = \frac{v_w k_o}{N+1}c_{H2O}$$

which is the result provided by the CTRW model in [12]. This analysis demonstrates the level of the assumptions involved in the CTRWM approach. As noted above, in a more general setting, the in-pore forward hopping frequency is not the same as at the pore mouth. Furthermore, the more general model of Eqn. 10 applies to wide (i.e. multi-pass) pore conditions of arbitrary cross-sectional area and it is also simple to extend Eqn. 10 to include rate processes at other junctions (e.g. local charged sites) within the pore and to relate the rate constants to the energy barriers associated with the particle movement along the channel.

3.2 The osmotic permeability, collective properties and friction in pore-arrays

In general, the flux of water molecules across a membrane subject to a chemical potential driving force, $\Delta\mu$, which in turn originates from an osmotic pressure gradient, may be expressed as

$$j_f = \frac{c_{H_2O}}{k_B T} p_f \Delta\mu \tag{11}$$

which is equivalent to Eqn. (3b). Here, c_{H_2O} is the molecular density of water while p_f is osmotic permeability as defined, *inter alia*, in [10, 14-16]. Using results from linear response theory[42,43] the permeability of membranes investigated in this work may be expressed as

$$c_{H_2O} p_f = \frac{c_{H_2O,M} D_M^{(z)} A}{L_p} = \left(\frac{N_T}{L_p}\right)^2 \int_0^\infty \langle u_z(t) u_z(0)\rangle dt \tag{12a}$$

$$= \left(\frac{N_T}{L_p}\right)^2 \lim_{t\to\infty} \frac{d\langle (z(t)-z(0))^2\rangle}{dt} \tag{12b}$$

where $c_{H_2O,M}$ is the water concentration within the membrane (which is related to c_{H_2O} via the partition coefficient[44] K between the bulk liquid phase on both sides of the membrane and the membrane pores), N_T is the total number of water molecules within the membrane at any given time and $D_M^{(z)}$ is the axial slip-diffusion (or viscous slip) coefficient of the water molecules within the CNT-pores of the membrane. The function in the integral of Eqn. 12a is the autocorrelation function (ACF) of the axial, collective centre-of-mass velocity of all of the water molecules present in the pores of the membrane, with the collective (number averaged) velocity given by

$$u_z(t) = \frac{1}{N_T}\sum_{i=1}^{N_T} v_{i,z}(t) \tag{13}$$

where $v_{i,z}(t)$ is the velocity of water molecule *i* along the z-axis of the CNT-pores. In Eqn. 12b, $z(t)$ is the corresponding collective, axial centre-of-mass position of the water molecules in the pore at any point in time:

$$z(t) = \frac{1}{N_T} \sum_{i=1}^{N_T} z_i(t) \tag{14}$$

For a membrane containing an array of N_p pores, each of which is accessible to water to the same extent, as studied in this work, we note that Eqn. 12a (and similarly Eqn. 12b) may be expressed as

$$c_{H_2O} p_f = \left(\frac{1}{L_p}\right)^2 \int_0^\infty \left\langle \sum_{j=1}^{N_p} \sum_{i=1}^{N(j)} v_{i,z}(t) \sum_{k=1}^{N_p} \sum_{i=1}^{N(k)} v_{i,z}(0) \right\rangle dt \tag{15a}$$

$$= \left(\frac{1}{L_p}\right)^2 \int_0^\infty \left\langle \sum_{j=1}^{N_p} N(j) u_z^{(j)}(t) \sum_{k=1}^{N_p} N(k) u_z^{(k)}(0) \right\rangle dt \tag{15b}$$

where $N(j)$ is the number of water molecules in pore *j*. The correlation function in Eqn. 15 may be considered a total one (TCF), in that it applies to all water molecules in the membrane (*i.e.* present in all of the CNT-pores). Considering Eqn. 15b, it is clear that this may be written in terms of the ACFs of the collective velocities in individual pores as well as cross-correlation functions (CCFs) of collective velocities between particular pores. Eqn. 15 demonstrates the possibility of significant cross-interactions between the individual pores in the array, unless they are well separated. Should this latter condition apply, then Eqn. 15 simplifies to

$$c_{H_2O} p_f \Big|_\infty = \left(\frac{1}{L_p}\right)^2 \int_0^\infty \left\langle \sum_{j=1}^{N_p} N^2(j) u_z^{(j)}(t) u_z^{(j)}(0) \right\rangle dt$$
$$= N_p c_{H_2O} p_f^*$$

where the subscript ∞ denotes the limit of infinite separation of all pores, and p_f^* is the permeability for a single, isolated pore.

In order to explore further the underlying nature of the collective pore-velocity correlations of Eqn. 15 which govern the distance-dependent friction observed in our previous work[27] it is appropriate to consider a simplified model of friction. For a given collective velocity within the pores, u_z, the friction may be approximated by a term due to isolated pores (the '∞' case), as well as an additional, distance-dependent term (a finite separation 'f' case), which will be zero in the case of the limit of infinite separation:

$$\frac{du_z}{dt} = -\left(1/\tau_\infty + 1/\tau_f\right) u_z \tag{16}$$

the solution of which is

$$u_z(t) = u_z(0)\exp\left(-\left(1/\tau_\infty + 1/\tau_f\right)t\right)$$

Here, the isolated-case $1/\tau_\infty$ friction term may be considered as a relaxation time which might possibly be expected in the decay of the ACF of the isolated-case collective velocity, while that of $1/\tau_f$ would be additional friction that might be manifest in the decay of the TCF of Eqn. 15, vis-à-vis that of the isolated case. At its simplest, one might model the distance-dependent friction term, $1/\tau_f$, to be inversely proportional to separation *d*, *i.e.* $1/\tau_f \sim 1/d$. Exploring this collective-velocity correlation further for the CNT-array:

$$\int_0^\infty \left\langle \left(\sum_{j=1}^{N_p} u_{j,z}(0) \right)^2 \right\rangle \exp\left(-\left(1/\tau_\infty + 1/\tau_f\right)t\right) dt = \left\langle \left(\sum_{j=1}^{N_p} u_{j,z}(0) \right)^2 \right\rangle \frac{1}{1/\tau_\infty + 1/\tau_f} \quad (17a)$$

$$= \frac{1}{N_T} \frac{k_B T}{m} \frac{1}{1/\tau_\infty + 1/\tau_f} \quad \text{where } 1/\tau_f \sim 1/d \quad (17b)$$

so that the permeability may be expressed in terms of the frictional terms:

$$\Rightarrow c_{H_2O} p_f = \frac{N_T}{L^2} \frac{k_B T}{m} \frac{1}{1/\tau_\infty + 1/\tau_f}$$

$$= c_{H_2O,M} \frac{A}{L} \frac{k_B T}{m} \frac{1}{1/\tau_\infty + 1/\tau_f}$$

leading to an effective slip-diffusion coefficient, D_{eff}:

$$p_f = \frac{K D_{eff} A}{L_p} \quad \text{where } \frac{1}{D_{\text{eff}}} = \frac{m}{k_B T}\left(1/\tau_\infty + 1/\tau_f\right) \quad (18)$$

4. Results and discussion

4.1 Static and alternating electric field effects and water self-diffusion

The permeation rates for water molecules through the SWCNT (in molecules per ns) embedded within the lipid layer at steady-state, in the absence and presence of axially-applied static and alternating electric fields, are provided in Fig. 3. Although the rate in the static field is marginally higher (notwithstanding overlapping error bars vis-à-vis the zero-field case), the alternating field results show no discernible trend when the electric field frequencies are incremented. It is important to note that each point represents the average fluxes of three independent simulations. Statistically, with the possible exception of the static-field case, we observe that there is no correlation between the field frequencies and the computed water permeation rates. This is in essential agreement with our previous study[27] although longer MD timescales from GPU-acceleration has led to reduced error bars in this work. In our previous work[26,27] monotonic decreases of water permeation rates with field frequencies were found for axial field application to (5,5) CNT-systems towards the zero-field rate at higher frequency, but not to (6,6) CNT-cases (or larger CNTs). For (5,5) CNT-systems, it was found that axial static electric fields enhanced water self-diffusion across them due to a decrease in the fluctuations of the number of water molecules within the pores, arising from the water molecules' dipole orientation with the field[26]. Although it is difficult to reach a conclusion for the (more statistically accurate vis-à-vis [27]) slightly larger static-field rate in the case of this work for (6,6) CNTs, due to overlapping error bars with respect to the zero-field result, the single-file nature of water transport in (6,6) CNTs, similar to the (5,5) case, suggests that a more minor decrease in the fluctuations of the number of water molecules inside is at play in the static field. This was observed in this study with results slightly lower than previous (6,6) data[26,27] although error bars for the size of the fluctuations were again overlapping vis-à-vis the zero-field case, so no firm statistical conclusion can be drawn in that regard. It would be necessary to carry out a greater number of longer runs to investigate this further, somewhat prohibitive even with ample GPU-accelerated computing resources.

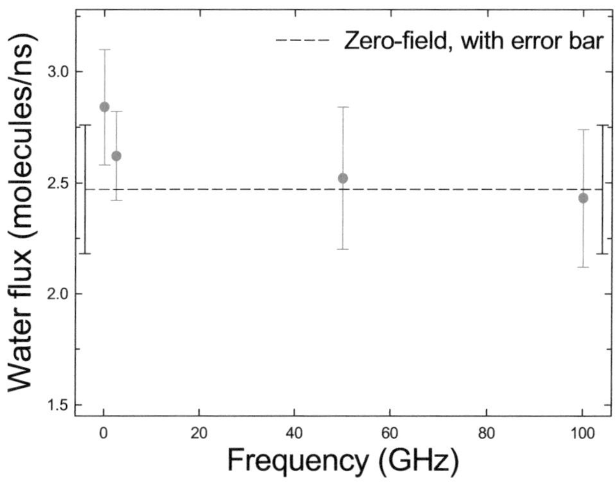

Fig. 3. Steady-state permeation rates under the influence of static and time-varying electrical fields. The dashed black line represents zero-field conditions with the corresponding error bars shown at either end thereof.

The water dipole orientations within the pore (see Fig. 4), illustrate more clearly the influence of the external field on the probability distributions of the cosine of the angle of the dipole with the positive *z*-axis, α. In agreement with our earlier results, the statistically more accurate results reported here illustrate that at high frequencies the distributions become more symmetric with peaks at -1 and 1, in very good agreement to the zero-field case. In the static field case the distribution is shifted almost entirely to the left side of the diagram.
From consideration of the fluxes and dipolar orientations, it is clearly evident that the results of the low-frequency electric field simulations are more similar to the static-field case, whereas high-frequency fields resemble zero field conditions. These results support the finding that the effects of low-frequency electric fields should be closer to those of static electric fields (if any), because the slower time variation of the field (*i.e.*, longer period) relative to permeation and loading timescales allows the field to influence the dynamics of these events. On the other hand, high-frequency electric fields do not readily permit enhancements in permeation or alterations in dipole alignment vis-à-vis the zero-field case, due to their faster oscillation and shorter period relative to permeation timescales in (6,6) CNTs.

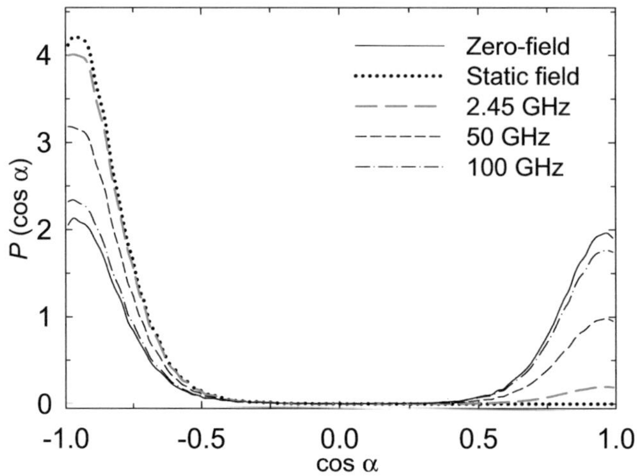

Fig. 4. Normalised probability distributions of the cosine of the angle which the dipole vector of the water molecules makes with the positive *z*-axis, α.

Considering single-molecule translational and librational modes in the isolated CNT system, the power spectra of the oxygen- and hydrogen-atom VACFs are shown in Figs. 5 and 6, respectively, under zero-field conditions, along with corresponding results for bulk water. The spectra are decomposed in terms of *x-y* and *z*-directions, and a translational 'bumping' mode vis-à-vis the CNT walls is clearly visible in the *x-y* plane at around 35 cm^{-1} (circa 0.8-0.9 ps) in the oxygen-atom spectra, along with corresponding librational modes of 35 and 160 cm^{-1} (around 0.18 and 0.8-0.9 ps). The axial-spectra, although somewhat more bulk-like, also exhibit important differences vis-à-vis the bulk state: notably a greater translational 'rattling' at 260 cm^{-1} (around 0.12 ps), and a marked reduction in higher-frequency librational motion, in accord with retarded rotational motion (evident in the preferred dipole distributions of Fig. 4).

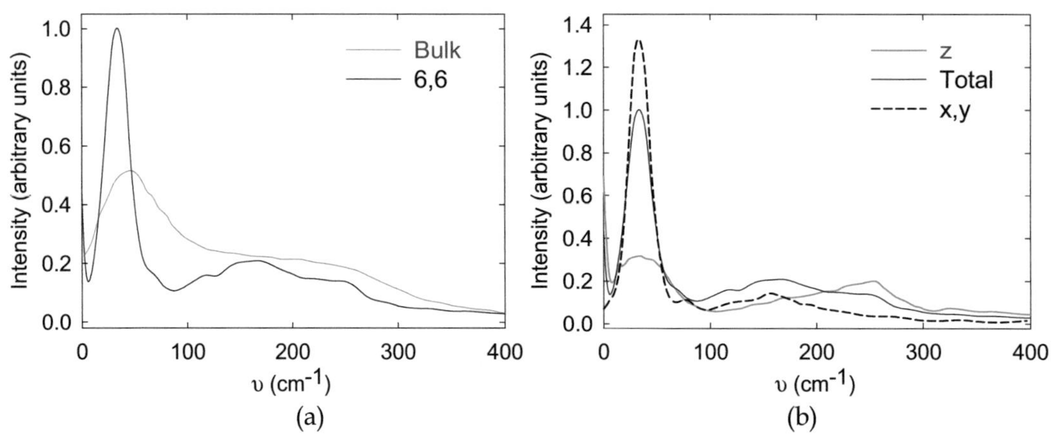

Fig. 5. Power spectra of single-water molecule VACFs for oxygen atoms in the isolated CNT under zero-field conditions, showing (a) the total spectrum along with corresponding results for bulk water, and (b) the *x-y* and *z*-direction spectra in the CNT.

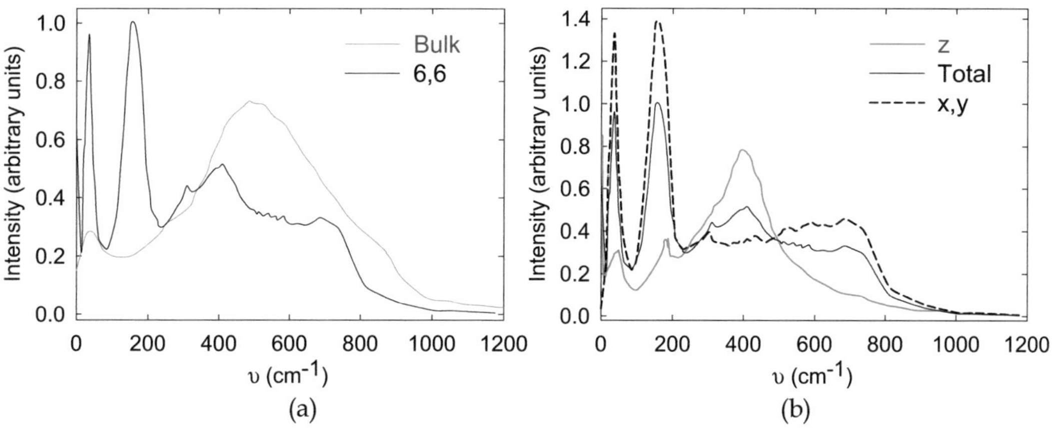

Fig. 6. Power spectra of single-water molecule VACFs for hydrogen atoms in the isolated CNT under zero-field conditions, showing (a) the total spectrum along with corresponding results for bulk water, and (b) the *x*-*y* and *z*-direction spectra in the CNT.

4.2 Water permeation through nanotube-arrays

Steady-state water diffusion rates, j_d, and hopping rates, k, were computed for the (6,6) SWCNT array simulations and are reported in Table I. The intra-pore water self-diffusion coefficients, D_p, provided in this Table were computed independently as described in [26] using the Einstein relation for self-diffusion (a special case of Eqn 12b)

$$D_p = \lim_{t\to\infty} \frac{1}{2} \frac{d\left\langle (\Delta z)^2 \right\rangle}{dt}$$

The results summarized here are essentially in agreement with previously reported values[25,26] albeit with reduced error bars due to the longer timescales rendered possible from GPU-acceleration. There is an incremental increase in j_d when the separation between the pores is augmented, which is a direct consequence of the incremental increase in D_p with the separation of the nanotubes for separations greater than 15 Å. The lower than anticipated diffusive flux for the 15 Å close packed structure when comparing the magnitude of j_d relative to D_p arises in part due to the distance-dependent friction effect[27] that retards water movement within the pores as demonstrated by the osmotic permeability results (the right-hand column of Table I) but also reflects a greater degree of interference between the pores at the pore mouths, namely a lowering of the forward (pore entry) rate constant k_f in the second term in Eqn. 10.

Estimates of the hopping rate constant, k, suggested by the CTRW model were also computed to compare our results with those reported in refs. [12-14], since, as previously shown[26,27] our results for the zero-field (6,6) CNT were in quite good agreement with those reported by Berezhkovskii and Hummer[12,13] but differed with the ones reported by Zhu and Schulten[14], even though both our results and the ones previously mentioned are based on the TIP3P/CHARMM potential. The hopping rate result, k = 57.4 ns^{-1}, reported here for simulations involving the 15 Å separation is more than a factor of 2 higher than the corresponding value reported by Zhu and Shulten (k = 26.9 ns^{-1}). As discussed in [27] only four tubes were employed in our simulations (embedded in a non-diffusing phospholipid membrane) while a close packed periodically imaged hexagonal array of twelve tubes was employed in the work conducted by Zhu and Schulten.[14] The much lower k observed in ref. 14 is explained by an increase in the friction effect generated by the twelve single files of water interacting via long range electrostatics, rather than the four nanotubes employed in our work.

Separation (Å)	j_d (ns^{-1})	p_d (nm^3.ns^{-1})	D_p (nm^2 ns^{-1})	^ak (ns^{-1})	^bp_f (nm^3.ns^{-1})	$^cp_f^{FM}$ (nm^3.ns^{-1})
15	1.69 (0.18)	0.050	1.94 (0.07)	57.40	0.896	1.74
20	2.12 (0.28)	0.063	2.00 (0.06)	59.17	1.156	1.85
25	2.32 (0.13)	0.069	2.14 (0.11)	63.31	1.268	1.93
∞	2.47 (0.29)	0.074	2.38 (0.12)	70.41	1.368	2.18

[a] The hopping rates defined by the CTRW model, k ($\equiv k_r$) = $2D_p/a^2$, where a = 0.26 nm is the projected length for a given water molecule along the pore axis (for more details see [12]).
[b] The osmotic permeability predicted by integration of the viscous slip total velocity-CF (TCF) (Eqn. 15).
[c] The osmotic permeability predicted by the simplified friction model (Eqn. 18).

Table I. Steady-state tracer permeation rates, tracer intrapore diffusion coefficients, hopping rates and osmotic permeabilities for the CNT arrays.

To demonstrate the existence of significant frictional effects between the aligned water files in [27] we estimated an effective viscosity along the axis of the water confined inside the CNTs. In summary, this was done by use of the appropriate Einstein relation[23] applied to tracking chains of six water molecules initially nearest the centre of each of the four nanotubes. The four hexamers of water so defined were treated as a fluid of four 'particles' and their motion was monitored until one or more of the water molecules in any given hexamer left its respective nanotube. The hexamers were then redefined and sampling continued until the next disruption in one of the hexamer chains. Averaging over each such occurrence in all four tubes in the zero-field simulations then enabled the determination of the effective viscosity of the files using:

$$\eta = \frac{V}{2k_B T}\lim_{t\to\infty}\frac{d}{dt}\left\langle \left(\mathrm{P}_{zz}(t)-\mathrm{P}_{zz}(0)\right)^2\right\rangle \text{ where } \mathrm{P}_{zz} = \frac{1}{V}\sum_{i=1}^{N_T} r_{iz}p_{iz} \tag{19}$$

The long-time slopes of the mean square displacement of the dynamical variable P_{zz} as functions of time (see Fig. 7) for each nanotube separation provided the effective viscosities 7.5 ± 0.45 (15 Å), 5.5 ± 0.4 (20 Å) and 4.4 ± 0.32 (25 Å) × 10^{-3} Nsm^{-2}. These are agreement with our previous study[27] although the standard deviations are reduced due to longer simulations. The significantly larger effective viscosity the closer the tube-tube separation confirms the strong interaction induced by the surroundings on the water molecules confined within each individual CNT pore. The dependence of η on the separation distance between the nanotubes is more clearly demonstrated in Fig. 8 where the fit to a simple $1/d$ dependence is shown to be excellent.

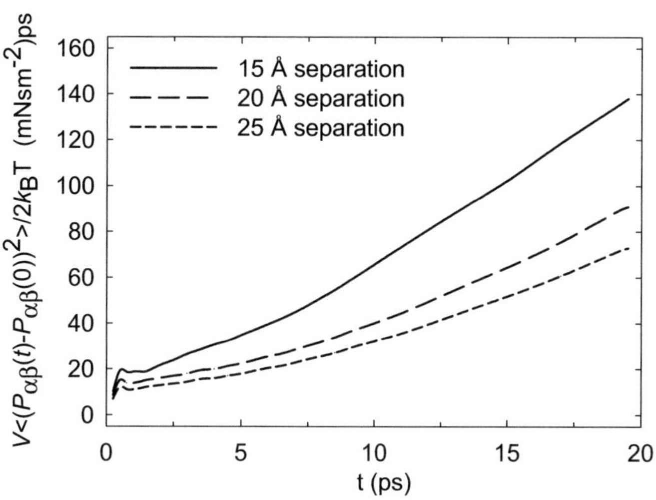

Fig. 7. Mean square displacement of the dynamical variable $P_{\alpha\beta}$ (with $a = z$ and $\beta = z$) for the four hexamer water files as functions of time and for three different nanotube-array separations.

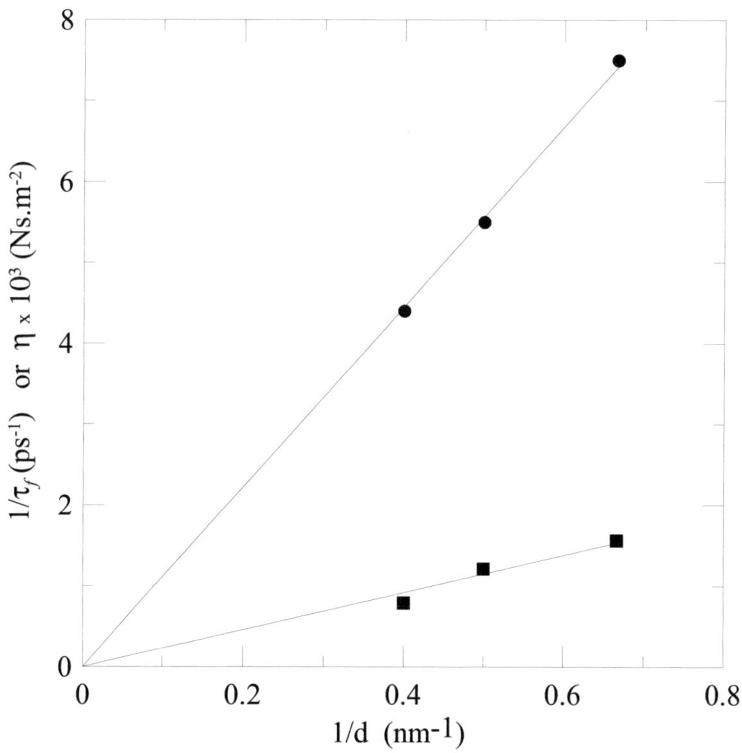

Fig. 8. Plots of the CNT array effective viscosity coefficient (from Eqn. 19), filled circles, and the $1/\tau_f$ distance-dependent friction parameter (inferred from the fitted peak-decay lines in Fig.9b), filled squares, versus the inverse tube-separations (i.e., 15, 20, 25 Å).

To investigate friction further, collective, axial VACFs of all of the water molecules in each individual CNT along the z-axis are shown in Fig. 9a, including the isolated tube case. A log-plot of this in Fig. 9b shows an essentially negative-exponential decay, along the general trend of an $A\exp(-t/\tau)\cos\omega t$ motif. Fitting to the successive peaks in the log-plot yields an isolated tube frictional relaxation time of 0.16 ps. From the other slopes, the distance-dependent frictional contribution may be determined for each of the four-CNT arrays (inferred from the fitted peak lines), and these are detailed in Table II. The $1/\tau_f$ distance-dependent friction parameter declines with increasing separation, and a plot of this versus the inverse separation as shown in Fig. 8, coupled with a regression line, shows an approximate $1/d$ relationship. Direct evaluation of the z-axis collective-velocity cross correlation function (VCCF) for each tube with the others in the three CNT-array cases confirms that the magnitude of the VCCF declines with increasing separation, and this is shown in Fig. 10.

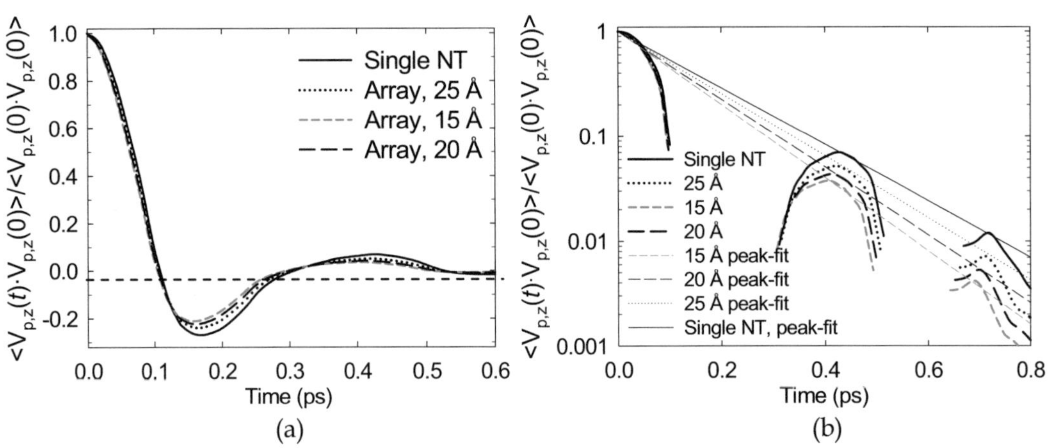

Fig. 9. (a) Normalised collective TCFs of all of the water molecules in each CNT along the z-axis; this is averaged over each of the four CNTs in the case of the arrays. (b) Log-plot of the normalised collective TCFs, showing negative-exponential fits to the peaks.

The osmotic permeabilities were evaluated in two independent ways, and the results are provided in the last two columns of Table II. Integration of the Green-Kubo-type representation of the viscous slip total velocity correlation function (TCF) (Eqn. 15) (see [42,43] and also [23, 45,46] for additional details on Green-Kubo analysis) leads to convergence within 10-15 ps and the results for the osmotic permeability are provided in the second-last column of Table II. These results are the exact values for the osmotic permeabilities. The larger values obtained from the approximate friction analysis (Eqn. 18) are to be expected given that the friction model does not take into account negative backscattering parts of the velocity correlation function clearly observed in Fig. 9a. Nonetheless, the general trend with nanotube separation is clearly evident. Also note that the ratio of the osmotic and diffusion permeabilities p_f/p_d in all cases (with p_f given by the exact results in the second last column of Table I) is approximately 18 in magnitude. This corresponds reasonably with the estimate of 15 obtained from an approximation suggested in reference [10].

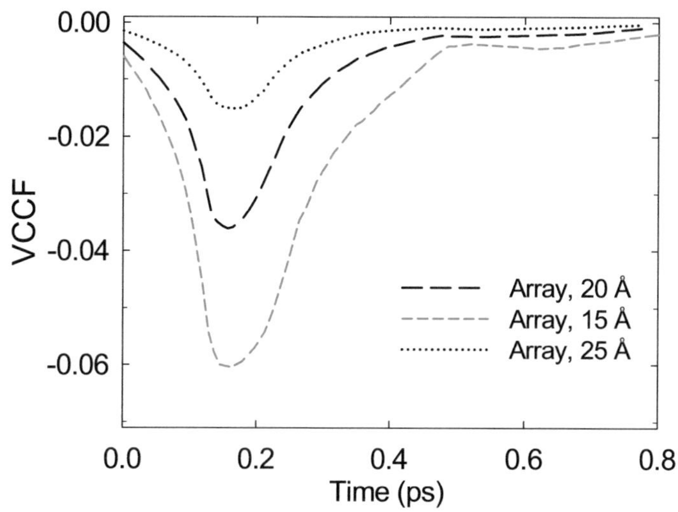

Fig. 10. Collective z-axis VCCFs of one pore with the other three in the CNT-array systems, averaged over all such cross-interactions and normalised vis-à-vis the four-averaged TCFs (shown in Fig. 9a).

Separation (Å)	τ_{eff} (ps)	τ_f (ps)*
15	0.128	0.64
20	0.134	0.825
25	0.142	1.26
∞	0.16	Infinite

* Inferred from $1/\tau_{eff} = 1/\tau_{eff,\infty} + 1/\tau_f$

Table II. Fitted values for effective frictional relaxation times, τ_{eff}, and values of the resultant separation-friction relaxation time, τ_f. See plot of $1/\tau_f$ versus $1/d$ (Fig. 8).

5. Conclusions

Single-file diffusion results through (6,6) SWCNTs, embedded in a lipid bilayer, under the influence of static and time-varying electric fields, have been reported. No substantial influence of alternating electric fields on water flux through SWCNTs of larger diameters was observed, although dipole orientations were affected markedly, with higher-frequency distributions resembling those of the zero-field case. In the case of static fields, however, the water flux appears somewhat higher than that of the zero-field case, owing possibly to a similar reduction in water-number fluctuations therein, as found in previous work for (5,5) CNTs[26,27]; an overlapping of error bars vis-à-vis the zero-field case, however, renders any firm statistical conclusion elusive in this respect. The (RMS) field intensities of 0.0065 V/Å are weak compared to previous molecular simulation studies employing static or time-varying electric fields[39] and also weak relative to fields generated by charge distributions[14], and are substantially more realistic with respect to experiment.

When arrays of CNTs are employed in the simulations the interpretation on the transport properties of the fluid within the individual nanotubes must take cognisance of inter-tube interactions. The (6,6) CNT-array simulations reported in this work have demonstrated the presence of a friction effect due to long-range electrostatic interactions between the moving water single-files within the pores, with better statistics than was possible in previous work.[14] It was shown that the generated friction is inversely proportional to tube separation, reaching values closer to the single-pore simulations (which should be considered as an infinitely-separated system) at the largest separation between the tubes, and estimation of an effective viscosity of the water molecules confined within the nanotubes confirmed this, along with consideration of the collective total velocity correlation functions (TCFs) for the water files within the arrays of CNTs .

A more general approach[42,43] to the concept of the osmotic permeability based on linear response theory leads to the recognition that p_f is simply related to the slip-diffusion (or viscous slip) coefficient of a fluid undergoing pressure-driven transport in a nanopore structure. The results reported in this work also support the approximate relationship $p_f \sim (N + 1)p_d$ suggested by Tajkhorshid *et al* [10] where N is the average number of water molecules within a single carbon nanotube.

6. References

[1] S. Iijima , Nature 354, 56 (1991).
[2] D.A. Britz and A.N. Khlobystov, Chem. Soc. Rev. 35, 637 (2006).
[3] N.W.S. Kam, M. O'Connell, J.A. Wisdom and H. Dai, Proc. Natl. Sci. (USA) 102, 11600 (2005).
[4] B.J. Landi, R.P. Raffaelle, S.L. Castro and S.G. Bailey, (2005). Progress in Photovoltaics: Research and Applications, 13, 165 (2005).
[5] A. Kongkanand, R.M. Dominguez and P.V. Kamat, Nano Letters, **7**, 676 (2007).
[6] M. Ø. Jensen, E. Tajkhorshid and K. Schulten. Biophys. J., 85, 2884 (2003).
[7] F. Zhu, E. Tajkhorshid and K. Schulten. FEBS Lett., 504, 212 (2003).
[8] F. Zhu, E. Tajkhorshid and K. Schulten., Biophys. J., 86, 50 (2004).
[9] M. Ø. Jensen and O. G. Mouritsen, Biophys. J., 90, 2270 (2006).
[10] E. Tajkhorshid, F. Zhu and K. Shulten, *Handbook of Material Modeling* (S. Yip ed., Springer: Netherlands 2005).
[11] M. Endo, M.S. Strano and P.M. Ajayan, *Carbon Nanotubes* (Springer- Verlag: Berlin, 2008; Vol. 111, pp 13-61).
[12] A. Berezhkovskii and G. Hummer, Phys. Rev. Lett., 89, 064503-1-4 (2002).
[13] G. Hummer, J. C. Rasaiah and J.P Noworyta, Nature 414 (6860), 188 (2001).
[14] F. Zhu and K. Schulten, Biophys. J., 85, 236 (2003).
[15] A. Kalra, S. Garde and G. Hummer, Proc. Natl. Acad. Sci. U.S.A., 100, 10175 (2004).

[16] F. Zhu, E. Tajkhorshid and K. Schulten, Phys. Rev. Lett., 93, 224501-1-4 (2004).
[17] J. Y Li,. X. J Gong, H. J. Lu, D.Li, H. P. Fang and R. H. Zhou, Proc. Natl. Acad. Sci. USA 104, 3687 (2007).
[18] X. Gong, J. Li, H. Lu, R. Wan, J. Li, J. Hu and H. Fang, Nat.Nanotechnol. 2, 709 (2007).
[19] B. Liu, X. Li, B. Li, B. Xu and Y. Zhao, Nano Lett. 9, 1386 (2009).
[20] S. Joseph and N.R. Aluru, Nano Lett., 8, 452 (2008).
[21] J.-Y. Li, Z.-X. Yang, H.-P. Fang, R.-H. Zhou and X.-W. Tang, Chinese Phys. Lett., 24, 2710 (2007).
[22] B. Mukherjee, P.K. Maiti, C. Dasgupta and A. K. Sood, J. Nanosci. Nanotechnol., 7, 1796 (2007).
[23] M.P. Allen and D.J. Tildesley, *Computer Simulation of Liquids*, Clarendon Press, Oxford (1987).
[24] D.C. Rapaport, 'The Art of Molecular Dynamics Simulation', 2nd ed., Cambridge (2004)
[25] D. Frenkel and B. Smit, 'Understanding Molecular Simulations: From Algorithms to Applications', 2nd ed., Academic Press (2001)
[26] J.-A. Garate, N.J. English and J.M.D. MacElroy. Molec. Sim. 35, 3 (2009).
[27] J.-A. Garate, N.J. English and J.M.D. MacElroy, J. Chem. Phys., 131, 114508-1 (2009).
[28] J.E. Stone, J.C. Phillips, P.L. Freddolino, D.J. Hardy, L.G. Trabuco and K. Schulten, J. Comput. Chem. 28(16), 2618–2640 (2007).
[29] J.C. Philips, R. Braun, W. Wang, J. Gumbart, E. Tajkhorshid, E. Villa, C. Chipot, R.D. Skeel, L. Kale and K. Shulten J. Comp. Chem. 26, 1781 (2005).
[30] A.D. Mackerell, Jr., D. Bashford, M. Bellot, R.L. Dunbrack. Jr., J.D Evanseck, M.J. Field, S. Fischer, J. Gao, H.Guo, S. Ha et al, J. Phys. Chem. B 102, 3586 (1998) .
[31] W.L. Jorgensen, J. Chandrasekhar, J.D. Madura, R.W. Impey and M.L. Klein, J. Chem. Phys. 79, 926 (1983)
[32] W. Humphrey, A.Dalke and K. Shulten, J. Mol. Graph.Model., 14, 33 (1996).
[33] T. Darden, D. York and L. Pedersen, J. Chem. Phys. 98, 10089 (1993).
[34] M. Tuckerman, B. J. Berne, and G. J. Martyna. R, J. Chem. Phys, 97, 1990 (1992).
[35] G.S. Grest and K. Kremer, Phys. Rev. A 33, 3628 (1986).
[36] J.P. Ryckaert, G. Ciccotti and H.J.C. Berendsen, J. Comp. Phys. 23, 327 (1977) .
[37] G.J. Martyna, D.J. Tobias and M.L. Klein, J. Chem. Phys 101, 1990 (1992).
[38] S.E Feller, Y. Zhang, R.W. Pastor and B.R. Brooks, J. Chem. Phys. 103, 4613 (1995).
[39] N.J. English and J.M.D MacElroy, J. Chem. Phys., 119, 11806 (2003).
[40] N.J. English and J.M.D MacElroy, J. Chem. Phys. 118, 1589 (2003).
[41] A.C. Metaxas and R.J. Meredith, 'Industrial Microwave Heating', Peter Peregrinus, London (1983).
[42] S.-H. Suh and J.M.D. MacElroy, Mol. Phys., 58, 445 (1986).
[43] J.M.D. MacElroy and S.-H.Suh, Mol. Phys., 60, 475 (1987).
[44] Note that the partition coefficient K relates the equilibrium concentrations of the water between the bulk liquid volume and the membrane volume as a whole (CNT pores and lipid layer) whereas K_r in Eqn. 6 is the equilibrium relation involving the CNT pore volume alone.

[45] D.J. Evans and G.P. Morriss, 'Statistical Mechanics of Non-equilibrium Liquids', 2nd ed., Cambridge (2008).
[46] D.A. McQuarrie, 'Statistical Mechanics', 2nd ed., University Science Books (2000).

22

Acute Toxicological Effects of Multi-Walled Carbon Nanotubes (MWCNT)

Balakrishna Murthy, P., Sairam Kishore, A. and Surekha, P.
International Institute of Biotechnology and Toxicology (IIBAT), Padappai, Kanchipuram district, Tamil Nadu, India

1. Introduction

Health effects of nanoparticles are attracting considerable and increasing concern of the public and government worldwide. Carbon nanotubes represent one of the fastest developing nanoparticle materials with production set to increase rapidly as a consequence of the useful properties of this material (Donaldson et al., 2006). The pulmonary toxicity of single-walled and multi-walled carbon nanotubes delivered at high doses and dose rates to the lower respiratory tract of rats and mice induced a high acute inflammatory response with granuloma formation and fibrosis as late effects (Warheit et al. 2004; Muller et al. 2005). However, the reports about toxicological research of multi-walled carbon nanotubes by the oral, dermal and ocular routes are not yet published. The oral, dermal and ocular routes represents an important portal of entries for MWCNT since several consumer which are on the market, already contain multi-walled carbon nanotubes.

Few studies provided only scanty insights into the interaction of MWCNT with the human body after entering via different portals. Hydroxylated single-walled carbon nanotubes (SWCNT) administered by gavage in mice (100 µL of a 15 µg/mL solution) are distributed to most of the organs and tissues, except the brain (Wang et al., 2004). The study by Wang et al., (2004) shows in mice that the radiomarked hydroxylated singled-walled carbon nanotubes administered intraperitoneally (100 µL of a 15 µg/mL solution) are distributed throughout the body, except the brain, pass through several compartments and are retained in the bones. Monteiro-Riviere et al. (2005) found multi-walled carbon nanotubes (MWCNT) in the cytoplasmic vacuoles of human epidermal keratocytes in vitro (up to 3.6 µm long), a decrease in cell viability and a significant increase in an inflammation marker (interleukin-8). This demonstrates the capability of MWCNT to penetrate the cell membrane. Huczko and Lange (2001) studied the effects on the skin and eyes of exposure to carbon nanotubes. The application of a saturated filter of a solution containing nanotubes did not cause irritation or allergy in volunteers. Ocular instillation of an aqueous suspension of nanotubes in rabbits did not cause irritation.

In the light of the increased production and proposed use of MWCNT in consumer products, there is a need for screening the potential toxicity of these nanoparticles. In the present study, we have made an attempt to investigate the acute oral, dermal, acute dermal irritation, eye irritation and skin sensitization potential of two different sizes (140±30, 10-15nm) of MWCNT.

2. Materials and methods

2.1 Nanomaterials

Two different sizes of multi-walled carbon nanotubes (MWCNT 1: 5–8 microns in length with 3–8nm inside diameter and outside diameter of 140±30 nm; Product No. 659258) and (MWCNT 2: 1–10 microns in length with 2–6nm inside diameter and outside diameter of 10–15 nm; Product No. 677248) purchased from Sigma–Aldrich, USA, were used to compare size dependent toxicity.

2.2 Particle-types and physicochemical characterization

The multi-walled carbon nanotubes used in the study was produced by catalytic chemical vapor deposition (CCVD). The MWCNT were composed of 99.9+% carbon with small amount (<0.1% Fe) of iron. The manufacturer specification for nanomaterial characterization was confirmed by the following techniques. MWCNT characteristics were assessed in the as-synthesized form prior to use for experiments or after dispersion in the vehicle (distilled water) for dosing. Solution characteristics were measured with dynamic light scattering (DLS).

2.2.1 Size, morphology

Multi-walled carbon nanotubes size was determined with scanning electron microscopy (SEM). In this study, Hitachi S-520 SEM was used at an accelerating voltage of 10,000V after depositing the samples onto aluminum stubs with double-sided carbon adhesive tape. Over 100 particles were counted and measured to determine average sizes and size distributions.

2.2.2 Dynamic light scattering (DLS)

Particle size measurements in distilled water were determined with dynamic light scattering as described by Murdock et al. (2008) on a Malvern Instruments
Zetasizer. Photon correlation spectroscopy or DLS is an analytical technique capable of measuring the size of very small particles, at low sample concentrations.

2.3 Animals and housing conditions

Experimental animals were obtained from in house animal facility. All procedures using animals were reviewed and approved by the institutional animal ethics committee.
The healthy Wistar (Crl:WI) rats of both sex, aged between 8 to10 weeks and body weights of 200 - 250 g were used for oral and dermal studies respectively. Females were nulliparous and non pregnant. The animal's parent stock was procured from Charles River, USA. Animals were housed in poly propylene cages with stainless steel grills, sieved and sterilized paddy husk used as bedding. Bedding material, cages, grills and water bottles were changed on alternate days. Animals were housed individually sex wise and group wise. Animals were acclimated for minimum period of five days in the controlled environment (temperature: 22±3 °C; relative humidity: 50±20% and light: 12 h light/dark cycle) and *adlibitum* supply of reverse osmosis water and a standard rodent pellet food (supplier: M/s. Tetragon Chemie Pvt. Ltd., Bangalore, India). In case of oral study, prior to the dosing over-night and following dosing, for a period of 3 hours food was withheld.

The following acute toxicity studies were conducted with two different sizes of MWCNT. A brief description of each of the test was provided in the methodology section below.

2.4 Acute oral toxicity test in Wistar rats

Acute oral toxicity -up and down procedure was conducted as per OECD 425 guidelines (OECD, 2006) with slight modifications in terms of usage of sexes and animal number. The limit test dose is selected (2000mg/kg b.w.) as per the OECD 425. A single dose of each test material (two different sizes of MWCNT) suspended in distilled water was administered by oral gavage to group of rats at a dose of 2000 mg/ kg b.w. with minimum of 48 h time interval. The test solution was prepared shortly prior to the administration. The dose volume maintained for all the groups was maximum 10 ml/kg b.w. Similarly, control group of animals (5 males and 5 females) were dosed in a sequential fashion with distilled water alone. Animals were observed for mortality/morbidity, clinical signs of toxicity and weekly body weight during the experimental period. Gross pathology was performed at the end of experimental period (day 14).

2.5 Acute dermal toxicity study in Wistar rats

The acute dermal toxicity test was conducted according to OECD 402 guidelines (OECD, 1987). A limit dose of 2000 mg/ kg b.w. of each test material (two different sizes of MWCNT) mixed with minimum volume (0.25ml) of distilled water was applied uniformly to a clipped 10% body surface area of different groups of rats comprising of 5 males and 5 females/ group. The test substance was held in contact with skin with a porous gauze dressing and bandaged with non-irritating adhesive tape up to 24 h. Following this, the residual test substance was wiped off gently from the skin using cotton soaked in water. Restrainer was used to prevent the ingestion of the test substance from the application site. Control group of animals (5 males and 5 females) were handled similarly without any treatment. Animals were observed for mortality/morbidity, clinical signs of toxicity and weekly body weight during the experimental period. Gross pathology was performed at the end of experimental period (day 14).

2.6 Acute dermal and eye irritation studies in New Zealand white rabbits

2.6.1 Animals and housing conditions

The female New Zealand white rabbits, 2.0 to 3.0 kilograms body weight ranges were used for dermal and eye irritation experiments. Females were nulliparous and non pregnant. Animals were housed individually in standard stainless steel cages and sterilized paddy husk as bedding. Bedding material was changed daily, whereas water bottles were changed on alternate days. Animals were acclimated for minimum period of five days in the controlled environment (temperature: 20±3 ◦C; relative humidity: 50±20% and light: 12 h light/dark cycle) and *adlibitum* supply of UV treated water and standard rabbit pellet food (supplier: M/s. Amrut Laboratory Animal Feed, Pune).

2.7 Acute dermal irritation study in New Zealand white rabbits

The acute dermal irritation test was conducted according to OECD 404 guidelines (OECD, 2002). A single dermal dose of 0.5 gram of each test material (two different sizes of MWCNT) moistened with minimum volume (0.25 ml) of distilled water was applied to a 6 cm^2 clipped area of skin. The application area was covered with 2-ply gauze square which was held in place with non-irritating tape and covered with porous tape for a semi-occlusive dressing. The rabbits were exposed to the test substance for 4 h after which the test substance was removed. Test sites were evaluated by Draize score for signs of dermal irritation approximately 60 min, 24, 48, and 72 h after test substance removal. Initially test was carried out using one animal. The negative response was confirmed using two additional animals. After 72h of observation the animals were euthanized using carbon dioxide.

2.8 Acute eye irritation study in New Zealand white rabbits

The acute eye irritation/corrosion test was conducted according to OECD 405 guidelines (OECD, 2002). A single dose of 0.1 gram of each test material (two different sizes of MWCNT) was applied into the conjunctival sac of left eye of each animal after gently pulling the lower lid away from the eyeball. The lids were then gently held together for about one second in order to prevent loss of the material. The right eye, which remains untreated, served as control. Initially test was carried out using one animal. The eyes of the test animals were washed with distilled water at 24 hours following the application of test materials and the responses obtained in the initial test were confirmed with two additional animals in the confirmatory test. The conjunctiva, iris, and cornea of both treated and control eyes were evaluated and scored according to Draize method at the end of 1, 24, 48, and 72 h following application of the test materials.

2.9 Skin sensitization potential test in Guinea pigs

The Skin sensitization potential test in Guinea pigs was conducted according to OCED 406 guidelines (OECD, 1992). The Dunkin Hartley guinea pigs, 300 to 400 grams body weight ranges were used in the experiments. The treatment group consists of 20 males whereas control group consist 10 males. The test was conducted in two different sizes of MWCNT. Animals were housed individually in standard stainless steel cages and sterilized paddy husk as bedding. Bedding material was changed daily, whereas water bottles were changed on alternate days. Animals were acclimated for minimum period of five days in the controlled environment (temperature: 20±3 ◦C; relative humidity: 50±20% and light: 12 h light/dark cycle) and *ad libitum* supply of UV treated water and standard rabbit pellet food (supplier: M/s. Amrut Laboratory Animal Feed, Pune). The diet was supplemented with ascorbic acid. The food and water was routinely analyzed and are considered not to contain any contaminants that could reasonably be expected to affect the purpose or integrity of the study. One flank of each animal was closely clipped free of hair, without any abrasion, 24 hours before the induction exposure. A cotton pad about 4 - 6 cm^2 in size was loaded with the test substance at a dose of 400 mg to all the animals of treatment group. The cotton pad was applied to the clipped area of animals of treatment group and held in contact with the skin by an occlusive patch and bandage dressing for a period of 6 h. The same treatment was repeated on day 7 and 14. Similarly, the animals in control group were treated with distilled water. On day 28, to the closely clipped posterior flank of both treatment and control groups, the nanomaterials were applied in cotton pad at a dose of 200 mg. As in induction exposure, the cotton pads were held in contact with the skin for 6 h by an occlusive patch and bandage dressing. The skin reactions were observed and scored at 30 h after application of the challenge Patch and 54 h after application of the challenge patch. The skin reactions were graded as per the Buehler sensitization scoring scale. On termination, the experiment animals were euthanized using carbon dioxide.

3. Results

3.1 Physicochemical characterization

3.1.1 Size

The average size of the MWCNT 1 is 166nm and MWCNT 2 is 100nm in SEM analysis. The morphology of the MWCNT (1 and 2) is open or closed tube. The MWCNT formed compact aggregates, making it difficult to determine individual particle size by SEM; however, a rough estimate for average size of the MWCNT 1 is 166nm and MWCNT 2 is 100nm **(Figures 1(a)-(d))**

Fig. 1. a) Scanning Electron Micrograph (SEM) of MWCNT 1

Fig. 1. b) Scanning Electron Micrograph (SEM) of MWCNT 1

Fig. 1. c) Scanning electron micrograph (SEM) of MWCNT 2

Fig. 1. d) Scanning electron micrograph (SEM) of MWCNT 2

3.1.2 Dynamic light scattering

The solution properties of nanomaterials (two different sizes of MWCNT) in distilled water were examined for changes in size due to agglomeration with dynamic light scattering (Murdock et al., 2008). Average size was calculated by the software from the intensity, volume and number distributions measured (**Table 1**). The polydispersity index (PDI) given is a measure of the size ranges present in the solution with a scale ranging from 0 to 1, with 0 being monodisperse and 1 being polydisperse (Wagner et al., 2007). The DLS results illustrate that depending on the material, once the nanomaterials are in solution they do not necessarily retain their "nano-size"(Murdock et al., 2008). The average size values(in solution) of 901nm for MWCNT 1, 554nm for MWCNT 2 were found along with high PDI readings. The formation of very strong nanomaterials aggregates due to vander waals forces is expected to occur for unmodified carbon nanomaterials in solution, which is shown for the nano-sized carbon black nanoparticles, which have much larger sizes than their primary size (Murdock et al., 2008).

Nanomaterial	Average size in water [a]	Polydispersity index (PDI)	PH	Specific surface area (SSA)
MWCNT 1 (O.D.X I.D X L: 110-170 nm × 3-8nm X 5-9 μm)	901 nm	1.000	7.2	10-15 m2/g
MWCNT 2 (O.D. × I.D. × L 10-15 nm × 2-6 nm × 0.1-10 μm)	554 nm	1.000	7.1	30-45 m2/g

[a] Measured before dosing in distilled water

Table 1. Multi-walled carbon nanotubes (MWCNT) Size in solution

3.2 Acute oral toxicity test in Wistar rats

There was no mortality evident in the study. Animals of treated groups showed normal and consistent gain in body weight when compared to the control group of animals **(Figure 2a and 2b).** No gross lesions were observed. Based on the above findings, the oral LD_{50} for MWCNT of two different sizes was greater then 2000 mg/kg b.w for Wistar rats.

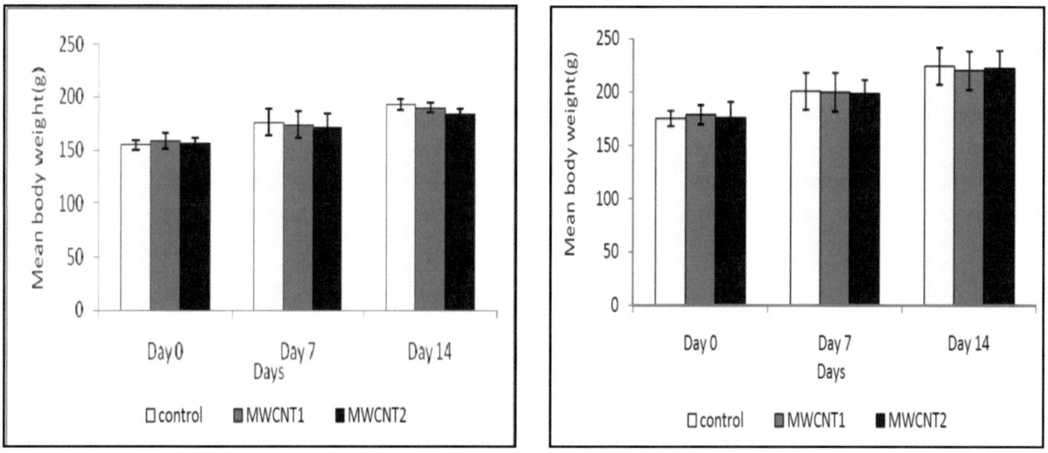

Fig. 2. a) & b) Acute oral toxicity test in rats - Mean body weight data (Males&Females)

3.3 Acute dermal toxicity study in Wistar rats

No mortality was evident in the study. Animals of treated groups showed normal and consistent gain in body weight when compared to the control group of animals **(Figure 3a and 3b).** No gross lesions were observed. Based on the above findings, the dermal LD_{50} for MWCNT of two different sizes was greater then 2000 mg/kg b.w for Wistar rats.

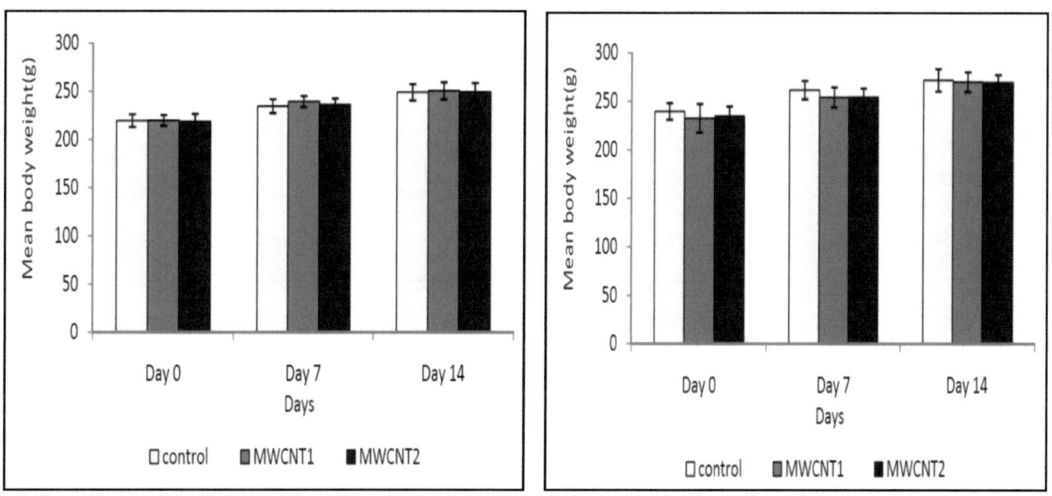

Fig. 3. a) & b) Acute dermal toxicity test in rats - Mean body weight data (Males & Females)

3.4 Acute dermal irritation study in New Zealand white rabbits

None of the rabbits treated with test material (two different sizes of MWCNT) showed any dermal reactions at 60 min, 24, 48 and 72 h after patch removal. None of the animals showed clinical signs of toxicity. Body weight gain of the individual animal was normal during the study period. Thus the Primary Irritation Index (PII) of two different sizes of MWCNT's is zero. Hence based on the above findings MWCNT 1 and 2 were considered as non irritant to the skin of the New Zealand white rabbits.

3.5 Acute eye irritation study in New Zealand white rabbits

The eyes of the rabbits were examined at 1, 24, 48, 72, 96 hours and 5th day after treating with test material. The maximum mean score for ocular lesions observed was 8.0 and 10.0 for MWCNT 1 and MWCNT 2 respectively (**Table 2 and 3**). Animals exhibited conjunctival redness, chemosis and discharge (score 1 or 2) from 1hour onwards. All the animals recovered from above ocular lesions on 5th day. Hence the MWCNT 1 and 2 were considered as minimally irritating to the eye of the New Zealand white rabbits.

Time of Observation	MWCNT 1 - Ocular Lesions			
	Total Score		Mean ± S.D.	
	Treated (left eye)	Control (right eye)	Treated (left eye)	Control (right eye)
1 hour	12	0	4.0 ± 0.00	0
24 hour	24	0	8.0 ± 0.00	0
48 hour	14	0	4.7 ± 1.15	0
72 hour	8	0	2.7 ± 1.15	0
96 hour	4	0	1.3 ± 1.15	0
5th Day	0	0	0	0

Table 2. Acute eye irritation test in New Zealand White rabbits - Mean ocular lesions – MWCNT 1

Time of Observation	MWCNT 2 - Ocular Lesions			
	Total Score		Mean ± S.D.	
	Treated (left eye)	Control (right eye)	Treated (left eye)	Control (right eye)
1 hour	16	0	4.7 ± 1.15	0
24 hour	30	0	10.0 ± 0.00	0
48 hour	18	0	6.0 ± 2.00	0
72 hour	10	0	3.3 ± 1.15	0
96 hour	4	0	1.3 ± 1.15	0
5th Day	0	0	0	0

Table 3. Acute eye irritation test in New Zealand White rabbits - Mean ocular lesions – MWCNT 2

3.7 Skin sensitization potential test in Guinea pigs

None of the animals treated with both the sizes of MWCNT did not exhibited any dermal reactions. Hence both the sizes of MWCNT could be considered as non sensitizer.

4. Discussion

Among 580 consumer nanotechnology-based products, one of the most common materials mentioned in the product descriptions is MWCNT based nanoparticles (Woodrow Wilson

International Center for Scholars, 2007). However, MWCNT remain one of the most controversial research areas regarding their toxicity to biological systems.

However, the effect of MWCNT *in vivo* has not been studied extensively. Therefore, we have made an attempt to scrutinize the acute toxicological potential of MWCNT. The adequate characterization of nanomaterials represents one of the key aspects of toxicity screening strategies. Adequate NM characterization is needed prior to the initiation of toxicological experimentation to ascertain the possible cause for the toxicity of NM and in absence of this would have limited significance (Warheit, 2008). Hence, we performed prior to toxicology experiments the physicochemical characterization of MWCNT like, size in dry state and distilled water, surface area to ascertain the possible cause for the toxicity. The two different sizes of MWCNT in distilled water showed increase in the size. Murdock et al., (2008) reported that once the nanomaterials are in water they do not necessarily retain their nano-size. Both the sizes (140±30, 10-15 nm) of MWCNT exhibited similar kind of results in acute oral, dermal and ocular irritation tests. Hence, there is no size dependant toxicity occurred in current acute toxicity experiments.

In our present acute oral and dermal toxicity studies in Wistar rats of two different sizes with MWCNT did not show any toxicity. Based on these results the acute oral and dermal LD_{50} is greater than 2000 mg/kg b.w. for both the sizes of MWCNT. The acute dermal irritation study in New Zealand white rabbits with two different sizes of MWCNT, exhibited no dermal reactions. Under the conditions of this test, both the sizes of MWCNT were considered to be non-irritating to the skin. However, acute eye irritation study with both the sizes of MWCNT produced conjunctival redness and discharge in the treated eye of the rabbits and the lesions were reversible. No skin reactions were observed in Guinea Pigs treated with both the sizes of MWCNT. Based on these findings, both the sizes of MWCNT could be classified as minimally-irritating to the eye of the rabbits.

In conclusion, the results of these acute toxicity studies demonstrated low hazard potential in mammals following acute exposures to the MWCNT in oral and dermal routes. The MWCNT screened in these studies may serve as a base-line data for risk assessment. Further studies are required to investigate the impact of multi-walled carbon nanotubes to cellular components, including genetic material.

5. Conclusion and future remarks

Despite of the wide usage of Multi-walled carbon nanotubes the toxicological information and potential *in vivo* health hazards are still fragmentary. Hence we have made an attempt to understand the toxicological potential of these nanomaterials in four different invivo models since most of the nanotoxicology studies were focused on *in vitro* models and only few research groups have dealt with *in vivo* systems. Toxicology studies with *in vivo* systems carry greater significance pertaining to their diversity in physiology and anatomy.

The current research work describes the results of four different *in vivo* toxicity studies conducted on multi-walled carbon nanotubes (MWCNT). The studies included are acute oral and dermal toxicity in Wistar rats, acute dermal and eye irritation study in New Zealand White rabbits and skin sensitization in Guinea pigs. Justification for these particular tests rests on the following criteria: potential routes of exposures (i.e., oral, dermal, and/or ocular) and the results of these acute studies demonstrated low hazard potential in mammals following acute exposures to the multi-walled carbon nanotubes (MWCNT) and there is no significant size dependent toxicity.

6. Acknowledgements

The authors are grateful to the IIBAT management for providing support for this work. We are thankful to Prof. P.V. Ramachandran, Head, Pathology and Dr. G. Selvam for their strong support in pathology work. We are also thankful to Mr. A. Goparaju, statistician for helping in analysis of data.

7. Conflict of interest

The authors declare no conflict of interest

8. References

Donaldson, K., Aitken, R., Tran, L., Stone, V., Duffin, R., Forest, G., & Alexander, A. (2006). Carbon nanotubes: A review of their properties in relation to pulmonarytoxicology and workplace safety, *Toxicological Sci* 92, 1, pp. 5-22.

Huczko, A., & Lange, H. (2001). Carbon nanotubes: Experimental evidence for a null risk of skin irritation and allergy, *Fullerene Sci Technol* 9, 2, pp.247-250.

Monteiro-Riviere, N.A., Namanich, R., Inman, A., Wang, Y., & Riviere, J. (2005). Multi-walled carbon nanotube interactions with human epidermal keratinocytes. *Toxicol Lett*, 155, pp. 377-384.

Muller, J., Huaux, F., Moreau, N., Misson, P., Heilier, J-F., Delos, M., Arras, M., Fonseca, A., Nagy, J.B., & Lison, D. (2005). Respiratory toxicity of multi-all carbon nanotubes. *Toxicol Appl Pharmacol*, 207, pp.221 231.

Murdock, R.C., Braydich-Stolle, L., Schrand, A.M., Schlager, J.J., & Hussain, S.M., (2008). Characterization of Nanomaterial Dispersion in Solution Prior to In Vitro Exposure Using Dynamic Light Scattering Technique. *Toxicol.Sci.* 101,2, pp.239-253.

Organisation for Economic Co-operation and Development (OECD). Guideline for the Testing of Chemicals Section 4 (Part 425), 2006.

Organisation for Economic Co-operation and Development (OECD). Guideline for the Testing of Chemicals Section 4 (Part 402), 1987

Organisation for Economic Co-operation and Development (OECD). Guideline for the Testing of Chemicals Section 4 (Part 404), 2002.

Organisation for Economic Co-operation and Development (OECD). Guideline for the Testing of Chemicals Section 4 (Part 405), 2002.

Organisation for Economic Co-operation and Development (OECD). Guideline for the Testing of Chemicals Section 4 (Part 406), 1992.

Wagner, A. J., Bleckmann, C. A., Murdock, R. C., Schrand, A. M., Schlager, J. J. & Hussain, S. M. (2007) Cellular interaction of different forms of aluminium nanoparticles in rat alveolar macrophages. *J. Phys. Chem. Biol*, 111, pp.7353–7359.

Wang, H., Wang, J., Deng, X., Sun, H., Shi, Z., Gu, Z., Liu, Y., & Zhaoc, Y. (2004). Biodistribution of carbon singlewall carbon nanotubes in mice. *J Nanosci Nanotech* 4, 8, pp.1019-1024.

Warheit, D.B., Laurence, B.R., Reed, K.L., Roach, D.H., Reynolds, G.A.M., & Webb, T.R. (2004). Comparative pulmonary toxicity assessment of single-wall carbon nanotubes in rats. *Toxicological Sci*, 77,pp.117-125.

Warheit, D. B. (2008) How meaningful are the results of nanotoxicity studies in the absence of adequate material characterization? *Toxicol. Sci*, 101, pp. 183–185.

Woodrow Wilson International Center for Scholars., 2007. A nanotechnology consumer products inventory.

23

Nanotoxicity: Exploring the Interactions Between Carbon Nanotubes and Proteins

Guanghong Zuo[1,2], Haiping Fang[1,2] and Ruhong Zhou[3,4]
[1]*Shanghai Institute of Applied Physics, Chinese Academy of Sciences, Shanghai,*
[2]*T-Life Research Center, Department of Physics, Fudan University, Shanghai,*
[3]*IBM Thomas J. Watson Research Center, Yorktown Heights, NY,*
[4]*Department of Chemistry, Columbia University, New York,*
[1,2]*China*
[3,4]*USA*

1. Introduction

Carbon nanotubes (CNTs), which are entirely composed of sp^2 carbons, can be viewed as graphite sheets that have been rolled into seamless cylinders. Since their discovery in 1991 [1], CNTs have become the "superstar" in the field of nanoscience and nanotechnology because of their unique structural, mechanical, and electronic properties [2]. Various promising biomedical applications have also been proposed, such as drug design [3], drug delivery [4], tumor therapy [5], tissue engineering [6], DNA recognition [7], and biosensor design [8]. It was predicted that the global market for CNTs could grow to $2 billion by 2014 [9].

On the other hand, with the fast advance of the nanotechnology and its widespread applications in various industries, there is a growing concern on the biosafety of these nanomaterials to human health [10-17]. Human skin, lungs, and gastrointestinal tract are in constant contact with the environment. Despite the skin is an effective barrier to foreign substances, the lungs and gastrointestinal tract are vulnerable. Recent experiments showed that unrefined CNTs might be aerosolized and release fine particles into air [18]. When the concentration of CNTs in air is sufficiently high, they can be inhaled and then migrate into other parts of body [19]. These inhaled CNTs can enter cells and accumulate in cytoplasms [20, 21], which may lead to the lung insult [22-24], immunologic toxicity [25], and adverse cardiovascular effects (on mice) [26]. Injections and implants, which are required for some biomedical applications, are the other possible routes of the exposure of human cells to nanoparticles. The experiments on mice showed that the tissue exposing directly to the injected CNTs may lead to asbestos-like pathogenic behaviors, including inflammation and the formation of lesions known as granulomas [27]. In addition, the injecting CNTs into the tail vein of male mice may cause the reversible testis damage [28]. Therefore, there is an urgent need for the understanding of nanotoxicity and the development of biocompatible nanomaterials.

Recently, the interactions between nanoparticles, such as CNTs, and biological molecules, such as proteins, have received great attention due to the aforementioned biosafety concerns. Proteins are the functional units of life. Studies on the interaction of CNTs and proteins may provide a key to understand the basic questions in the nanotoxicology and nanopharmacology. In 1999, Mioskowski and coworkers observed that the protein streptavidin can grow into helical crystals on the surface of CNTs in experiment [29]. In 2003, Park et al. showed experimentally that the CNT is effective in blocking some biological membrane ion channels due to its structural match in size and shape [30]. Using atomic force microscope (AFM) and Fourier transform infrared spectroscopy (FTIR), Karajanagi *et al.* investigated the conformational changes of two enzymes, α-chymotrypsin (CT) and soybean peroxidase (SBP), after adsorbing onto the SWCNT [31]. They argued that the distribution of the hydrophobic residues on the surface of the enzymes may determine the protein conformational changes when adsorbed onto the CNT. The important role of hydrophobic residues during the interactions between proteins and CNTs was also found by the experiments of Goldberg-Oppenheimer and Regev [32], in which they explored the conformational changes of bovine serum albumin (BSA) bound with CNTs via Cryogenic temperature transmission electron microscopy (Cryo-TEM). And the experiments on the interactions between small peptides and CNTs also showed that the tryptophan residues play a key role in the binding of the peptides on CNTs [33, 34].

In addition to these studies mentioned above, various experimental techniques have also been developed to shed light on this challenging problem [35-37]. However, the mechanism, particularly the effect of CNTs on the structure and function of biomolecules which may lead to the loss of the original function of proteins, is far from being understood. The inherent difficulties involved in these complex settings can limit their applications due to too many factors involved ("side effects") and experimental results may vary for the same proteins interacting with the same nanoparticles (such as same chemical composition, same shape, length, and aggregation property, etc.) under same conditions (such as same concentration, temperature, and pressure, etc). The computational simulations, on the other hand, might mimic and model this kind of interactions and avoid some complicated side effects. Molecular dynamics (MD) is one of these computational techniques, which is widely used in the studies of biomolecules [38-55], and nanoscale systems [56-60]. It is found effective in providing insights to the interactions between proteins and CNTs. For example, a recent MD simulation revealed some stepwise conformational changes of the sub-domain of human serum albumin (HSA) due to its adsorption onto the CNT surfaces [61].

In this chapter, we will review some of our recent works on the interactions between single-wall carbon nanotubes (SWCNTs) and proteins, and their effects on protein structure and function, using large scale molecular dynamics simulations. It is widely accepted that specific interactions between proteins and other molecules, often based on the structures of active sites, play a vital role in the function of proteins. Thus, if a SWCNT can occupy or disrupt the active site of a protein, it will very likely affect the function of the protein. First, we investigated the probability of the SWCNT disrupting the active site by using a signaling and regulatory protein WW domain (a relatively small protein) as an example. It was found that the SWCNT could plug into the hydrophobic cores of WW domains. Two key residues forming the functional scaffold in the native structure were separated by the SWCNT. This plugging of the SWCNT seriously broke the structure of the active site and reduced the possibility of the WW domain in binding with its partners. Second, we investigated the three-way binding competition among a SWCNT, protein SH3 domain (another small signaling and regulatory protein domain), and a proline rich motif ligand. It was found that the SWCNT had a very high probability occupying the active site of the SH3 domain, which prevented the ligand from binding to the active site. In these two cases, the effects of the SWCNT on the proteins would lead to the loss of the original function of the proteins. Finally, the interactions between the SWCNT and human blood serum proteins, including BSA (bovine serum albumin), γ-Ig (gamma-globulin), Tf (transferring), and BFg (fibrinogen), were also investigated by MD simulations, to confirm that the mechanisms identified with these small protein domains, such as WW domains and SH3 domains, also apply to much larger proteins. It is found that the hydrophobic interactions between the SWCNTs and nonpolar residues, particularly those aromatic ones through so-called π–π stacking, play a key role in the interaction between the SWCNTs and proteins.

2. Results

2.1 Disruption of protein active sites

Proteins are molecular machines, building blocks, and arms of living cells. The enormous variety of protein functions is based on their high specificity in partnering with other molecules, with which they interact -- a relationship that resembles a key and lock. This specific relationship demands a fairly unique spatial structure of the protein, in which the hydrophobic residues are buried in the core (named hydrophobic core). The pristine CNTs are extremely hydrophobic. Thus it is possible for the pristine CNTs to plug into the hydrophobic core due to the strong affinity between the CNT and the hydrophobic residues in the core. This plugging will lead to the loss of the original function of the protein, and thus is a potential mechanism for the toxicity of the CNTs (and maybe other hydrophobic nanoparticles).

We take a WW domain interacting with various sized SWCNTs as an example to illustrate this idea [62]. The WW domain is a protein module in signaling and regulatory proteins as the functional module to identify and bind the proline-rich motifs (PRMs) of its binding partner [63-67]. They exist as a triple stranded anti-parallel β-sheet structure [68-70]. Two highly conserved aromatic residues, a tryptophan residue in the third β-strand and an aromatic residue in the second β-strand, are together in the native structure to form the scaffold to bind the proline residue in the PRM. For the WW domain used in our simulation, which is a module of the human Yes-associated protein [71], the two key residues are W39 at the third β-strand and Y28 in the second β-strand. The SWCNTs used in the simulation are armchair of (m, n), where m = n = (4, 5, 6), corresponding to the tube diameters of (5.38 Å, 6.73 Å, 8.08 Å), respectively. The SWCNT and protein are initially well separated, with a distance between the geometric centre of protein domain and the SWCNT of 15.0 Å. Each system was solvated with TIP3P model water. We performed 36 independent simulations for the YAP65 WW domain and these three sizes of SWCNTs with different initial velocities, each with 200 ns (see more detail in Computational Methods).

Interaction of YAP65 WW domain with SWCNT

In our simulations, we found several cases that the SWCNTs plugged into the YAP65 WW domain between the second and third β-strands, as one representative structure shown in Fig. 1(a). Fig. 1(b) shows the superposition of the structure of the WW domain in the complex and in the native state. It was found that the main change of the protein conformation in the complex was its third β-strand. It unfolds into a loop and wraps on the SWCNT. Most of the contacts between the second and third β-strand are broken. Essentially, these conformational changes of the WW domain are pre-requirements for the formation of this complex of the protein domain and the SWCNT. In the complex, nine residues in the second and third β-strands contact with the SWCNT. Most of them are hydrophobic residues, such as W39. This implies that the hydrophobic interaction, the dominant force for protein folding [72], is the key to this phenomenon, in which the SWCNT plugged into the hydrophobic core of the WW protein domain to form a stable protein-SWCNT complex.

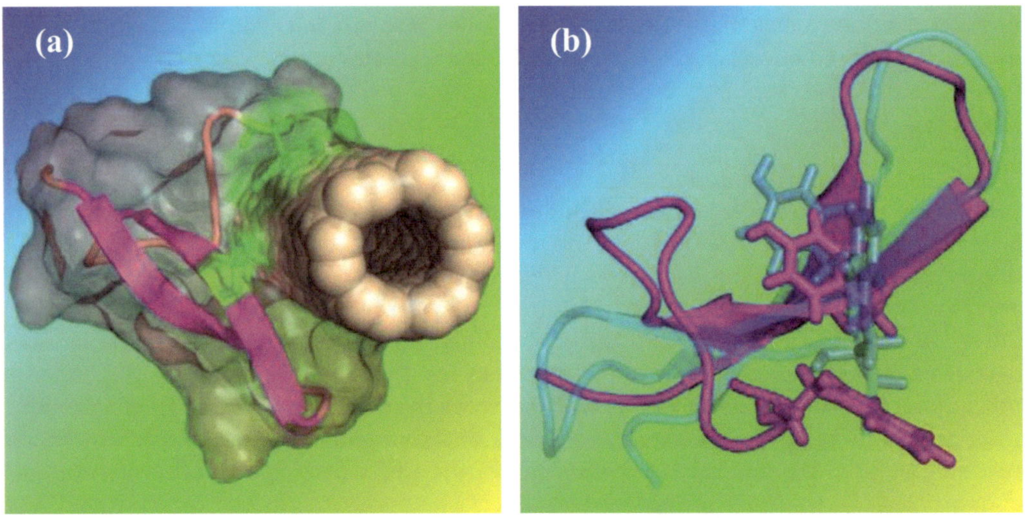

Fig. 1. Typical structures of (a) a complex of the SWCNT plugging into the YAP65 WW domain and (b) The WW domain in the complex (magenta and solid) superposed with its native state (cyan and half semitransparent). In the complex, the YAP65 WW domain is shown in cartoon with magenta strand and salmon loop, and the key residues (W39 and Y28) are identified by green sticks. The (6, 6) armchair SWCNT has a diameter of 8.08 Å, and its atoms are shown in wheat spheres. The solvated surfaces of the complex are shown in semitransparent.

Fig. 2(a) displays some representative snapshots of these complexes at different times to show the plugging process of the SWCNT. The interface area between the protein domain and the SWCNT (denoted by S, shown in Fig. 2(b)) is used to illustrate this process. Here S is defined as half of the difference between the solvent accessible surface area of the complex and the sum of solvent accessible surface areas of the protein and the SWCNT [73]. Thus at $t = 0$, S is small (~50 Å^2) since the YAP65 WW domain and the SWCNT are initially separated. The YAP65 WW domain and the SWCNT approached each other very quickly. As a result, S rose to about 200 Å^2 within only 1 ns. This indicated that the SWCNT was adsorbed onto the protein surface (see the snapshot of $t = 1$ ns in Fig. 2(a)). In the next 45 ns, S gradually increased from ~200 Å^2 to ~250 Å^2, with significant fluctuations along the way. It was found that the contacting surface region of the domain kept changing during the process of adsorption. That is, the SWCNT was constantly seeking for a more stable binding site. Around $t = 45$ ns, there is a negative impulse in the interface area S. A careful examination on the trajectory showed that the SWCNT began to plug into the second and third β-strands and the negative impulse was a result of the opening of the second and the third β-strands before the SWCNT was "swallowed" by the protein. S reached its maximal value of ~300 Å^2 at around t = 80 ns, meaning that the SWCNT was finally successful in plugging into the protein domain. In these structures, the SWCNT was basically wrapped by the second and the third β-strands, forming a complex of the WW domain with the SWCNT. From t = 80 ns up to 200 ns, there were only minor fluctuations of S at ~300 Å^2, showing the stability of this protein-SWCNT complex (data after 100 ns were not shown since the plugging was mainly done in the first 100 ns).

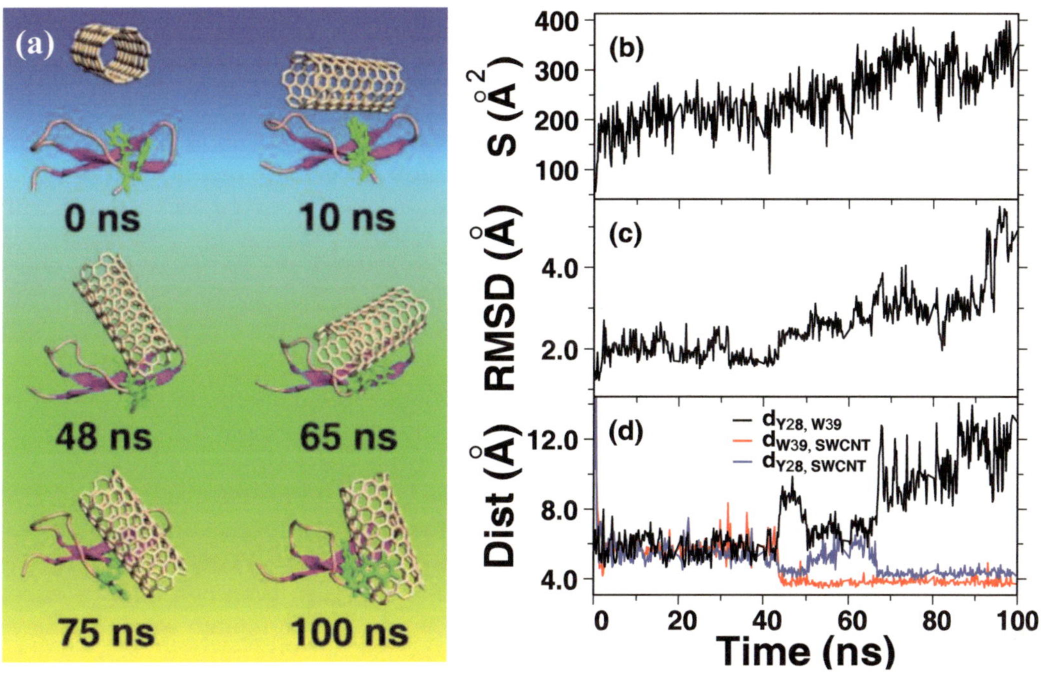

Fig. 2. A typical trajectory of the YAP65 WW domain together with a SWCNT.
(a) Representative snapshots at various time points. The proteins are shown in cartoon with magenta strand and salmon loop. The binding scaffold residues are shown in green sticks and the SWCNT are shown in wheat sticks. (b) The interface area of the WW domain and the SWCNT, (c) The RMSD of the YAP65 WW domain from its native structure, and (d) The distances between the SWCNT and the key residues of YAP65 WW domain, W39 (red) and Y28 (blue), and distance between these two residues (black), denoted by $d_{W39, SWCNT}$, $d_{Y28, SWCNT}$, and $d_{W39, Y28}$, respectively, as a function of time.

In order to investigate the conformational change of the YAP65 WW domain, we computed the RMSD from its native structure in the MD trajectory, as shown in Fig. 2(c). As a comparison, we had also performed simulations for the YAP65 WW domain only (without any SWCNT or ligand). It was found that without the disruption of the SWCNT, the RMSD of the WW domain remained at a value of around 2.0 Å. From the RMSD shown in Fig. 2(c), the conformational change of the YAP65 WW domain can be described as four stages. In the first stage, from 0 to ~45 ns, the RMSD maintained at roughly 2.0 Å. That is, there was little impact of the SWCNT on the conformation of the YAP65 WW domain despite its adsorption on the SWCNT surface. In the second stage, from ~45 to ~70 ns, the RMSD increased gradually from ~2.0 Å to ~3.5 Å, which was a notable conformational change and corresponded to the opening of the second and third β-strands due to the insertion of the SWCNT. The third stage is from ~70 to ~90 ns. In this stage, a small plateau for the RMSD, around 3.0 Å, appeared. It was due to the partial recovery of the contacts between the second and third β-strands. Around 90 ns, there was a significant jump in the RMSD, from ~3.0 to ~4.0 Å, which initiates the fourth stage. Detailed studies showed that there was a change in both the orientation and binding site of the indole ring of W39 residue on the SWCNT during this jumping process. From t = ~95 ns, the RMSD kept roughly constant at ~4.0 Å, indicating the final complex of the YAP65 WW domain and SWCNT was quite stable. This relatively large RMSD indicates a significant conformational alteration of the YAP65 WW domain caused by the insertion of the SWCNT.

The W39 and Y28 are the key residues for the function of the YAP65 WW domain. The interaction between the SWCNT and these two key residues were also investigated. Fig. 2(d) shows the distances between these two residues, and distances between each residue with the SWCNT as a function of time, denoted by $d_{W39,Y28}$, $d_{W39,SWCNT}$, and $d_{Y28,SWCNT}$, respectively. Corresponding to the adsorption of the SWCNT on the YAP65 WW, there is a drop of $d_{W39,SWCNT}$ and $d_{Y28,SWCNT}$ in the first ns. And then $d_{W39,Y28}$, $d_{W39,SWCNT}$, and $d_{Y28,SWCNT}$ all kept almost as a constant at ~6.0 Å with small fluctuations until t = 45 ns, showing that the two key residues were "confined" in their native state, and the initial adsorption of the SWCNT does not impact much on their structures and orientations. In the period of t = 45 - 50 ns, there were an impulse jump in $d_{W39,Y28}$ to a value around 9 Å and a drop in $d_{Y28,SWCNT}$ to a valley around 4 Å. These changes indicate the detachment of W39 with Y28 and the approach of Y28 to the SWCNT. However, $d_{W39,SWCNT}$ decreased sharply to < 4 Å at the same time and remained nearly unchanged for the rest of the simulation times, which indicates a strong and favorable interaction between W39 and the SWCNT. It is interesting to note that this favorable tryptophan-CNT interaction has also been observed in recent experiments for CNT-peptide interactions [33, 34]. Therefore, even though we do not have direct experimental evidence at this point for this plugging of SWCNTs, it might not be totally impossible due to the strong and favorable interactions between CNTs and hydrophobic residues with aromatic rings such as tryptophan residues. It was very interesting to find that both $d_{Y28,W39}$ and $d_{Y28,SWCNT}$ returned to their previous values (before t = 45 ns) at ~50 ns. A careful examination of the trajectory movie showed that there was a rotation of the SWCNT orientation, with the SWCNT firmly seizing the W39 in the period of t = 50 -70 ns. After t = 70 ns, $d_{Y28,SWCNT}$ decreased to a small value again (~4 Å), and the $d_{W39,Y28}$ turned to a significantly larger value (8-12 Å). This indicates a significantly larger separation of the two active-site residues, which corresponds to a wide opening of the second and third β-strands, indicating a potential loss of its function.

Stability and hydrophobic interaction

We had computed various interaction energies of the system as functions of the interface area between YAP65 WW domain and SWCNT (shown in Fig. 3). For the interaction energy of the total system (using the initial state as reference, see Fig. 3(a)), there is a basin at about S = 280 Å^2, with the interaction energy about 30.0 kcal/mol lower than the reference state, the state where YAP65 WW domain and the SWCNT are well separated. From Fig. 2(b), we know that S ~ 280 Å^2 corresponds to the state in which SWCNT plugs into the YAP65 WW domain. Therefore, the complex with SWCNT inserted into the core of YAP65 WW domain is the energetically more favored state.

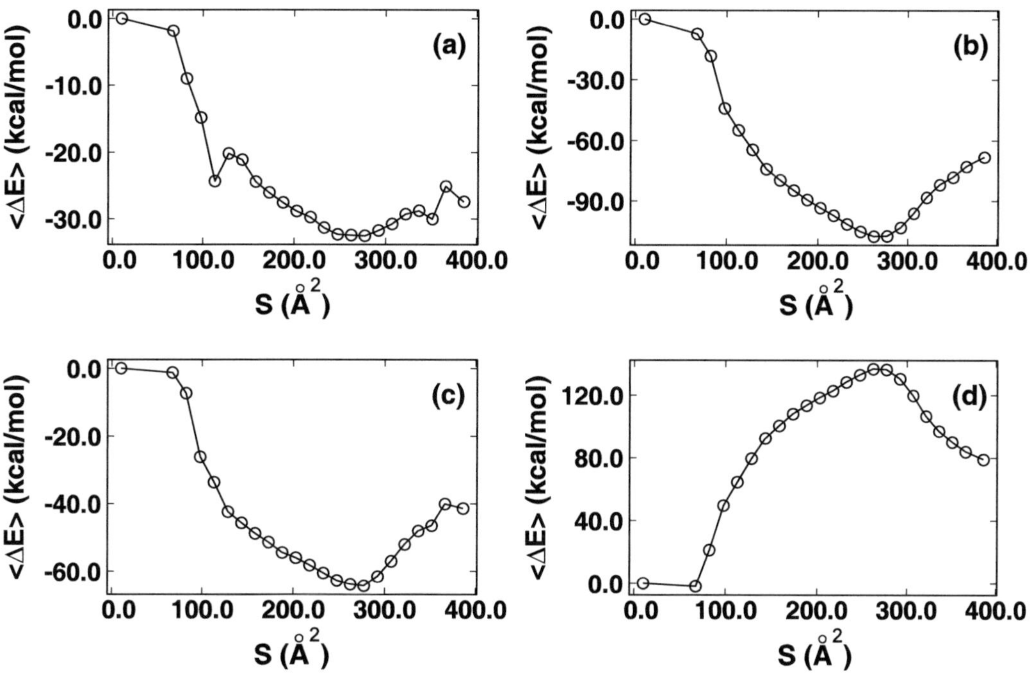

Fig. 3. Interaction energy from the initial state (a) of the total system, (b) of solutes, (c) of solvent, and (d) between solute and solvent, as a function of the interface area S between the YAP65 WW domain and the SWCNT, averaged from the six trajectories of the MD simulations. The interaction energy of the initial state, with the SWCNT and WW domain well separated, is set as 0.

Fig. 3(b-d) show the interaction energies of the solute (YAP65 WW domain and SWCNT), the solvent (the water molecules), and between solute and solvent, as functions of the interface area, respectively. As shown in these figures, the internal interaction energies of the solute and the solvent (see fig. 3(b and c)) change in the opposite direction of the interaction energy between the solute and the solvent (see fig. 3(d)). Clearly, in our current system, the internal interaction energies of the solute and the solvent dominate the total energy of the system, which both have an energetic basin around S=280 $Å^2$. This implies that the solute and the solvent (water) are favored to interact with themselves and not favored to interact with each other, i.e., a separation of the two is energetically more favored, which is best described as so-called hydrophobic interactions. It is the hydrophobic interaction between the YAP65 WW, SWCNT and water that makes the plugging of SWCNT into YAP65 WW domain.

Binding site and structural flexibility

In each case of a SWCNT plugging into the YAP65 WW domain, we have found that the YAP65 WW domain displays the same binding site after relentless search. It was shown above that the hydrophobic interaction dominated the binding and plugging of the SWCNT. However, there are also hydrophobic residues, including a tryptophan residue, on the first β-strand of the YAP65 WW. Thus the question was why the SWCNT plugged into the gap between the second and third β-strand instead of that between the first and second β-strand. The flexibility of different part of the YAP65 WW domain may be a factor of selection of the plugging site. To illustrate this idea, Fig. 4 shows the root mean square fluctuations (RMSF) of the average of non-hydrogen atoms of each residue for the YAP65 WW domain (the simulation was performed without any SWCNT or ligand). Generally, the loops and turns have larger fluctuations. The notable point of this figure is the difference between the three β-strands. Clearly, the fluctuations of the third β-strand residues are significantly higher than those of the other two β-strands. This indicated that the third β-strand was more flexible than the other two β-strands. That is, the gap between the second and third β-strand is easier to be opened, and thus the SWCNT is easier to penetrate it.

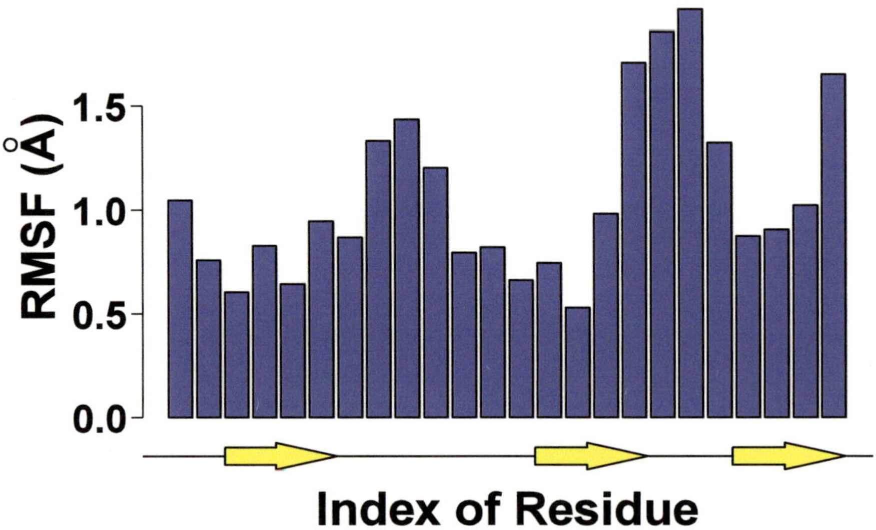

Fig. 4. RMSF of the non-hydrogen atoms for the simulation of the YAP65 WW domain without SWCNT. The yellow arrows show the three β-strands of the WW domain. The secondary structure information of the domain is obtained by DSSP [74].

Effect on protein function

As shown above, the SWCNT plugged into the second and third β-stands and formed a stable complex. In the protein-SWCNT complex, two active residues of YAP65 WW domain (W39 and Y28) are separated, thus the primary functional unit, the scaffold, for binding with prolines in the proline rich motifs (PRM) is disrupted. In general, the SWCNT insertion of the binding site is much more disruptive than the typical surface adsorption, which may result in the loss of the original function of the WW domain. To demonstrate this, we performed simulations for the system with an additional PRM ligand (sequence GTPPPPYTVG), which binds to both the YAP65 WW domain in its native state [71], and to the protein-SWCNT complex. In the PRM binding simulations, the PRM was initially placed at 25 Å away from the center of the protein-SWCNT complex. There were 7045 TIP3P water molecules and one Cl^- (which was used to neutralize the system), and all other MD setups and parameters were similar to the previous simulations. Ten independent runs were performed for each system in NVT ensemble at 298 K.

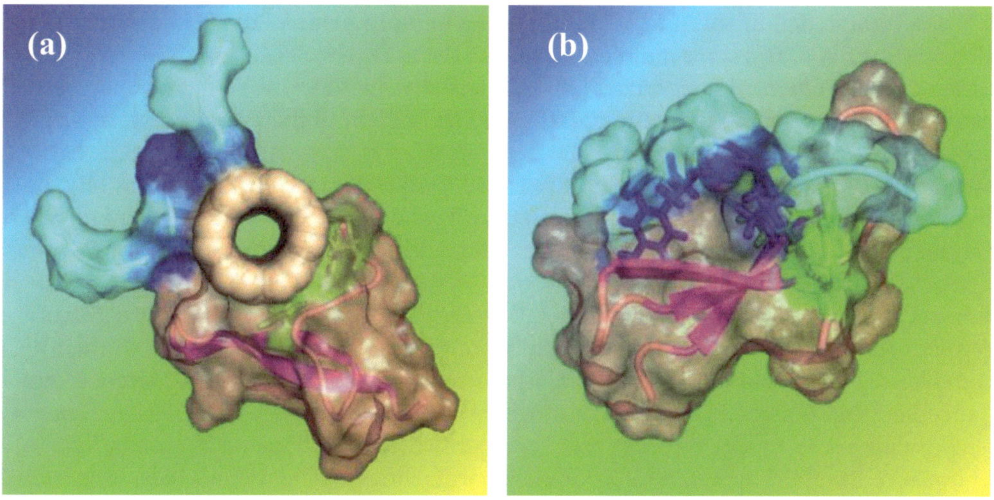

Fig. 5. A typical structure of (a) the protein-SWCNT complex with the adsorbed proline-rich motif (PRM) and (b) the domain binding with the PRM (PDB code: 1JMQ). Here the YAP65 WW protein domain is shown in cartoon with magenta strand and salmon loop, and the key residues (W39 and Y28) are identified by green sticks. The atoms of the (6, 6) armchair SWCNT are shown in wheat spheres. The PRM ligand with the sequence GTPPPPYTVG is shown in cyan cartoon with the key residues identified by blue sticks. The solvated surfaces of the complex are shown in semitransparent.

Our simulations showed that the protein-SWNT complex was stable, and the PRM always bound to the SWCNT instead of the binding site of the YAP65 WW, as shown in one representative structure in Fig. 5(a). As for comparison, we also showed the native structure with the PRM in Fig. 5(b). Clearly, the binding site (labeled by the green stick in these two figures) had been blocked by the SWCNT in the YAP65 WW domain-SWCNT complex. These findings indicate that the insertion of the SWCNT disrupted and blocked the binding site of the YAP65 WW domain, and disabled its function of identifying and binding with the PRM. To validate this prediction from our *in silico* approach, *in vitro* and/or *in vivo* experiments are highly needed. Interestingly, there is recent evidence from both experiment and theory showing that some antibody from mouse immune repertoire can "absorb" the C60 nanoparticle through similar hydrophobic interactions [75, 76].

2.2 Competitive binding with ligands to receptors

The scheme of protein functioning can be roughly described into three steps: binding, transforming, and releasing. The binding, especially the specific binding, between proteins and ligands is the primary step for the protein functioning. Again, hydrophobic interactions often play a vital role in the binding of proteins with ligands. If pristine CNTs (which are extremely hydrophobic) pass through cell membranes and possess stronger binding affinity than ligands with hydrophobic binding sites of some key proteins, these CNTs may hold their hydrophobic binding sites, obstruct their binding processes, and disrupt the functions of these proteins. This competitive binding may be another potential mechanism of toxicity of hydrophobic nanoparticles, including CNTs. All the aforementioned studies focus on the direct interactions of nanoparticles with proteins. The effect of nanoscale particles on the ligand-receptor binding, i.e., a three way nanoparticle-ligand-protein competitive binding, is much less studied.

We take a proline-rich ligand (PPPVPPRR) and its binding module, the SH3 domain, together with a pristine SWCNT as an example to illustrate the idea. The SH3 domain is one of the most abundant protein interaction domains [77], and usually found in the signaling and regulatory proteins as the functional module to identify and bind PRM of their binding partners [78-84]. Fig. 6(a) shows the native structure of the SH3 domain binding with the ligand. It exists as a characteristic β-barrel fold which consists of five β-strands arranged as two tightly packed anti-parallel β-sheets [85]. The residues, including F8, W36, P50, and Y53, are the key residues to bind ligands [78, 86]; and the hydrophobic interaction is the dominant force for the binding process [87]. The SWCNTs are armchair of (m, n), where m = n = 3, corresponding to the tube diameters of 4.04 Å with a length of 19.54 Å. The ligand, SWCNT and protein are initially well separated from each other. The initial distances of the geometric centers, of both the SWCNT and the ligand, from that of the SH3 domain are set at 30 Å, but in different orientations. The distance between the geometric centers of the SWCNT and ligand is about 42 Å. They were solvated with TIP3P model water as well. We performed 20 independent simulations for the SH3 domain, the PRM ligand, and the SWCNT with different initial velocities, each with 200 ns (see more detail in Computational Methods).

Interaction of SH3 domain, ligand and SWCNT

We find that the SWCNT occupies the binding pocket and contacts directly with the key residues of the SH3 domain in 13 of the 20 simulation trajectories. In this way, the SWCNT prevents the ligand from forming the native contacts with the SH3 domain binding pocket.

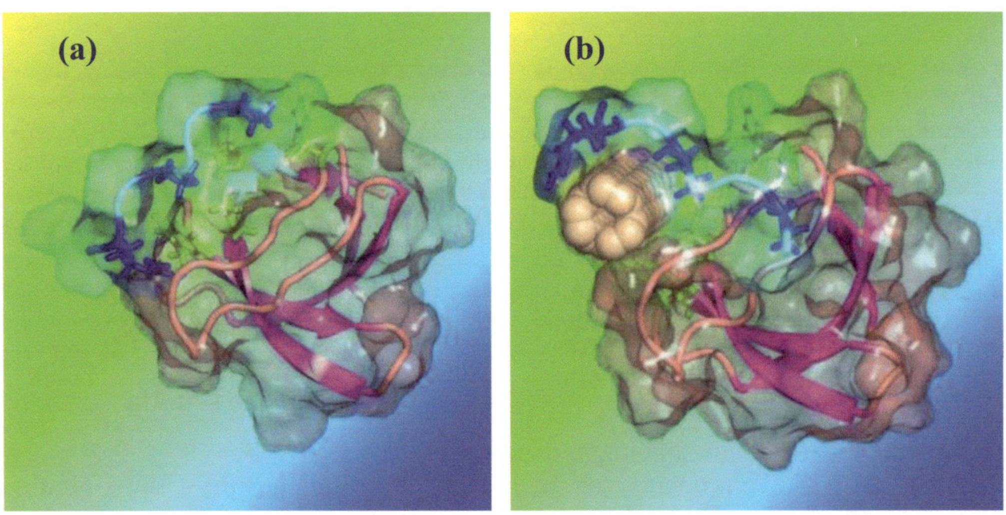

Fig. 6. Structures of (a) the SH3 domain binding with the PRM ligand (PDB: 1CKB) in its native state; and (b) a typical snapshot in which the SWCNT occupies the binding pocket and contacts with the key residues of the SH3 domain, while the ligand binds partly to the SWCNT and partly to the SH3 domain; and Here the SH3 domains are shown in cartoon with magenta strand and salmon loop, and the key residues (F8, W36, P50 and Y53) in the pocket are identified by green sticks. The atoms of the SWCNT are shown in wheat spheres. The ligands with sequence PPPVPPRR are shown in cyan cartoon with the key residues identified by blue sticks. The solvated surface of complexes is shown semitransparent.

Fig. 6(b) shows a typical structure for the case that SWCNT occupies the binding pocket of SH3 and the ligand binds to the SWCNT. The SWCNT directly contacts with nine residues of the SH3 domain in the pocket, including the key residues W36, P50, and Y53. In the other 12 cases, the SWCNT contacts with at least one of the key residues of the binding pocket of the SH3 domain. And in 11 of those 13 trajectories, the ligand binds partly to the SWCNT and partly to the surface of the SH3 domain (near the SWCNT). In other two cases, the ligand only binds to the surface of the SH3 domain (but not the active site) and does not bind to the SWCNT. Even though the other seven trajectories do not show a direct binding of the SWCNT with the SH3 domain binding pocket, the SWCNT does display some tendency of the binding. For example, in three of the seven trajectories, the SWCNT is getting close to the key W37 residue in the binding pocket; and in the other two trajectories, it is getting close to the key Y03 residue. The incomplete binding might be caused by the fact that our simulation lengths, 200+ ns, are still not long enough.

The binding process is illustrated by representative snapshots as well as the interface areas between the SH3 domain, the ligand, and the SWCNT, using a typical trajectory (see Fig. 7), in which the SWCNT first binds onto the SH3 and then the ligand binds onto the SWCNT.

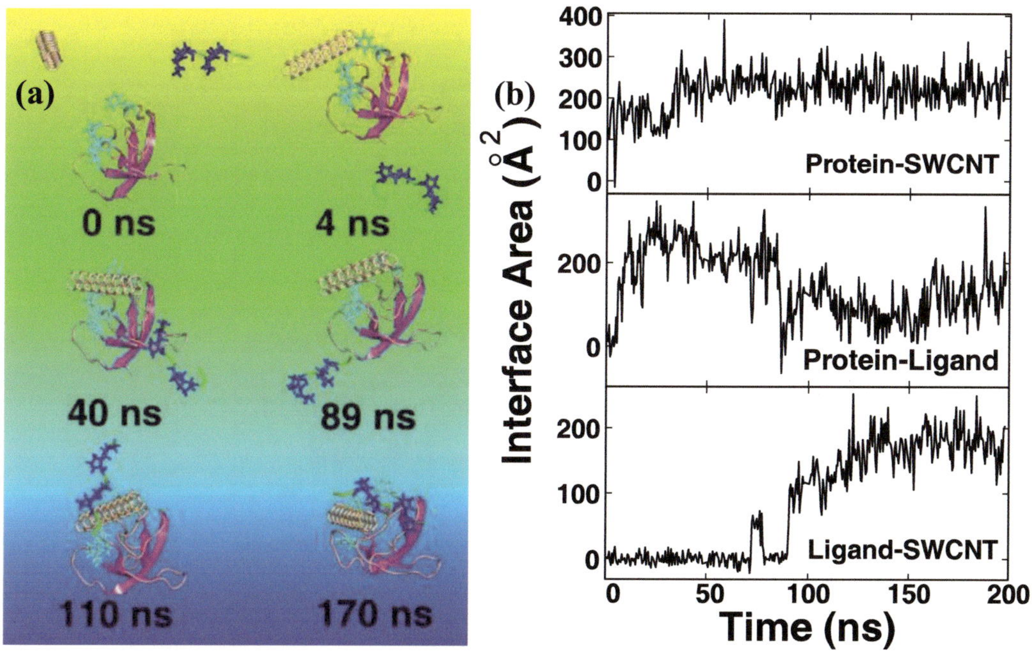

Fig. 7. One representative trajectory in which the SWCNT adsorbs onto the binding pocket and makes direct contacts with the key residues of the SH3 domain, while the ligand binds partly with the SWCNT and partly with the SH3 domain. (a) Some representative snapshots. Here the proteins are shown in cartoons with magenta strand and salmon loop. The key residues of the protein domain are noted by red sticks, the ligands with the sequence PPPVPPRR are shown in cyan cartoon with the key residues identified by blue sticks. The SWCNTs are shown in wheat sticks. (b) The protein-SWCNT, protein-ligand, and ligand-SWCNT interface areas as function of time.

To illustrate the binding process of the ligand, we also show the interface areas of the ligand with the SH3 domain (protein-ligand), and the ligand with the SWCNT (ligand-SWCNT), denoted by S_{PL} and S_{LC}, respectively (see Fig. 7(b)). At t = 0, S_{PL} = S_{LC} = 0 since the ligand is separated from both the SH3 domain and the SWCNT initially. In the first 5 ns, S_{PL} increases to about 200 Å^2 very quickly, and maintains that value until t = ~ 90 ns. At the snapshot of t = 40 ns, the ligand is adsorbed onto the SH3 domain (but not at the binding pocket). On the other hand, S_{LC} still keeps at about 0. The most notable event happened at about 90 ns. After that, S_{PL} decreases quickly from 200 Å^2 to a very small value, and later recovers back to ~150 Å^2. Meanwhile, S_{LC} increases to about 100 Å^2. At t = 89 ns, the ligand has temporarily separated from the SH3 domain, diffuses to the other side of the SH3 domain, and then binds to both the SWCNT and the SH3 domain (the structure is shown in the snapshot at 100 ns). From t = 90 ns to t = 200 ns, both S_{PL} and S_{LC} remain at these values with normal fluctuations, indicating a stable three-way bound structure among the SWCNT, the ligand, and the SH3 domain.

From the interface areas, we showed the main picture of the binding process of the SWCNT, the ligand and the SH3 domain. To give a more detailed picture about the binding process at the residue level, we had computed the contact number of the SWCNT and the ligand with the residues of SH3 domain. Here we used the protein residue as a unit base in accounting (see Fig. 8), and a contact is counted if the distance between a non-hydrogen atom in the

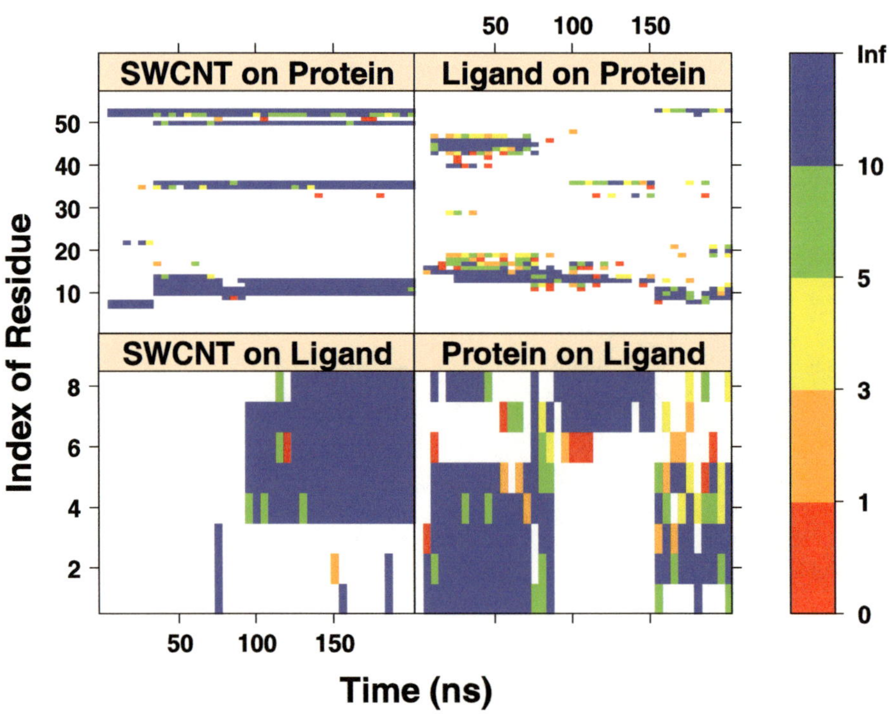

Fig. 8. Numbers of the contacts of the SWCNT and the ligand with the residues of SH3 domain, together with those of the SH3 domain and the SWCNT with the residues of the ligand. The number denoted by the color and the " blank" meaning no contact. There is a contact if the distance between a non-hydrogen atom in the bound object and a non-hydrogen atom in the residue is less than 5 Å.

bound object and a non-hydrogen atom in a residue is less than 5 Å. Generally speaking, the closer the object is to the protein residue, the larger the contact number is between the protein and the object, and the stronger the object binds to the protein residue. Similarly, from the contacts of the SWCNT on the SH3 domain, we found that there were three stages in the binding process. In the initial stage, from 0 to ~3 ns, the SH3 domain and the SWCNT were separated; in the second stage, from ~3 to ~35 ns, the SWCNT was adsorbed onto four residues of the SH3 domain; in the third stage, from ~35 ns to the end of the simulation, the SWCNT bound closely with many residues (up to 9) of the SH3 domain, including the key residues W36, P50, and Y53.

Furthermore, the contacts between the SWCNT and the ligand are found to be quite stable, *i.e.*, these contacts remain in contact once they appear (see Fig. 8). On the contrary, the contacts between the ligand and the protein (both ligand on protein and protein on ligand) vary with time, indicating that the relative positions between the ligand and the SH3 domain change with time much more often. Further, the great changes of the ligand-protein contacts happen at t = ~88 ns (Fig. 8), corresponding to the previous observation that the ligand temporarily separates from the SH3 domain, then diffuses to the other side of the SH3 domain, and finally re-adsorbs onto both the SWCNT and the SH3 domain. Conformational fluctuations of the ligand on the SWCNT and SH3 domain are observed after the final adsorption. For example, from 90 ns to 150 ns, the ligand binds both the SH3 domain (on the residues 12-13, 16, and 36) and the SWCNT by the C-terminal residues; while from 150 ns to 200 ns, the ligand binds the SH3 domain (on the residues 9~11, and 53) by the N-terminal and the SWCNT by C-terminal (see also Fig. 8).

Free energy landscape

Fig. 9(a) shows the free energy landscape as a function of two reaction coordinates: the interface area between the SH3 domain and the SWCNT (S_{PC}), and the minimal distance between the SWCNT and the key residues of the SH3 domain (including F08, W36, P50, and Y53), denoted by D_{KC}. Clearly, the SWCNT will prevent the regular binding of the ligand to the SH3 domain if it binds to one of the key residues. Thus, the reaction coordinate D_{KC} could be a good reaction coordinate to describe the binding free energy landscape. To obtain a reasonable result, we sampled the structures with a time interval of 20 ps in the last 180 ns of all the 20 trajectories. The global minimum in this landscape is found at S_{PC} ~ 220 Å^2 and D_{KC} < 5.0 Å (see Fig. 9(a)) with a binding free energy of -6.08 kcal/mol, which corresponds

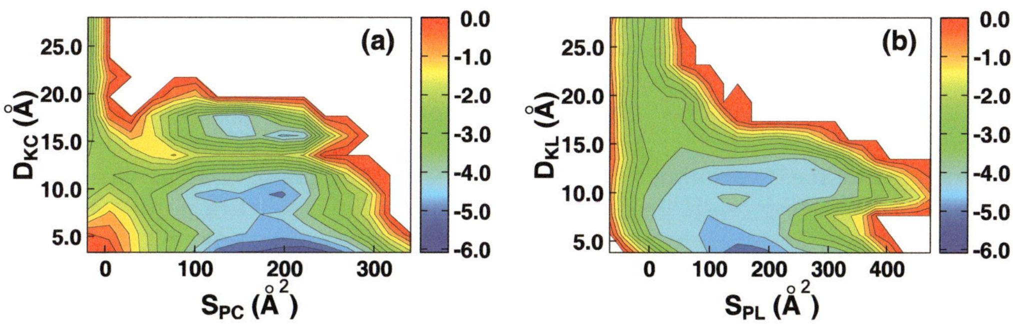

Fig. 9. The binding free energy landscapes of the SWCNT (a) and the ligand (b) with the SH3 domain. "S_{PC}" and "S_{PL}" denote the interface area of the SWCNT and ligand with the SH3 domain, respectively. "D_{KC}" and "D_{KL}" are the minimal distance of the SWCNT and the ligand from these key residues, respectively. The unit of the free energy is kcal/mol.

to the state where the SWCNT binds to the SH3 domain binding pocket. Two local minima, in which S_{PC} ~ 150 Å^2 & D_{KC} ~ 8.0 Å, and S_{PC} ~ 180 Å^2 & D_{KC} ~ 18.0 Å, are also observed. Detail studies show that these two states correspond to the SWCNT binding to the side (the region around the residue W37) and the opposite (the region around the residue Y03) of the binding pocket of the SH3 domain, respectively.

The binding of the ligand to the SH3 domain is also described by the free energy landscape as a function of the two equivalent reaction coordinates: the interface area between the ligand and the SH3 domain (S_{PL}), and the distance between the ligand and the key residues of the SH3 domain (D_{KC}) (shown in Fig. 9(b)). The lowest binding affinity for the ligand is -5.51 kcal/mol, which is 0.57 kcal/mol less than the SWCNT shown above. This explains why the PRM ligand loses the competition to the SWCNT in the binding to the SH3 domain. Interestingly, the binding free energy landscape also shows that there are two basins, with the free energy of -5.51 kcal/mol (the lowest for the ligand binding) and -5.16 kcal/mol, respectively. Both states have similar values of S_{PL} ~ 180 Å^2, i.e., adsorbing onto the SH3 domain with similar contacting surface areas. The difference lies in the D_{KL}, i.e., the distance to the key residues of the SH3 domain binding pocket. The deeper one has a very small value of D_{KL} ~ 5.0 Å, which indicates that the ligand is around the binding pocket. The other basin has a larger value of D_{KL} ~ 12.0 Å, indicating that the ligand is far away the binding pocket. We note that the same D_{KL} (~12 Å) does not mean that there is only one binding site for the ligand in this case. A detailed study shows that the ligand binds to a broad region of the SH3 domain with residues 11-19, 23-25, and 43-47 involved.

π–π stacking interaction

From Fig. 9(a), there are three basins in the free energy landscape of the SWCNT binding on the SWCNT. When the SWCNT was in the global minimum (binding on the binding pocket), it contacted with at least one of the aromatic residues, including F08, W36 and Y53. The upper panel of Fig. 10 shows the typical binding mode of the SWCNT contacting with these three aromatic residues, in which their side chains of are almost parallel with the SWCNT. For the two local minima, the SWCNT contacted with the Y03 and W37, respectively (shown in the lower panel of Fig. 10). Overall, the SWCNT always contacts with aromatic residues when it binds onto the SH3 domain. In the SH3 domain native structure, there are a total of 47 residues exposed to environment (solvent, ligand, or SWCNT), but only ten of them are hydrophobic and five of them aromatic residues. This indicates the SWCNT is favored to contact with the hydrophobic and aromatic residues (more below).

In order to describe quantitatively the degree by which the SWCNT favors to contact with the hydrophobic and aromatic residues, we estimated the p-values of distribution of the hydrophobic and aromatic residues in contact with the SWCNT based on the χ^2 hypothesis [88] (see Fig. 11). It was found that the SWCNT is remarkably favored in binding with the hydrophobic residues of the SH3 domain. The p-value for the SWCNT in contacting with hydrophobic residues (Ile, Val, Leu, Met, Ala, Trp, and Phe, following the hydrophobicity classifications of this NCBI website: http://www.ncbi.nlm.nih.gov/Class/Structure/aa/aa_explorer.cgi), aromatic residues (His, Tyr, Phe, and Trp), and the hydrophobic aromatic residues (Phe and Trp; here we do not include Tyr residue, even though many people might classify Tyr to be hydrophobic aromatic residue as well) were 6.51×10^{-3}, 2.42×10^{-12}, and 2.00×10^{-19}, respectively. This indicates that the interactions between the SWCNT and the hydrophobic residues, particularly the aromatic residues (π–π stacking interactions), play an important role in the binding of SWCNT and proteins. We note that favorable interactions of the carbon nanotube with the hydrophobic and aromatic residues had also been observed in recent carbon nanotube-peptide experiments [33, 34, 89]. That is, these interactions are independent of the sizes of proteins and CNTs. Therefore, even though the SWCNT used in our simulation is quite small, the observation of the high probability to occupy the binding pocket of the protein by the SWCNT is extendable to the hydrophobic nanoscale particles of larger lengths, which is expected to be observed with more extensive computations with the development of the supercomputers as well as advanced simulation techniques. It should also be noted that the π–π stacking interactions might be underestimated in standard force fields due to the lack of polarization, and more advanced techniques such as quantum mechanics simulations might be needed to fully catch the effect; however, a recent study shows that classical simulations can be sometimes even better to capture this π–π stacking interactions than the quantum mechanics simulations due to the facts that the classical force fields are often directly fitting from experimental data, while the quantum mechanics simulations, on the other hand, might suffer from the limited size and boundary conditions [90].

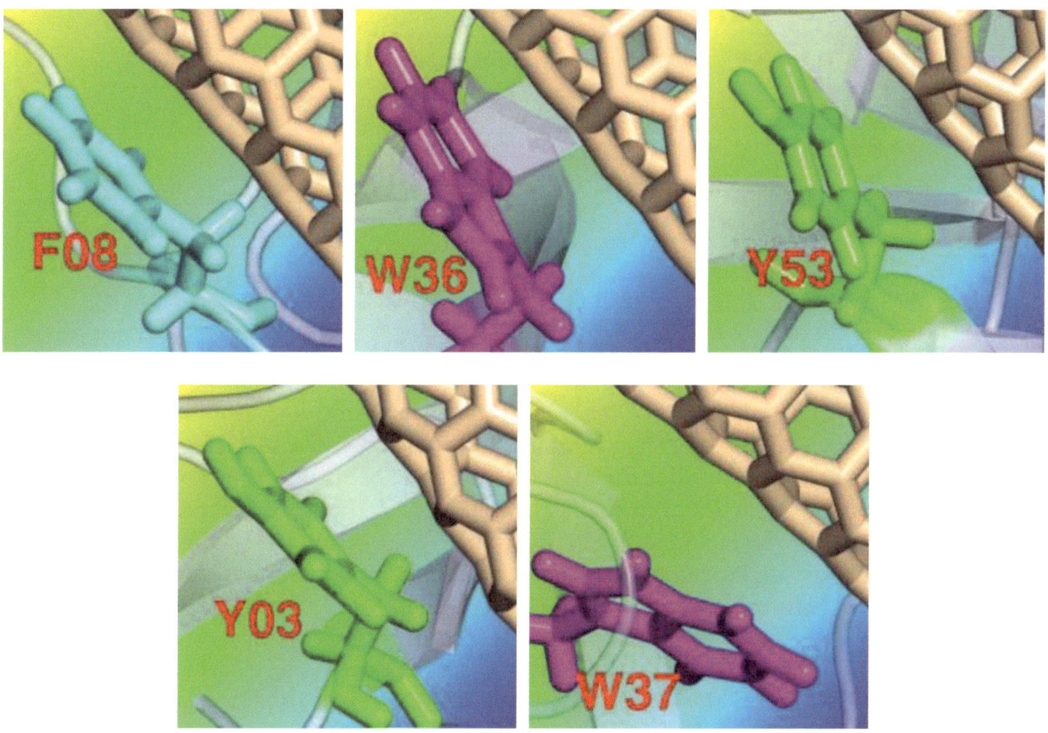

Fig. 10. Interaction between the SWCNT and the aromatic residues of SH3 domain. The top panel shows the key residues of the binding pocket, including F08, W36 and Y53. The bottom panel shows the aromatic residues which are exposed but not belong to the binding pocket, including Y03 and W37. Here the SH3 domains are shown in semitransparent sky blue cartoon. The aromatic residues are shown in sticks (magenta tryptophan, cyan phenylalanine, and green tyrosine) and labeled by red text. The SWCNTs are shown in wheat sticks.

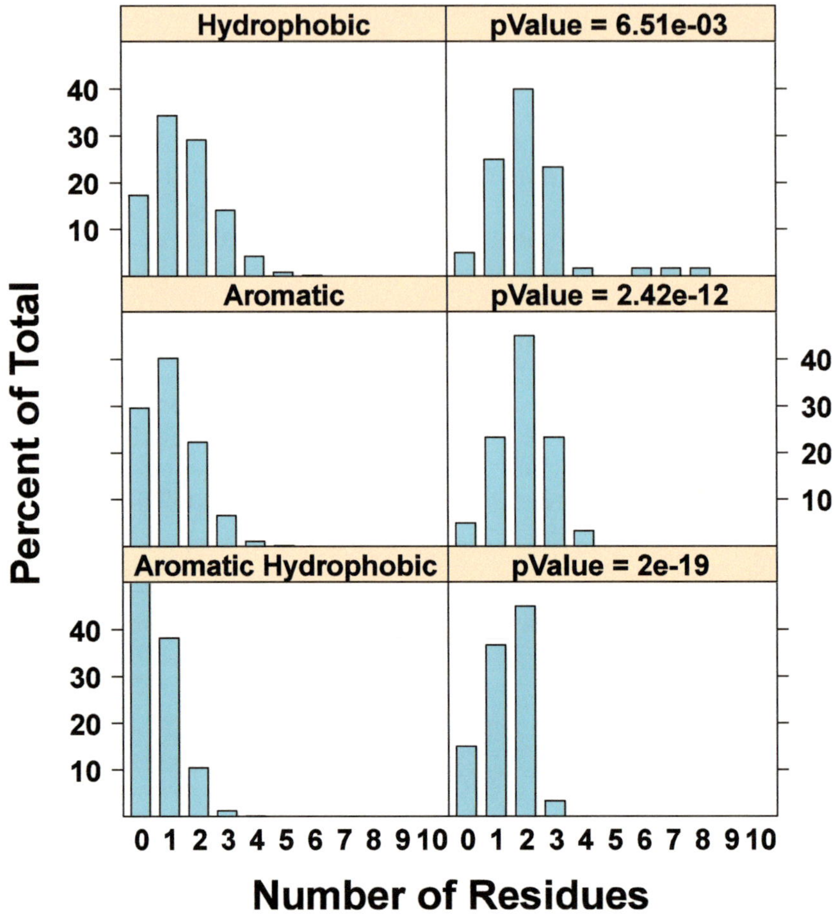

Fig. 11. Distribution of the number of the hydrophobic, aromatic, and aromatic hydrophobic residues contacted with SWCNT. Here a residue is contacted with the SWCNT when the distance between one of its non-hydrogen atoms and SWCNT is less than 5 Å. The left column shows the expected distribution based on the number of the residues and the total number of the residues contacted with SWCNT. The right column shows the practical distribution. The p-values for the distribution are estimated on the χ^2 hypothesis test, and shown on the right column.

Effect on the binding event

It is noteworthy that a previous MD simulation has shown that the ligand could recognize the binding pocket and form the native binding mode with the SH3 domain, in the absence of nanoscale particles [87]. The long-range electrostatic interactions are found to play an important role in the ligand *recognition* of the binding pocket, through its two arginine residues on the C-terminal [87]. The PRM ligand's R7 and R8 residues have favorable long-range electrostatic interactions with SH3-domain's acidic residues, D14, E16, D17, and E33 on RT and n-sCr loops, near one end of the binding pocket, which provides guidance to an accelerated ligand binding mode search. This electrostatically accelerated association of proteins was also previously observed in experiments [91, 92]. And studies had confirmed that the arginine residue is conserved in the PRM to this SH3 domain [78, 79]. The stability of the final *binding*, on the other hand, is driven by the structural match and hydrophobic interactions between the ligand (proline residues and V4) and the SH3 domain (F8, F10, W36, P50, P52, and Y53), in addition to the salt-bridges formed by R7/R8 (see Fig. 6(a)). In our current complex system with the presence of SWCNT, the SWCNT wins the competition over the ligand for the binding pocket even without any guidance from the long-range electrostatic interactions (since SWCNT is uncharged in our simulations), indicating a very strong hydrophobic interaction between the SWCNT and the SH3 domain. Interestingly, the ligand still manages to *locate* the area of the active site and bind with both the SWCNT and SH3 domain (though no longer in the binding pocket), which implies the PRM ligand still maintains some *recognition* capability due to the long-range electrostatic interactions through its two C-terminal arginine residues.

The structural match between the ligand and active site stabilized the native binding mode. However, it raised the demand for an extensive conformational space mapping even with the electrostatic acceleration. And the hydrophilic residues in the ligand (the two terminal arginines) often shield the hydrophobic ones in the water environment to "inhibit" their exposure to the active site. On the other hand, the hydrophobic interactions between the SWCNT and the hydrophobic residues of the binding pocket, particularly those aromatic ones, are strong and nonspecific. This makes the SWCNT more straightforward to be adsorbed onto the binding pocket. Thus it takes only about 35 ns for the SWCNT to reach a stable binding state at the active site, while it takes ~130 ns for the ligand to come near the SH3 domain in the previous MD simulations [87]. This is also consistent with the above free energy landscape analysis, which reveals that the SWCNT is more favorable than the PRM ligand to be bound to the active site of the SH3 domain. That is, in the competition for the binding pocket of the SH3 domain, the SWCNT has advantages over the PRM ligand in both kinetics and thermodynamics. The SWCNT essentially occupies the binding pocket of the SH3 domain and interrupts its native binding with the PRM ligand.

2.3 Interaction with human blood serum proteins

All the above simulations were done with relatively short CNTs (~20 Å) and small proteins (SH3 domain has 57 amino acids and WW 26 amino acids) due to the enormous computational resources needed. In order to see whether longer SWCNTs and larger proteins might exhibit similar behaviors, we have recently performed simulations with IBM BlueGene supercomputers on much larger systems – the SWCNTs interacting with human blood serum proteins BSA (bovine serum albumin), γ-Ig (gamma-globulin), Tf (transferring), and BFg (fibrinogen), also in an attempt to compare with experiments directly. The total number of atoms in the solvated systems is up to 1.06 million. Figure 12 shows one example of the system setup with BFg dimmer interacting with SWCNT in solution. Since this is still an ongoing research, only a very brief description of the current results will be given. More simulation results and detailed comparisons with experiments will be reported elsewhere [93].

To elucidate the mechanism of protein adsorption by the SWCNT, up to 150 ns molecular dynamics simulations have been performed for each protein-SWCNT complex in explicit solvent. In our experiments, including AFM images, fluorescence spectroscopy, and SDS-PAGE, we found that these serum proteins display a competitive binding/adsorption with the SWCNTs, with BFg showing the largest adsorption capacity. In addition, the adsorption capacity displays a general trend of BFg > γ-Ig > Tf > BSA [93].

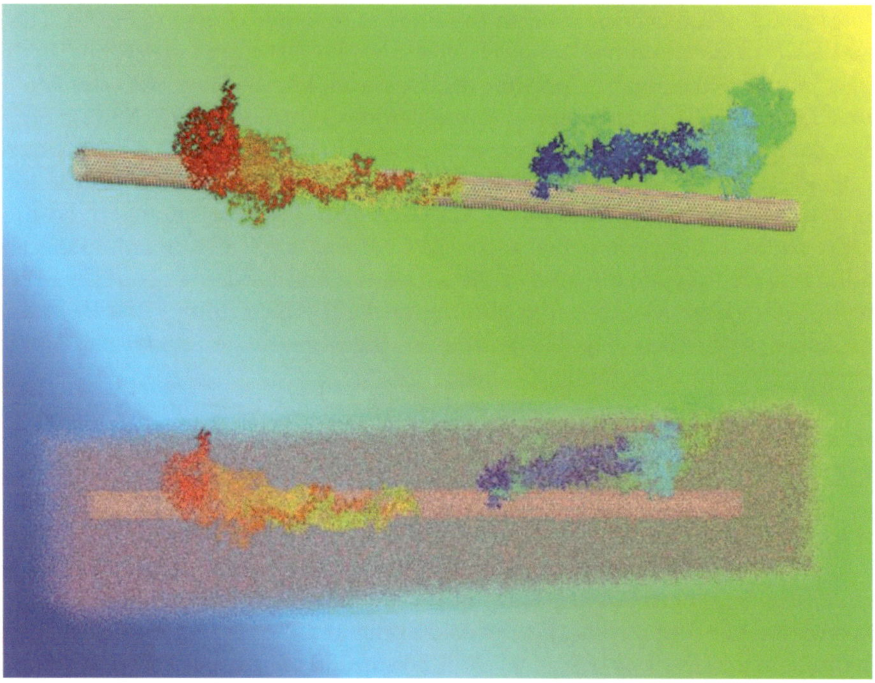

Fig. 12. Representative snapshot from the simulation trajectory of the human blood serum protein BFg interacting with a SWCNT. The solvated complex system has 1,058,598 atoms. The top panel shows the BFg-SWCNT complex alone, while the bottom panel also includes the solvation box. Here the protein BFg is shown in ribbons colored by chain. The SWCNT is shown in wheat sticks, and water in dots (with only O atoms shown).

As mentioned above, the aromatic residues, tryptophan (Trp), tyrosine (Tyr) and phenylalanine (Phe), could produce π-π stacking interactions between aromatic rings of amino acids and six-member rings of the SWCNT. As a first analysis, the numbers of these three kinds of amino-acid residues on the surface of initial conformation for each protein molecule were counted. The approximate numbers of (Trp, Tyr, Phe) for BFg, γ-Ig, Tf and BSA were in sequence of (14, 47, 27); (2, 25, 10); (1, 13, 15); and (1, 9, 10); respectively. This implies that the aromatic residues on the surface of these blood serum proteins particularly consist of Tyr and Phe, and lack of Trp, except for BFg which has 14 surface Trp residues. Most of the Trp residues are wrapped in the core of proteins, and play a lesser role in these serum protein-SWCNT complex adsorptions, which seems to be different from the above WW domain and SH3 domain cases where Trp residues play a significant role.

The snapshots of MD simulation results describe the preferred binding sites on proteins for SWCNT. If we highlight the residues within 5 Å distance from the surface of SWCNT, the aromatic residues Tyr and Phe stand out as the key residues involved in the adsorption. Interestingly, only Tyr and Phe were observed in the adsorption on the surface of SWCNT (at least in the first 150 ns simulations, longer runs might reveal larger protein conformational changes and exposure of Trp residues to the SWCNT surface as well), and Trp residues made little contribution for adsorption in the present MD simulations, which are still very short in time scale as compared to experiments (5 min to 1 hour). Nevertheless, our simulations show that there is a positive correlation between the total numbers of Trp,

Tyr, and Phe residues and the binding surface areas, thus the adsorption content, of the proteins, respectively.

In order to illuminate the above positive correlation relationship more clearly, we calculated the contact residue number and contact surface area of various proteins, in adsorption of SWCNT versus time. We used the average contact residue number (ACRN) of the protein to quantify its adsorption ability with SWCNT. The ACRN for BFg during 140 to 150 ns is 94. The ACRN for γ-Ig after its saturation during 70 to 150 ns is 28. The ACRN for Tf after its saturation during 80 to 150 ns is 13. The ACRN for BSA after its saturation during 65 to 150 ns is 12. The relationship of ACRN of various protein complex systems shows an order of BFg > γ-Ig > Tf > BSA, which is in good agreement with the experimental findings [93] (and also compatible with the above simple analysis using aromatic hydrophobic residues). Similarly, we computed the average contact surface area (ACSA) for these blood serum proteins (see Fig. 13). The ACSA for BFg, γ-Ig, Tf and BSA during the responding time ranges are 4360 Å^2, 1240 Å^2, 580 Å^2 and 470 Å^2 respectively, which again shows an order of BFg > γ-Ig > Tf > BSA, in good agreement with the experimental findings [93].

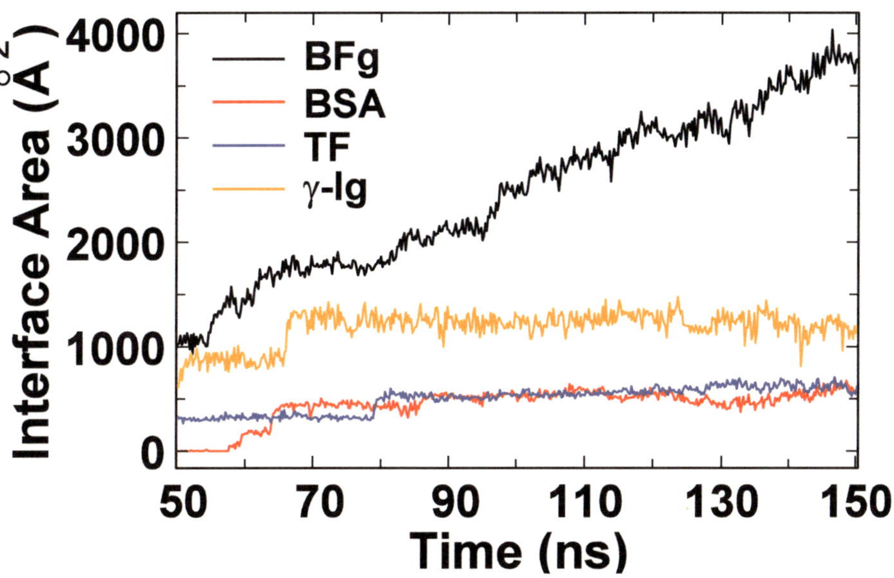

Fig. 13. The protein-SWCNT interface area as a function of time for various human blood serum proteins. Here the interface area is defined as half of the difference between the solvent accessible surface area of the complex and the sum of solvent accessible surface areas of the protein and the SWCNT.

The most abundant Tyr and Phe residues on the surface of BFg molecule can be attributed to the highest protein quantities binding to SWCNT. Interestingly, the π-cation interactions are also found to play a role, though smaller, in the BFg-SWCNT binding. The average Arg and Lys residues in contact with SWCNT (ACRN-Arg, ACRN-Lys) are found to be about ~1.3 and ~3.5, respectively (out of the total ~94 ACRN above). The other three proteins γ-Ig, Tf, and BSA show similar results, with the (ACRN-Arg, ACRN-Lys) having values of (0.0, 0.2), (1.4, 3.2), (0.1, 2.4), respectively. These findings are consistent with a previous report by Kagan and coauthors, where they found that the interactions between human

myeloperoxidase (hMPO) and the carboxylated nanotubes involved both arginine (π–π-cation interaction) and tyrosine (π–π interaction) residues [94]. Our current simulations show that the π–π stacking interactions are the driving forces for the competitive binding of human serum proteins onto SWCNTs.
Finally, it should be noted that carbon nanotubes often exist in much longer lengths in *in vitro* experiments as compared to our simulation lengths, which are typically a few to a few tens of nanometers. However, there are recent studies showing that the enzyme horseradish peroxidase (HRP) can biodegrade SWCNTs into very small fragments [95]. Therefore short SWCNTs may exist in the living body and studying SWCNTs with relatively short lengths might be of significant importance to human health as well.

3. Conclusion

We had investigated the effect of the CNTs on the conformation and function of proteins by using the large scale molecular dynamics simulations. WW domain, SH3 domains were used as examples to illustrate our ideas that the CNTs can affect the function of the proteins by either disrupting the structure of active site or shielding the active site from competing ligands. In the simulation of the SWCNT and the WW domain, it was found that the SWCNT can plug into the hydrophobic core of WW domain to form a stable protein-SWCNT complex. This insertion of the SWCNT broke the scaffold which used to bind the proline rich motifs (PRM), and thus reduced the possibility of the direct binding between the PRM and the WW domain. In the case of SH3 domain, we studied a three-way binding competition among the SWCNT, the PRM ligand, and the SH3 domain, and found that the SWCNT had a very high probability occupying the binding pocket of the ligand in the SH3 domain, with about 0.6 kcal/mol more favorable binding affinity than the original PRM ligand. Moreover, in most of the simulation trajectories, the ligand will be adsorbed onto the SWCNT, which may further reduce the possibility of the correct binding for the ligand onto the active site of the SH3 domain. These adverse effects of the SWCNTs on proteins, including both the disruption of the active sites and the competitive binding with the ligands, might seriously damage the original functions of proteins, suggesting the potential nanotoxicity of the SWCNT. The interactions between the SWCNTs and the hydrophobic residues, particularly the aromatic residues (π–π stacking interactions), are found to play a key role in both the insertion of the SWCNT and the competitive binding with ligands. Moreover, we also simulated much larger systems, with human serum proteins interacting with SWCNTs, and found a competitive binding among all these serum proteins, with the adsorption capacity in an order of BFg > γ-Ig > Tf > BSA, in good agreement with experimental findings. Again, the π–π stacking interactions are found to be the driving forces for the competitive binding of human serum proteins onto SWCNTs. These findings might have shed some light to the general mechanism of the interactions between hydrophobic nanoparticles and proteins.

4. Computational method

Modeling WW domain and SWCNT

The protein YAP65 WW domain was prepared from the Protein Data Bank (PDB code: 1JMQ, truncated to include residues 15-40), and modeled by AMBER03 force field [96]. The lengths of all those SWCNTs are 19.54 Å. The carbon atoms of the SWCNTs were modeled as uncharged Lennard–Jones particles with a cross-section of σ_{cc} = 3.40 Å, and a depth of the potential well of ε_{cc}= 0.36 kJ/mol [56, 57]. The interactions between these carbon atoms of SWCNT and other atoms were generated by the AMBER03 force field [96]. The combined systems were then solvated in a rhombic dodecahedral periodic box with the distance between the solutes and box boundary at least 8 Å. The numbers of water molecules were 2694, 2709 and 2735 for the system with the SWCNT of m = 4, 5, and 6, respectively. And a Cl^- was added into solution to neutralize the system.

Modeling SH3 domain, Ligand and SWCNT

The SH3 domain and the ligand were prepared from the Protein Data Bank (PDB code: 1CKB [85], residues 134-190 for the SH3 domain, and residue 1-8 for the ligand), and modeled by AMBER03 force field [96]. The carbon atoms of the SWCNT were also modeled as uncharged Lennard–Jones particles with a cross-section of σ_{cc} = 3.40 Å, and a depth of the potential well of ε_{cc} = 0.36 kJ/mol [56, 57]. Two three-way (SWCNT + ligand + protein) binding complex systems were set up. In the first system, the initial separations of the geometric centre of the SWCNT and the ligand from that of the SH3 domain were both set as 30 Å, a distance long enough to avoid the starting point bias. The initial orientations of the SWCNT and the RPM ligand versus the SH3 domain were set at different directions. The resulting complex was then solvated in a rhombic dodecahedral periodic box, with the distance between the solutes and the boundary of the box at least 10 Å. The final complex system size is 12,378 atoms. For the second complex system, a similar procedure was followed, but with the SWCNT and ligand swapped in space (thus in the opposite orientations when compared to the first system) for effective sampling. The final system size is 13,367 atoms. The TIP3P [97] water model was used for the salvation, and three Na^+ were added into solution to neutralize each system.

WW domain and SH3 domain MD simulations

The MD simulations were performed by using the Gromacs package 4.0 [98]. In the simulations, the covalent bonds involving H atoms were constrained by the LINCS algorithm [99], allowed a time step of 2 fs. The long range electrostatic interactions were treated with the particle-mesh Ewald method (PME) with a grid spacing of 1.2 Å [100]. The cutoff for the van der Waals interaction was set to 10 Å. After energy minimization, all the systems were equilibrated by MD simulations for 200 ps at a constant pressure of 1 bar and temperature of 298 K using Berendsen coupling [101]. Then the production simulations were performed in the NVT ensemble at 298 K.

Human serum proteins interacting with SWCNT

The human blood serum proteins simulated in this work were taken from the Protein Data Bank and Swiss-Model Repository as BFg with PDB ID: 1DEQ, γ-Ig with PDB ID: 3HR5, Tf with PDB ID: 2HAV, and BSA with SWISS-MODEL ID: 432779d395a52bfc9f6574bc3e98afcd_1. All SWCNTs used in the molecular dynamics (MD) simulations for the adsorption of CNT for proteins BFg, Ig, Tf, and BSA were armchair CNT (14, 14) with diameter as around 2.0 nm, which were consistent with the experimental ones. The geometrical coordinate parameters of CNT were generated by using Nanotube Modeler software (www.jcrystal.com/products/wincnt/).

MD simulations were performed by NAMD [102] of version 2.6. The various blood proteins were modeled by CHARMM 32b1 force field [103]. Carbon atoms of SWCNTs were assumed to be type CA with Lennard-Jones parameters $\varepsilon_{cc} = -0.0700\,\text{kcal/mol}$,

$R_{min}/2 = 1.9924$ Å. The interaction between these carbon atoms of SWCNTs and other atoms were generated by CHARMM 32b1 force field. The various blood proteins were well-separated from SWCNT with the minimum distances up to 40.0 Å initially. All systems were solvated in periodic TIP3P modeled water box, and neutralized by adding sodium and chloride ions with 0.2 M salt concentration as physiological condition, with system sizes up to 1,058,598 atoms (see Fig. 12 for one example of BFg).

Simulations were performed at constant temperature of 310 K, and pressure of 1 atm, with the PME method for long-range electrostatic interactions and time step of 2 fs. The cutoff for the Van der Waals interaction was set to 12 Å. All systems were simulated in the NPT ensemble for more than 150 ns. Visual Molecular Dynamics (VMD 1.8.7.) graphics viewer software was utilized to illustrate the adsorption of proteins and SWCNT in different representations.

5. Acknowledgement

We thank Dr. Peng Xiu, Dr. Chunlei Wang, Dr. Payel Das, Dr. Seung-gu Kang, and Prof. Bruce Berne for helpful discussions. This research is supported in part by grants from NNSFC (10825520), NBRPC (973 Program: 2007CB936000 and 2007CB814800), Shanghai Leading Academic Discipline Project (B111), and Shanghai Supercomputer Center of China. RZ acknowledges the support from the IBM BlueGene Science Program.

6. References

[1] Iijima, S., *Helical microtubules of graphitic carbon.* Nature, 1991. 354: p. 56-58.

[2] Li, W.Z., et al., *Large-scale synthesis of aligned carbon nanotubes.* Science, 1996. 274: p. 1701-1703.

[3] Prato, M., K. Kostarelos, and A. Bianco, *Functionalized carbon nanotubes in drug design and discovery.* Accounts of Chemical Research, 2008. 41(1): p. 60-68.

[4] Bhirde, A.A., et al., *Targeted killing of cancer cells in vivo and in vitro with EGF-directed carbon nanotube-based drug delivery.* ACS Nano, 2009. 3(2): p. 307-316.

[5] Thakare, V.S., et al., *Carbon nanotubes in cancer theragnosis.* Nanomedicine, 2010. 5(8): p. 1277-1301.

[6] Zanello, L.P., et al., *Bone cell proliferation on carbon nanotubes.* Nano Letters, 2006. 6(3): p. 562-567.

[7] Tu, X., et al., *DNA sequence motifs for structure-specific recognition and separation of carbon nanotubes.* Nature, 2009. 460(7252): p. 250-253.

[8] Liu, N., et al., *Carbon Nanotubes for Electrochemical and Electronic Biosensing Applications.* NanoScience in Biomedicine, 2009: p. 205-246.

[9] Thayer, A.M., *Carbon nanotubes by the metric ton.* Chemical & Engineering News, 2007. 85(46): p. 29-35.

[10] Service, R.F., *Is nanotechnology dangerous?* Science, 2000. 290: p. 1526-1527.

[11] Gilbert, N., *Nanoparticle safety in doubt.* Nature, 2009. 460: p. 937-937.

[12] Nel, A., et al., *Toxic potential of materials at the nanolevel.* Science, 2006. 311: p. 622-627.

[13] Zhao, Y., G. Xing, and Z. Chai, *Nanotoxicology: Are carbon nanotubes safe?* Nature Nanotechnology, 2008. 3(4): p. 191-192.

[14] Maynard, A.D., et al., *Safe handling of nanotechnology.* Nature, 2006. 444(7117): p. 267-269.

[15] Bi, S.P., J. Zhang, and J.J. Cheng, *Call from China for joint nanotech toxicity-testing effort.* Nature, 2009. 461(7264): p. 593-593.

[16] Donaldson, K. and C.A. Poland, *Nanotexicology: New insights into nanotubes.* Nature Nanotechnology, 2009. 4(11): p. 708-710.
[17] Buzea, C., I.I. Pacheco, and K. Robbie, *Nanomaterials and nanoparticles: Sources and toxicity.* Biointerphases, 2007. 2(4): p. MR17-MR71.
[18] Maynard, A.D., et al., *Exposure to carbon nanotube material: Aerosol release during the handling of unrefined single-walled carbon nanotube material.* Journal of Toxicology and Environmental Health-Part A, 2004. 67: p. 87-107.
[19] Ryman-Rasmussen, J.P., et al., *Inhaled carbon nanotubes reach the subpleural tissue in mice.* Nature Nanotechnology, 2009. 4: p. 747-751.
[20] Kolosnjaj, J., H. Szwarc, and F. Moussa, *Toxicity studies of carbon nanotubes.* Advances in Experimental Medicine and Biology, 2007. 620: p. 181-204.
[21] Porter, A.E., et al., *Direct imaging of single-walled carbon nanotubes in cells.* Nature Nanotechnology, 2007. 2: p. 713-717.
[22] Ma-Hock, L., et al., *Inhalation toxicity of multiwall carbon nanotubes in rats exposed for 3 months.* Toxicological Sciences, 2009. 112: p. 468-481.
[23] Schipper, M.L., et al., *A pilot toxicology study of single-walled carbon nanotubes in a small sample of mice.* Nature Nanotechnology, 2008. 3(4): p. 216-221.
[24] Shvedova, A.A., et al., *Unusual inflammatory and fibrogenic pulmonary responses to single-walled carbon nanotubes in mice.* American Journal of Physiology - Lung Cellular and Molecular Physiology, 2005. 289: p. L698-708.
[25] Mitchell, L.A., et al., *Mechanisms for how inhaled multiwalled carbon nanotubes suppress systemic immune function in mice.* Nature Nanotechnology, 2009. 4: p. 451-456.
[26] Li, Z., et al., *Cardiovascular effects of pulmonary exposure to single-wall carbon nanotubes.* Environmental Health Perspectives, 2007. 115: p. 377-382.
[27] Poland, C.A., et al., *Carbon nanotubes introduced into the abdominal cavity of mice show asbestos-like pathogenicity in a pilot study.* Nature Nanotechnology, 2008. 3: p. 423-428.
[28] Bai, Y.H., et al., *Repeated administrations of carbon nanotubes in male mice cause reversible testis damage without affecting fertility.* Nature Nanotechnology, 2010. 5(9): p. 683-689.
[29] Balavoine, F., et al., *Helical crystallization of proteins on carbon nanotubes: A first step towards the development of new biosensors.* Angewandte Chemie International Edition, 1999. 38(13-14): p. 1912-1915.
[30] Park, K.H., et al., *Single-walled carbon nanotubes are a new class of ion channel blockers.* Journal of Biological Chemistry, 2003. 278: p. 50212-50216.
[31] Karajanagi, S.S., et al., *Structure and function of enzymes adsorbed onto single-walled carbon nanotubes.* Langmuir, 2004. 20(26): p. 11594-11599.
[32] Goldberg-Oppenheimer, P. and O. Regev, *Exploring a nanotube dispersion mechanism with gold-labeled proteins via Cryo-TEM imaging.* Small, 2007. 3(11): p. 1894-1899.
[33] Wang, S., et al., *Peptides with selective affinity for carbon nanotubes.* Nature Materials, 2003. 2: p. 196-200.
[34] Zheng, L.F., D. Jain, and P. Burke, *Nanotube-peptide interactions on a silicon chip.* Journal of Physical Chemistry C, 2009. 113: p. 3978-3985.
[35] Cedervall, T., et al., *Understanding the nanoparticle–protein corona using methods to quantify exchange rates and affinities of proteins for nanoparticles.* Proceedings of the National Academy of Sciences of the United States of America, 2007. 104: p. 2050-2055.
[36] Klein, J., *Probing the interactions of proteins and nanoparticles.* Proceedings of the National Academy of Sciences of the United States of America, 2007. 104: p. 2029-2030.
[37] Rocker, C., et al., *A quantitative fluorescence study of protein monolayer formation on colloidal nanoparticles.* Nature Nanotechnology, 2009. 4: p. 577-580.

[38] Duan, Y. and P.A. Kollman, *Pathways to a protein folding intermediate observed in a 1-microsecond simulation in aqueous solution.* Science, 1998. 282: p. 740-744.
[39] Levy, Y., P.G. Wolynes, and J.N. Onuchic, *Protein topology determines binding mechanism.* Proceedings of the National Academy of Sciences of the United States of America, 2004. 101: p. 511-516.
[40] Snow, C.D., et al., *Absolute comparison of simulated and experimental protein-folding dynamics.* Nature, 2002. 420(6911): p. 102-106.
[41] Zhou, R.H., et al., *Hydrophobic collapse in multidomain protein folding.* Science, 2004. 305(5690): p. 1605-1609.
[42] Liu, P., et al., *Observation of a dewetting transition in the collapse of the melittin tetramer.* Nature, 2005. 437(7055): p. 159-162.
[43] Zuo, G., J. Hu, and H. Fang, *Effect of the ordered water on protein folding: an off-lattice Gō-like model study.* Physical Review E (Statistical, Nonlinear, and Soft Matter Physics), 2009. 79(3 Pt 1): p. 031925.
[44] Mayor, U., et al., *The complete folding pathway of a protein from nanoseconds to microseconds.* Nature, 2003. 421: p. 863-867.
[45] Mirny, L. and E. Shakhnovich, *Protein folding theory: From lattice to all-atom models.* Annual Review of Biophysics and Biomolecular Structure, 2001. 30: p. 361-396.
[46] Garcia, A.E. and D. Paschek, *Simulation of the pressure and temperature folding/unfolding equilibrium of a small RNA hairpin.* Journal of the American Chemical Society, 2008. 130: p. 815-816.
[47] Miyashita, N., J.E. Straub, and D. Thirumalai, *Structures of beta-Amyloid Peptide 1-40, 1-42, and 1-55-the 672-726 Fragment of APP-in a Membrane Environment with Implications for Interactions with gamma-Secretase.* Journal of the American Chemical Society, 2009. 131: p. 17843-17852.
[48] Hua, L., et al., *Dynamics of water confined in the interdomain region of a multidomain protein.* Journal of Physical Chemistry B, 2006. 110: p. 3704-3711.
[49] Liu, P., et al., *Hydrophobic aided replica exchange: an efficient algorithm for protein folding in explicit solvent.* Journal of Physical Chemistry B, 2006. 110: p. 19018-19022.
[50] Zhou, R., *Exploring the protein folding free energy landscape: coupling replica exchange method with P3ME/RESPA algorithm.* Journal of Molecular Graphics & Modelling 2004. 22: p. 451-463.
[51] Zhou, R.H., et al., *Destruction of long-range interactions by a single mutation in lysozyme.* Proceedings of the National Academy of Sciences of the United States of America, 2007. 104(14): p. 5824-5829.
[52] Gao, Y.Q., W. Yang, and M. Karplus, *A structure-based model for the synthesis and hydrolysis of ATP by F1-ATPase.* Cell, 2005. 123: p. 195-205.
[53] Karplus, M., et al., *Protein structural transitions and their functional role.* Philos Transact A Math Phys Eng Sci, 2005. 363: p. 331-355; discussion 355-356.
[54] Roitberg, A.E., A. Okur, and C. Simmerling, *Coupling of replica exchange simulations to a non-Boltzmann structure reservoir.* Journal of Physical Chemistry B, 2007. 111: p. 2415-2418.
[55] Li, W.F., et al., *Metal-coupled folding of cys2his2 zinc-finger.* Journal of the American Chemical Society, 2008. 130(3): p. 892-900.
[56] Gong, X.J., et al., *A charge-driven molecular water pump.* Nature Nanotechnology, 2007. 2: p. 709-712.
[57] Hummer, G., J.C. Rasaiah, and J.P. Noworyta, *Water conduction through the hydrophobic channel of a carbon nanotube.* Nature, 2001. 414: p. 188-190.
[58] Tu, Y.S., et al., *Water-mediated signal multiplication with Y-shaped carbon nanotubes.* Proceedings of the National Academy of Sciences of the United States of America, 2009. 106: p. 18120-18124.

[59] Li, J.Y., et al., *Electrostatic gating of a nanometer water channel.* Proceedings of the National Academy of Sciences of the United States of America, 2007. 104: p. 3687-3692.
[60] Giovambattista, N., et al., *Hydrophobicity of protein surfaces: Separating geometry from chemistry.* Proceedings of the National Academy of Sciences of the United States of America, 2008. 105: p. 2274-2279.
[61] Shen, J.-W., et al., *Induced stepwise conformational change of human serum albumin on carbon nanotube surfaces.* Biomaterials, 2008. 29(28): p. 3847-3855.
[62] Zuo, G.H., et al., *Plugging into proteins: Poisoning protein function by a hydrophobic nanoparticle.* ACS Nano, 2010. 4(12): p. 7508-7514.
[63] Chen, H.I. and M. Sudol, *The WW domain of Yes-associated protein binds a proline-rich ligand that differs from the consensus established for Src homology 3-binding modules.* Proceedings of the National Academy of Sciences of the United States of America, 1995. 92(17): p. 7819–7823.
[64] Garrus, J.E., et al., *Tsg101 and the vacuolar protein sorting pathway are essential for HIV-1 budding.* Cell, 2001. 107: p. 55-65.
[65] Lu, P.J., et al., *The prolyl isomerase Pin1 restores the function of Alzheimer-associated phosphorylated tau protein.* Nature, 1999. 399(6738): p. 784-788.
[66] Sudol, M., et al., *Characterization of the mammalian YAP (yes-associated protein) gene and Its role in defining a novel protein module, the WW domain.* Journal of Biological Chemistry, 1995. 270(24): p. 14733-14741.
[67] Zheng, H.W., et al., *The prolyl isomerase Pin1 is a regulator of p53 in genotoxic response.* Nature, 2002. 419(6909): p. 849-853.
[68] Macias, M.J., et al., *Structural analysis of WW domains and design of a WW prototype.* Nature Structural Molecular Biology, 2000. 7(5): p. 375-379.
[69] Huang, X., et al., *Structure of a WW domain containing fragment of dystrophin in complex with beta-dystroglycan.* Nature Structural Biology, 2000. 7(8): p. 634-638.
[70] Sudol, M., *Structure and function of the WW domain.* Progress in Biophysics and Molecular Biology, 1996. 65(1-2): p. 113-132.
[71] Macias, M.J., et al., *Structure of the WW domain of a kinase-associated protein complexed with a proline-rich peptide.* Nature, 1996. 382: p. 646-649.
[72] Dill, K., *Dominant forces in protein folding.* Biochemistry, 1990. 29: p. 7133-7155.
[73] Nadassy, K., S.J. Wodak, and J. Janin, *Structural features of protein–nucleic acid recognition sites.* Biochemistry, 1999. 38: p. 1999-2017.
[74] Kabsch, W. and C. Sander, *Dictionary of protein secondary structure: pattern recognition of hydrogen-bonded and geometrical features.* Biopolymers, 1983. 22: p. 2577-2637.
[75] Braden, B.C., et al., *X-ray crystal structure of an anti-Buckminsterfullerene antibody fab fragment: biomolecular recognition of C(60).* Proceedings of the National Academy of Sciences of the United States of America, 2000. 97: p. 12193-12197.
[76] Noon, W.H., Y. Kong, and J. Ma, *Molecular dynamics analysis of a buckyball-antibody complex.* Proceedings of the National Academy of Sciences of the United States of America, 2002. 99 Suppl 2: p. 6466-6470.
[77] Venter, J.C., et al., *The sequence of the human genome.* Science, 2001. 291(5507): p. 1304-1351.
[78] Ball, L.J., et al., *Recognition of proline-rich motifs by protein-protein-interaction domains.* Angewandte Chemie International Edition, 2005. 44(19): p. 2852-2869.
[79] Macias, M.J., S. Wiesner, and M. Sudol, *WW and SH3 domains, two different scaffolds to recognize proline-rich ligands.* FEBS Letters, 2002. 513(1): p. 30-37.
[80] Bucciantini, M., et al., *Inherent toxicity of aggregates implies a common mechanism for protein misfolding diseases.* Nature, 2002. 416(6880): p. 507-511.
[81] Gorina, S. and N.P. Pavletich, *Structure of the p53 tumor suppressor bound to the ankyrin and SH3 domains of 53BP2.* Science, 1996. 274(5289): p. 1001-1005.

[82] Zarrinpar, A., R.P. Bhattacharyya, and W.A. Lim, *The structure and function of proline recognition domains.* Sci STKE, 2003. 2003(179): p. RE8.

[83] Li, X.F., et al., *Structural basis of robo proline-rich motif recognition by the srGAP1 Src homology 3 domain in the slit-robo signaling pathway.* Journal of Biological Chemistry, 2006. 281(38): p. 28430-28437.

[84] Kobashigawa, Y., et al., *Structural basis for the transforming activity of human cancer-related signaling adaptor protein CRK.* Nature Structural & Molecular Biology, 2007. 14(6): p. 503-510.

[85] Wu, X.D., et al., *Structural basis for the specific interaction of lysine-containing proline-rich peptides with the N-terminal SH3 domain of c-Crk.* Structure, 1995. 3: p. 215-226.

[86] Nguyen, J.T., et al., *Exploiting the basis of proline recognition by SH3 and WW domains: Design of n-substituted inhibitors.* Science, 1998. 282: p. 2088-2092.

[87] Ahmad, M., W. Gu, and V. Helms, *Mechanism of fast peptide recognition by SH3 domains13.* Angewandte Chemie International Edition, 2008. 47(40): p. 7626-7630.

[88] Freedman, D., R. Pisani, and R. Purves, *Statistics.* 2007: W.W. Norton & Co.

[89] Brown, S., T.S. Jespersen, and J. Nygard, *A genetic analysis of carbon-nanotube-binding proteins.* Small, 2008. 4(4): p. 416-420.

[90] Tomasio, S.D. and T.R. Walsh, *Atomistic modelling of the interaction between peptides and carbon nanotubes.* Molecular Physics, 2007. 105(2-3): p. 221-229.

[91] Selzer, T., S. Albeck, and G. Schreiber, *Rational design of faster associating and tighter binding protein complexes.* Nature Structural Biology, 2000. 7(7): p. 537-541.

[92] Schreiber, G. and A.R. Fersht, *Rapid, electrostatically assisted association of proteins.* Nature Structural Biology, 1996. 3(5): p. 427-431.

[93] Ge, C., et al., *π–π stacking interaction driven competitive adsorption of human blood proteins on single-wall carbon nanotubes and its implications for resulting cytotoxicity.* Nature Nanotechnology, 2011: p. submitted.

[94] Kagan, V.E., et al., *Carbon nanotubes degraded by neutrophil myeloperoxidase induce less pulmonary inflammation.* Nature Nanotechnology, 2010. 5(5): p. 354-359.

[95] Allen, B.L., et al., *Biodegradation of single-walled carbon nanotubes through enzymatic catalysis.* Nano Letters, 2008. 8: p. 3899-2903.

[96] Duan, Y., et al., *A point-charge force field for molecular mechanics simulations of proteins based on condensed-phase quantum mechanical calculations.* Journal of Computational Chemistry, 2003. 24: p. 1999-2012.

[97] Jorgensen, W.L., et al., *Comparison of Simple Potential Functions for Simulating Liquid Water.* Journal of Chemical Physics, 1983. 79(2): p. 926-935.

[98] Hess, B., et al., *GROMACS 4: Algorithms for highly efficient, load-balanced, and scalable molecular simulation.* Journal of Chemical Theory and Computation, 2008. 4: p. 435-447.

[99] Hess, B., *P-LINCS: A parallel linear constraint solver for molecular simulation.* Journal of Chemical Theory and Computation, 2008. 4(1): p. 116-122.

[100] Darden, T., D. York, and L. Petersen, *Particle mesh Ewald: an N log(N) method for Ewald sums in large systems.* The Journal of Chemical Physics, 1993. 98: p. 10089–10092.

[101] Berendsen, H.J.C., et al., *Molecular dynamics with coupling to an external bath.* The Journal of Chemical Physcis, 1984. 81(8): p. 3684–3690.

[102] Phillips, J.C., et al., *Scalable molecular dynamics with NAMD.* Journal of Computational Chemistry, 2005. 26(16): p. 1781-1802.

[103] Feller, S.E. and A.D. MacKerell Jr, *An improved empirical potential energy function for molecular simulations of phospholipids.* Journal of Physical Chemistry B, 2000. 104(31): p. 7510-7515.

Part 5

CNTs for Energy Related Applications

24

Carbon Nanotubes Supported Metal Nanoparticles for the Applications in Proton Exchange Membrane Fuel Cells (PEMFCs)

Zhongqing Jiang[1] and Zhong-Jie Jiang[2]
[1]*Department of Chemical Engineering, Ningbo University of Technology, Ningbo, Zhejiang,*
[2]*Department of Nature and Sciences, University of California, California,*
[1]*China*
[2]*USA*

1. Introduction

Proton exchange membrane fuel cells (PEMFCs), as an important alternative energy source with great potentials for use in applications ranging from cellular phones, laptops, digital camera and other conventionally battery driven devices to long-term stationary monitoring electronics, have been extensively investigated (K. Matsumoto et al., 2009; P. Xing et al., 2004). In a PEMFC, the conversion of chemical energy to electrical energy occurs via a direct electrochemical reaction, and its efficiency is directly dependent upon the catalysts used (X. Wang et al., 2004). To be commercially viable, it is generally required that these catalysts have high durability, low cost and higher activities in oxygen reduction and/or fuel oxidation reaction (Kundu et al., 2009). Currently, the most widely used catalysts in the PEMFCs are metal nanoparticles (NPs), mainly Pt and/or Pt based alloys (W.Z. Li et al., 2002, 2003, 2004; Prabhuram et al., 2006; X. Wang et al., 2005). These metal NPs are usually characterized with high activities in oxygen reduction and/or fuel oxidation reaction due to their high surface to volume ratio and improved Fermi levels for redox reactions (W.Z. Li et al., 2006; S.Y. Wang et al., 2009; B.H. Wu et al., 2009; Y. Zhao et al., 2007). However, individual metal NPs are usually unstable and prone to loss of their catalytic activity due to their irreversible aggregation during the electrochemical processes. For practical uses, therefore, specific supports are mostly used to mobilize and prevent these metal NPs from aggregation.

Among various types of supports, carbon nanotubes (CNTs) are undoubtedly the most widely used. CNTs are allotropes of carbon with a cylindrical nanostructure, and are characterized as elongated fullerenes with diameters ranging from 1-100 nm (Wunderlich, 2007) and lengths of up to several microns. They can be single walled (called as single walled CNTs, SWCNTs) or multiple walled (called as multiple walled CNTs, MWCNTs) (Cassell et al., 1999; Iijima, 1991; Ijima & Ichihashi, 1993; Journet et al., 1997; Thess et al., 1996). These cylindrical carbon molecules have unique properties, such as high-surface area, good electronic conductivity, strong mechanical properties and high-chemical stability, which make them potentially useful in many applications in nanotechnology, electronics, optics, and other fields of materials science, as well as potential uses in architectural fields (Baughman et al., 2002; Pham-Huu et al., 2002). Studies have shown that the deposition of metal NPs onto the surface of CNTs can not only protect these particles from aggregation, but also improve their catalytic activities, and even produce properties that are not accessible to CNTs and metal NPs alone, which are important for their use in PEMFCs (W.Z. Li et al., 2006; S.Y. Wang et al., 2009; B.H. Wu et al., 2009; Y. Zhao et al., 2007). However, due to the inertness of the CNT walls, the effective attachment of metal catalysts remains a challenge, especially for the solution-based methods for the preparation of metal deposited CNTs (metal/CNTs) (Balasubramanian & Burghard, 2005). Surface functionalization of CNTs is, therefore, generally required to enable a homogeneous and uniform deposition of metal NPs (J. Chen et al., 1998; Yu et al., 1998). It is, however, demonstrated that the surface functionalization methods have great influence on the structure and properties of CNTs. In some cases, harsh chemical or electrochemical oxidations applied with concentrated strong acid could lead to a structural destruction to CNTs, resulting in decrease in their electrical conductivity and correspondingly a possible reduction in the catalytic activity of the obtained metal/CNTs when used in PEMFCs (Koshio et al., 2001; J. Li et al., 1998; Qu et al., 2005, 2006). For rational design of metal/CNTs catalysts, it is therefore important to well understand the various methods used for the CNT functionalization.

Except for the methods used for the CNT functionalization, the catalytic activity of the metal/CNTs is also affected by the size and distribution of deposited metal NPs. Since the dispersion and particle size of metal NPs largely determine the utilization and catalytic activity of metal/CNTs, the synthesis of metal NPs supported by CNTs with a controlled manner is of fundamental and practical importance. Indeed, researches have demonstrated that the deposition, distribution, and crystalline size of metal NPs supported on CNTs are significantly dependent upon the method used to synthesize metal/CNTs, the types of functional groups on the surface of metal NPs, and the way that metal NPs are adsorbed. It is therefore necessary to well know about the various methods used to synthesize metal/CNTs for the preparation of catalysts of high efficiency.

In the chapter, we will first provide brief recapitulations of the concepts of various surface functionalization methods of CNTs, and some possible shortcomings associated with these methods. This is followed by descriptions of the various methods used for the preparation of metal/CNTs, and the way that the metal ions and metal NPs are adsorbed onto CNTs is also elucidated. For the activity validation of the synthesized catalysts, it is essential to directly use them in fuel cells. It is demonstrated that the performance of the catalysts in the fuel cell is also affected by the methods for synthesis of membrane electrode assembly (MEA), which is the core of a fuel cell. Thus, the activity validation of the synthesized catalysts and the methods used for the synthesis of MEAs are also described and discussed in this chapter.

2. Methods for functionalization of CNTs

Over the years, a great deal of research has been conducted on the surface modification of CNTs (Hirsch, 2002; Y. Lin et al., 2004; Tasis et al., 2006). The modification of these quasi one-dimensional structures can be carried out by the covalent attachment of chemical groups through reactions onto the π-conjugated skeleton of CNT or by the noncovalent adsorption or wrapping of various functional molecules (Saha & Kundu, 2010). Covalent surface modification of the CNTs leads to a permanent change to the CNT surface. In this case, the CNTs are functionalized with reactive groups which can later form covalent bonds with another molecule (Hirsch, 2002; Tasis et al., 2006). Non-covalent surface modification does not involve the formal chemical bond formation between a molecule and the surface of CNT. The functionalization of CNTs is through adsorption of functional molecules via van der Waals forces, electrostatic forces, hydrogen bonding, or other attractive forces (Y. Lin et al., 2004). Studies have shown that the stability and catalytic activity of metal/CNT composites are strongly dependent upon the way that the CNTs are functionalized. In this section, various CNT functionalization methods based on covalent and noncovalent surface modifications will be discussed. Specifically, a novel CNT functionalization method based on a plasma treatment is also presented. The plasma surface modification is a newly reported method for the CNT functionalization, which is characterized to be a mild surface modification approach and effectively prevent CNTs from the structural destructions possibly caused by other surface functionalization methods (Jiang et al., 2009, 2011). The obtained metal/CNTs are reported to be with higher catalytic activity in a methanol oxidation. Therefore the plasma surface functionalization method has great potentials for the preparation of metal/CNTs of high efficiency as catalysts. Additionally, nitrogen dopped CNTs (N-CNTs) which show great promises as supports of metal NPs for the PEMFC applications, are also introduced. Due to the presence of N, the N-CNTs are reported to exhibit a strong binding to metals NPs. It can therefore avoid using functionalization processes that might be detrimental to the catalytic properties of the obtained metal/N-CNTs (Maiyalagan et al., 2005).

2.1 Covalent functionalization

2.1.1 Oxidative treatment of CNT surfaces

Among various surface functionalization methods, oxidative treatment of CNT surfaces is probably the most widely employed, which has been used to remove amorphous carbon for purification purposes, to open CNT ends for metal NP insertion and to create functional groups on their surface favoring the anchorage of metallic NPs. It can be performed by mixing CNTs with different oxidants such as HNO_3, H_2SO_4, $KMnO_4$, OsO_4, HNO_3/H_2SO_4, RuO_4, $K_2Cr_2O_7$, polyphosphoric acid, citric acid, ozone gas and H_2O_2 (Hernadi et al., 2001; Hirsch et al., 2002; Hwang et al., 1995; Kannan et al., 2008; Kyotani et al., 2001; Mawhinney et al., 2000a, 2000b; Y.C. Xing et al., 2004; Yu et al., 1998; J. Zhang et al., 2003; Zheng et al., 2006). The oxidized CNTs are usually functionalized with hydroxyl (-OH), carboxyl (-COOH), carbonyl (-C=O) and sulfate (-OSO_3H) groups, which can serve as metal-anchoring sites facilitating metal nuclei formation and subsequent particle growth (Georgakilas et al., 2002; Kongkanand et al., 2006a; Michelson et al., 2001).

Oxidation with nitric acid solutions is a simple, effective and commonly used approach to functionalize CNTs, which can lead to the formation of CNTs functionalized with carboxylic acid functions, as well as of lactones, phenols, carbonyls, anhydrides, ethers and quinones (Bambagioni et al., 2009). The oxidation of CNTs occurs primarily on the CH_n groups (MWCNT defects), giving rise to the formation of alcohols -OH, then C=O and finally carboxylic acid groups, and the density of the functional groups and subsequently deposited metal NPs are strongly dependent upon the concentration of HNO_3 used to treat the CNTs (Bambagioni et al., 2009; Reddy & Ramaprabhu, 2007). Since the oxidation with HNO_3 solutions can provide CNTs with a large amount of anchoring sites facilitating the deposition of metal NPs of smaller size with homogeneous dispersion, the subsequent fabricated metal/CNT composites usually exhibit a high efficiency in the PEMFC application. As reported by Han et al. (K.I. Han et al., 2004), the electrocatalysts supported on CNTs functionalized by the HNO_3 treatment showed improved activity over a commercially available electrocatalyst, E-TEK.

For a better functionalization of CNTs, some oxidants mixtures are also employed to functionalize CNTs with suitable surface for the deposition of metal NPs. For example, Wei et al. (Wei et al., 2008) used a H_2SO_4/H_2O_2 solution to functionalize CNTs. The functionalized MWCNTs were characterized to be terminated with carboxyl groups. Halder et al. (Halder et al., 2009) tried a surface treatment of CNTs by its mixing with a combination of concentrated HNO_3 and H_2SO_4 which gave very good surface functionalization on the wall of CNTs. Liu et al. (Liu et al., 2002) reported a high density of oxygen containing species on the CNT surface by a $K_2Cr_2O_7/H_2SO_4$ oxidative treatment.

In general, the oxidative treatment technique can functionalize CNTs with oxygen-containing functional groups on their walls, which could increase the surface binding sites of CNTs, avoid the aggregation of the subsequent deposited metal NPs, improve the dispersion of metal NPs, and reduce the average size of metal NPs deposited. The surface functional groups (e.g., carboxyl, hydroxyl, carbonyl groups) on the oxidized CNTs are mostly concentrated at defects sites or at the end tips of CNTs, where the strain and/or the chemical reactivity are higher. However, such functionalization method is inevitably accompanied with a few problems, such as uneven distribution of the surface functional groups, structural damage, and thus partial loss in electrical conductivity of the CNTs. Additionally, due to the hydrophobic surface of CNTs which makes them tend to form aggregates in polar solvents, the surface oxidation of the CNTs is mostly incomplete. That is because during the functionalization process, the CNTs inside these aggregates may not be attacked by the oxidative agents but remain unmodified. However, to use CNTs as a heterogeneous catalyst support, the entire surface of CNTs needs to be oxidized for functionalization, so that highly dispersed catalysts could be achieved. Although prolonged acid oxidation at higher temperatures could lead to an improvement in the quality of CNT functionalization, this might result in more oxidative damage on the graphene structures, leading to potentially more severe loss in the electric conductivity of the carbon nanomaterials. Therefore, in an effort to prepare highly dispersed, high-loading Pt NPs on CNTs, an effective method of CNT functionalization should be sought.

2.1.2 Photochemical oxidation of CNT surfaces

Compared to the oxidation technique mentioned above, the functionalization of CNTs by a photochemical oxidation of surfaces is a more facile and eco-friendly surface functionalization method. The reaction can be conducted in a gas phase and dry process with zero-emission of liquid wastes, providing CNTs with a large amount of carbonyl and carboxyl groups in a very short period of time. Its high efficiency and adjustability in the CNT functionalization provides an additional advantage to control the chemical and physical properties of CNTs. As reported by Asano et al. (Asano et al., 2006), the functionalization of the CNTs by the photochemical oxidation with a short-wavelength ultraviolet irradiation could produce the CNT surface with a high density of oxygen-containing functional groups. An enhancement of the active surface area and the performance of methanol oxidation for the Pt NPs deposited on the photochemical oxidized CNTs, which was attributed to the high dispersion and dense deposition of Pt NPs on the oxygen groups-rich surface, was demonstrated by Jang et al (Jang et al., 2009).

2.1.3 Sonochemical treatment

Sonochemical treatment is reported to be a relatively mild surface modification technique, which can alleviate the damage of CNTs possibly caused by the higher temperature oxidation to a certain extent. During a sonochemical surface modification process, the ultrasonic waves can produce microscopic bubbles in the liquid. A collapsing of these microscopic bubbles results in shock waves, which produce dangling bonds on the surface of CNTs that undergo further chemical reactions and provide the oxidative power to incorporate acidic sites. It is found that the sonochemical treatment of CNTs under acidic aqueous solution (HNO_3 and/or H_2SO_4) conditions can provide CNTs functionalized with –C=O, -C-O-C-, -COO-, and –C-OH groups, which is important for the deposition of uniformly dispersed Pt NPs. The ability to produce CNTs with high densities of functional groups for high loading deposition of metal NPs on CNTs using the sonochemical technique makes it a promising way for the CNT functionalization. Reddy and Ramaprabhu (Reddy & Ramaprabhu, 2007) functionalized the purified SWCNTs by an ultrasonication of CNTs in concentrated nitric acid. The lower power was used to reduce the damage to CNTs during the ultrasonication. The treated CNTs with less structural damage were characterized to be functionalized with high concentrations of -OH and -COOH (Rajalakshmi et al., 2005). Xing et al. (Y.C. Xing et al., 2004, 2005) have shown that the Pt NPs deposited on sonochemically treated CNTs exhibited a much higher catalytic activity than those supported on the carbon black when used in the PEMFCs. This enhancement of electrochemical activity is attributed to the unique structures of CNTs and the strong interactions between the Pt NPs and the CNT support (Y.C. Xing, 2005).

2.1.4 Silane-assisted method

Several recent reports show that CNTs can be chemically functionalized by silane coupling agents (Kamavarama et al., 2009; Ma et al., 2006; X. Sun et al., 2003; Villers et al., 2006). In the silane assisted functionalization approach, CNTs are mixed with a solution containing a silane derivative and water in ethanol. Upon hydrolysis, the silane derivative form an acid silicate on the surface of CNTs, permitting the exchange of H^+ by the metal ions for the subsequent deposition of metal NPs by a reduction of the adsorbed metal ions. For example, Sun et al. (X. Sun et al., 2003) used a silane derivative to functionalize the CNT surface with -SO_3H group for the deposition of Pt NPs. The deposition of the Pt NPs was carried out by immersing the CNTs in a solution containing $PtCl_2$, a silane derivative and water in ethanol (X. Sun et al., 2003), which produced a Pt^{2+} adsorbed CNTs. The CNT supported Pt NPs were formed in a flow of H_2 and Ar. Fig. 1 shows the transmission electron microscopy (TEM) images of Pt NPs deposited on CNTs in the absence and the presence of the silane precursor. It shows that the functionalization with silane derivative could facilitate the uniform deposition of Pt on the CNT's surface, producing well dispersed Pt particles with a smaller size.

A major drawback of the silane-assisted functionalization method is that the electrocatalytic activity of the obtained metal/CNTs catalysts is improved not as largely as it is expected. In this method, an electrochemical insulating layer, organosilane, is inserted between metal and CNTs and thus decrease the interactions between them, which is unforvable for the improvement of the activity of the obtained catalysts. As reported by Ma et al. (Ma et al., 2006), Pt NPs deposited on a silane modified CNTs showed only slightly better electrocatalytic activity in the PEMFC than the commercial electrodes.

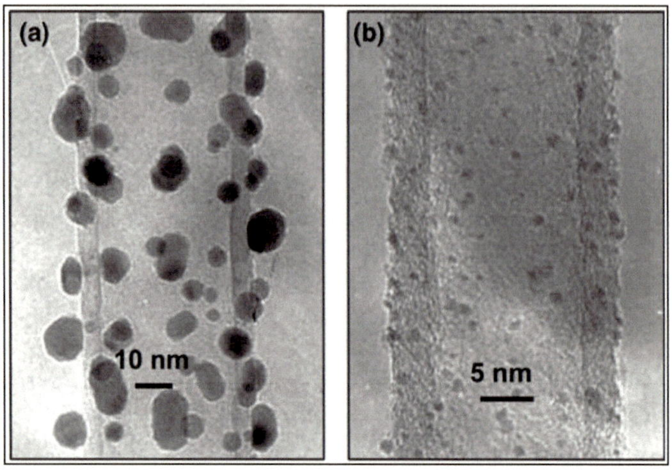

Fig. 1. TEM images of Pt NPs deposited on CNTs (a) in the absence and (b) in the presence of the silane precursor. Reprinted with permission from X. Sun et al., 2003. Copyright 2003 Elsevier Science Ltd.

2.1.5 Ionic liquids treatment

Ionic liquids (ILs) represent a group of solvents that consist only of ions existing in the liquid state at low temperatures. Due to their high thermal and chemical stability, high ionic conductivity, wide electrochemical windows, and negligible vapor pressure, ILs are expected to be superior solvents for many chemical processes (Parvulescu & Hardacre, 2007; Welton, 1999) and therefore attract considerable technological and scientific interests in recent years. Advanced progress in the development of catalysts for the PEMFCs applications indicates that an introduction of ILs into the reaction systems for preparing Pt/CNT composites provides a possible way to obtain composites with excellent catalytic and electrocatalytic performance (Park et al., 2009; B.H. Wu et al., 2009).

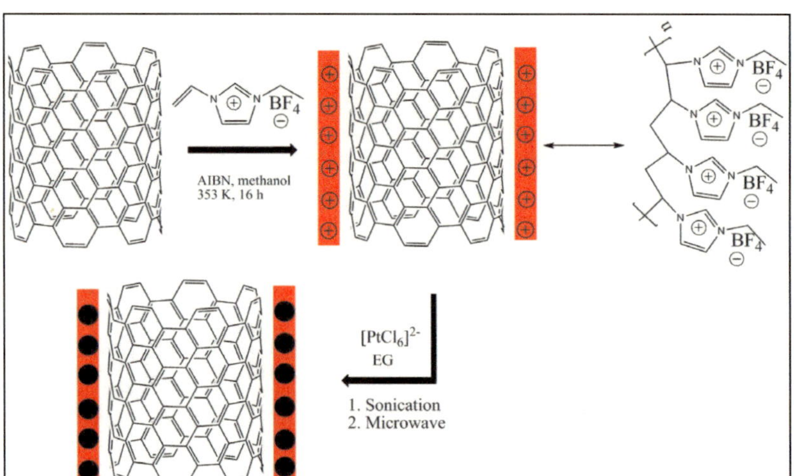

Fig. 2. Schematic of the modification of CNTs with PIL and the preparation of Pt/PIL-CNTs nanohybrids. EG: ethylene glycol, AIBN: 2,2′-azobisisobutyronitrile.

In an IL based method for the CNT surface functionalization, the CNTs are usually functionalized first to produce a suitable surface for the grafting of the IL molecules. As reported by Zhao et al. (Z.W. Zhao et al., 2006), MWCNTs used for the surface modification were pretreated in concentrated HNO_3 before functionalized with ILs, 1-ethyl-3-methylimidazolium bis(trifluoromethylsulfonyl) imide and 1-butyl-3-methylimidazolium bis (trifluoromethylsulfonyl) imide. It was reported that the functionalization with the ILs made a big contribution to the formation of small, homogeneous Pt NPs and to the suppression of agglomeration of CNTs. The deposited Pt NPs exhibited an electrochemical surface area with 21% higher than the commercial Pt/C. For the better deposition of metal NPs, Wu et al. (B.H. Wu et al., 2009) reported a novel method based on the thermal-initiation free radical polymerization of the IL monomer, 3-ethyl-1-vinylimidazolium tetrafluoroborate ([VEIM]BF_4), to form an ionic-liquid polymer (PIL) on the CNT surface, which introduced a large number of surface functional groups on the CNTs with uniform distribution to anchor and grow metal NPs (Fig. 2). The process of modification by PIL would lead to less structural damage to CNTs than the typical acid-oxidation treatment because of the mild polymerization of the IL monomer. The subsequently deposited Pt and PtRu NPs therefore exhibited a smaller particle size, a better dispersion, a higher electrochemical active area and correspondingly better performance in the direct electrooxidation of methanol than those on the CNTs without the PIL modification.

The disadvantage of the IL based functionalized method is that this approach requires an initial step of the CNT functionalization for the subsequent IL or PIL functionalization. It does not only increase the complexity of process, but possibly causes a structural damage to CNTs although it is alleviated compared to oxidative treatment techniques.

2.1.6 Electrochemical modification

The surface modification of CNTs by an electrochemical method is an attractive approach for functionalization of CNTs. In comparison to other functionalization methods, the electrochemical modification of CNTs can produce CNTs with a C-C covalent bond, which is strong and suitable as a substrate for the deposition of NPs, provides a uniform functional surface, which can effectively prevent the undesired nucleation processes on the CNTs surface, and facilitate the formation of metal NPs with a narrower size distribution due to the specific electrostatic interaction between the substrate and the adsorbed metals. The surface functionalization of CNTs by an electrochemical coupling of aromatic diazonium salts and phenyl residues have been reported (Bahr et al., 2001; Balasubramanian et al., 2003). For example, Guo and co-workers (Guo & H.-L. Li, 2004) covalently functionalized CNTs by grafting an ordered 4-aminobenzene monolayer onto the CNT surface via electro-reduction of 4-nitrobenzenediazonium tetrafluoroborate through cyclic voltammetry (CV), as shown in Fig. 3.

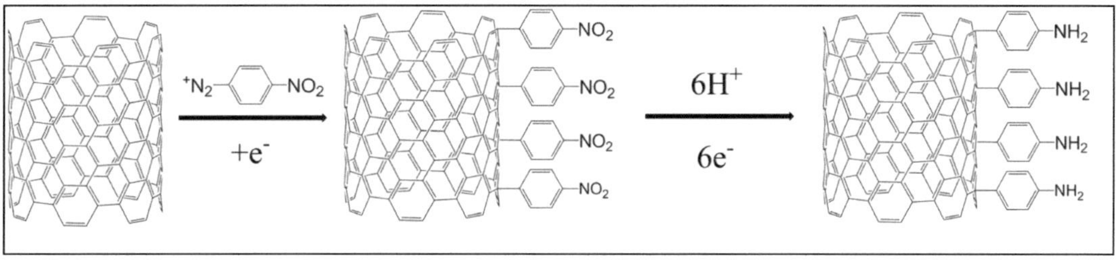

Fig. 3. Schematic for the electrochemical modification of CNTs.

The electrocatalytic properties of Pt/MWCNT composites for methanol oxidation have been investigated by the CV and the high electrocatalytic activity was observed. This might be attributed to the small particle size, high dispersion of platinum particles and the particular properties of the MWCNT supports.

2.2 Non-covalent functionalization

Surface modification of CNTs by non-covalent functionalization is particularly attractive due to its possibility to preserve the electronic network of CNTs (Dyke & Tour, 2004; Moghaddam et al., 2004). The non-covalent functionalization of CNTs is mainly through adsorption of functional molecules via their interactions with CNTs by van der Waals forces, electrostatic forces, hydrogen bonding, and other attractive forces (Y. Lin et al., 2004). Various species of polymers (Andrews et al., 2002; G. Wu et al., 2006), polyelectrolytes (S.W. Lee et al., 2009; S.Y. Wang et al., 2009; W. Yang et al., 2008), surfactants (J.F. Lin et al., 2010), proton-conducting polymers (Tian et al., 2007; Sarma et al., 2005; Scibioh et al., 2008) and other functional molecules (Du et al., 2008) can non-specifically bind with the external surface of CNTs without using non-covalent coupling.

The wrapping of CNTs by polymer molecules was developed by Connel and co-workers (Connel et al., 2001) to prepare individual, well dispersed CNTs in aqueous solution based on the non-covalent attachment of macromolecules on CNTs. The method relies on the thermodynamic favorability of the interactions of CNT–polymer with respect to that of CNT–water, which leads to the hiding of the hydrophobic surface of the CNTs by the attachment of the polymer molecules. When mixing with polyelectrolytes, the energy balance favors CNT wrapping, yielding a high density of charged surface sites which can then serve as a good starting point for the alternating monolayer adsorption of the oppositely charged components through a layer by layer (LBL) technique, driven by electrostatic and van der Waals interactions (Correa-Duarte et al., 2006; Ostrander et al., 2001). As schematically shown in Fig. 4, upon the wrapping of CNTs with a negatively charged polyelectrolyte, a positively charged monolayer can then be deposited, which serves as the real template for the NP adsorption via the electrostatic interactions.

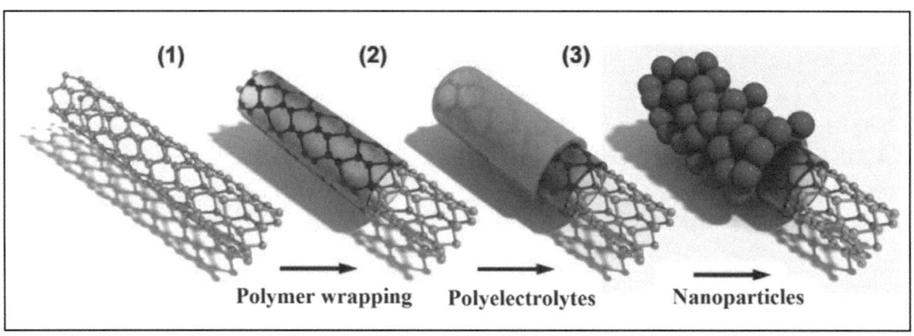

Fig. 4. Schematic illustration of a non-covalent functionalization of CNTs, involving (1) polymer wrapping, (2) self-assembly of polyelectrolytes and (3) NP deposition. Reprinted with permission from Correa-Duarte & Liz-Marzán, 2006. Copyright 2006 Royal Society of Chemistry.

Polybenzimidazole (PBI) (Chemical structure of PBI, Fig. 5a) and its derivatives are some of the most promising candidates for high temperature polymer electrolytes since the proton transfer occurs not only by the vehicle mechanism but also by the hopping mechanism (Q. Li et al., 2004; J.T. Wang et al., 2004), which is an good feature for increasing the rate of the proton transfer and shows great promises in improving the performance of PEMFCs (Okamoto et al., 2008). From the point of view of materials science, PBI can act as a proton-conducting material for PEMFCs that can be operative even under dry conditions above 100°C (the PEMFC operations at higher temperatures afford many benefits such as decreased carbon monoxide poisoning of the catalyst metal NPs, increased catalytic reaction rate, easy removal of generated water, and so on (Q. Li et al., 2003), and therefore is a promising candidate as a substitution for Nafion, which is a widely used proton exchange membrane (PEM) in low-temperature PEMFC systems (Deluca et al., 2006; Heitner-Wirguin, 1996; Kerres, 2001). Especially, PBI is expected to act as i) a metal adsorbing material via the coordination of the metal ion with the aromatic nitrogen on PBI, ii) a MWCNT-solubilizing material, and iii) a proton conductor. Therefore, studies on the applications of PBI in PEMFCs are of great interests. Indeed, it has already reported that the aromatic compounds have a great potential to individually dissolve SWCNTs through a physical adsorption mechanism based on the π–π interactions (Okamoto et al., 2008). A MWCNT/PBI/Pt nanocomposite has been developed by Okamoto et al. (Okamoto et al., 2009). They reported that as a result of the PBI wrapping, the loading efficiency of the Pt NPs onto the MWCNTs was dramatically improved up to 58.8% compared to that of the pristine MWCNTs (41.0%). The process also allows homogeneous immobilization of Pt NPs onto the surface of MWCNTs. Far-Fourier transform infrared spectroscopy shows the existence of a peak from the Pt–N bonding, indicating that these improvements were derived from the coordination of the Pt ion with the PBI molecules. The CV measurements revealed that the Pt NPs deposited on the MWCNT/PBI showed higher utilization efficiency (74%) as electrocatalysts in the PEMFC application compared to those on the pristine MWCNT (39%).

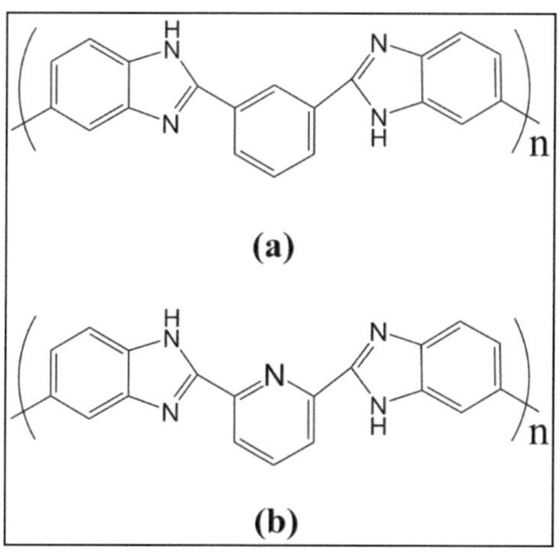

Fig. 5. Chemical structures of (a) PBI and (b) PyPBI.

A PBI derivative (pyridine-containing polybenzimidazole (PyPBI))-wrapped MWCNTs (PyPBI/MWCNTs) was fabricated by Fujigaya et al. (Fujigaya et al., 2009). Among various types of PBI derivatives reported to date, the pyridine-containing polybenzimidazoles (PyPBI, Fig. 5b) is known to possess significantly higher proton conductivity due to its higher acid doping ability and better mechanical properties compared to the conventional PBIs, which endow them with higher capability of proton transfer and improved stability under the electrochemical process, and therefore is expected to exhibit improved properties in the applications of PEMFCs. The Pt ion can be efficiently adsorbed onto the obtained PyPBI-wrapped MWCNTs via the coordination reaction, and the successive reduction of the Pt ion forms rather uniform Pt NPs on the surfaces of the MWCNT/PyPBI. The CV measurements for the hybrids (MWCNT/PyPBI/Pt) showed a high electrochemical surface area (Fujigaya et al., 2009), which was due to the formation of the "ideal triple-phase boundary nanostructure" that was demonstrated by the high resolution TEM (HRTEM) observations as shown in Fig. 6. This result provides useful information for the design and fabrication of highly efficient CNT-based electrocatalysts for the PEFC systems.

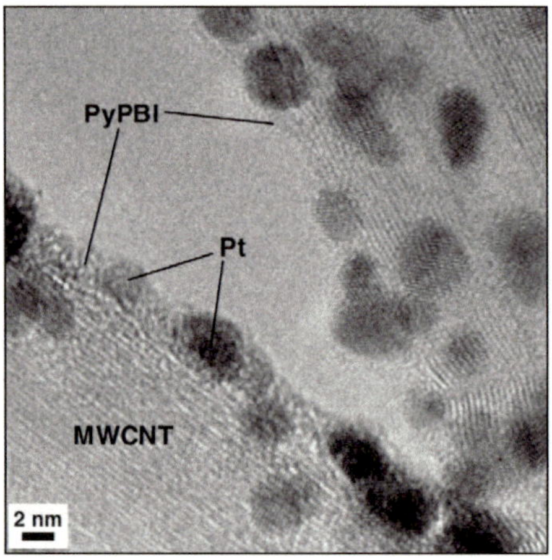

Fig. 6. Typical HRTEM image of the MWCNT/PyPBI/Pt. The Pt NPs are penetrated into the thin PyPBI-coating layer to contact closely with the MWCNT surfaces. Reprinted with permission from Fujigaya et al., 2009. Copyright 2009 Elsevier Science Ltd.

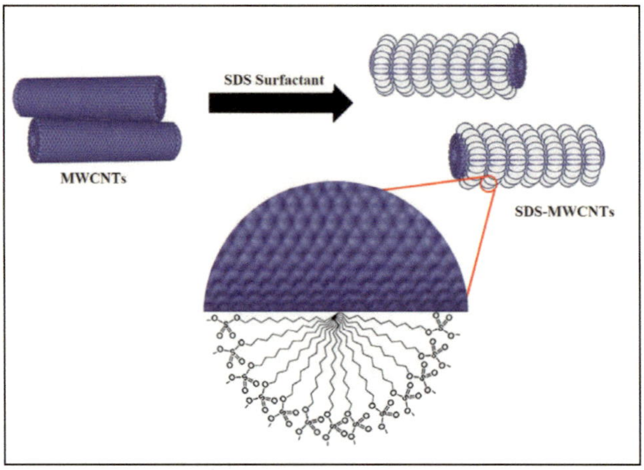

Fig. 7. Schematic representation of SDS-MWCNT micelles. Reprinted with permission from J.F. Lin et al., 2010. Copyright 2010 Elsevier Science Ltd.

Surfactants are a class of amphiphilic organic compounds, which contain both hydrophobic groups (their *tails*) and hydrophilic groups (their *heads*). Therefore, it is possible to functionalize CNTs with surfactant molecules by non-covalent adsorption of hydrophobic groups onto the surface of CNTs and extension of hydrophilic groups to the adsorption of metal ions and/or metal NPs. The adsorption of surfactants enables homogenous suspension of CNTs as individual tubes by decreasing the interfacial surface tension (Moore et al., 2003). As reported by Richard et al. (Richard et al., 2003) and Islam et al. (Islam et al., 2003), the SDS molecules could be chemically adsorbed on the surface of CNTs with the formation of hemimicelles along the graphite network of CNTs. As shown in Fig. 7 for the schematic structure of SDS-MWCNTs, the hydrophobic tails of the micelles enable attachment to the inert surface of MWCNTs and the hydrophilic heads with negative charge enable separation/dispersion of MWCNTs into individual tubes. Lin et al. (J.F. Lin et al., 2010a) used micelle-encapsulated MWCNTs with SDS as a catalyst support to deposit Pt NPs. The HRTEM images revealed the crystalline nature of Pt NPs with a diameter of ~4 nm on the surface of MWCNTs. A single PEMFC with total catalyst loading of 0.2 mg Pt cm^{-2} (anode 0.1 and cathode 0.1mg Pt cm^{-2}, respectively) has been evaluated at 80 °C with H_2 and O_2 gases using Nafion-212 electrolyte. The Pt/MWCNTs synthesized by using modified SDS-MWCNTs showed a peak power density of 950 mW cm^{-2}. Accelerated durability evaluation was carried out by conducting 1500 potential cycles between 0.1 and 1.2 V with 50 mV s^{-1} scan rate, H_2/N_2 at 80 °C. The PEMFC with Pt/SDS-MWCNTs as catalysts showed superior stability in performance compared to the commercial Pt/C composites.

Another promising and intriguing area of developing science is the surface modification of CNTs by the proton-conducting polymers owing to their novel applications in electronic and electro-optical devices. Innovative attempts have been developed to design and synthesize conducting polymer/CNT composite materials for various target applications such as electrochemical devices, light-emitting diodes, chromatography, electrostatic discharge protection, corrosion protecting paint and electrocatalysts. As reported by Selvaraj et al. (Selvaraj & Alagar, 2008), the combination of conducting polymers with CNTs would offer an attractive composite support for electrocatalysts in ethylene glycol (EG) oxidation to enhance the activity and stability based on the morphological modification or electronic interaction between the two components. In that work, polythiophene (PTh) was chosen as the conducting polymer matrix due to its relatively wide potential stability, reproducible synthesis and good electronic conducting properties. The prepared PTh/CNT composites were further decorated with Pt and PtRu NPs by the chemical reduction of the corresponding metal salts using HCHO as the reducing agent. The presence of CNTs in conjugation with a conducting polymer produced a good supports for the catalyst deposition, which allowed the formation of Pt and PtRu NPs with higher dispersion and thereby a better catalytic behavior towards EG oxidation. Results showed that the Pt/PTh–CNT and PtRu/PTh–CNT modified electrodes show enhanced electrocatalytic activity and stability towards the electro-oxidation of EG than the Pt/PTh electrodes.

Treatment of CNTs with surfactants, polymers and other capping agents, are generally tedious and in most cases, additional heat treatment steps are needed to get rid of the non-conducting polymer and surfactants attached to the Pt or Pt alloy NPs. In this respect, Wang et al. (D. Wang et al., 2010) reported a simple and novel method to functionalize the MWCNTs by using tetrahydrofuran (THF) solvent. To demonstrate the effectiveness of the method, they selected the syntheses of Pt and binary PtSn NPs on THF-functionalized

MWCNTs due to their importance for the electrooxidation reactions of methanol and ethanol in low temperature fuel cells. THF is an oxygen-containing heterocycle with five-membered rings. The presence of a σ–π attractive force between the π bonds of CNTs and the σ bonds of cyclopentanes of THF enabled the surface functionalization of CNTs due to a π–π stacking (D.Q. Yang et al., 2005a). Such interaction also makes the MWCNTs easily dispersible. The electronegativity difference between carbon and oxygen makes the C–O bond moderately polar with a sterically accessible oxygen atom. In a chloroplantic acid solution, THF adsorbed CNTs could be protonated, which makes the adsorption of $PtCl_6^{2-}$ and Sn^{4+} ions to the sterically accessible oxygen atoms by an electrostatic self-assembly. The formation of Pt and PtSn NPs on the MWCNTs could be realized by a H_2 treatment as shown in Fig. 8. The TEM image showed that the well-dispersed Pt and PtSn NPs can be directly deposited onto the THF-functionalized MWCNTs without any prior chemical oxidation treatments and the as-prepared Pt/MWCNT and PtSn/MWCNT catalysts show a high activity and stability for the ethanol oxidation in acid solutions. The advantages of the THF-functionalized CNT catalyst support are its simplicity and effectiveness in the deposition of highly dispersed Pt and Pt alloys on CNTs.

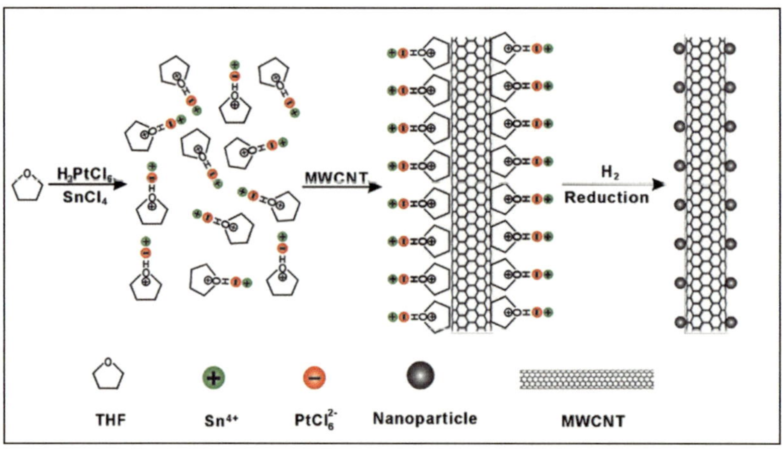

Fig. 8. Schematic of the synthesis of PtSn NP catalysts on the THF-functionalized MWCNTs. Reprinted with permission from D. Wang et al., 2010. Copyright 2010 Elsevier Science Ltd.

2.3 Plasma surface modification

The surface modification and functionalization methods described above, such as the addition of polyelectrolytes, supramolecular complexation with surfactants, functional organics, or polymers, could effectively increase the surface binding sites on the surface of CNTs for the subsequent deposition of metal NPs, avoid the aggregation of metal NPs, improve the dispersion of metal NPs, and simultaneously reduce the average size of metal NPs deposited (R.J. Chen et al., 2001, 2003; Holzinger et al., 2001; Star et al., 2001). In most cases, however, some severe problems accompanied with such surface modification and functionalization methods, such as uneven distribution of the surface functional groups, structural damage, blockage of the direct touch between metal NPs and MWCNTs, could lead to partial lose in the electrical conductivity of the carbon supports, reduce the interactions between metal and CNTs, and thereby the performance of the obtained electrocatalysts (Anderson et al., 2002; Hsin et al., 2007). Additionally, these methods usually require the use of a large amount of chemicals, toxic solvents and/or extreme conditions, which is easier to cause environmental pollution. In order to minimize the above disadvantages during the preparation, it is highly desired to develop a mild surface functionalization technique to introduce homogeneous distributed functional groups with a high density onto the surface of CNTs, but cause less structural damage to the CNTs (and thus retain good electrical conductivity) and no pollution to the environment. Various dry processes, including both nonreactive and reactive plasmas (Brunetti et al., 2008; Q. Chen et al., 2000, 2001; Khare et al., 2004; Plank et al., 2003, 2004; Yan et al., 2005) and low-energy ion beam bombardment in a vacuum (D.Q. Yang et al., 2005), have been found as good candidates. Compared to wet approaches, dry plasma processing may be easier to control, with relatively less contamination. Plasma treatment is an efficient method in the field of surface modifications. The excited species, radicals, electrons, ions, and UV light within plasma strongly interact with the surfaces of CNTs breaking the C=C bonds and creating active sites for binding of functional groups, as a result, chemical and physical modifications occur on the surfaces. Compared to other chemical modification methods, the plasma treatment method has the advantages of shorter reaction time, nonpollution, and providing a wide range of different functional groups depending on plasma parameters such as power, used gases, treatment time, and pressure. Thus, this method offers the possibility of scaling up to produce large quantities necessary for commercial use. Plank et al. (Plank et al., 2003) reported the surface functionalization of CNTs by CF_4 plasma. The reaction was conducted at the room temperature in a short duration of time. Scanning electron microscopy (SEM) images indicates the dimension and morphology of CNTs have been preserved after a CF_4 plasma exposure. X-ray photoelectron spectroscopy (XPS) demonstrates the prevalence of covalent C–F bonds on the CNTs after CF_4 exposure. Recently, Yang et al. (D.Q. Yang & Sacher, 2006; G.X. Zhang et al., 2007) studied the effect of plasmas on highly oriented pyrolytic graphite (HOPG), where they found that Ar, O_2, N_2, and H_2O plasmas could break C-C bonds, producing -C· free radical defects that, on atmospheric exposure, reacted with components of air (H_2O and O_2) to produce oxidized carbon species (C-OH, C=O, and COOH). These oxidized carbon species could facilitate the deposition of metal NPs due to hydrogen bonding between the hydroxyl groups on the NP surface and these species on the HOPG that are introduced upon atmospheric exposure of the free radicals produced during the plasma treatment. Similar results were reported to a plasma modification of CNTs (D.Q. Yang & Sacher, 2008). It showed that the exposure of CNTs to Ar plasma or O_2 plasma produced surface defects on the surface of CNTs which could act as both nucleation and binding sites for the deposition of Pt NPs. The XPS and TEM analyses showed that the interactions between Pt NPs and CNTs were enhanced by the Ar or O_2 plasma treatment.

The possible mechanisms associated with the plasma treatment of CNTs include the generation of the C-O, C=O, and O-C=O bonds, as shown in Fig. 9 (C. Chen et al., 2009). Since the π bonds in C=C are active and are the most susceptible to the plasma attacks, it is believed that the radicals are first generated on the dissociated π bonds in C=C, which further react with active oxygen atoms (Fig. 9A). This explains the decrease in the C=C fraction after a plasma treatment. This process may produce C-O bonds, and then the C-OH bonds are formed through stabilization by hydrogen atom transfer from the same or a neighboring chain. The hydrogen atoms can also be introduced during the synthesis phase of MWCNTs or via atmosphere exposure. Oxygen radicals are considered to be generated on the surfaces of MWCNTs, which could lead to the formation of the new C=O bonds through intramolecular reorganization on the C-C bonds, as shown in Fig. 9B. The formation of O-C=O bonds is believed to be due to the C=O bonds through the combination of the plasma-generated radicals on the C=O bonds with the active oxygen atoms. After stabilization with proton transfer, HO-C=O can be formed, as shown in Fig. 9C. Compared to pure O_2 plasma treatment, Ar/O_2 plasma treatment enhances the C-O and O-C=O fractions, and the C-O and O-C=O fractions increase with increasing plasma power and the treatment time. The efficiency of Ar/O_2 mixture gas plasma treatment is higher than that of pure O_2 plasma treatment, since the content of active oxygen in Ar/O_2 mixture gas plasma is higher than that in pure O_2 plasma. Ar atoms and/or ions present in the plasma can also interact with the surfaces of MWCNTs creating active sites for further oxygen functionalization. Indeed, in some case, N-containing groups can also be formed on the surface of CNTs (Ruelle et al., 2008 and C. Chen et al., 2010). A plasma discharge can create enough electron energy to fractionize NH_3, forming metastable ions of NH_2, NH, N, and H as well as radicals, which can be incorporated into the surface of CNTs during the plasma treament.

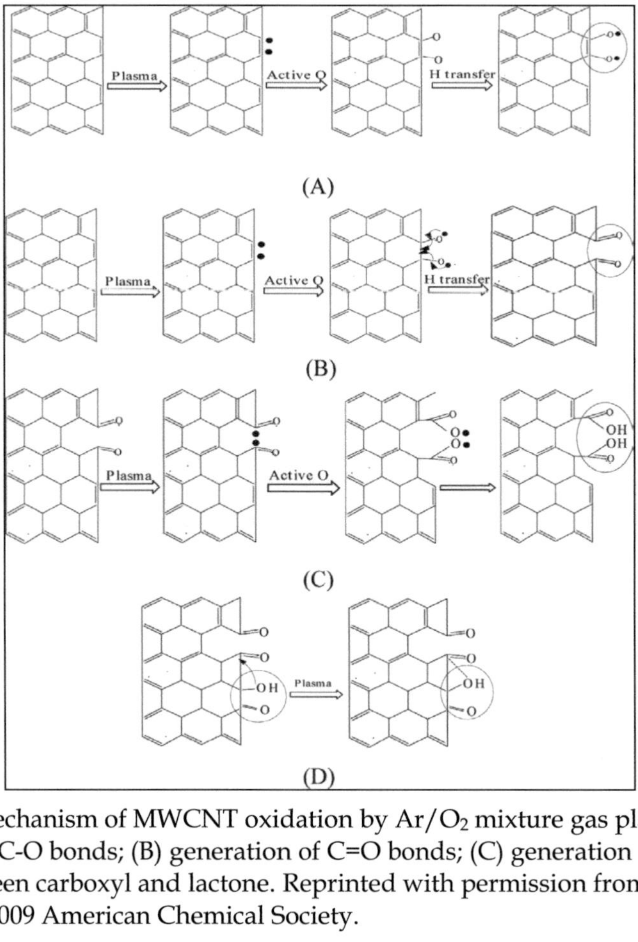

Fig. 9. Possible mechanism of MWCNT oxidation by Ar/O_2 mixture gas plasma treatment: (A) generation of C-O bonds; (B) generation of C=O bonds; (C) generation of O-C=O bonds; (D) transfer between carboxyl and lactone. Reprinted with permission from C. Chen et al., 2009. Copyright 2009 American Chemical Society.

The investigation of the electrocatalytic activity of Pt NPs on the plasma treated CNTs has been conducted by our group (Jiang et al., 2009, 2011). The results showed that the Pt NPs on the N_2 plasma treated CNTs exhibited a significantly higher electrochemical activity towards the methanol oxidation in an acid solution, in comparison to those on the CNTs functionalized by other modification methods. The structures of CNTs and the catalyst/CNTs interactions were found to play important roles in determining the performance of the catalysts. The Pt NPs deposited on the MWCNTs functionalized by a strong acid treatment which easily leads to the structural damage of MWCNTs showed a much lower performance of Pt/MWCNTs in methanol oxidation reaction, due to the decrease in the conductivity of MWCNTs caused by the structural damage. An insertion of impurities between Pt NPs and MWCNTs could also result in a decrease in electron migration from metal to MWCNTs and give rise to the decrease of electrochemical performance of Pt/MWCNTs in methanol oxidation reaction. It therefore suggests that to obtain Pt/MWCNTs with higher catalytic activities, it is necessary to adopt a mild surface modification approach and to make metal NPs directly deposit onto the CNT surface.

2.4 Nitrogen-doped CNTs

As mentioned above, a chemical modification of the surface of CNTs by covalent functionalization could reduce considerably the mechanical and electronic performance of CNTs due to the introduction of large numbers of defects, and in some cases, the electrocatalysts on the non-covalently functionalized CNTs shows low activity over fuel oxidation due to the poor conductivity of functional molecules and low conduction between the metal NPs and the CNTs and between the neighboring CNTs, which require us to seek for new approaches for the preparation of electrocatalysts. It is recently found that the use of nitrogen doped CNTs (N-CNTs) could be considered as one of promising options. The introduction of N can lead to the formation of CNTs with high surface areas (Feng et al., 2008), high densities of defects (Tao et al., 2007), chemically active impurity sites (Nxumalo et al., 2008; Tao et al., 2007) and narrow widths (the numbers of walls decrease with N inclusion) (Nxumalo et al., 2008). The N-doped nanotubes are found to be either metallic or narrow energy gap semiconductors (Huang et al., 2000; Miyamoto et al., 1997), thus offering the possibility of greater electrical conductivity as compared to the pure CNTs. Studies have shown that the N-doped CNTs and their composites can be used as support materials and have great potentiality in the PEMFC catalyst applications. Due to the presence of chemically active nitrogenated sites (substitutional and pyridinic nitrogen), the N-doped CNTs are reported to have enhanced activity and selectivity in many catalytic applications (Shao et al., 2008), and exhibit a strong binding to metals, leading to excellent metal dispersion in the metal/N-CNT materials (Droppa Jr. et al., 2002). It can therefore avoid using functionalization processes that might be detrimental to the catalytic properties of the obtained metal/N-CNT composites (Maiyalagan et al., 2005; C.L. Sun et al., 2005; C.-H. Wang et al., 2006, 2007; Zamudio et al., 2006). For example, Maiyalagan et al. (Maiyalagan et al., 2005) studied the electrocatalytic properties of Pt/N-CNTs synthesized by a reduction of Pt^{2+} adsorbed on the surface of unfunctionalized N-CNTs. The obtained N-CNT-supported Pt NPs were reported to be homogeneously dispersed on the nanotubes. An enhanced catalytic activity and stability toward methanol oxidation was observed with Pt/N-CNTs in comparison with commercial Pt/C catalyst supplied by E-TEK. The authors of that work attributed the enhanced catalytic activity and stability of Pt/N-CNTs to the factors, such as, the higher dispersion of Pt NPs which increased the availability of electrochemically active surface area, the appearance of the specific active sites at the metal–support boundary and strong and specific metal–support interaction. The investigation of the microstructure and electrochemical activity of the PtRu supported by N-CNTs was reported by Wang et al. (C.-H. Wang et al., 2006). These N-CNTs were directly grown on the carbon cloth (N-CNTs–carbon cloth composite electrode) using a microwave plasma enhanced chemical vapour deposition, and then used as the template to support the subsequently sputtered PtRu nanoclusters. The ferricyanide/ferrocyanide redox reaction in CV measurements showed a faster electron transfer on the N-CNTs–carbon cloth composite electrode than the one with carbon cloth alone. In addition, it was found that the PtRu nanoclusters supported by the N-CNTs–carbon cloth composite electrode had considerably higher electrocatalytic activity in the methanol oxidation than the carbon cloth counterpart, which suggested a high performance of the N-CNTs/carbon cloth composite electrode, its suitability for direct methanol fuel cell applications.

3. Synthesis and characterization of metal NPs supported on CNTs

It is know that the electrocatalytic activities of catalysts on the CNTs are greatly dependent upon their size, morphology, composition and dispersion, which are determined by the way that they are produced and the way they are adsorbed onto the surface of CNTs. Since the dispersion and particle size of metal NPs largely determine the utilization and catalytic activity of metal/CNTs, the synthesis of metal NPs supported by CNTs with a controlled manner is of great importance for the design of catalysts of higher efficiency. Up to now, the most widely used catalysts for the application of PEMFCs are Pt and Pt-based alloys due to their large surface to volume ratio, improved catalytic activities relative to their bulk material. The synthesis of metal NPs/CNT composites can be performed either by nucleation and growth of metal NPs on the surface of CNTs or by attachment of preformed NPs in the bulk solution onto the surface of CNTs. In the following sections, the methods used to prepare NPs/CNTs will be reviewed and discussed.

3.1 Formation of metal NPs directly on CNTs

Nucleation and growth of metal NPs directly on the surface of CNTs is the mostly used method to prepare metal NPs/CNTs catalysts. The preparation of such NPs/CNT composites can be conducted either physically or chemically. Metal NPs are absorbed on the surface of MWCNTs mainly based on van der Waals interactions, electrostatic interactions and coordination interactions between metal particles and functional groups, which in some cases seem to be sufficiently strong to guarantee meaningful adhesion (K.C. Lee et al., 2006; X. Sun & Saha, 2008).

3.1.1 Physical methods

In a physical method, bulk metals are thermally vaporized followed by a sputtering of metal gases onto the surface of CNTs. The sputtering-deposition method is a recently developed approach to prepare the PEMFC cathode catalysts, aiming at metal loading reduction and metal utilization improvement. It has been demonstrated that the sputter-deposition technique is a good way to deposit small and uniform metal NPs on CNTs with sizes well controlled by the sputtering time and current. This method can also generate a thinner catalyst layer that could give a higher fuel cell cathode performance and, at the same time, considerably reduce the metal loading. The physical deposition of Pt NPs on the surface of CNTs was reported by Tang et al. (Z. Tang et al., 2010), who produced Pt NPs with 4 nm in diameter and a narrow size distribution. A high maximum power density of 595 mW cm^{-2} was observed for a low Pt loading of 0.04 mg cm^{-2} at the cathode for the PEMFC application. The deposition of Pt NPs on nitrogen-dopped MWCNTs (N-MWCNTs) was done by Sun et al. (C.L. Sun et al., 2005). The well-separated Pt NPs with an average diameter of 2 nm were formed on the side-walls of N-MWCNTs. In that work, the nitrogen incorporation in the MWCNTs might play a critical role in the self-limited growth of the Pt NPs. The CV results showed that the Pt/N-MWCNT catalyst had improved electrochemical activity towards methanol oxidation and showed great promises for a future μDMFC device. However, because the preparation of Pt/CNT composites usually requires the use of extremely high temperatures, this technique may face some technical challenges with respect to the electrode mass production.

3.1.2 Chemical methods

Compared with the physical methods, the chemical methods have the significant advantage of being able to easily control the primary structures of NPs, such as size, shape, and composition, as well as to achieve mass production. A large variety of chemical methods such as impregnation method, colloidal method, ion-exchange method, electrochemical method, microwave heated polyol method, have been reported for the preparation of metal/CNTs composites as a catalyst for the DMFC applications. Different growth control mechanisms and strategies are used in each of the different chemical deposition methods.

3.1.2.1 Impregnation method

The impregnation method is the most widely used wet-chemical method, which is a simple and straightforward for depositing metal NPs on the CNTs for the preparation of the PEMFC catalysts and is thus an attractive choice for a large-scale synthesis. The method involves the impregnation of the support material with a salt solution containing the metal to be deposited, followed by a reduction step (Asano et al., 2006; Liao et al., 2006; Y. Lin et al., 2005; Lordi et al., 2001). During an impregnation process, metal ions are initially adsorbed to the surface of functionalized CNTs by homogeneously mixing CNTs with the metal precursors in a solution. The chemical reduction of the metal ions on the surface of the CNTs can either be carried out by a liquid phase reduction using borohydride, formic acid or hydrazine as a reductive agent, or by a gas phase reduction using a flowing hydrogen gas as a reductive agent under elevated temperature. For the impregnation method, the size and distribution of Pt NPs are affected by many factors, in which the chemical modifications of the surface of CNTs will play a major role since the pristine surface of CNTs is relatively inert unfavorable for the deposition of metal NPs. A desired way in this case is to functionalize the surface of CNTs first through a chemical reaction as discussed in Section 2. As one example, Li and coworkers (W.Z. Li et al., 2004) reported the synthesis of Pt/MWCNT nanocomposites by using the impregnation method, which was then used as electrocatalyst applied in a direct methanol fuel cell (DMFC). In that work, the Pt NPs were deposited on the pre-functionalized MWCNTs by reduction of Pt precursor with EG, which produced a Pt/MWCNT composite with a homogeneously dispersed spherical Pt NPs of a narrow particle size distribution. The obtained Pt/MWCNTs were characterized to exhibit significantly higher performance than the Pt loaded on the commercial XC-72 carbon when used in the DMFC. This improvement in catalytic performance was attributed to the greater dispersion of the supported Pt particles.

3.1.2.2 Electrochemical method

The electrochemical method for the preparation of metal/CNTs is very similar to the process of the impregnation method, except for an electrochemical reduction of the adsorbed metal ions rather than the chemical reduction. In this process, functionalized CNTs are first mixed with the metal precursors in aqueous solution to produce a homogeneous solution. A pulse current, such as direct current or CV, is then added for the reduction of metal ions promoting the deposition of metal NPs on CNTs, which usually produces metal NPs/CNTs with high efficiency in PEMFCs as compared to those prepared by the conventional deposition techniques (Choi et al., 1998; Taylor et al., 1992; Thompson et al., 2001). An approach for the electrochemical deposition of Pt particles with a narrow size distribution on CNTs was reported by Tsai et al. (Tsai et al., 2006), who successfully electrodeposited Pt and PtRu NPs on the dense CNTs directly grown on carbon cloths in EG containing H_2SO_4 aqueous solutions. Prior to the electrodeposition of Pt or PtRu NPs, all the specimens with CNTs directly grown on carbon cloths (CNT/CC) underwent a hydrophilic treatment at 50 mV s^{-1} for 100 cycles with potential ranged from –0.25 to +1.25 V_{SCE} (V_{SCE} means that the potential was quoted against a saturated calomel electrode (SCE)) in an O_2 saturated 2 M H_2SO_4 aqueous solution at ambient condition. To achieve a larger driving force for dechlorination of the Pt and Ru precursors, more negative deposition potentials are usually favorable. EG acted as a stabilizing surfactant to prevent the particles from agglomeration during the electrodeposition processes and could also enhance the dechlorination of Pt and Ru precursor salts and led to the formation of NPs. In the meantime, nano-sized Pt and PtRu particles were also found in specimens treated at two pre-selected negative potentials. The particle sizes of Pt on CNTs ranging from ~4.5 to ~9.5 nm and PtRu on CNTs (PtRu/CNTs) ranging from ~4.8 to ~5.2 nm were obtained and was reported to exhibit improved electrocatalytic activity in methanol oxidation in comparison to the corresponding commercially available catalysts.

3.1.2.3 Colloidal method

Colloidal method involves the nucleation of metal clusters on the surface of CNTs, followed by growth of these clusters, or involves the formation of a metal oxide colloid, followed by simultaneous reduction and adsorption, or adsorption followed by chemical reduction. In this method, the size of the metal NPs is largely controlled or stabilized by the protecting agents, such as ligands, surfactants or polymers (Kuo et al., 2005). The colloidal metal NPs are stabilized by either steric hindrance or by electrostatic charges. In recent years, there have been considerable interests in the development of colloidal methods to prepare Pt catalysts supported on the CNTs with a narrow particle size distribution (Kongkanand et al., 2006b; C. Lee et al., 2005; W.Z. Li et al., 2003; X. Li et al., 2004, 2006; Yoshitake et al., 2002). For example, Li and co-worker (X. Li et al., 2004) used the surfactant 3-(N,N-dimethyldodecylammonio) propanesulfonate (SB12) as a stabilizer to prepare Pt NPs supported on the CNTs by reduction of H_2PtCl_6 with methanol (X. Li et al., 2004, 2006). The Pt NPs deposited on the functionalized CNTs were well-dispersed with an average size of 2.2 nm (X. Li et al., 2004).

Though the colloidal method can provide a narrow size distribution of metal NPs, the major drawback is the presence of a protecting agent, which may decrease the catalytic performance of the NPs. As a result, the organic protecting layers used for the protection of the electrocatalysts prepared by the colloid method must be removed prior to their use in the PEMFCs. The desired way is to prepare colloidal NPs without the use of protecting agents. Such fascinating way has been recently reported by Yoshitake et al. (Yoshitake et al., 2002), who synthesized a Pt/CNT catalyst for the use of PEMFCs by the colloid method. In the preparation, a colloidal Pt oxide was first prepared by adding $NaHSO_3$ and H_2O_2 into an aqueous solution of H_2PtCl_6 without using an organic stablizer. The adsorption of the Pt oxide colloids was done through its mixing with the single-wall carbon nanohorns (SWNH) powder. The reduction of Pt oxides was carried out by a H_2 gas. The produced Pt/SWNH catalyst showed very homogeneous dispersion of Pt NPs with an average size of 2 nm and exhibited higher electrocatalytic activity in a PEMFC.

3.1.2.4 Ion-exchange method

An ion-exchange method is an effective technique for depositing metal NPs on the CNTs without using protecting agents, reducing agents or precursor complexes. In this technique, a metal cation complex, such as $[Pt(NH_3)_4]^{2+}$, is ion-exchanged with hydrogen ions of the acid functional groups on the surface of the CNTs. After the ion-exchange process, the Pt cation complex is reduced to the Pt NPs in an H_2 atmosphere. The interaction between the acid functional groups and the Pt precursor determines the dispersion and size of the metal particles. The preparation of a Pt/CNT composite by the ion-exchange method has been reported by Yin and co-workers (Y. Shao et al., 2006; J. Wang et al., 2007). In their work, an electrochemically functionalized CNT electrode was immersed in a solution of the platinum cation-complex salt for 48 h, which resulted in the ion-exchange of the hydrogen ions of the functional group on the surface of the CNTs with the Pt cation complex. The immersed CNTs was then filtrated and washed thoroughly with distilled water. The reduction of the adsorbed platinum complex precursor to its metallic form was carried out by the treatment with hydrogen gas at 190 °C. It was reported the Pt NPs were highly dispersed on the CNTs with dispersion much better than those prepared by the borohydride method (J. Wang et al., 2007).

3.1.2.5 Microwave heated polyol method

In a microwave heated polyol method, a polyol (ethylene glycol) solution containing catalyst metal precursor salts is refluxed at high temperature by a microwave heating in order to homogeneously decompose EG and create an active reducing agent for metal ions (W. Chen et al., 2005; Z. Liu et al., 2005). A metal support could be optionally present to capture the depositing metal particles in this process. Unlike the conventional conductive heating strategy used to thermally activate the polyol, which has a heterogeneous temperature distribution, the fast heating by microwaves can accelerate the reduction of the metal precursor ions and the nucleation of the metal NPs. In addition, the homogeneous microwave heating could reduce the temperature and concentration gradients in the reacting sample solution, resulting in a more uniform environment for the nucleation and growth of metal particles (W. Chen et al., 2005; Z. Liu et al., 2005). The synthesis of Pt/CNT or PtRu/CNT catalysts using a microwave heated polylol process (W. Chen et al., 2005; Z. Liu et al., 2004, 2005) has been reported by several research groups. The obtained Pt/CNT or PtRu/CNT catalysts are usually characterized with greater catalytic activity towards oxygen reduction than the catalysts fabricated by some other techniques although both catalysts had almost the same Pt particle sizes.

3.2 Connecting metal NPs and CNTs

In the methods presented above for the preparation of metal/CNT composites, the formation of metal NPs directly occurs on CNTs. However, in some cases, the deposition of metal NPs is realized by adsorption of preformed metal NPs in the solution through either the formation of covalent bonds between the functional groups on metal NPs and the functional groups present on the CNT surface or the intermolecular interactions such as hydrophobic, π–π stacking or electrostatic attractions between them (Georgakilas et al., 2007).

3.2.1 Covalent linkage

The metal NPs anchored on CNTs by a covalent linkage usually exhibit a long life and high catalytic performance when used in PEMFCs. Due to the high strength of the covalent interactions, the metal NPs are usually well separated on the CNTs and exhibit a higher stability during the electrochemical process. The higher degree of dispersity increases electrocatalytically active surface areas of metal NPs which therefore exhibit a high activity in methanol oxidation and oxygen reduction. For example, Yang et al. (D.Q. Yang et al., 2006b) reported the fabrication of Pt/CNTs by the covalent attaching of the Pt NPs onto the CNTs using benzyl mercaptan as an interlinker. In their work, the CNTs were first functionalized with benzyl mercaptan by a π-Stacking. The functionalized CNT surface interacted strongly with the Pt NPs through the formation of Pt-S bonds and resulted in a very high Pt NP loading both high dispersion and a narrow size distribution, as schematically illustrated in Fig. 10.

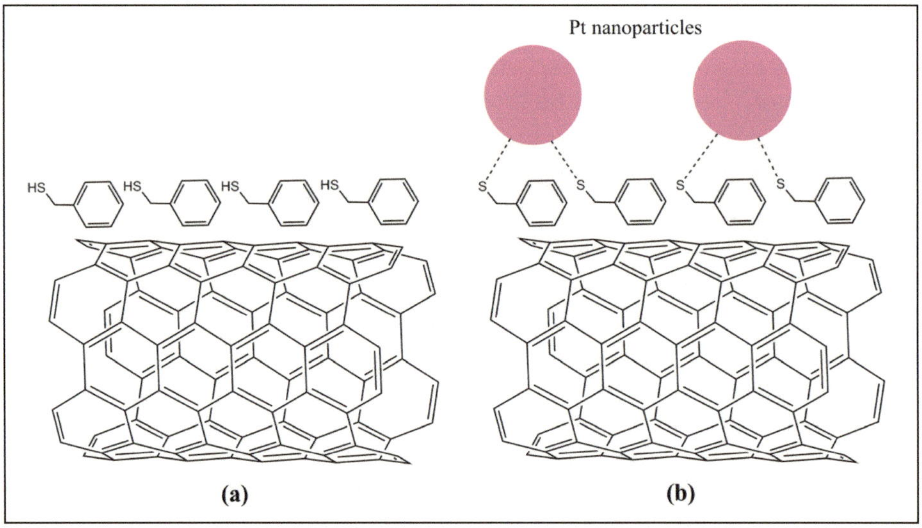

Fig. 10. Schematic of (a) a CNT functionalized with benzyl mercaptan via π-π bonding and (b) the bonding of Pt NPs to the functionalized CNT via covalent S-Pt bond formation. Reprinted with permission from D.Q. Yang et al., 2006b. Copyright 2006 American Chemical Society.

3.2.2 Hydrophobic interactions and hydrogen bonds

The interactions between hydrophobic ligands forming the monolayer passivating the metal surface with hydrophobic molecules adsorbed on CNTs and hydrogen bonds between the molecules on NPs and CNTs can be used to immobilize the metal NPs onto CNTs. A combination of hydrophobic and hydrogen bond interactions will make the attachment of metal NPs on CNTs more tightly, and therefore be employed as a promising way to the preparation of metal/CNT composites as catalysts for PEMFC applications (L. Han et al., 2004). As reported by Han et al. (L. Han et al., 2004), a Au/CNT composite has been synthesized by the hydrophobic interactions of the alkyl chains on decanethiol and mercaptoundecanoic acid adsorbed Au NPs with the CNT surface and hydrogen bonds between carboxylic groups of CNTs and those present on the NP surface. Due to the strong interactions provided by the hydrophobic interactions and the hydrogen bonds between ligated Au NPs and CNTs, the obtained Au/CNTs showed a very stable structure, cannot be disassembled by sonication in hydrophobic solvents. However, thermal treatment of the sample, for example, at 300 ℃ would induce aggregation of NPs that stick strongly to the nanotube surface due to the removal of the capping shells of the Au NPs, which decreased the interactions between Au NPs and CNTs (Georgakilas et al., 2007).

3.2.3 π-Stacking

The deposition of metal NPs through the π-Stacking involves a cover of metal NPs with phenyl-containing molecule and the adsorption onto the surface of CNTs by a π-Stacking interaction. As recently reported by Mu et al. (Mu et al., 2005), they have applied this method to prepare the Pt/CNTs by the surface modification of Pt NPs with triphenylphosphine (PPh_3) and the subsequent deposition of the modified Pt NPs onto the CNTs through the π-Stacking. The main disadvantage of the method is that a special annealing process is usually required to remove PPh_3 molecules, which might lead to some degree of aggregation of Pt NPs. Indeed, it is demonstrated that even though the thermal treatment would result in aggregation of Pt NPs, the mean size of the deposited Pt NPs remained small and showed higher electrocatalytic activity and better tolerance to poisoning species in the methanol oxidation than the commercial E-TEK catalyst.

3.2.4 Electrostatic interactions

Electrostatic interaction is a commonly used method to anchor metal colloids to CNTs. Because the functionalized CNTs with an ionic polyelectrolyte are charged, which can serve as an anchor for metal NPs oppositely charged. In typical examples, the oxidized CNTs are modified with a cationic polyelectrolyte and exposed to the negatively charged metal NPs. By choosing different kinds of polyelectrolytes the surface of the CNT can be also negatively charged in order to the deposit of positively charged NPs. The electrostatic approach has been used to build mixed Au/MWCNT layers using the LBL (layer-by-layer) methodology (Kim & Sigmund, 2004). In the work, the acid-functionalized MWCNTs were first covered with a layer of a positively charged polymer [PDDA, poly(diallyldimethylammonium chloride)] and then with a layer of a negatively charged polymer [PSS, poly(sodium 4-styrenesulfonate)]. Subsequently, the positively charged Au NPs were anchored through electrostatic interactions to the PSS layer. The positively charged Au NPs may also interact directly with the MWCNTs presenting carboxylate groups on their surface. However, the direct binding happened with a much lower density of Au deposition than that observed for the PDDA/PSS LBL method.

4. Activity validation of the synthesized catalysts in a fuel cell operation

The most direct and effective method for the activity validation of the synthesized catalysts is to directly use them in a single fuel cell. For fuel cells, the activity of a catalyst can be deduced from their performance. The most commonly used way to reflect the performance of the fuel cells is the polarization (or current-voltage) curve of the MEA which is the core of the PEM fuel cell, composed of an anode gas diffusion layer (GDL), an anode catalyst layer, a membrane (the PEM), a cathode catalyst layer, and a cathode gas diffusion layer, as shown in Fig. 11 which schematically shows a single typical PEMFC. Two data collection modes are frequently used in obtaining the polarization curve, conducted either by adjusting the cell voltage then recording the current density, or by adjusting the current density then recording the cell voltage, with the latter being the most popularly used in the fuel cell performance data collection. A typical polarization curve of a cell obtained by collecting the cell voltage as a function of current density is shown in Fig. 12, which can then used to yield the power density of the MEA (cell voltage × current density) plotted as a function of current density. From the power density curve, the maximum power density of the fuel cell MEA can be then known as well as the maximum volume power density and the mass power density of a fuel cell stack. In principal, beside the catalysts, the performance of a fuel cell (polarization curve) is also affected by the quality and property of MEA and the operating conditions, such as temperature, pressure, relative humidity (RH), gas flow rates, etc. Therefore, for the sake of systematical improvements in the efficiency of the fuel cell, a better understanding of the effects of the quality and property of the MEA on the performance of the fuel cell is essential.

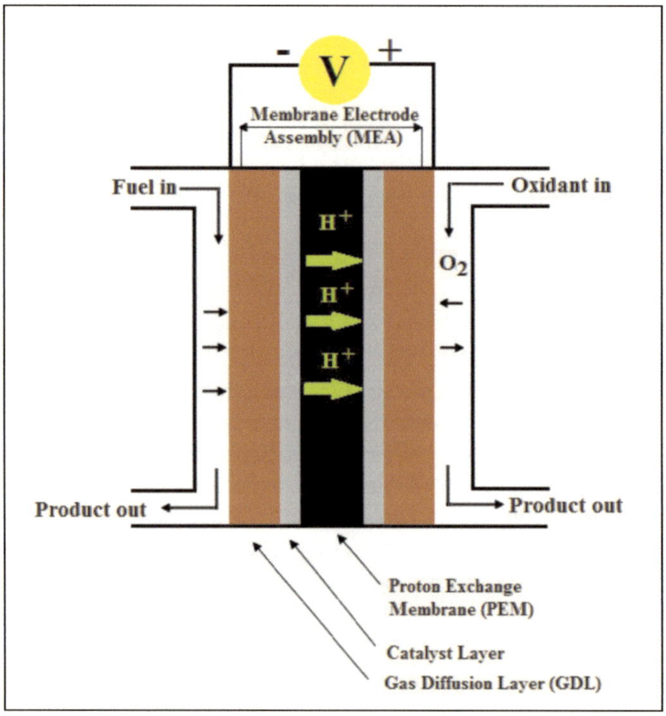

Fig. 11. Schematic of a single typical PEMFC.

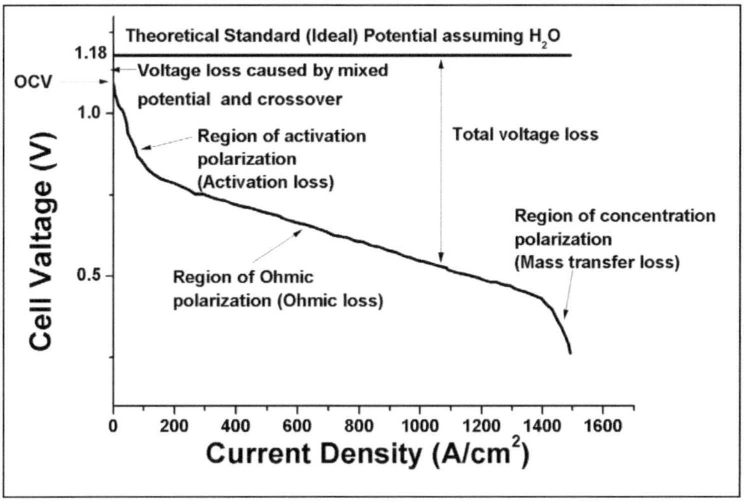

Fig. 12. Typical polarization curve of PEMFCs.

4.1 Synthesis of metal/CNT based MEA

As the core of the PEM fuel cell, the MEA conducts the conversion of the chemical energy of the fuel (i.e., hydrogen) into electricity through the electrochemical oxidation of fuel at the anode and the electrochemical reduction of oxygen at the cathode. Therefore, the MEA component materials, structure, and fabrication technologies largely determine the performance of a PEMFC. An optimization of MEA is of great importance for the improvement of the PEMFC performance (Shen, 2008). An ideal MEA allows all active sites of catalysts in the catalyst layer to be accessible to the reactant, protons and electrons, and can effectively remove produced water from the catalyst layer (CL) and gas diffusion layers (GDL). As mentioned above, a typical MEA for a single PEMFC (J.M. Tang et al., 2007), is composed of a PEM, anode and cathode electrodes, and anode and cathode GDL (schematically shown in Fig. 11). According to differences in preparation processes and structures, hydrophilic catalyst layers can be prepared either by a membrane-based or a GDL-based method, as shown in Fig. 13. For the membrane-based method, the MEA is fabricated by depositing the catalyst ink directly onto a dry and fixed membrane or by coating catalyst ink onto a blank substrate (e.g., PTFE film) and then transferring the coating catalyst ink onto the membrane (Wilson & Gottesfeld, 1992), which is then sandwiched between two GDLs and followed by a hot pressing step, while in the GDL based method, the catalyst ink is directly painted or sprayed onto the pre-treated GDL and then hot pressed onto the membrane. In these two methods, the catalyst ink used for coating the membrane and a blank substrate and the GDLs can be prepared by mixing the metal/CNT catalyst with ionomer firstly, which can improve the contact between the catalyst particles and the ionomer, and thus help to improve catalyst utilization. It has been reported that an ionomer-bonded hydrophilic catalyst layer could improve Pt utilization by up to 45.4% (Cheng et al., 1999). The notable advantages of such an ionomer-bonded hydrophilic electrode include (Girishkumar et al., 2005):

1. improved bonding between the membrane and the catalyst layer;
2. uniform continuity of the electronic and ionic paths for all catalyst sites due to the uniform dispersion of catalyst in the ionomer;

3. high metal NPs utilization resulting from good contact between the catalyst and the protonic conductor;
4. relatively low catalyst loading without performance losses.

However, there are still some inevitable drawbacks associated with this kind of catalyst layers. For example, due to the lack of hydrophobic passages, gas transportation from the GDL to the reaction sites becomes difficult, and the produced water tends to accumulate in the electrode and block the gas transport paths, leading to a decrease in fuel cell performance. In addition, due to the degradation of the ionomers, its ability to bind with the catalyst particles will decrease, causing lowered reliability as well as durability problems. Recent efforts, therefore, turn to the preparation of the catalyst layers with reduced thicknesses. A thin catalyst layer can minimize the shortcomings associated with an ionomer-bonded hydrophilic catalyst of thick layer and improve the efficiency of the mass transfer at the interface, such as the efficient movement of protons, electrons, and dissolved reactants in the reaction zone, which is beneficial to reduce catalyst loading and increase mass power density of a MEA. However, this requires the use of catalysts of higher efficiency.

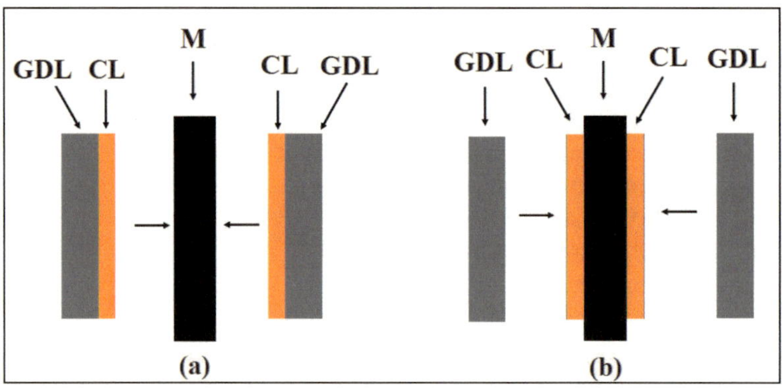

Fig. 13. Configuration of MEAs for (a) the CL on GDL mode and (b) the CL on membrane mode. GDL - gas diffusion layer; CL - catalyst layer; M - membrane.

4.2 Performance of metal/CNT based MEA

The performance of fuel cells with metal/CNT composites as catalysts in PEMFCs has been reported by many groups (Hernández-Fernández et al., 2010; W.Z. Li et al., 2005; J.F. Lin et al., 2010b; T. Matsumoto et al., 2004a, 2004b; Prabhuram et al., 2006; Saha et al., 2008; X. Wang et al., 2005), it is generally believed that the performance of a CNT-based MEA is better than that of conventional metal/supports MEA (T. Matsumoto et al., 2004a, 2004b; Prabhuram et al., 2006; Saha et al., 2008; X. Wang et al., 2005). Although the detailed mechanism how the CNTs affect the catalytic activity of the metal/CNTs are still not well understood, it is generally believed that the enhancement of the electro-catalytic activity of the metal/CNT composites may arise from the following reasons:

1. Good dispersion of metal NPs. The well-dispersed NPs on the surface of CNTs make them more accessible to the fuel oxidation reaction.
2. Unique structure of CNTs. Because of the novel morphology and electrical properties of CNTs, a fast transfer of charges through the composites is possible, which results in their high electrocatalytic activity.

3. Small sizes of deposited metal NPs. The surface-to-volume ratio becomes larger when the size of NPs decreases, which increases the percentage of atoms at the surface accessible to the fuel oxidation reaction. In addition, with decreasing the size of metal NPs, their Fermi level improves, which make the electron transfer easier, favoring the subsequent redox reaction.
4. High purification of CNTs. The metal NPs are reported to be sensitive to some elements, such as sulfur. Trace amount of sulfur adsorption would lead to a decrease in the catalytic activity of metal NPs. The CNT supports contain fewer organic impurities, unlike other carbon materials such as the XC-72 carbon (contain ca. 0.2 at. % sulfur), which is important to maintain the high catalytic activity of metal/CNTs.
5. Porous structures of CNTs. Porous structures influence the reactant-product mass transport and therefore have a big effect on the activities of catalysts.

5. Conclusions and outlooks

In the PEMFCs, the conversion of chemical energy of the fuel (i.e., hydrogen) into electricity is carried out by the catalysts, which is of great importance in determining the performance of the PEMFCs. Many years of studies give strong evidences that metal/CNTs are more active in the fuel catalytic oxidation and provide better performance than other catalyst systems when used in the PEMFCs and thus attract tremendous attentions in recent years. However, for the preparation of metal/CNTs, surface functionalization of the CNTs is mostly required to produce the CNTs with suitable surface properties for metal deposition. The surface functionalization might lead to the structural destruction of the CNTs, which is detrimental to prepare the metal/CNTs of high efficiency in catalysis. It is generally believed that a mild surface modification method is desired for the functionalization of CNTs, which can effectively prevent the CNTs from the structural destruction and has great promises to synthesize the high efficient metal/CNT catalysts. Additionally, besides the functionalization methods, the catalytic activity of the metal/CNTs is also affected by the size and dispersion of the deposited metal NPs. For rational design of catalysts of high efficiency, it is therefore essentially important to know exactly the factors that affect the activity of the catalysts.

Currently, although the development of PEMFCs is moving toward commercialization due to the impressive research effort in recent years, significant challenges including detailed mechanism how the CNTs affect the catalytic activity of the metal/CNTs and high materials cost remain to be solved. It is clear that this research in these areas would be one of important on-going topics in the development of more highly efficient catalysts with low cost to meet the requirements of fuel cell commercialization.

6. References

Anderson, M. L.; Stroud, R. M. & Rolison, D. R. (2002). Enhancing the activity of fuel-cell reactions by designing three-dimensional nanostructured architectures: Catalyst-modified carbon−silica composite aerogels. *Nano Letters*, Vol. 2, No. 3, 235-240, ISSN 1530-6984

Andrews, R.; Jacques, D.; Qian, D. & Rantell, T. (2002). Multiwall carbon nanotubes: synthesis and application. *Accounts of Chemical Research*, Vol. 35, No. 12, 1008-1017, ISSN 0001-4842

Asano, K.; Kondo, D.; Kawabata, A.; Takei, F.; Nihei, M. & Awano, Y. (2006). Chemical modification of multiwalled carbon nanotubes by vacuum ultraviolet irradiation dry process. *Japanese Journal of Applied Physics,* Vol. 45, No. 4B, 3573-3576, ISSN 0021-4922

Bahr, J. L.; Yang, J.; Kosynkin, D. V.; Bronikowski, M. J.; Smalley, R. E. & Tour, J. M. (2001). Functionalization of carbon nanotubes by electrochemical reduction of aryl diazonium Salts: a bucky paper electrode. *Journal of the American Chemical Society,* Vol. 123, No. 27, 6536-6542, ISSN 0002-7863

Balasubramanian, K. & Burghard, M. (2005). Chemically functionalized carbon nanotubes. *Small,* Vol. 1, No. 2, 180-192, ISSN 1613-6810

Balasubramanian, K.; Friedrich, M.; Jiang, C.; Fan, Y.; Mews, A.; Burghard, M. & Kern, K. (2003). Electrical transport and confocal raman studies of electrochemically modified individual carbon nanotubes. *Advanced Materials,* Vol. 15, No. 18, 1515-1518, ISSN 0935-9648

Bambagioni, V.; Bianchini, C.; Marchionni, A.; Filippi, J.; Vizza, F.; Teddy, J.; Serp, P. & Zhiani, M. (2009). Pd and Pt–Ru anode electrocatalysts supported on multi-walled carbon nanotubes and their use in passive and active direct alcohol fuel cells with an anion-exchange membrane (alcohol =methanol, ethanol, glycerol). *Journal of Power Sources,* Vol. 190, No. 2, 241-251, ISSN 0378-7753

Baughman, R. H.; Zakhidov, A. A. & Heer, W. A. (2002). Carbon nanotubes-the route toward applications. *Science,* Vol. 297, No. 5582, 787-793, ISSN 0036-8075

Brunetti, F. G.; Herrero, M. A.; Muñoz, J. M.; Díaz-Ortiz, A.; Alfonsi, J.; Meneghetti, M.; Prato, M. & Vázquez, E. (2008). *Journal of the American Chemical Society,* Vol. 130, No. 25, 8094-8100, ISSN 0002-7863

Cassell, A. M.; Raymakers, J. A.; Kong, J. & Dai, H. J. (1999). Large scale CVD synthesis of single-walled carbon nanotubes. *Journal of Physical Chemistry B,* Vol. 103, No. 31, 6484-6492, ISSN 1520-6106

Chen, C.; Liang, B.; Ogino, A.; Wang, X. & Nagatsu, M. (2009). Oxygen functionalization of multiwall carbon nanotubes by microwave-excited surface-wave plasma treatment. *Journal of Physical Chemistry C,* Vol. 113, No. 18, 7659-7665, ISSN 1932-7447

Chen, C.; Liang, B.; Lu, D.; Ogino, A.; Wang, X. & Nagatsu, M. (2010). Amino group introduction onto multiwall carbon nanotubes by NH_3/Ar plasma treatment. *Carbon,* Vol. 48, No. 4, 939-948, ISSN 0008-6223

Chen, J.; Hamon, M. A.; Hu, H.; Chen, Y. S.; Rao, A. M.; Eklund, P. C. & Haddon, R. C. (1998). Solution properties of single-walled carbon nanotubes. *Science,* Vol. 282, No. 5386, 95-98, ISSN 0036-8075

Chen, Q. & Dai, L. (2000). Plasma patterning of carbon nanotubes. *Applied Physics Letters,* Vol. 76, No. 19, 2719-2721, ISSN 0003-6951

Chen, Q.; Dai, L.; Gao, M.; Huang, S. & Mau, A. (2001). *Journal of Physical Chemistry B,* Vol. 105, No. 3, 618-622, ISSN 1520-6106

Chen, R. J.; Bangsaruntip, S.; Drouvalakis, K. A.; Kam, W. S.; Shim, M.; Li, Y.; Kim, W.; Utz, P. J. & Dai, H. (2003). Noncovalent functionalization of carbon nanotubes for highly specific electronic biosensors. *Proceedings of the National Academy of Sciences of the United States of America,* Vol. 100, No. 9, 4984-4989, ISSN 0027-8424

Chen, R. J.; Zhang, Y.; Wang, D. & Dai, H. (2001). Noncovalent sidewall functionalization of single-walled carbon nanotubes for protein immobilization. *Journal of the American Chemical Society*, Vol. 123, No. 16, 3838-3839, ISSN 0002-7863

Chen, W.; Zhao, J.; Lee, J. Y. & Liu, Z. (2005). Microwave heated polyol synthesis of carbon nanotubes supported Pt nanoparticles for methanol electrooxidation. *Materials Chemistry and Physics*, Vol. 91, No. 1, 124-129, ISSN 0254-0584

Cheng, X.; Yi, B.; Han, M.; Zhang, J.; Qiao, Y. & Yu, J. (1999). Investigation of platinum utilization and morphology in catalyst layer of polymer electrolyte fuel cells. *Journal of Power Sources*, Vol. 79, No. 1, 75-81, ISSN 0378-7753

Choi, K. H.; Kim, H. S. & Lee, T. H. (1998). Electrode fabrication for proton exchange membrane fuel cells by pulse electrodeposition. *Journal of Power Sources*, Vol. 75, No. 2, 230-235, ISSN 0378-7753

Connel, M. J. O.; Boul, P.; Ericson, L. M.; Huffman, C.; Y. Wang, E. H.; Kuper, C.; Tour, J.; Ausman, K. D. & Smalley, R. E. (2001). Reversible water-solubilization of single-walled carbon nanotubes by polymer wrapping. *Chemical Physics Letters*, Vol. 342, No. 3-4, 265-271, ISSN 0009-2614

Correa-Duarte, M. A. & Liz-Marzán, L. M. (2006). Carbon nanotubes as templates for one dimensional nanoparticle assemblies. *Journal of Materials Chemistry*, Vol. 16, No. 1, 22-25, ISSN 0959-9428

Deluca, N. W. & Elabd, Y. A. (2006). Polymer electrolyte membranes for the direct methanol fuel cell: A review. *Journal of Polymer Science Part B: Polymer Physics*, Vol. 44, No. 16, 2201-2225, ISSN 0887-6266

Droppa Jr., R.; Hammer, P.; Carvalho, A. C. M.; Santos, M. C. D. & Alvarez, F. (2002). Incorporation of nitrogen in carbon nanotubes. *Journal of Non-Crystalline Solids*, Vol. 299-302, No. 2, 874-879, ISSN 0022-3093

Du, C. Y. & Zhao, T. S. & Liang, Z. X. (2008). Sulfonation of carbon-nanotube supported platinum catalysts for polymer electrolyte fuel cells. *Journal of Power Sources*, Vol. 176, No. 1, 9-15, ISSN 0378-7753

Dyke, C. A. & Tour, J. M. (2004). Overcoming the insolubility of carbon nanotubes through high degrees of sidewall functionalization. *Chemistry – A European Journal*, Vol. 10, No. 4, 812-817, ISSN 0947-6539

Feng, J. -M.; Li, Y. -L.; Hou, F. & Zhong, X. -H. (2008). Controlled growth of high quality bamboo carbon nanotube arrays by the double injection chemical vapor deposition process. *Materials Science and Engineering A*, Vol. 473, No. 1-2, 238-243, ISSN 0921-5093

Fujigaya, T.; Okamoto, M. & Nakashima, N. (2009). Design of an assembly of pyridine-containing polybenzimidazole, carbon nanotubes and Pt nanoparticles for a fuel cell electrocatalyst with a high electrochemically active surface area. *Carbon*, Vol. 47, No. 14, 3227-3232, ISSN 0008-6223

Gamez, A.; Richard, D.; Gallezot, P.; Gloaguen, F.; Faure, R. & Durand, R. (1996). Oxygen reduction on well-defined platinum nanoparticles inside recast ionomer. *Electrochimica Acta*, Vol. 41, No. 2, 307-314, ISSN 0013-4686

Georgakilas, V.; Gournis, D.; Tzitzios, V.; Pasquato, L.; Guldi, D. M. & Prato, M. (2007). Decorating carbon nanotubes with metal or semiconductor nanoparticles. *Journal of Materials Chemistry*, Vol. 17, No. 26, 2679–2694, ISSN 0959-9428

Georgakilas, V.; Kordatos, K.; Prato, M.; Guldi, D. M.; Holzinger, M. & Hirsch, A. (2002). Organic functionalization of carbon nanotubes. *Journal of the American Chemical Society*, Vol. 124, No. 5, 760-761, ISSN 0002-7863

Girishkumar, G.; Rettker, M.; Underhile, R.; Binz, D.; Vinodgopal, K.; McGinn, P. & Kamat, P. (2005). Single-wall carbon nanotube-based proton exchange membrane assembly for hydrogen fuel cells. *Langmuir*, Vol. 21, No. 18, 8487-8494, ISSN 0743-7463

Guo, D.-J. & Li, H.-L. (2005). High dispersion and electrocatalytic properties of platinum on functional multi-walled carbon nanotubes. *Electroanalysis*, Vol. 17, No. 10, 869-872, ISSN 1040-0397

Halder, A.; Sharma, S.; Hegde, M. S. & Ravishankar, N. (2009). Controlled attachment of ultrafine platinum nanoparticles on functionalized carbon nanotubes with high electrocatalytic activity for methanol oxidation. *Journal of Physical Chemistry C*, Vol. 113, No. 4, 1466-1473, ISSN 1932-7447

Han, K. I.; Lee, J. S.; Park, S. O.; Lee, S. W.; Park, Y. W. & Kim, H. (2004). Studies on the anode catalysts of carbon nanotube for DMFC. *Electrochimica Acta*, Vol. 50, No. 2-3, 791-794, ISSN 0013-4686

Han, L.; Wu, W.; Kirk, F. L.; Luo, J.; Maye, M. M.; Kariuki, N. N.; Lin, Y.; Wang, C. & Zhong, C. J. (2004). A direct route toward assembly of nanoparticle−carbon nanotube composite materials. *Langmuir*, Vol. 20, No. 14, 6019-6025, ISSN 0743-7463

Heitner-Wirguin, C. (1996). Recent advances in perfluorinated ionomer membranes: structure, properties and applications. *Journal of Membrane Science*, Vol. 120, No. 1, 1-33, ISSN 0376-7388

Hernadi, K.; Siska, A. & Thien-Nga, L. (2001). Reactivity of different kinds of carbon during oxidative purification of catalytically prepared carbon nanotubes. *Solid State Ionics*, Vol. 141-142, 203-209, ISSN 0167-2738

Hernández-Fernández, P.; Montiel, M.; Ocón, P.; de la Fuente, J. L. G.; García-Rodríguez, S.; Rojas, S. & Fierro, J. L. G. (2010). Functionalization of multi-walled carbon nanotubes and application as supports for electrocatalysts in proton-exchange membrane fuel cell. *Applied Catalysis B: Environmental*, Vol. 99, No. 1-2, 343-352, ISSN 0926-3373

Hirsch, A. (2002). Functionalization of single-walled carbon nanotubes. *Angewandte Chemie International Edition*, Vol. 41, No. 11, 1853-1959, ISSN 1433-7851

Holzinger, M.; Vostrowskv, O.; Hirsch, A.; Hennrich, F.; Kappes, M.; Weiss, R. & Jellen, F. (2001). Sidewall functionalization of carbon nanotubes. *Angewandte Chemie International Edition*, Vol. 40, No. 21, 4002-4005, ISSN 1433-7851

Hsin, Y. L.; Hwang, K. C. & Yeh, C. T. (2007). Poly(vinylpyrrolidone)-modified graphite carbon nanofibers as promising supports for PtRu catalysts in direct methanol fuel cells. *Journal of the American Chemical Society*, Vol. 129, No. 32, 9999-10010, ISSN 0002-7863

Huang, Y.; Gao, J. & Liu, R. (2000). Structure and electronic properties of nitrogen-containing carbon nanotubes. *Synthetic Metals*, Vol. 113, No. 3, 251-255, ISSN 0379-6779

Hwang, K. C. (1995). Efficient cleavage of carbon graphene layers by oxidants. *Journal of the Chemical Society, Chemical Communications*, Vol. 2, No. 24, 173-174, ISSN 0022-4936

Iijima, S. (1991). Helical microtubules of graphitic carbon. *Nature*, Vol. 354, 56-58, ISSN 0028-0836

Ijima, S. & Ichihashi, T. (1993). Single-shell carbon nanotubes of 1-nm diameter. *Nature*, Vol. 363, 603-605, ISSN 0028-0836

Islam, M. F.; Rojas, E.; Bergey, D. M.; Johnson, A. T. & Yodh, A. G. (2003). High weight fraction surfactant solubilization of single-wall carbon nanotubes in water. *Nano Letters*, Vol. 3, No. 2, 269-273, ISSN 1530-6984

Jang, I. Y.; Lee, S. H.; Park, K. C.; Wongwiriyapan, W.; Kim, C.; Teshima, K.; Oishi, S.; Kim, Y. J. & Endo, M. (2009). Effect of photochemically oxidized carbon nanotubes on the deposition of platinum nanoparticles for fuel cell catalysts. *Electrochemistry Communications*, Vol. 11, No. 7, 1472-1475, ISSN 1388-2481

Jiang, Z. Q.; Jiang, Z.-J. & Meng, Y. D. (2011). High catalytic performance of Pt nanoparticles on plasma treated carbon nanotubes for electrooxidation of ethanol in a basic solution. *Applied Surface Science*, Vol. 257, No. 7, 2923-2928, ISSN 0169-4332

Jiang, Z. Q.; Yu, X. Y.; Jiang, Z.-J.; Meng, Y. D. & Shi, Y. C. (2009). Synthesis of monodispersed Pt nanoparticles on plasma processed carbon nanotubes for methanol electro-oxidation reaction. *Journal of Materials Chemistry*, Vol. 19, No. 37, 6720-6726, ISSN 0959-9428

Journet, C.; Maser, W. K.; Bernier, P.; Loiseau, A.; Chappelle, M. L.; Lefrant, S.; Deniard, P.; Lee, R. & Fischer, J. E. (1997). Large-scale production of single-walled carbon nanotubes by the electric-arc technique. *Nature*, Vol. 388, No.756-758, ISSN 0028-0836

Kamavarama, V.; Veedub, V. & Kannana, A. M. (2009). Synthesis and characterization of platinum nanoparticles on in situ grown carbon nanotubes based carbon paper for proton exchange membrane fuel cell cathode. *Journal of Power Sources*, Vol. 188, No. 1, 51-56, ISSN 0378-7753

Kannan, R.; Kakade, B. A. & Pillai, V. K. (2008). Polymer electrolyte fuel cells using Nafion-based composite membranes with functionalized carbon nanotubes. *Angewandte Chemie International Edition*, Vol. 47, No.14, 2653–2656, ISSN 1433-7851

Kerres, J. A. (2001). Development of ionomer membranes for fuel cells. *Journal of Membrane Science*, Vol. 185, No. 1, 3-27, ISSN 0376-7388

Khare, B. N.; Wilhite, P.; Quinn, R. C.; Chen, B.; Schingler, R. H.; Tran, B.; Imanaka, H.; So, C. R.; Bauschlicher, C. W. & Meyyappan, M. (2004). Functionalization of carbon nanotubes by ammonia glow-discharge: experiments and modeling. *Journal of Physical Chemistry B*, Vol. 108, No. 24, 8166-8172, ISSN 1520-6106

Kim, B. & Sigmund, W. M. (2004). Functionalized multiwall carbon nanotube/gold nanoparticle composites. *Langmuir*, Vol. 20, No. 19, 8239-8242, ISSN 0743-7463

Kongkanand, A.; Kuwabata, S.; Girishkumar, G. & Kamat, P. V. (2006). Single-wall carbon nanotubes supported platinum nanoparticles with improved electrocatalytic activity for oxygen reduction reaction. *Langmuir*, Vol. 22, No. 5, 2392-2396, ISSN 0743-7463

Kongkanand, A.; Vinodgopal, K.; Kuwabata, S. & Kamat, P. V. (2006). Highly dispersed Pt catalysts on single-walled carbon nanotubes and their role in methanol oxidation. *Journal of Physical Chemistry B*, Vol. 110, No. 23, 16185-16188, ISSN 1520-6106

Koshio, A.; Yudasaka, M.; Zhang, M. & Iijima, S. (2001). A simple way to chemically react single-wall carbon nanotubes with organic materials using ultrasonication. *Nano Letters*, Vol. 1, No. 7, 361-363, ISSN 1530-6984

Kundu, S.; Nagaiah, T. C.; Xia, W.; Wang, Y.; Van Dommele, S.; Bitter, J. H.; Santa, M.; Grundmeier, G.; Bron, M.; Schuhmann, W. & Muhler, M. (2009). Electrocatalytic activity and stability of nitrogen-containing carbon nanotubes in the oxygen reduction reaction. *Journal of Physical Chemistry C*, Vol. 113, No. 32, 14302-14310, ISSN 1932-7447

Kuo, P. L.; Chen, C. C. & Jao, M. W. (2005). Effects of polymer micelles of alkylated polyethylenimines on generation of gold nanoparticles. *Journal of Physical Chemistry B*, Vol. 109, No. 19, 9445-9450, ISSN 1520-6106

Kyotani, T.; Nakazaki, S.; Xu, W. H. & Tomita, A. (2001). Chemical modification of the inner walls of carbon nanotubes by HNO_3 oxidation. *Carbon*, Vol. 39, No. 5, 782-785, ISSN 0008-6223

Lee, C.; Ju, Y.; Chou, P.; Huang, Y.; Kuo, L. & Oung, J. (2005). Preparation of Pt nanoparticles on carbon nanotubes and graphite nanofibers via self-regulated reduction of surfactants and their application as electrochemical catalyst. *Electrochemistry Communications*, Vol. 7, No. 4, 453-458, ISSN 1388-2481

Lee, K. C.; Zhang, J. J.; Wang, H. J. & Wilkinson D. P. (2006). Progress in the synthesis of carbon nanotube- and nanofiber-supported Pt electrocatalysts for PEM fuel cell catalysis. *Journal of Applied Electrochemistry*, Vol. 36, No. 5, 507–522, ISSN 0021-891X

Lee, S. W.; Kim, B.; Chen, S.; Shao-Horn, Y. & Hammond, P. T. (2009). Layer-by-Layer assembly of all carbon nanotube ultrathin films for electrochemical applications. *Journal of the American Chemical Society*, Vol. 131, No. 2, 671-679, ISSN 0002-7863

Li, J.; Moskovits, M. & Haslett, T. L. (1998). Nanoscale electroless metal deposition in aligned carbon nanotubes. *Chemistry of Materials*, Vol. 10, No. 7, 1963-1967, ISSN 0897-4756

Li, Q.; He, R.; Gao, J. A.; Jensen, J. O. & Bjerrum, N. J. (2003). The CO poisoning effect in PEMFCs operational at temperatures up to 200°C. *Journal of the Electrochemical Society*, Vol. 150, No. 12, A1599-A1605, ISSN 0013-4651

Li, Q.; He, R.; Jensen, J. O. & Bjerrum, N. J. (2004). PBI-based polymer membranes for high temperature fuel cells–preparation, characterization and fuel cell demonstration. *Fuel cells*, Vol. 4, No. 3, 147-159, ISSN 1615-6846

Li, X.; Ge, S.; Hui, C. L. & Hsing, I. -M. (2004). Well-dispersed multiwalled carbon nanotubes supported platinum nanocatalysts for oxygen reduction. *Electrochemical and solid state letters*, Vol. 7, No. 9, A286-A289, ISSN 1099-0062

Li, X. & Hsing, I. -M. (2006). The effect of the Pt deposition method and the support on Pt dispersion on carbon nanotubes. *Electrochimica Acta,* Vol. 51, No. 25, 5250-5258, ISSN 0013-4686

Li, W. Z.; Liang, C. H.; Qiu, J. S.; Zhou, W. J.; Han, H. M.; Wei, Z. B.; Sun, G. Q. & Xin, Q (2002). Carbon nanotubes as support for cathode catalyst of a direct methanol fuel cell. *Carbon*, Vol. 40, No. 5, 791-794, ISSN 0008-6223

Li, W. Z.; Liang, C. H.; Zhou, W. J.; Qiu, J. S.; Li, H. Q.; Sun, G. Q. & Xin, Q. (2004). Homogeneous and controllable Pt particles deposited on multi-wall carbon nanotubes as cathode catalyst for direct methanol fuel cells. *Carbon*, Vol. 42, No. 2, 436-439, ISSN 0008-6223

Li, W. Z.; Liang, C. H.; Zhou, W. J.; Qiu, J. S.; Zhou, Z. H. & Sun, G. Q. (2003). Preparation and characterization of multi-walled carbon nanotube supported platinum for cathode catalysts of direct ethanol fuel cells. *Journal of Physical Chemistry B*, Vol. 107, No. 26, 6292-6299, ISSN 1520-6106

Li, W. Z.; Wang, X.; Chen, Z. W.; Waje, M. & Yan, Y. S. (2005). Carbon nanotube film by filtration as cathode catalyst support for proton-exchange membrane fuel cell. *Langmuir*, Vol. 21, No. 21, 9386-9389, ISSN 0743-7463

Li, W. Z.; Wang, X.; Chen, Z. W.; Waje, M. & Yan, Y. S. (2006). Pt-Ru supported on double-walled carbon nanotubes as high-performance anode catalysts for direct methanol fuel cells. *Journal of Physical Chemistry B*, Vol. 110, No. 31, 15353-15358, ISSN 1520-6106

Liao, S. J.; Holmes, K. A.; Tsaprailis, H. & Birss, V. I. (2006). High performance PtRuIr catalysts supported on carbon nanotubes for the anodic oxidation of methanol. *Journal of the American Chemical Society*, Vol. 128, No. 11, 3504-3505, ISSN 0002-7863

Lin, J. F.; Kamavaram, V. & Kannan, A. M. (2010). Synthesis and characterization of carbon nanotubes supported platinum nanocatalyst for proton exchange membrane fuel cells. *Journal of Power Sources*, Vol. 195, No. 2, 466-470, ISSN 0378-7753

Lin, J. F.; Mason, C. W.; Adame, A.; Liu, X.; Peng, X. H. & Kannan, A. M. (2010). Synthesis of Pt nanocatalyst with micelle-encapsulated multi-walled carbon nanotubes as support for proton exchange membrane fuel cells. *Electrochimica Acta*, Vol. 55, No. 22, 6496-6500, ISSN 0013-4686

Lin, Y.; Cui, X.; Yen, C. & Wai, C. M. (2005). Platinum/carbon nanotube nanocomposite synthesized in supercritical fluid as electrocatalysts for low-temperature fuel cells. *Journal of Physical Chemistry B*, Vol. 109, No. 30, 14410–14415, ISSN 1520-6106

Lin, Y.; Taylor,S.; Li, H.; Fernando, K. A. S.; Qu, L.; Wang,W.; Gu,L.; Zhou, B. & Sun, Y. -P. (2004). Advances toward bioapplications of carbon nanotubes. *Journal of Materials Chemistry*, Vol. 14, No. 4, 527-541, ISSN 0959-9428

Liu, Z.; Gan, L. M.; Hong, L.; Chen, W. & Lee, J. Y. (2005). Carbon-supported Pt nanoparticles as catalysts for proton exchange membrane fuel cells. *Journal of Power Sources*, Vol. 139, No. 1-2, 73-78, ISSN 0378-7753

Liu, Z.; Lee, J. Y.; Chen, W.; Han, M. & Gan, L. M. (2004). Physical and electrochemical characterizations of microwave-assisted polyol preparation of carbon-supported PtRu nanoparticles. *Langmuir*, Vol. 20, No. 1, 181-187, ISSN 0743-7463

Liu, Z.; Lin, X.; Lee, J. Y.; Zhang, W.; Han, M. & Gan, L. M. (2002). Preparation and characterization of platinum-based electrocatalysts on multiwalled carbon nanotubes for proton exchange membrane fuel cells. *Langmuir*, Vol. 18, No. 10, 4054-4060, ISSN 0743-7463

Lordi, V.; Yao, N. & Wei, J. (2001). Method for supporting platinum on single-walled carbon nanotubes for a selective hydrogenation catalyst. *Chemistry of Materials*, Vol. 13, No. 3, 733-737, ISSN 0897-4756

Ma, P. C.; Kim, J. & Tang, B. Z. (2006). Functionalization of carbon nanotubes using a silane coupling agent. *Carbon*, Vol. 44, No. 15, 3232-3238, ISSN 0008-6223

Maiyalagan, T.; Viswanathan, B. & Varadaraju, U. V. (2005). Nitrogen containing carbon nanotubes as supports for Pt – Alternate anodes for fuel cell applications. *Electrochemistry Communications*, Vol. 7, No. 9, 905-912, ISSN 1388-2481

Matsumoto, K.; Higashihara, T. & Ueda, M. (2009). Locally and densely sulfonated poly(ether sulfone)s as proton exchange membrane. *Macromolecules*, Vol. 42, No.4, 1161-1166, ISSN 0024-9297

Matsumoto, T.; Komatsu, T.; Arai, K.; Yamazaki, T.; Kijima, M.; Shimizu, H.; Takasawa, Y. & Nakamura, J. (2004). Reduction of Pt usage in fuel cell electrocatalysts with carbon nanotube electrodes. *Chemical Communications*, Vol. 7, 840-841, 1359-7345

Matsumoto, T.; Komatsu, T.; Nakano, H.; Arai, K.; Nagashima, Y.; Yoo, E.; Yamazaki, T.; Kijima, M.; Shimizu, H.; Takasawa, Y. & Nakamura, J. (2004). Efficient usage of highly dispersed Pt on carbon nanotubes for electrode catalysts of polymer electrolyte fuel cells. *Catalysis Today*, Vol. 90, No. 3-4, 277-281, ISSN 0920-5861

Mawhinney, D. B.; Naumenko, V.; Kuznetsova, A.; Jr. Yates, J. T.; Liu, J. & Smalley, R. E. (2000). Infrared spectral evidence for the etching of carbon nanotubes: ozone oxidation at 298 K. *Journal of the American Chemical Society*, Vol. 122, No. 10, 2383-2384, ISSN 0002-7863

Mawhinney, D. B.; Naumenko, V.; Kuznetsova, A.; Jr. Yates, J. T.; Liu, J. & Smalley, R. E. (2000). Surface defect site density on single walled carbon nanotubes by titration. *Chemical Physics Letters*, Vol. 324, No.1-3, 213-216, ISSN 0009-2614

Michelson, E. T.; Chiang, I. W.; Zimmermann, J. L.; Boul, P. J.; Lozano, J.; Liu, J.; Smalley, R. E.; Hauge, R. H. & Margrave, J. L. (1999). Solvation of fluorinated single-walled carbon nanotubes in alcohol solvents. *Journal of Physical Chemistry B*, Vol. 103, No. 21, 4318-4322, ISSN 1520-6106

Miyamoto, Y.; Cohen, M. L. & Louie, S. G. (1997). Theoretical investigation of graphitic carbon nitride and possible tubule forms. *Solid State Communications*, Vol. 102, No. 8, 605-608, ISSN 0038-1098

Moghaddam, M. J.; Taylor, S.; Gao, M.; Huang, S. M.; Dai, L. M. & McCall, M. J. (2004). Highly efficient binding of DNA on the sidewalls and tips of carbon nanotubes using photochemistry. *Nano Letters*, Vol. 4, No. 1, 89-93, ISSN 1530-6984

Moore, V. C.; Strano, M. S.; Haroz, E. H.; Hauge, R. H. & Smalley, R. E. (2003). Individually suspended single-walled carbon nanotubes in various surfactants. *Nano Letters*, Vol. 3, No. 10, 1379-1382, ISSN 1530-6984

Mu, Y.; Liang, H.; Hu, J.; Jiang, L. & Wan, L. (2005). Controllable Pt nanoparticle deposition on carbon nanotubes as an anode catalyst for direct methanol fuel cells. *Journal of Physical Chemistry B*, Vol. 109, No. 47, 22212-22216, ISSN 1520-6106

Nxumalo, E. N.; Nyamori, V. O. & Coville, N. J. (2008). CVD synthesis of nitrogen doped carbon nanotubes using ferrocene/aniline mixtures. *Journal of Organometallic Chemistry*, Vol. 693, No. 17, 2942-2948, ISSN 0022-328X

Okamoto, M.; Fujigaya, T. & Nakashima, N. (2008). Individual dissolution of single-walled carbon nanotubes by using polybenzimidazole, and highly effective reinforcement of their composite films. *Advanced Functional Materials*, Vol. 18, No. 12, 1776-1782, ISSN 1616-301X

Okamoto, M.; Fujigaya, T. & Nakashima, N. (2009). Design of an assembly of poly(benzimidazole), carbon nanotubes, and Pt nanoparticles for a fuel-cell electrocatalyst with an ideal interfacial nanostructure. *Small*, Vol. 5, No. 6, 735-740, ISSN 1613-6810

Ostrander, J. W.; Mamedov, A. A. & Kotov, N. A. (2001). Two modes of linear Layer-by-Layer growth of nanoparticle–polylectrolyte multilayers and different interactions in the Layer-by-layer deposition. *Journal of the American Chemical Society*, Vol. 123, No. 6, 1101-1110, ISSN 0002-7863

Park, H. S.; Choi, B. G.; Yang, S. H.; Shin, W. H.; Kang, J. K.; Jung, D. & Hong, W. H. (2009). Ionic-liquid-assisted sonochemical synthesis of carbon-nanotube-based nanohybrids: control in the structures and interfacial characteristics. *Small*, Vol. 5, No. 15, 1754-1760, ISSN 1613-6810

Parvulescu, V. I. & Hardacre, C. (2007). Catalysis in ionic liquids. *Chemical Reviews*, Vol. 107, No. 6, 2615-2665, ISSN 0009-2665

Pham-Huu, C.; Keller, N.; Roddatis, V. V.; Mestl, G.; Schloegl, R. & Ledoux, M. J. (2002). Large scale synthesis of carbon nanofibers by catalytic decomposition of ethane on nickel nanoclusters decorating carbon nanotubes. *Physical Chemistry Chemical Physics*, Vol. 4, No. 3, 514-521, ISSN 1463-9076

Plank, N. O. V.; Cheung, R. & Andrews, R. J. (2004). Thiolation of single-wall carbon nanotubes and their self-assembly. *Applied Physics Letters*, Vol. 85, No. 15, 3229-3231, ISSN 0003-6951

Plank, N. O. V.; Jiang, L. & Cheung, R. (2003). Fluorination of carbon nanotubes in CF_4 plasma. *Applied Physics Letters*, Vol. 83, No. 12, 2426-2428, ISSN 0003-6951

Prabhuram, J.; Zhao, T. S.; Tang, Z. K.; Chen, R. & Liang, Z. X. (2006). Multiwalled carbon nanotube supported PtRu for the anode of direct methanol fuel cells. *Journal of Physical Chemistry B*, Vol. 110, No. 11, 5245-5252, ISSN 1520-6106

Qu, L. T. & Dai, L. M. (2005). Substrate-enhanced electroless deposition of metal nanoparticles on carbon nanotubes. *Journal of the American Chemical Society*, Vol. 127, No. 31, 10806-10807, ISSN 0002-7863

Qu, L. T.; Dai, L. M. & Osawa, E. (2006). Shape/size-control led syntheses of metal nanoparticles for site-selective modification of carbon nanotubes. *Journal of the American Chemical Society*, Vol. 128, No. 16, 5523-5532, ISSN 0002-7863

Rajalakshmi, N.; Ryu, H.; Shaijumon, M. M. & Ramaprabhu, S. (2005). Performance of polymer electrolyte membrane fuel cells with carbon nanotubes as oxygen reduction catalyst support material. *Journal of Power Sources*, Vol. 140, No. 2, 250-257, ISSN 0378-7753

Reddy, A. L. M. & Ramaprabhu, S. (2007). Pt/SWNT-Pt/C Nanocomposite Electrocatalysts for Proton-Exchange Membrane Fuel Cells. *Journal of Physical Chemistry C*, Vol. 111, No. 44, 16138-16146, ISSN 1932-7447

Richard, C.; Balavoine, F.; Schultz, P.; Ebbesen, T. W. & Mioskowski, C. (2003). Supramolecular self-assembly of lipid derivatives on carbon nanotubes. *Science*, Vol. 300, No. 5620, 775-778, ISSN 0036-8075

Ruelle, B.; Felten, A.; Ghijsen, J.; Drube, W.; Johnson, R. L. & Liang, D. (2008). Functionalization of MWCNTs with atomic nitrogen: electronic structure. *Journal of Physics D: Applied Physics*, Vol. 41, No. 4, 045202 (1-4), ISSN 0022-3727

Saha, M. S. & Kundu, A. (2010). Functionalizing carbon nanotubes for proton exchange membrane fuel cells electrode. *Journal of Power Sources*, Vol. 195, No. 19, 6255–6261, ISSN 0378-7753

Saha, M. S.; Li, R. & Sun, X. (2008). High loading and monodispersed Pt nanoparticles on multiwalled carbon nanotubes for high performance proton exchange membrane fuel cells. *Journal of Power Sources*, Vol. 177, No. 2, 314-322, ISSN 0378-7753

Sarma, L. S.; Lin, T. D.; Tsai, Y. W.; Chen, J. M. & Hwang, B. J. (2005). Carbon supported Pt–Ru catalysts prepared by the Nafion stabilized alcohol-reduction method for application in direct methanol fuel cells. *Journal of Power Sources*, Vol. 139, No. 1-2, 44-54, ISSN 0378-7753

Scibioh, M. A.; Oh. I. H.; Lim, H. T.; Hong, S. A. & Ha, H. Y. (2008). Investigation of various ionomer-coated carbon supports for direct methanol fuel cell applications. *Applied Catalysis B: Environmental*, Vol. 77, No. 3-4, 373-385, ISSN 0926-3373

Selvaraj, V. & Alagar, M. (2008). Ethylene glycol oxidation on Pt and Pt–Ru nanoparticle decorated polythiophene/multiwalled carbon nanotube composites for fuel cell applications. *Nanotechnology*, Vol. 19, No. 4, 045504 (1-8), ISSN 0957-4484

Shao, Y.; Sui, J.; Yin, G. & Gao, Y. (2008). Nitrogen-doped carbon nanostructures and their composites as catalytic materials for proton exchange membrane fuel cell. *Applied Catalysis B: Environmental*, Vol. 79, No. 1, 89-99, ISSN 0926-3373

Shao, Y.; Yin, G.; Wang, J.; Gao, Y. & Shi, P. (2006). Multi-walled carbon nanotubes based Pt electrodes prepared with in situ ion exchange method for oxygen reduction. *Journal of Power Sources*, Vol. 161, No. 1, 47-53, ISSN 0378-7753

Shen, P. K. (2008). PEM Fuel Cell Catalyst Layers and MEAs, In: *PEM fuel cell electrocatalysts and catalyst layers*, J. J. Zhang, (Ed.), 355-380, Springer, ISBN 978-1-84800-935-6, London, United Kingdom

Star, A.; Stoddart, J. F.; Steuerman, D.; Diehl, M.; Boukai, A.; Wong, W.; Yang, X.; Chung, S.; Choi, H. & Heath, J. R. (2001). Preparation and properties of polymer-wrapped single-walled carbon nanotubes. *Angewandte Chemie International Edition*, Vol. 40, No. 9, 1721-1725, ISSN 1433-7851

Sun, C. L.; Chen, L. C.; Su, M. C.; Hong, L. S.; Chyan, O.; Hsu, C. Y.; Chen, K. H.; Chang, T. F. & Chang, L. (2005). Ultrafine platinum nanoparticles uniformly dispersed on arrayed CNx nanotubes with high electrochemical activity. *Chemistry of Materials*, Vol. 17, No. 14, 3749-3753, ISSN 0897-4756

Sun, X.; Li, R.; Villers, D.; Dodelet, J. P. & Desilets, S. (2003). Composite electrodes made of Pt nanoparticles deposited on carbon nanotubes grown on fuel cell backings. *Chemical Physics Letters*, Vol. 379, No. 1-2, 99-104, ISSN 0009-2614

Sun, X. & Saha, M. S. (2008). Nanotubes, nanofibers and nanowires as supports for catalysts, In: *PEM fuel cell electrocatalysts and catalyst layers*, J. J. Zhang, (Ed.), 665-714, Springer, ISBN 978-1-84800-935-6, London, United Kingdom

Tang, J. M.; Jensen, K.; Waje, M.; Wenzhen; Larsen, P.; Pauley, K.; Chen, Z.; Ramesh, P.; Itkis, M. E.; Yan, Y. & Haddon, R. C. (2007). High performance hydrogen fuel cells with ultralow Pt loading carbon nanotube thin film catalysts. *Journal of Physical Chemistry C*, Vol. 111, No. 48, 17901-17904, ISSN 1932-7447

Tang, Z.; Poh, C. K.; Lee, K. K.; Tian, Z. Q.; Chua, D. & Lin, J. Y. (2010). Enhanced catalytic properties from platinum nanodots covered carbon nanotubes for proton-exchange membrane fuel cells. *Journal of Power Sources*, Vol. 195, No. 1, 155-159, ISSN 0378-7753

Tao, X. Y.; Zhang, X. B.; Sun, F. Y.; Cheng, J. P.; Liu, F. & Luo, Z. Q. (2007). Large-scale CVD synthesis of nitrogen-doped multi-walled carbon nanotubes with controllable nitrogen content on a $Co_xMg_{1-x}MoO_4$ catalyst. *Diamond and Related Materials*, Vol. 16, No. 3, 425-430, ISSN 0925-9635

Tasis, D.; Tagmatarchis, N.; Bianco, A. & Prato, M. (2006). Chemistry of carbon nanotubes. *Chemical Reviews*, Vol. 106, No.1105-1136, ISSN 0009-2665

Tauster, S. J.; Fung, S. C.; Baker, R. T. K. & Horsley, J. A. (1981). Strong interactions in supported-metal catalysts. *Science*, Vol. 211, No. 4487, 1121-1125, ISSN 0036-8075

Taylor, E. J.; Anerson, E. B. & Vilambi, N. R. K. (1992). Preparation of high-platinum-utilization gas diffusion electrodes for proton-exchange-membrane fuel cells. *Journal of the Electrochemical Society*, Vol. 139, No. 5, L45-L46, ISSN 0920-5861

Thess, A.; Lee, R.; Nikolaev, P.; Dai, H. J.; Petit, P.; Robert, J.; Xu, C. H.; Lee, Y. H.; Kim, S. G.; Rinzler, A. G.; Colbert, D. T.; Scuseria, G. E.; Tomanek, D.; Fischer, J. E. & Smalley, R. E. (1996). Crystalline ropes of metallic carbon nanotubes. *Science*, Vol. 273, No. 5274, 483-487, ISSN 0036-8075

Thompson, S. D.; Jordan, L. R. & Forsyth, M. (2001). Platinum electrodeposition for polymer electrolyte membrane fuel cells. *Electrochimica Acta*, Vol. 46, No. 10-11, 1657-1663, ISSN 0013-4686

Tian, Z. Q.; Jiang, S. P.; Liu, Z. C. & Li, L. (2007). Polyelectrolyte-stabilized Pt nanoparticles as new electrocatalysts for low temperature fuel cells. *Electrochemistry Communications*, Vol. 9, No. 7, 1613-1618, ISSN 1388-2481

Tsai, M.; Yeh, T. & Tsai, C. (2006). An improved electrodeposition technique for preparing platinum and platinum–ruthenium nanoparticles on carbon nanotubes directly grown on carbon cloth for methanol oxidation. *Electrochemistry Communications*, Vol. 8, No. 9, 1445-1452, ISSN 1388-2481

Villers, D.; Sun, S. H.; Serventi, A. M. & Dodelet, J. P. (2006). Characterization of Pt nanoparticles deposited onto carbon nanotubes grown on carbon paper and evaluation of this electrode for the reduction of oxygen. *Journal of Physical Chemistry B*, Vol. 110, No. 51, 25916-25925, ISSN 1520-6106

Wang, C. H.; Du, H. Y.; Tsai, Y. T.; Chen, C. P.; Huang, C. J.; Chen, L. C.; Chen, K. H. & Shih, H. C. (2007). High performance of low electrocatalysts loading on CNT directly grown on carbon cloth for DMFC. *Journal of Power Sources*, Vol. 171, No. 1, 55-62, ISSN 0378-7753

Wang, C. -H.; Shih, H. -C.; Tsai, Y. -T. Du, H. -Y. Chen, L. -C. & Chen, K. -H. (2006). High methanol oxidation activity of electrocatalysts supported by directly grown nitrogen-containing carbon nanotubes on carbon cloth. *Electrochimica Acta*, Vol. 52, No. 4, 1612-1617, ISSN 0013-4686

Wang, D.; Lu, S. & Jiang, S. P. (2010). Tetrahydrofuran-functionalized multi-walled carbon nanotubes as effective support for Pt and PtSn electrocatalysts of fuel cells. *Electrochimica Acta*, Vol. 55, No. 8, 2964-2971, ISSN 0013-4686

Wang, J.; Yin, G.; Shao, Y.; Wang, Z. & Gao, Y. (2007). Platinum deposition on multiwalled carbon nanotubes by ion-exchange method as electrocatalysts for oxygen reduction. *Journal of the Electrochemical Society*, Vol. 154, No. 7, B687-B693, ISSN 0013-4651

Wang, J. T.; Savinell, R. F.; Wainright, J.; Litt, M. & Yu. H. (1996). A H_2/O_2 fuel cell using acid doped polybenzimidazole as polymer electrolyte. *Electrochimica Acta*, Vol. 41, No. 2, 193-197, ISSN 0013-4686

Wang, S. Y.; Jiang, S. P.; White, T. J.; Guo, J. & Wang, X. (2009). Electrocatalytic activity and interconnectivity of Pt nanoparticles on multiwalled carbon nanotubes for fuel cells. *Journal of Physical Chemistry C*, Vol. 113, No. 43, 18935-18945, ISSN 1932-7447

Wang, X.; Li, W.; Chen, Z.; Waje, M. & Yan, Y. (2006). Durability investigation of carbon nanotube as catalyst support for proton exchange membrane fuel cell. *Journal of Power Sources*, Vol. 158, No. 1, 154-159, ISSN 0378-7753

Wang, X.; Waje, M. & Yan, Y. (2005). CNT-based electrodes with high efficiency for PEMFCs. *Electrochemical and Solid State Letters*, Vol. 8, No. 1, A42-A44, ISSN 1099-0062

Wang, Y.; Xu, X.; Tian, Z.; Zong, Y.; Cheng, H. & Lin, C. (2006). Selective heterogeneous nucleation and growth of size-controlled metal nanoparticles on carbon nanotubes in solution. *Chemistry – A European Journal*, Vol. 12, No. 9, 2542-2549, ISSN 0947-6539

Wei, Z. D.; Yan, C.; Tan, Y.; Li, L.; Sun, C. X.; Shao, Z. G.; Shen, P. K. & Dong, H. W. (2008). Spontaneous reduction of Pt(IV) onto the sidewalls of functionalized multiwalled carbon nanotubes as catalysts for oxygen reduction reaction in PEMFCs. *Journal of Physical Chemistry C*, Vol. 112, No. 7, 2671-2677, ISSN 1932-7447

Welton, T. (1999). Room-temperature ionic liquids. solvents for synthesis and catalysis. *Chemical Reviews*, Vol. 99, No. 8, 2071-2084, ISSN 0009-2665

Wilson, M. S. & Gottesfeld, S. (1992). High performance catalyzed membranes of ultra-low Pt loadings for polymer electrolyte fuel cells. *Journal of the Electrochemical Society*, Vol. 139, No. 2, L28-L30, ISSN 0013-4651

Wu, B. H.; Hu, D.; Kuang, Y. J.; Liu, B.; Zhang, X. H. & Chen, J. H. (2009). Functionalization of carbon nanotubes by an ionic-liquid polymer: dispersion of Pt and PtRu nanoparticles on carbon nanotubes and their electrocatalytic oxidation of methanol. *Angewandte Chemie International Edition*, Vol. 48, No. 26, 4751-4574, ISSN 1433-7851

Wu, G.; Li, L.; Li, J. H. & Xu, B. Q. (2006). Methanol electrooxidation on Pt particles dispersed into PANI/SWNT composite films. *Journal of Power Sources*, Vol. 155, No. 2, 118-127, ISSN 0378-7753

Wunderlich, W. (2007). Growth model for plasma-CVD growth of carbon nano-tubes on Ni-sheets. *Diamond and Related Materials*, Vol. 16, No. 2, 369-378, ISSN 0925-9635

Xing, P.; Robertson, G. P.; Guiver, M. D.; Mikhailenko, S. D. & Kaliaguine, S. (2004). Sulfonated poly(aryl ether ketone)s containing the hexafluoroisopropylidene diphenyl moiety prepared by direct copolymerization, as proton exchange membranes for fuel cell application. *Macromolecules*, Vol. 37, No. 21, 7960-7967, ISSN 0024-9297

Xing, Y. C.; Li, L.; Chusuei, C. C. & Hull, R. V. (2005). Sonochemical oxidation of multiwalled carbon nanotubes. *Langmuir*, Vol. 21, No. 9, 4185-4190, ISSN 0743-7463

Xing, Y. C. (2004). Synthesis and electrochemical characterization of uniformly-dispersed high Loading Pt nanoparticles on sonochemically-treated carbon nanotubes. *Journal of Physical Chemistry B*, Vol. 108, No. 50, 19255-19259, ISSN 1520-6106

Yan, Y. H.; Chan-Park, M. B.; Zhao, Q.; Li, C. M. & Yue, C. Y. (2005). Functionalization of carbon nanotubes by argon plasma-assisted ultraviolet grafting. *Applied Physics Letters*, Vol. 87, No. 21, 213101-213103, ISSN 0003-6951

Yang, D. Q.; Hennequin, B. & Sacher, E. (2006). XPS demonstration of π−π interaction between benzyl mercaptan and multiwalled carbon nanotubes and their use in the adhesion of Pt nanoparticles. *Chemistry of Materials*, Vol. 18, No. 21, 5033-5038, ISSN 0897-4756

Yang, D. Q.; Rochette, J. F. & Sacher, E. (2005). Controlled chemical functionalization of multiwalled carbon nanotubes by kiloelectronvolt argon ion treatment and air exposure. *Langmuir*, Vol. 21, No. 18, 8539-8545, ISSN 0743-7463

Yang, D. Q.; Rochette, J. F. & Sacher, E. (2005). Spectroscopic evidence for π−π interaction between poly(diallyl dimethylammonium) chloride and multiwalled carbon nanotubes. *Journal of Physical Chemistry B*, Vol. 109, No. 10, 4481-4484, ISSN 1520-6106

Yang, D. Q. & Sacher, E. (2006). Platinum nanoparticle interaction with chemically modified highly oriented pyrolytic graphite surfaces. *Chemistry of Materials*, Vol. 18, No. 7, 1811-1816, ISSN 0897-4756

Yang, D. Q. & Sacher, E. (2008). Strongly enhanced interaction between evaporated Pt nanoparticles and functionalized multiwalled carbon nanotubes via plasma surface modifications: effects of physical and chemical defects. *Journal of Physical Chemistry C*, Vol. 112, No. 11, 4075-4082, ISSN 1932-7447

Yang, W.; Wang, X.; Yang, F.; Yang, C. & Yang, X. (2008). Carbon nanotubes decorated with Pt nanocubes by a noncovalent functionalization method and their role in oxygen reduction. *Advanced Materials*, Vol. 20, No. 13, 2579–2587, ISSN 0935-9648

Yoshitake., T.; Shimakawa., Y.; Kuroshima., S.; Kimura., H.; Ichihashi., T. & Kudo., Y. (2002). Preparation of fine platinum catalyst supported on single-wall carbon nanohorns for fuel cell application. *Physica B*, Vol. 323, No. 1-4, 124-126, ISSN 0921-4526

Yu, R. Q.; Chen, L. W.; Liu, Q. P.; Lin, J. Y.; Tan, K. L.; Ng, S. C.; Chan, H.; Xu, G. Q. & Hor, T. (1998). Platinum deposition on carbon nanotubes via chemical modification. *Chemistry of Materials*, Vol. 10, No. 3, 718-722, ISSN 0897-4756

Zamudio, A.; Elias, A. L.; Rodriguez-Manzo, J. A.; Lopez-Urias, F.; Rodriguez-Gattorno, G.; Lupo, F.; Ruhle, M.; Smith, D. J.; Terrones, H.; Diaz, D. & Terrones, M. (2006). Efficient anchoring of silver nanoparticles on N-Doped carbon nanotubes. *Small*, Vol. 2, No. 3, 346-350, ISSN 1613-6810

Zhang, G. X.; Yang, D. Q. & Sacher, E. (2007). X-ray photoelectron spectroscopic analysis of Pt nanoparticles on highly oriented pyrolytic graphite, using symmetric component line shapes. *Journal of Physical Chemistry C*, Vol. 111, No. 2, 565-570, ISSN 1932-7447

Zhang, J.; Zou, H.; Qing, Q.; Yang, Y.; Li, Q.; Liu, Z. & Guo, Z. D. (2003). Effect of chemical oxidation on the structure of single-walled carbon nanotubes. *Journal of Physical Chemistry B*, Vol. 107, No. 16, 3712-3718, ISSN 1520-6106

Zhao, Y.; Fan, L. Z.; Zhong, H. Z.; Li, Y. F. & Yang, S. H. (2007). Platinum nanoparticle clusters immobilized on multiwalled carbon nanotubes: Electrodeposition and enhanced electrocatalytic activity for methanol oxidation. *Advanced Functional Materials*, Vol. 17, No. 9, 1537-1541, ISSN 1616-301X

Zhao, Z. W.; Guo, Z. P.; Ding, J.; Wexler, D.; Ma, Z. F.; Zhang, D. Y. & Liu, H. K. (2006). Novel ionic liquid supported synthesis of platinum-based electrocatalysts on multiwalled carbon nanotubes. *Electrochemistry Communications*, Vol. 8, No. 2, 245-250, ISSN 1388-2481

Zheng, H. T.; Li, Y. L.; Chen, S. X. & Shen, P. K. (2006). Effect of support on the activity of Pd electrocatalyst for ethanol oxidation. *Journal of Power Sources*, Vol. 163, No. 1, 371-375, ISSN 0378-7753